Annual Reports
on the Progress of Chemistry
Volume 74, 1977

SECTION B
Organic Chemistry

The Chemical Society
Burlington House, London WIV OBN

British Library Cataloguing in Publication Data

Annual reports on the progress of chemistry.
Section B: Organic chemistry.
Vol. 74: 1977.
1. Chemistry, Organic
I. Chemical Society
547 QD251.2

ISBN 0-85186-101-6
ISSN 0069-3030

Set in Times on Linotron and printed offset by
J. W. Arrowsmith Ltd., Bristol, England

Made in Great Britain

Contributors

P. Andreozzi, B.S., Ph.D., *Case Western Reserve University, Cleveland, Ohio, U.S.A.*
R. Baker, B.Sc., Ph.D., *University of Southampton*
A. J. Boulton, B.A., Ph.D., *University of East Anglia*
D. G. Buckley, M.Sc., Ph.D., A.R.A.C.I., *Queen Mary College, London*
A. T. Bullock, B.Sc., Ph.D., *University of Aberdeen*
R. Bushby, D.Phil., B.Sc., A.R.C.S., *University of Leeds*
H. A. J. Carless, M.A., D.Phil., *Birkbeck College, London*
P. M. Collins, B.Sc., Ph.D., *Birkbeck College, London*
R. G. Coombes, B.Sc., Ph.D., *The City University, London*
A. Cox, M.A., B.Sc., Ph.D., D.I.C., *University of Warwick*
D. A. Evans, B.Sc., Ph.D., *University of Southampton*
H. Hudson, Ph.D., D.Sc., C.Chem., F.R.I.C., *Polytechnic of North London*
B. M. Hursthouse, B.Sc., Ph.D., *Queen Mary College, London*
M. G. Hutchings, M.Sc., Ph.D., *I.C.I. Organics, Manchester*
G. Klopman, D.Sc., *Case Western Reserve University, Cleveland, Ohio, U.S.A.*
R. Lines, B.Sc., Ph.D., *The Norwegian Institute of Technology, The University of Trondheim, Norway*
S. A. Matlin, B.Sc., Ph.D., A.R.C.S., D.I.C., C.Chem., M.R.I.C., *The City University, London*
G. P. Moss, B.Sc., Ph.D., A.R.C.S., D.I.C., *Queen Mary College, London*
S. Neidle, B.Sc., Ph.D., A.R.C.S., D.I.C., *King's College, London*
G. Pattenden, Ph.D., C.Chem., F.R.I.C., *University of Nottingham*
Pill-Soon Song, Ph.D., *Texas Tech. University, U.S.A.*
A. Stewart, Ph.D., *I.C.I. Corporate Laboratory, Runcorn*
M. C. Summers, Ph.D., *University of Cambridge*
D. R. Taylor, M.A., Ph.D., *University of Manchester*
D. J. Thompson, Ph.D., *I.C.I. Corporate Laboratory, Runcorn*
T. P. Toube, B.Sc., M.Sc., Ph.D., *Queen Mary College, London*
M. V. Twigg, Ph.D., *I.C.I. Corporate Laboratory, Runcorn*
J. H. P. Utley, B.Sc., Ph.D., *Queen Mary College, London*
R. S. Ward, M.A., Ph.D., *University College, Swansea*
D. C. Williams, Ph.D., *Trinity College, Dublin*

Preface

The 1977 *Annual Reports* exhibit the same kind of coverage as seen in the 1976 volumes. However, in view of the suspension of the non-viable *Specialist Periodical Reports* titles covering much of inorganic chemistry, it is planned to increase the treatment of those areas in Section A of the 1978 Reports. The Society's Books and Reviews Committee has also recognised that the present coverage of physical chemistry in *Annual Reports* is somewhat irregular and uneven and that possibly the literature needs of practising physical chemists are therefore not being met. Hence an investigation is being undertaken into the type of Reports desired by physical chemists, in particular the areas to be covered and the depth of the treatment, bearing in my mind the Society's existing *Specialist Periodical Reports* titles in this area. It is thus hoped to be able to determine in time for the 1979 *Annual Reports* whether a separate volume for physical chemistry would be feasible.

With these 1977 Reports Professor M. F. Lappert has completed his term of office as Senior Reporter for the Inorganic Chemistry part of Section A, and we thank him for his great efforts over the past three years. We welcome as a new Senior Reporter Dr. G. P. Moss who collaborates with his colleague Dr. J. H. P. Utley on the coverage and organisation of Section B, Organic Chemistry.

Contents

Chapter 1 Introduction 1
By J. H. P. Utley and G. P. Moss

Chapter 2 Physical Methods and Techniques 3
Part (i) Mass Spectrometry 3
By T. P. Toube

1 Ion Structures: Aliphatic 4

2 Ion Structures: Aromatic 6

3 Fragmentation Processes 8

4 Techniques 9

Part (ii) X-Ray Crystallography 11
By M. B. Hursthouse and S. Neidle

1 Direct Methods 11

2 Electron Density Distribution 12

3 Molecular Conformations 12

Part (iii) Ultraviolet and Visible Spectroscopy of Bio-organic Molecules 18
By P.-S. Song

1 Introduction 18

2 Cyclic Polyenes 18
Porphyrins 18
Chlorophylls 22
Corrins and Tetrapyrroles 24

3 Linear Polyenes 26
Simple Polyenes 26
Retinyl Polyenes 30
Carotenoids 32

4 Heterocyclic Compounds 34
Indoles and Tryptophan 34

Nucleic Acid Bases 35
Purine Bases 35
Pyrimidine Bases 37
Flavins 38

Chapter 3 Theoretical Chemistry 41
By G. Klopman and P. Andreozzi

1 **Introduction** 41

2 **Theoretical Aspects of Solvation** 41
Quantum Mechanical Methods 41
Solvation of Monoatomic Ions 46
Solvation of Organic Cations 47
Association of Monoatomic Ions with Organic Molecules 49

3 **Theoretical Aspects of Hydrogen Bonding** 51
Studies on its Origin and Nature 51
Structural Studies 52
Theoretical Aspects of Proton Transfer 54

4 **Other Molecular Associations** 56

Chapter 4 Reaction Mechanisms 59
Part (i) Pericyclic Reactions 59
By R. J. Bushby

1 **Cycloadditions and Cycloreversions** 59

2 **Sigmatropic Reactions** 63

3 **Ene Reactions** 68

4 **Electrocyclic Reactions** 68

Part (ii) Polar Reactions 71
By H. R. Hudson

1 **Introduction** 71

2 **Nucleophilic Substitution at Saturated Carbon** 71

3 **Nucleophilic Substitution at sp^2 Carbon** 75
Vinylic Substitution 75
Nucleophilic Attack on C=N Double Bonds 78

4 **Carbocations** 78
Preparation, and Rearrangements in Super-acid Media 78
Other Aspects 81

5 **β-Elimination** 82

6 Electrophilic Addition 83

7 Ion–Molecule Reactions in the Gaseous Phase 84
Proton Transfer 85
Gas-phase Acidities 85
Gas-phase Basicities 86
Nucleophilic Substitution 87
Carbocations 88
Stabilities 88
Condensation Reactions 88
Ester Cleavage 89

Part (iii) Electron Spin Resonance Spectroscopy and Free Radical Reactions 90
By A. T. Bullock

1 Kinetics and Mechanism 90
2 Chemically Induced Dynamic Electron Polarization 103

Chapter 5 Arynes, Carbenes, Nitrenes, and Related Species 105
By S. A. Matlin

1 Arynes 105

2 Nitrenes 106

3 Carbenes 110
Generation 110
Reactions 111

Chapter 6 Organometallic Chemistry 119
Part (i) The Transition Elements 119
By A. Stewart, D. J. Thompson, and M. V. Twigg

1 Introduction 119

2 Metal-catalysed Hydrogenation 119

3 Isomerization 124

4 Dimerization, Oligomerization, and Polymerization 126

5 Carbonylation 127

6 Reaction of Co-ordinated Ligands 128

7 Olefin Metathesis 132
Mechanistic Studies 132
Applications 133

8 Use of Metal Cluster Complexes in Catalysis 134

Part (ii) Main-Group Elements 136
By M. G. Hutchings

1 Introduction 136

2 Group I 137

3 Group II 141
Magnesium 141
Zinc, Cadmium, and Mercury 141

4 Group III 142
Boron 142
Aluminium, Gallium, and Thallium 145

5 Group IV 147
Silicon 147
Germanium, Tin, and Lead 150

6 Group V 151

Chapter 7 Electro-organic Chemistry 153
By R. Lines

1 Introduction 153

2 Anodic Processes 153
The Anodic Oxidation of Carboxylates 153
The Anodic Oxidation of Neutral Organic Compounds 154

3 Cathodic Processes 158
The Cathodic Reduction of Organic Cations 158
The Cathodic Reduction of Neutral Organic Compounds 158

4 Miscellaneous 162

Chapter 8 Photochemistry 165
By H. A. J. Carless

Chapter 9 Aliphatic Compounds 175

Part (i) Hydrocarbons 175
By D. R. Taylor

1 Acetylenes 175

2 Alkanes 180

3 Allenes 181

4 Olefins and Dienes 184

Part (ii) Other Aliphatic Compounds 194
By R. S. Ward

1 **Amines and Imines** 194

2 **Nitriles and Isocyanides** 198

3 **Nitro-compounds** 199

4 **Azo- and Diazo-compounds** 200

5 **Alcohols and Ethers** 201

6 **Aldehydes and Ketones** 202

7 **Carboxylic Acids** 205

8 **Esters and Lactones** 206

9 **Amides and Lactams** 208

10 **Anhydrides and Imides** 209

11 **Phosphorus Compounds** 209

12 **Sulphur Compounds** 211

13 **Halogen Compounds** 213

Chapter 10 Aromatic Compounds 215
By R. G. Coombes

1 **Introduction** 215

2 **Benzene Isomers, Oxides, and Homobenzenes** 217

3 **Benzene and its Derivatives** 219
General 219
Electrophilic Substitution 223
Nucleophilic Substitution 227
Biaryls 229
Quinones and Related Compounds 230

4 **Cyclophanes** 232

5 **Molecular Rearrangements** 235

6 **Condensed Systems** 237

7 **Non-benzene Systems** 241
Three- and Four-membered Rings 241
Five- and Seven-membered Rings 243
Annulenes 246

Chapter 11 Heterocyclic Chemistry 251
By A. J. Boulton

1 Heterocycles in Functional Group Transformations 251

2 General Heterocyclic Synthesis 252

3 Three-membered Rings 252

4 Four-membered Rings 257

5 Five-membered Rings 259

6 Six-membered Rings 272

7 Seven-membered and Larger Rings 278

8 Reviews 282

Chapter 12 Alicyclic Chemistry 285
By A. Cox

1 Introduction 285

2 Synthesis 285
Three- and Four-membered Rings 285
Five- and Six-membered Rings 287
Seven- and Eight-membered Rings 289
Annulenes 290
Polycyclic Systems 291

3 Stereochemistry 295

4 Structural Properties and Orbital Interactions 297

5 Reactions 301
Metal-promoted Reactions 301
Thermally Induced Reactions 305

Chapter 13 Synthetic Methods 309
By G. Pattenden

1 Introduction 309

2 Alkanes 309

3 Alkenes 310

4 Alkynes, Enynes, and Allenes 314

5 Halogeno-compounds 318

6 Alcohols 319

7 Ethers 320

8 Amines 321

9 Aldehydes and Ketones 322
Functionalized Aldehydes and Ketones 325

10 Carboxylic Acids and Anhydrides 332

11 Esters and Functionalized Esters 334
Lactones 335
Macrolides 336

12 Nitriles 338

13 Alkylation 338

14 Ring Synthesis 340

Chapter 14 Biological Chemistry 343
Part (i) Monosaccharides 343
By P. M. Collins

1 Introduction 343

2 Glycosides 343

3 Unsaturated Sugars 349

4 Dicarbonyl Sugars 351

5 Esters 354

6 Ethers and Acetals 356

7 Halo- and Thio-sugars, and other Inorganic Sugar Derivatives 358

8 Deoxy-sugars and Branched-chain Sugars 361

9 Amino-sugars 364

10 Synthesis of Non-carbohydrate Compounds 366

Part (ii) Insect Chemistry 367
By R. Baker and D. A. Evans

1 Introduction 367

2 Sex Attraction and Related Phenomena 367

3 Aggregation Pheromones and Population Attractants 372

4 Pheromones of Social Insects and Related Species 372

5 Alarm Behaviour in Aphids 373

6 Defence Secretions 373

7 Host Food, Host Prey, and Oviposition Attractants 375

8 Antifeedants and Repellants 376

9 Biosynthesis and Biotransformation 377

10 Perception of Stimuli 378

11 Techniques of Microscale Structure Elucidation 379

12 Behaviour-modifying Chemicals in Pest Control 379

13 Synthetic Studies 380
Acyclic Derivatives 380
Alicyclic Derivatives 387

Part (iii) Tetrapyrroles and their Biosynthesis 392
By. D. G. Buckley

1 Introduction 392

2 Biosynthesis of Tetrapyrroles: General 393

3 Biosynthesis of Uro'gen-III: the 'Type-III Problem' 394
The Nature of the Rearrangement 396
The Role of Cosynthetase: Timing of the Rearrangement 398
Dipyrrolic Intermediates in Uro'gen-III Biosynthesis 398
Proof that Rearrangement Occurs at the Tetrapyrrole Level 404

4 The Pathway from Uro'gen-III to Protoporphyrin-IX 406
Intermediates between Uro'gen-III and Copro'gen-III 407
Biosynthesis of Protoporphyrin-IX from Copro'gen-III 411
Oxidative Decarboxylation of the Propionic Acid Side-chains 411
Aromatization of Proto'gen-IX 416

5 The Iron and Magnesium Branches 416
The Haemoproteins 416
The Chlorophylls 417
The Bacteriochlorophylls 419

6 The Biosynthesis of Vitamin B_{12} 423
Sirohydrochlorin 426

7 Miscellaneous 428
New Techniques 428
Bile Pigments 428
Haemoglobin/Myoglobin Model Studies 431

Part (iv) Enzyme Chemistry 432
By M. C. Summers and D. C. Williams

1 **Irreversible Enzyme Inhibitors** 432

2 **Acetylenic Irreversible Enzyme Inhibitors** 433

3 **Steroid Isomerases** 437

4 **Biotin-requiring Enzymes** 440

5 **Adenylate Cyclase: Receptor Complexes** 442

6 **Multienzyme Complexes** 443

Author Index 447

Section A, Physical and Inorganic Chemistry, contains the following items

Part I PHYSICAL CHEMISTRY
1 Introduction. By P. A. H. Wyatt
2 Electrolyte Solutions. By A. K. Covington and A. D. Pethybridge
3 Solid Electrolytes. By M. D. Ingram and C. A. Vincent
4 Luminescence of Organic Solids. By J. O. Williams
5 Studies of Molecular Motion in Liquids and Solids using Low Frequency Dielectric Relaxation and related Techniques. By G. Williams and J. Crossley

Part II INORGANIC CHEMISTRY
6 Introduction. By M. F. Lappert
7 The Typical Elements,
Part I: Groups I and II. By R. H. Cragg
Part II: Group III. By G. E. Toogood
Part III: Groups IV and V. By J. D. Smith
Part IV: Groups VI–VIII. By R. H. Cragg
8 Chemistry of the *d*- and *f*-Block Metals.
Part I: The Transition Metals. By J. R. Dilworth, G. J. Leigh, and R. L. Richards
Part II: Scandium, Yttrium, the Lanthanides, and the Actinides. By K. W. Bagnall
9 Organometallic Compounds. By C. J. Cardin, D. J. Cardin, R. J. Norton, and K. R. Dixon

ORGANIC CHEMISTRY

1 Introduction

By J. H. P. UTLEY and G. P. MOSS

Department of Chemistry, Queen Mary College, University of London, Mile End Road, London E1 4NS

Contributors have again been asked to restrict the number of papers referred to in their articles. This policy has a number of advantages: the volume is kept to a manageable size, is speedily published, and critical and selective reviewing is encouraged. The unavoidable subjectiveness of the articles is balanced over the years by changes in reviewers (and Senior Reporters!). We hope that the result is, for most organic chemists, the provision of broad and up to date coverage. Fuller accounts of specialist topics are provided in the Society's series of *Specialist Periodical Reports*.

Chapters on mass spectrometry and *X*-ray crystallography are included; these topics are now reviewed every second year. Other topics are reviewed less frequently (see Table). Chapters on (i) u.v. and visible spectroscopy of bio-organic molecules and (ii) tetrapyrroles and their biosynthesis are included for the first time. The former topic has not been reviewed for many years and the latter article is particularly timely because, it seems, many problems in the field have now been solved. Attention is also drawn to the chapter on polar reactions in which Dr. Hudson covers recent work on ion–molecule reactions in the gas phase. Each year we ask contributors to comment on highlights in their subject. This year most of the authors felt that 1977 had seen a steady advance but no startling innovations.

Finally, we wish to thank all of the contributors for producing and processing material promptly and the Editorial Staff of the Chemical Society for so efficiently converting a mass of typescript into this volume.

Table *Synopsis of Reports 1972–1977*

Subject	1972	1973	1974	1975	1976	1977
Physical Methods:						
Mass spectrometry	√	√	√	√	×	√
Optical rotatory dispersion/circular dichroism	√	×	√	×	√	×
X-ray crystallography	√	√	×	√	×	√
N.m.r. spectroscopy	√	√	√	√	×	×
Theoretical chemistry	√	×	×	√	√	√
Reaction Mechanisms:						
Aromatic[a]	√	√	×	×	×	×
Pericyclic	√	√	√	√	√	√
Polar reactions	√	√	√	√	√	√
Free radical reactions	√	√	√	√	√	√

Table *Synopsis of Reports 1972–1977*—cont.

	1972	1973	1974	1975	1976	1977
Arynes, carbenes, nitrenes, and related species	√	√	√	√	√	√
Molecular rearrangements[b]	√	√	×	×	×	×
Organic electrochemistry	√	√	√	√	√	√
Photochemistry	√	√	√	√	√	√
Aliphatic compounds	√	√	√	√	√	√
Alicyclic compounds	√	√	√	√	√	√
Aromatic compounds	√	√	√	√	√	√
Heterocyclic compounds	√	√	√	√	√	√
Organometallic Compounds:						
Transitional elements	√	√	√	√	√	√
Main-group elements	×	×	√	√	√	√
Synthetic (general) methods	√	√	√	√	√	√
Alkaloids	√	√	×	×	√	×
Nucleic acids	√	√	√	√	√	×
Enzymes	√	√	√	√	√	√
Biosynthesis	√	√	√	√	√	×
Other chapters	*c*	*c, d, e, f*	*g, h, i, j, k*	*l, m, n*	*k, o, p*	*f, n, q, r*

[a] From 1974 aromatic reaction mechanisms have been included with aromatic compounds. [b] From 1974 molecular rearrangements have been included in appropriate sections, such as aromatic compounds. [c] Terpenes and steroids (from 1974 terpenes have been included with alicyclic compounds). [d] Post reactions of polymers. [e] Organophosphorus compounds. [f] Monosaccharides. [g] HPLC. [h] Affinity chromatography. [i] Histamines and cholinergic agonists. [j] Porphyrins and related compounds. [k] Peptides and proteins. [l] Laser Raman and infrared spectroscopy. [m] Steroids. [n] Insect chemistry. [o] Gas chromatography–mass spectrometry. [p] Neurochemistry. [q] U.v. and visible spectroscopy. [r] Tetrapyrroles.

2 Physical Methods and Techniques
Part (i) Mass Spectrometry

By T. P. TOUBE
Chemistry Department, Queen Mary College, London E1 4NS

In this Report the policy of selection which governed previous Reports[1,2] is continued but the volume of material in two years' literature has led to an even greater need for selectivity than before. Fortunately, part of the period under review has been admirably covered in the fourth Specialist Periodical Report.[3] A number of relevant books have appeared, dealing with physical aspects,[4] applications in the field of natural products,[5,6] conference proceedings,[7—9] compilations of data,[10,11] and ion cyclotron resonance (ICR) spectroscopy.[12] Other publications contain chapters on aspects of mass spectrometry.[13—15] Topics of recent reviews include: charge localization,[16] newer methods of structure determination,[17] collisional activation (CA),[18] and peptide sequencing,[19] as well as other aspects of mass spectrometry.[20]

[1] T. P. Toube, *Ann. Reports* (*B*), 1974, **71**, 7.
[2] T. P. Toube, *Ann. Reports* (*B*), 1975, **72**, 3.
[3] 'Mass Spectrometry', ed. R. A. W. Johnstone, (Specialist Periodical Reports), The Chemical Society, London, 1977, Vol. 4.
[4] 'Mass Spectrometry', ed. A. Maccoll, MTP International Review of Science, Physical Chemistry Series Two, Vol. 5, Butterworths, London, 1975.
[5] 'Progress in Mass Spectrometry', ed. H. Budzikiewicz, Vol. 3; M. Hesse and H. P. Bernhard, 'Alkaloide ausser Indol-, Triterpen- und Steroidalkaloide', Verlag Chemie, Weinheim, 1975.
[6] Z. V. Zaretskii, 'Mass Spectrometry of Steroids', Wiley, New York, 1976.
[7] 'Advances in Mass Spectrometry', ed. N. R. Daly, Vol. 7, Heyden, London, 1977.
[8] 'Advances in Mass Spectrometry in Biochemistry and Medicine', ed. A. Frigerio and N. Castagnoli, Wiley, New York, 1976, Vol. 1; ed. A. Frigerio, Wiley, New York, 1977, Vol. 2.
[9] 'Quantitative Mass Spectrometry in Life Sciences', ed. A. P. Leenheer and R. R. Roncucci, Elsevier, Amsterdam, 1977.
[10] H. M. Rosenstock, K. Draxl, B. W. Steiner, and J. T. Herron, 'Energetics of Gaseous Ions', *J. Phys. and Chem. Ref. Data*, 1977, **6**, Suppl. 1.
[11] H. Cornu and R. Massot, 'Compilation of Mass Spectral Data', 2nd Edn., Heyden, London, 1975.
[12] T. A. Lehman and M. M. Bursey, 'Ion Cyclotron Resonance Spectroscopy', Wiley, New York, 1976.
[13] J. B. Lambert, H. F. Schurvell, L. Verbit, R. G. Cooks, and G. H. Stout, 'Organic Structural Analysis', Macmillan, New York, 1976.
[14] 'Determination of Organic Structures by Physical Methods', ed. F. C. Nachod, J. J. Zuckerman, and E. W. Randall, Vol. 6, Academic Press, New York, 1976.
[15] J. M. Wilson in 'Structure Determination in Organic Chemistry', ed. L. M. Jackman, M. T. P. International Review of Science, Organic Chemistry Series Two, Vol. 1, Butterworths, London, 1976.
[16] D. H. Williams and J. H. Beynon, *Org. Mass Spectrometry*, 1976, **11**, 103.
[17] D. H. Williams, *Israel J. Chem.*, 1975, **14**, 33.
[18] K. Levsen and H. Schwarz, *Angew. Chem. Internat. Edn.*, 1976, **15**, 509.
[19] H. Nau, *Angew. Chem. Internat. Edn.*, 1976, **15**, 75.
[20] A. L. Burlingame, B. J. Kimble, and P. J. Derrick, *Analyt. Chem.*, 1976, **48**, 368R.

The appearance of draft IUPAC recommendations on symbolism and nomenclature for mass spectrometry[21] is particularly welcome; the present Report, however, will continue to use the more common m/e symbol rather than the recommended m/z.

1 Ion Structures: Aliphatic

The examination of the actual structures of ions in the vapour phase continues to be an active area of research. The dangers of inferring molecular structures from fragmentation patterns augmented by deuterium-labelling results are amusingly illustrated in the mass spectra of the dioxolans (1), (2), and (3).[22] There is a loss of $^2H\cdot$ from the molecular ion of (1), whereas that of (2) loses only $^1H\cdot$, implying that the loss is from the 2-position. However, ionized (3) loses only $^1H\cdot$, not $^2H\cdot$, suggesting that an isotope effect forces a loss from the 5-position. Interestingly, (4) loses only ^2H.

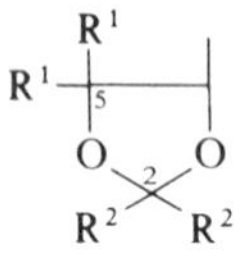

(1) $R^1 = R^2 = D$
(2) $R^1 = D, R^2 = H$
(3) $R^1 = H, R^2 = D$

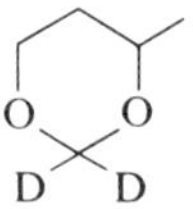

(4)

Symmetry-forbidden elimination reactions can be detected by the release of kinetic energy, derived from the reverse activation energy for the process, from the metastable ion,[1,23] and this effect is observed for the loss of H_2 from $CH_2{=}\overset{+}{O}H$. However, the loss of H_2 from CH_3O^+ shows no kinetic energy release,[24] indicating that it does not decompose *via* rearrangement to the isomeric protonated aldehyde structure (1,1-elimination adjacent to a vacant p-orbital is a permitted process).

The shapes of metastable peaks can be used to give information on the potential energy surfaces over which ions of low excess internal energy decompose and hence contribute to our knowledge of the structures of fragmenting ions.[25] Appearance potential (AP) measurements and analysis of metastable peak shapes suggest that CH_4N^+ and $C_2H_6N^+$ ions from a variety of sources decompose from common structures in each case.[26] On the other hand $C_2H_3N^{+\cdot}$ ions have been shown[27] by CA, metastable analysis, and mass-analysed ion kinetic energy spectroscopy (MIKES) to exist as the discrete structures $:CHCH{=}\overset{+\cdot}{N}H$, $CH_3C\overset{+\cdot}{N}$, $CH_3\overset{+\cdot}{N}C:$, and $CH_2{=}C{=}\overset{+\cdot}{N}H$. The CA spectra of $C_2H_4O_2^{+\cdot}$ ions formed by the McLafferty rearrangement of aliphatic acids[28] indicate that they do not tautomerize to the acetic

21 *Org. Mass Spectrometry*, 1977, **12**, 115.
22 J. U. R. Nielsen, S. E. Jørgensen, S. Fredericksen, R. B. Jensen, G. Schroll, and D. H. Williams, *Acta Chem. Scand. (B)*, 1977, **31**, 227.
23 D. H. Williams and G. Hvistendahl, *J. Amer. Chem. Soc.*, 1974, **96**, 6753.
24 R. D. Bowen and D. H. Williams, *J.C.S. Chem. Comm.*, 1977, 378.
25 D. H. Williams, *Accounts Chem. Res.*, 1977, **10**, 280.
26 A. G. Loudon and K. S. Webb, *Org. Mass Spectrometry*, 1977, **12**, 283.
27 J. van Thuijl, J. J. van Houte, A. Maquestiau, R. Flammang, and C. De Meyer, *Org. Mass Spectrometry*, 1977, **12**, 196.
28 K. Levsen and H. Schwarz, *J.C.S. Perkin II*, 1976, 1231.

acid molecular ion structure. Gaseous $C_2H_5S^+$ ions exhibit by CA a variety of stable structures,[29] including $CH_3SCH_2^+$, $CH_3\overset{+}{C}HSH$, and $\overline{CH_2CH_2\overset{+}{S}}H$, but not $HSCH_2CH_2^+$ nor $CH_2{=}CH\overset{+}{S}H_2$. An even greater number of discrete structures are found for $C_3H_7S^+$.[30] Metastable ion characteristics for $C_2H_5N^{+\cdot}$ show[31] separate structures for $CH_3CH{=}\overset{+\cdot}{N}H$, $CH_2{=}CH\overset{+\cdot}{N}H_2$, and $CH_2{=}NCH_3$, but interconversion between $CH_2NHCH_2^{+\cdot}$ and $\overline{CH_2CH_2\overset{+\cdot}{N}}H$. Results for $CH_3CH{=}\overset{+}{N}H_2$ and $CH_2{=}\overset{+}{N}HCH_3$ show[32] that their decomposition by loss of H_2 or C_2H_2 occurs over the same potential energy surfaces whereas for the analogous reactions of oxygen-containing species the ions $CH_3CH{=}\overset{+}{O}H$ and $CH_2{=}\overset{+}{O}CH_3$ remain discrete.

Results from CA studies suggest that protonation of propene leads[33] within 10^{-5} s to the same ion as protonation of cyclopropane. The decomposition of $C_3H_7O^+$ has been shown[34,35] to follow the sequence (5)→(6)→(7) prior to the loss of ethene. The rate-determining step is the 1,2-hydrogen shift (5)→(6). Loss of ^{13}C labels from such ions from various precursors can be rationalized in terms of the potential energy surfaces for the reactions. It has been suggested that $C_3H_7S^+$ ions[36] from thiols and thioethers decompose from a common structure.

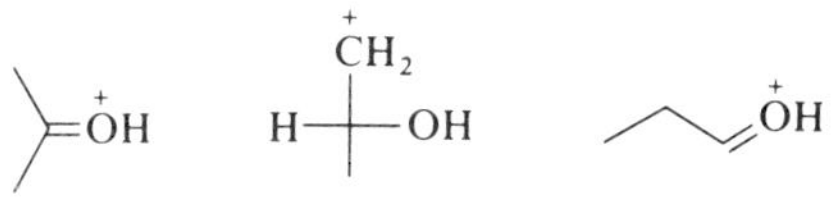

(5) (6) (7)

Examination of the electron-impact (EI) mass spectra of a number of C_8H_{16} cycloalkanes[37] with various ring sizes indicates that four-, five-, and six-membered rings fragment after ring opening to the isomeric 1-alkene with subsequent double-bond migration, while the decomposition of the molecular ions of larger-ring compounds is from the structure with the ring intact. However, CA studies on $C_4H_8^{+\cdot}$ ions[38] show that cyclobutane itself is stable on the 10^{-5} s time-scale and can be distinguished from but-1-ene, but-2-ene, and 2-methylpropene. Methylcyclopropane may rearrange to a mixture of the isomeric alkenes, whereas the $C_4H_8^{+\cdot}$ ion from cyclohexane appears to be a mixture of the normal alkenes. The thermochemistry of these ions has been studied[39] as well as their fragmentation after EI or field ionization (FI). Potential energy surfaces for $C_4H_9O^+$ ions[40] show

[29] B. van de Graaf and F. W. McLafferty, *J. Amer. Chem. Soc.*, 1977, **99**, 6806.
[30] B. van de Graaf and F. W. McLafferty, *J. Amer. Chem. Soc.*, 1977, **99**, 6810.
[31] J. L. Holmes and J. K. Terlouw, *Canad. J. Chem.*, 1976, **54**, 1007.
[32] R. D. Bowen, D. H. Williams, and G. Hvistendahl, *J. Amer. Chem. Soc.*, 1977, **99**, 7509.
[33] P. P. Dymerski, R. M. Prinstein, P. F. Bente, and F. W. McLafferty, *J. Amer. Chem. Soc.*, 1976, **98**, 6834.
[34] G. Hvistendahl, R. D. Bowen, and D. H. Williams, *J.C.S. Chem. Comm.*, 1976, 294.
[35] R. D. Bowen, J. R. Kalman, and D. H. Williams, *J. Amer. Chem. Soc.*, 1977, **99**, 5481.
[36] W. J. Broer and W. D. Weringa, *Org. Mass Spectrometry*, 1977, **12**, 326.
[37] F. Borchers, K. Levsen, H. Schwarz, C. Wesdemiotis, and R. Wolfschütz, *J. Amer. Chem. Soc.*, 1977, **99**, 1716.
[38] T. Nishishita, F. M. Bockhoff, and F. W. McLafferty, *Org. Mass Spectrometry*, 1977, **12**, 16.
[39] J. L. Holmes, G. W. Weese, A. S. Blair, and J. K. Terlouw, *Org. Mass Spectrometry*, 1977, **12**, 424.
[40] F. Borchers, K. Levsen, H. Schwarz, C. Wesdemiotis, and H. U. Winkler, *J. Amer. Chem. Soc.*, 1977, **99**, 6359.

that for the decomposition of (8) the rate-determining step is isomerization to (9) and for (10) isomerization to (11). Predictions consistent with observed results may be deduced from the potential energy curves.

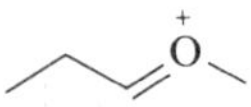

(8)

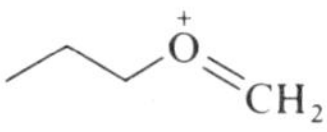

(9)

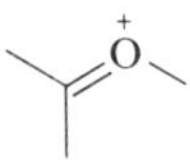

(10)

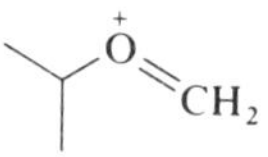

(11)

An extensive investigation of the isomeric octenes,[40] unsaturated in the 1-, 2-, 3-, and 4-positions, by CA, field ionization kinetics, ^{2}H and ^{13}C labelling, and examination of their metastable ions has shown that the decomposing ions isomerize to a mixture of interconverting structures within 10^{-9} s. The ions which do not decompose, however, are only partially isomerized.

Evidence has been presented for a number of cyclic oxonium ions in the vapour phase. Using a combined scan of the accelerating voltage (V) and the potential across the electrostatic analyser (E), where V/E^2 is kept constant, a spectrum equivalent to MIKES is produced.[41] Using this technique it has been shown[42] that $ArO(CH_2)_nOAr$ may give rise to ions of structure (12) ($n = 4$ or 5) after 10^{-6}—10^{-5} s. The analogous system (12; $n = 2$) isomerizes[43] to $Ph\overset{+}{O}{=}CHCH_3$. On the other hand the corresponding sulphonium ion systems (13) ($n = 2$,[43] 3, 4, or 5[44]) appear to be more stable, although the ion with $n = 2$ isomerizes after 10^{-5} s in a manner similar to its oxonium counterpart. Labelling experiments (^{2}H, ^{13}C) suggest that the molecular ions of (14) ($n = 1$ or 2) form the bridged oxonium ions (15) ($n = 1$ or 2) by loss of PhO· in the source or first drift region of a mass spectrometer.[45]

(12) X = O
(13) X = S

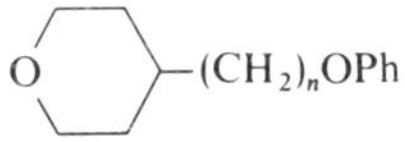

(14)

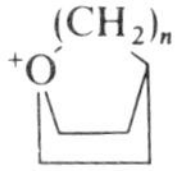

(15)

2 Ion Structures: Aromatic

A comparison of the kinetic energies of decomposing $C_6H_6O^{+\cdot}$ ions[46] suggests that there are at least two isomeric structures, the particular structure found depending on the precursor of the ion. Collisionally induced decompositions and ^{2}H labelling show[47] that on the 10^{-5} s time-scale the phenol molecular ion does not interconvert with the ketonic structure (16). Although ion kinetic energy studies of the loss of

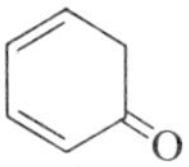

(16)

[41] S. Evans and R. Graham, *Adv. Mass Spectrometry*, 1974, **6**, 429.
[42] C. C. Van de Sande, *Org. Mass Spectrometry*, 1976, **11**, 121.
[43] K. Levsen, H. Heimbach, C. C. Van de Sande, and J. Monstrey, *Tetrahedron*, 1977, **33**, 1785.
[44] C. C. Van de Sande, *Org. Mass Spectrometry*, 1976, **11**, 130.
[45] C. C. Van de Sande, M. Vanhooren, and F. Van Gaever, *Org. Mass Spectrometry*, 1976, **11**, 1206.
[46] M. K. Hoffman, M. D. Friesen, and G. Richmond, *Org. Mass Spectrometry*, 1977, **12**, 150.
[47] F. Borchers, K. Levsen, C. B. Theissling, and N. M. M. Nibbering, *Org. Mass Spectrometry*, 1977, **12**, 746.

CO from $C_6H_6O^{+\cdot}$ ions do not exclude fragmentation from a common structure, the CA results[48] show a number of non-decomposing forms and suggest that the decomposing ion may be an acyclic $\alpha\beta$-unsaturated aldehyde.

The $C_7H_7^+$ ion[1] continues to attract a considerable amount of interest. An earlier result[49] indicated that whereas benzyl methyl ether and 7-methylcycloheptatriene produced *m/e* 91 ($C_7H_7^+$) ions that reacted with dimethylamine in an ICR spectrometer, the *m/e* 91 ion from 7-methoxycycloheptatriene did not. This has been amplified by an ICR study[50] of $C_7H_7^+$ ions from toluene and halogenated toluenes (*o*-, *m*-, *p*-, and benzyl; F, Cl, Br, and I) and the reactions of such ions with the parent compounds. There appear to be two species of ions, reactive (benzyl) ions and unreactive (tropylium) ones. The evidence suggests that ring expansion, when it occurs, takes place after the formation of the *m/e* 91 species, except in the case of toluene itself; the fluorinated derivatives produce $C_7H_6F^+$ instead, and here again ring expansion may take place in the parent compound rather than in the fragment. These results are similar to, but not identical to those from CA studies.[51] Calculations[52] of heats of formation and other thermodynamic properties for these ions support the view[49,50,53,54] that the reactive species in the ICR reactions is benzyl and the unreactive one tropylium. Theoretical chemists have entered the fray. Calculations (MINDO/3) have been made to try to evaluate the interconversion[55] of benzyl and tropylium and of their substituted analogues[56] in the mass spectrometer, the mechanism of the toluene–cycloheptatriene interconversion,[57] and the stabilities and fragmentation routes of these and other aromatic cations (C_3–C_8).[58—60]

Benzoyl ions have been investigated by MIKES.[61] The kinetic energy released in the metastable decompositions can be used to estimate the excess energy in the parent ion giving rise to the $PhCO^+$ fragment and hence may help to characterize the parent ion. The apparent heat of formation of the fragment will be a function of the excess energy. All the $PhCO^+$ ions studied appeared identical, except for those derived from $PhCOCF_3$ and, possibly, from PhCOPh. A study of the distribution[62] of such kinetic energy release may give information about the mechanism of fragmentation. Benzoyl ions from alkyl benzoates[63] have excess energies of *ca.* 8 kcal mol^{-1} for methyl and ethyl benzoates, and *ca.* 18 kcal mol^{-1} for the benzoates of some higher alcohols. If these excess energies are taken into account the

[48] A. Maquestiau, Y. van Haverbeke, R. Flammang, C. De Meyer, K. G. Das, and G. S. Reddy, *Org. Mass Spectrometry*, 1977, **12**, 631.
[49] A. Venema and N. M. M. Nibbering, *Tetrahedron Letters*, 1974, 3013.
[50] J.-A. A. Jackson, S. G. Lias, and P. Ausloos, *J. Amer. Chem. Soc.*, 1977, **99**, 7515.
[51] F. W. McLafferty and J. Winkler, *J. Amer. Chem. Soc.*, 1974, **96**, 5182.
[52] J.-L. M. Abbond, W. J. Hehre, and R. W. Taft, *J. Amer. Chem. Soc.*, 1976, **98**, 6072.
[53] J. Shaw, R. C. Dunbar, and G. Olah, *J. Amer. Chem. Soc.*, 1974, **96**, 6227.
[54] R. C. Dunbar, *J. Amer. Chem. Soc.*, 1975, **97**, 1382.
[55] C. Cone, M. J. S. Dewar, and D. Landman, *J. Amer. Chem. Soc.*, 1977, **99**, 372.
[56] M. J. S. Dewar and D. Landman, *J. Amer. Chem. Soc.*, 1977, **99**, 4633.
[57] M. J. S. Dewar and D. Landman, *J. Amer. Chem. Soc.*, 1977, **99**, 2446.
[58] M. J. S. Dewar and H. S. Rzepa, *J. Amer. Chem. Soc.*, 1977, **99**, 7432.
[59] M. J. S. Dewar and D. Landman, *J. Amer. Chem. Soc.*, 1977, **99**, 7439.
[60] L. Radom and H. F. Schaefer, *J. Amer. Chem. Soc.*, 1977, **99**, 7522.
[61] J. F. Elder, J. H. Beynon, and R. G. Cooks, *Org. Mass Spectrometry*, 1976, **11**, 415.
[62] J. F. Elder, R. G. Cooks, and J. H. Beynon, *Org. Mass Spectrometry*, 1976, **11**, 423.
[63] S. Tajima, T. Azami, and T. Tsuchiya, *Org. Mass Spectrometry*, 1977, **12**, 24.

heats of formation of $C_6H_5^+$ by loss of CO from these benzoyl ions have a linear relationship to the number of degrees of freedom in the precursor ester.

Gaseous $C_8H_9^+$ ions exhibit a number of discrete structures after 10^{-5} s. By CA[64] at least thirteen such structures, many of them substituted benzyl or tropylium ions, have been detected. By contrast, $C_{11}H_9^+$ ions[65] from six aromatic precursors all appear to have a common (possibly benztropylium) structure or to be a mixture of rapidly interconverting isomeric forms.

3 Fragmentation Processes

Interest in the interaction of remote functional groups (see previous Reports[1,2]) continues. The fragmentation of $PhCH_2NH(CH_2)_n OCH_2Ph$ ($n = 2$—7)[66] after EI may involve the loss of $PhCH_2\cdot$ or PhCHO, apparently *via* hydrogen transfer to the nitrogen atom. The ring size in the transition state appears to be a major factor determining the fragmentation pathway; the loss of $PhCH_2\cdot$ is absent when $n = 1$, while the expulsion of PhCHO is favoured at low values of n, a six-membered transition state for hydrogen transfer in each reaction being favoured. Loss of water from the molecular ion of (17) involves interaction between the two hydroxy-groups,[67] a process requiring considerable deformation of the diyne system. On the other hand, the chemical ionization (CI) spectra of some C_4 1,4-diols[68] show hydrogen-bridged $(M+H)^+$ peaks for the saturated and (2*Z*)-but-2-ene compounds (18), but not for the (2*E*)-but-2-ene or but-2-yne diols. In the CI mass

(17)

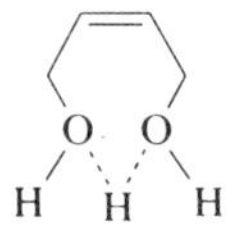

(18)

spectra of some dicarboxylic acids $HO_2C(CH_2)_nCO_2H$ and their methyl esters,[69] variations in their behaviour (*e.g.* ease of loss of water from the acids, isobutane reagent gas; ease of protonation of the esters, NH_3 reagent gas) can be correlated with n.

The McLafferty rearrangement is perhaps the best-known mass spectrometric reaction. Recent studies of carbonyl compounds have drawn attention to competing fragmentation processes which have tended to be overshadowed by the McLafferty rearrangement. One of the most interesting of these fragmentations is the loss of an alkane fragment from ketones at low energy. The expulsion of propane from the molecular ion of di-isopropyl ketone[70] leads to the fourth most abundant ion in the mass spectrum. At low energies MIKES shows that it constitutes the main primary fragmentation pathway. Similar reactions are observed in

[64] C. Köppel, C. C. Van de Sande, N. M. M. Nibbering, T. Nishishita, and F. W. McLafferty, *J. Amer. Chem. Soc.*, 1977, **99**, 2883.
[65] H. Schwarz, C. Wesdemiotis, H. Heimbach, and K. Levsen, *Org. Mass Spectrometry*, 1977, **12**, 213.
[66] A. P. Bruins and N. M. M. Nibbering, *Org. Mass Spectrometry*, 1976, **11**, 271.
[67] H. Schwarz, *Org. Mass Spectrometry*, 1976, **11**, 313.
[68] H. Schwarz, U. Rapp, and B. Windel, *Org. Mass Spectrometry*, 1976, **11**, 219.
[69] R. J. Weinkam and J. Gal, *Org. Mass Spectrometry*, 1976, **11**, 188.
[70] J. F. Litton, T. L. Kruger, and R. G. Cooks, *J. Amer. Chem. Soc.*, 1976, **98**, 2011.

other ketones and in amines and alcohols (propan-2-ol loses methane), but not in McLafferty ions, which are enolic. Another reaction which competes with the McLafferty rearrangement has been observed using ^{2}H labelling of the acids (19) and (20), which cleave[71] by rearrangement of a β-hydrogen atom to the carbonyl oxygen followed by the shift of a hydrogen atom from the α- to the β-carbon before fragmentation. The McLafferty and double McLafferty ions from (21) do

(19)

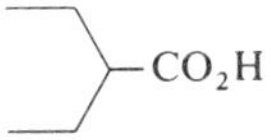

(20)

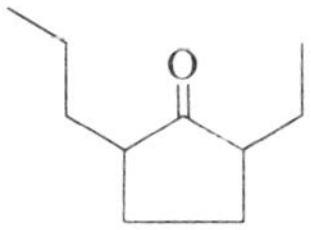

(21)

not behave in an ICR spectrometer[72] as either keto or enol species. By CA they have been shown not to be ketones, although the metastable ions appear to be enolic and behave like 'normal' McLafferty ions. This behaviour has been rationalized in terms of fragmentation following the facile α-cleavage of the cyclic ketone, with subsequent slow ring closures to yield the metastable ions which have low internal energies.

The fragmentation of allenes and acetylenes[73] involves analogues of the McLafferty rearrangement, with hydrogen migration to the central carbon of the allene system as well as sometimes to the terminal carbon. The elimination of ethene from nona-1,2-diene involves rupture of the C-2—C-3 bond.

The loss of acetic acid from 2-butyl acetate[74] has been shown to involve the loss of a hydrogen from C-3 in a low activation energy process which is insensitive to the energy of the ionizing electrons but sensitive to changes in the temperature of the ion source.

4 Techniques

The use of linked scans of electrostatic sector voltages and accelerating potentials has already been mentioned.[41] A number of linked-scan methods have been developed, including scans for producing CA results[75—77] using instruments of conventional geometry. Linked scans involving the magnetic sector field (B) can also be carried out. A critical evaluation of the various methods for observing metastable ions has been published.[78] The advantages and disadvantages of the techniques are discussed: the scan with B^2/E constant is the only one of the linked-scan techniques which preserves information on releases of kinetic energy.

[71] D. J. McAdoo, D. N. Witiak, and F. W. McLafferty, *J. Amer. Chem. Soc.*, 1977, **99**, 7265.
[72] J. R. Hass, R. G. Cooks, J. F. Elder, M. M. Bursey, and D. G. I. Kingston, *Org. Mass Spectrometry*, 1976, **11**, 697.
[73] J. R. Wiersig, A. N. Yeo, and C. Djerassi, *J. Amer. Chem. Soc.*, 1977, **99**, 532.
[74] M. M. Green, T. J. Mangner, S. P. Turner, and F. J. Brown, *J. Amer. Chem. Soc.*, 1976, **98**, 7082.
[75] A. F. Weston, K. R. Jennings, S. Evans, and R. M. Elliot, *Internat. J. Mass Spectrometry Ion Physics*, 1976, **20**, 317.
[76] D. L. Kemp, R. G. Cooks, and J. H. Beynon, *Internat. J. Mass Spectrometry Ion Physics*, 1976, **21**, 93.
[77] R. Robbiani, T. Kuster, and J. Seibl, *Angew. Chem. Internat. Edn.*, 1977, **16**, 120.
[78] R. K. Boyd and J. H. Beynon, *Org. Mass Spectrometry*, 1977, **12**, 163.

The use of MIKES (employing either EI or CI for the generation of the ions) for the direct identification of the components of crude mixtures,[79] for example of alkaloids,[80] has been shown to be effective.

A thin tungsten rod, freshly broken at abraded notches for each sample, has been suggested[81] as a simple emitter for FI and field desorption (FD) sources. Comparisons of the utility of FD spectra relative to other techniques (*e.g.* EI, FI) continue to be made, for example in the study of natural products.[82]

The CA spectra of the isometric pentenes[83] cannot be used to distinguish them, although the isomeric hexenes do show some differentiation. However, an elegant application of the ion–molecule reaction[84] between an alkene and vinyl methyl ether as a probe for the localization of double bonds has been demonstrated.

A novel consequence of the widespread use of mass spectrometry in the determination of the structures of complex organic molecules has been the development of synthetic schemes based on the combination of fragments observed in the mass spectra of the target molecules. Such 'retro mass spectrometric syntheses'[85,86] generally involve cycloaddition reactions.

79 T. L. Kruger, J. F. Litton, R. W. Kondrat, and R. G. Cooks, *Analyt. Chem.*, 1976, **48**, 2113.
80 T. L. Kruger, R. G. Cooks, J. L. McLaughlin, and R. L. Ranieri, *J. Org. Chem.*, 1977, **42**, 4161.
81 M. Anbar and G. A. St. John, *Analyt. Chem.*, 1976, **48**, 198.
82 G. W. Wood, N. Mak, and A. M. Hogg, *Analyt. Chem.*, 1976, **48**, 981.
83 T. Nishishita and F. W. McLafferty, *Org. Mass Spectrometry*, 1977, **12**, 75.
84 A. J. V. Ferrer-Correia, K. R. Jennings, and D. K. S. Sharma, *J.C.S. Chem. Comm.*, 1975, 973; *Org. Mass Spectrometry*, 1976, **11**, 867.
85 T. Kametani, T. Higa, C. V. Loc, M. Ihara, M. Koizumi, and K. Fukumoto, *J. Amer. Chem. Soc.*, 1976, **98**, 6186.
86 T. Kametani and K. Fukumoto, *Accounts Chem. Res.*, 1976, **9**, 316.

2 Physical Methods and Techniques Part (ii) *X*-Ray Crystallography

By M. B. HURSTHOUSE

Chemistry Department, Queen Mary College, Mile End Road, London E1 4NS

and

S. NEIDLE

Biophysics Department, King's College, 26—29 Drury Lane, London WC2B 5RL

Since 1975, when this topic was last reviewed, there have been many publications concerned with *X*-ray structure determination of organic compounds. Again, readers are referred to the extensive (and almost comprehensive) bibliographies available.[1,2] In addition the Cambridge Crystallographic Data Centre is developing a computerized information retrieval service, which will give scientists interested in molecular structure and conformation fast access to the wealth of information in the literature. There are already signs that intelligent use of this facility can reveal hitherto unsuspected features of, and variations in, the geometry of molecular fragments in series of related compounds.

1 Direct Methods

Direct methods of structure analysis which enable one to proceed from the diffraction data to a structure without either structural information or a heavy atom, continue to be routinely applied to most 'small' problems. The further development of these methods, for those structures that do not succumb to straightforward analysis, has received much attention in the period under review. Several series of papers have described various (at present) theoretical approaches to solving the phase problem *via* relationships between phases of considerable degrees of complexity.[3] Applications of quartet relationships to real structures have met with some success.[4,5] An alternative approach has been to develop the concept of 'magic integers',[6] which enable many possible phase sets to be examined using only a straightforward modification of the well-established

1 'Molecular Structures and Dimensions', Crystallographic Data Centre, Cambridge University, 1977 Vol. 8.

2 'Molecular Structures by Diffraction Methods', ed. L. E. Sutton and G. A. Sim, (Specialist Periodical Reports), The Chemical Society, London, 1976, Vol. 4.

3 See for example the sextet relationships of S. Fortier and H. Hauptmann, *Acta Cryst.*, 1977, **A33**, 694, 697; also the quintet relationships of G. Giacovazzo, *ibid.*, p. 944.

4 A. A. Freer, C. J. Gilmore, D. M. Mant, and J. McConwick, *J.C.S. Chem. Comm.*, 1977, 296; C. J. Gilmore, *Acta Cryst.*, 1977, **A33**, 712.

5 See for example, R. Bonnett, J. E. Davies and M. B. Hursthouse, *Nature*, 1976, **262**, 326.

6 P. Main, *Acta Cryst.*, 1977, **A33**, 750.

MULTAN procedure. It seems that these methods have the attraction of relative simplicity, and do actually solve difficult structures.[7] Of especial interest to the non-theoretical crystallographer is an analysis of a number of 'failures' of crystal-structure analysis;[8] it is apparent that exercise of a non 'black-box' mentality in the application of the direct methods computer programs available can often pay dividends in the solution of such structures. Some, even small structures, are, it seems, inherently impossible to solve by simple approaches. A procedure to overcome these problems has been suggested,[9] which 'should be able routinely to solve structures with 100 atoms'.

2 Electron Density Distribution

More papers have described the determination of electron density distributions in molecules. X-N maps for sulphamic acid, using *X*-ray and neutron data collected at 78 K, show strong bond populations in all bonds, with peaks in the S—N and S—O bonds similar to peaks in bonds between first row atoms only.[10] Broad peaks for the lone pairs on oxygen atoms suggest a situation intermediate between *sp* and sp^2 hybridization, and the absence of any build-up of electron density in the H···O bonds supports the idea of an electrostatic model. For the molecule of [$^2H_{12}$]-2,5-dimethylhex-3-yne-2,5-diol, the deformation density maxima are 0.58 and 0.36 e $Å^{-3}$ for the triple and single C—C bonds but only 0.09 for the C—O single bonds.[11] In the molecule of tetraphenylbutatriene, the central C=C bond length is 1.260 Å, in contrast to the outer bonds whose lengths are 1.348 Å.[12a] The deformation densities in the butadiene skeleton show that the π-lobes in adjacent C=C bonds are mutually perpendicular. The deformation densities in allene, carboxyl, and amide groups have also been determined for a complex between allenedicarboxylic acid and acetamide.[12b] In the molecule of *sym*-dibenzo-cyclo-octa-1,5-diene-3,7-diyne (1), the electron density around the centre of the triple bond is asymmetric, with a slight accumulation of charge inside the ring.[13]

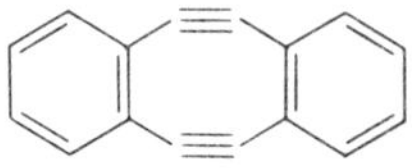

(1)

3 Molecular Conformations

The incorporation of potential energy calculations into structural studies is also attracting more and more attention. One of the main problems in this kind of work is the reliability and suitability of the interatomic potential functions used in the calculations and attempts are being made to determine these experimentally. An example of this approach is the calculation of such functions *via* an analysis of

[7] P. Main, *Acta Cryst.*, 1978, **A34**, 31.
[8] L. Lessinger, *Acta Cryst.*, 1976, **A32**, 538.
[9] M. M. Woolfson, *Acta Cryst.*, 1977, **A33**, 219.
[10] J. W. Bats, P. Coppens, and T. F. Koetzle, *Acta Cryst.*, 1977, **B33**, 37.
[11] R. B. Helmhold and A. Vos, *Acta Cryst.*, 1977, **A33**, 456.
[12] (*a*) Z. Berkovitch-Yellin and L. Leiserowitz, *Acta Cryst.*, 1977, **B33**, 3657; (*b*) Z. Berkovitch-Yellin, L. Leiserowitz, and F. Nader, *ibid.*, p. 3670.
[13] R. Destro, T. Pilati, and M. Simonetta, *Acta Cryst.*, 1977, **B33**, 447.

crystal lattice vibrational frequencies in some hydrocarbons.[14] Applications of potential energy calculations have been described for the system 1-ethyl-1-methyl-4-piperidinium perchlorate, where deformations of some interbond angles in the cation have been correlated with particular van der Waals' contacts,[15] and for the adrenaline molecule.[16] In this latter case, the conformation of the molecule found in some structures does not correspond to a minimum in the free-molecule conformational energy map and it has been shown that the lattice energy compensates for the loss in conformation energy. Structure analyses of some polyphenyls have also raised some interesting conformational points. At room temperature the crystal structure of *p*-terphenyl apparently contains only one molecule per asymmetric unit, in which some of the atoms show very large thermal motion. On cooling, two of the unit cell edges double up to give a superstructure where the end rings of the molecule are stabilized into one of the two potential wells associated with each C—C single bond.[17a] Similar effects are found for *p*-quaterphenyl[17b] and biphenyl.[17c]

A topic in many ways related to the study of molecular conformation is that of the geometry of strained molecules, particularly those containing ring systems. Classic examples of strained aromatic molecules are the cyclophanes, and studies of meta-[18] and para-cyclophane derivatives have been described. In the centrosymmetric, quadruple-layered cyclophane (2),[19] the two outer benzene rings are boat-shaped and the inner rings adopt a novel twist shape.

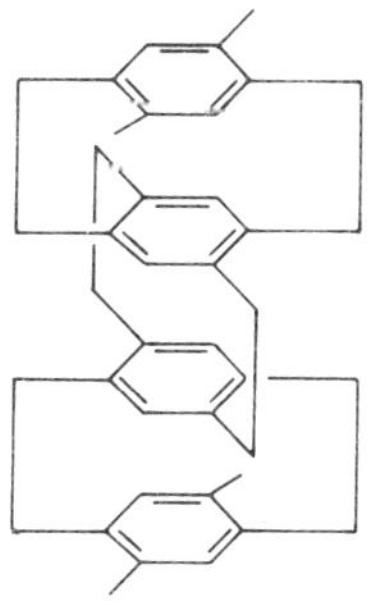

(2)

In the crowded fulvalene 1-methyl-2-(9′-fluorenylidene)-1,2-dihydropyridine,[20] the twist about the linkage bond is 35.7°, indicating a significant contribution from the form (3). Other examples of strained-ring molecules which show bond lengthening and/or angle strain are the *A-C′-C* stereoisomer of dodecahydrotrypticene,[21] the triterpene oxide companulin (4),[22] and 6,9-diaza-5,10-

[14] T. L. Starr and D. E. Williams, *Acta Cryst.*, 1977, **A33**, 771.
[15] W. Fedeli, F. Mazza, E. Giglio, C. Quagliata, and N. Scarcelli, *Acta Cryst.*, 1976, **B32**, 878.
[16] J. Caillet, P. Claveni and B. Pullmann, *Acta Cryst.*, 1976, **B32**, 2740.
[17] (*a*) J. L. Baudoir, Y. Delugeard, and H. Caillean, *Acta Cryst.*, 1976, **B32**, 150; (*b*) Y. Delugeard, J. Desucher and J. L. Baudoir, *ibid.*, p. 702; (*c*) G. P. Charbonneau and Y. Delugeard, *ibid.*, p. 1420.
[18] Y. Kai, N. Yasuoka, and N. Kasai, *Acta Cryst.*, 1971, **B33**, 754.
[19] H. Mizuno, K. Nishiguchi, T. Toyoda, T. Otsubo, S. Misumi, and N. Morimoto, *Acta Cryst.*, 1977, **B33**, 329.
[20] H. L. Ammon, *Acta Cryst.*, 1976, **B32**, 2693.
[21] A. Albinati, J. Brückner, and G. Allegra, *Acta Cryst.*, 1977, **B33**, 229.
[22] F. Mo, *Acta Cryst.*, 1977, **B33**, 641.

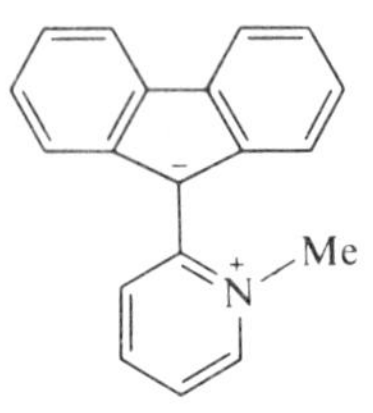

(3)

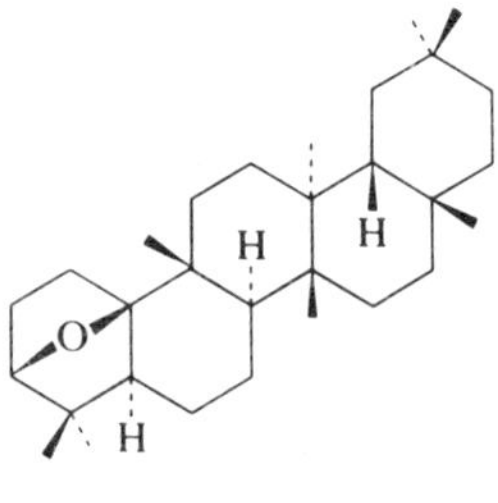

(4)

dioxotricyclo[7,3,0,0^{1,6}]dodecane (5), a molecule designed to contain non-planar amide bonds.[23] In the overcrowded, but planar, tetramethylcytoxine molecule (6)

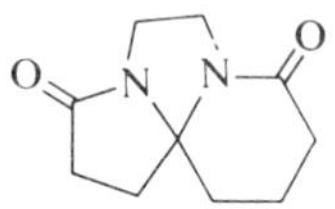

(5)

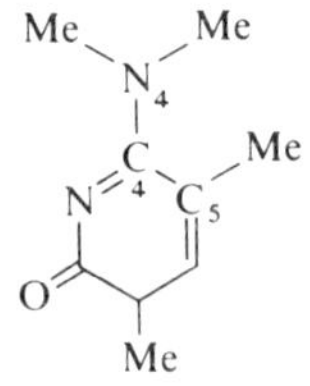

(6)

the steric strain arising out of the retention of the sp^2 geometry for the dimethylamino nitrogen atom and the Me···Me clash is taken up by the spreading of the angles at N-4, C-4, and C-5.[24] In our last Report, we mentioned three papers which described the correlation of crystal structure of some carboxylic acids with their reaction with amines and ammonia, and another example of this has been given. In the structures of (*S*)-(+)-2,2-diphenylcyclopropanecarboxylic acid and (*R*)-(+)-2,2-diphenyl-1-methylcyclopropanecarboxylic acid, the crystal packing, particularly the closed environment of the polar groups, explains the observed ditropic attack by NH_3 gas on single crystals.[25] In a similar study[26] the crystal structure analyses of dimethyl *meso*-(2*R*,3*S*,4*R*,5*S*)- and *meso*-(2*R*,3*R*,4*S*,5*S*)-α,α'-dimethyl-β,β'-dibromoadipate have assisted in the interpretation of the reaction of single crystals of the two compounds with gaseous NH_3 and amines to yield the corresponding diesters of (*E,E*)- and (*E,Z*)-butadiene by double dehydrobromination. In both cases, the reaction is topochemically controlled in that the configurations of the products correlate directly with the conformation of the starting molecules in their crystals. Other examples of this very interesting kind of study are the solid-state dehydration of 2-hydroxy-2-(β-benzoyl-β-phenylhydrazyl)indane-1,3-dione [(7a)→(7b)],[27] where compound (7a) also appears to be the first example of the type of intermediate found in reactions of carbonyl

[23] S. E. Ealick and R. Vander Helm, *Acta Cryst.*, 1977, **B33**, 76.
[24] J. K. Dattaguysta, W. Saenger, K. Bolewska, and I. Kulakowska, *Acta Cryst.*, 1977, **B33**, 85.
[25] C. C. Chiang, C.-T. Lin, A. H.-J. Wand, D. Y. Curtin and I. C. Paul, *J. Amer. Chem. Soc.*, 1977, **99**, 6303.
[26] D. Rabinovich and Z. Shakked, *Acta Cryst.*, 1977, **B33**, 809.
[27] S. A. Puckett, I. C. Paul, and D. Y. Curtin, *J. Amer. Chem. Soc.*, 1976, **98**, 787, 2371.

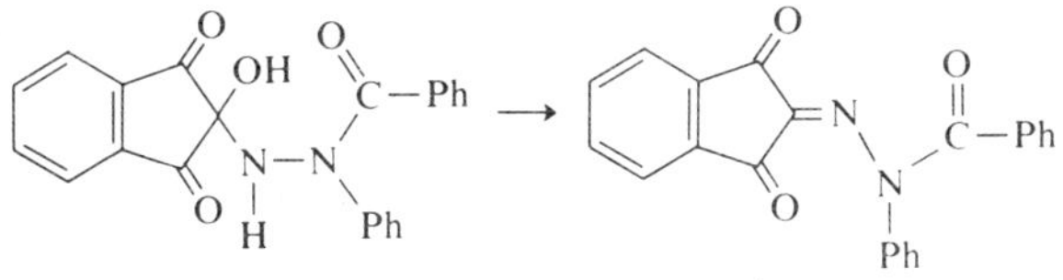

(7a) (7b)

compounds with arylhydrazines and related compounds, and the conversion in the solid state of the yellow form of 2-(4′-methoxyphenyl)-1,4-benzoquinone into the red form.[28] In this latter reaction, the process begins at a limited number of nucleation sites in the crystal and spreads by migration in well-defined fronts and the colours of the two forms correlate with the different types of packing adopted, particularly in relation to intermolecular π-interactions.

Peptide-containing antibiotics continue to be of major structural interest, partly on account of their interesting conformational and hydrogen-bonding properties. [Phe^4, Val^6]antamanide is a synthetic analogue of the cyclic decapeptide antamanide that possesses biological activity in respect of ion selectivity. It appears to have a novel type of (5 → 1) NH···O=C hydrogen bond.[29] An *X*-ray study of a degradation product has finally confirmed the structure of the complex antibiotic vancomycin (8);[30] suggestions have been made for a possible binding mode to

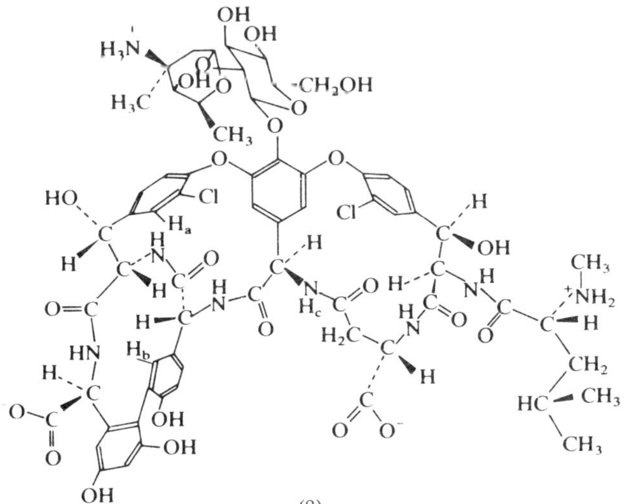

(8)

peptides terminating in the residues D-alanyl-D-alanine, a crucial process in its biological action. An α-helical conformation has been observed[31] for the amide units in the crystal structure of *N*-acetylactinobolin.

[28] G. R. Desirajin, I. C. Paul and D. Y. Curtin, *J. Amer. Chem. Soc.*, 1977, **99**, 1594.
[29] I. L. Karle, *J. Amer. Chem. Soc.*, 1977, **99**, 5152.
[30] G. M. Sheldrick, P. G. Jones, O. Kennard, D. H. Williams, and G. A. Smith, *Nature*, 1978, **271**, 223.
[31] R. B. Von Dreele, *Acta Cryst.*, 1976, **B32**, 2852.

The linear tetrapyrrole bilirub has a compact conformation in the solid state which is stabilized by intramolecular hydrogen bonds,[5] a feature shown by both independent molecules in the asymmetric unit. This structure is not dissimilar to that of a biladiene analogue (9)[32]—both have the two planar pyrromethane groups hinged together at right angles.

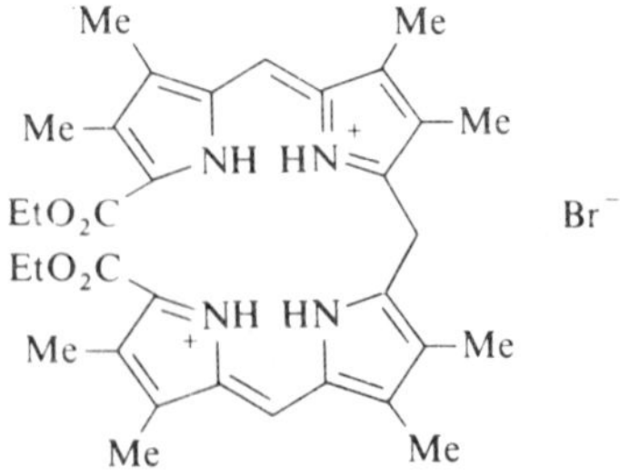

(9)

The remarkable crystal structure of anhydrous cholesterol contains eight independent molecules in the asymmetric unit,[33] a total of 224 non-hydrogen atoms. Although the skeletal frameworks are all the same, there are considerable variations in the tail conformations. The packing of molecules in the crystal lattice differs from that in cholesterol monohydrate,[34] which has a bilayer-type structure with the rings almost parallel to one another. An X-ray analysis of 3β-acetoxy-17aα-(2-acetoxyethoxy)-17α,17aβ-dimethyl-D-homoandrost-5-en-17β-ol (10) has disproved the notion of isolating rotamers that have restricted rotation about the steroidal C-17—C-20 bond.[35] The structure of prostaglandin $F_{2\alpha}$(11) has two

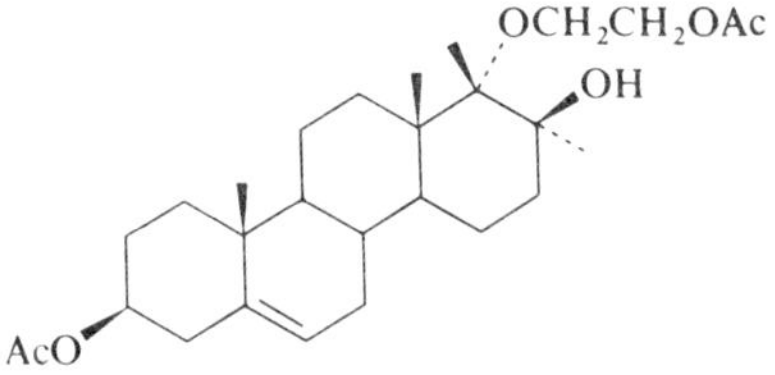

(10)

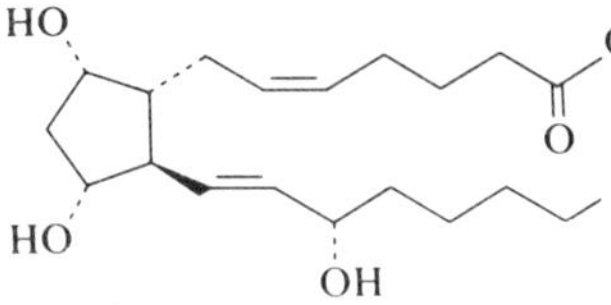

(11)

independent molecules in the asymmetric unit: both have the side-chains aligned roughly parallel.[36] A model for the biological recognition by prostaglandin receptors has been suggested on the basis of this structure, which explains much of the binding data.

The structures of a number of nucleic acid constituents have been reported, some of which are notable for their broader biological significance and interest. A new pattern of hydrogen-bonding has been observed in the structure of 5-nitro-6-methyluracil,[37] which has adjacent bases perpendicular to one another, in contrast

[32] G. Struckmeier, U. Thewalt, and J. Engel, *J.C.S. Chem. Comm.*, 1976, 963.
[33] H. S. Shieh, L. G. Hoard, and C. E. Nordman, *Nature*, 1977, **267**, 287.
[34] B. M. Craven, *Nature*, 1976, **260**, 727.
[35] Y. Osawa, T. Makino, and C. M. Weeks, *J.C.S. Chem. Comm.*, 1976, 990.
[36] D. A. Langs, M. Erman, and G. T. DeTitta, *Science*, 1977, **197**, 1003.
[37] R. Parthasarathy and T. Srikrishnan, *Acta Cryst.*, 1977, **B33**, 1749.

to the normal parallel arrangements. The structures of the hypermodified nucleosides t^6A and g^6A (the former occurs in the anticodon loop of tRNA), have suggested[38] conformational alterations in this loop that depend on the codon being read. The interactions of drugs with nucleic acids have been investigated *via* the structures of several model systems such as ethidium-dinucleoside[39,40] and 9-aminoacridine-dinucleoside[41] complexes, and an extrapolation from these models to the polymers has been attempted.[42] The crystal structure[43] of a proflavine-dinucleoside phosphate (CpG) has, however, cast some doubt on the general validity of these conclusions; in particular the conformational changes induced in the miniature double helix can be related to even small alterations in the base pair geometry.[44]

[38] R. Parthasarathy, J. M. Ohrt, and G. B. Chheda, *Biochemistry*, 1977, **16**, 4999.
[39] C. C. Tsai, S. C. Jain, and H. M. Sobell, *J. Mol. Biol.*, 1977, **144**, 301.
[40] S. C. Jain, C. C. Tsai, and H. M. Sobell, *J. Mol. Biol.*, 1977, **144**, 317.
[41] T. D. Sakore, S. C. Jain, C. C. Tsai, and H. M. Sobell, *Proc. Nat. Acad. Sci. U.S.A.*, 1977, **74**, 188.
[42] H. M. Sobell, C. C. Tsai, S. C. Jain and S. G. Gilbert, *J. Mol. Biol.*, 1977, **114**, 333.
[43] S. Neidle, A. Achari, G. L. Taylor, H. M. Berman, H. L. Carrell, J. P. Glusker, and W. C. Stallings, *Nature*, 1977, **269**, 304.
[44] H. M. Berman, S. Neidle, and R. K. Stodola, *Proc. Nat. Acad. Sci. U.S.A.*, 1978, **75**, 828.

2 Physical Methods and Techniques Part (iii) Ultraviolet and Visible Spectroscopy of Bio-organic Molecules

By P.-S. SONG

Department of Chemistry, Texas Tech University
Lubbock, Texas 79409, U.S.A.

1 Introduction

Although recent developments in various spectroscopic properties of organic molecules have been reviewed in these Reports, there has not been a specific review of the literature on the u.v. and visible spectroscopy of bio-organic molecules in recent years. The present Report attempts to review selected literature in this field. In reviewing the literature, luminescence spectroscopy of biomolecules will be largely excluded, unless it is directly relevant to the discussion of the u.v.–visible spectroscopy of molecules under consideration. Furthermore, intermolecular interactions (*e.g.* chlorophyll aggregates, excitons, complexes, *etc.*) and effects of proteins on u.v.–visible spectra of chromophores (*e.g.* retinal in rhodopsin) will not be discussed.

2 Cyclic Polyenes

Porphyrins.—Porphyrins, including metalloporphyrins, can be treated as cyclic conjugated polyenes. Chlorophylls (chl) and bacteriochlorophylls (Bchl) are structurally and spectroscopically derived from the chlorin and bacteriochlorin chromophores, respectively. The absorption spectra of these compounds are characterized by visible bands in the 500—800 nm region and near-u.v. bands (Soret) in the region of 350—450 nm. The latter are considerably more intense ($\varepsilon \approx 1—3 \times 10^5, f = 1—2$) than the former ($\varepsilon \approx 1—8 \times 10^4, f = 0.01—0.3$), particularly in symmetrical porphyrins. This spectral shape contrasts with the well known spectral feature (*i.e.* intense visible band with weak or no near-u.v. bands) of linear polyenes such as carotenoids.

The genealogy of theoretical models for explaining the visible (and u.v.) spectral characteristics of porphyrins is as follows:

Free-electron perimeter model of Simpson (1949) and Kuhn (1949)

↓

Hückel molecular orbital (MO) model of Longuet-Higgins, Rector, and Platt (1950)

↓

Four-orbital, configuration interaction (CI) model of Gouterman (1959)

↓

SCF MO CI Pariser–Parr–Pople model of Weiss, Kobayashi, and Gouterman (1965)

↓

All-valence-electron MO model of Schaffer and Gouterman (1972)

↓

Ab initio SCF CI model of Christoffersen and his co-workers (1977)[1]

In the free-electron perimeter model, porphyrin of D_{4h} symmetry can be viewed as an 18π-electron cyclic polyene with an approximately circular field around the perimeter. The highest occupied (orbital angular momentum quantum number $q = \pm 4$) and lowest empty orbital ($q = \pm 5$) levels are doubly degenerate. Between these degenerate levels, four excited electron configurations are possible, accounting for $\pi \rightarrow \pi^*$ transitions to doubly degenerate $B_{x,y}$ (from $\Delta q = \pm 1$ configurations) and $Q_{x,y}$ (from $\Delta q = \pm 9$ configurations) states. These transitions are assigned to the Soret and visible bands, respectively.

The degenerate Q and B bands split into Q_x, Q_y and B_x, B_y, respectively, as the symmetry of the metalloporphyrin-like system descends from D_{4h} to D_{2h} (*e.g.* free-base porphyrin) and C_{2v} (*e.g.* chlorin). The splits between Q_x and Q_y and B_x and B_y are about 3000 cm^{-1} and 250 cm^{-1}, respectively, as predicted from semi-empirical SCF MO CI calculations including electron correlation.[2] Previous MO calculations and *ab initio* CI[1] tended to overestimate the B_x–B_y split.

In addition to the four-state ($Q_{x,y}$ and $B_{x,y}$) absorption spectrum, one prominent vibronic satellite appears for each Q_x 0–0 and Q_y 0–0 origin, thus producing the well known visible bands I(Q_x, 0–0), II(Q_x, 1–0), III(Q_y, 0–0), and IV(Q_y, 1–0) for free-base porphyrins. Four similar bands (I—IV) appear in the spectra of chlorins and bacteriochlorins, except that Bands I and III are designated as Q_y(0–0) and Q_x(0–0), respectively, in accordance with the polarization axes. Qualitatively speaking, $Q_{x,y}$ and $B_{x,y}$ bands of porphyrins may be compared with $L_{a,b}$ (from $\Delta q = \pm 5$ configurations) and $B_{a,b}$ (from $\Delta q = \pm 1$ configurations) bands of naphthalene and other polyacenes, respectively.

Although the above model is qualitatively satisfactory in describing the major spectral features of Soret and visible bands in porphyrins, the weak- and near-u.v. ($N_{x,y}$, $L_{x,y}$, $M_{x,y}$, *etc.*) bands, $n \rightarrow \pi^*$ transitions, and effects of metal ions are not at all represented. The assignment of bands II and IV as vibronic satellites of the $Q_{x,y}$ transitions is by no means definitive.[3—5] Alternatively, band IV may be assigned to a third $\pi \rightarrow \pi^*$ transition (*e.g.* a band at 485 nm with $\varepsilon = 3.4 \times 10^4$ for meso-tetraphenylporphin and at 495 nm for octaethylporphin in the vapour phase).[6] The SCF MO CI theory also predicts such a transition,[7] but *ab initio* CI theory (*see*

[1] (*a*) J. D. Petke, G. M. Maggiora, L. L. Shipman, and R. E. Christoffersen, *J. Mol. Spectroscopy.*, in press; (*b*) D. Spangler, G. M. Maggiora, L. L. Shipman, and R. E. Christoffersen, *J. Amer. Chem. Soc.*, 1977, **99**, 7470, 7478.

[2] K. Tomono and K. Nishimoto, *Bull. Chem. Soc.* Japan. 1976, **49**, 1179.

[3] R. Plus and M. Lutz, *Spectroscopy Letters*, 1974, **7**, 73.

[4] J. Bohandy, B. F. Kim, and C. K. Jen, *J. Mol. Spectroscopy*, 1973, **43**, 199.

[5] A. H. Corwin, A. B. Chivois, R. W. Poor, D. G. Whitten, and E. W. Baker, *J. Amer. Chem. Soc.*, 1968, **90**, 6577.

[6] J. B. Kim, J. J. Leonard, and F. R. Longo, *J. Amer. Chem. Soc.*, 1974, **94**, 3986.

[7] A. J. McHugh, M. Gouterman, and C. Weiss, jun. *Theor. Chim. Acta*, 1972, **24**, 346.

below) does not support this. Furthermore, this assignment is not likely in view of the low-temperature Shpol'skii spectroscopic analysis of the vibronic origin for this band.[8]

The *ab initio* CI calculations[1] predict that one electronic transition (to the doubly degenerate 2^1E state) in Mg^{2+}-porphyrin accounts for the major Soret intensity, while the higher-lying states and several forbidden $\pi \rightarrow \pi^*$ transitions make relatively minor contributions to the Soret band at 387 nm. The *N* band at *ca.* 327 nm is identified as arising from $\pi \rightarrow \pi^*$ transition to another doubly degenerate 4^1E state. In free-base porphyrin, five $\pi \rightarrow \pi^*$ transitions (to 2^1B_{2u}, 2^1B_{3u}, 1^1B_{1u}, 3^1B_{2u}, and 3^1B_{3u}) are predicted to contribute to the Soret band intensity with a maximum at 385 nm. The *N* band at 315 nm may consist of two allowed transitions, 4^1B_{3u} and 4^1B_{2u}. Absorption bands of porphyrins in the vacuum-u.v. region are apparently diffuse.[9] No evidence for Rydberg states was revealed in the vacuum-u.v. spectra of tetraphenylporphin and its metalloporphyrin derivatives.[10] The vapour-phase absorption bands of porphins from 640–210 nm have been assigned to Q_x, Q_y, *B*, *N*, *L*, and *M*(*Q*, *B*, *N*, and *L* for Mg^{2+}-porphin).

The $n \rightarrow \pi^*$ transitions in porphyrins including chlorins can originate from the pyridine-like nitrogen *n*-orbitals and from formyl or aza substituents at the cyclopentanone ring and methine bridge positions. A different definition of $n \rightarrow \pi^*$ assignment for porphyrins and chlorins has also been described.[11] A PPP MO variant taking into account *n*-bonding orbital electrons predicts an $n \rightarrow \pi^*$ transition of free-base porphin at 316 nm.[12] Similarly, the extended HMO method predicts an $n \rightarrow \pi^*$ transition in the near u.v.[7] in contrast to the CNDO/S prediction at 269 nm for porphins and chlorins.[13] There is indirect evidence that the Soret band broadening in tetra-azaporphin and phthalocyanine is attributable to the underlying n, π^* state which is localized at the methine bridge.[14] Linear dichroism of free-base and Mn^{2+}-porphyrins shows all transitions in the range 350—900 nm to be in-plane polarized,[15] in contrast to an earlier report which showed a negative dichroism at *ca.* 400 nm in tetraphenylporphin. It now appears, however, that dichroism at 400 nm is definitely negative[16] and that it represents either $n \rightarrow \pi^*$ or a vibronic component of $\pi \rightarrow \pi^*$ origin. Various studies do not generally support the $n \rightarrow \pi^*$ assignment for the visible bands I and III, which has been made by some workers.

It is clear that information concerning the location and intensity of $n \rightarrow \pi^*$ bands in porphyrins (and chlorophylls, *see below*) is either lacking or only tentative. However, it is quite likely that $^1n, \pi^*$ state(s) lie in the vicinity of the Soret and near u.v. region, resulting in band broadening of the $\pi \rightarrow \pi^*$ transitions.[17] *Ab initio* CI calculation also predicts an $n \rightarrow \pi^*$ (1^1B_{1u}) transition under the Soret band.[1]

[8] A. T. Grudyushko, K. N. Solov'ev, and A. S. Starukhin, *Optika i Spectroskopiya*, 1976, **40**, 267.
[9] B. H. Schechtman and W. E. Spicer, *J. Mol. Spectroscopy*, 1970, **33**, 28.
[10] L. Edwards, D. H. Dolphin, M. Gouterman, and A. E. Adler, *J. Mol. Spectroscopy*, 1970, **35**, 90; *ibid.*, 1971, **38**, 16.
[11] S. G. Boxer, G. L. Closs, and J. J. Katz, *J. Amer. Chem. Soc.*, 1974, **96**, 7058.
[12] M. Sundbom, *Acta Chem. Scand.*, 1968, **22**, 1317.
[13] G. M. Maggiora and L. J. Weimann, *Chem. Phys. Letters*, 1973, **22**, 297.
[14] M. Gouterman, in 'Excited States of Matter', ed. C. W. Shoppee, Texas Tech. University Press, Lubbock, Texas, 1973, p. 63.
[15] R. Gale, R. D. Peacock, and B. Samori, *Chem. Phys. Letters*, 1976, **37**, 430.
[16] B. Norden and A. Davidson, *Chem. Phys. Letters*, 1976, **37**, 433.
[17] K. N. Solov'ev, A. T. Gradushko, and M. P. Tsvirko, *J. Luminescence*, 1976, **14**, 365.

The visible spectrum of porphin with its characteristic four-band (I—IV) system changes to the typical D_{4h}-porphyrin spectrum of two-band (α and β bands) upon metal substitution,[18] as indicated earlier. The effects of different metal substitutions on the porphyrin absorption spectrum are partly attributed to perturbations on orbital energy levels of b_1 (or a_{2u} in symmetry notation), $c_1(e_g)$ and $c_2(e_g)$ molecular orbitals which are substantially localized on the central N atoms. Effects of different metal ions on the D_{4h} porphyrin spectrum are generally not drastic, with some exceptions such as Mn^{2+}-, Fe^{2+}-, and Fe^{3+}-porphyrins. Band I (Q_x, 0–0) usually shifts to the red with increasing atomic number of the central metal ion, but spectral shifts of $Q \leftarrow A$ and $B \leftarrow A$ bands in both direction and magnitude are determined by ligand (L)–porphyrin interactions in L-Me-porphyrin-type complexes (for example, see ref. 5). Also, the B_x–B_y band split by fifth and sixth ligands (*e.g.* cytochome *c*) could be larger than the Q_x–Q_y split, in contrast to the case of free-base porphyrin.

In general, high-spin metalloporphyrins show stronger charge-transfer (CT) bands than low-spin complexes, as has been found for Fe^{2+}-tetraphenylporphyrin.[19] CT bands arising from the occupied porphyrin MO (a_{2u} in particular) to the empty metal *d*-orbitals (e_g) may occur throughout the near-u.v.–near-i.r. region, depending on the strength of ligand field and the nature of ligands. For example, porphyrin $\rightarrow Fe^{3+}$ CT bands have been assigned for Fe^{3+}-myoglobin using polarized absorption spectroscopy and the weak band at 695 nm in Fe^{3+}-cytochrome *c* (low spin) was found to be out-of-plane polarized, consistent with the ligand (a_{2u}) $\rightarrow$ metal (d_{z^2}; a_{1g}) CT transition,[20] whereas ligand (a_{2u} or a_{1u}) $\rightarrow$ metal (d_π; e_g) CT transitions are predicted to occur in the near-u.v. region (320—400 nm) in Cr^{3+}-tetraphenylporphyrin in chloroform.[21]

The $d \rightarrow d$ transitions are generally so weak that their resolution is usually masked by the strong porphyrin bands. Metalloporphyrins of d^3, d^4, d^6, and d^8 configurations show $d \rightarrow d$ transitions of low frequency in the near-i.r. or i.r. region (*e.g.* Cr^{3+}-tetraphenylporphyrin[21]), as can be predicted from porphyrin being a weak-field ligand. The $d \rightarrow d$ transitions for various metalloporphyrin complexes have been assigned, but the assignment of these transitions are often not unambiguous owing to CT and porphyrin bands which are more intense.

It is well known that the visible band, $Q \leftarrow A$, of porphyrins with D_{4h} symmetry can be described as a planar oscillator as a consequence of the two-fold degeneracy for $Q_{x,y}$ states. Thus, the degree of fluorescence polarization with respect to the visible absorption in porphyrins is independent of excitation wavelengths within the visible band and is in the neighbourhood of $\frac{1}{7}$, a value characteristic of a planar oscillator.[22] The fluorescence polarization degree increases substantially as the porphyrin symmetry is lowered from D_{4h} to D_{2h} or C_{2v}, as in free-base porphins and chlorins, respectively. The $S_2 \leftarrow S_0$ transition (Q_y in free-base porphyrins and Q_x in hydrogenated porphyrins) is then polarized perpendicular to the $S_1 \leftarrow S_0$ absorption or fluorescence polarization axis, yielding a negative degree of

18 M. Gouterman, L. K. McCaffery, and M. D. Rowe, *J.C.S. Dalton*, 1972, 596.
19 H. Kobayashi and Y. Yanagawa, *Bull. Chem. Soc.*, Japan, 1972, **45**, 450.
20 W. A. Eaton and R. M. Hochstrasser, *J. Chem. Phys.*, 1967, **46**, 2533; *ibid.*, 1968, **49**, 985.
21 M. Gouterman, L. K. Hanson, G. E. Khalil, and W. R. Leenstra, *J. Chem. Phys.*, 1975, **62**, 2343.
22 G. D. Yegorova, V. A. Mashenkov, K. N. Solov'ev, and N. A. Yushkevich, *Biofizika*, 1973, **18**, 40.

fluorescence polarization. However, fluorescence polarization does not yield absolute orientations of these transition moment dipoles.

The lowest transition, $Q_x \leftarrow A$, in free-base porphins is theoretically predicted to be polarized along the NH—NH axis (x) and the $Q_y \leftarrow A$ transition is perpendicular to the former. This is consistent with the effects of substitution along the x- and y-axes on the hyperchromicity of the Q_x (band I) and Q_y (band III) bands, respectively, and with the polarized reflectance spectroscopy of free-base tetraphenylporphin single crystal[23] and the polarized quasi-line absorption spectrum of Zn^{2+}-porphin in triphenylene.[24,25]

Vibronic coupling between electronic states (*e.g.* Q state with higher electronic states) can be invoked to explain mixed polarization characteristics of the porphyrin absorption bands. According to the vibronic theory, the Q_x 1–0 band gains its intensity from the y-component of the Soret transitions through b_{1g} vibration-induced mixing, whereas the a_{1g} vibration mixes the x-component of the Soret band with the Q_x 1–0 band.[26] High-resolution vibrational analysis of the absorption spectra,[25–27] site-selection fluorescence spectroscopy,[28] and resonance Raman spectroscopy[29] provide powerful tools to carry out vibronic analysis of the visible and Soret bands of metalloporphyrins.

Fluorescence polarization degree varies over the Soret band of porphins (*e.g.* octaethylporphin), suggesting that there are two more or less perpendicularly polarized transitions to B_x and B_y states,[17] but other electronic transitions which are not readily resolved by the fluorescence polarization technique are also likely to contribute to the Soret band, as mentioned earlier.

Chlorophylls.—Chls and Bchls show absorption spectra similar to those of chlorin and bacteriochlorin, respectively. Chl-*a* shows the characteristic four visible bands system [band I at 660 nm = Q_y (0–0); II at 615 nm = Q_y (1–0); III at 575 nm = Q_x (0–0); IV at *ca.* 530 nm = Q_x (1–0)] and the Soret band (429- and 410 nm) with absorbance peak ratios of 1.0:0.16:0.1:0.06:1.31:0.87 in ether. The oscillator strength, *f*, of the Q_y (0–0), averaged from absorption spectra in different solvents, is 0.151.[30] Attempts have been made to assign the u..v. bands of chls in the 200—410 nm region,[31,32] but they remain to be established. The $Q_y \leftarrow A$ band of Bchl-*a* occurs at 769 nm and the Soret bands at 392 and 357 nm with absorbance peak ratios of 1:0.55:0.8 in ether.

Protochl-*a*, chl-*a*, and Bchl-*a* assume approximately circular, triangular, and rectangular π-electron fields, respectively. Thus, the $Q_y \leftarrow A$ band shows a progressive red shift, whereas the $Q_x \leftarrow A$ band shows a relatively small blue shift. The Q_y–Q_x split is also expected to increase in going from protochl-*a* to Bchl-*a*

[23] B. G. Anex and R. S. Umans, *J. Amer. Chem. Soc.*, 1964, **86**, 5026.
[24] B. F. Kim, J. Bohandy, and C. K. Jen, *Spectrochim. Acta*, 1974, **30A**, 2031.
[25] B. F. Kim and J. Bohandy, *J. Mol. Spectroscopy*, 1977, **65**, 90.
[26] Y. J. Aronowitz and M. Gouterman, *J. Mol. Spectroscopy*, 1977, **64**, 267.
[27] J. A. Shelnutt, D. C. O'Shea, N. Y. Yu, L. D. Cheng, and R. H. Felton, *J. Chem. Phys.*, 1977, **66**, 3387.
[28] J. Fünfschilling and D. F. Williams, *Photochem. and Photobiol.*, 1975, **22**, 151; *ibid.*, 1977, **26**, 109.
[29] T. G. Spiro, *Accounts Chem. Res.*, 1974, **7**, 339.
[30] L. L. Shipman, *Photochem. and Photobiol.*, 1977, **26**, 287.
[31] C. Weiss, jun. *J. Mol. Spectroscopy*, 1972. **44**, 37.
[32] P.-S. Song, T. A. Moore, and M. Sun, in 'The Chemistry of Plant Pigments' ed. C. O. Chichester, Academic Press, New York, 1972, pp. 33—74.

Configuration analysis of wavefunctions for chl-*a* and Bchl-*a* suggests that the Q_y states of these molecules are very similar, as the peripheral C=C (3b—4b) bond in chl-*a* is not locally excited to any appreciable extent. The isocyclic ring C=O group is also unimportant in substantially perturbing the $Q_y \leftarrow A$ transition of chlorin and bacteriochlorin macrocycles. Furthermore, the vinyl substituent does not significantly contribute to the Q_y state functions of chls. These analyses are, therefore, consistent with the use of the four-otbital model of porphyrins for chls in that the low-energy electronic states $Q_{x,y}$ and $B_{x,y}$ can be described by two degenerate MOs of the macrocylic ring.[33]

It is possible that a shallow hump at 325 nm in the absorption spectrum of chl-*a* is an $n \rightarrow \pi^*$ band. Since cyclopentanone shows the $n \rightarrow \pi^*$ band in this general region, this possibility is energetically feasible. However, the intensity at 325 nm is too high to be attributed to an $n \rightarrow \pi^*$ band of chl-*a*, although perturbations generated by nearby states can enhance its intensity considerably. It should also be noted that in deuterioporphyrins $N \leftarrow A$ ($\pi \rightarrow \pi^*$) transitions with $\varepsilon \leqslant 10^4$ occur in the same region (320—350 nm). ^{15}N N.m.r. studies[11] suggest that the lowest $n \rightarrow \pi^*$ transition is localized in ring IV of chl-*a* and pheophytin-*a*, but the suggestion that the $n \rightarrow \pi^*$ transition might occur near the Q_y band with relatively strong intensity has not been directly confirmed.

The absorption band maximum (I, Q_y, 0–0) of chl-*a* is polarized ($P_0 > 0.4$) parallel to the fluorescence emission oscillator and its polarization approximately corresponds to the y-axis according to the PPP MO calculation.[34] Band II (Q_y, 1–0) at 615 nm is made up of two components, one at *ca.* 632 nm with $P_0 \sim 0$ and another at 613 nm with $P_0 \sim 0.28$. These bands may be assigned to Q_y (1–0) and Q_y (2–0), respectively. If this assignment is correct, the 575 nm band with $P_0 \sim 0.03$ is most likely assigned to Q_x (0–0) (band III). M.c.d. results are consistent with the assignment here.[35] Pheophytin-*a*, chlorin, and porphins at 77 K also show negative polarization for this band. The $Q_x \leftarrow A$ transition is thus polarized (approximately along the x-axis) nearly perpendicular to the $Q_y \leftarrow A$ transition.[34] However, Shipman *et al.*[36] assign the 635 nm component (or *ca.* 632 nm[34]) of chl-*a* in ethanol at 77 K to the Q_x(0–0) on the basis of spectral analogies with Bchl-*a* and an *apparent* shift of the room-temperature absorption band III (Q_x, 0–0) to *ca.* 635 nm at 77 K. Although this is possibly due to the effect of an anisotropic polarizability of the medium with increased refractive index at 77 K, preferentially red-shifting the Q_x (0–0) band, the Q_y (1–0) assignment for this band with a negative polarization degree is more likely on the basis of the following observation.

High-resolution fluorescence excitation and polarization spectra of chl-*a* in ether at 15 K revealed a well-resolved peak at 632 nm.[37] If this peak were Q_x(0–0), one would not see its mirror image in the fluorescence spectrum. However, the 723 cm^{-1} vibrational mode of the 632 nm excitation band shows its mirror image (729 cm^{-1}) in the fluorescence emission spectrum at 15 K.

[33] P.-S. Song, C. A. Chin, I. Yamazaki, and H. Baba, *Internat. J. Quantum Chem. Quantum Biol. Symp.*, 1977, **4**, 305; *ibid.*, 1976, **3**, 89.
[34] P. Koka and P.-S. Song, *Biochim. Biophys. Acta*, 1977, **495**, 220.
[35] C. Houssier and K. Sauer, *J. Amer. Chem. Soc.*, 1970, **92**, 779.
[36] L. L. Shipman, T. M. Cotton, J. R. Norris, and J. J. Katz, *J. Amer. Chem. Soc.*, 1976, **98**, 8222.
[37] W. W. Mantulin and P.-S. Song, unpublished results.

The minimum polarization at 435 nm in chl-*a* is attributable to the $B_x \leftarrow A$ transition, in agreement with theoretical predictions and m.c.d. results.[35] Absorption and polarization characteristics of protochl-*a*, chl-*b*, and pheophytins are qualitively similar to those of chl-*a*. In particular, chl-*b* shows a fluorescence excitation polarization spectrum closely resembling that of chl-*a* in ether at 77 K, except that the Q_y (0–0) band shows a lower P_0 value (0.35) than that of chl-*a*.[38] The large red shift of Q_y relative to that of chl-*a* is accompanied by increase in the Q_y–Q_x gap, which contributes to improved resolution of fluorescence polarization in Bchl-*a*. Thus, P_0 for Q_y (0–0) of Bchl-*a* is 0.4 in ethanol at 77 K, whereas P_0 for Q_x (0–0) at 625 nm is 0.28.[39]

The above discussion is based on the simplifying assumption that each of the resolved vibronic bands, Q_y(0–0) and (1–0), Q_x (0–0) and (1–0), contains no strong contribution due to other overlapping vibronic transitions. However, low-temperature site-selection spectroscopy of chl-*a* and -*b* in which the fluorescence spectra are resolved into a number of vibronic components[28] clearly indicates complexity in interpreting the polarization spectra of chlorophyls and porphyrins.

Corrins and Tetrapyrroles.—Corrins show two visible bands designated as α- and β-bands, along with the intense near-u.v. γ-band ($\approx$Soret) and moderately intense u.v. bands. The α- and β-bands are the 0–0 and 1–0 components of the first allowed electronic transition. Kuhn (1959) was the first to describe the corrin spectrum in terms of the free electron model. The four-orbital model equivalent to the Gouterman model for porphyrins was proposed by Day[40] within the framework of the HMO and PPP MO approximations. The latter was also applied to corrins by others.[41,42] The HMO method was found to be inadequate for describing the spectral intensities of corrins. On the basis of these results, it is now well established that the visible and near-u.v. band intensities in corrins are mainly of the $\pi \rightarrow \pi^*$ type. This conclusion is generally valid for corrinoids such as vitamin B_{12}, since 'free-base' corrins (*e.g.* descobalt-B_{12}) show the characteristic visible and near-u.v. spectral intensities. The $\pi \rightarrow \pi^*$ assignment for the entire u.v.–visible spectrum of free-base corrins is clearly confirmed by the polarized phosphorescence excitation spectrum of descobalt-B_{12} which shows $P_0 < 0.1$ for all visible and u.v. peaks with respect to the out-of-plane polarized phosphorescence of $^3\pi, \pi^*$ symmetry.[43]

The low-temperature absorption spectrum of descobalt-B_{12} in ethanol shows well resolved peaks at 525, 493, 465(sh), 435(sh), 395, 375, 358(sh), 326, 314, 300(sh), 285, 280(sh), 268, 260(sh), and 238 nm with peak ratios of 1: 0.76: 0.36: 0.12: 0.16: 0.17: 0.12: 2.09: 1.04: 0.64: 0.48: 0.54: 1.39: 0.76: 0.43. Although calculated oscillator strengths are invariably greater than experimental *f* values, the *f* ratios produced by the PPP SCF MO CI calculation are 1(Q_y):0.07(Q_x):6.8 (near u.v. $B_{x,y}$) which agrees with the observed ratios of 1:0.07:6.5 in descobalt B_{12}. In analogy to the porphyrin notation, we have assigned

[38] C. A. Chin, Ph.D. Thesis Texas Tech University, Lubbock, Texas, 1975.
[39] T. G. Ebrey and R. K. Clayton, *Photochem. and Photobiol.*, 1969, **10**, 109.
[40] P. Day, *Theor. Chim. Acta*, 1967, **7**, 328.
[41] H. Johansen and L. L. Ingraham, *J. Theor. Biol.*, 1969, **23**, 191.
[42] P. O. Offenhartz, B. H. Offenhartz, and M. M. Fung, *J. Amer. Chem. Soc.*, 1970, **92**, 2966.
[43] R. D. Fugate, C. A. Chin, and P.-S. Song, *Biochim. Biophys. Acta*, 1976, **421**, 1.

the 525 nm peak and its vibrational satellites at 493, 465, and 435 nm to Q_y 0–0, 1–0, 2–0, and 3–0, respectively, while the 395, 375, and 358 nm peaks are designated as Q_x 0–0, 1–0, and 2-0, respectively. The subscripts *x* and *y* correspond to the polarization axes.[43] The 362 nm maximum may then be designated as a B_x equivalent. These designations are consistent with the relative polarization data discussed below.

The lowest-energy transition Q_y, arises mainly from the 7(HOMO)$\rightarrow$8(LEMO) configuration, while Q_x and B_x transitions are contributed to mainly by 6$\rightarrow$8 and 7$\rightarrow$9 configuration mixing.[41,42] The fourth ($S_4 \leftarrow S_0$) and fifth ($S_5 \leftarrow S_0$) transitions are mainly due to 7$\rightarrow$10 and 6$\rightarrow$9 configurations, respectively. The latter is polarized nearly parallel to the Q_y-axis in descobalt B_{12}, whereas the S_4 transition is polarized between the *x*- and *y*-axes.[43] The predicted polarization axes for Q_y and Q_x bands are the *y*- and *x*-axis, respectively. The fluorescence polarization spectrum of descobalt-B_{12} shows that Q_x and B_x transitions are polarized perpendicular to the Q_y band,[43,44] reminiscent of the chl-*a* spectrum.[33] Although the linear dichroism of vitamin B_{12} in stretched poly(vinyl alcohol) film is too small to deduce the absolute polarization axes of the visible and near-u.v. bands,[45] the negative dichroism at the near-u.v. band is indicative of the perpendicular polarization of the B_x transition with respect to the Q_y polarization direction. Both c.d.[43,44,46] and m.c.d.[47] spectra also suggest this polarization picture. The absolute polarizations of the absorption spectrum of a single-crystal Ni^{2+}-corrin (nirrin) indicate that the visible band is polarized along the *y*-axis and the near-u.v. band is oppositely polarized.[48] Thus, the polarized reflection spectroscopy result is consistent with the PPP predictions.[43]

Biliverdin, an open tetrapyrrole bile pigment derived from haems, shows an absorption spectrum similar to those of porphyrins with the visible band maximum at 670 nm ($\varepsilon \sim 1.1 \times 10^4$) and a near u.v. band at 376 nm ($\varepsilon \sim 3.8 \times 10^4$) in water (pH 11.8) at room temperature.[49] The absorption bands are not resolved. The low-temperature absorption spectrum of biliverdin in ethanol at 77 K resolves shoulders at 660 nm ($f = 0.14$) and 435 nm ($f \sim 0.09$), along with maxima at 707 nm ($f = 0.19$) and 380 nm ($f = 1.06$).[50] These spectral characteristics are reminiscent of the spectra of porphyrins. For, descriptive purposes, we will use the porphyrin notations, $Q_{x,y}$, for the visible and near-u.v. ('Soret') bands of biliverdin.[50] Thus, the long-wavelength band at 707 nm is designated as Q_y which is predominantly due to the 16(HOMO)$\rightarrow$17 (LEMO) configuration. The second band at 660 nm resolved in the low-temperature ethanol glass is largely contributed by the 14$\rightarrow$17 configuration, while the strong near-u.v. transition arises from the 16$\rightarrow$18 configuration is predicted to occur between the Q_x and near-u.v. regions.

It is generally agreed that the conformation of biliverdin is neither fully linear nor fully cyclic.[49,50] The most likely conformation is probably 'semi-circular', since

[44] A. J. Thompson, *J. Amer. Chem. Soc.*, 1969, **91**, 2780.
[45] R. Eckert and H. Kuhn, *Z. Elektrochem.*, 1960, **64**, 356.
[46] R. Bonnett, D. M. Godfrey, V. B. Math, P. M. Scopes, and R. N. Thomas, *J.C.S. Perkin I*, 1973, 252.
[47] B. Briat and C. Djerassi, *Nature*, 1968, **217**, 918.
[48] B. G. Anex and G. J. Eckhardt, *Abstracts, Symposium on Molecular Structure and Spectroscopy, Columbus*, Ohio, 1966.
[49] G. Blauer and G. Wagniere, *J. Amer. Chem. Soc.*, 1975, **97**, 1949; *ibid.*, 1976, **98**, 7806.
[50] Q. Chae and P.-S. Song, *J. Amer. Chem. Soc.*, 1975, **97**, 4176.

the oscillator strength ratio, $f_{\text{visible}}/f_{\text{near u.v.}}$, is inconsistent with either extreme conformation. It should be noted that this ratio is a sensitive measure of the chromophore conformation, with the maximum ratio that for linear polyenes and the minimum ratio that for cyclic polyenes (*see above*). In addition, the fluorescence excitation polarization spectrum of biliverdin yields a value of *ca.* 50° for the angle between the polarization axes of the visible and near-u.v. bands, consistent with a semi-circular conformation.[50] The spectroscopic assignments and conformational analysis of other phycobilin chromophores are now in progress using the methods described for biliverdin (J. Jung and P.-S. Song, to be published).

The absolute polarization directions of the visible and near-u.v. absorption bands are not known. However, approximate polarization directions for these bands have been deduced from the linear dichroic data which show negative and positive dichroism for the 675 and 370 nm bands, respectively, of biliverdin in stretched PVA film.[50] By assigning the PVA stretching axis to the long axis of the semi-circular conformation of biliverdin, the polarization of the visible band at 675 nm has been determined to be along the A–C ring axis, while the 380 nm band is polarized perpendicular to the former. It should be noted that a number of simplifying assumptions are involved in these deductions, although the relative polarizations of these two bands are consistent with the polarized fluorescence excitation spectrum mentioned earlier. Further refinements in the data analysis of linear dichroic spectra and a polarized single-crystal spectroscopic study are needed for definitive assignments of the polarization axes.

3 Linear Polyenes

Simple Polyenes.—In contrast to cyclic polyenes, linear polyenes show strong absorption at the first absorption band, with considerably weaker absorption at shorter wavelengths. Until recently, the electronic spectra of simple linear polyenes of conjugated C=C bonds ($N = 2$—6, where N is the number of conjugated C=C bonds) have been 'satisfactorily' described in terms of free-electron and simple MO theories.

For a polyene within the 'four-orbital' framework (*e.g.* butadiene), the electronic transitions arise from electron configurations corresponding to the change in orbital quantum number, $\Delta n = 1$ for N(HOMO)$\rightarrow(N+1)$(LEMO) and $\Delta n = 2$ for $(N-1)\rightarrow(N+1)$ and $N\rightarrow(N+2)$ configurations. The degeneracy of the latter excitations (*c.f.* Coulson–Rushbrooke theorem) is lifted *via* configuration interactions between them. It is well known that the first transition, ${}^1B_u \leftarrow {}^1A_g$ (or ${}^1B \leftarrow {}^1A$ in Platt notation), is strongly electric-dipole allowed with its polarization direction along the long molecular axis. Both oscillator strength [$f \sim 0.74$ for all-*trans*-butadiene, ~ 3.4 for all-*trans*-β-carotene] and $\lambda_{\max}$ increase with conjugation. The latter converges to an asymptotic value of 610 nm for 'infinitely' long polyenes as a result of C—C and C=C bond-order alternations, which restrict the electron motion in a one-dimensional periodic potential, and/or π-electron correlation effects.[51]

The ${}^1A_g^+ \leftarrow A_g^-$ and ${}^1A_g^- \leftarrow A_g^-$ transitions are symmetry-forbidden, but the former can be readily observable even in *trans*-polyenes and it becomes strongly

[51] A. Ovchinnikov, I. I. Ukrainskii, and G. V. Kventsel, *Sov. Phys. Uspekhi*, 1973, **15**, 575.

allowed in *cis*-polyenes ('cis peak') due to the loss of the centre of symmetry. The doubly forbidden ($g \rightarrow g, - \rightarrow -$) $^1A_g^- \leftarrow A_g^-$ transition is mainly the result of a doubly excited configuration. In contrast to the ^1B_u state, which is ionic in character, the $^1A_g^-$ excited state is covalent with a high degree of double-bond character for the C—C single bond relative to the ^1B_u or A_g^- ground state.

The long held traditional assignment of polyene spectra described above has been questioned on both experimental and theoretical grounds in recent years.[52] The new assignment puts the $^1A_g^-$ excited state below the strongly allowed ^1B_u state. The lowering of the $^1A_g^-$ state arises from configuration interactions among singly and multiply excited configurations.[53,54] The quantum mechanical prediction for the location of $^1A_g^-$ is by no means definitive at the present state of computation, since both *ab initio* and semi-empirical MO–CI (singly plus doubly excited configurations included) used involve several approximations that may affect the accuracy of the predicted location of the $^1A_g^-$ state; some of these approximations include neglect of the σ-core and its redistribution, restriction of basis set, *etc.* It is also possible that the order of the $^1A_g^-$ state reverses upon inclusion of triplyexcited configurations in the CI calculations. Nonetheless, the new order for the $^1A_g^-$ state is strongly supported on both theoretical and experimental grounds, as mentioned below.

The gas-phase study of *cis* and *trans*-hexa-1, 3, 5-triene showed no detectable absorption on the red edge of the main band and that results of PPP-type calculations including only singly excited configurations matched more closely the experimental u.v. spectra than the singly + doubly excited configuration interaction calculations (see Gavin and Rice, in refs. 52 and 55). Karplus *et al.*[55] subsequently argued that the failure to detect the ^1A_g absorption band in the gas-phase spectrum of the hexatriene is not inconsistent with the original PPP (single + double CI) prediction for the low-lying $^1A_g^-$ state, since the transition to $^1A_g^-$ occurs too close to the allowed ^1B_u band and it is symmetry-forbidden.[55] Apparently, such a 'masking' is not a problem for diphenylpolyenes at low temperature where the weak absorption has been seen.[52]

The low-lying $^1A_g^-$ state is consistent with the valence-bond picture of polyenes, in contrast to the free-electron and simple LCAO MO methods. The valence-bond picture indicates that the $^1A_g^-$-state is lowered in energy, as the conjugation of polyenes is extended, and that the $^1A_g^- \leftarrow A_g^-$ transition may occur below the $^1B_u \leftarrow A_g^-$ transition.[52]

The assignment of $^1A_g^-$ to the lowest excited $^1\pi, \pi^*$ state has a profound implication in re-interpreting spectroscopic and photochemical properties of linear polyenes. The former includes several anomalies (*e.g.* almost no overlap between absorption and emission bands) which had been largely explained on the basis of Franck–Condon arguments (*e.g.* distortion of the polyene molecule by a low-frequency mode). In addition to the near 'non-overlap' of the $^1B_u \leftarrow A_g^-$ absorption and the fluorescence bands, the spectroscopic anomalies of polyenes include

52 (*a*) B. Hudson and B. Kohler, *Ann. Rev. Phys. Chem.*, 1974, **25**, 437; see original references therein (1972–1974); (*b*) R. L. Christensen and B. E. Kohler, *J. Phys. Chem.*, 1976, **80**, 2197.
53 R. J. Buenker and J. L. Whitten, *J. Chem. Phys.*, 1968, **49**, 5381.
54 K. Schulten and M. Karplus, *Chem. Phys. Letters*, 1972, **14**, 305.
55 M. Karplus, R. M. Gavin, jun., and S. A. Rice, *J. Chem. Phys.*, 1975, **63**, 5507.

(*a*) anomalously long τ_F and (*b*) apparent differential solvent shifts of the main ^1B_u absorption and fluorescence bands. Thus, the former shifts to the red as the solvent polarizability, $(n^2-1)/(n^2+2)$, increases, but the fluorescence maximum of diphenyloctatetraene shifts to the red only slightly.[52] As a result, the Stokes shift ranges from 2000 to 6000 cm^{-1}. In general, polyenes show a large Stokes shift of 3000—7000 cm^{-1}.[56] Along with such a large Stokes shift, there is an apparent separation between the $^1B_u \leftarrow A_g^-$ 0–0 and the first observed line of the fluorescence. This separation is understandable in mixed-crystal and Shpol'skii quasi-line spectra of diphenyloctatetraene at 1.8—4.2 K.[52] A series of new quasi-lines can be observed between the red edge of the absorption band and the blue edge of fluorescence under high-resolution conditions. Such weak quasi-lines are not attributable to the $^1B_u \leftarrow A_g^-$ transition.[52] These new absorption lines are rich in low-frequency vibrations (<500 cm^{-1}). The weak, sharp absorption shows an apparent 0–0 frequency at 452.24 nm for diphenyloctatetraene in a biphenyl host, coincident with the first emission line of the fluorescence spectrum. The weak absorption and fluorescence show an approximate mirror image. In contrast to the strong $^1B_u \leftarrow A_g^-$ absorption band, the weak band ($f \approx 0.05 \pm 0.02$) is relatively insensitive to the solvent matrix. On the basis of this and other less direct evidence, the weak absorption has been assigned to the $^1A_g^- \leftarrow A_g^-$ transition, consistent with the single + double CI PPP calculations. It is concluded that this new assignment is generally valid for all polyenes, particularly for diphenylpolyenes of 3–6 conjugated C=C bonds, decapentaene carboxylic acid, 2,10-dimethylundecapentaene, parinaric acid,[57] and retinyl polyenes (*see below*).[52] Furthermore, it is predicted that the $^1A_g^-$ state is lowered in energy with increasing conjugation, in contrast to the allowed ^1B_u state which converges to the asymptotic level.

The $^1A_g^- \leftarrow A_g^-$ absorption and its fluorescence spectrum of 2, 10-dimethylundecapentaene in *n*-nonane matrices at 4.2 K have been recorded at high resolution, revealing several predominant low frequency, non-totally symmetric vibrations (*see above*). These vibrations apparently induce the forbidden $^1A_g^- \leftarrow A_g^-$ transition in agreement with the Herzberg-Teller vibronic theory which predicts the mixing of the closely lying $^1A_g^-$ and ^1B_u states.[52,58]

The fluorescence polarization at the weak edge of the absorption in diphenyloctatetraene remains high (>0.4), indicating that the transition moment is parallel to the emission and ^1B_u transition moments.[59] Earlier, this was used as an argument against the $^1A_g^-$ assignments.[59] However, the polarization data can be accommodated in terms of the Herzberg–Teller-type vibronic mixing of the two states, $^1A_g^-$ and ^1B_u. The vibronic mixing is expected to decrease with the extent of conjugation, as the gap between the two states increases, resulting in a quadratic decrease in the oscillator strength of the former transition.[52,60]

In an attempt to obtain an independent evidence for the $^1A_g^-$ state assignment in polyene, electron scattering (ES) spectroscopy has been applied to hexatriene. A shoulder at 282 nm in the ES spectrum increases its intensity as the momentum transfer is raised. This is taken as evidence for the $^1A_g^-$ assignment of the lowest

[56] M. Kovner, *Acta Physicochim.* U.R.S.S., 1944, **19**, 385.
[57] L. A. Sklar, B. S. Hudson, M. Peterson, and J, Diamond, *Biochemistry*, 1977, **16**, 813.
[58] R. L. Christensen and B. E. Kohler, *J. Chem. Phys.*, 1975, **63**, 1837.
[59] T. A. Moore and P.-S. Song, *Chem. Phys. Letters*, 1973, **19**, 128.
[60] J. B. Birks and D. J. S. Birch, *Chem. Phys. Letters*, 1975, **31**, 608.

singlet excited state in polyenes.[61] Several attempts to resolve the $^1A_g^-$ state of polyenes have also been made without conclusive evidence for or against the assignment by using two-photon absorption for which the $^1A_g^- \leftarrow A_g^-$ transition is allowed. In 1, 4-diphenylbutadiene, it is possible that the two-photon absorption occurs at *ca.* 500 cm^{-1} below the ^1B_u state at 347.2 nm.[62] The $^1A_g^-$ state in diphenylhexatriene and diphenyloctatetraene has also been located at 386 and 442.5 nm, respectively.[62] However, two-photon absorption with the thermal blooming technique did not reveal the $^1A_g^-$ state for hexa-1, 3, 5-triene, and more recent studies of electron-impact and multiphoton ionization spectra of hexatriene did not provide conclusive evidence for the low-lying $^1A_g^-$ state in the 270—300 nm region.[63] Nonetheless, the failure to observe such a low-lying state in hexatriene does not necessarily entail rejection of the new state ordering in polyenes, since the two-photon absorption cross-section could be too small to be measured and/or masked by nearby intense two-photon absorption bands (*e.g.* Rydberg band).

Another line of evidence for the low-lying $^1A_g^-$ state has been explored using c.d. and m.c.d. A weak absorption at the red edge of the main absorption band of cycloheptatriene shows an m.c.d. sign opposite to that of the latter. This weak band has been assigned to the $^1A_g^-$-type band.[52] However, m.c.d. has not yielded unambiguous conclusions in several polyene systems including those complexed at inducible optical active sites on proteins. The induced c.d. spectra of α- and β-parinaric acid octadeca-9,11,13,15-tetraenoic acid)–bovine serum albumin complexes showed a negative c.d. band at the long-wavelength edge and a positive c.d. band at λ_{max}.[64] In addition, no mirror image between the main absorption and fluorescence spectra was observed. These results are consistent with the $^1A_g^-$ assignment for the weak absorption. However, interpretation of the induced c.d. spectra is complicated by the exciton contribution arising from inter-chromophore interactions.[64]

It appears that the $^1A_g^-$ assignment as the lowest $\pi \rightarrow \pi^*$ transition in polyenes is an attractive proposal which is able to accommodate characteristic anomalies (*see above*) of polyene spectra. There still seem to be some spectroscopic features not readily resolved at present, however. These include the apparent absence of the $^1A_g^-$ band intensity in unsubstituted hexatriene in the gas phase, uncertainties as to accurate values of the $^1A_g^- \leftarrow A_g^-$ oscillator strength, and the location of the $^1A_g^-$ state in longer polyenes with six or more conjugated C=C bonds. Apparently, the end aryl group substitution can affect the relative ordering of $^1A_g^-$ and ^1B_u states.[64] It is concluded that $^1A_g^-$ lies below ^1B_u for all orientations of the phenyl group in diphenylhexatriene and diphenyloctatetraene.[60] Accepting the $^1A_g^-$ nature for the weak red edge absorption of polyenes (*e.g.* diphenyloctatetraene) and assuming the mirror image between the $^1A_g^-$ absorption and fluroescence which overlap just slightly, one would then expect the intensity ratio of the absorption at λ_{max} to that

[61] D. E. Post, W. M. Hetherington, and B. Hudson, *J. Chem. Phys.*, 1976, **64**, 4020.
[62] (*a*) R. L. Swofford and W. M. McClain, *J. Chem. Phys.*, 1973, **59**, 5740; (*b*) G. R. Holtom and W. M. McClain, *Chem. Phys. Letters*, 1976, **44**, 436.
[63] (*a*) A. J. Towarowski and D. S. Kliger, *Chem. Phys. Letters*, 1977, **50**, 36; (*b*) W. M. Flicker, O. A. Mosher, and A. Kupperman, *ibid.*, 1977, **45**, 492; (*c*) D. H. Parker, S. J. Sheng, and M. A. El-Sayed, *J. Chem. Phys.*, 1976, **65**, 5534.
[64] (*a*) L. A. Sklar, B. S. Hudson, and R. D. Simoni, *Proc Nat. Acad. Sci. U.S.A.*, 1975, **72**, 1649; (*b*) L. A. Sklar, B. S. Hudson, and R. D. Simoni, *Biochemistry*, 1977, **16**, 5100.

of $^1A_g^-$ 0–0 to be roughly comparable to that of the fluorescence at $\lambda_{F,max}$ and $^1A_g^-$ 0–0.

The long-wavelength absorption and fluorescence of β-apo-8′-carotenal and astacene in the solid state have been assigned to the $^1A_g^-$ state.[65] However, these are likely to be due to impurities[32] or reaction products.

Retinyl Polyenes.—Retinyl polyenes such as retinol and retinal show broad, structureless absorption spectra at room temperature and at 77 K. The main absorption band, $^1B_u \leftarrow A_g^-$, $\lambda_{max} \approx$ 330—335 nm, $f \approx 1.36$, of all-*trans*-retinol is due to the ^1B_u-type (or ^1B in Platt notation) transition. (Hereafter, we will retain the C_{2h} or free-electron notation for genealogical correlation with the previous section.) The *cis* band, $^1A_g^+ \leftarrow A_g^-$ (or $^1C \leftarrow A$), occurs at 240—250 nm. and its intensity is substantially enhanced in *cis*-isomers (f ≈ 0.1—0.2). The ^1B_u transition is most likely polarized along the long molecular axis, while the former is polarized nearly perpendicular to the latter (~55° from the fluorescence polarization[32,59] and 65° from PPP–CI calculation). Because of the lack of symmetry and distortion of the planarity at the C-6—C-7 bond, extensive configuration mixing of various excited states is expected. Except for the above two states, $^1B_u^-$ and two $^1A_g^-$ states of retinol that may lie in the 210—290 nm region have not been resolved. One of the $^1A_g^-$ states has recently been assigned as the lowest π, π^* state (352 nm; *i.e.* 1600 cm^{-1} lower than ^1B_u) in all-*trans*-retinol on the basis of two-photon absorption.[66] This assignment is consistent with the earlier assignment and with spectroscopic anomalies of retinol which are shared by other polyenes (*see above*) including 2,10-dimethylundecapentaene, axerophtene, anhydro-vitamin A, and all-*trans*-retinal.[52,67] The $^1A_g^-$ level in all-*trans*-retinol is *ca.* 1600 cm^{-1} below the ^1B_u state, as mentioned above. This gap is less than in 2,10-dimethylundecapentaene (3250 cm^{-1} for the latter).[67]

The absorption bands of retinyl polyenes containing β-ionone are usually diffuse and structureless even at low temperature. The diffuseness is expected from vibronic interactions between close-lying states, but the torsional potential of the C-6—C-7 bond in the ground and excited states appears to play the most important role.[68] Thus, it is noted that anhydrovitamin A at 77 K[67] and all-*trans*-retinol bound specifically to β-lactoglobulin (unpublished results) show structured absorption spectra, possibly as the result of a fixed torsional angle about the C-6—C-7 bond.

Retinals exhibit absorption spectra similar to that of retinol, with λ_{max} at *ca.* 360—380 nm ($f \approx 1$) which is due to the $^1B_u \leftarrow A_g^-$ genealogy. The ^1B_u transition is nearly maximally polarized with respect to the fluorescence oscillator under photoselection conditions at 77 K.[69] This transition is most likely polarized along the

[65] (*a*) K. Mandal and T. N. Misra, *Bull. Chem. Soc.*, Japan, 1976, **49**, 198, 975; (*b*) B. Mallik, K. M. Jain, K. Mandal, and T. N. Misra, *Indian J. Pure Appl. Phys.*, 1975, **13**, 699.

[66] (*a*) R. R. Birge, J. A. Bennett, B. M. Pierce, and T. M. Thomas, *J. Amer. Chem. Soc.*, 1978, in press; (*b*) R. R. Birge, K. Schulten, and M. Karplus, *Chem. Phys. Letters*, 1975, **31**, 451.

[67] (*a*) R. L. Christensen and B. E. Kohler, *Photochem. and Photobiol.*, 1973, **18**, 293; (*b*) *ibid.*, 1974, **19**, 401.

[68] B. Honig, A. Warshel, and M. Karplus, *Accounts Chem. Res.*, 1975, **8**, 92, and references therein.

[69] (*a*) P.-S. Song, Q. Chae, M. Fujita, and H. Baba, *J. Amer. Chem. Soc.*, 1976, **98**, 819; (*b*) T. A. Moore, Ph.D. Thesis Texas Tech. University, Lubbock, Texas 1975;(*c*) T. A. Moore and P. S. Song, *Nature (New Biol.)*, 1973, **242**, 30.

long molecular axis and shorter-wavelength transitions are expected to be polarized at some angle to the former on the basis of the declining degrees of fluorescence polarization at $\lambda < 320$ nm. The PPP calculation predicts an angle of 52° between ^1B_u and $^1A_g^+$ (*cis* peak) polarization axes.[69]

The lowest excited state is suggested to be of $^1A_g^-$ type in all-*trans*-retinal and 11-*cis*-12-*s*-*trans*-retinal.[66,67] Although this suggestion is consistent with the spectral anomalies (long τ_F^0, large Stokes shift, *etc.*) shared by other polyenes and retinol, its frequency and intensity are not accurately known at the present, partly owing to the lack of highly resolved absorption and fluorescence spectra. Furthermore, the fluorescence of retinal is λ_{ex}-dependent,[67,69,70] making it difficult to utilize fluorescence as a probe for the $^1A_g^-$ state.

Retinal (*e.g.* all-*trans*-isomer) does not fluoresce in dry non-polar solvents, but the H-bonded fraction of retinal in H-bonding solvents is apparently responsible for the fluorescence. This accounts for the λ_{ex}-dependence of fluorescence in the solvent.[71] The lowest excited singlet state of all-*trans*-retinal in dry hydrocarbon solvents[71] and of 11-*cis*-retinal[67] is then assigned to the $^1n,\pi^*$ state.

The $^1A_g^- \leftarrow A_g^-$ transition in retinal is no longer symmetry-forbidden. For example, the PPP calculation of f values ranges from 0.16 for the planar 11-*cis*-*s*-*trans*-retinal ($^1A_g^- = S_1$) to 0.22 for the planar 11-*cis*-*s*-*cis*-retinal ($^1A_g^- = S_2$). The calculated f values for other retinal isomers range from 0.176 for 13-*cis*- and 0.288 for 11-*cis*-12-*s*-(150°)-*trans*-retinal. The half-bandwidth of the main absorption band of the *trans*- and *cis*-isomers remains approximately the same.[67,72] Thus, the predicted strong absorption due to the $^1A_g^-$ transition is not readily apparent in the low-temperature absorption spectra.

Without identifying the symmetry ($^1A_g^-$ *vs.* ^1B_u), we have argued for the lowest singlet state to be a non-discrete π,π^* state admixed with a nearby n,π^* state on the basis of the observation that retinal carbonyl–cation binding enhances the fluorescence quantum yield in a manner analogous to the H-bonding effect noted above.[69] However, $^1\pi,\pi^*$–n,π^* perturbation is apparently not responsible for the band broadness, since the torsional potential about the C-6—C-7 bond plays an important role in the spectral diffuseness in retinals.[68]

Cation binding to the all-*trans*-retinal carbonyl results in less than a 3 nm red shift.[69] However, the H-bonded retinal–phenol complex shows a substantial spectral shift ($\lambda_{max} \approx 385$ nm in 3-methylpentane and 420 nm in 3-methylpentane-1mM-phenol; $\lambda_{F,max} \approx 520$ and 600 nm, respectively).[71] Furthermore, a significant overlap between the main absorption and fluorescence, uncharacteristic of other polyenes, is observed. Thus, it is possible that the strong H-bonding has either lowered the ^1B_u state below the $^1A_g^-$ and $^1n,\pi^*$ states or introduced a charge-transfer character into the lowest singlet state.

The 77 K absorption spectra of retinals show structured band(s) over the 260—320 nm region, which have been assigned to an $n \rightarrow \pi^*$ transition.[72] The intensity of this band is enhanced for *cis*-isomers. However, other assignments (second B_u^-, A_g^-, σ, π^*, *etc.*) for this region cannot be ruled out. The *cis* peak (of $^1A_g^+$

[70] R. S. Becker, K. Inuzuka, J. King, and D. E. Balke, *J. Amer. Chem. Soc.*, 1971, **93**, 43.

[71] (*a*) T. Takemura, P. K. Das, G. Hug, and R. S. Becker, *J. Amer. Chem. Soc.*, **98**, 7099; (*b*) R. S. Becker, G. Hug, P. K. Das, A. M. Schaffer, T. Takemura, N. Yamamoto, and W. Waddell, *J. Phys. Chem.*, 1976, **80**, 2265.

[72] R. R. Birge, M. J. Sullivan, and B. E. Kohler, *J. Amer. Chem. Soc.*, 1976, **98**, 358.

genealogy) is located at *ca.* 250 nm, and is strongly enhanced in 11-*cis*-retinal at 77 K.[72] The PPP double-CI calculation of 11-*cis*-retinal predicts two additional ^1A_g states (3^1A_g and 4^1A_g or ^1B_u in 11-*cis*-12-*s*-*cis*) and the $^1A_g^+$ state.[66] Thus, the structured absorption band at the 260—320 nm region ($\lambda_{max} \approx 280$ nm)[73] may be assigned to one or both of these ^1A_g transitions.

From the foregoing review, it is clear that the assignments of various bands other than the main absorption band (B_u) are still tentative. The spectroscopy of retinal Schiff's bases is even less characterized than that of retinals, although general features of the absorption spectra of Schiff's bases at room termperature and 77 K are similar to those of corresponding retinals.[73] Recently, a unusually large red shift of protonated retinal Schiff's base has been observed at 77 K, showing absorption maximum at *ca.* 540 nm. It is suggested that the red shift results from a lowering of the transition energy of the protonated Schiff's base as a result of the preferential lowering of the excited singlet state due to its interaction with the hydrogen halide solvent cage.

Carotenoids.—Carotenoid aldehydes containing one β-ionone end are called β-apo-carotenals, of which retinal (five C=C bonds) may be regarded as the shortest carotenal, with λ_{max} ~360 nm. The main band maximum shifts to the red with increasing conjugation (*e.g.* 414 nm for β-apo-12'-carotenal with seven C=C bonds and 508 nm for torularhodinaldehyde or 3', 4'-dehydro-β,ψ-caroten-17'-al with thirteen C=C bonds). This shift is usually accompanied by enhanced vibrational resolution in the visible absorption band. This trend is also observed with carotenoids with two β-ionone ends (*e.g.* β-carotene). The vibrational structure is even more prominent at 77 K.[74,75] Thus, the torsional potential around the C-6—C-7 *s*-bond is no longer the dominant factor for the spectral diffuseness, as the spectral diffuseness decreases with the number of C=C bonds in carotenoids.[76] The total bandwidth ($\delta\bar{\nu}/\text{cm}^{-1}$) is empirically given by $\delta\bar{\nu} = \Delta + 64000N \times \delta/n^2$, where n, N_1 and $\delta(=0.244 \pm 0.021\ \text{cm}^{-1})$ are the number of double bonds, the number of β-ionone rings, and the reduction in the effective number of double bonds per β-ionone ring contributing to the spectral width, respectively. Thus, the second term gives the width contribution due to the torsional potential around the C-6—C-7 *s*-bond, while the first term, $\Delta = 549 \pm 14\ \text{cm}^{-1}$, is of all contributions to the bandwidth from factors other than the distribution of torsional angles.

The main absorption band, ^1B_u ($f \approx 3$), is probably polarized along the long axis of the carotenoid conjugation in β-carotene and related carotenoids. The polarized absorption and reflectance spectra of all-*trans*-β-carotene single crystals have been measured. The ^1B_u band is polarized at an angle of between 0 and 40° to the long molecular axis.[77] Linear dichroic results of carotenoids in stretched

[73] (*a*) A. M. Schaffer, W. H. Waddell, and R. S. Becker, *J. Amer. Chem.* Soc., 1974, **96**, 2063; (*b*) W. H. Waddell, A. M. Schaffer, and R. S. Becker, *ibid.*, 1977, **99**, 8456.

[74] (*a*) P.-S. Song and T. A. Moore, *Photochem. and Photobiol.*, 1974, **19**, 435; (*b*) Q. Chae, P.-S. Song, J. E. Johansen, and S. Liaaen-Jensen, *J. Amer. Chem. Soc.*, 1977, **99**, 5609; (*c*) P. Koka and P.-S. Song, *Biochim. Biophys. Acta*, 1977, **495**, 220.

[75] B. Ke, F. Imsgard, H. Kjøsen, and S. Liaaen-Jensen, *Biochim. Biophys. Acta*, 1970, **210**, 139.

[76] R. Hemley and B. E. Kohler, *Biophys. J.* 1978, in press.

[77] (*a*) D. Chapman, R. J. Cherry, and A. Morrison, *Proc. Roy. Soc.* 1967, **A301**, 173; (*b*) L. J. Parkhurst and B. G. Anex, *J. Chem. Phys.*, 1966, **45**, 862.

polyethylene films and in liquid crystals are interpretable in terms of the long molecular axis polarization of the main band.[74,78] The main absorption polarization in *cis*-carotenoids is also expected to be along the long molecular axis.

According to the new assignment,[52,76] the lowest singlet state in carotenoids is also of the A_g^- symmetry. While the frequencies of the transitions to the allowed excited singlet state, ^1B_u, can be fitted to the equation, $\bar{\nu}/\text{cm}^{-1} = 16500 + 64\,100/n$, where n is the effective number of double bonds in carotenoids and $\bar{\nu}$ reaches an assymptotic value for long carotenoids, it is predicted that the lowest singlet ($^1A_g^-$) state continues to be lowered in energy with n. Unfortunately, carotenoids are non-fluorescent, thus making it difficult to confirm this prediction directly. At present, there are no theoretical (*e.g.* double CI) or experimental data on the exact location of $^1A_g^-$ in carotenoids. Not surprisingly, the PPP singly excited CI predicts the $^1A_g^-$ state to be at an energy above $^1A_g^+$. The linear dichroism remains constant over the main absorption band and its red edge in *cis*-carotenal spectra.[74] Since the oscillator strength of the $^1A_g^-$ band is not expected to be negligible, as is the case for retinal,[66] the dichroic ratio measurements may have detected unique polarization characteristics. However, the failure to resolve different dichroic ratios does not necessarily rule out the presence of a $^1A_g^-$ band. Thus, the weak absorption at $\lambda > \lambda_{max}$ observed for *cis*-carotenals at 77 K may be attributable to either a $^1A_g^-$ or another $\pi \rightarrow \pi^*$ transition, strongly perturbed by the carbonyl group or by exciton interactions (*e.g.* see ref. 69), since carotenols do not show such long-wavelength absorption.[74] An $n \rightarrow \pi^*$ transition strongly admixed with π,π^* character is also a possibility.[79] Thus, no definite assignment for the weak long-wavelength band in carotenals can be made at present.

The nature of the weak absorption bands at wavelengths shorter than λ_{max} in carotenoids is not well established, except for the '*cis* peak' (*e.g.* 337 nm for 15, 15′-*cis*-β-carotene in hexane at room remperature; 350 nm at 77 K), which becomes strongly allowed in *cis*-carotenoids. The *cis* peak can be safely assigned to the $^1A_g^+$ state ($^1C \leftarrow {}^1A$ in Platt notation), consistent with the PPP MO prediction.[32,78]

The polarization of the *cis*-band has been deduced from linear dichroic spectroscopy of β-carotene and several other carotenals,[74,75,78] and its polarization is perpendicular to the long molecular axis. The 375 nm band in 13-*cis*-rhodopinal at 77 K is highly structured, analogous to the structured band at *ca.* 280 nm of retinal[71] and renierapurpurin-20-al ($\lambda \approx 376$ nm and 320 nm, both having some structure and essentially identical dichroic ratios). It is not certain which of these bands is of the $^1A_g^+$ genealogy; the former is tentatively identified as the $^1C(^1A_g^+)$ transition. The shorter-wavelength band ($^1D \leftarrow {}^1A$) is then assignable to the second $^1B_u \leftarrow A_g^-$ transition. Carotenoids with chiral centres usually yield three distinct c.d. bands of alternating signs in the region of 200—300 nm. It is clear that several weakly absorbing transitions occur in u.v. spectra of carotenoids, and more detailed studies are needed to delineate the number and identity of these transitions in carotenoids. In contrast to carotenals, spectra of the Schiff's bases of carotenals and carotenones lack any vibrational structure.

[78] T. A. Moore and P.-S. Song, *J. Mol. Spectroscopy*, 1974, **52**, 209, 216, 224, and references therein.

[79] (*a*) V. R. Salares, R. Mendelsohn, P. R. Carex, and H. J. Bernstein, *J. Phys. Chem.*, 1976, **80**, 1137; (*b*) V. R. Salares, N. M. Young, H. J. Bernstein, and P. R. Carey, *Biochemistry*, 1977, **16**, 4751.

The wavelength maxima of near-u.v. (including *cis*-peak) and u.v. bands arising from excited configurations involving higher empty MOs in carotenoids can be predicted by Dales's rule which states that λ_x (x = 2nd, 3rd transition, *etc.*) lies very close to the λ_{max} (the main long-wavelength band) of a polyene with n/x, where n is the number of double bonds. This empirical rule is useful, but it fails in longer and substituted carotenoids, particularly at low temperature. Although several PPP-type calculations have been made, the spectroscopic characterization of shorter-wavelength bands of carotenoids remains to be carried out, as mentioned above. As it stands now, the state ordering in a typical carotenoid (*e.g.* all-*trans*-β-carotene) is $^1B_u < {^1A_g^+} < 2^1B_u$, with the location of the symmetry-forbidden $^1A_g^-$ state yet to be spectroscopically resolved. However, the validity of the low-lying $^1A_g^-$ state for organic vapour–absorbed solid samples of β-apo-8′-carotenal and other polyenes[65] is questionable. The recent theoretical calculations of absorption and fluorescence spectra of β-carotene suggest that the ^1B_u state relaxes prior to emission by changing bond alternation along the relaxation co-ordinate in the excited state. This implies that the relaxed ^1B_u state is the lowest and emitting state.[80] However, the fluorescence spectrum used is probably in error, since pure β-carotene does not fluoresce.

4 Heterocyclic Compounds

Indoles and Tryptophan.—The $\pi \rightarrow \pi^*$ transitions, ^1L_b and ^1L_a, are not well resolved in indoles, owing to considerable overlap and vibronic mixing between these two transitions.[81,82] The u.v. absorption in the region of 260—275 nm is assigned to the ^1L_a transition, although the ^1L_b intensity (maximum at 287—290 nm) contributes to the overall extinction and spectral shape in this region. It has been possible to resolve the ^1L_a and ^1L_b bands of indole and tryptophan from fluorescence excitation polarization (P) measurements in propylene glycol at −58 °C, by monitoring the long-wavelength fluorescence where the $^1L_a \rightarrow {^1A}$ emission predominates (*cf.* indole shows dual fluorescence due to $^1L_b \rightarrow {^1A}$ and $^1L_a \rightarrow {^1A}$).[82] The long-wavelength edge at 305—310 nm with $P \sim 0.4$ is assigned to the $^1L_a \leftarrow {^1A}$ transition, in agreement with previous studies. The $^1L_b \leftarrow {^1A}$ 0–0 maximum in indole is located at 289.5 nm where the fluorescence polarization degree is minimal ($P \approx 0.1$), with its vibrational satellites at 286 ($0 + 538.3$ cm^{-1}) and 282.5 nm ($0 + 717.6$ or 735.7 cm^{-1}).[83] The P value then reaches a maximum ($P \approx 0.3$) at 260 nm. Both ^1L_a and ^1L_b contribute more or less equally to the absorption at $\lambda \lesssim 250$ nm.

Tryptophan shows essentially similar polarization characteristics. The $^1L_b\,\lambda_{max}$ (0–0) is resolved at 291 nm, whereas the $^1L_a\,\lambda_{max}$ is at 267 nm.

The angle θ between ^1L_a and ^1L_b transition moments can be calculated from the PPP MO method and it ranges from 54° with CI to 78° without CI.[81] If we assign the long-wavelength edge of the absorption to the ^1L_a (0–0) transition, which possesses a maximum P with respect to the $^1L_a \rightarrow {^1A}$ emission, an estimate of θ can

80 T. Kakitani and H. Kakitani, *J. Phys. Soc.* Japan, 1977, **42**, 1287.

81 (*a*) P.-S. Song and W. E. Kurtin, *J. Amer. Chem. Soc.*, 1969, **91**, 4892; (*b*) M. Sun and P.-S. Song, *Photochem. and Photobiol.*, 1977, **25**, 3.

82 B. Valeur and G. Weber, *Photochem. and Photobiol.*, 1977, **25**, 441.

83 E. H. Strickland, J. Horwitz, and C. Billups, *Biochemistry*, 1970, **9**, 4914.

be anywhere between 45 and 90°, depending on the assumptions made concerning the linearity of the oscillators and the resolution of ^1L_a and ^1L_b intensities in the excitation and emission spectra. The lower value results if the 290 nm absorption with minimum P with respect to the $^1L_a \rightarrow {}^1A$ fluorescence is assumed to be exclusively due to ^1L_b, whereas the higher value results if both ^1L_a and ^1L_b contribute equally to this absorption. The actual composition of extinction at this peak is probably somewhere between these assumptions.

According to a polarized single-crystal absorption study,[84] the ^1L_a and ^1L_b transition moments are oriented at −38° and 54° to the long molecular axis, respectively. On the other hand, the PPP CI method[81] yields the ^1L_a polarization in agreement with the above determination (polarization along N-1—C-4 axis), but the ^1L_b transition is nearly short-axis polarized.

Nucleic Acid Bases.—In spite of the obvious simplicity, the genealogy of u.v. bands of purine and pyrimidine in terms of the benzene notation (^1L_b, ^1L_a, and $^1B_{a,b}$ or $^1B_{2u}$, $^1B_{1u}$, and $^1E_{1u}$)[85] has been heuristically useful. Thus, the u.v. bands of purine in trimethyl phosphate (TMP) at 265 nm ($\varepsilon \approx 6.9 \times 10^3$), 240 nm ($\varepsilon \approx 3 \times 10^3$), and 200 ($\varepsilon \approx 18.1 \times 10^3$)—188($\varepsilon \approx 21.1 \times 10^3$) nm are correlated to $^1L_b(B_{2u})$, $^1L_a(B_{1u})$, and $^1B_{a,b}$ (E_{1u}) bands, respectively, and the u.v. bands of pyrimidine in methylcyclohexane at 242 nm ($\varepsilon \approx 2 \times 10^3$), 210 nm ($\varepsilon \approx 1 \times 10^3$), and 190 nm ($\varepsilon \approx 6 \times 10^3$) are also correlated respectively. According to this genealogical correlation, for example, the u.v. bands of 9-ethyl guanine in TMP at 275 nm ($\varepsilon \approx 9.4 \times 10^3$), 256 nm ($\varepsilon \approx 15.4 \times 10^3$), and 203($\varepsilon \approx 20 \times 10^3$)—190($\varepsilon \approx 27.4 \times 10^3$) nm are correlated to $L_b(B_{2u})$, $L_a(B_{1u})$, and $B_{a,b}$ (E_{1u}), respectively, and similar correlations are made to the absorption bands of cytosine at 277 nm ($\varepsilon \approx 7.5 \times 10^3$), 237 nm ($\varepsilon \approx 3.5 \times 10^3$), and 204 ($\varepsilon \approx 11.9 \times 10^3$)—185 ($\varepsilon \approx 12.2 \times 10^3$) nm. In addition, polarization directions of these bands can be correlated with those of benzene. For example, the 260 nm band (^1L_b) of 1-methyl-thymine is polarized along the axis intersecting N-1—C-2 and C-4—C-5 bonds (inclined by 11 ± 2° toward N-1).[86] With the above description of the gross features of the u.v. spectra of bases, more recent studies will be selectively reviewed in the following sections.

Purine Bases. Analysis of polarized reflection spectra of the *(100)* face of purine crystals yields the transition moment directions of both $\pi \rightarrow \pi^*$ and $n \rightarrow \pi^*$ transitions.[87] The first band, at 294 nm ($f \approx 0.0057$ in the crystal, 0.003—0.005 in solution), is out-of-plane polarized and is due to an $n \rightarrow \pi^*$ transition, which probably is highly localized at the N-3 position of purine.[85] The second band, at 263 nm (^1L_b), is in-plane polarized at +48° from the C-4—C-5 axis (toward C-6). The weak band resolved at 250 nm in the crystal is out-of-plane polarized and is assigned to the second $n \rightarrow \pi^*$ transition, which is probably localized at N-1. The strong absorption at 200 nm is probably due to B_b, as it is polarized nearly parallel to the 263 nm band, but the identity of the out-of-plane polarized 190 nm band remains to be established. An $n \rightarrow \pi^*$ assignment for this band is unlikely owing to

[84] Y. Yamamoto and J. Tanaka, *Bull. Chem. Soc.*, Japan, 1972, **45**, 1362.
[85] (*a*) L. B. Clark and I. Tinoco, jun., *J. Amer. Chem. Soc.*, 1965, **87**, 11; (*b*) W. Hug and I. Tinoco, jun., *ibid.*, 1973, **95**, 2803; (*c*) W. Hug and I. Tinoco, jun., *ibid.*, 1974, **96**, 665.
[86] R. F. Stewart and N. Davidson, *J. Chem. Phys.*, 1963, **39**, 255.
[87] (*a*) H. H. Chen and L. B. Clark, *J. Chem. Phys.*, 1969, **51**, 1862; (*b*) L. B. Clark, *J. Amer. Chem. Soc.*, 1977, **99**, 3934.

its relatively strong intensity ($\varepsilon \approx 6 \times 10^3$). The experimental polarization directions reported here are in good agreement with PPP-type results, but agreement is less satisfactory in purine bases containing additional functional groups such as C=O or NH_2.

The $n \rightarrow \pi^*$ transitions in purine bases have been calculated by the CNDO-CI method.[85,88] Agreement with available experimental $n \rightarrow \pi^*$ transition energies is not sufficiently quantitative, but it is possible to use theoretical data in aiding the resolution of weak $n \rightarrow \pi^*$ bands in purines. The lowest $n \rightarrow \pi^*$ transition is predicted at 256 nm for purine, 228 nm for adenine, and 239 nm for guanine-N-1, which are theoretically above the lowest $\pi \rightarrow \pi^*$ (^1L_b) transition in the respective bases.[85] The calculations based on the same CNDO method predict an $n \rightarrow \pi^*$ transition to be the lowest in adenine, thymine, and cytosine, but not in guanine.[88] At least for purine, there is no doubt that the lowest singlet excited state is an n,π^* type, as the high-resolution spectra of purine in crystal and in inert-gas matrices at 4—20 K exhibit vibrational progression characteristic of an $n \rightarrow \pi^*$ transition (318.5 nm in low-temperature matrix).[89]

The assigned wavelengths of the lowest $n \rightarrow \pi^*$ transitions in adenine and guanine are 280 and 264 nm, respectively, by means of polarized absorption and reflectance spectra of crystals, solvent effects, and c.d. spectroscopy. Poly(dA) shows a positive c.d. peak at 280 nm which can be assigned to an $n \rightarrow \pi^*$ transition of adenine.[90] Similarly, the 270 nm absorption in deoxyadenosine is assigned to an $n \rightarrow \pi^*$ transition. However, the identification of an $n \rightarrow \pi^*$ transition at 264 nm is not definitive, since this region is significantly masked by $\pi \rightarrow \pi^*$ transitions. The SCrRPA method predicts the second $n \rightarrow \pi^*$ at 259 nm, however. It is reasonable to conclude that the lowest singlet transition is of an $n \rightarrow \pi^*$ type in adenine.[91] The lowest $n \rightarrow \pi^*$ transition appears to be strongly localized on N-3.[85] The location of the lowest $n \rightarrow \pi^*$ transition in guanine remains to be resolved. Recent polarized single-crystal reflectance data suggest an $n \rightarrow \pi^*$ transition at 300.3 nm for 9-ethylguanine at 77 K.[87] However, the possibility of an exciton and matrix-shifted $^1B_{2u}$ (0–0) band at 77 K cannot be ruled out.

The long-wavelength band ($\lambda_{max} \approx 274$ nm) of single crystals of 9-methyladenine:1-methylthymine (AT dimer) is due to the lowest $\pi \rightarrow \pi^*$ transition of the former. Its polarization is along the short molecular axis C-5—C-6), while the weaker band with $\lambda_{max} \sim 255$ nm is long-axis polarized.[86] According to the commonly used convention, the transition moment directions (θ) for purines are measured as positive towards C-6 (clockwise) with respect to the C-4—C-5 axis. The first $\pi \rightarrow \pi^*$ transition moment of adenine is then at an angle of $-12 \pm 3°$. Thus, the two $\pi \rightarrow \pi^*$ transitions are polarized at least 45° from each other. This conclusion is further confirmed by the dichroic analysis of the spectra of adenine and 9-methyladenine.[91]

88 N. V. Zheltovskii and V. I. Danilov, *Biofizika*, 1974, **19**, 784.

89 (*a*) J. J. Smith, *Photochem. and Photobiol.*, 1976, **23**, 365; (*b*) M. J. Robey and I. G. Ross, *ibid.*, 1975, **21**, 363.

90 (*a*) C. A. Bush and H. A. Scheraga, *Biopolymers*, 1969, **7**, 395; (*b*) C. A. Bush, *J. Amer. Chem. Soc.*, 1973, **95**, 214.

91 (*a*) A. F. Fucaloro and L. S. Forster, *J. Amer. Chem. Soc.*, 1971, **93**, 6443; (*b*) *Spectrochim. Acta*, 1974, **30A**, 883.

The 260 nm band of adenine in solution is correlated to $^1B_{2u}(L_b)$, but it is apparently composed of two electronic transitions $(L_a + L_b)$, as predicted by theoretical treatments and spectroscopic measurements such as m.c.d. The linear dichroism along the short-wavelength tail of the 260 nm band decreases, indicating the presence of a transition corresponding to the second $\pi \rightarrow \pi^*$ transition (L_a) which is polarized at some angle to the L_b transition moment direction. The difficulty with the L_a assignment is that its f value appears to be weaker than that of the L_b band, in contrast to theory and to what is expected from correlation with the benzene and naphthalene spectra.

9-Ethylguanine in TMP shows absorption bands at 275 nm (shoulder, $\varepsilon \approx 9.4 \times 10^3$, L_b or B_{2u}) 256 nm ($\varepsilon \approx 15.4 \times 10^3$, L_a or B_{1u}), and 203 ($\varepsilon \approx 20 \times 10^3$)—190 ($\varepsilon \approx 27.4 \times 10^3$, $B_{a,b}$ or E_{1u}) nm. In aqueous solution, pH 6.8, is the 256 nm band of 9-ethylguanine shifts to 249 nm.[87] Based on combined information from polarized specular reflectance measurements on single crystals of 9-ethylguanine and from fluorescence polarization data, the first band (B_{2u}) is either polarized nearly short-axis or close to the N-1—N-3 direction (*i.e.* $44 \pm 5°$ or $-14 \pm 5°$ with the N-3—C-6 direction), while the second band (B_{1u}) is polarized close to the C-2—C-8 axis (*i.e.* $115 \pm 10°$ or $95 \pm 10°$ with the N-3—C-6 direction).[92] Newer results from polarized single-crystal reflection spectra of 9-ethylguanine yield polarization angles of $-4°$ or $+35°$ for B_{2u} and $-75°$ for B_{1u},[87] in general agreement with the above results. Polarization directions for guanine, HCl, $2H_2O$ are also available.[87] Although PPP generates reasonable $\pi \rightarrow \pi$ transition energies (*e.g.* 277 nm for B_{2u} and 246 nm for B_{1u}), the calculated polarization directions are not satisfactory, as their directions are reversed relative to the observed directions. Both experiments based on polarized reflectance, c.d., m.c.d., and fluorescence polarization and theories confirm that the first two $\pi \rightarrow \pi^*$ transitions in guanine and guanosine are polarized nearly perpendicular to each other.

Pyrimidine Bases. Uracil in TMP shows $^1B_{2u}(L_b)$ at 258 nm ($\varepsilon \approx 7.8 \times 10^3$, $f = 0.175$) and $^1E_{1u}$ $(B_{a,b})$ at 203 ($\varepsilon \approx 8.2 \times 10^3$)—181 ($\varepsilon \approx 11.8 \times 10^3$) nm. The $^1n,\pi^*$ state probably is at higher energy (250 nm)[85] than $^1B_{2u}$. The c.d. of uridine in water (pH 7) resolves a band at 240 nm with negative ellipticity which may be assigned to $^1B_{1u}$[93], in agreement with some theoretical treatments (*e.g.* B_{1u} at 224 nm by PPP, but 195 nm by CNDO-CI or 210 nm estimated by analogy with the 237 nm band of cytosine[85]). The 258 nm band of uracil may be of composite nature, as revealed by (*a*) the electric field effect[94] and (*b*) the polarized single-crystal reflectance spectra of its model, 6-azauracil, which shows resolved bands at 256 and 278 nm, in addition to the weak band.[95] The latter appears to correspond to the weak absorption at *ca.* 230 nm in solution. 1,3-Dimethyluracil and thymines show similar absorption spectra. However, the nature of the weak band hidden under the strong absorption band is yet to be elucidated.

Several attempts using c.d. solvent perturbation, and other spectroscopic methods have not been successful in resolving $n \rightarrow \pi^*$ transitions in uracil.

[92] (*a*) P. R. Callis, B. Fanconi, and W. T. Simpson, *J. Amer. Chem. Soc.*, 1971, **93**, 6679; (*b*) P. R. Callis and W. T. Simpson, *ibid.*, 1970, **92**, 3593.
[93] K. K. Cheong, Y. C. Fu, R. K. Robins, and H. Eyring, *J. Phys. Chem.*, 1969, **73**, 4219.
[94] K. Seibold and H. Labhart, *Biopolymers*, 1971, **10**, 2063.
[95] J. N. Brown, L. M. Trefonas, A. F. Fucaloro, and B. G. Anex, *J. Amer. Chem. Soc.*, 1974, **96**, 1597.

However, the polarized single-crystal absorption spectrum of 1-methyluracil suggests a weak $n \rightarrow \pi^*$ transition ($\varepsilon \approx 247$) at 264 nm which is polarized perpendicular to the molecular plane; the first $\pi \rightarrow \pi^*$ transition (B_{2u}) in the crystal is at 275.5 nm[96]

The longest-wavelength band in 1-methyluracil and 1-methylthymine is polarized almost parallel to the N-1—C-4 direction (0° or 7° and 19°, respectively), with respect to the N-1—C-4 direction with clockwise rotation denoted as +). Semi-empirical MO calculations[97] are in satisfactory agreement with the experimental directions. Another polarized absorption study reports a polarization angle of $-38 \pm 14°$ for the main band at 260 nm for thymine itself, and the second absorption band at 220 nm is polarized along the short molecular axis (−86°).[98] The angle of 48° between these two transition moment directions is substantially lower than the value for uracil and thymine estimated from the linear dichroic data.[91]

Cytosine in TMP shows λ_{max} at ~277 nm ($^1B_{2u}$, $\varepsilon \approx 7.5 \times 10^3$), a shoulder at 237 nm ($^1B_{1u}$, $\varepsilon \approx 3.5 \times 10^3$), and 204 nm ($\varepsilon \approx 11.9 \times 10^3$)—185 nm ($\varepsilon \approx 12.2 \times 10^3$, $^1E_{1u}$). The first two $\pi \rightarrow \pi^*$ transitions are polarized nearly parallel, according to the polarized reflectance study on cytosine,[91] cytosine monohydrate, and 1-methylcytosine.[92] The $^1B_{2u}$ band is polarized at an angle of $14 \pm 1°$, and $^1B_{1u}$ is polarized at $-5 \pm 3°$.[96] Although agreement between observed and MO-calculated polarization angles is good (within ~10°) for thymine and uracil, this is not the case with cytosine. However, CNDO-CI seems to yield better angles.[85]

From polarized absorption measurements on oriented films of polycytidylic acid, an $n \rightarrow \pi^*$ transition was located at 278 nm.[99] Such an $n \rightarrow \pi^*$ transition may be strongly localized on N-3.[89] However, both CNDO–CI (calc. $n \rightarrow \pi^*$ at 253 nm) and polarized single crystal absorption[96] studies fail to resolve an $n \rightarrow \pi^*$ at λ longer than the first $\pi^* \rightarrow \pi^*$ band.

Flavins.—Flavins (*e.g.* riboflavin) show four major absorption bands at 445 nm ($\varepsilon \approx 12 \times 10^3$), 360 nm ($\varepsilon \approx 10 \times 10^3$), 270 nm ($\varepsilon \approx 32 \times 10^3$), and 230 nm. All four bands are attributable to the $\pi \rightarrow \pi^*$-type transition, although other types of transition, particularly $n \rightarrow \pi^*$', may contribute to some of these band intensities. The PPP calculations on the isoalloxazine nucleus support the $\pi \rightarrow \pi^*$ nature of the major absorption bands in flavins.[100] Calculated transition energies and relative oscillator strengths of the flavin nucleus, 7,8-dimethylisoalloxazine, are in agreement with observed absorption spectra of flavins.[101]

The PPP theory predicts a weak $\pi \rightarrow \pi^*$ transition (S_3) at *ca.* 300 nm, where a fiuorescence polarization minimum is also recorded, indicating a separate transition in this region.[101] M.c.d. provides indirect evidence for this transition, as the m.c.d.

[96] (*a*) W. Eaton and T. P. Lewis, *J. Chem. Phys.*, 1970, **53**, 2164; (*b*) T. P. Lewis and W. Eaton, *J. Amer. Chem. Soc.*, 1971, **93**, 2054.

[97] J. S. Kwiatkowski and B. Pullman, *Adv. Heterocyclic Chem.*, 1975, **18**, 199; for theoretical results see references therein.

[98] M. Tanaka and J. Tanaka, *Bull. Chem. Soc.* Japan, 1971, **44**, 672, 938.

[99] A. Rich and M. Kasha, *J. Amer. Chem. Soc.*, 1960, **82**, 6197.

[100] (*a*) J. L. Fox, S. P. Laberge, K. Nishimoto, and L. S. Forster, *Biochim. Biophys. Acta*, 1967, **136**, 544; (*b*) K. Nishimoto, *Bull. Chem. Soc.* Japan, 1967, **40**, 2493.

[101] (*a*) P.-S. Song, *Ann. New York Acad. Sci.*, 1969, **158**, 410; (*b*) P.-S. Song, *Internat. J. Quantum Chem.*, 1969, **3**, 303; (*c*) P.-S. Song, T. A. Moore, and W. E. Kurtin, *Z. Naturforsch.*, 1972, **B27**, 1011,

maximum is at $\lambda < \lambda_{max} \approx 370$ nm for riboflavin.[102] C.d.[103] and linear dichroic[104] studies support this possibility.

The polarized single-crystal absorption of FMN in flavodoxin indicates that both S_1 and S_2 bands are polarized along the long molecular axis.[105] Defining the polarization angle θ as the clockwise angle measured from the long axis (N-10—N-1 direction), the S_1 and S_2 bands at 450 and 370 nm are then polarized at 16° and −4°, respectively, making $\Delta\theta$ of 20° for the angle between the two transition moments. The Kronig–Kramers transformation of specular reflection spectra of crystalline bis(10-methylisoalloxazine)copper(II) perchlorate tetrahydrate yields the corresponding absorption spectra.[106] Again, S_1 ($f \approx 0.16$) and S_2 ($f \approx 0.15$) transitions are found to be polarized along the long molecular axis, with $\theta = 26°$ and 1°, respectively, and $\Delta\theta$ of 25°. The rotation of the two transition moments by 5–10° clockwise relative to those of the polarized single-crystal absorption of FMN may be attributed to substituent or crystal environmental effects. Considering this, agreement between the two sets of polarization data is excellent. Thus, earlier theoretical transition moment directions calculated from PPP (S_1 along C-8—N-3 and S_2 along C-7—C-2, with $\Delta\theta = 25$—30° depending on several parameters in the calculation)[101] are experimentally confirmed. The polarization of both transitions is along the C-7—C-2 axis, according to MINDO/3.[107] Both fluorescence polarization[101,108] and linear dichroism[104] data are qualitatively consistent with the polarization directions from the single-crystal reflection and absorption studies and the PPP prediction. Polarization directions of S_1 and S_2 bands are not significantly altered by substituting N-5 with carbon, as in deazaflavin, yielding $\Delta\theta = 25°$ (from fluorescence polarization) and 27° (from PPP[108]).

Higher-energy $\pi \rightarrow \pi^*$ transitions of flavins in the 200—300 nm region are not well understood. It is likely that the u.v. band at 270 nm is composed of more than one $\pi \rightarrow \pi^*$ transition. Thus, specular reflectance spectroscopy of 10-methylisoalloxazine reveals bands at 284 nm ($f = 0.2$), 257 nm ($f = 0.39$), and 221 nm ($f = 0.11$), whereas the solution spectrum shows only two peaks at 269 nm ($f = 0.55$) and 225 nm ($f = 0.11$). Their polarization angles are 3°, −19°, and 42°, respectively.[106]

Four possible $n \rightarrow \pi^*$ transitions localized at N-1, C-2=O, C-4=O, and N-5 are not spectrally resolved, as they are masked under the intense $\pi \rightarrow \pi^*$ bands. It is likely that the lowest $n \rightarrow \pi^*$ transition is localized at N-5 and that its low-intensity band is hidden under the second $\pi \rightarrow \pi^*$ band at 370 nm.[108] C.d. and specular reflection spectra of flavins do not reveal the location of an $n \rightarrow \pi^*$ band, indicating that such a transition, if present, is probably too weak ($\varepsilon < 500$) to be detectable. The $n \rightarrow \pi^*$ transition assigned to the solvent-sensitive shoulders of the 450 and 370 nm bands by various authors are not experimentally compatible, and its location remains obscure. However, it appears most likely that the lowest $n \rightarrow \pi^*$

[102] (*a*) D. E. Edmondson and G. Tollin, *Biochemistry*, 1971, **10**, 113; (*b*) G. Tollin, *ibid.*, 1968, **7**, 1720.
[103] H. Harders, S. Förster, W. Voelter, and A. Bacher, *Biochemistry*, 1974, **13**, 3360.
[104] J. Siodmiak and D. Frackowiak, *Photochem. and Photobiol.* 1972, **16**, 173.
[105] W. A. Eaton, J. Hofrichter, M. W. Makinen, R. D. Anderson, and M. L. Ludwig, *Biochemistry*, 1975, **14**, 2146.
[106] M. W. Yu, C. J. Fritchie, A. F. Fucaloro, and B. G. Anex, *J. Amer. Chem. Soc.*, 1976, **98**, 6496.
[107] M. F. Teitell, S. H. Suck, and J. L. Fox, *J. Amer. Chem. Soc.*, 1978, in press.
[108] (*a*) M. Sun, T. A. Moore, and P.-S. Song, *J. Amer. Chem. Soc.* 1972, **94**, 1730; (*b*) M. Sun and P.-S. Song, *Biochemistry*, 1973, **12**, 4663.

transition occurs near the 370 nm band, as dimethylalloxazines, the tautomers of dimethylisoalloxazines, with the $S_1 \leftarrow S_0$ λ_{max} at ~380—390 nm are considerably less fluorescent than the latter, suggesting that an n,π^* state lies only slightly above the fluorescent π,π^* state.[108] In this argument, it is assumed that the tautomerism does not affect $n \rightarrow \pi^*$ energy as much as it does $\pi \rightarrow \pi^*$ energy.

The fluorescence polarization degree in the red region of the first absorption band of alloxazine and lumichromes is >0.4. The angle between the two near-u.v. absorption bands (378 and 318 nm) of alloxazine is estimated to be about 30° from the fluorescence polarization data. This angle is somewhat larger than in isoalloxazines. PPP also predicts a larger angle (43°).[108]

The predominant configuration for the 580 nm band of flavin semiquinone arises from the same orbital pairs as in oxidized flavins, *i.e.* HOMO → LEMO (half-occupied).[105] However, this band cannot be unambiguously identified as the S_1 transition which may arise from an orbital promotion other than HOMO → LEMO.[109] Both experimental and theoretical investigations are called for before satisfactory spectroscopic assignments including polarizations can be developed for flavin semiquinone.

Work derived from this laboratory has been supported by the Robert A. Welch Foundation (D–182). I also wish to thank Professors R. Birge, R. Christofferson, J. L. Fox, and B. Kohler for making their preprints available.

[109] B. Grabe, *Acta Chem. Scand.*, 1974, **28**, 363.

3 Theoretical Chemistry

By G. KLOPMAN and P. ANDREOZZI

Department of Chemistry, Case Western Reserve University Cleveland, Ohio 44106, U.S.A.

1 Introduction

The use of quantum mechanics as a probe into the origin and nature of molecular associations has recently become the subject of much attention. Included in this area are the semi-empirical and *ab initio* studies of molecular interactions in solution, particularly in water and other hydrogen-bonding solvents. Despite the inherent difficulties in such studies, quantum mechanics has become an extremely powerful tool for the determination of the geometry and properties of solvated ions and hydrogen-bonded complexes. In many cases it is now possible to provide a detailed analysis of the components of molecular interaction energies, precisely defining those terms responsible for the stability of these associated species. Recognizing the importance of such investigations to the field of theoretical chemistry, we have devoted this chapter to a review of the major contributions in this area during the past year.

2 Theoretical Aspects of Solvation

Quantum Mechanical Methods.—Since the majority of interactions of interest to chemists occur in solution, it is not surprising that quantum mechanical techniques are frequently used to study solute–solvent interactions. Ideally, the solute–solvent aggregate is treated as a supersystem in which all the electrons of the aggregate are considered in the field of all the nuclei fixed in appropriate positions. A review of such 'supermolecular' methods was recently published by Pullman *et al.*[1] Two problems remain associated with the supermolecule technique preventing its widespread and exclusive use within the framework of normal molecular orbital calculations. First, since the number of orbitals involved in the solute–solvent aggregate is large, only a few solvent molecules may actually be considered. Secondly, the size of the system also limits the number of solvent conformations and orientations around the solute which are actually calculated.

Despite these limitations, however, the supermolecule approach can provide valuable insight into the structure and bonding of solutions often not obtainable from experiment. For example, an STO-3G calculation has been made in an attempt to simulate the hydrogen-bonding of formamide in the liquid state and in

[1] A. Pullman, Proceedings of the Second International Congress of Quantum Chemistry, ed. R. Parr and B. Pullman, Reidel, Dordrecht, Holland, 1976.

dilute aqueous solutions.[2] Although the relationship between the model system [formamide + $(H_2O)_5$ molecules (1) representing an aqueous formamide solution and $(formamide)_5$ molecules (2) representing pure liquid formamide] and the liquid

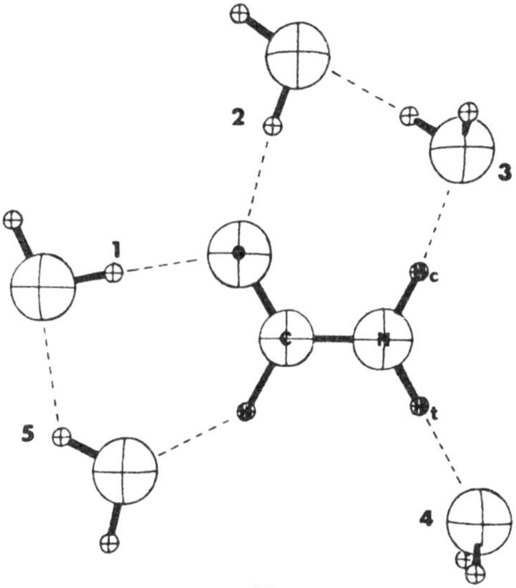

(1)

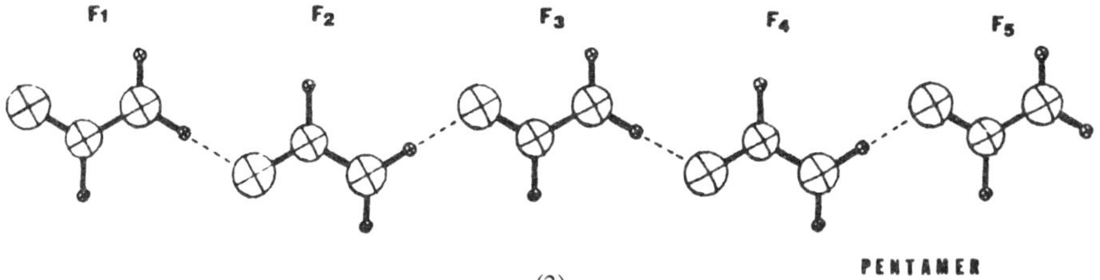

(2)

state was crude, some valuable structural information was obtained. A comparison of the interaction energies of the linear formamide tetramer with that of the mixed linear–cyclic tetramers indicated that linear chains are more stable than mixed linear–cyclic systems for the pure liquid. Calculations also showed that the interaction energy of $H_2O(5)$ with the CH proton is very small, so that a totally hydrated formamide molecule might be sufficiently accurately represented by considering only four H_2O molecules in the first hydration shell.

In some cases, attempts are being made to include many molecules of solvent in the calculation of solute–solvent supersystems. For this purpose, Monte Carlo techniques have proven to be particularly useful. As they are usually applied to calculations on liquids, Monte Carlo methods are used to generate a series of configurations and orientations of the solute–solvent aggregate which are then weighted appropriately to calculate average thermodynamic properties of the

[2] J. F. Hinton and R. D. Harpool, *J. Amer. Chem. Soc.*, 1977, **99**, 349.

system. As Clementi[3] points out, these statistical properties are important factors to be considered for a realistic description of an ensemble of interacting molecules.

In one Monte Carlo study, Owicki and Scheraga investigated the structure and properties of liquid water[4] and dilute aqueous methane[5] in the isothermal–isobaric ensemble at 298 K and atmospheric pressure. The method provided reasonable estimates of the internal energy, heat capacity, and mean molar volume for each system. Energy probability distribution functions were used to help describe the hydrogen-bonding interactions in the liquids. It was found that the hydrogen-bonding environment in water is better described as a continuum of bent or stretched bonds rather than relatively discrete sets of bonded and non-bonded interactions. Interactions between water molecules in the solvation shell of methane are more stable and more sharply distributed than those in the bulk liquid. The size of this effect diminishes with increasing distance from methane, supporting the view that non-polar solutes increase the degree of hydrogen-bonding or 'structure' in their hydration shell.

Clementi[6] also used a Monte Carlo technique to study the structure of a number of alkali-metal halides surrounded by a cluster of 200 water molecules. Co-ordination numbers and hydration-shell radii for the cations and anions were computed, and information about the cluster size and shape was also determined.

Another quantum mechanical model which has been used quite extensively to study solute–solvent interactions is based on the development of pair or higher-order potentials. The basic idea is to perform thorough *ab initio* calculations for small systems in which only one or two solvent molecules are explicitly included. An analytical many-body potential is then fitted to the data. Assuming that the pair potentials are transferable, one can then include many solvent molecules interacting with the solute. This procedure is much easier to use than straightforward quantum mechanical calculations of a supermolecule with hundreds of solvent molecules.

One excellent example of this approach to the treatment of solvation effects was a series of papers by Clementi and co-workers. In an introductory paper Clementi[3] outlines a theoretical and computational model used to describe complex chemical systems of interacting species. His methodology operationally links both quantum chemistry and statistical thermodynamics and is subsequently tested by considering the interaction of water with various biomolecules. In the first two papers of the series, SCF-LCAO-MO computations were presented for 21 amino-acids[7] and the four bases of DNA[8] interacting with one molecule of water at a large number of positions and orientations around each biomolecule. The 1690 computed total energies for the H_2O–amino complexes and the 368 computed total energies for the H_2O–base complexes were then fitted with an analytical potential of the form

$$E=\sum_{i}\sum_{j\neq i}(-A_{ij}^{ab}/r_{ij}^{6}+B_{ij}^{ab}/r_{ij}^{12}+C_{ij}^{ab}/r_{ij})+E(\text{biomolecule})+E(\text{water})$$

where i and j designate two atoms, one in the biomolecule and the second in the

[3] E. Clementi, *Bull. Soc. chim. belges*, 1976, **85**, 969.
[4] J. C. Owicki and H. A. Scheraga, *J. Amer. Chem. Soc.*, 1977, **99**, 7403.
[5] J. C. Owicki and H. A. Scheraga, *J. Amer. Chem. Soc.*, 1977, **99**, 7413.
[6] E. Clementi, R. Barsotti, J. Fromm, and R. O. Watts, *Theor. Chim. Acta*, 1976, **43**, 101.
[7] E. Clementi, F. Cavallone, and R. Scordamaglia, *J. Amer. Chem. Soc.*, 1977, **99**, 5531.
[8] R. Scordamaglia, F. Cavallone, and E. Clementi, *J. Amer. Chem. Soc.*, 1977, **99**, 5545.

water molecule, a is an index that distinguishes the electronic environment of an atom in the biomolecules, b is an index that distinguishes between either a hydrogen or the oxygen atom in the H_2O molecule, A, B, and C are fitting constants, and E(biomolecule) and E(water) are the total energies of the biomolecule and of the water molecule, respectively, at infinite separation from the other.

Contour energy maps can then be used to illustrate the structural organization of water around the various biomolecules. Some of these are shown in Figure 1. In

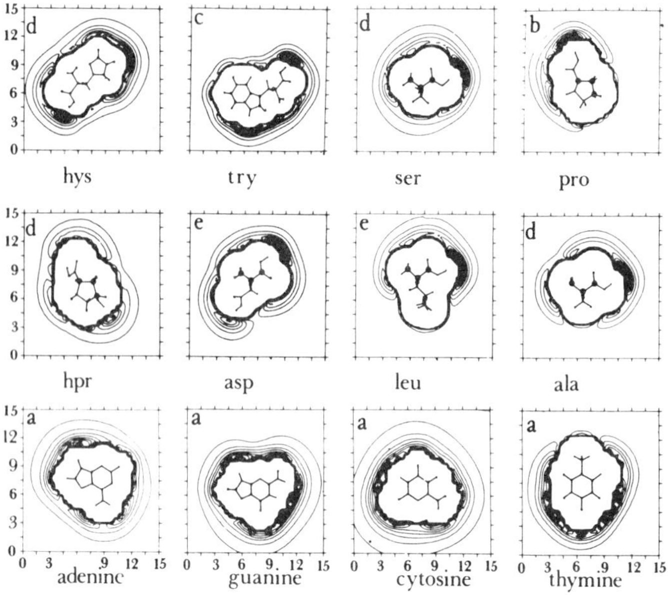

Figure 1 *Structural organization of water around various biomolecules*
(Reproduced by permission from *Bull. Soc. chim. belges*, 1976, **85**, 969)

the concluding paper of the series, Bolis and Clementi[9] investigate the reliability and transferability of their library of pair potentials, using phenylalanine as a test case. They computed the interaction with a molecule of water at 75 different positions relative to phenylalanine in the SCF-LCAO-MO approximation. The interaction energies for the complex using the fitted pair potentials displayed a reasonable agreement with those computed in the *ab initio* framework, with an error of about 1 kcal mol^{-1}.

Attempts have also been made to simulate the solvent molecules of solute–solvent clusters by point charges or point dipoles. One successful attempt in this area was the fractional-charge model developed by Noell and Morokuma.[10] This model consists of placing fractional point charges at the location of solvent atomic centres. Neither electrons nor atomic basic functions are explicitly associated with such centres. If desired, however, a specified number of solvent molecules may be

[9] G. Bolis, and E. Clementi, *J. Amer. Chem. Soc.*, 1977, **99**, 5550.
[10] J. O. Noell and K. Morokuma, *J. Phys. Chem.*, 1976, **80**, 2675.

included explicitly in the usual *ab initio* framework. One may consider this as a solute core to which additional solvent molecules may be added as fractionally charged centres. The fractional charges chosen are those which reproduce the calculated or experimental dipole moment of the solvent molecule.

Extensive testing of the model for the hydration of Li^+ and F^- was performed to assess the applicability of the method. As expected, it was found that the interaction energies between the ions and their solvent hosts were nearly independent of whether the first-shell water molecules are explicitly included. The fractional-charge model was also applied to the electronic structure of H_3N–BH_3 in the crystal state. In this case, H_3N–BH_3 in the centre of a unit cell acted as a solute and ten nearest neighbours as the solvent systems. Finally, the method was used to study changes in the nature of the complex H_3N–HF associated with hydration. Here, it was found that addition of the first hydration shell has a significant effect on the general characteristic of the potential energy surface for the complex. The minimum on the potential energy surface has shifted to a longer H—F distance and shorter N—H distance. In addition, one has a levelling of the potential energy surface, and a conformation representative of a proton-transferred-type structure ($NH_4^+\cdots F^-$) now has an energy *ca.* 5 kcal mol^{-1} above the minimum (compared with 27 kcal mol^{-1}) in the gas phase.

Despite the reported success of the above method, it must be noted that the model lacks the proper short-range repulsive behaviour owing to the fact that orbitals are not explicitly associated with fractionally charged solvent molecules. The method does not include short-range exchange repulsion, which prevents the meaningful optimization of solvent conformations. Furthermore, although the method accurately depicts the electrostatic interaction between solute and solvent and the polarization of solute by solvent, it is not capable of representing charge-transfer or exchange interactions nor of assessing the polarization of solvent by solute.

A related method which has been used to study interactions in solution considers the solvent molecules as point dipoles. *Ab initio* calculations are then performed on the solute in the presence of dipolar fields. Simons,[11] for example, reported on some initial attempts at developing such a model for anion–solvent interactions and the hydration energy of solvated electrons. The dipole representation of water molecules was a rather poor approximation for the study of anion–solvent interactions. Simons's fixed finite dipole model also was not capable of providing an accurate description for the properties of the hydrated electron.

A somewhat different approach for studying solute–solvent interactions is based on modifying the hamiltonian of the system to include the solvent field effect. One recent example involved the use of CNDO/2 with a modified shell repulsion potential for calculation of the properties of alkali halide salts in water.[12] The shell interaction potential was approximated by a simple function containing several empirically chosen parameters characteristic of each atom. These parameters were selected to obtain the correct dependence of the total energy on the interatomic distance for certain simple systems. The formation energies of the hydrates of alkali-metal ions and halide ions calculated in this manner agreed with experimen-

[11] J. Simons, *Internat. J. Quantum Chem.*, 1977, **11**, 971.
[12] A. V. Bandura, N. P. Novoselov, and R. A. Evaresto, *Theor. and Exp. Chem.*, 1977, **12**, 463.

tal values within limits of 1—3 kcal mol^{-1} whereas with the normal CNDO/2 procedure they were 1.5—3 times larger than the experimental ones.

Electrostatic molecular potentials[13] have often been used to represent the potential around molecules. The resulting diagrams allow one to visualize the potential wells where solvent molecules are likely to be found. It has now been shown that to generate such potential maps, it is not necessary to determine accurate molecular wavefunctions.[14] Indeed, a model where the molecule is composed of completely localized electron pairs is often sufficient to generate very accurate potential maps.

This model was recently extended by Bonaccorsi, Scrocco, and Tomasi[15] to include the polarization of the solute by the solvent. These authors derived an analytical expression of the polarization term in the interaction energy between a molecule and a point-like charge. The molecule is partitioned into a number of groups defined with respect to each bond and each lone pair, and suitable group polarizabilities are associated with the groups. The polarization energy is calculated with similar procedures on the same basis set as the electrostatic term, permitting one to compare numerical values of similar accuracy. The method was used to study the approach of a point charge along two opposite directions of the molecular axis of HCN and along the nitrogen lone-pair axis for several amines.

Some attempts have also been made to develop quantum mechanical models for dipole–dipole and van der Waals interactions in molecular systems. Gavezzotti and Simonetta[16] developed a model to account for intermolecular co-operation and dipole–dipole interactions during molecular rearrangements in organic crystals. They found that the simplest way to study the energetic changes accompanying the reorientational motion of crystals was to consider a molecular cluster, made up of a finite number of molecules, and to calculate the pairwise interactions of each molecule with all others in the cluster. A given rearrangement was visualized as the displacement of one or more internal co-ordinates of one fundamental molecule. Then one or more other molecules in the cluster were allowed appropriate motions with respect to which the cluster energy was minimized. The model was successfully applied to the study of the role of intermolecular co-operation in lowering rotational barriers of some naphthalene and benzene derivatives.

For van der Waals systems, a valence-bond-type method has been used to calculate the long-range interaction potentials between molecules.[17] Calculations showed that the anisotropic electrostatic interactions, even for compounds with no net dipole, were of the same order of magnitude as the dispersion terms. The anisotropic forces are believed to have important effects on some crystal properties.

Solvation of Monoatomic Ions.—A great number of quantum mechanical studies have been made to investigate the solvation of various cations and anions in water and other solvents. One extensive Monte Carlo study by Clementi[6] dealt with the hydration of alkali-metal ions in clusters of up to 200 water molecules. The study produced valuable information on the cluster size and shape around the ions and

[13] R. Bonaccorsi, E. Scrocco, and J. Tomasi, *J. Chem. Phys.*, 1970, **52**, 5270.
[14] R. Bonaccorsi, E. Scrocco, and J. Tomasi, *J. Amer. Chem. Soc.*, 1977, **99**, 4546.
[15] R. Bonaccorsi, E. Scrocco, and J. Tomasi, *Theor. Chim. Acta*, 1977, **43**, 63.
[16] A. Gavezzotti and M. Simonetta, *Acta Cryst.*, 1976, **A32**, 997.
[17] P. E. S. Wormer, F. Mulder, and A. Van Der Avoird, *Internat. J. Quantum Chem.*, 1977, **11**, 959.

gave reasonable estimates of the co-ordination numbers and hydration-shell radii of alkali-metal cations.

Using the supermolecule technique, Pullman[18] also studied the solvation and fixation of the metal cations of Groups I and II. She found that the bond energies for the cations interacting with H_2O increased in the following order: $Be^{2+} > Mg^{2+} > Ca^{2+} > Li^{+} > Na^{+} > K^{+}$. This order was rationalized on the basis of two observations: first, the electrostatic attraction for dications should be larger than for monocations and, secondly, within each group of metals the exchange repulsion term favours the ion of smaller ionic radius. The same order of bond energies was determined in another non-empirical investigation of these cations interacting with HF and NH_3 as solvents.[19] On going from HF to H_2O to NH_3 the bond energy of the ion–solvent complex increased. These results were also explained in terms of the electronegativity and effective size of the F, O, and N atoms involved in the donor–acceptor interactions.

Ab initio calculations[20] for the solvation of Al^{3+} and Cu^{2+} failed to reproduce the experimental hydration numbers and hydration enthalpies when solvation shells beyond the first were neglected. However, a model of Al^{3+} with six water molecules in the first hydration shell and twelve in the second shell yielded a computed hydration energy of 1041 kcal mol^{-1}, in reasonable agreement with the experimental enthalpy of 1116 kcal mol^{-1}. This model also accounts for the S_N1 mechanism proposed for the exchange of water molecules between the hydration sphere of Al^{3+} and the bulk solvent. Provided that the exchange is considered to take place between the first and second hydration shells, the activation energy for the S_N1 process (1) is 21 kcal mol^{-1}, compared with 38 kcal mol^{-1} for the S_N2 process (2).

$$[Al(H_2O)_6]^{3+} \xrightarrow{S_N1} [Al(H_2O)_5(H_2O^{2s})]^{3+} \qquad (1)$$

$$[Al(H_2O)_6(H_2O^{2s})]^{3+} \xrightarrow{S_N2} [Al(H_2O)_7]^{3+} \qquad (2)$$

In general, quantum mechanical studies on the solvation of anions are inherently more difficult than those for cations owing to the size and polarizability of the anions. We previously mentioned some difficulties encountered in the calculations of anion–solvent systems using point dipoles to represent the solvent.[4] One successful anion–solvent study was a recent CNDO/2 investigation of the optimal co-ordination number (n) of the aquo-complexes of Br^- and I^-. It was found that the optimal co-ordination occurs when these ions are surrounded by six water molecules.[21]

Solvation of Organic Cations.—A number of attempts have been made to investigate the influence of a solvent on the relative stability of organic cations. For example, McManus and Worley[22] studied halonium ion–carbocation equilibria (3) using MINDO/3 to evaluate the importance of carbocation solvation. All of the cyclic chloronium ions which they studied were 10—30 kcal mol^{-1} more stable than

[18] A. Pullman, *Bull. Soc. chim. belges.*, 1976, **85**, 963.
[19] V. M. Pinchuk, Y. A. Kruglyak, and M. D. Dolgushin, *Theor. and Exp. Chem.*, 1977, **12**, 116.
[20] H. Veillard, *J. Amer. Chem. Soc.*, 1977, **99**, 7194.
[21] V. B. Volkov and D. A. Zhogolev, *Chem. Phys. Letters*, 1977, **49**, 591.
[22] S. P. McManus and S. D. Worley, *Tetrahedron Letters*, 1977, 555.

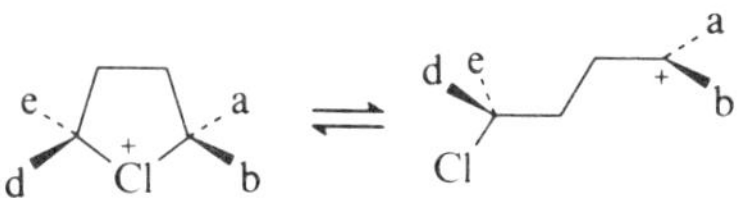

(3)

their open chloroalkyl carbocation isomers, suggesting that chlorine is quite effective in internally solvating the carbocations in the absence of solvent. MINDO/3 predicted that a tertiary carbocation is more than 10 kcal mol^{-1} less stable than its isomeric chloronium ion. Since the experimental value is known to be between 0 and 2 kcal mol^{-1} in solution in SO_2ClF they concluded that the solvent must supply a substantial amount of stabilization to the open carbocation.

Jorgensen[23] reported MINDO/3 and perturbation theory calculations for complexes of carbocations with an HCl molecule to provide insight into the relative extent of specific solvation at carbocation centers. He found that specific solvation at positively charged carbons becomes less favourable with increasing charge delocalization. For example, the relatively localized homocubyl cation (3) was better stabilized[24] by the solvated leaving group than the more delocalized pyramidal ion (4) and the bishomoaromatic ion (5), having the relative energies 0.0, 3.2, and 6.6 kcal mol^{-1} respectively. Interestingly enough, the gas-phase energies are in a distinctly different order, (4)<(5)<(3). Thus, it is concluded that the relative energies of isomeric carbonium ions may vary appreciably from solution to the gas phase, and caution must be exercised in using relative energies of carbonium ions in the gas phase as a gauge for the relative energies of intermediates in solution.

(3) (4) (5)

Since Jorgensen's model considered explicitly only interactions with one solvent molecule, it was believed that introducing additional solvent molecules could eliminate the discrepancy between the solvation energy of the delocalized and localized isomers. To probe this issue more thoroughly, Jorgensen and Munroe[25] studied the influence of increasing solvation on the relative energies of bisected and bridged ethyl cations. Their MINDO/3 results revealed that the solvation energy difference for these cations showed no sign of being reduced even after addition of up to five solvent molecules, supporting their view that additional solvation will not return the energetic order to that for the isolated ions in the gas phase.

Another study of interactions between solvent and organic ions involved an investigation of the nature of solvation and hydrogen-bonding of pyridinium ions in H_2O at the STO-3G level.[26] Relative solvation enthalpies for the transfer of a series of pyridinium ions from the gas phase to water were calculated by combining

[23] W. L. Jorgensen, *J. Amer. Chem. Soc.*, 1977, **99**, 280.
[24] W. L. Jorgensen, *J. Amer. Chem. Soc.*, 1977, **99**, 4272.
[25] W. L. Jorgensen and J. E. Munroe, *Tetrahedron Letters*, 1977, 581.
[26] E. M. Arnett, B. Chawla, L. Bell, M. Taagepera, W. J. Hehre, and R. W. Taft, *J. Amer. Chem. Soc.*, 1977, **99**, 5729.

relative heats of ionization in the gas phase and in water with relative heats of solvation of the corresponding neutral pyridines. Both the hydrogen bonds from water to the pyridines (6) and from the pyridinium ions to water (7) were investigated. As expected, the hydrogen bond between the pyridinium ions and water was very strong and accounted for most of the difference between the gas-phase basicities and those in water.

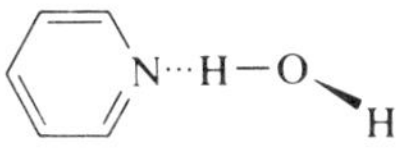

(6)

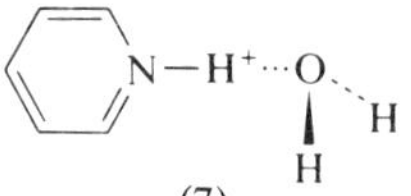

(7)

Theoretical calculations have also been used to study the influence of solvent on the isomer distribution of electrophilic reactions. For example, Rayez and Dannenberg[27] carried out INDO calculations for *ortho*-, *meta*-, and *para*-protonated toluene in the gas phase and in the presence of one or two trifluoroacetic acid solvent molecules. The calculated percentage of *ortho*, *meta*-, and *para*-products was found to be in agreement with experiment. The *para*/*ortho* ratio significantly decreased upon solvation whereas the *para*/*meta* ratio remained large, supporting the idea that relative cationic solvation might be quite important in consideration of normal organic reactivities.

Association of Monoatomic Ions with Organic Molecules.—The stability of various conformers of a molecule can be affected by co-ordination with cations. For example, CNDO/2 calculations have shown that the non-planar conformers of but-1-ene,[28] which are less stable in the free state than the planar *cis* form, have larger stabilization energies when co-ordinated with Na^+ ions These changes are believed to be particularly important in biochemical molecules, where the presence of a cation may favour a structure susceptible to producing specific enzymatic action. It is thus not surprising to find that a number of such studies have found their way into a theoretical investigation.

Pullman and co-workers have written an interesting series of papers on the binding of cations to biomolecules. *Ab initio* computations were performed on the binding of the Na^+ ion to the purine and pyrimidine bases of the nucleic acids.[29] The bases uracil (8), cytosine (9), guanine (10), and adenine (11) are shown, along

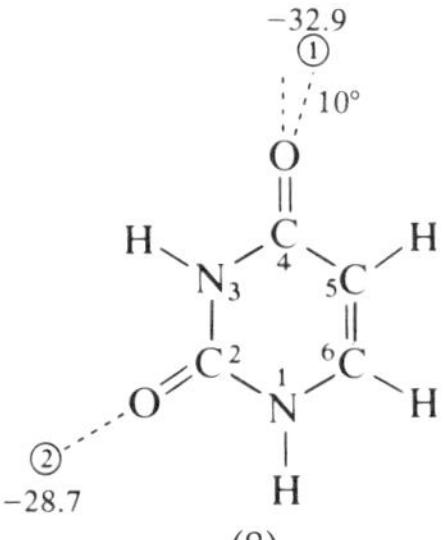

(8)

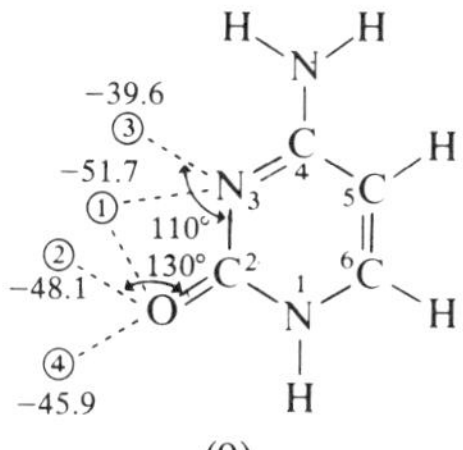

(9)

[27] J. C. Rayez and J. J. Dannenberg, *Tetrahedron Letters*, 1977, 671.
[28] V. W. Meiler, D. Deininger, and D. Michel, *Z. phys. Chem. (Leipzig)*, 1977, **258**, 139.
[29] D. Perahia, A. Pullman, and B. Pullman, *Theor. Chim. Acta*, 1977, **43**, 207.

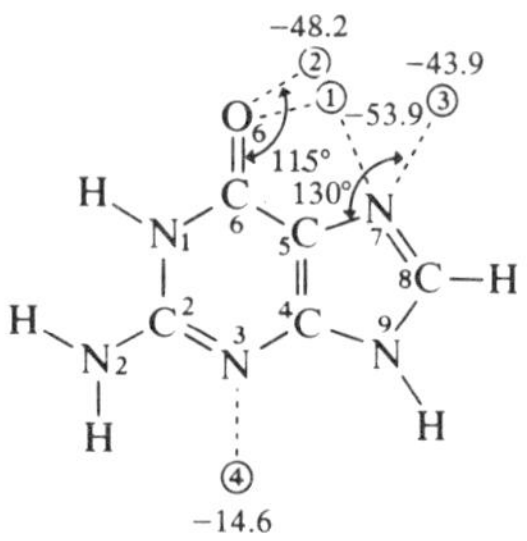

(10)

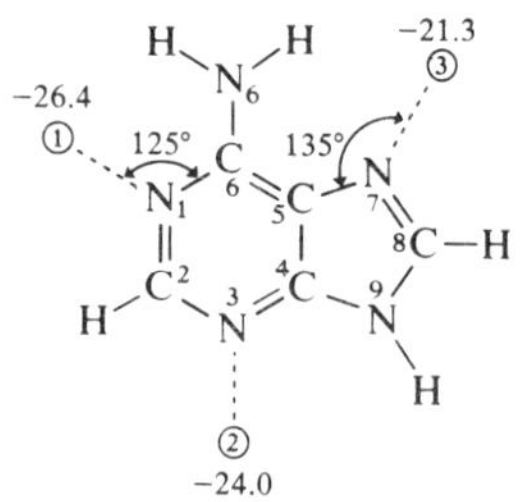

(11)

with the favoured binding site of the cation. The calculated binding energies, in kcal mol^{-1}, were −32.9, −51.73, −53.89, and −26.35 for (8), (9), (10), and (11) respectively. These computations indicate two principal differences between these interactions. First, for bases containing both oxygen and nitrogen, cation binding at an individual oxygen binding site is favoured over binding at the nitrogen atom. This inversion in the intrinsic binding ability of Na^+ compared with H^+ is due to the exchange repulsive component of the binding energy which, for the same distance, is larger for nitrogen than for oxygen. Secondly, we note the appearance in cytosine and guanine of bridged positions between a nitrogen and a carbonyl oxygen as the preferred binding sites. In the electrostatic potential energy maps for free cytosine and guanine, the nitrogen and oxygen atoms were associated with separate minima and the bridge position was less favourable. However, the equilibrium distance for a proton is smaller than for Na^+, so that the proton can move into the potential minima close to the heteroatoms.

In a second paper of the series, Pullman *et al.*[30] reported *ab initio* computations on the binding of alkali-metal and alkaline-earth cations to the phosphate group to determine the effects of binding on the conformational properties of the phosphodiester linkage and the polar head of phospholipids. Using an extended STO-3G basis set, they studied the interactions of Na^+, Mg^{2+}, K^+, and Ca^{2+} with the dimethyl phosphate anion [(12) as the model for the phosphodiester linkage] at a bridged (B) and an external (E) site with three fundamental conformations. In all cases, strongest binding corresponded to the B-site, the two bivalent cations binding more strongly than the univalent ones. Fixation of the cations at the B-site did not influence the order of conformational preference with respect to the torsion about the $P{-}O_{ester}$ bonds existing in the free molecule. However, binding of cations to the external E-site was able to produce substantial modification in the order of conformational stabilities. Thus, depending on the site of binding, cations may or may not perturb the intrinsic conformational preference of the phosphodiester linkage.

The authors also studied the interaction of Na^+ at the B- and E-sites of ethanolamine phosphate (EP) (13) in an attempt to investigate the effects of cation binding on the conformational properties of phospholipids. The intrinsically preferred conformation for the free EP was a highly folded structure associated with $(\alpha_4, \alpha_5) = (270°, 30°)$. The most stable form of the Na^+,EP adduct was substantially

[30] B. Pullman, N. Gresh, H. Berthod, and A. Pullman, *Theor. Chim. Acta*, 1977, **44**, 151.

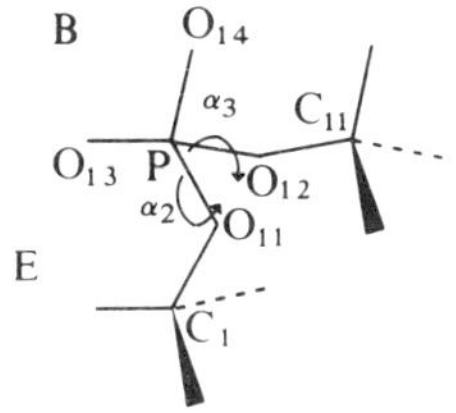

(12)

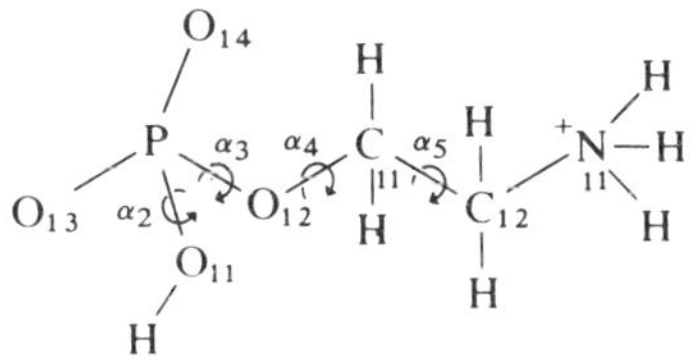

(13)

different however, with $(\alpha_4, \alpha_5) = (-30°, 180°)$. Thus, fixation of the cation at B had the effect of extending the structure of the polar head with respect to α_4. On the other hand, considering Na^+ binding with the E-site, the most stable conformation is very nearly the same as that for free EP. Again, it was found that the site of binding may have a profound influence on the conformation of a ligand.

3 Theoretical Aspects of Hydrogen Bonding

Studies on its Origin and Nature.—Among the various factors involved in molecular association, hydrogen-bonding is probably the most important. It is thus not surprising to find that the study of its origin and nature continues to generate considerable theoretical interest. Various energy-decomposition schemes have been used in the past to describe the factors responsible for its occurrence. Most of them attempt to separate the total interaction energy into electrostatic, exchange, and polarization energies. Morokuma *et al.*[31] now propose a new energy-decomposition scheme where the influence of charge transfer is evaluated as well. The procedure involves the manipulation of a matrix generated by the Hartree–Fock molecular orbitals of the isolated molecules. In their article, the authors describe the specific steps for modifying the Gaussian 70 package for the calculation of these new energy components.

This energy-decomposition analysis was applied to the study of $(H_2O)_2$, $(HF)_2$, $H_3N\cdots HF$, and other complexes in which the proton donor is HF, H_2O, NH_3, or CH_4 and the proton acceptor is HF, H_2O, or NH_3.[32] The essential energy components in hydrogen-bonding include electrostatic, polarization, exchange repulsion, charge transfer, and their coupling. For 'normal' hydrogen bonds in which the proton acceptor is F, O, or N and the proton donor is a polar bond F—H, O—H, or N—H, the hydrogen bond is strongly electrostatic in nature, with a small but significant contribution from charge transfer. Although charge transfer plays only a minor role in strong, more electrostatic hydrogen bonds such as $H_3N\cdots HF$, it is essential for the stabilization of weaker bonds such as $H_3N\cdots HCH_3$ and $H_2O\cdots HCH_3$. During hydrogen-bond formation, the electron density on the hydrogen-bonding proton decreases owing to exchange repulsion and polarization, while its concentration on the acceptor atom is principally due to charge transfer and polarization. Charge-redistribution effects in the non-interacting parts of the donor and acceptor are controlled by the polarization term and generally follow an alternating charge pattern. The energy components of some lithium complexes

[31] K. Kitaura and K. Morokuma, *Internat. J. Quantum Chem.*, 1977, **10**, 325.
[32] H. Umeyama and K. Morokuma, *J. Amer. Chem. Soc.*, 1977, **99**, 1316.

$(LiF)_2$ and $(LiH)_2$ were compared with the hydrogen-bonded $(HF)_2$ complex. Based on these comparative findings, it appears that hydrogen-bonding is a unique type of association which always involves a short, polar, and strong H—X bond as the proton donor. Since the polarity is moderate, the electrostatic energy is not too great, and exchange repulsion prevents the hydrogen from approaching too closely to the proton acceptor. Because of all these effects, hydrogen bonds are always intermediate to weak interactions.

A similar conclusion was reached by Dolgushin *et al.*[33] These authors investigated the nature of the hydrogen bond by a comparative analysis of the interactions of Li^+, F^-, OH^-, and H_2O with H_2O. In agreement with Morokuma's results they found the exchange terms to be repulsive but largely offset by the larger electrostatic attraction. In addition, they found that the total charge of the proton-donor or proton-acceptor system is the critical point for hydrogen-bond formation. If the donor system is negatively charged or if the acceptor system is positively charged, no hydrogen bond results. If, on the other hand, the acceptor system is negatively charged, a strong hydrogen bond can exist.

Clark *et al.*[34] investigated the electronic reorganizations accompanying core and valence-shell ionization in some simple hydrogen-bonded systems, including $H_3N\cdots H_2O$, $(H_2O)_2$, $HF\cdots H_2O$, and $(H_2O)_3$, using a 4-31G and double-zeta basis set. The relaxation energies for a given core hole were found to increase on going from monomer to dimer, irrespective of whether the binding energy for the core level increased or decreased. Core holes in the dimers caused substantial changes in the hydrogen-bond energies compared with the neutral systems.

Structural Studies.—A number of studies were aimed at investigating the role of intramolecular and intermolecular hydrogen-bonding in determining the stereochemistry of molecular systems. For example, Newton *et al.*[35] attempted to determine the role of intramolecular factors in the planarity of the carbon–oxygen framework in the α-hydroxycarbonyl moiety of many α-hydroxy-acids and carboxylates. STO 4-31G calculations on three α-hydroxycarbonyl systems, glycolic acid (14), the glycolate anion (15), and glycoaldehyde (16), gave equilibrium

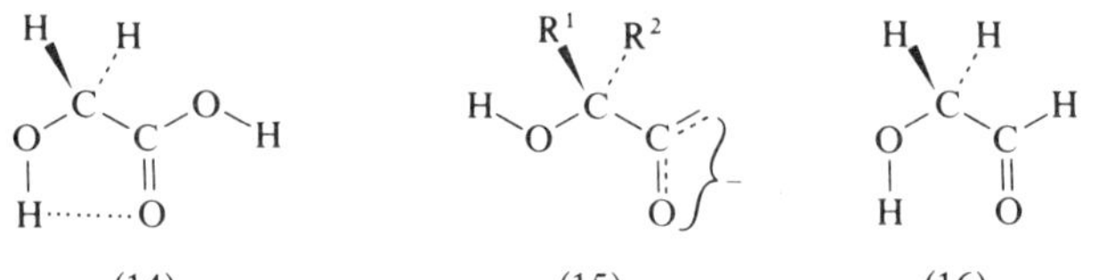

(14) (15) (16)

conformations corresponding to internally hydrogen-bonded structures. Rotations of the α-hydroxy-group were calculated to cost much less energy than distortions which destroyed the planarity of the carbon–oxygen framework. These results were consistent with the crystal-state structures of the molecules where intermolecular hydrogen-bonding takes precedence over internal hydrogen-bonding. Although the relative stabilities of these structures may be partly explained in

33 M. D. Dolgushin and V. M. Pinchuk, *Theor. Chim. Acta*, 1977, **45**, 157.
34 D. T. Clark and B. J. Cromarty, *Theor. Chim. Acta*, 1977, **44**, 181.
35 M. D. Newton and G. A. Jeffrey, *J. Amer. Chem. Soc.*, 1977, **99**, 2413.

terms of intramolecular hydrogen bonds, analysis of the charge distribution indicated that the short-range O—H···O effects are smaller than those generally found in intermolecular hydrogen-bonding. Furthermore no weakening of the O—H bond of the α-hydroxy-group was found when the acid and the aldehyde were compared with non-hydrogen-bonded conformations.

In a related study, Vishveshwara and Pople[36] investigated the stereochemistry of the α-amino counterparts of glycolic acid (14) and glycoaldehyde (16). Using a 4-31G basis set, they calculated the equilibrium geometry of glycine (17) and α-aminoacetaldehyde (18) as well as the effect of rotation of the α-amino-group on the relative stabilities of these compounds. The most stable conformations corresponded to structures in which the amine group forms two bifurcated hydrogen bonds with the carbonyl oxygen. The rotational potential energy surfaces of the acid and the aldehyde also showed two other stable conformations. For glycine, this corresponded to a structure with a hydrogen bond between the hydroxyl proton and nitrogen (19), lying only 2.2 kcal mol^{-1} above (17). For α-aminoacetaldehyde, the second local minimum lay only 1.4 kcal mol^{-1} above (18) and was stabilized by a strong interaction between the aldehyde proton and the nitrogen lone-pair electrons (20).

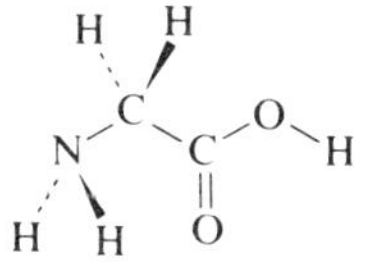

(17)

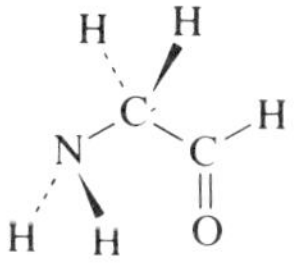

(18)

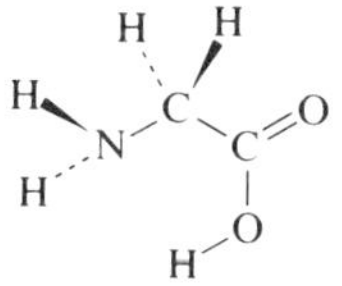

(19)

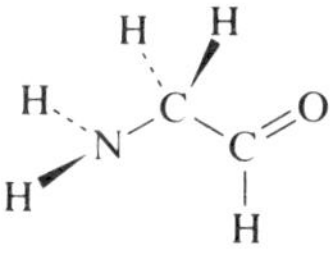

(20)

Quantum mechanical methods were also used to study the effects of hydrogen-bonding on peptide bonds. In one such study,[37] the peptide unit was modelled by *trans-N*-methylacetamide (NMA) (21), which was allowed to interact with various hydrogen-bonding species such as water, formamide, imidazole, $CH_3NH_3^+$, and $HCOO^-$. PRDDO calculations indicated that all of the species made non-planar deformations of the peptide unit more difficult. The effects of these species on the flexibility and electronic charge distribution of NMA can be understood by recognizing that hydrogen bonds to either the carbonyl or the amine group of the peptide linkage will stabilize (22) more than (21). Partial and full hydration of the peptide

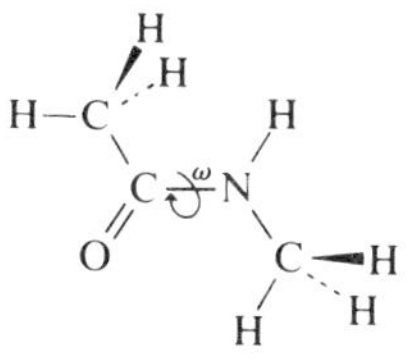

(21)

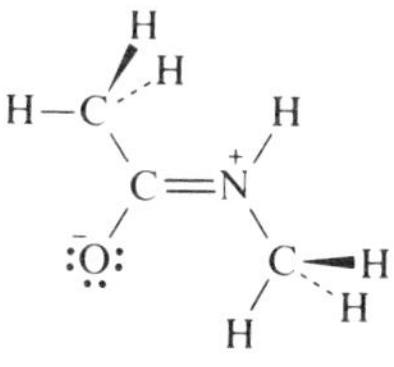

(22)

unit were also studied. It was found that many of the effects of a full hydration shell could be simulated by the interaction of only two water molecules.

[36] S. Vishveshwara and J. Pople, *J. Amer. Chem. Soc.*, 1977, **98**, 2422.
[37] S. Scheiner and C. W. Kern, *J. Amer. Chem. Soc.*, 1977, **99**, 7042.

In a somewhat related work, molecular orbital calculations were used to study the hydrolysis of formamide as a model for the hydrolysis of peptides by carboxypeptidase A.[38] A proton donor H_3O^+ was positioned near the nitrogen atom of the model substrate. Nucleophilic attack by a water dimer, a model for the basic form of Glu-270, led to hydrolysis of the peptide (23). A Li cation positioned proximate

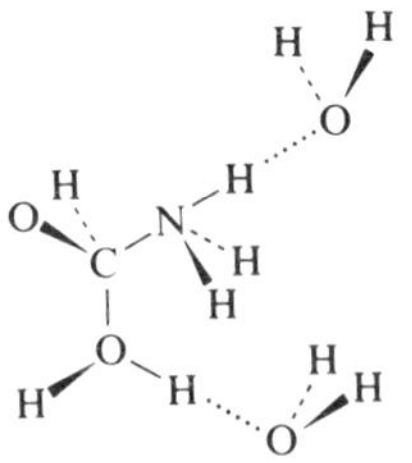

(23)

to the carbonyl oxygen was seen to facilitate hydrolysis. The electrophile first polarizes the carbonyl bond of the substrate, reducing the negative charge on the oxygen and making the carbon more susceptible to nucleophilic attack. Li^+ and $[Be(OH)(NH_3)_2]^+$, believed to be good models for Zn^{2+} and its ligands, were found to be more effective at catalysing the hydrolysis than were various hydrogen-bonding species like NH_4^+.

Theoretical Aspects of Proton Transfer.—A number of theoretical studies have been related to the question of proton transfer in hydrogen-bonded systems. Iwata and Morokuma[39] examined the ground and various excited states of the formic acid monomer and dimer (24) with an STO-3G basis set. The two-configuration elec-

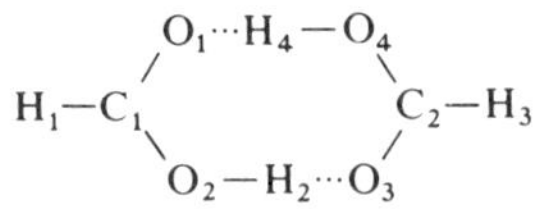

(24)

tron hole potential method was used for the calculation of the excited state of the dimers. The potential energy curves for the symmetrical simultaneous movement of two bridging protons for all of the states were studied, and it was found that proton transfer in the ground state was the one with the lowest barrier. The authors did not find an excited state with a low barrier for the hydrogen exchange. However, they note that since the transitions being considered, $n-\pi^*$ and $\pi-\pi^*$, are not directly involved in weakening the O-2—H-2 bond or strengthening the H-2···O-3 bond, the barrier should not change drastically upon excitation. As a result, states which should have low barriers for proton exchange are those involving an excitation from the O-2—H-2 σ-orbital or to the O-2—H-2 σ^*-orbital.

The potential energy surface for the proton transfer in the H_2S dimer[40] was also calculated. Despite the limited basis set used, the calculated energy of

[38] S. Scheiner and W. N. Lipscomb, *J. Amer. Chem. Soc.*, 1977, **99**, 3466.
[39] S. Iwata and K. Morokuma, *Theor. Chim. Acta*, 1977, **44**, 323.
[40] K. Pecul, *Theor. Chim. Acta*, 1977, **44**, 77.

207 kcal mol^{-1} was in fairly good agreement with the experimental value of 151—199 kcal mol^{-1} for the energy of proton transfer. On the basis of their results, the authors suggest that the H_2S dimer can be stable at any orientation of the monomers.

The behaviour of the hydrogen-bonded proton in malonaldehyde [skeleton as in (25)] continues to stir the interest of theoreticians, and its dynamics have been

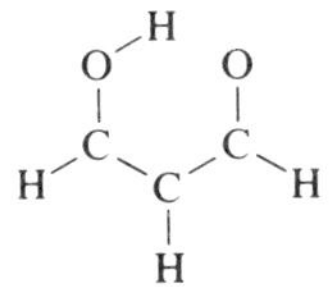

(25)

studied by determining the reaction co-ordinates from the gradients of potential obtained with a CNDO/2 routine.[41] The motion along the reaction co-ordinate was calculated, considering all vibrational motions perpendicular to the reaction co-ordinate. It was found that a rapid transformation occurs between the two asymmetric conformers through a low potential barrier. One of the interesting features of this work was the finding that the distribution of the hydrogen-bonded proton on the reaction co-ordinate varied with the increase of temperature. The mean position of the proton was symmetric in the low-temperature range and became asymmetrically oriented only at high temperature. The existence of an out-of-plane vibrational motion with a negative force constant led the investigators to suggest that the motion of the proton proceeds along an ellipse-like orbit.

The possibility of tunnelling in the proton-exchange reactions between methyl-oxonium ion and methyl alcohol, methyl alcohol and methoxide ion, hydronium ion and water, and water and hydroxide ion was investigated by STO 4-31G calculations on the proton transfer in the above systems.[42] The calculated tunnelling frequencies were about two orders of magnitude larger than the experimental values for the four systems considered. Small perturbations, such as the rotation of methyl groups, were found to destroy the symmetry of the profiles and were believed to be responsible for the discrepancies between calculated and experimental results.

In a recent paper, attempts were made to study proton-transfer mechanisms in the water dimer and trimer coupled to an environment.[43] The environment was simulated by reaction fields together with uniform external electric fields. The reaction fields were varied with respect to the direction of the dipole moment of the supermolecule, and g, the solute–surrounding coupling tensor, was used to gauge the reaction field strength. For negative g values, the proton potential curve dimer has only one minimum. For positive g values, however, a second minimum is apparent in the region corresponding to the ion-pair structure $H_3O^+\cdots OH^-$ as g increases in value (Figure 2). A potential barrier against the proton displacement from H_3O^+ toward OH^- can thus be calculated, but was seen to depend on the magnitude of g.

[41] S. Kato, K. Kato, and K. Fukui, *J. Amer. Chem. Soc.*, 1977, **99**, 684.
[42] J. H. Busch and J. R. de la Vega, *J. Amer. Chem. Soc.*, 1977, **99**, 2397.
[43] O. Tapia and E. Poulain, *Internat. J. Quantum Chem.*, 1977, **11**, 473.

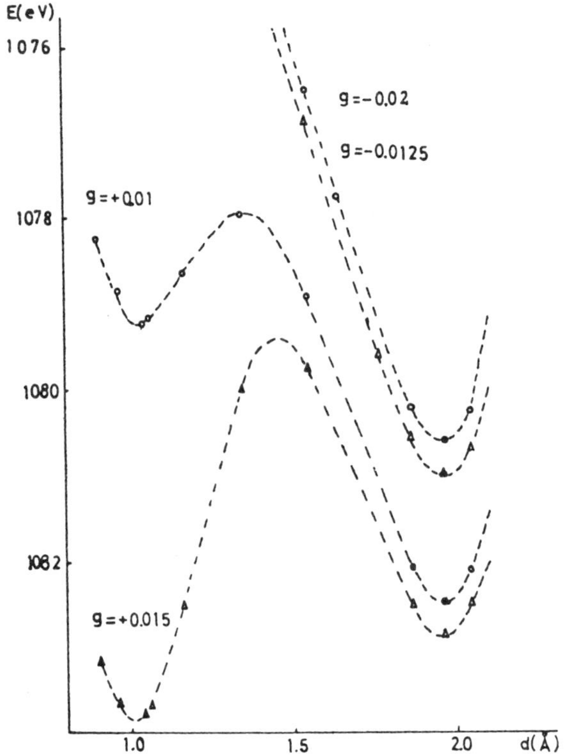

Figure 2 *Proton potential curves for the water dimer model as a function of the orientation and reaction field strength*
(Reproduced by permission from *Internat. J. Quantum Chem.*, 1977, **11**, 473)

4 Other Molecular Associations

Quantum mechanical calculations have been used to investigate van der Waals, charge-transfer, and ionic associations. For example, Lochmann *et al.*[44] used the PCILO method to analyse the individual energy contributions responsible for the stabilization of van der Waals, hydrogen-bonding, and charge-transfer complexes. The nonane dimer, representing a van der Waals type interaction, was found to be stabilized by -0.85 kcal mol^{-1} while the water dimer and a Li^+–*cis*-but-2-ene complex were stabilized by -4.15 and -3.60 kcal mol^{-1} respectively. The zeroth-order energy involving an electrostatic interaction was destabilizing for all systems. Stabilization of the nonane dimer was seen to arise from inter-bond correlation as well as delocalization, but the delocalization contribution was significantly more important in the other two complexes considered. In a second paper, Lochmann calculated intermolecular interactions, again using the supermolecule approach, within the PCILO method but this time using fixed polarities.[45] By employing fixed

[44] R. Lochmann and T. Weller, *Internat. J. Quantum Chem.*, 1976, **10**, 909.
[45] R. Lochmann, *Internat. J. Quantum Chem.*, 1977, **11**, 293.

polarities, determined by minimization of the polarization energy of the isolated subsystems of the complex for all distances between them, his PCILO calculations were without any iteration cycle. Both the stabilization energies and equilibrium distances for the nonane dimer, water dimer, and a benzene–tetracyanoethylene complex were in good agreement with the calculated values using optimized wave-functions.

A more sophisticated analysis of non-covalent interactions was presented by Kollman,[46] using a STO 4-31G basis set and the Morokuma energy-decomposition scheme. The interaction energies and equilibrium structures for a wide variety of intermolecular complexes were presented, including van der Waals molecules, hydrogen-bonded complexes, charge-transfer complexes, radical complexes, and three-body interactions. It was found that the electrostatic energy and electrostatic potential are the most useful guides to qualitative and semi-quantitative predictions of the energies of non-covalent interactions. The electrostatic energy also correctly predicts the directionality of many interactions, including $(Cl_2)_2$, $(HF)_2$, $(HCl)_2$, $F^- \cdots H_2O$, and $Cl^- \cdots H_2O$. Consequently, it was suggested that a general equation can be used to calculate the interaction energy for many non-covalent interactions involving Lewis acids and bases. The procedure was found to be successful in a host of test cases, including $(Cl_2)_2$, $CO_2 \cdots H_2O$, $SO_2 \cdots NH_3$, $Li^+ \cdots OH_2$, $F^- \cdots H_2O$, $NH_4^+ \cdots F^-$, and $F^- \cdots HF$.

In a related paper, Kollman *et al.*[47] calculated the proton affinities, Li^+ affinities, and hydrogen-bond affinities for some simple bases. As before, they found that the electrostatic potential is a good guide to predicting the relative H^+, HF, and Li^+ affinity for basic sites. However, the order of the basicities of various compounds varied significantly, depending on the nature of the acid and on the substituent on the basic site. Specifically, the methyl substituent effect was found to be quite different for fluoro and amine bases. In the case of fluoro bases, electrostatic and charge-redistribution effects reinforce each other; in amine bases they oppose each other.

STO-3G and 4-31G calculations have been used to determine the relative proton affinities of pyridine and the diazines.[48] The experimentally observed order (pyridine > 1,2-diazine > 1,3-diazine > 1,4-diazine) was reproduced only at the 4-31G level. In the protonated ions, increasing stability correlated with increasing an orbital energy of the base. This contrasts with the behaviour of the hydrogen-bonded complexes, where increasing stability correlated with increasing *p*-character of the nitrogen lone-pair orbital.

A series of electron donor–acceptor complexes of halogens was studied by Umeyama *et al.*[49] Based on the energy components essential for binding, the complexes were qualitatively classified as weak electrostatic–charge-transfer complexes, intermediate electrostatic complexes, weak electrostatic complexes, and very weak dispersion–charge-transfer complexes. *N*-Methyl substituent effects were studied in a number of donor–NH_2R systems. It was found that methyl substitution in H_2N–ClF resulted in a small destabilization due to cancellation of

[46] P. Kollman, *J. Amer. Chem. Soc.*, 1977, **99**, 4875.
[47] P. Kollman and S. Rothenberg, *J. Amer. Chem. Soc.*, 1977, **99**, 1333.
[48] J. E. Del Bene, *J. Amer. Chem. Soc.*, 1977, **99**, 3617.
[49] H. Umeyama, K. Morokuma, and S. Yamabe, *J. Amer. Chem. Soc.*, 1977, **99**, 330.

exchange repulsion and charge-transfer terms (see Table). This contrasted with a small stabilization effect for H_3N–BH_3 due to polarization–exchange repulsion stabilization and large stabilization for $H_3N\cdots H^+$, which was found to result from a polarization effect.

Table *Comparison of interaction energy components and* N-*methyl substituent effects between various complexes at equilibrium geometry in* kcal mol^{-1}.

		H_3N–ClF	H_3N–BH_3	H_3N–H^+
	R/Å	2.717	1.705	1.02
Total interaction energy	ΔE	−8.23 (0.29)	−44.7 (−0.8)	−221.9 (−8.5)
Electrostatic	ES	−11.18 (0.32)	−92.9 (−1.2)	−99.8 (3.3)
Exchange repulsion	EX	7.41 (0.53)	86.9 (4.4)	0.0 (0.0)
Polarization	PL	−1.05 (0.02)	−17.2 (−5.0)	−27.4 (−12.8)
Charge transfer	CT	−3.59 (−0.57)	−27.1 (−1.4)	−88.3 (−3.4)
Coupling	MIX	0.19 (−0.01)	5.6 (2.4)	−6.5 (4.4)

The numbers in parentheses are the difference between the value for the CH_3H_2N complex and that for the H_3N complex. A negative number indicates that the CH_3N_2N complex is more stable and *vice versa*.

LaGrange and co-workers[50] studied a similar series of charge-transfer complexes with a 4-31G basis set. For the complexes studied, H_3N—Cl_2, H_2O—Cl_2, and HF—Cl_2, it was found that the intermolecular distance is smaller than the sum of the van der Waals radii. It was also found that the energy of formation of the complexes diminished regularly as the electronegativity of the donor atom increased.

Finally, SCF studies have been carried out on a novel class of molecular complexes involving neutral alkali-metal and halogen atoms with dipolar molecules, in particular interactions between Li, Na, or F with NH_3, H_2O, HF, PH_3, H_2S, or HCl.[51] The minimum-energy structures for the alkali-metal atom–dipolar molecule interactions are M–BH_n, where the negatively charged end of the hydride approaches the neutral atom. There is transfer of charge from the hydrides to the metal, and this leads to a rather substantial dipole moment for the complexes. It was found that the interaction energies of a lithium atom with different hydrides decrease in the order $NH_3 > H_2O > HF > PH_3 > H_2S > HCl$. The interaction of Na with the hydrides follows the same relative order as the Li complexes but the interaction energies are smaller.

[50] J. LaGrange, G. Leroy, and G. Louterman-Leloup, *Bull. Soc. chim. belges.*, 1976, **86**, 241.
[51] M. Trenary, H. F. Shaefer, and P. Kollman, *J. Amer. Chem. Soc.*, 1977, **99**, 3885.

4 Reaction Mechanisms
Part (i) Pericyclic Reactions

By R. J. BUSHBY
Department of Organic Chemistry, The University, Leeds LS2 9JT

A new, two-volume work on pericyclic reactions has been published[1] which contains a number of useful articles, particularly those where a fresh view of the literature has been obtained by dividing according to substrate type (carbanions, carbonium ions, *etc.*) rather than the more normal division by reaction type.

1 Cycloadditions and Cycloreversions

More energy surface calculations on cycloaddition reactions have been reported[2] and in an interesting paper Houk[3] has compared the results of semi-empirical and of *ab initio* MO methods. For the Diels–Alder reaction and the addition of 1,3-dipoles to olefins he concludes that most NDO semi-empirical methods favour an unsymmetrical (biradical) mode of addition whereas *ab initio* methods favour a symmetrical (concerted) transition state (TS) because neglect of overlap, even in the MINDO method, results in the filled-orbital–filled-orbital repulsions being under-estimated. In particular these NDO methods under-estimate the repulsive interaction between the two highest occupied molecular orbitals, which for an allowed cycloaddition contributes more to the destabilization of the unsymmetrical TS than to the symmetrical TS. Although this provides a simple explanation of why the results obtained by the two methods differ it does not enable one to judge which is closer to the truth as the effect of correlation on the balance between symmetric and unsymmetric TS remains largely a matter for conjecture. However, the symmetric TS seems the more likely as Salem's work on the Diels–Alder reaction had previously shown that a parallelism exists between the results of STO-3G and 4-31G + 3 × 3CI calculations.[4]

As far as the experimental evidence on [2 + 4] cycloaddition reactions is concerned, Firestone has produced another lively article[5] which contends that this points towards biradical intermediates, but most authors seem to favour the more balanced arguments forwarded by Huisgen in favour of the concerted mechanism.[6]

1 'Pericyclic Reactions', ed. A. P. Marchand and R. E. Lehr, Academic Press, New York, 1977, Vols. I and II.
2 (*a*) L. A. Burke and G. Leroy, *Theor. Chim. Acta*, 1977, **44**, 219; *cf. ibid.*, 1975, **40**, 313; (*b*) M. V. Basilevsky, A. G. Shamov, and V. A. Tikhomirov, *J. Amer. Chem. Soc.*, 1977, **99**, 1369.
3 P. Caramella, K. N. Houk, and L. N. Domelsmith, *J. Amer. Chem. Soc.*, 1977, **99**, 4511.
4 R. E. Townsend, G. Ramunni, G. Segal, W. J. Hehre, and L. Salem, *J. Amer. Chem. Soc.*, 1976, **98**, 2190.
5 R. A. Firestone, *Tetrahedron*, 1977, **33**, 3009.
6 R. Huisgen, *J. Org. Chem.*, 1976, **41**, 403.

There can be no doubt, however, that suitable substitution can divert most of these reactions into a two-step pathway and also that intermediates are involved in many thermal [2+2] cycloaddition reactions. In an interesting discussion of the nature of these intermediates and of the problems involved in their interception[7] it has once again been stressed that an over-rigid distinction cannot be drawn between the biradical (1) and the zwitterion (2). However, the reaction of tetracyanoethylene with enol ethers has been taken as a model of one involving an intermediate of high zwitterionic type.[7b]

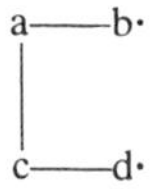

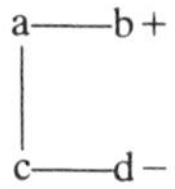

(1) (2)

Most treatments of orientation in cycloaddition reactions have been in terms of PMO theory[8] and Trong Anh and co-workers have pointed out the advantages of their simple approach to the problem.[8a,b] This concentrates on selecting the site at which bonding is first likely to occur by looking for the most important HOMO–LUMO *two-centre* interaction (rather than the *four-centre* interaction normally employed) which then dictates the course of the reaction. This method produces reliable predictions in most cases, even the troublesome case of acrolein dimerization,[8c] and the results are, furthermore, independent of the degree of concertedness of the reaction.

As well as the more traditional σ–ρ correlations,[9] localization[10] and FMO[11] approaches have also proved their value in rationalizing relative reactivities. Hence, the energy barriers for the cycloaddition of maleic anhydride to various aromatic systems may be correlated with the corresponding changes in resonance energy.[10] On the other hand, as predicted by FMO theory, values of log k for the Diels–Alder reaction of various 1-aryl-4-benzylidene-5-pyrazolones[11a] (3)

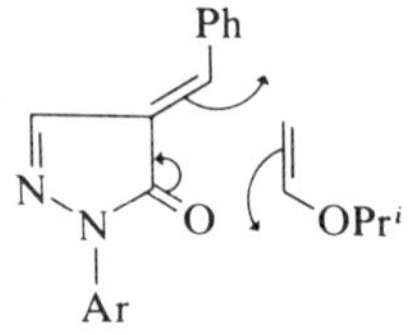

(3)

correlate well with their measured reduction potentials, and the rates of addition of diazomethane to double bonds correlate with experimental estimates of the energy difference E(HOMO, diazomethane)$-E$(LUMO, dipolarophile.[11b] When

[7] (*a*) R. Huisgen, *Accounts Chem. Res.*, 1977, **10**, 199; (*b*) *ibid.*, p. 117.

[8] (*a*) O. Eisenstein, J. M. Lefour, N. Trong Anh, and R. F. Hudson, *Tetrahedron*, 1977, **33**, 523; (*b*) C. Minot and N. Trong Anh, *ibid.*, p. 533; (*c*) V. Bachler and F. Mark, *ibid.*, p. 2857; (*d*) V. Bachler and F. Mark, *Theor. Chim. Acta*, 1976, **43**, 121.

[9] C. K. Bradsher, T. G. Wallis, I. J. Westerman, and N. A. Porter, *J. Amer. Chem. Soc.*, 1977, **99**, 2588; S. Ito and I. Saito, *Tetrahedron Letters*, 1977, 1203.

[10] W. C. Herndon, *J.C.S. Chem. Comm.*, 1977, 817; *cf.* R. D. Brown, *J. Chem. Soc.*, 1950, 691, 2730.

[11] (*a*) G. Desimoni, P. P. Righetti, E. Selva, G. Tacconi, V. Riganti, and M. Specchiarello, *Tetrahedron*, 1977, **33**, 2829; (*b*) J. Geittner, R. Huisgen, and R. Sustmann, *Tetrahedron Letters*, 1977, 881; (*c*) *ibid.*, p. 877.

however, CNDO calculated energies are employed more than just the simple FMO interactions have to be included.[11c]

Further kinetic evidence has been provided for a twisted, quasi-zwitterionic TS in the thermal retro-[2+2] reactions[12] of cyclobutanones, and among photochemical [2+2] reactions a particularly neat example has been reported by Paquette[13] in the conversion of compounds (4) into (5) *via* the caged intermediate (6).

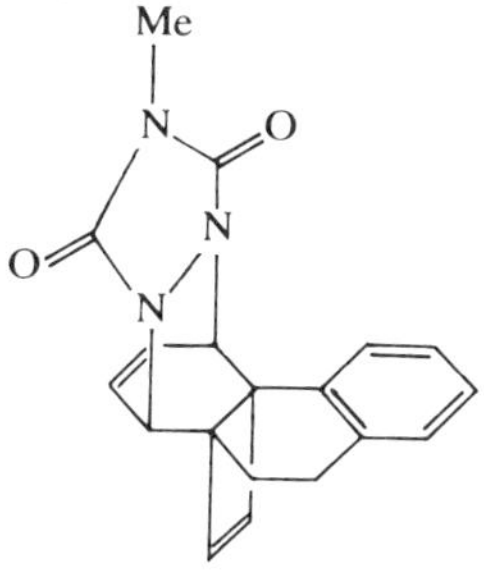

(4)

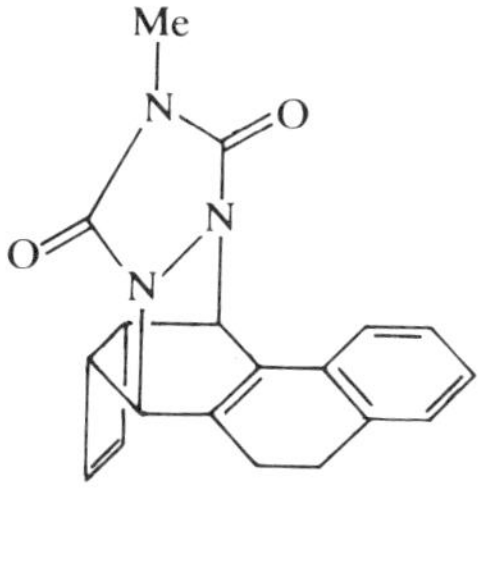

(5)

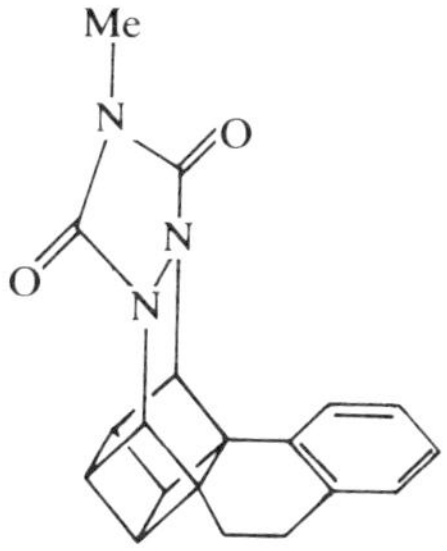

(6)

Studies of the Diels–Alder reaction have once again emphasized that factors besides secondary orbital interactions, particularly steric factors, can influence *endo/exo* ratios[14] and *X*-ray crystallography has confirmed that the adduct formed between 2 cyano-1,3-diphenylallyl anion and *trans*-stilbene also has the expected geometry (7) for a concerted [2+4] suprafacial addition process.[15] Among the related [2+4] addition reactions of 1,3-dipoles Huisgen has reported details of work on azomethine imines[16] and Padwa has continued his studies of nitrillium ylides.[17] He has suggested[17a] that the observation that ylide (8; R = H, CF_3, or aryl) gives adduct (9), whereas the corresponding ylide in which R is alkyl gives adduct (10), can be rationalized by consideration of the geometry of the intermediate.[18] Hence the intermediate with R = H or aryl is thought to be linear

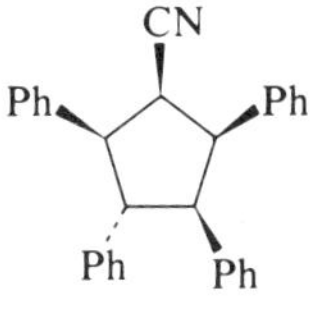

(7)

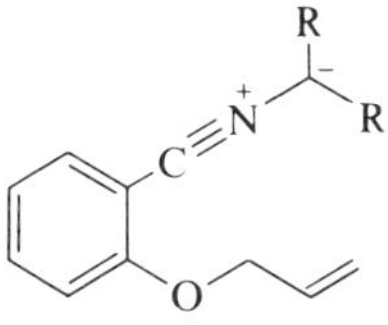

(8)

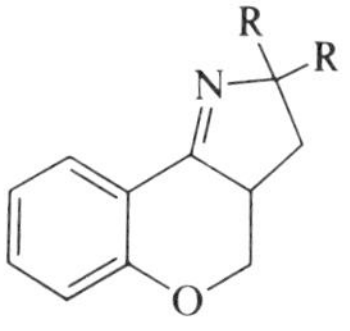

(9)

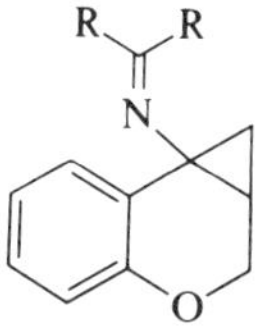

(10)

[12] H. M. Frey and R. A. Smith, *J.C.S. Perkin II*, 1977, 752; *ibid.*, p. 2082.
[13] L. A. Paquette, T. G. Wallis, K. Hirotsu, and J. Clardy, *J. Amer. Chem. Soc.*, 1977, **99**, 2815.
[14] T. Sasaki, K. Kanematsu, K. Iizuka, and N. Izumichi, *Tetrahedron*, 1976, **32**, 2879; D. W. Jones, *J.C.S. Perkin I*, 1977, 980; K. Seguchi, A. Sera, and K. Maruyama, *Bull. Chem. Soc. Japan*, 1976, **49**, 3558.
[15] W. T. Ford and G. F. Luteri, *J. Amer. Chem. Soc.*, 1977, **99**, 5330.
[16] R. Huisgen and A. Eckell, *Chem. Ber.*, 1977, **110**, 522, 540, 559, 571; R. Huisgen, M. V. George, A. S. Kende, and A. Eckell, *ibid.*, p. 578.
[17] (*a*) A. Padwa, P. H. J. Carlsen, and A. Ku, *J. Amer. Chem. Soc.*, 1977, **99**, 2798; (*b*) A. Padwa and P. H. J. Carlsen, *ibid.*, p. 1514; A. Padwa and N. Kamigata, *ibid.*, p. 1871.
[18] P. Caramella, R. W. Gandour, J. A. Hall, C. S. Deville, and K. N. Houk, *J. Amer. Chem. Soc.*, 1977, **99**, 385.

and to behave as a 1,3-dipole but the intermediate with R = alkyl is thought to be bent, with high resultant carbene character.

An independent synthesis and study of the reactions of compound (11)[19] has shown that of the two adducts formed between tropone and cyclopentadiene, (12) and (13), the first probably arises by initial formation of (11) followed by a Cope

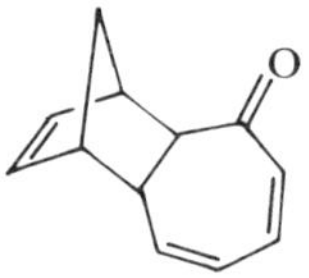

(11)

(12)

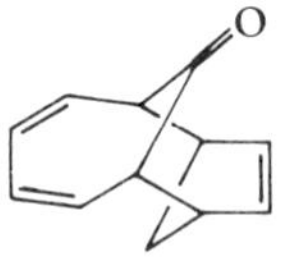

(13)

rearrangement but that the second cannot be formed from this intermediate. It is probably the result of a concerted [4+6] cycloaddition. This completes an interesting analogy between this reaction and the reaction of cycloheptatriene and 2,5-dimethyl-3,4-diphenylcyclopentadienone.[20]

Whereas it has now been proved that the reported [2+6] adduct between *N*-ethoxyazepine and diethyl azodicarboxylate is in fact a normal Diels–Alder product[21a] the formation of an analogous adduct (14) with nitrosobenzene has been confirmed by *X*-ray crystallography.[21b]

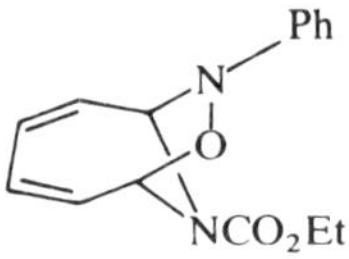

(14)

The reaction of electron-deficient acetylenic dienophiles with the pentalenes (15) (X = H, CHO, CN, or CO_2Me) has been shown to involve attack on the electron-rich ring to give the [2+8] adducts (16) whereas cyclopentadiene attacks the electron-poor ring to give the [2+4] adducts (17).[22] It is also interesting to note

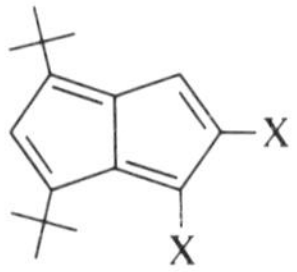

(15)

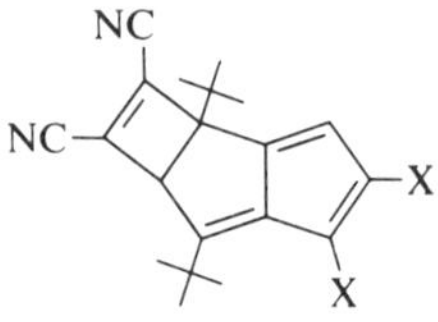

(16)

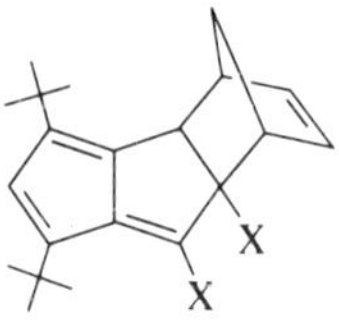

(17)

that although 8-phenylheptafulvene seems to give with tetracyanoethylene the expected [2+8] adduct (18), the 8,8-diphenyl compound gives a [2+4] adduct

[19] M. Franck-Neumann and D. Martina, *Tetrahedron Letters*, 1977, 2293.
[20] K. N. Houk and R. B. Woodward, *J. Amer. Chem. Soc.*, 1970, **92**, 4143.
[21] (*a*) W. S. Murphy and K. P. Raman, *J.C.S. Perkin I*, 1977, 1824; (*b*) W. S. Murphy, K. P. Raman, and B. Hathaway, *ibid.*, 2521.
[22] M. Suda and K. Hafner, *Tetrahedron Letters*, 1977, 2449, 2453; see also B. Kitschke and H. J. Lindner, *ibid.*, p. 2511.

(19).[23] In this case, presumably, steric rather than electronic factors are responsible for the change in selectivity. A remarkable formal [2+8] adduct (20), whose structure has been confirmed by *X*-ray crystallography, is also formed in the reaction of 4-phenyl-1,2,4-triazoline-3,5-dione with octalene.[24]

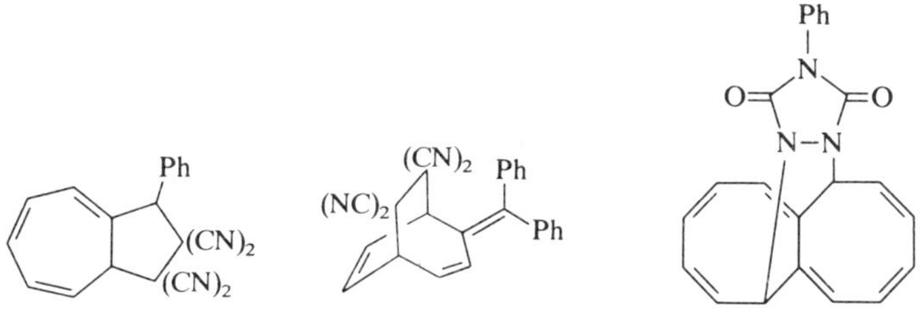

(18) (19) (20)

Aspects of the elimination of nitrogen[25] and of sulphur dioxide[26] have been reviewed and a detailed study of the elimination of nitrogen from optically active *cis*- and *trans*-3-ethyl-5-methylpyrazolines[27] has shown that the results for the two isomers differ, so that no universal stereochemical course can be assigned to these reactions. It has also been shown that elimination of nitrogen is enormously accelerated by the introduction of a trimethylsilyl group in the 4-position.[28] This has been rationalized in terms of concerted nitrogen loss and silyl migration (21). Hammett LFER studies of the elimination of sulphur dioxide from the thiiren dioxides (22) have provided evidence for a stepwise process[29] and it has been shown that elimination of sulphur monoxide from the *cis*- and *trans*-thiiran oxide (23) proceeds with 95% retention of stereochemistry,[30] in contrast to earlier results for related systems.[30b]

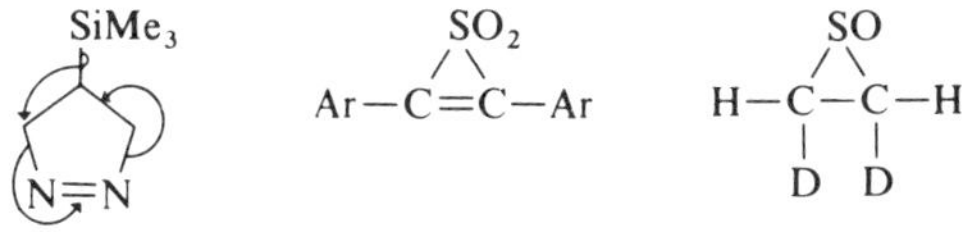

(21) (22) (23)

2 Sigmatropic Reactions

STO-3G and 4-31G *ab initio* MO calculations on the 1,3-sigmatropic rearrangement of vinyl alcohol suggest that the hydrogen follows an antarafacial route but that the energy barrier to this is high (calculated as 85 kcal mol^{-1}).[31] This apparent

[23] K. Komatsu, M. Fujimori, and K. Okamoto, *Tetrahedron*, 1977, **33**, 2791.
[24] E. Vogel, H.-V. Runzheimer, F. Hogrefe, B. Baasner, and J. Lex, *Angew. Chem. Internat. Edn.*, 1977, **16**, 871.
[25] H. Meier and K.-P. Zeller, *Angew. Chem. Internat. Edn.*, 1977, **16**, 835.
[26] W. L. Mock, in Vol. II of ref. 1.
[27] T. C. Clarke, L. A. Wendling, and R. G. Bergman, *J. Amer. Chem. Soc.*, 1977, **99**, 2740.
[28] R. F. Cunico and H. M. Lee, *J. Amer. Chem. Soc.*, 1977, **99**, 7613.
[29] J. C. Philips and O. Morales, *J.C.S. Chem. Comm.*, 1977, 713.
[30] (*a*) W. G. L. Aalbersberg and K. P. C. Vollhardt, *J. Amer. Chem. Soc.*, 1977, **99**, 2792; (*b*) G. E. Hartzell and J. N. Paige, *J. Org. Chem.*, 1967, **32**, 459; K. Kondo, M. Matsumoto, and A. Negishi, *Tetrahedron Letters*, 1972, 2131.
[31] W. J. Bouma, D. Poppinger, and L. Radom, *J. Amer. Chem. Soc.*, 1977, **99**, 6443.

resistance to unimolecular tautomerism has been advanced as an explanation of the stability of this species in the vapour phase. In the liquid form tautomerism is rapid but here bimolecular mechanisms are far more likely. In contrast semi-empirical MO calculations on other 1,3-sigmatropic shifts predict a suprafacial route, probably with high biradical/zwitterionic character,[32] and experimental evidence in support of this latter prediction has been reported.[33]

1,5-Sigmatropic shifts in indenyl systems continue to attract attention,[34] and for the rearrangement of the vinyl-substituted indenes (24) it has been shown that a good LFER exists between the rate of the sigmatropic shift and that for the corresponding attack of morpholine on $CH_2{=}CHX$.[34a] This emphasizes the importance of the secondary orbital interaction between the HOMO(indenyl) and LUMO(migrating vinyl group). A similar secondary orbital interaction (25) has been used to explain the high rate of rearrangement of the spirononatriene (26) relative to its isomer (27) (rate ratio > 1000:1).[35] Studies of conformationally constrained indenyl systems[34a] suggest that the migrating vinyl group prefers an *exo* orientation, as in compound (28), rather than an *endo* orientation, as in compound (29). An *exo* orientation avoids the unfavourable secondary orbital interactions (marked *) found in the *endo* TS (30).

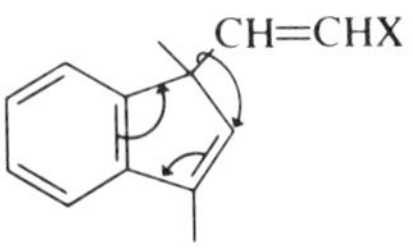

(24)

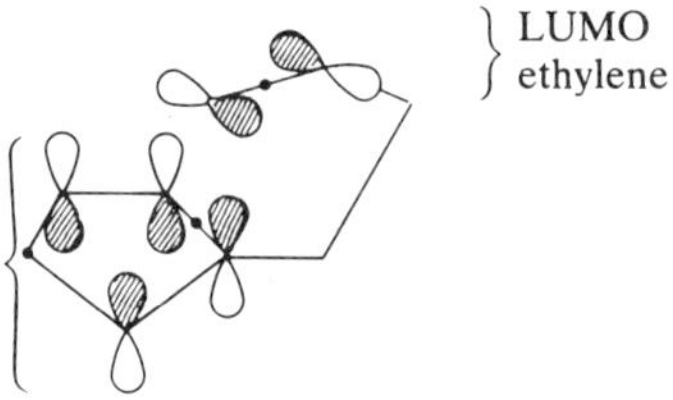

(25)

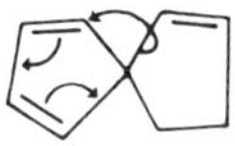

(26)

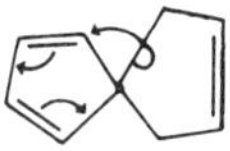

(27)

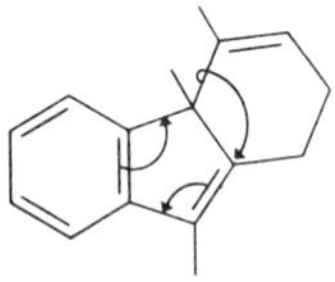

(28)

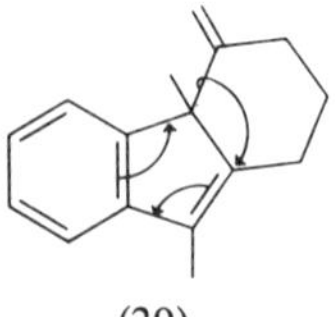

(29)

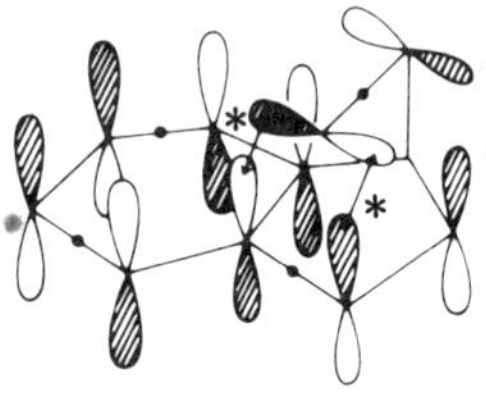

[32] W. W. Schoeller, *J. Amer. Chem. Soc.*, 1977, **99**, 5919; J. P. Grima, F. Choplin, and G. Kaufmann, *J. Organometallic Chem.*, 1977, **124**, 315; T. Minato, S. Inagaki, H. Fujimoto, and K. Fukui, *Bull. Chem. Soc. Japan*, 1977, **50**, 1651.

[33] J. E. Baldwin, S. E. Branz, and J. A. Walker, *J. Org. Chem.*, 1977, **42**, 4142; W. Kirmse and H.-R. Murawski, *J.C.S. Chem. Comm.*, 1977, 122.

[34] (*a*) D. J. Field and D. W. Jones, *J.C.S. Chem. Comm.*, 1977, 688; (*b*) D. W. Jones and G. Kneen, *J.C.S. Perkin I*, 1977, 1313; W. A. Pettit and J. W. Wilson, *J. Amer. Chem. Soc.*, 1977, **99**, 6372; K. K. de Fonseka, C. Manning, J. J. McCullough, and A. J. Yarwood, *ibid.*, p. 8257.

[35] M. F. Semmelhack, H. N. Weller, and J. S. Foos, *J. Amer. Chem. Soc.*, 1977, **99**, 292.

Schmid and co-workers[36] have provided impressive evidence for potential 2,5-biradical character (31) in the TS of the Cope rearrangement by studying the quantitative effect of radical-stabilizing substituents at these positions, *viz.* (32;

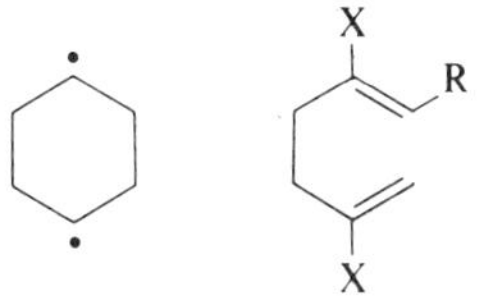

(31) (32)

R = Me, X = H, CN, or CO_2Me). They point out, however, that this biradical mechanism is unable to explain all of the phenomena associated with these rearrangements and suggest that they are normally concerted and only diverted into the biradical pathway when such 2,5-substituents are present. Similar studies have been reported by Dewar[37] who, however, no longer[37b] favours this flexible mechanism hypothesis but proposes a 2,5-biradicaloid even in the rearrangement of hexa-1,5-diene. His MINDO/3 calculations[38] suggest that this is a true intermediate and that it prefers a chair conformation. The balance between boat and chair TS (or intermediate!) in other 3,3-sigmatropic rearrangements has been investigated by several groups.[39] For 3,3′-dimethyl-3,3′-bicyclopropenyl[39a] the chair-like TS (33) is preferred over the boat-like TS (34) by 4.3 kcal mol^{-1} and a chair-like TS also seems to be preferred in the thermal (35a) and acid-catalysed (35b) amino-Claisen rearrangements.[39b] These latter reactions have been studied in some detail[39b] and appear to involve a concerted pericyclic mechanism. The large acceleration brought about by protonation of the nitrogen is attributed to charge delocalization in the reaction (36). In contrast the Claisen rearrangement

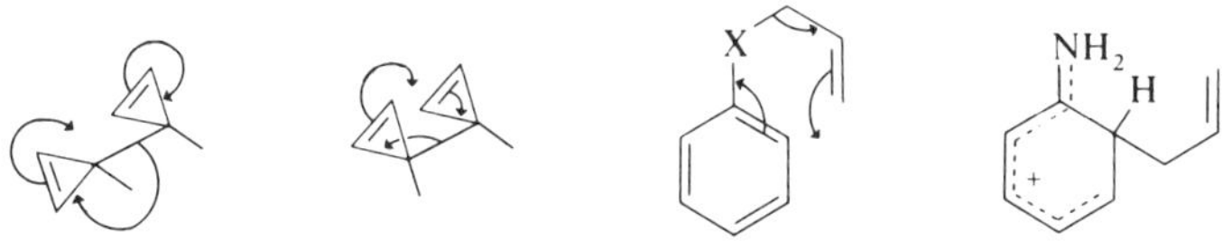

(33) (34) (35) a; X = NH
b; X = NH_2^+ (36)

(37) implicated in the reaction of allylic alcohols with cyclic orthoesters[39c] may proceed either predominantly through a chair-like TS or a boat-like TS, depending largely on steric factors, and the two TS's also seem to be finely balanced in the rearrangements of *ortho*-dienones such as compound (38).[39d] In this particular

[36] (*a*) R. Wehrli, H. Schmid, D. Bellus, and H.-J. Hansen, *Helv. Chim. Acta*, 1977, **60**, 1325; full English summary, *Chimia (Switz.)*, 1976, **30**, 416.

[37] (*a*) M. J. S. Dewar and L. E. Wade, *J. Amer. Chem. Soc.*, 1977, **99**, 4417; (*b*) cf. *J. Amer. Chem. Soc.*, 1973, **95**, 290.

[38] M. J. S. Dewar, G. P. Ford, M. L. McKee, H. S. Rzepa, and L. E. Wade, *J. Amer. Chem. Soc.*, 1977, **99**, 5069.

[39] (*a*) J. H. Davis, K. J. Shea, and R. G. Bergman, *J. Amer. Chem. Soc.*, 1977, **99**, 1499; (*b*) S. Jolidon and H.-J. Hansen, *Helv. Chim. Acta*, 1977, **60**, 978; (*c*) R. J. Cave, B. Lythgoe, D. A. Metcalfe, and I. Waterhouse, *J.C.S. Perkin I*, 1977, 1218; (*d*) A. Wunderli, T. Winkler, and H.-J. Hansen, *Helv. Chim. Acta*, 1977, **60**, 2436.

case the *threo*-form rearranges mainly *via* a chair-like TS (39) whereas the *erythro*-form prefers a boat-like TS (40).

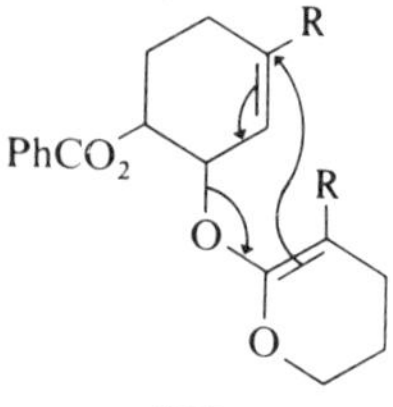

(37)

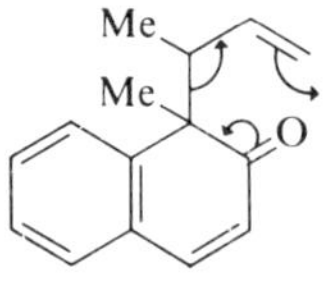

(38)

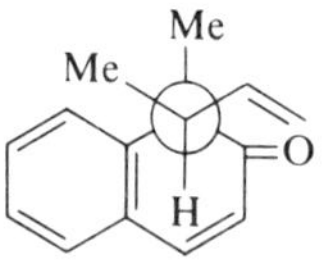

(39)

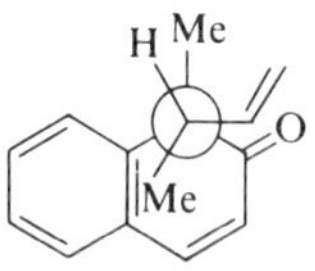

(40)

Several studies have been reported which bear on the possible involvement of radical pairs in the sigmatropic rearrangements of anionic and zwitterionic compounds.[40] Hence the 2,3-sigmatropic rearrangement of the anion (41)

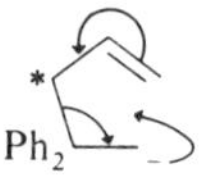

(41)

normally proceeds with inversion of the allyl group but as the temperature of the reaction is raised there is some scrambling of the isotopic label (marked *), pointing to a dissociation–recombination mechanism.[40a] Studies of pentadienyl-*N*-acyl-ammonioamidates (42)[40b] show that the proportions of the 1,2- and 5,2-rearrangement products (43) and (44) are characteristic only of the substituents R^1, R^2, and R^3 and not the group X, and hence these rearrangements are formulated as involving the radical intermediate (45), whereas the proportion of the 3,2-rearrangement product (46) is a function of the nature of X, and this product must be formed, at least in part, by a different mechanism, possibly *via* the concerted TS (47). Studies of the Wittig rearrangement of the optically active ether (48)[40c] show similar degrees of racemization in the products of both 1,2- and 1,4-shifts, in contrast to the corresponding Stevens rearrangement, where the product of the

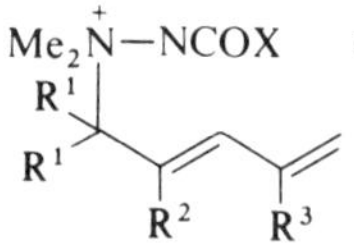

(42)

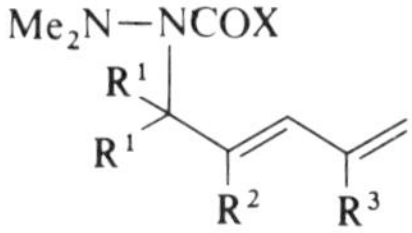

(43)

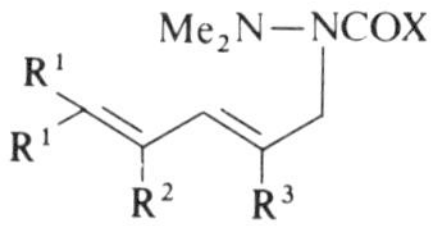

(44)

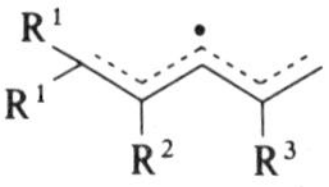

(45)

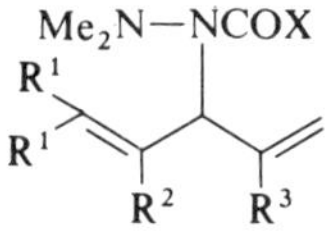

(46)

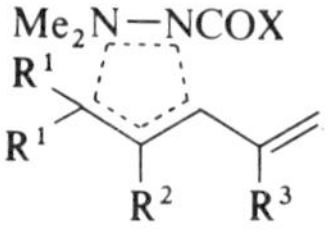

(47)

[40] (*a*) E. Grovenstein and A. B. Cottingham, *J. Amer. Chem. Soc.*, 1977, **99**, 1881; (*b*) K. Chantrapromma, W. D. Ollis, and I. O. Sutherland, *J.C.S. Chem. Comm.*, 1977, 97; (*c*) H. Felkin and C. Frajerman, *Tetrahedron Letters*, 1977, 3485; (*d*) E. F. Jenny and J. Druey, *Angew. Chem. Internat. Edn.*, 1962, **1**, 155.

1,4-shift is the more highly racemized of the two[40d] (Scheme). An attractive, simple explanation of this observation is based on the assumption that the anion (48) adopts a *cis* configuration and the anion (50) an all-*trans* configuration; then the radical pair (49) is presumably generated in the configuration shown, in which the α-phenylethyl radical has an equal distance to travel to form either product, whereas the radical pair (51) is generated in an extended configuration where it has much further to travel in order to produce the 1,4-rearrangement product, with consequently greater chances for racemization.

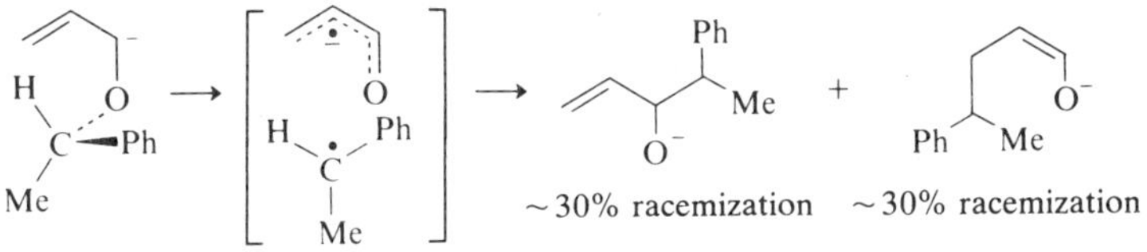

(48) (49)

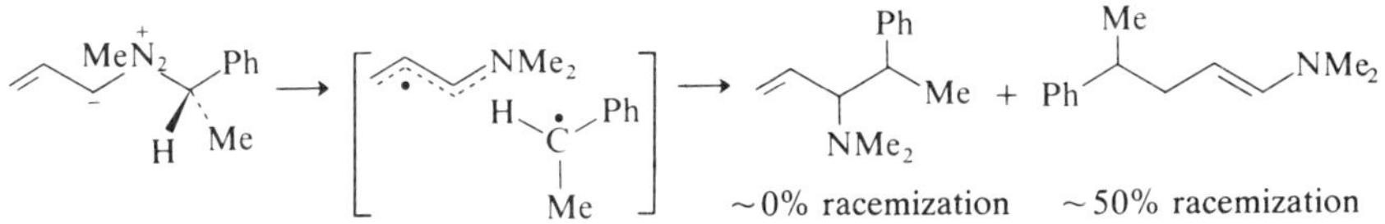

(50) (51)

Scheme

Various mechanisms have been demonstrated for dyotropic rearrangements of silicon compounds, all of which appear to involve some degree of nucleophilic attack on silicon.[41] Hence the rearrangement of the allyl system (52) is thought to involve initial co-ordination between the oxygen and the silicon, followed by migration of the allyl group as shown (53),[41a] and the fluoride-ion-catalysed rearrangement of the same compound to proceed *via* initial displacement of the

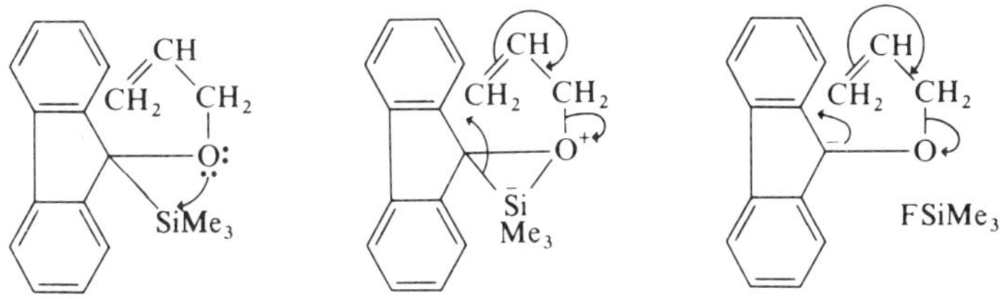

(52) (53) (54)

trimethylsilyl group to give the intermediate (54).[41b] Similarly the rearrangement of the ester (55) is formulated as involving the intermediate (56).[41c]

[41] (*a*) M. T. Reetz, *Chem. Ber.*, 1977, **110**, 965; (*b*) M. T. Reetz and N. Greif, *ibid.*, p. 2958; (*c*) M. T. Reetz and N. Greif, *Angew. Chem. Internat. Edn.*, 1977, **16**, 712.

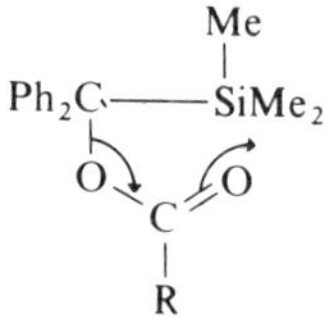

(55)

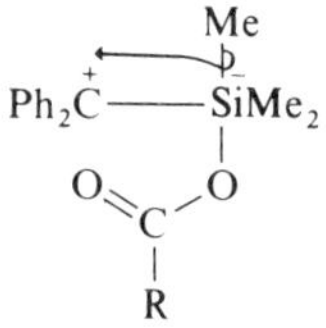

(56)

3 Ene Reactions

Detailed studies of the ene reaction of terminal alkenes with maleic anhydride have been interpreted in terms of a concerted mechanism and a preferred *exo* TS,[42] and the stereochemistry of the dimer formed from the tricycloundecene (57) has also been advanced as evidence for a concerted process.[43] Lewis-acid catalysis of the ene reaction of chloral allows the reaction to be performed under mild, synthetically useful conditions and added ferric chloride has also been shown to reverse the preferred stereochemistry in the reaction with α-pinene.[44] This can be understood in terms of the simple steric model illustrated in formulae (58) and (59),

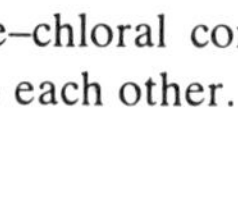

(57)

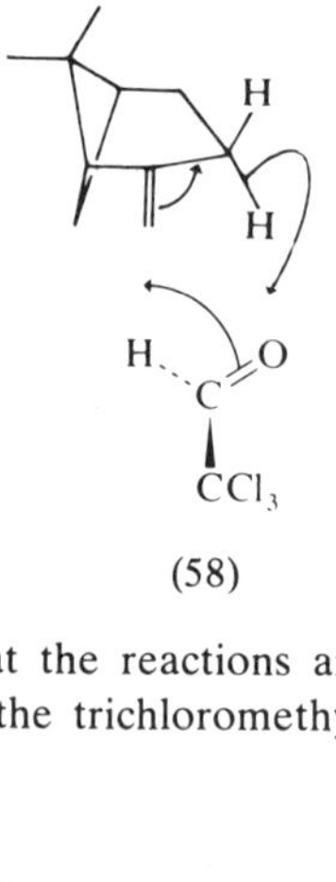

(58)

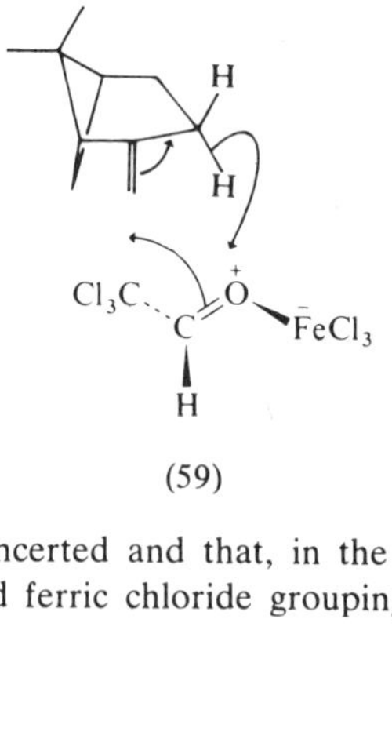

(59)

in which it is assumed that the reactions are concerted and that, in the ferric chloride–chloral complex, the trichloromethyl and ferric chloride groupings are *trans* to each other.

4 Electrocyclic Reactions

Several theoretical studies of electrocyclic reactions have been reported[45] and calculations of the energy surface for the opening of cyclopropylidenes to allenes have revealed a complex combination of disrotatory and conrotatory motions and provided an explanation of how chirality is retained.[45a]

[42] F. R. Benn, J. Dwyer, and I. Chappell, *J.C.S. Perkin II*, 1977, 533.
[43] R. Greenhouse, W. T. Borden, T. Ravindranathan, K. Hirotsu, and J. Clardy, *J. Amer. Chem. Soc.*, 1977, **99**, 6955.
[44] G. B. Gill and B. Wallace, *J.C.S. Chem. Comm.*, 1977, 380, 382.
[45] (*a*) P. W. Dillon and G. R. Underwood, *J. Amer. Chem. Soc.*, 1977, **99**, 2435; (*b*) N. L. Bauld and J. Cessac, *ibid.*, p. 23; Y. Jean and A. Devaquet, *ibid.*, p. 1949; M. J. S. Dewar, G. P. Ford, and H. S. Rzepa, *J.C.S. Chem. Comm.*, 1977, 728; R. Cimiraglia, M. Persico, and J. Tomasi, *J. Phys. Chem.*, 1977, **81**, 1876.

The ring-opening reactions of Dewar-benzene[46] and of bicyclopentene[47] have been shown to proceed in a straightforward manner, even though formally disallowed. In the case of 5-methylbicyclopentene Baldwin[47] has provided evidence that products (60) and (61) arise by a 'hot molecule' rearrangement of the

(60) (61)

initial 1,4-bond cleavage product, as a result of which he claims that the controversy which has surrounded these rearrangements may be finally 'retired'.

The electrocyclic ring-opening reactions of epoxides[48] have been reviewed[48a] and trapping of the resultant carbonyl ylides has shown that, as expected, the thermal reaction proceeds mainly in a conrotatory[48b,c] and the photochemical reaction mainly in a disrotatory manner.[48d,e] In both cases, however, some cross-over products are obtained and these seem to arise not only by isomerization of the intermediate but also by a competing ring opening in the disallowed sense. In the case of vinyl-substituted systems,[48c,d] such as compound (62),[48c] the initial product (63) of conrotatory ring opening can be trapped with maleic anhydride or may undergo a disrotatory closure to form a five-membered ring (64).

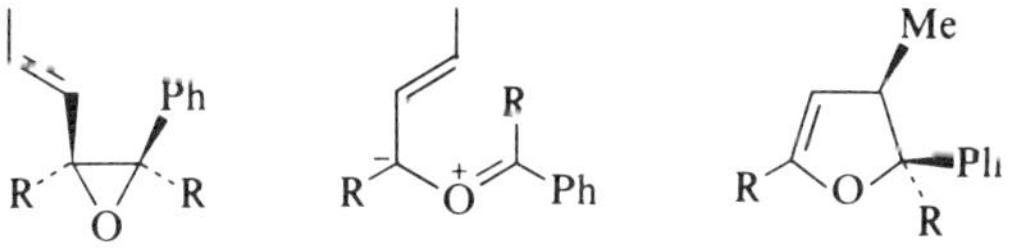

(62) (63) (64)

It is interesting to note that n.m.r. observation of the reduction of the cyclopropyl bromide (65) with lithium in THF at −60 °C shows initial formation of the *Z,E*-anion (66)[49] which then, more slowly, isomerizes to the equilibrium *Z,E*/*E,E*

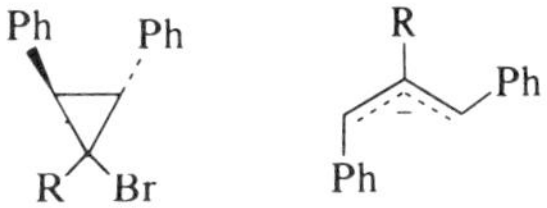

(65) (66)

mixture. This is consistent with the expected disrotatory opening of an intermediate cyclopropyl radical.[49b] Attempts to probe the stereochemistry of the

46 M. J. Goldstein and R. S. Leight, *J. Amer. Chem. Soc.*, 1977, **99**, 8112.

47 G. D. Andrews and J. E. Baldwin, *J. Amer. Chem. Soc.*, 1977, **99**, 4853.

48 (*a*) R. Huisgen, *Angew. Chem. Internat. Edn.*, 1977, **16**, 572; (*b*) V. Markowski and R. Huisgen, *J.C.S. Chem. Comm.*, 1977, 439, 440; (*c*) M. S. Medimagh and J. Chuche, *Tetrahedron Letters*, 1977, 793; (*d*) W. Eberbach and U. Trostman, *ibid.*, p. 3569; (*e*) V. Markowski and R. Huisgen, *ibid.*, 1976, p. 4643.

49 (*a*) G. Boche and D. R. Schneider, *Tetrahedron Letters*, 1976, 3657; (*b*) *cf.* M. Newcomb and W. T. Ford, *J. Amer. Chem. Soc.*, 1974, **96**, 2968; and S. Sustmann and C. Rüchardt, *Chem. Ber.*, 1975, **108**, 3043.

related pentadienyl anion–cyclopentenyl anion interconversion[50] have been largely frustrated by uncertainties which surround the geometry of the reacting species[50a] or by isomerization of the reaction product.[50b]

Thermal cyclization of the 12π-electron system (67) gives the expected product of conrotatory closure (68).[51] This cannot, however, be claimed as a great success for the predictive power of the orbital symmetry rules, as the reaction is reversed photochemically!

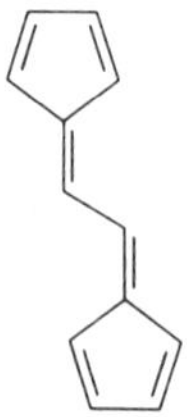

(67)

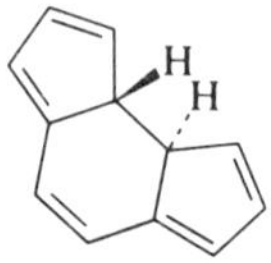

(68)

[50] (*a*) D. H. Hunter, S. K. Sim, and P. R. Steiner, *Canad. J. Chem.*, 1977, **55**, 1229; (*b*) C. W. Schoppee and G. N. Henderson, *J.C.S. Perkin I*, 1977, 1028.

[51] H. Sauter, B. Gallenkamp, and H. Prinzbach, *Chem. Ber.*, 1977, **110**, 1382.

4 Reaction Mechanisms
Part (ii) Polar Reactions

By H. R. HUDSON
Department of Chemistry, The Polytechnic of North London, Holloway Road, London N7 8DB

1 Introduction

Nucleophilic substitution at saturated carbon, particularly in solvolysis reactions, continues to account for a substantial proportion of the research publications coming under this heading. The first section of this year's Report deals with some recent studies on structural effects in this area, whilst a critical account of medium effects on the rate and mechanism of solvolysis reactions is given elsewhere.[1] Also included below are sections on nucleophilic substitution at unsaturated carbon in compounds containing carbon–carbon (vinylic) or carbon–nitrogen double bonds, carbocations, addition, and elimination. Carbanions, reactions of carbonyl compounds, and other relevant topics are deferred to make way for a new section on ion–molecule reactions in the gas phase, in so far as these throw light on the mechanisms of polar organic reactions. Full coverage of all aspects of organic reaction mechanisms is given in an annual review of this subject.[2]

2 Nucleophilic Substitution at Saturated Carbon

The tool of increasing electron demand, which was introduced to study participation and the non-classical carbonium-ion problem,[3] has become a powerful means for the investigation of solvolytic processes. The method involves the attachment of a series of substituted aryl groups to the potential carbocation centre and a determination of the ρ^+ value of the system as a measure of its sensitivity to electronic influences.

In the solvolysis of aryldialkylcarbinyl *p*-nitrobenzoates (1) in 80% aqueous acetone,[4] the similarity in ρ^+ values for t-cumyl, 3-methyl-2-butyl, and 3-pentyl derivatives indicates that the stabilizing effect of the alkyl groups (Me, Et, and Pr^i) must be nearly the same in each case. High negative values for the cyclopropyl and cyclobutyl members of the 1-aryl-1-cycloalkyl *p*-nitrobenzoate series (2) have been attributed to the effect of *I*-strain, which destabilizes the carbocation and results in

[1] T. W. Bentley and P. von R. Schleyer, *Adv. Phys. Org. Chem.*, 1977, **14**, 1.
[2] 'Organic Reaction Mechanisms, 1977', ed. A. R. Butler and M. J. Perkins, Wiley, London, 1978.
[3] H. C. Brown, 'The Non-Classical Ion Problem', Plenum, New York, 1977.
[4] H. C. Brown, M. Ravindranathan, E. N. Peters, C. G. Rao, and M. M. Rho, *J. Amer. Chem. Soc.*, 1977, **99**, 5373.

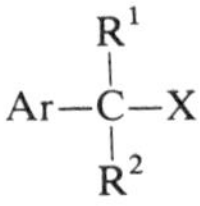

(1)

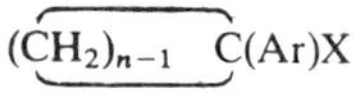

(2)

X = p-$NO_2C_6H_4CO_2$-; Ar = p-XC_6H_4 (X = MeO, Me, H, or CF_3) or 3,5-$(CF_3)_2C_6H_3$

an increased electron demand on the aryl system. Cyclohexyl also gives rise to a relatively high negative ρ^+ value, which has been attributed to the resistance of this conformationally stable system to the introduction of an sp^2 centre.

Unambiguous evidence for participation of carbon has been obtained by the use of a similar procedure in the solvolysis of a series of 9-aryl-9-pentacyclo-$[4,3,0,0^{2,4}\,0^{3,8}\,0^{5,7}]$nonyl p-nitrobenzoates (3).[5] The ρ^+ value for this system (−2.05) is comparable to that observed in the solvolysis of 7-norbornenyl derivatives in which π-participation is significant. Comparison of solvolysis rates with those for the corresponding 7-norbornyl derivatives (4) shows also that with increasing electron demand at the cationic centre, the rates of solvolysis of the pentacyclic derivatives increase markedly relative to those for the 7-norbornyl compounds. Solvolysis of the parent substrate (3; R = H) is 10^{10}—10^{12} times faster than for the related 7-norbornyl derivative. The results, which are also supported by low 9-Me/9-H and 9-Ph/9-Me rate ratios in the pentacyclic series, are interpreted in terms of ($\pi\sigma$)-participation by the remote cyclopropane bond, leading to the trishomocyclopropenyl cation (5).

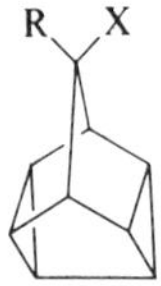

(3)

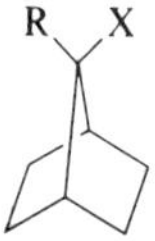

(4)

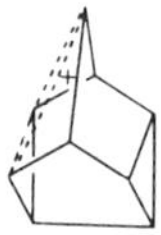

(5)

X = p-$NO_2C_6H_4CO_2$-; R = p-XC_6H_4 (X = MeO, H, or CF_3) or 3,5-$(CF_3)_2C_6H_3$

Absence of significant σ-participation has been demonstrated in the solvolyses of tertiary 2-aryl-2-norbornyl (6) and 2-aryl-2-camphenilyl p-nitrobenzoates (7).[6] In

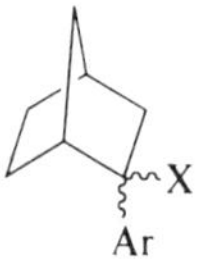

(6)

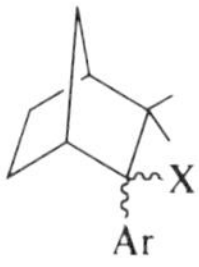

(7)

X = p-$NO_2C_6H_4CO_2$-; Ar = p-XC_6H_4 (X = MeO, Me, H, or CF_3) or 3,5-$(CF_3)_2C_6H_3$

these series, the effect of deactivating the aromatic ring, and thereby increasing electron demand at the cationic centre, should be to increase the *exo*/*endo* rate ratios if σ-participation is an important factor. Essentially constant *exo*/*endo* rate

5 H. C. Brown and M. Ravindranathan, *J. Amer. Chem. Soc.*, 1977, **99**, 299.
6 H. C. Brown, K. Takeuchi, and M. Ravindranathan, *J. Amer. Chem. Soc.*, 1977, **99**, 2684.

ratios are, however, obtained within each series. The results imply that the high *exo*/*endo* rate ratios observed in the tertiary 2-norbornyl systems are due mainly to decreased rates of reaction in the sterically hindered *endo* direction of the norbornane structure; and by extrapolation, a similar conclusion has been drawn for the corresponding secondary systems.

A fundamental difference in the behaviour of tertiary and secondary systems has been reported in the solvolysis of 5-norbornen-*exo*-2-yl substrates.[7] Whereas the secondary ester (8) undergoes assisted ionization during acetolysis to yield the symmetrical norbornenyl–nortricyclyl cation (9) and thence racemic products (10) and (11) (Scheme 1), solvolysis of optically active 1,2-dimethyl-5-norbornen-*exo*-2-yl *p*-nitrobenzoate (12) in 90% aqueous acetone yields 2-methylene-1-methyl-5-norbornene (16) (with 62% retention of configuration), 1,2-dimethyl-5-norbornen-*exo*-2-ol (15) (with 15% retention of configuration), and racemic 1,6-dimethyl-3-nortricyclanol (17) (Scheme 2). The results, which are also similar for methanolysis, are consistent with the initial formation of a symmetrical tertiary cation (13).

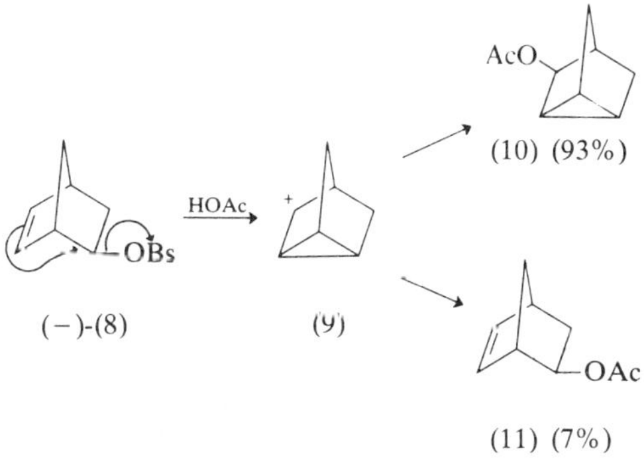

Scheme 1

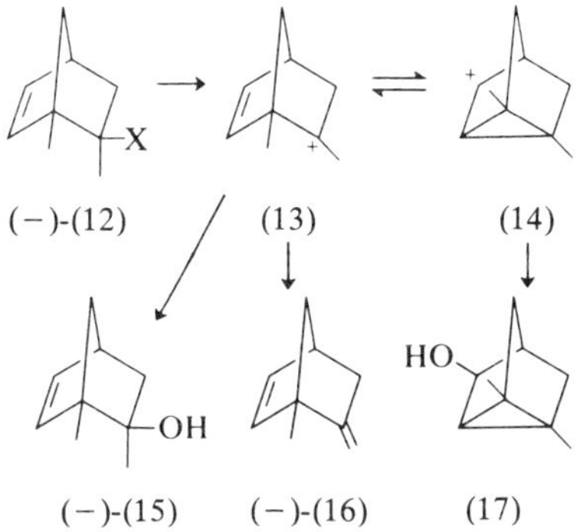

(X = p-$NO_2C_6H_4CO_2$-)

Scheme 2

[7] H. L. Goering and C.-S. Chang, *J. Amer. Chem. Soc.*, 1977, **99**, 1547.

Evidence for C—C σ-delocalization in simple secondary carbocations has been obtained in the acetolysis of some *N*-alkyl-*N*-nitroso-acetamides, which give small yields of primary acetates.[8] For example, the cyclohexyl derivative gives cyclohexyl acetate and cyclopentylmethyl acetate in the ratio 99.05 : 0.95. Rearrangement of 1-ethylpropyl to 2-methylbutyl (0.3%) and of s-butyl to isobutyl (0.2%) similarly occurs. It has been argued that the equilibrium coefficient between a secondary and a primary cation (70—80 kJ mol^{-1} apart in energy) will be *ca.* 10^{12}, and would require a physically unrealistic difference of 10^9 or more in the reaction rates of two assumed classical intermediates to account for the product ratios observed. A non-classical ion or protonated cyclopropane (19), generated only when the R—C and C—A bonds of the substrate (18) are anti-parallel, has been proposed to account for these strongly endothermal secondary to primary rearrangements (Scheme 3), and it has been suggested that C—C σ-delocalization in simple carbocations may be more widespread than has frequently been implied.

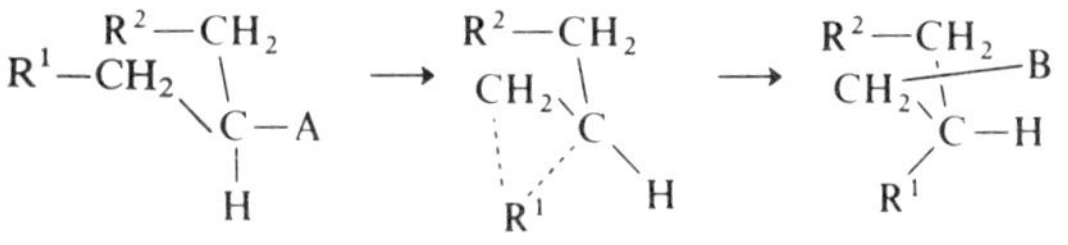

(18) (19)

Scheme 3

Evidence has been obtained for the involvement of protonated cyclopropane intermediates in the trifluoroacetolysis of n-[1-^{14}C]butyl mercuric perchlorate at 50 or 72 °C.[9] Under these conditions the ^{14}C label appears at all positions of the major solvolysis product (2-butyl trifluoroacetate), at positions 1 and 2 of the n-butyl ester, and is divided in a 50 : 50 ratio between C-1 and the rest of the molecule in the small amount of isobutyl ester (*ca.* 2%) which is formed. The results are consistent with 8—14% of the products being formed from a series of equilibrating

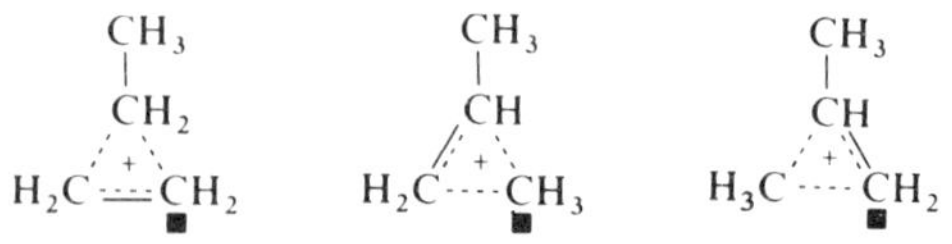

(20) (21) (22)

■ = ^{14}C

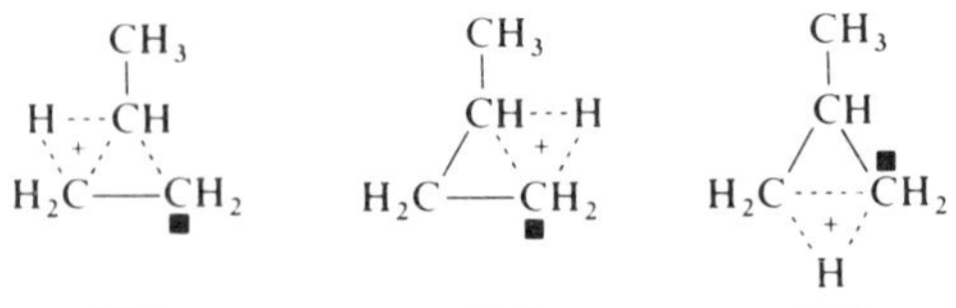

(23) (24) (25)

[8] C. M. Cooper, P. J. Jenner, N. Perry, H. Storesund, J. Russell-King, and M. C. Whiting, *J.C.S. Chem. Comm.*, 1977, 668.

[9] C. C. Lee and R. Reichle, *J. Org. Chem.*, 1977, **42**, 2058.

edge- or corner-protonated cyclopropanes [(20)—(25)]. At 35 °C, only successive 1, 2-hydride shifts appear to be involved (Scheme 4).

$$CH_3CH_2CH_2{}^{14}\overset{+}{C}H_2 \rightarrow CH_3CH_2\overset{+}{C}H{}^{14}CH_3 \rightleftharpoons CH_3\overset{+}{C}HCH_2{}^{14}CH_3$$

Scheme 4

The description of secondary solvolysis in terms of competing neighbouring-group (k_Δ) and nucleophilic solvent-assisted (k_s) processes is now well established [equation (1)].[10] In limiting or unassisted solvolyses, the rate constant is represen-

$$k_1 = k_\Delta + k_s \qquad (1)$$

ted by k_c, and there has been much debate on the possible importance of such a process in secondary systems.[11] Recent studies on the solvolysis of cyclo-octyl tosylate have now shown that the solvolysis of a secondary substrate by the k_c process may not be unusual, even in a rather nucleophilic solvent and where there are no obvious barriers to back-side approach by the nucleophile, provided that relief of ground-state strain is sufficient to provide a competitive pathway.[12] The kinetic evidence is based on comparisons with adamantyl tosylate in aqueous ethanol or aqueous 2,2,2-trifluoroethanol and is supported by theoretical calculations on the ionization of methylcyclo-octane (Scheme 5) (the methyl group

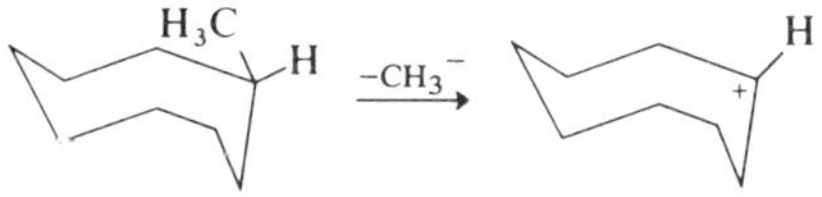

Scheme 5

being considered sterically similar to a leaving group such as chloride or *p* nitrobenzoate). The calculations indicate a value of -3.06 kcal mol^{-1}, for the δ-strain whereas other acyclic or monocyclic substrates show positive δ-strain values.

In another context, new light has been thrown on the importance of steric acceleration in solvolytic reactions by a determination of the *X*-ray crystal structure of tris(t-butyl)methyl *p*-nitrobenzoate.[13] Both B-strain, which results from crowding of the bulky t-butyl groups, and F-strain, which results from steric interference between the t-butyl groups and the *p*-nitrobenzoate moiety, are clearly demonstrated by distortions of bond lengths and bond angles. The relief of both types of strain as the transition state is formed must therefore be assumed to have significant effects on reaction rate.

3 Nucleophilic Substitution at sp^2 Carbon

Vinylic Substitution.—In spite of considerable work in recent years on vinyl cations as reaction intermediates,[14] a number of fundamental questions related to the

10 B. Capon and S. P. McManus, 'Neighbouring Group Participation', Vol. 1, Plenum, New York, 1976.
11 J. R. Pritt and M. C. Whiting, *J.C.S. Perkin II*, 1975, 1458, and references cited therein.
12 J. M. Harris, D. L. Mount, M. R. Smith, and S. P. McManus, *J. Amer. Chem. Soc.*, 1977, **99**, 1283.
13 P.-T. Cheng, S. C. Nyburg, C. Thankachan, and T. T. Tidwell, *Angew. Chem. Internat. Edn.*, 1977, **16**, 654.
14 Z. Rappoport, *Accounts Chem. Res.*, 1976, **9**, 265; M. Hanack, *ibid.*, p. 364; *Ann. Reports (B)*, 1975, **72**, 71.

extent of bond-breaking in the transition state and to substituent effects have remained unanswered. Recent studies on the solvolysis of ring-substituted β-styryl trifluoromethanesulphonates (triflates) (26) and their dideuterio-analogues (27)

$$H_2C{=}C(Ar)OSO_2CF_3 \qquad D_2C{=}C(Ar)OSO_2CF_3$$

(26) (27)

$Ar = XC_6H_4$ (X = H, *p*-Cl, *m*-Cl, *p*-CF_3, or *p*-NO_2)

have gone some way towards elucidating these factors.[15] For both series, excellent Hammett plots were obtained with $\rho = -4.1$, which is indicative of considerable charge development in the transition state, even with the strongly deactivating *p*-nitro-substituent present. A β-deuterium isotope effect of $k_H/k_D = 1.45$ for the parent compounds agrees well with earlier results. Taken with other literature data, a full range of k_H/k_D values from 1.21 for *p*-methoxy to 1.71 for *p*-nitro is obtained, and shows a linear correlation with substituent σ^+ values. The results indicate that rotation of the aryl group to deconjugate with the C=C double bond and to conjugate with the nascent empty *p*-orbital occurs early, and must be complete, or nearly so, at the transition state leading to the vinyl cation.

Novel rearrangements of the vinyl cation have been reported in the solvolyses of the spiro-triflate (28) and the primary vinyl triflate (30).[16] At 130 °C in aqueous ethanol or aqueous trifluoroethanol, buffered with pyridine, the spiro substrate gives a yield of 90—98% of diene (29). It has been concluded that migration of the adjacent cyclohexyl bond *to* the vinyl cation centre occurs, probably in a concerted fashion, with anchimeric assistance (Scheme 6). Cycloheptanone is the principal

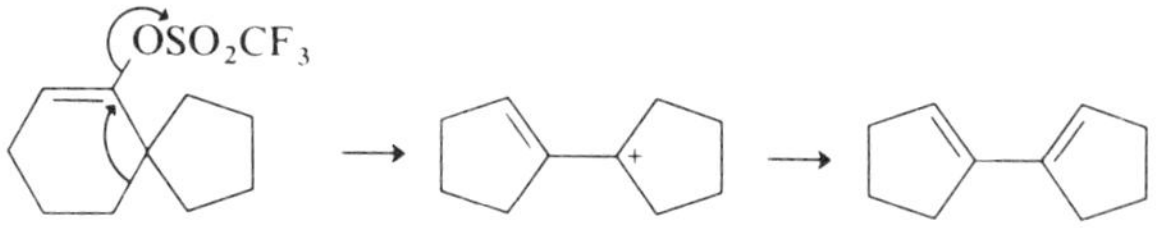

(28) (29)

Scheme 6

product obtained under similar conditions from the primary vinyl substrate and it has been suggested that in this case concerted ionization and bond migration *across* the vinyl group occurs, to give the cyclic vinyl cation (31) as the first intermediate (Scheme 7). The rearrangement suggests that a 'bent' secondary vinyl cation of this type is more stable than the corresponding linear primary vinyl cation.

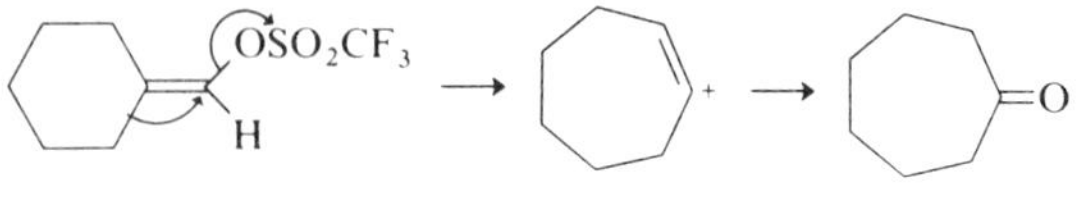

(30) (31)

Scheme 7

[15] P. J. Strang, R. J. Hargrove, and T. E. Dueber, *J.C.S. Perkin II*, 1977, 1486.
[16] P. J. Strang and T. E. Dueber, *Tetrahedron Letters*, 1977, 563.

acetolysis of triaryl[2-^{13}C]vinyl bromides (Scheme 8) in the presence of silver
shown that the relative extents of isotopic scrambling which occur during the
acetolysis of triaryl[2-^{13}C]vinyl bromides (Scheme 8) in the presence of silver

$$Ar_2{}^{13}C{=}C(Ar)Br \xrightarrow{HOAc} Ar_2{}^{13}C{=}C(Ar)OAc + AcO^{13}C(Ar){=}CAr_2$$

Scheme 8

acetate are in the approximate ratio 1:2:3 for the phenyl, *p*-tolyl, and *p*-anisyl derivatives, respectively.[17] The differences are much smaller than would be expected if migratory aptitude was the predominant factor in determining the extent of rearrangement, and it has been suggested that the stability of the initial cation and the effect of the particular aryl group on the electrophilic character of the migration terminus are also important. Higher percentages of rearrangement occur in trifluoroacetic acid. Kinetic solvent isotope effect studies show that an addition–elimination sequence can occur, although it becomes less significant if a salt such as silver or sodium acetate is added.

The acid-catalysed hydrolysis of vinyl sulphides has been shown to proceed by a mechanism analogous to that of vinyl ethers.[18] Slow proton transfer to the C=C double bond gives a sulphur-stabilized carbenium ion which reacts rapidly with water to form products (Scheme 9).

$$CH_2{=}CH{-}SR \xrightarrow[slow]{H^+} CH_3{-}\overset{+}{C}H{-}SR \xrightarrow[fast]{H_2O} CH_3CHO + HSR$$

Scheme 9

An interesting controversy has arisen over the stereochemistry of nucleophilic substitution of (*E*)- and (*Z*)-3-chloro-2-phenylpropenonitriles [(32) and (33); X = Cl].[19] Whilst retention of configuration is confirmed in reactions of these isomers

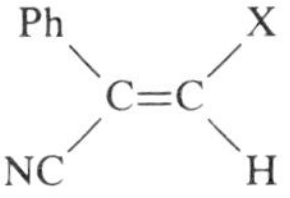

(*E*)-isomer
(32)

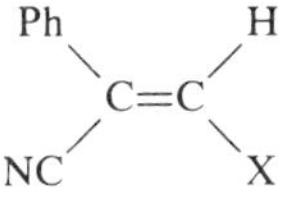

(*Z*)-isomer
(33)

with weak bases such as morpholine, piperidine, and phenol–triethylamine, it is claimed that 1:1 mixtures of the corresponding (*E*)- and (*Z*)-vinyl ethers are obtained in reactions with strong bases such as sodium ethoxide in ethanol or sodium phenoxide in THF.[20] Such a result is explicable by initial retention, followed by a number of 'racemization' mechanisms,[21] and could possibly be due to the use of higher concentrations of reactants in the later work. Of more fundamental importance is the view[20] that the previous configurational assignments for (*E*)- and (*Z*)-3-ethoxy-2-phenylpropenonitriles [(32) and (33); X = OEt][19] should be

[17] C. C. Lee, A. J. Paine, and E. C. F. Ko, *Canad. J. Chem.*, 1977, **55**, 2310.
[18] R. A. McClelland, *Canad. J. Chem.*, 1977, **55**, 548.
[19] Z. Rappoport and A. Topol, *J.C.S. Perkin II*, 1975, 982.
[20] G. Le Guillanton and M. Cariou, *J.C.S. Perkin II*, 1977, 997.
[21] Z. Rappoport, *J.C.S. Perkin II*, 1977, 1000.

reversed. This would imply that an unprecedented inversion of configuration had occurred in the substitution reaction with ethoxide, and the result should be viewed with caution at this stage.

Nucleophilic Attack on C=N Double Bonds.—Nucleophilic substitutions of diarylimidoyl chlorides (34; X = Cl) by secondary amines in benzene or acetonitrile show complex mechanistic behaviour. In benzene, both first- and second-order terms in [amine] are observed, and the Hammett plots show minima which are indicative of two competitive processes having opposite electronic demands.[22] The results have been ascribed to a combination of an S_N2(ion-pair) process for electron-donating substituents (Scheme 10) and a nucleophilic addition–elimination process (Ad_N-*E*)

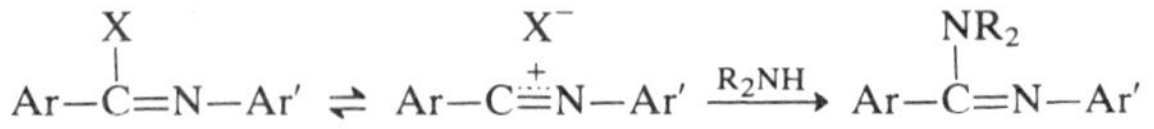

(34)

Scheme 10

for electron-attracting substituents (Scheme 11). Superimposed on these schemes are less important third-order amine-catalysed routes in which the amine may serve as a base, electrophile, or bifunctional catalyst. In acetonitrile the amine-catalysed route disappears, and it has been suggested that the Ad_N-*E* process occurs, together with reaction either *via* ion-pairs or free nitrilium ions.

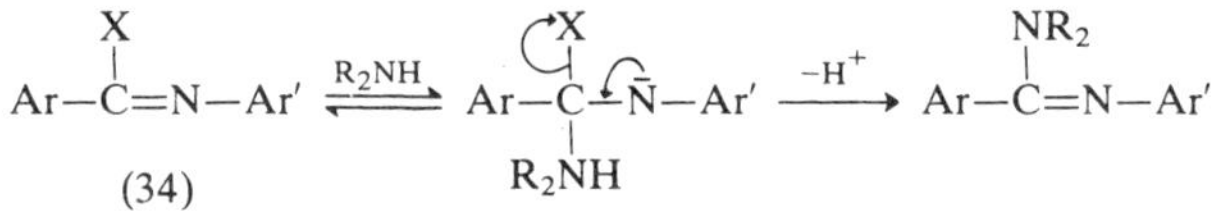

(34)

Scheme 11

The less reactive imidoyl cyanides (34; X = CN) undergo nucleophilic substitution by amines in acetonitrile or by alkoxides in alcohols, only by an addition–elimination sequence.[23] It has been suggested that a common initial step involves nucleophilic attack on the imidoyl cyanide by the amine, alkoxide ion, or alcohol and that the subsequent expulsion of the leaving group may be uncatalysed or amine-catalysed in acetonitrile, or solvent-assisted in the alcohol.

4 Carbocations

Preparation, and Rearrangements in Super-acid Media.—The first direct observation of a long-lived cyclopropyl cation has been made by dissolving the highly strained 11-methyl-11-bromotricyclo[4,4,1,$0^{1,6}$]undecane (35) in SO_2ClF at −60 °C and adding the solution slowly to SbF_5 in SO_2ClF at −120 °C (Scheme 12).[24] The ^{1}H n.m.r. spectrum at −90 °C reveals substantial deshielding of the

[22] R. Ta-Shma and Z. Rappoport, *J. Amer. Chem. Soc.*, 1977, **99**, 1845.
[23] R. Ta-Shma and Z. Rappoport, *J.C.S. Perkin II*, 1977, 659.
[24] G. A. Olah, G. Laing, D. B. Ledlie, and M. G. Costopoulos, *J. Amer. Chem. Soc.*, 1977, **99**, 4196.

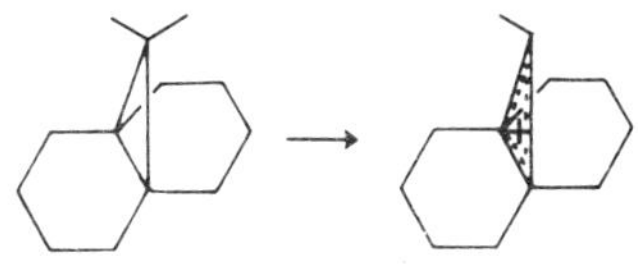

(35) (36)

Scheme 12

methyl and methylene protons, which is indicative of a discrete carbocation structure. The ^{13}C n.m.r. spectrum shows that the ion is unsymmetrical, and is best described as a 'bent' cyclopropane cation (36). Deshielding at C-1, C-6, and C-11 suggests that the positive charge is delocalized by interaction of the C-1—C-6 σ-bond with the p-orbital at C-11.

Some new examples of substituted pyramidal dications have been prepared by dissolution of the precursors (37) and (38) in $HFSO_3$–SbF_5 (1:1) in SO_2ClF at −60 °C.[25] Those containing two ethyl or two isopropyl groups are found to have these substituents in the *basal* positions, and their formation is rationalized as shown (Scheme 13).

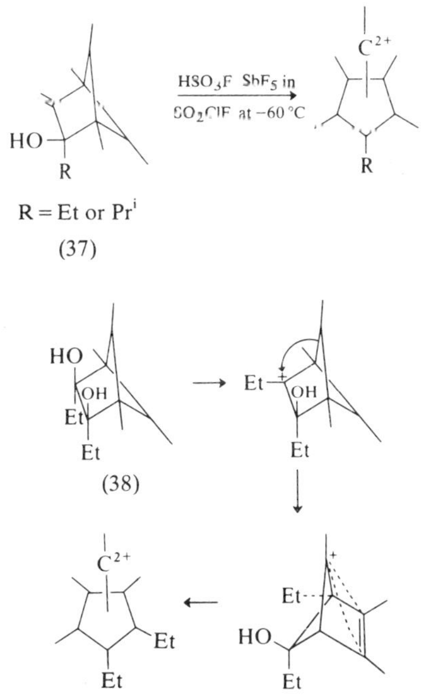

Scheme 13

[25] C. Giordano, R. F. Heldeweg, and H. Hogeveen, *J. Amer. Chem. Soc.*, 1977, **99**, 5181.

Ring closure of allyl to cyclopropylcarbinyl has been systematically studied by treatment of 4-methylpent-1-en-3-ol (39; $R^1 = Me$, $R^2 = H$), 2,4-dimethylpent-1-en-3-ol (39; $R^1 = R^2 = Me$), and pent-1-en-3-ol (39; $R^1 = R^2 = H$). with FSO_3H–SbF_5 in SO_2ClF at −78 or −120 °C.[26] The corresponding cyclopropylcarbinyl cations (41), which are thought to be formed *via* the homoallylic ions (40), rearrange further at higher temperatures to the more highly substituted allylic structures (42), (Scheme 14). The general order of thermodynamic stability for ions in which $R^1 = R^2 = H$ is $(RCH{\cdots}CH{\cdots}CH_2)^+ <$ cyclo-$C_3H_5C^+R_2 <$ $(RCH{\cdots}CH{\cdots}CHR)^+$, and the stabilizing effect of a cyclopropyl group on a carbenium centre has been estimated as 11—17 kcal mol^{-1} larger than that of a vinyl group.

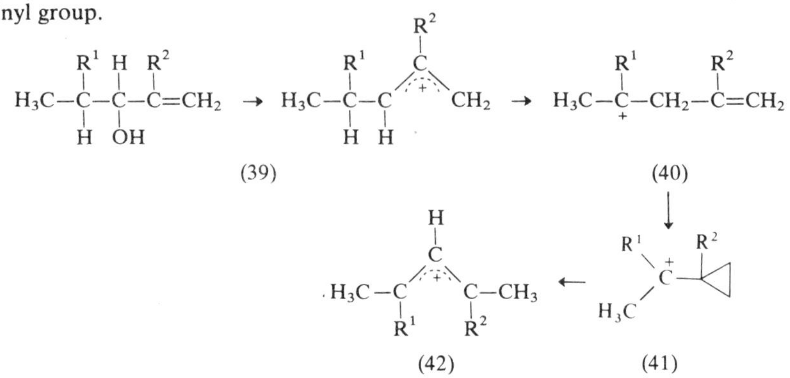

Scheme 14

A study of rearrangements and equilibria in the ions formed under stable ion conditions from β-arylethyl chlorides and their side-chain-substituted derivatives shows that σ-bridged ethylenearenium (phenonium) ions are formed only from 2-chloroethylbenzene and its ring-substituted derivatives.[27] The ethylenebenzenium ion (43) is formed quantitatively in HF–SbF_5 at −60 °C, and subsequently rearranges at higher temperatures *via* the highly unstable 2-phenylethyl cation (44) to give the benzylic ion (45) (Scheme 15). Substitution of methyl

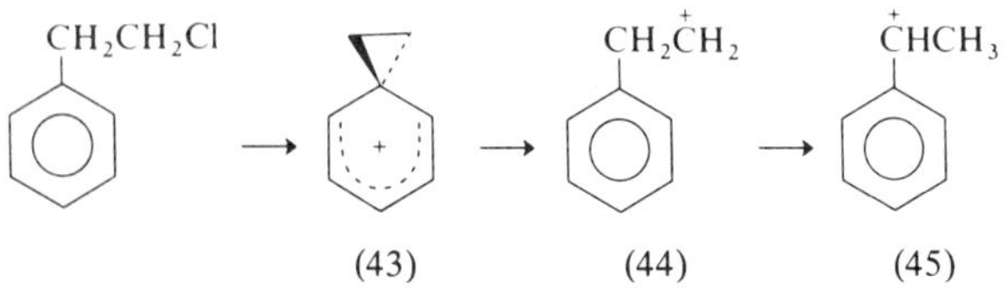

Scheme 15

groups at C_α or C_β leads, upon ionization, to rearranged benzylic and/or equilibrating β-phenylethyl cations, but the experimental determination of the energy difference between these two types is possible only in the case of the ions derived

[26] H. Mayr and G. A. Olah, *J. Amer. Chem. Soc.*, 1977, **99**, 510.
[27] G. A. Olah, R. J. Spear, and D. A. Forsyth, *J. Amer. Chem. Soc.*, 1977, **99**, 2615.

from 1-(*p*-methoxyphenyl)-2-chloropropane (Scheme 16). The relatively small energy difference (8—10 kcal mol^{-1}) has been interpreted in terms of stabilization of the β-phenylethyl cation by π-bridging (46), although the concentration of this species at equilibrium is thought to be exceedingly low.

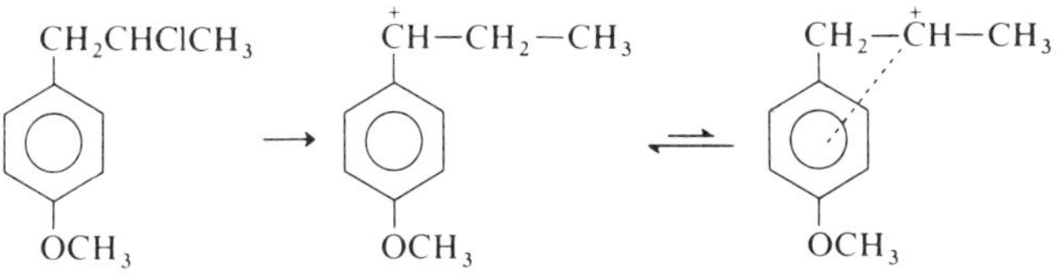

(46)

Scheme 16

The proton-decoupled and proton-coupled ^{13}C n.m.r. spectra of all C_3 to C_8 alkyl cations show the methyl substituent effects to be constant and additive.[28] They can therefore be used to estimate ^{13}C chemical shifts of static alkyl cations as well as of symmetrical or unsymmetrical equilibrating carbocations. There is little or no contribution from hydrogen- or methyl-bridged structures in equilibrating t-cations, and the experimental data allow calculations to be made of ΔG, ΔH, and ΔS for the differences between them. Application of the additive substituent effect method to equilibrating secondary carbocations, *e.g.* s-butyl, shows significant deviations of the estimated from the experimental values and indicates that there is a contribution from partial hydrogen-bridging.

Other Aspects.—A critical examination of the use of ^{13}C chemical shifts for establishing electron densities in carbocations has been made on the basis of solvolysis rate studies for the *p*-nitrobenzoates shown [(47)—(49)].[29] The large

	$(CH_3)_3C—X$	$Ph(CH_3)_2C—X$	$—C(CH_3)_2—X$
	(47)	(48)	(49)
($X = p$-$NO_2C_6H_4CO_2$-)			
Relative k_1^*	1.0	10^3	$10^{2.7}$
^{13}C shift/p.p.m (from TMS)	329.2	254.4	280.6

(*80% aqueous acetone, 25 °C)

rate enhancements observed with phenyl or cyclopropyl substituents on the α-carbon atom appear to result from major electron supply from the ring system rather than from the relief of B-strain. Nevertheless, the relative rates show no correlation with the ^{13}C chemical shifts for the corresponding carbocations, and it has been questioned whether such shifts can be used in their present state of understanding to prove unequivocally the structures of ions such as 2-norbornyl or 2-bicyclo[2,1,1]hexyl.

A '*bona fide* free methyl cation' has been generated in the liquid phase by allowing tritiated methane to undergo β-decay in solution in benzene and toluene

28 G. A. Olah and D. J. Donovan, *J. Amer. Chem. Soc.*, 1977, **99**, 5026.
29 H. C. Brown and E. N. Peters, *J. Amer. Chem. Soc.*, 1977, **99**, 1712.

[equation (2)].[30] Alkylation of the aromatic ring ensues. This novel approach appears to provide a means for observing the intrinsic reactivity in solution or even

$$CT_4 \xrightarrow{\beta\text{-decay}} CT_3^+ + {}^3He + \beta^- \qquad (2)$$

in the solid state of a simple carbocation, completely unperturbed by the effects of solvent, or counterion, *etc.*

5 β-Elimination

The range of mechanisms available for β-elimination has been discussed previously [*cf. Ann. Reports* (*B*), 1976, **73**, 61; 1975, **72**, 76]. Substrates with good leaving groups attached to a s- or t-carbon atom and having no activation of the proton on C_β react with weak bases which are also strong nucleophiles by a mechanism designated *E*2*C*. The precise nature of the 'loose' transition state involved is still a topic for active discussion, which centres on whether there is a covalent, 'S_N2-like,' component to the interaction of the base with C_α[(50) or (51)] or whether such interaction is electrostatic in nature (52). What is said to be compelling evidence

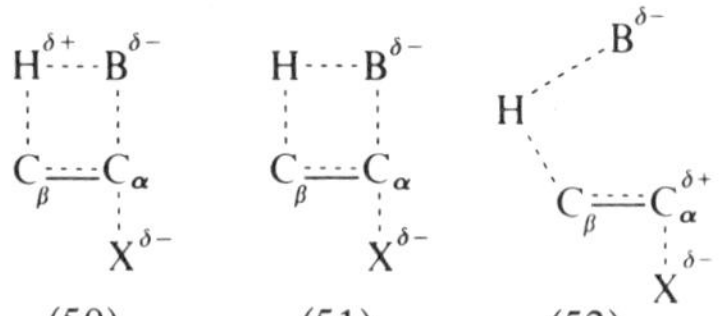

(50) (51) (52)

against the extensive development of positive charge on C_α has been obtained by comparison of the secondary deuterium isotope effects associated with the β'C—H and γC—H bonds of cyclohexyl tosylate in its elimination reaction with tetrabutylammonium hydroxide in acetone.[31] The similarity in magnitude of these effects ($k_{\beta'\text{-d}}/k_{\gamma\text{-d}} = 0.98$) is taken to indicate a substantial degree of formation of a double bond in the transition state. It has, however, been pointed out[32] that 1,3-diaxial interactions between the γ-hydrogens and the leaving group will be relieved as the transition state is formed, whatever the mechanism, and that the changes in associated vibrational frequencies will induce an isotope effect at C_γ of the type observed. An examination of the elimination (*E*2) and substitution (S_N2) reactions of cyclohexyl tosylate with triphenylphosphine shows that this neutral weak base, which is also a good carbon nucleophile, is a poor reagent for elimination compared to anionic weak bases which promote the *E*2*C* process (PhS^-, OAc^-, Cl^-, PhO^-, Br^-, *etc*), or compared to strong neutral bases such as triethylamine or 1,5-diazabicyclo[4,3,0]non-5-ene (DBN), whose reactions are more *E*2*H*-like in character.[32] The results suggest that whilst interaction between an anionic base and C_α is important in the *E*2*C* transition state, this interaction is primarily electrostatic.

The methoxide-induced eliminations of 1,1-diaryl-2,2-dichloroethanes [$(p\text{-}XC_6H_4)_2CHCHCl_2$] and their β-deuteriated analogues provide what is claimed to

[30] F. Cacace and P. Giacomello, *J. Amer. Chem. Soc.*, 1977, **99**, 5477.
[31] D. Cook, *J. Org. Chem.*, 1976, **41**, 2173.
[32] D. J. McLennan, *J.C.S. Perkin II*, 1977, 293.

be the first example of mechanistic changeover from *E*2 (for X = H or Cl) to *E*1cB (for X = NO_2) as a result of changing the substituent in the β-bound aromatic ring.[33] Similar conclusions have been reached as a result of studying chlorine isotope effects in the same system (X = MeO, Cl, H, or NO_2), and it has been shown that the C—Cl bond must be very nearly intact in the *E*2 transition state[34] [*cf. Ann. Reports* (*B*), 1976, **73**, 63 for a preliminary communication]. The influence of changes in substituents on C_α on the mechanism followed is exemplified in the reactions of methanolic methoxide with 2,2-di-(*p*-nitrophenyl)-1,1,1-trifluoroethane, 2,2-di-(*p*-nitrophenyl)-1-fluoroethane, and their β-deuteriated analogues.[35] The presence of three fluorine atoms on C_α leads to a clear example of reversible formation of the carbanion intermediate, *i.e.* the $(E1cB)_R$ mechanism is in operation. In contrast, the analogous trichloro-substrate has been shown to undergo elimination *via* the irreversible carbanion mechanism, designated $(E1cB)_I$.[36] With only one fluorine (*i.e.* the leaving group) on C_α, the reaction follows the *E*2 mechanism with a carbanion-like transition state.

An interesting example of alkene formation with hydride ion as the leaving group is provided by the elimination reactions of organolithium and organomagnesium compounds in the presence of hydride acceptors, *e.g.* tetraphenylcarbonium tetrafluoroborate, tricyclohexylborane, or tri-s-butylborane [equation (3)].[37]

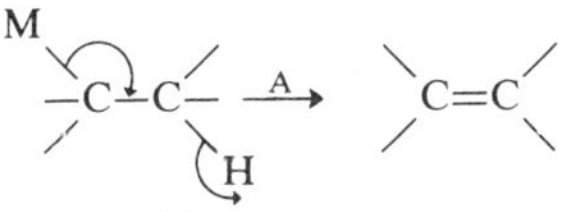

(3)

M = Li or Mg; A = $Ph_4C^+ BF_4^-$, $(cyclo\text{-}C_6H_{11})_3B$, or $(s\text{-}C_4H_9)_3B$

Preferential formation of the Hofmann product has been attributed to steric influences.

6 Electrophilic Addition

Two possible mechanisms have generally been considered to be available for the acid-catalysed hydration of olefins. One involves the rate-determining protonation of carbon (the $A_{SE}2$ mechanism) (Scheme 17), whilst the alternative route, which

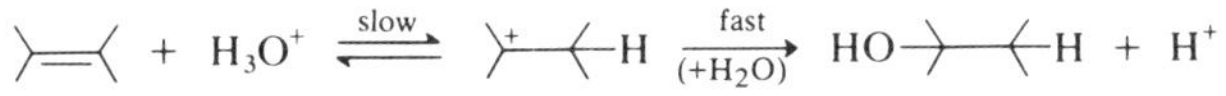

Scheme 17

reflects the dependence of reaction rate on acidity function rather than pH, involves the initial formation of a π-complex, which subsequently undergoes rate-determining conversion into the solvated carbocation (Scheme 18). A review of the literature data on olefin hydration, together with numerous new experimental results, now provides evidence that all olefin hydrations proceed by the $A_{SE}2$

33 D. J. McLennan, *J.C.S. Perkin II*, 1977, 1753.
34 A. Grout, D. J. McLennan, and I. H. Spackman, *J.C.S. Perkin II*, 1977, 1758.
35 J. Kurzawa and K. T. Leffek, *Canad. J. Chem.*, 1977, **55**, 1696.
36 D. J. McLennan and R. J. Wong, *J.C.S. Perkin II*, 1974, 1373.
37 M. T. Reetz and W. Stephan, *Angew. Chem. Internat. Edn.*, 1977, **16**, 44.

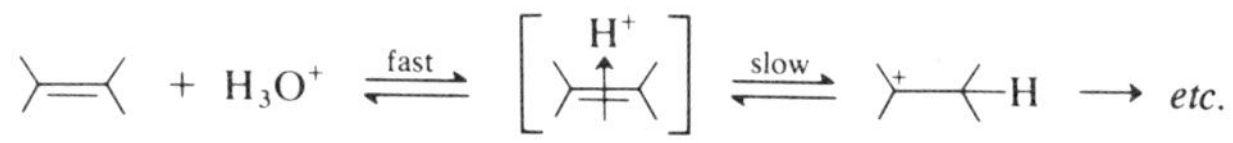

Scheme 18

process.[38] This conclusion is based on the extremely good correlation [equation (4)] which is obtained by plotting the second-order constants for olefin hydration in

$$\log k_2 = \rho \Sigma \sigma_p^+ + C \qquad (4)$$

aqueous acid at 25 °C against the sum of the relevant σ_p^+ parameters for the alkyl substituents attached to the carbocation centre of the intermediate, $R^1R^2R^3C^+$. For a range of ninety-six olefins, including 1,1- and 1,2-disubstituted alkenes, 2-substituted buta-1,3-dienes, substituted styrenes, vinyl esters, *N*-vinylacetamide, and 2-bromopropene, a correlation coefficient of 0.938 is obtained, with $\rho = -10.5$ and $C = -8.92$. A comparison of ethylene with *p*-nitrostyrene at H_0 −7.37 shows that ethylene is the less reactive, as predicted by equation (4), and indicates that ethylene itself protonates through an $A_{SE}2$ transition state, to give the open ethyl cation. This quite remarkable conclusion is at variance with many currently held views on the role of carbocation intermediates in nucleophilic solvents. Indeed, it has so far proved impossible to detect a free ethyl cation in super-acid media, let alone in water, and it seems likely that the interpretation of this result will come under careful scrutiny.

Two sets of workers have now concluded that, contrary to previous reports, solvent effects do not cause significant changes in the relative rates of bromination of alkenes and alkynes, although the absolute rates may change considerably.[39,40] Structural effects are, however, significant in determining the rate-constant ratio $k_{C=C}/k_{C\equiv C}$, which is *ca.* 10^3 for styrene and phenylacetylene, but *ca.* 1.0 for cinnamic acid (or its 4-nitro-derivative) and the corresponding acetylene, in a range of hydroxylic solvents.[39] The role of structural effects is greatly enhanced in chlorinated hydrocarbon solvents, in which the reaction is also seen to be second-order in bromine.[41] The importance of electrophilic solvent assistance to the leaving group (Br^-) is clear, but there is a difference of opinion as to whether the cationic intermediate is subject to specific nucleophilic solvation,[41] or simply to a medium effect.[40] It has been proposed that the accepted mechanism of olefin bromination (Scheme 19) is also valid for acetylenes.[40]

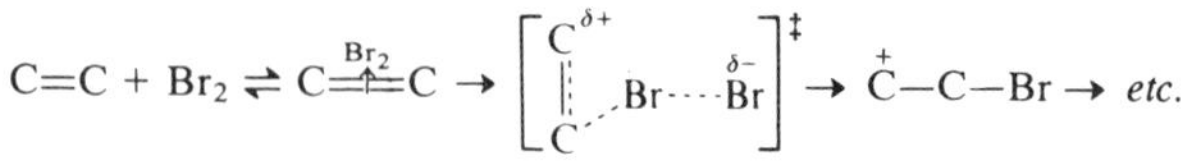

Scheme 19

7 Ion–Molecule Reactions in the Gaseous Phase

Increasing attention has been paid in recent years to the 'intrinsic' (*i.e.* solvent-free) properties of basic processes such as proton transfer, nucleophilic substitution,

[38] V. J. Nowlan and T. T. Tidwell, *Accounts Chem. Res.*, 1977, **10**, 252 and references cited therein.
[39] G. H. Schmid, A. Modro, and K. Yates, *J. Org. Chem.*, 1977, **42**, 2021.
[40] M.-F. Ruasse and J.-E. Dubois, *J. Org. Chem.*, 1977, **42**, 2689.
[41] A. Modro, G. H. Schmid, and K. Yates, *J. Org. Chem.*, 1977, **42**, 3673.

carbocation formation, *etc.*, as revealed by studies of ion–molecule reactions in the gaseous phase. Investigations have been made mainly with techniques such as pulsed ion cyclotron resonance (ICR) mass spectrometry,[42] pulsed electron beam high-pressure mass spectrometry,[43] photoionization mass spectrometry,[44] or that involving the use of flowing afterglow apparatus.[45] Apart from the inherent interest of gas-phase reactions, the results can throw considerable light on the nature of solvation processes and can enable a clearer insight to be obtained into many aspects of solution chemistry.

Proton Transfer.—Earlier work has shown that the presence of alkyl substituents in a molecule can increase both acidity and basicity in the gaseous phase, an effect said to be due primarily to polarization of the alkyl group by the nearby ionic centre.[46] Distant alkyl groups, however, appear to destabilize a negative charge, so that gas-phase acidities tend to be decreased by their influence in phenols, carboxylic acids, and acetylenes.[47]

Gas-phase Acidities. The order of acidities for a series of alkanethiols has now been shown,[48] like that for alcohols,[47a] to be exactly the reverse in the gaseous phase to that in aqueous solution, *i.e.* for the gas-phase reaction shown [equation (5) (X = O

$$RX^- + Bu^tXH \rightarrow Bu^tX^- + RXH \qquad (5)$$
$$(X = O \text{ or } S)$$

or S)] $\Delta G^\ominus$ increases for various R groups in the order Me < Et < Pr^n (given for X = S only) < Pr^i < Bu^t. Whereas the reversal of acidity order for alcohols in dimethyl sulphoxide has been shown to result primarily from the effect of the size of the alkyl substituent on the heat of solution of the alkoxide anion,[49] a similar analysis for alcohols in water cannot be made, as only the free energies of ionization are available. A consideration of the effect of alkyl groups on the relative values of enthalpy and entropy of ionization for alkanethiols in water shows, however, that the main factor causing the inversion of order in this case is the entropy of ionization.[48] It has further been concluded that the main factor controlling the relative gas-phase acidities of alkanethiols, hydrogen sulphide, and the alcohols is the electron affinity of the corresponding $RS^\cdot$ or $RO^\cdot$ radicals, bond energies being nearly constant in each series. Water is anomalous.

The intrinsic acidities of some fifty substituted phenols and benzoic acids have been determined from equilibrium constants and $\Delta G^\ominus$ values for their gas-phase proton-transfer reactions with a range of standard aliphatic acids of known acidities [equation (6)].[50] The results show that phenol and acetic acid have similar acid

42 J. L. Beauchamp, *Ann. Rev. Phys. Chem.*, 1971, **22**, 527.
43 A. J. Cunningham, J. D. Payzant, and P. Kebarle, *J. Amer. Chem. Soc.*, 1972, **94**, 7627.
44 A. D. Williamson, P. Le Breton, and J. L. Beauchamp, *J. Amer. Chem. Soc.*, 1976, **98**, 2705; P. R. Le Breton, A. D. Williamson, J. L. Beauchamp, and W. T. Huntress, *J. Chem. Phys.*, 1975, **62**, 1623; M. S. Foster, A. D. Williamson, and J. L. Beauchamp, *Internat. J. Mass Spectrometry Ion Phys.*, 1974, **15**, 429.
45 D. K. Bohme, R. S. Hemsworth, J. W. Rundle, and H. I. Schiff, *J. Chem. Phys.*, 1973, **58**, 3504.
46 J. I. Brauman and L. K. Blair, *J. Amer. Chem. Soc.*, 1968, **90**, 6561; 1970, **92**, 5986.
47 (*a*) R. T. McIver, jun., and J. S. Miller, *J. Amer. Chem. Soc.*, 1974, **96**, 4323; (*b*) R. T. McIver, jun., and J. H. Silvers, *ibid.*, 1973, **95**, 8462; (*c*) K. Hiraoka, R. Yamdagni, and P. Kebarle, *ibid.*, 1973, **95**, 6833.
48 J. E. Bartmess and R. T. McIver, jun., *J. Amer. Chem. Soc.*, 1977, **99**, 4163.
49 E. M. Arnett, L. E. Small, R. T. McIver, jun., and J. S. Miller, *J. Amer. Chem. Soc.*, 1974, **96**, 5638.
50 T. B. McMahon and P. Kebarle, *J. Amer. Chem. Soc.*, 1977, **99**, 2222.

$$A_1H + A_2^- \rightarrow A_1^- + A_2H \qquad (6)$$

strengths in the gaseous phase, whereas benzoic acid is stronger by *ca.* 8.7 kcal mol^{-1}. Solvation must account for the differences in water, in which acetic and benzoic acids have similar strengths and are stronger than phenol by nearly 6 pK_a units. It can be visualized that the acetate anion, with its relatively small methyl group and with its negative charge localized on the two oxygen atoms, will be more strongly solvated than will the phenoxide anion, which is larger and which has the charge delocalized over the aromatic ring. Ionization of acetic acid in water is therefore relatively more favourable. The relative acid strengths of benzoic acid in the gaseous phase and in aqueous solution have been interpreted in terms of contributions from charge-separated canonical forms, for which solvation is more favourable than for the uncharged species, and which are more important in the case of the acid than the anion.

Substituent effects for *meta*- and *para*-substituted benzoic acids in the gaseous phase are revealed through linear plots of free-energy changes for the gas-phase reaction (7) against Taft's σ^0 or Brown's σ values, the former giving a slightly

$$XC_6H_4CO_2^- + PhCO_2H \rightarrow XC_6H_4CO_2H + PhCO_2^- \qquad (7)$$

better correlation. *o*- and *p*-Hydroxybenzoic acids have anomalously high gas-phase acidities, but it can be shown that in the gaseous phase the phenolic proton is in each case the more acidic, and that the compounds are better regarded as substituted phenols in this context. Enhanced acidity of the phenolic proton can be attributed to the $-M$ effect of the *ortho*- or *para*-carboxy-group. On the other hand, the carboxy-group itself is made less acidic by the $+M$ effect of a hydroxy-group in the *ortho*- or *para*-position. The strength of the *ortho*-hydroxy-acid is further increased by stabilization of the anion through internal hydrogen-bonding between O^- and the CO_2H group.

A good linear correlation is obtained between the gas-phase and aqueous solution acidities of *meta*- and *para*-substituted phenols, as measured by $\Delta G^\ominus$ under these two conditions, for the same proton-transfer reaction [equation (8)]. It is

$$XC_6H_4O^- + PhOH \rightarrow XC_6H_4OH + PhO^- \qquad (8)$$

noteworthy that $\Delta G^\ominus$ is 6.8 times larger in the gas phase than in aqueous solution for phenols and 10.6 times larger for the acids. Although attenuations of substituent effects from gaseous phase to solution are not uncommon, those for the phenols and benzoic acids in water are unusually large, and are attributed to hydrogen-bonding in this solvent, in which substituent effects are largely reflected in the entropy term. In keeping with a previous study of gas-phase acidities and solution acidities in dimethyl sulphoxide,[51] the attenuation of substituent effects in phenols in this medium is small (factor of *ca.* 2) and is contained in the enthalpy term.

Gas-phase Basicities. It has been shown that the observed strengths of amines in the gaseous phase are consistent with a simple electrostatic model.[52] An alternative

[51] F. G. Bordwell, J. E. Bartmess, G. E. Drucker, Z. Margolis, and W. G. Matthews, *J. Amer. Chem. Soc.* 1975, **97**, 3226.
[52] D. H. Aue, H. M. Webb, and M. T. Bowers, *J. Amer. Chem. Soc.*, 1976, **98**, 311, 318.

approach[53] shows that an excellent linear correlation is obtained by plotting $\Delta G^{\ominus}$ or $\Delta H^{\ominus}$ values for the gas-phase proton-transfer reaction [equation (9)] against

$$RCH_2NH_2 + CH_3\overset{+}{N}H_3 \rightarrow RCH_2\overset{+}{N}H_3 + CH_3NH_2 \qquad (9)$$

new 'intrinsic' substituent constants, σ_I, obtained from gas-phase ionization-potential data and polarizability models.[54] Two interpretations have been offered.[53] Either (*a*) the observed effects of alkyl substituents on gas-phase base strengths and the σ_I parameters are both determined by polarizability effects, and involve no internal inductive effects, or (*b*) the σ_I parameters are a measure of internal inductive effects of alkyl groups, but the observed effects of the alkyl substituents are a combination of polarizability and internal inductive effects, both having very nearly the same structural dependencies on chain length and branching. The second interpretation is probably correct in view of recent comparisons of the quantitative effects of alkyl groups on gas-phase acidities of alcohols with those of gas-phase basicities of amines.[55]

Nucleophilic Substitution.—Studies of gas-phase S_N2 reactions [equation (10)] have revealed a wide variety of reaction rates.[56] In certain cases, intermediate

$$X^- + RY \rightarrow Y^- + RX \qquad (10)$$

adducts between anions and alkyl halides have been obtained (*e.g.* $ClCH_3Br^-$), with stabilities ranging from 8.6 to 14.4 kcal mol^{-1}.[57] The possibility of detecting the neutral product, RX, has also been demonstrated and it has been shown that the interaction of chloride ion with either *cis*- or *trans*-4-bromocyclohexan-1-ol proceeds with preponderant inversion of configuration.[58] Reaction rates and efficiencies (*i.e.* the fraction in each case of collisions resulting in reaction) for a wide range of systems (X = OH, Cl, Br, F, CN, MeO, MeS, or Bu^{t}O; RY = MeCl, MeBr, CF_3CO_2Me, PhOMe, MeOMe, Me_3CCH_2Cl, Me_3CCO_2Et, MeSH, or $MeCO_2H$), lead to the conclusion that the reaction is best described in terms of a 'double-well' potential [curve (*a*), Figure 1], the energy barrier in many cases being lower than the energy of the reactants.[59] The increase in the barrier in solution results solely from the differential solvation of the reactants and the transition state, the more localized charge on the reactant anion making this species better solvated. In a polar aprotic solvent such as dimethyl sulphoxide the barrier is raised less than in a protic solvent, which affords the possibility of specific hydrogen-bonding (Figure 1). The intrinsic nucleophilicities of anions in the gaseous phase follow the reverse order of the polarizabilities (*e.g.* $MeO^- > MeS^-$ and $F^- > Cl^- > Br^-$), a result which may be due to the stronger interaction which will occur between the more

[53] R. W. Taft and L. S. Levitt, *J. Org. Chem.*, 1977, **42**, 916.
[54] L. S. Levitt and H. F. Wilding, *Progr. Phys. Org. Chem.*, 1976, **12**, 119.
[55] R. W. Taft, in 'Proton Transfer Reactions', ed. E. F. Caldin and V. Gold, Chapman and Hall, London, 1974, Ch. 2, p. 66.
[56] J. I. Brauman, W. N. Olmstead, and C. A. Lieder, *J. Amer. Chem. Soc.*, 1974, **96**, 4030; K. Tanaka, G. I. Mackay, J. D. Payzant, and D. K. Bohme, *Canad. J. Chem.*, 1976, **54**, 1634 and references cited therein.
[57] R. C. Dougherty, J. Dalton, and J. D. Roberts, *Org. Mass Spectrometry*, 1974, **8**, 77; R. C. Dougherty and J. D. Roberts, *ibid.*, p. 81; R. C. Dougherty, *ibid.*, p. 85.
[58] C. A. Lieder and J. I. Brauman, *J. Amer. Chem. Soc.*, 1974, **96**, 4028; *Internat. J. Mass Spectrometry Ion Phys.*, 1975, **16**, 307.
[59] W. N. Olmstead and J. I. Brauman, *J. Amer. Chem. Soc.*, 1977, **99**, 4219.

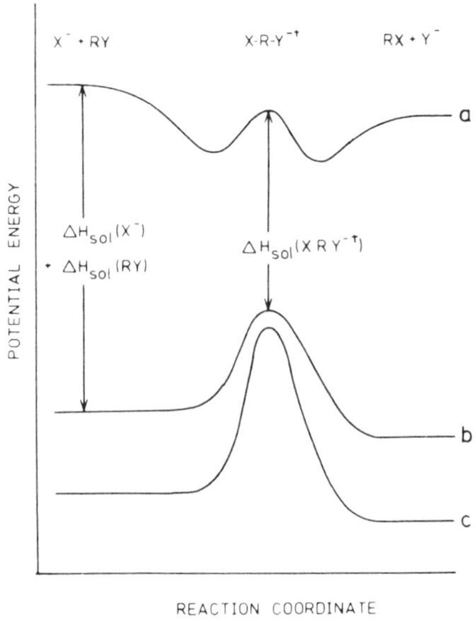

Figure 1 *Representative diagrams of the reaction co-ordinates for a nucleophilic displacement reaction* (a) *in the gaseous phase, and* (b) *in dipolar aprotic and* (c) *in protic solvents*
(Reproduced by permission from *J. Amer. Chem. Soc.*, 1977, **99**, 4219)

concentrated charge of the smaller anions and the carbon centre. The parallel between polarizability and nucleophilicity of anions in protic solvents is thus seen to be purely an artefact of solvation effects; indeed, the order may be reversed in aprotic solvents.

Carbocations.—*Stabilities.* Heterolytic bond-dissociation energies $D(R^+—Br^-)$ have been determined for a range of alkyl carbenium ions, acyl cations, and cyclic halonium ions, by examining the equilibria shown [equation (11)].[60] A combina-

$$R_1^+ + R_2X \rightleftharpoons [R_1XR_2{}^+]^* \rightleftharpoons R_1X + R_2^+ \quad (11)$$

tion of the gas-phase data with heats of solution in $HSO_3F–SbF_5$ gives, *via* appropriate chemical cycles, the relative enthalpies of solvation, which have been shown to be related to ion size, the smaller ions being better solvated. Relative stabilities (measured as bromide affinities) for the cyclic bromonium ions are the same in the gaseous phase as in solution. In both cases, stability increases with ring size and, in the three-membered rings, with methyl substitution. The range of stabilities is, however, attenuated in solution. The adamantyl cation is found to be more stable than t-butyl in the gaseous phase, indicating that the strain energy due to non-planarity of the adamantyl system is smaller than the stabilization afforded by interaction of the charge with the hydrocarbon framework.

Condensation Reactions. A number of gas-phase condensation reactions of carbenium ions, of possible relevance to prebiotic synthesis, have been shown to be

[60] R. H. Staley, R. D. Wieting, and J. L. Beauchamp, *J. Amer. Chem. Soc.*, 1977, **99**, 5964.

analogous to the corresponding reactions in solution. With water, the isopropyl and t-butyl cations yield protonated alcohols, which can undergo further hydration (Scheme 20), or elimination to yield alkene.[61] Other n-donors (MeOH, NH_3,

$$C_4H_9^+ \xrightarrow{H_2O} [C_4H_9(OH_2)]^+ \xrightarrow{H_2O} [C_4H_9(OH_2)_2]^+ \xrightarrow{H_2O} [C_4H_9(OH_2)_3]^+ \rightarrow \textit{etc.}$$

Scheme 20

$MeNH_2$, *etc.*) probably condense in a similar fashion.[61] The gas-phase equivalent of the Koch–Haaf synthesis has likewise been shown to occur with carbenium ions generated in methane containing small amounts of carbon monoxide and water (Scheme 21), the species produced being identified as the protonated acid.[62]

$$R^+ \xrightarrow{CO} R\overset{+}{C}O \xrightarrow{H_2O} RCOOH_2^+$$

Scheme 21

Formation of formic acid was also noticed, but this underwent decarbonylation, possibly as shown (Scheme 22).

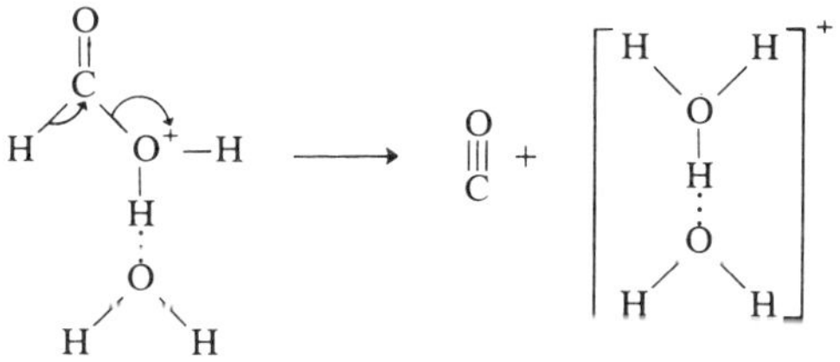

Scheme 22

Ester Cleavage.—An interesting divergence between behaviour in the gaseous phase and in solution is seen in the mechanism of ester cleavage by alkoxide ion. Whereas the $B_{AC}2$ mechanism is almost universal in base-catalysed hydrolyses in aqueous media, the gas-phase reaction of deuteriomethoxide ion with methyl trifluoroacetate or methyl benzoate yields the carboxylate anion as the only detectable product [equation (12)].[63] A possible mechanism involves attack on the

$$CD_3O^- + RCO_2CH_3 \rightarrow RCO_2^- + CD_3OCH_3 \qquad (12)$$

ester alkyl group *via* an S_N2 transition state (54). In protic solvents, on the other hand, a tetrahedral intermediate (55) is more likely to be favoured by stabilization through hydrogen-bonding to the carbonyl oxygen atom.

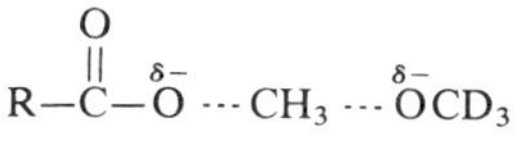

(54)

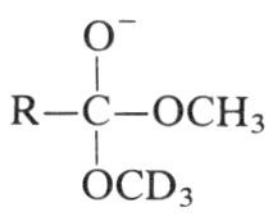

(55)

[61] K. Hiraoka and P. Kebarle, *J. Amer. Chem. Soc.*, 1977, **99**, 360.
[62] K. Hiraoka and P. Kebarle, *J. Amer. Chem. Soc.*, 1977, **99**, 366.
[63] M. Comisarow, *Canad. J. Chem.*, 1977, **55**, 171.

4 Reaction Mechanisms

Part (iii) Electron Spin Resonance Spectroscopy and Free Radical Reactions

By A. T. BULLOCK

Department of Chemistry, University of Aberdeen, Old Aberdeen, Scotland AB9 2UE

Of several review articles and books published during the year, attention is especially drawn to a book devoted to applications of e.s.r. to polymer research[1] and an admirable, albeit wholly theoretical, article on theories of chemically induced dynamic electron polarization (CIDEP).[2] Wan and Elliott[3] have briefly reviewed CIDEP mechanisms and, more importantly, have given illustrations of how CIDEP and CIDNP may be used together to study photochemical reactions. E.s.r. studies of triplet states have been reviewed[4] with attention drawn to cyclic 4π-electron systems, CH_2, and environmental effects. Especial emphasis was put onto ground state triplets. Finally, of relevance to the general field of this report, although not specifically about e.s.r. studies, is a review of free-radical rearrangements in telomerization.[5]

1 Kinetics and Mechanism

Much of the work described in this section deals with radicals having the unpaired electron centred largely on atoms other than carbon. There have been detailed studies of the generation and subsequent reactions of alkyloxy and aryloxy radicals. A possible general route to alkyloxy radicals has been described which involves the photolysis ($\lambda = 365$ nm) of alkoxy-vanadium(V) chelates.[6] The primary step is

$$\underset{\text{(1)}}{\mathrm{Q}\overset{\mathrm{O}}{\overset{\|}{\underset{\underset{\mathrm{OR}}{|}}{\mathrm{V}}}}\mathrm{Q}} + h\nu \rightarrow \underset{\text{(2)}}{\mathrm{Q}\overset{\mathrm{O}}{\overset{\|}{\mathrm{V}}}\mathrm{Q}} + \cdot\mathrm{OR} \qquad (1)$$

where Q is the 8-quinolyloxo ligand and R represents a series of nineteen alkyl groups. The production of paramagnetic compound (2), containing V^{IV}, was

1 B. Rånby and J. F. Rabek, 'Polymers/Properties and Applications. Vol. 1. E.S.R. Spectroscopy in Polymer Research', Springer-Verlag, Berlin, Heidelberg, and New York, 1977.
2 P. W. Atkins and G. T. Evans, 'Advances in Chemical Physics', XXXV, p. 1 *et. seq.*, ed. I. Prigogine and S. A. Rice, J. Wiley and Sons, New York, London, Sydney, and Toronto, 1976.
3 J. K. S. Wan and A. J. Elliot, *Accounts Chem. Res.*, 1977, **10**, 161.
4 E. Wasserman and R. S. Hutton, *Accounts Chem. Res.*, 1977, **10**, 27.
5 R. Kh. Freidlina and A. B. Terent'ev, *Accounts Chem. Res.*, 1977, **10**, 9.
6 S. M. Aliwi and C. H. Bamford, *J.C.S. Faraday I*, 1977, **73**, 776.

demonstrated directly by e.s.r. whilst comparison of the quantum yields for photo-initiated free radical polymerization of methyl methacrylate in the presence of (1) showed that each molecule of chelate decomposing gave rise to one initiating radical. Subsequent reactions of the alkyloxy radicals were studied by spin-trapping experiments using phenyl t-butyl nitrone (PBN) at various concentrations. In the case of primary alkyloxy radicals the e.s.r. spectrum depended on [PBN]. For example, for $R = Bu^n$, a simple spectrum was obtained at [PBN] = 0.008 M which was attributable to the adduct from the isomerized hydroxyalkyl radical (4).

$$\underset{(3)}{Pr^nCH_2O\cdot} \rightarrow \underset{(4)}{Pr^n\dot{C}HOH} \qquad (2)$$

At high [PBN] (*ca.* 0.3 M) both (3) and (4) were trapped. Most secondary alkyloxy radicals gave the same spectra at high and low concentrations of PBN. The coupling constants $a^{\beta\text{-H}}$ and a^N of the resultant nitroxides were characteristic of alkyl adducts. Clearly β-scission occurs according to the general equation

$$R'(R'')CHO\cdot \rightarrow R'CHO + \cdot R'' \qquad (3)$$

However, β-scission was found to be less important for $R = Pr^i$, cyclopentyl, and cyclohexyl, where the behaviour was similar to that found for the primary alkyloxy radicals, *i.e.* complex spectra at high [PBN] with $a^{\beta\text{-H}}$ and a^N suggesting alkyloxy and hydroxyalkyl adducts. Only two derivatives with t-alkyl groups were studied, namely $R = Bu^t$ and t-pentyl. Both gave simple spectra at high and low [PBN] with splittings characteristic of alkyl adducts. Again this was thought to be a consequence of β-scission, *viz.*

$$Me_3CO\cdot \rightarrow \cdot CH_3 + Me_2CO \qquad (4a)$$

$$EtMe_2CO\cdot \rightarrow \cdot C_2H_5 + Me_2CO \qquad (4b)$$

Product analyses confirmed this, the yield of Me_2CO being 60% for reaction (4a) and 100% for (4b).

The reactions of secondary alkyloxy radicals have also been studied in a flow system.[7] They were produced by Ti^{III} reduction of secondary alkyl hydroperoxides and cyclic hydroperoxides. No RO· radicals were observed directly but some were trapped by the *aci*-anion of nitromethane and characterized by the e.s.r. parameters of the $ROCH_2NP_2^{\cdot-}$ adducts. In all cases, carbon-centred radicals were observed to be derived from the parent alkyloxy radicals and were explained in terms of the following four reaction pathways.

(i) 1,2-hydrogen shift [*cf.* equation (2) for n-alkyloxy radicals]

$$R^1R^2CHO\cdot \xrightarrow[H_2O]{1,2\text{-H}} \cdot CR^1R^2OH \qquad (5)$$

(ii) 1,5-hydrogen shift

$$\textit{e.g. } Me(CH_2)_2CH(O\cdot)Me \xrightarrow{1,5\text{-H}} \cdot CH_2(CH_2)_2CHMeOH \qquad (6)$$

[7] B. C. Gilbert, R. G. G. Holmes, and R. O. C. Norman, *J. Chem. Research*, 1977, (S)1; (M)0101.

(iii*a*) C—C fragmentation

e.g. 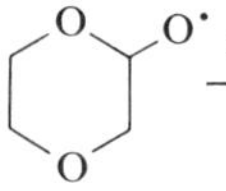 $\xrightarrow{F_{C-C}}$ $\cdot CH_2O(CH_2)_2OCHO$ (7)

(iii*b*) C—Me fragmentation

e.g. $MeCH_2OCH(O\cdot)Me \xrightarrow{F_{C-Me}} \cdot CH_3$ (8)

(iv) C—O fragmentation

e.g. $Me_2CHOC(O\cdot)Me, \xrightarrow[(ii) 1,2\text{-}H]{(i) F_{C-O}} Me_2\dot{C}OH$ (9)

The authors suggest that the 1,2-H shift [equation (5)] is not intramolecular but probably involves the solvent.[8] It was concluded that alkyloxy radicals from secondary hydroperoxides show an ease of reaction in the order 1,5-H $> F_{C-C} >$ 1,2-H $> F_{C-Me}$. This contrasts with the order for primary alkyloxy radicals, *viz.* 1,2-H $\approx$ 1,5-H $> F_{C-C}$.

Aryloxy radicals have been obtained by photorearrangement of nitro-compounds.[9] Photochemical rearrangements of nitro-groups are often postulated to proceed *via* the nitrite[10] and it is claimed[9] that the detection of aryloxy radicals during the photolysis of several hindered nitro-compounds is strong evidence for the intermediate formation of nitrite groups. As an example, photolysis of 6-nitrobenzo[*a*]pyrene (5) gave the spectrum of the aryloxy radical (6) [Scheme (1)]. The radical (6) had previously been prepared by H-atom abstraction from benzo[*a*]pyren-6-ol.[11]

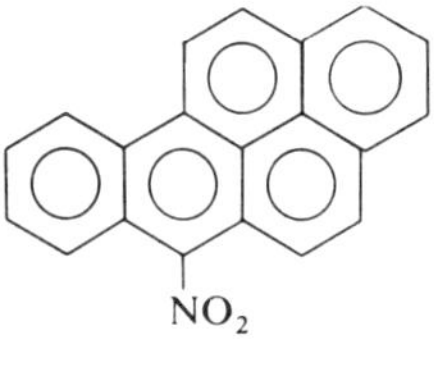

$\xrightarrow{h\nu}$ ArONO $\longrightarrow$ Ar$\dot{O}$
(6)

(5) = $ArNO_2$

Scheme 1

Other oxygen-centred radicals have been studied, notably the decay kinetics of cumylperoxy radicals.[12] Previous e.s.r. determinations of the decay kinetics of peroxy radicals have given first, second, and intermediate orders. The authors point out that in most previous reports the radicals were produced photolytically

[8] B. C. Gilbert, R. G. G. Holmes, H. A. H. Laue, and R. O. C. Norman, *J.C.S. Perkin II*, 1976, 1047.
[9] Y. Ioki, *J.C.S. Perkin II*, 1977, 1240.
[10] O. L. Chapman, D. C. Heckert, J. W. Reasoner, and S. P. Thackaberry, *J. Amer. Chem. Soc.*, 1966, **88**, 5550.
[11] C. Nagata, M. Inomata, M. Kodama, and Y. Tagashira, *Gann*, 1968, **59**, 289.
[12] S. Fukuzumi and Y. Ono, *J.C.S. Perkin II*, 1977, 622.

and suggest that the measured kinetics may have been complicated by the presence of other products of the photolysis. To avoid this, the cumylperoxy radicals were produced by flowing a solution (CCl_4, benzene, and cumene solvents) of cumyl hydroperoxide (0.74—4.70 M) through a tube packed with 20—30 mesh manganese dioxide or cobalt oxide supported on silica. On stopping the flow, decay of the cumylperoxy radicals was found to be strictly first-order ($k_t = 0.15 \pm 0.015\ s^{-1}$ at 300 K) irrespective of the initial hydroperoxide and cumylperoxy concentrations and of the nature of the solvent and catalyst. The authors subsequently[13] confirmed the first-order decay by measuring the radical concentrations during the steady-state decomposition of cumene hydroperoxide with PbO_2, the latter being held in suspension in the cavity. The decomposition was found to be a radical chain-reaction. The hydroperoxide formed a complex on the surface of PbO_2, the complex subsequently dissociated to the peroxy radical which then desorbed into solution. From the kinetic studies it was found necessary to postulate the formation of a 2:1 complex between the hydroperoxide and the peroxy radical. Two pieces of spectral data supported this, namely the g-value and the observed line-width both varied monotonically with the initial concentration of cumene hydroperoxide. The kinetic, g-value, and linewidth results were all accommodated by $K = 0.50\ l^2\ mol^{-2}$ for the equilibrium

$$RO_2\cdot + 2RO_2H \rightleftharpoons (RO_2\cdots 2RO_2H) \qquad (10)$$

The same authors have used similar techniques, *i.e.* the heterogeneously catalysed production of radicals, to study the liquid-phase autoxidation of cumene with PbO_2,[14] the determination of cross-propagation rate coefficients in the autoxidation of hydrocarbons,[15] and the mechanism of the formation of the p-benzosemiquinone anion over manganese dioxide.[16] In the autoxidation of cumene (with PbO_2), the kinetics were studied over the temperature range 291—393 K.[14] E.s.r. was used to monitor the concentration of the chain-propagating cumylperoxy radical and the rate of oxygen consumption was measured under the same conditions. Both the radical concentration and the rate of oxygen consumption were constant with time and independent of the catalyst weight: liquid volume ratio. Experimentally, it was found that,

$$[RO_2\cdot] = k_a[RH][PbO_2]^0 \qquad (11)$$

and

$$-d[O_2]/dt = k_b[RH]^2[PbO_2]^0 \qquad (12)$$

These results, together with the product distribution, were shown to be consistent with the following mechanism.

$$\textit{Initiation} \qquad RO_2H + PbO_2 \xrightarrow{R_i} RO_2\cdot \qquad (13)$$

$$\textit{Propagation} \qquad RO_2\cdot + RH \xrightarrow{k_{14}} R\cdot + RO_2H \qquad (14)$$

[13] S. Fukuzumi and Y. Ono, *J.C.S. Perkin II*, 1977, 625.
[14] S. Fukuzumi and Y. Ono, *J.C.S. Perkin II*, 1977, 784.
[15] S. Fukuzumi and Y. Ono, *J. Phys. Chem.*, 1977, **81**, 1895.
[16] Y. Ono, T. Matsumura, and S. Fukuzumi, *J.C.S. Perkin II*, 1977, 1421.

$$R\cdot + O_2 \xrightarrow{k_{15}} RO_2\cdot \quad (15)$$

$$RO\cdot + RO_2H \xrightarrow{k_{16}} ROH + RO_2\cdot \quad (16)$$

$$RO_2\cdot + RO_2\cdot \xrightarrow{k_{17}} 2RO\cdot + O_2 \quad (17)$$

Termination

$$RO_2\cdot \xrightarrow{k_{18}} PhCOMe + MeO\cdot \quad (18)$$

$$MeO\cdot + RO_2\cdot \xrightarrow{\text{fast}} \text{stable products} \quad (19)$$

The value of $2k_{17}$ was found to be $5.0 \times 10^5\ \text{l mol}^{-1}\ \text{s}^{-1}$ (291 K), in agreement with that found earlier.[13] The following Arrhenius parameters were found:

$$\log_{10}(k_{14}/\text{l mol}^{-1}\ \text{s}^{-1}) = 5.3 - 30/\theta \quad (20a)$$

and

$$\log_{10}(k_{17}/\text{l mol}^{-1}\ \text{s}^{-1}) = 10.9 - 30/\theta \quad (20b)$$

where $\theta = 2.303\ RT\ \text{kJ mol}^{-1}$.

The PbO_2-catalysed decomposition of t-butyl hydroperoxide in the presence of hydrocarbons has been studied by the same method.[15] It has been shown that simultaneous measurements of $[Bu^tO_2\cdot]$ (by e.s.r.) and the rate of oxygen generation lead to a direct determination of the rate coefficients for cross-propagation reactions (*e.g.* hydrogen abstraction from hydrocarbons by $Bu^tO_2\cdot$) and for self-reaction of $Bu^tO_2\cdot$. Rate coefficients for termination, chain transfer, and cross termination involving $Bu^tO_2\cdot$ were also determined by e.s.r. It was suggested that the technique provides a method for determining absolute rate coefficients for reactions involving $Bu^tO_2\cdot$ at various temperatures that is more accurate and simple than the classic hydroperoxide method.[17]

Before leaving this technique, we note that the mechanism of the formation of the *p*-benzosemiquinone anion over manganese dioxide has been studied.[16] The kinetic results could be fitted to the equation

$$d[SA]/dt = k(b[HQ]_0 - [SA])(c[MnO_2]_0 - [SA]) \quad (21)$$

where k, b, and c are constants, $[MnO_2]_0$ is the MnO_2 weight to liquid volume ratio, and $[HQ]_0$ and $[SA]$ are respectively the initial concentrations of hydroquinone and semiquinone anion radical. It was proposed that the active surface species $MnO_2{}^*$ abstracts a hydrogen atom from hydroquinone to give the neutral radical, which then desorbs and is converted into the semiquinone anion radical. The number of surface species of $MnO_2{}^*$ was determined to be $7.7 \times 10^{18}\ \text{m}^{-2}$, in good agreement with previous values obtained for the oxidations of 1,1-diphenyl-2-picrylhydrazine[18] and cinnamyl alcohol.[19]

[17] J. A. Howard, W. A. Schwalm, and K. U. Ingold, *Adv. Chem. Ser.*, 1968, No. 75, p. 6.
[18] A. T. T. Oei and J. L. Garnett, *J. Catalysis*, 1970, **19**, 176.
[19] D. Dollimore and K. H. Tonge, *J. Chem. Soc. (B)*, 1967, 1380.

The reactions of esters and anhydrides in a Ti^{III}/H_2O_2 flow system have been studied in some detail[20] following an earlier observation that vinyl acetate in this flow system gave mainly the spectrum of the methyl radical.[21] It has now been found that the alkyl radicals are formed by the route indicated in equations (22)—(24), *viz.*,

$$RCO_2R' + HO_2^- \rightarrow RCO_2O^- + R'OH \quad (22)$$

$$RCO_2OH + Ti^{III} \rightarrow RCO_2\cdot + OH^- + Ti^{IV} \quad (23)$$

$$RCO_2\cdot \rightarrow R\cdot + CO_2 \quad (24)$$

The first step is perhydrolysis followed by one-electron reduction of the resultant peroxo-acid. Decarboxylation of the acyloxy radical $RCO_2\cdot$ then gives the alkyl radical. In some cases, the spin trap $CH_2{:}NO_2^-$ enabled the intermediate acyloxy radicals to be detected in addition to the alkyl radicals. From the relative concentrations of the spin adducts $RCO_2CH_2NO_2^{\cdot -}$ and $RCH_2NO_2^{\cdot -}$ it was found that the rates of decarboxylation decrease in the series shown in Scheme 2. A lower limit

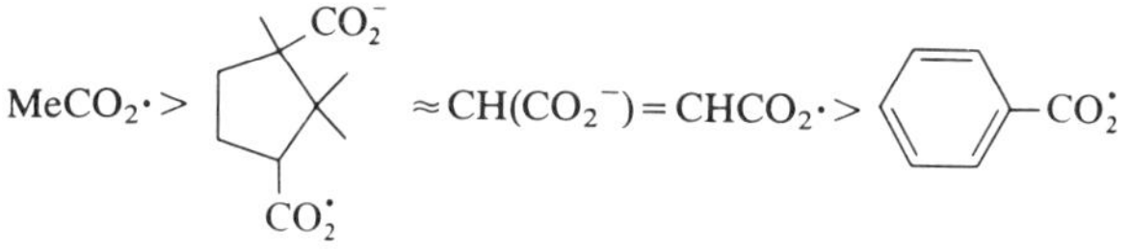

Scheme 2

for the rate of decarboxylation of $CH_3CO_2\cdot$ was estimated to be $2 \times 10^7\ s^{-1}$ whilst an upper limit for $PhCO_2\cdot$ is $2.5 \times 10^5\ s^{-1}$. Decarboxylation reactions have also been observed in the adduct radicals formed by the reaction of photochemically produced acyl radicals with the fumarate dianion.[22] The acyl adducts (7) underwent decarboxylation to give the 1-carboxy-3-oxyallyl radical dianions (8) [Scheme 3]. For R = Me and Et, the e.s.r. spectra of (7) and (8) were observed simul-

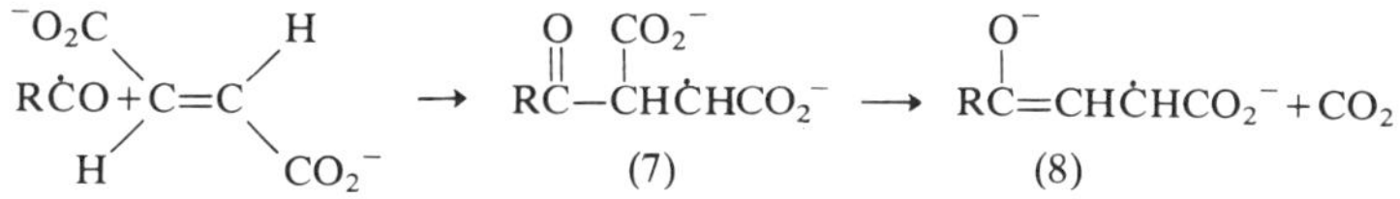

Scheme 3

taneously. Plots of $\log_{10}\{[RC(O^-){=}CH\dot{C}HO_2^-]/[RCOCH(CO_2^-)\dot{C}HCO_2^-]\}$ *vs* $1/T$ showed that the activation energies for the decarboxylation step were equal for R = Me and Et to within experimental error and had the value $40 \pm 2\ kJ\ mol^{-1}$. The acyl radicals were readily identifiable since hyperfine coupling to protons bonded to

[20] B. C. Gilbert, R. G. G. Holmes, P. D. R. Marshall, and R. O. C. Norman, *J. Chem. Research*, 1977, (S)172; (M)1949.
[21] D. J. Edge, B. C. Gilbert, R. O. C. Norman, and P. R. West, *J. Chem. Soc. (B)*, 1971, 189.
[22] S. Steenken and M. Lyda, *J. Phys. Chem.*, 1977, **81**, 2201.

C_4 and C_5 (C_β and C_α of the acyl precursor) of the allyl radical (8) could be measured. In contrast to these examples[20,22] of ready decarboxylations, a study of the radical zwitterions formed by the one-electron oxidation of a series of methoxylated benzoic acids by Tl^{2+}, Ag^{2+}, and $SO_4^{\cdot-}$ has shown that the presence of even one methoxy-group is sufficient to stabilize the radicals with respect to decarboxylation.[23] The radical zwitterions could also be produced from the addition of ·OH to mono-, di-, and tri-methoxylated benzoic acids.[24] The adducts reacted with H^+ ($k = 10^8$—10^9 l mol^{-1} s^{-1}) to give the zwitterions *via* elimination of water from the protonated adduct.

There have been several reports of sulphur-centred radicals. Thus, a series of aromatic and aliphatic sulphinyl radicals have been observed and characterized.[25] The aromatic sulphinyls were produced by one or other of the three routes shown in Scheme 4, where the ·OH radicals were generated in a conventional Ti^{III}/H_2O_2

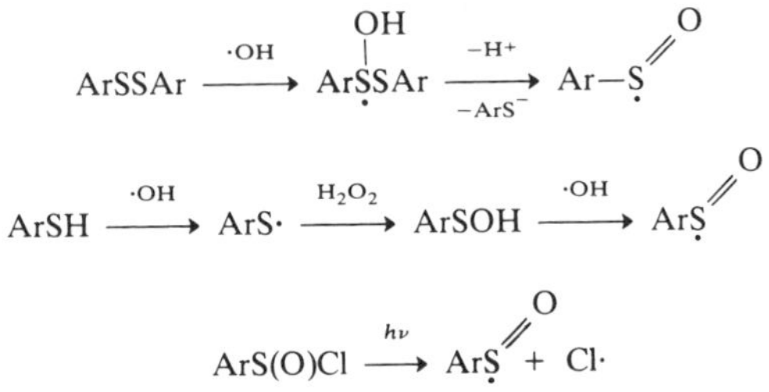

Scheme 4

flow system. The aromatic radicals were all π-radicals with extensive delocalization on to the aromatic ring. The aliphatic sulphinyl radicals were generated by photolysis of sulphinyl chlorides at low temperatures (*ca.* −100 °C) and were characterized by $g \approx 2.011$ and $a^{\beta\text{-H}}$ in the range 0.8—1.1 mT. Particular attention was paid to Me$\dot{S}$O, for which $g = 2.0100$, a(3H) = 1.15 mT, and ΔH, the peak-to-peak derivative linewidth, was 1.10 mT. The authors suggest that a previous report[26] of the spectrum of Me$\dot{S}$O, obtained by the photolysis of t-butyl-methanesulphenate, refers to some other sulphur radical [a(3H) = 0.657 mT, g = 2.00965, ΔH = 0.06 mT]. Previous failures to detect Me$\dot{S}$O and Et$\dot{S}$O were attributed to efficient spin–rotation relaxation for which $T_1 \propto \eta/T$. Sulphinyl radicals have also been observed in a study of the photochemical decomposition of organic sulphites and reactions of the sulphites with alkyloxy radicals.[27] Dialkyl sulphites $(R^1R^2CHO)_2S{=}O$ reacted with Bu^{t}O· in two ways. Addition at the sulphur atom gave sulphuranyloxy radicals $(R^1R^2CHO)_2\dot{S}(O)OBu^t$ while hydrogen abstraction also occurred, giving the carbon-centred radical $\cdot CR^1R^2OS(O)OCHR^1R^2$. This then fragmented to give alkoxysulphinyl radicals $R^1R^2CHO\dot{S}O$. Photolysis of the dialkyl sulphites in the absence of di-t-butyl peroxide resulted in fission of the S—O bond to give alkoxysulphinyl and alkyloxy ($\cdot OCHR^1R^2$) radicals; the latter

[23] P. O'Neill, S. Steenken, and D. Schulte-Frohlinde, *J. Phys. Chem.*, 1977, **81**, 26.
[24] P. O'Neill, S. Steenken, and D. Schulte-Frohlinde, *J. Phys. Chem.*, 1977, **81**, 31.
[25] B. C. Gilbert, C. M. Kirk, R. O. C. Norman, and H. A. H. Laue, *J.C.S. Perkin II*, 1977, 497.
[26] T. Kawamura, P. J. Krusic, and J. K. Kochi, *Tetrahedron Letters*, 1972, 4075.
[27] B. C. Gilbert, C. M. Kirk, and R. O. C. Norman, *J. Chem. Research*, 1977, (S)173; (M)1974.

subsequently rearranged to $\cdot CR^1R^2OH$. The spectrum of $HO\dot{S}O$ was also observed in this work and possible schemes for its generation were given.

The structure and formation of some dialkoxysulphuranyl radicals (10) have been described.[28] Two routes seemed to involve sulphenates (9), namely the photolysis of disulphides, R^1SSR^1, and thioethers, R^1SR^1, in the presence of peroxides. These are shown in Scheme 5. Support for the intermediacy of sulphenates in routes (a) and (b) of Scheme 5 comes from the fact that sulphenates

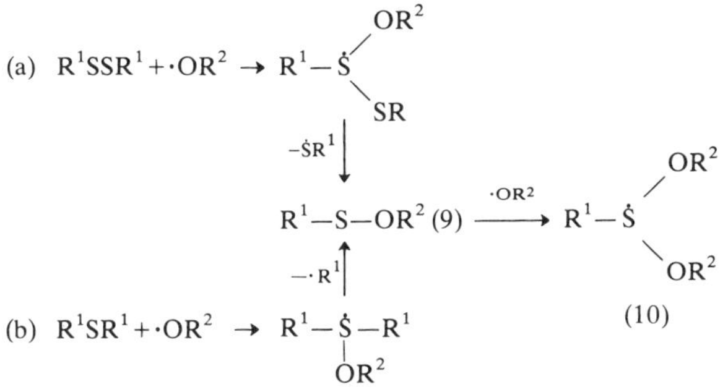

(10)

Scheme 5

themselves give dialkoxysulphuranyl radicals on photolysis. Evidence for the loss of $\cdot R^1$ in Scheme 5(b) was provided by the result for theitan, from which the ring-opened radical $\cdot CH_2CH_2CH_2SOBu^t$ was detected. The third route to dialkoxysulphuranyl radicals was shown to be the direct addition of alkyl radicals to dialkylsulphoxylates $(RO)_2S$. Except for $R^1 = CF_3$ ($g = 2.0079$) all g values were in the range 2.0090—2.0096. Radicals of the type $RCH_2\dot{S}(OBu^t)_2$ show spectra consistent with structure (11) whilst $Ph\dot{S}(OBu^t)_2$ is essentially a π-radical of structure (12). Deuterium labelling experiments, together with e.s.r., have shown

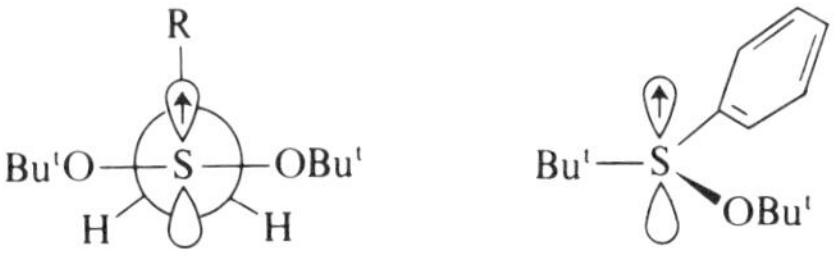

(11) (12)

that the addition of MeO· radicals to $(MeO)_2S$, to give the trimethoxysulphuranyl radical $(MeO)_3S\cdot$, has a high degree of stereoselectivity.[29] The incoming radical was found to take up an apical site in the adduct. Furthermore, neither intra- nor inter-molecular ligand exchange took place to an appreciable extent during the average lifetime of the sulphuranyl radicals.

Of radicals carrying a significant spin density on nitrogen atoms, nitroxides continue to attract attention. The syntheses, e.s.r. spectra, and stabilities of two

[28] W. B. Gara, B. P. Roberts, B. C. Gilbert, C. M. Kirk, and R. O. C. Norman, *J. Chem. Research*, 1977, (S)152; (M)1748.
[29] J. W. Cooper and B. P. Roberts, *J.C.S. Chem. Comm.*, 1977, 228.

3-oxy-1,3-diazacyclohexene-1-oxide radicals [(13a), (13b)] have been described.[30] They were found to be significantly more stable than comparable α-unsubstituted nitroxides.[31] Thus, (13a) decomposed with first-order kinetics with $k = 6 \times 10^{-4}\ s^{-1}$ (298 K). The authors discounted the possibility of a monomer–dimer equilibrium, which would show first-order kinetics for decomposition involving either slow bimolecular disproportionation of monomer or unimolecular decomposition of dimer, and suggest the possibility of H-atom transfer from C-4 to O-1 producing a reactive carbon atom. From the coupling constant of $a^{H\text{-}4}$ (1.09 mT) the appropriate dihedral angle was estimated to be 10° in (13a). This is

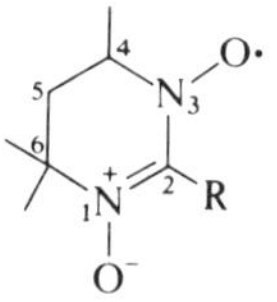

(13) a; R = H
b; R = Me

consistent with H-4 having a pseudo-axial conformation, two of the three methyl groups being pseudo-equatorial. Kinetic studies have also been carried out on the nitroxides (14), (15), and (16), produced by photolysis of Et_3SiH solutions of 2-, 3-,

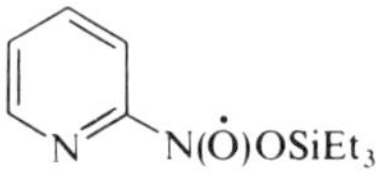

(14)

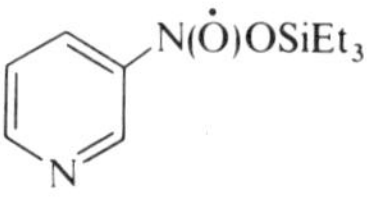

(15)

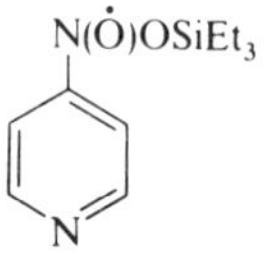

(16)

and 4-nitropyridines *in situ*.[32] Again first-order decays were observed but temperature-jump experiments clearly showed that dimer lay in the reaction pathway. Equation (25) shows the mechanism, although a clear choice between decay of monomer and decay of dimer could not be made

$$\text{Products} \leftarrow 2\text{ArN}(\dot{\text{O}})\text{OR} \rightleftharpoons \text{Dimer} \rightarrow \text{Products} \qquad (25)$$

A literature misassignment has been corrected. The e.s.r. signal obtained on the reaction of pentyl nitrite with aniline in benzene had been attributed to the σ-radical PhN—NO·.[33] However, it has now been shown[34] that this reaction and the reaction of pentyl nitrite with 1,3-diphenyl-triazine give signals consisting of overlapping spectra. The species responsible were shown to be diphenylnitroxide and phenylpentyloxynitroxide. The reactions between trifluoronitrosomethane and some 1,3-diketones have been shown to generate the nitroxide radicals $CF_3N(\dot{O})CH(COR^1)COR^2$ and their tautomers $CF_3N(\dot{O})C(COR^1){:}\,C(OH)R^2$

[30] S. N. Ghriofa, R. Daray, and M. Conlon, *J.C.S. Perkin I*, 1977, 651.
[31] D. F. Bowman, T. Gillan, and K. U. Ingold, *J. Amer. Chem. Soc.*, 1971, **93**, 6555.
[32] L. Lunazzi, G. Placucci, and N. Ronchi, *J.C.S. Perkin II*, 1977, 1132.
[33] A. F. Levit,and I. P. Gragerov, *Zhur. org. Khim.*, 1969, **5**, 31.
[34] J. I. G. Cadogan, R. G. M. Landells, R. M. Paton, J. T. Sharp, and R. U. Weber, *J. Chem. Research*, 1977 (S)108.

together with the iminoxy radicals $(R^2CO)(R^1CO)C{:}\,N\dot{O}$.[35] Under similar reaction conditions, however, diethyl 2-methyl-3-oxosuccinate afforded only the nitroxide $CF_3N(\dot{O})CMe(CO_2Et)COCO_2Et$. A reaction scheme was proposed which involved hydroxylamine anions $CF_3N(\bar{O})CH(COR^2)COR^1$ as intermediates.

A series of iminyl and triazenyl radicals have been derived by radical attack on some organic azides.[36] Bu^{t}O· radicals were found to react with a series of primary and secondary alkyl azides to give spectra characteristic of the iminyl radicals ($g = 2.0028$—2.0029, $a^N = 0.95$—0.98 mT), *viz.*

$$Bu^tO\cdot + HC(R^1)(R^2)\bar{N}{-}\overset{+}{N}{\equiv}N \xrightarrow{-H} Bu^tOH + N_2 + R^1(R^2)C{=}N\cdot \qquad (26)$$

The intermediate α-azidoalkyl radicals were not observed. In contrast, $Et_3Si\cdot$ and $Ph_3Si\cdot$ added to primary alkyl azides to give either $R^2{-}N{=}N{-}\dot{N}SiR^1_3$ (17) or $R^2(R^1_3Si)N{-}N{=}N\cdot$ (18). From a consideration of g-values (2.0010—2.0012) and the three values of a^N (1.76—2.0 mT, 0.34—0.40 mT, and 0.12—0.19 mT) it was argued that (18) was the probable structure, *i.e.* the 3,3-disubstituted triazenyl radical.

The chain process involved in the decomposition of diphenyldiazomethane induced by copper perchlorate has been studied kinetically and product analyses have been carried out. The e.s.r. spectra observed suggested that the radical cations $Ph_2\overset{+}{\dot{C}}N_2$ and $Ph_2\overset{+}{\dot{C}}N{:}\,NCPh_2$ are key intermediates in the chain reaction.[37]

Following earlier studies of the σ^*–π^* orbital crossover in a series of fluorinated benzene radical anions,[38] the isotropic spectra of several fluorinated pyridine anions have been observed in an adamantane matrix.[39] Very large couplings to ^{19}F in penta- and 2,3,4,6-tetrafluoropyridine anions together with small values of a^F for the more lightly fluorinated species indicate the former to be σ-radicals whilst the latter are π-radicals. The σ^*–π^* orbital crossover was rationalized in terms of the stabilization of σ^* and destabilization of π^* orbitals due to the inductive effects by back donation from fluorine. Its relevance to the present report lies in the suggestion by the authors that the availability of low-lying σ^* states should be carefully considered when assessing the chemistry of polyfluoro-aromatics. This is especially true for their reactions with nucleophiles.

Phosphorus-centred radicals are represented by studies of some reactions of phosphate radicals[40] and of the stereochemical non-rigidity and relative ligand apicophilicities of some phosphoranyl radicals.[41,42] The phosphate radical $P\dot{O}_4^{2-}$ and its protonated forms $HP\dot{O}_4^-$ and $H_2P\dot{O}_4$ were prepared *in situ* by photolysis of potassium peroxodiphosphate ($P_2O_8^{4-} \xrightarrow{h\nu} 2P\dot{O}_4^{2-}$).[40] While the phosphate radicals themselves were not detected by e.s.r., their adducts to fumaric and maleic acids and to the *aci*-anion of nitromethane were observed. In general, the reactions were similar to the related species $S\dot{O}_4^-$, *i.e.* adduct radicals were obtained

35 B. L. Booth, D. J. Edge, R. N. Haszeldine, and R. G. G. Holmes, *J.C.S. Perkin II*, 1977, 7.
36 J. W. Cooper, B. P. Roberts, and J. N. Winter, *J.C.S. Chem. Comm.*, 1977, 320.
37 D. Bethell, K. L. Handoo, S. A. Fairhurst, and L. H. Sutcliffe, *J.C.S. Chem. Comm.*, 1977, 326.
38 M. B. Yim and D. E. Wood, *J. Amer. Chem. Soc.*, 1976, **98**, 2053.
39 M. B. Yim, S. DiGregorio, and D. E. Wood, *J. Amer. Chem. Soc.*, 1977, **99**, 4260.
40 P. Maruthamuthu and H. Taniguchi, *J. Phys. Chem.*, 1977, **81**, 1944.
41 J. W. Cooper, M. J. Parrott, and B. P. Roberts, *J.C.S. Perkin II*, 1977, 730.
42 R. W. Dennis, I. H. Elson, B. P. Roberts, and R. C. Dobbie, *J.C.S. Perkin II*, 1977, 889.

with unsaturated substrates, hydroxyalkyl radicals from aliphatic alcohols, and inorganic radicals such as $\dot{C}O_3^-$ from HCO_3^- and $\dot{P}O_3^{2-}$ from HPO_3^{2-}. However, they differed in their reactions with aliphatic and aromatic carboxylic acids. At neutral pH, phosphate radicals caused hydrogen abstraction from saturated aliphatic mono- and dicarboxylic acids giving rise to α-carbon radicals, whereas $S\dot{O}_4^-$ gave mainly radicals produced by decarboxylation. On the other hand, phthalic acid gave a substituted phenyl radical on reaction with the peroxodisulphate system but did not do so with the phosphate radical. The authors concluded that $HP\dot{O}_4^-$ is a milder oxidant than the structurally related $S\dot{O}_4^-$ radical.

A series of phosphoranyl radicals have been found to exhibit linewidth effects which were interpreted in terms of intramolecular ligand exchange at the phosphorus atom.[41] In general, it was concluded that the following order of ligand apicophilicity holds:

$$F, Cl, RCO_2 > RC(O)NR, OCN > RO, R_2N > H > R$$

Other line-shape changes have been observed in some fluoroalkoxy- and fluoroalkyl-phosphoranyl radicals.[42] It was concluded that the apicophilicity of $-CF_3$ was less than that of $-Cl$ but usually greater than RO·. Additions of trifluoromethyl radicals to trialkyl phosphites was found to be reversible[42] just as an earlier study had demonstrated the reversibility of the addition of methyl radicals.[43] Radical addition to tervalent phosphorus, when followed by β-scission of an existing ligand, has been noted to be the free-radical equivalent of the Arbuzov rearrangement.[44] $Me_2N\cdot$ and phenyl radicals were generated photolytically and allowed to react with a series of *cis/trans* isomeric five- and six-membered ring phosphites. The reactions were found to be nearly stereospecific and the authors concluded that for the phosphoranyl intermediates, permutational isomerization of the Berry or turnstile mechanisms could not compete kinetically with the product-forming β-scission.

In the field of carbon-centred radicals, a flow system has been designed to test for the intermediacy of free radicals in the currently accepted mechanistic scheme for the reaction between benzylic halides and aromatic radical anions.[45] The accepted scheme is

$$RX + NaC_{10}H_8 \rightarrow R\cdot \rightarrow \dot{R}^- \rightarrow \text{products} \qquad (27)$$

The bis(3,5-di-t-butylphenyl)methyl radical was observed in the reaction between its bromide precursor and sodium naphthalenide. The generation of phenyl radicals and their subsequent abstraction reactions with a series of aliphatic substrates have been reported.[46] The abstracted atoms were H, Br, and I. Competitive experiments confirmed the nucleophilicity of the phenyl radical and it has been suggested that the transition state contains a significant contribution from polar structures such as $Ph^+\cdots\dot{H}\cdots\bar{C}HXMe$. Radical addition to alkynes and the subsequent intramolecular reactions of the resultant vinyl radicals have been

[43] J. W. Cooper and B. P. Roberts, *J.C.S. Perkin II*, 1976, 808.
[44] W. G. Bentrude, W. Del Alley, N. A. Johnson, M. Murakami, K. Nishikida, and H-W. Tan, *J. Amer. Chem. Soc.*, 1977, **99**, 4383.
[45] K. Schreimer, H. Oehling, H. E. Ziegler, and I. Angres, *J. Amer. Chem. Soc.*, 1977, **99**, 2638.
[46] B. Ashworth, B. C. Gilbert, and R. O. C. Norman, *J. Chem. Research*, 1977 (S)94; (M)1101.

described.[47] The vinyl radicals were not detected but were clearly implicated. Intermolecular addition and intramolecular abstraction of the vinyl radicals gave rise to the observed radicals. In the detailed schemes deduced, several examples of 1,5-hydrogen atom shifts were observed, including an oxygen-to-carbon example.

In view of the wide range of reported values for the rate coefficient for mutual termination of t-butyl radicals, the kinetics of this reaction have been studied carefully in isobutane and cyclopentane over the temperature range 170—330 K.[48] The radicals were generated by several photolytic routes, using a rotating sector–digital signal averaging technique. The results were that $2k_t = 1.1 \times 10^{10}\ \mathrm{l\ mol^{-1}\ s^{-1}}$ (298 K), with $E_a = 4.3\ \mathrm{kJ\ mol^{-1}}$ in both solvents. These are close to recent gas-phase measurements at the same temperature ($2k_t = 9 \times 10^{9}\ \mathrm{l\ mol^{-1}\ s^{-1}}$)[49] and to those for termination in a series of alkanes[50] and were well fitted to the Smoluchowski equation, suitably modified by the microviscosity theory.[51] The concordance of the results from several different methods of radical production indicated that the observed rate coefficients were not significantly affected by geminate recombination in the solvent cage. It was pointed out[48] that recent gas- and liquid-phase results cast serious doubt on the validity of the currently accepted thermochemistry for alkyl radicals.[52]

In addition to examples cited earlier, there have been several interesting reports on spin-trapping studies. The traps used most widely remain 2-methyl-2-nitroso-propane (MNP) and phenyl-t-butyl nitrone (PBN). There have been two reports on spin-trapping in the γ-radiolysis of alcohols.[53,54] High concentrations of MNP were used to detect the intermediate methyl radicals in the radiolysis of Bu^tOH, MeCH(OH)Me, and $MeCH_2OH$ and the yields of nitroxide adducts were in the same order as the yields of methane.[53] However, no adduct was observed in the case of MeOH, which suggests either that the methyl radical is not a precursor of methane in this case or that it has excess energy and reacts with solvent rather than with the spin trap. PBN was the trap of choice in a study of the γ-radiolysis of fluorinated alcohols.[54] In general, both hydrogen-atom and radical adducts were observed. A particularly important observation was that in mixtures of $(CF_3)_2CHOH$ and Me_2CHOH, the ratio [H adduct]/[radical adduct] increased by more than one hundred-fold as the composition varied from the fully protiated to the pure fluorinated alcohol. This reflected a change in loss pathways for hydrogen atoms, especially hydrogen-abstraction reactions in the protiated case.

Both MNP and PBN have been used in a study of the thermolysis of aryldiazo alkyl ethers.[55] When carried out in aromatic solvents, the reaction yields biaryls and is thought to involve aryl radicals as intermediates. The reaction is summarized by the equation

$$Ar^1{-}N{=}N{-}O{-}R + Ar^2H \rightarrow Ar^1Ar^2 + N_2 + ROH \qquad (28)$$

[47] W. T. Dixon, J. Foxall, G. H. Williams, D. J. Edge, B. C. Gilbert, H. Kazarians-Moghaddam, and R. O. C. Norman, *J.C.S. Perkin II*, 1977, 827.
[48] J. E. Bennett and R. Summers, *J.C.S. Perkin II*, 1977, 1504.
[49] D. A. Parkes and C. P. Quinn, *J.C.S. Faraday I*, 1976, **73**, 1952.
[50] H. Schuh and H. Fischer, *Internat. J. Chem. Kinetics*, 1976, **8**, 341.
[51] A. Spernol and K. Wirtz, *Z. Naturforsch.*, 1953, **8a**, 522.
[52] R. Hiatt and S. W. Benson, *Internat. J. Chem. Kinetics*, 1973, **5**, 385.
[53] F. P. Sargent and E. M. Gardy, *J. Phys. Chem.*, 1977, **81**, 1215.
[54] A. C. Ling and L. Kevan, *J. Phys. Chem.*, 1977, **81**, 605.
[55] R. M. Paton and R. U. Weber, *J.C.S. Chem. Comm.*, 1977, 769.

Product analyses of isomer ratios, the observation of CIDNP, and the spin-trapping experiments all confirmed the intermediacy of aryl radicals in this reaction. The phenyl radical has been trapped both by MNP and PBN in the radiolysis of benzene.[56] Previous evidence for its presence as an intermediate had been rather indirect.

The apparatus and method for trapping radicals produced in a silent electric discharge in the gas phase has been described.[57a] One surprising feature was that trapping seemed to be quite selective, only one or two radicals being observed in any one experiment. For example, gas chromatographic analysis of the products from 2-methylpropane showed no less than seventeen products. However, only the 2-methylpropyl radical was trapped in observable quantities. A general review of spin trapping of photolytically-produced radicals in the gas phase has also appeared.[57b]

The kinetics of spin trapping continue to receive attention.[58,59] A combination of pulse radiolysis and spin-trapping data was used to determine the rate of addition of MeO· to MNP, for which a rate coefficient of $1.3 \times 10^8\ \mathrm{l\ mol^{-1}\ s^{-1}}$ (−45 °C) was obtained.[58] A more extensive report is concerned with the measurement of rate coefficients for spin-trapping primary alkyl radicals with MNP and PBN.[59] The method made use of the fact that the 5-hexenyl radical (19) isomerizes to cyclopentylmethyl (20) at a rate k_c which is reliably known. The spin adducts of the two isomers were distinguished by labelling (19) at the starred position with ^{13}C.

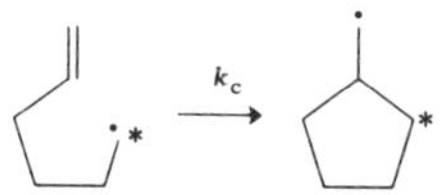

(19) (20)

Hyperfine coupling to this carbon was only detectable in the adduct of (19) with both traps. The same rate coefficient for trapping k_T was assumed for both isomers and was calculated from

$$k_T = k_c[19T\cdot]/[T][20T\cdot] \qquad (29)$$

where [19T·] represents the concentration of the adduct of (19) with the trap T and [20T·] is the concentration of the adduct of (20) with T. At 40 °C in benzene $k_T(\mathrm{PBN}) = 1.34 \times 10^5\ \mathrm{l\ mol^{-1}\ s^{-1}}$ and $k_T(\mathrm{MNP}) = 90.2 \times 10^5\ \mathrm{l\ mol^{-1}\ s^{-1}}$. The authors suggested extensions of the technique.

Other spin-trapping studies include investigations of hydrogen atom abstraction from polystyrene by Bu^{t}O· radicals,[60] micellar catalysis of radical reactions,[61] and superoxide–alkyl halide reactions.[62]

[56] F. P. Sargent and E. M. Gardy, *J. Chem. Phys.*, 1977, **67**, 1793.
[57] (*a*) D. B. Hibbert, A. J. B. Robertson, and M. J. Perkins, *J.C.S. Faraday I*, 1977, **73**, 1499; (*b*) E. G. Janzen, *Creat. Detect. Excited State*, 1976, **4**, 83.
[58] F. P. Sargent, *J. Phys. Chem.*, 1977, **81**, 89.
[59] P. Schmid and K. U. Ingold, *J. Amer. Chem. Soc.*, 1977, **99**, 6434.
[60] N. Ohto, E. Niki, and Y. Kamiya, *J.C.S. Perkin II*, 1977, 1416.
[61] D. P. Bakalik and J. K. Thomas, *J. Phys. Chem.*, 1977, **81**, 1905.
[62] M. V. Meritt and R. A. Johnson, *J. Amer. Chem. Soc.*, 1977, **99**, 3713.

2 Chemically Induced Dynamic Electron Polarization

There have been two reports of S–$T_{\pm1}$ polarization in CIDEP.[63,64] Both involve the use of time-resolved e.s.r., typically 2 μs after a radiolysis pulse. In the first,[63] $\cdot CH_2CO_2H$ was produced in H_2O/H_2SO_4 solutions of sodium acetate, the other important species being H·. At pH 1.3, the low-field line of the triplet from $\cdot CH_2CO_2H$ was in enhanced emission, the central line in emission, and the high-field line in enhanced absorption (but considerably less intense than the other two lines). However, in D_2O/H_2SO_4 solution an essentially 'normal' CIDEP spectrum was obtained, *i.e.* low-field enhanced emission, unpolarized central line, and the high-field line in enhanced absorption more intense than the low-field line. The other radical present was D·. These results were explained in the following way. Substantial hyperfine coupling in H· was responsible for making S–T_{-1} polarization feasible for radical pairs involving H· atoms. This coupling splits the T_{+1} and T_{-1} levels while T_0 is unaffected in first order. Thus, not only is the S–T_{-1} energy gap substantially reduced but differentiation of the $T_{\pm1}$ levels allows S–$T_{\pm1}$ polarization pathways to contribute to the usual S–T_0 polarization. Since D· has a much smaller hyperfine coupling (7.754 mT) than has H· (50.66 mT) there was no appreciable S–T_{-1} mixing observed from radical pairs involving D·. Similar mixing was observed by the same authors in an investigation of CIDEP in the pulse radiolysis of aqueous solutions of micelles.[64] N_2O-saturated solutions of several anionic and cationic surfactants were studied. Below the critical micelle concentration (CMC), the radical pair S–T_0 mixing was dominant in all cases. This mechanism still obtained above the CMC, since micelle formation is a dynamic equilibrium, but emission and enhanced absorption were observed, with emission in substantial excess. Especially noteworthy was the emission from central lines in the spectra (radicals were of the general type $R^1CH_2\dot{C}HCH_2R^2$). This behaviour, noted only above the CMC, was ascribed to S–T_{-1} mixing consequent upon restricted diffusion in the aggregate phase.

CIDEP has been observed in radicals (21) and (22) when solutions of maleimide, and the N-ethyl and N-methyl derivatives, in various alcohols, AH, were examined

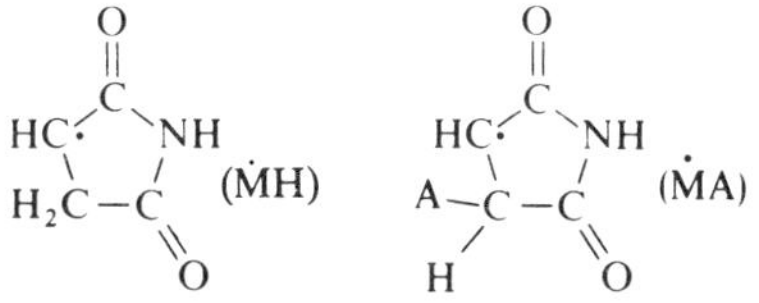

(21) (22)

using a pulse photolysis–signal averaging technique.[65] The observed spectra showed contributions from both radical pair and triplet mechanisms. The results were accommodated by the mechanism given in equations (30a—e).

$$M(S_0) \xrightarrow{h\nu} M^*(S) \xrightarrow{ISC} M^*(T) \quad (30a)$$

[63] A. D. Trifunac and D. J. Nelson, *J. Amer. Chem. Soc.*, 1977, **99**, 289.
[64] A. D. Trifunac and D. J. Nelson, *Chem. Phys. Letters*, 1977, **46**, 346.
[65] P. B. Ayscough, T. H. English, G. Lambert, and A. J. Elliot, *J.C.S. Faraday I*, 1977, **73**, 1302.

$$M^*(T)+AH \longrightarrow \dot{M}H+\dot{A} \qquad (30b)$$

$$\dot{A}+M(S_0) \longrightarrow \dot{M}A \qquad (30c)$$

$$\dot{M}A, \dot{M}H+M(S_0) \longrightarrow \text{polymer} \qquad (30d)$$

$$\dot{M}A, \dot{M}H \longrightarrow \text{products} \qquad (30e)$$

Evidence that the triplet state was involved in the H-abstraction step (30b) was inferred from the initial polarization of the $\dot{M}H$ radicals. The counter-radical $\dot{A}$ reacted within microseconds, adding to the double bond of $M(S_0)$, as evidenced by the transfer of its polarization to $\dot{M}A$. Kinetic measurements suggested that $\dot{M}A$ and $\dot{M}H$ have second-order termination rate coefficients $> 10^9\,l\,mol^{-1}\,s^{-1}$ in alcoholic solvents at room temperature and may also undergo further addition to maleimide [reaction (30d)].

Finally, a nitrogen laser coupled to a time-resolved e.s.r. spectrometer has been used to study the sensitized and unsensitized photoreductions of biacetyl with triethylamine.[66] It was found that the amine quenched both singlet and triplet biacetyl. The reaction of the triplet resulted in efficient radical production but was not rapid enough to produce significant electron polarization. The sensitizer used was benzophenone. In this case polarized spectra of biacetyl radical anions were observed and shown to arise in the following manner. Rapid reaction of the triplet benzophenone with the amine gave rise to polarization in the primary radical $Me\dot{C}HNEt_2$. This polarization was then transferred *via* reaction with ground state biacetyl to the secondary radical anions. Estimates of the initial polarization of both primary radicals $Me\dot{C}HNEt_2$ and $Ph_2\dot{C}OH$ showed these to be equal, thus providing the first quantitative evidence for this requirement of the triplet mechanism.

[66] K. A. McLaughlan, R. C. Sealy, and J. M. Wittman, *J.C.S. Faraday II*, 1977, **73**, 926.

5 Arynes, Carbenes, Nitrenes, and Related Species

By S. A. MATLIN

Department of Chemistry, The City University, St. John Street, London EC1V 4PB

1 Arynes

It has been a quiet year for aryne chemistry. The action of dialkylamines on *o*- and *p*-difluorobenzenes gives rise to fluorobenzyne intermediates from which isomeric mixtures of dialkylaminofluorobenzenes are formed.[1] Nucleophilic attack on the aryne triple bond by a side chain carbanion α to a nitrile group is the basis of a new synthesis of benzoheterocycles,[2] as illustrated by the high-yield formation of *N*-methylisoindole (Scheme 1).

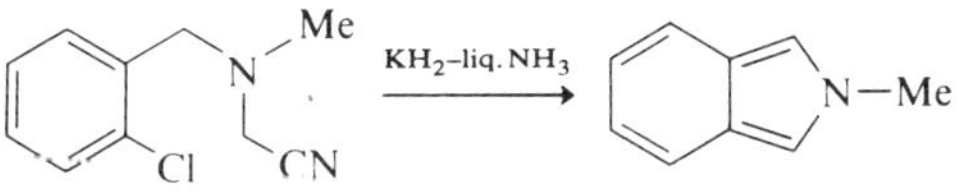

Scheme 1

The formation of biphenylenes substituted on one ring has been accomplished by the gas phase pyrolysis of mixtures of precursors of appropriate benzynes.[3] The mediocre yields are offset by the simplicity of this method.

The cyclobuta[*c*]benzyne (1) has been generated from the corresponding dihalide and undergoes the expected trapping reaction with furan.[4] Compound (1) does not appear to form a dimer, in marked contrast to the behaviour of its isomer (2) reported last year.[5] This presumably reflects the lower stability of (1) compared with (2).

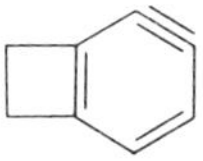

(1)

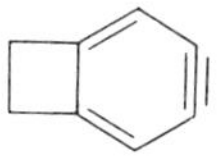

(2)

Competition between nucleophilic attack (route a) and [4+2] and [2+2]cyclo-addition pathways (routes b and c respectively) is seen in the reactions of allylic Grignard reagents with benzyne (Scheme 2).[6]

1 M. F. Moreau-Hochu and P. Caubere, *Tetrahedron*, 1977, **33**, 955.
2 B. Jaques and R. G. Wallace, *Tetrahedron*, 1977, **33**, 581.
3 A. Martineau and D. C. DeJongh, *Canad. J. Chem.*, 1977, **55**, 34.
4 R. L. Hillard and K. P. C. Vollhardt, *Angew. Chem. Internat. Edn.*, 1977, **16**, 399.
5 R. L. Hillard and K. P. C. Vollhardt, *J. Amer. Chem. Soc.*, 1976, **98**, 3579; S. A. Matlin, *Ann. Reports (B)*, 1976, **73**, 85.
6 J. G. Duboudin, B. Jousseaume, and M. Pinet, *J.C.S. Chem. Comm*,. 1977, 454.

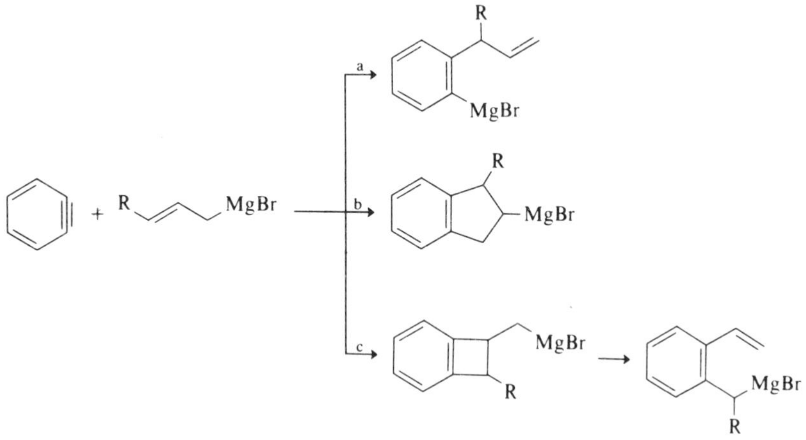

Scheme 2

1-Naphthyne reacts with the amine (3) by a [4+2] route, with subsequent loss of diethylamine giving phenanthrene. Other arynes behave analogously.[7]

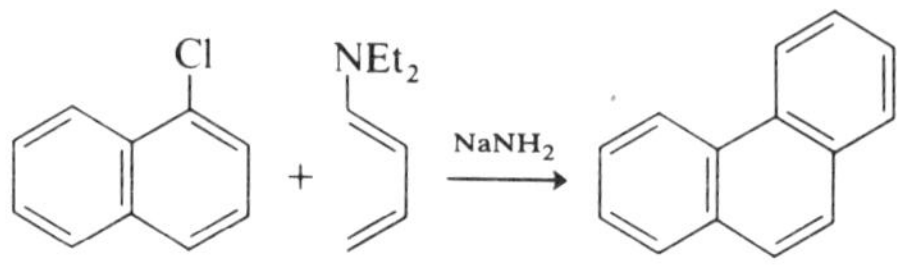

(3)

The initial adduct of 2-vinylnaphthalene and benzyne undergoes an unusual concerted ene reaction[8] with a further molecule of benzyne (Scheme 3; R = H or D).

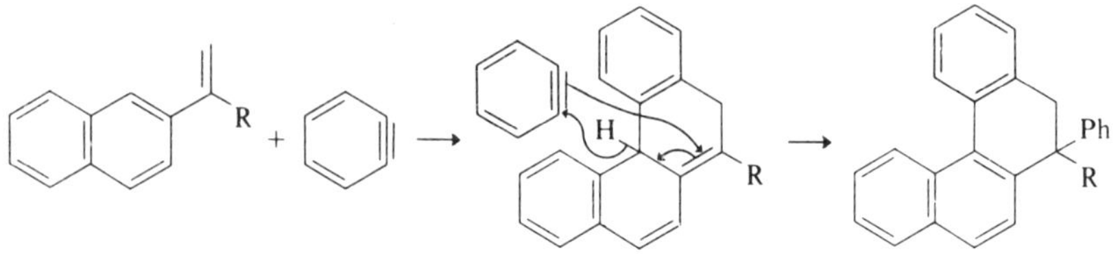

Scheme 3

2 Nitrenes

In contrast to previous *ab initio* studies, new calculations (generalized valence bond and configuration interaction) predict the ground state of aminonitrene (H_2NN) to be a singlet lying about 15 kcal mol^{-1} below the triplet state.[9]

[7] S. Tanimoto, *Tetrahedron Letters*, 1977, 2903.
[8] Y. Ittah, I. Shahak, J. Blum, and J. Klein, *Synthesis*, 1977, 678.
[9] J. H. Davis and W. A. Goddard, *J. Amer. Chem. Soc.*, 1977, **99**, 7111.

The low-temperature addition of the nitrenes formed by the oxidation of *N*-aminophthalimide (4) and *N*-aminobenzoxazolone (5) to olefins conjugated to

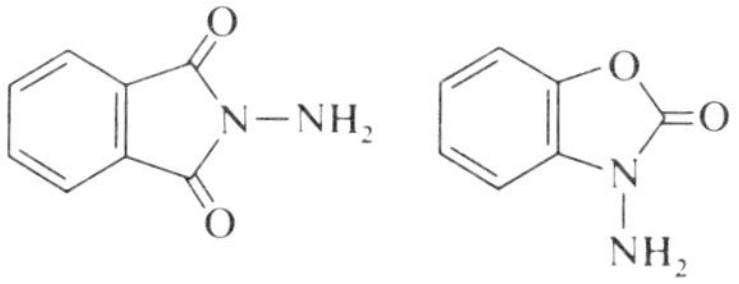

(4) (5)

ester, vinyl, or aryl functions is highly stereoselective.[10] For example, phthalimidonitrene (Z—N̈:) adds to methyl acrylate at −30 °C, affording exclusively the *cis*-aziridine (6), which largely inverts to the more stable *trans*-isomer (7) on

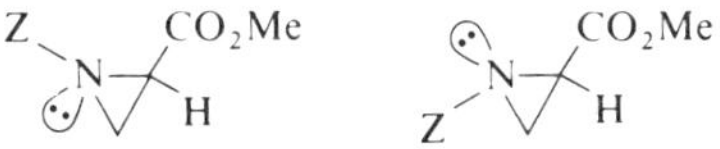

(6) (7)

warming. This strong preference for formation of the *cis*-isomer appears to be due to a secondary orbital interaction between the π electrons of the group conjugated to the olefin and those of the carbonyl group in the *N*-acyl-*N*-nitrene.

When phthalimidonitrene is generated in the presence of β-diketones, addition of the nitrene to the enol double bond is followed by opening of the azirane ring to give an amide (Scheme 4).[11]

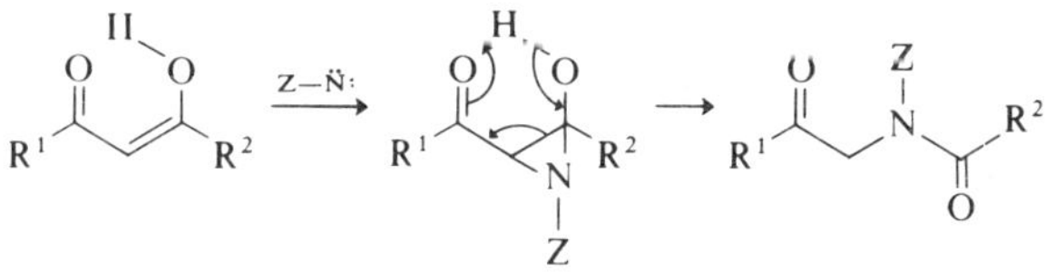

Scheme 4

Arylnitrenes are known to equilibrate with the corresponding 7-azanorcaradienes, which can be trapped by a variety of nucleophiles as azepines.[12]

The photolysis of α-azido-α, γ-dienoic esters (8) leads to pyrrole formation[13] *via* the azirines (9). Contrary to earlier reports, the thermolysis of 2-penta-1,3-dienyl-2*H*-azirines (10) also affords pyrroles and not azepines.[14]

However, azepine formation has been cleverly induced in one instance.[15] The azirine (11; R = H) cyclises to a pyrrole *via* δ C—H insertion by the nitrene (12; R = H) and the corresponding methyl-substituted case (11; R = Me) gives a

[10] R. S. Atkinson and J. R. Malpass, *J.C.S. Perkin I*, 1977, 2242.

[11] H. Person, K. Luanglath, and A. Foucaud, *Tetrahedron Letters*, 1977, 221.

[12] M. Masaki, K. Fukui, and J. Kita, *Bull. Soc. chim. France*, 1977, **50**, 2013; B. Nay, E. F. V. Scriven, H. Suschitzky, and Z. U. Khan, *Synthesis*, 1977, 757; S. E. Carroll, B. Nay, E. F. V. Scriven, H. Suschitzky, and D. R. Thomas, *Tetrahedron Letters*, 1977, 3175; S. E. Carroll, B. Nay, E. F. V. Scriven, and H. Suschitzky, *ibid.*, p. 943.

[13] H. Hemetsberger, I. Spira, and W. Schonfelder, *J. Chem. Res. (S)*, 1977, 247.

[14] K. Isomura, T. Tanaka, and H. Taniguchi, *Chem. Letters*, 1977, 397.

[15] K. Isomura, H. Taguchi, T. Tanaka, and H. Taniguchi, *Chem. Letters*, 1977, 401.

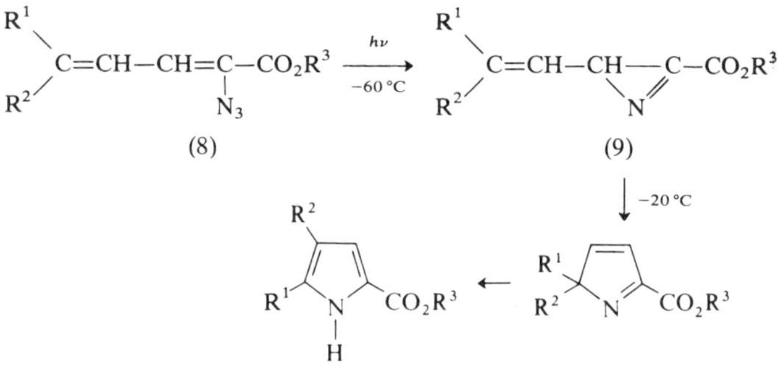

(8) (9)

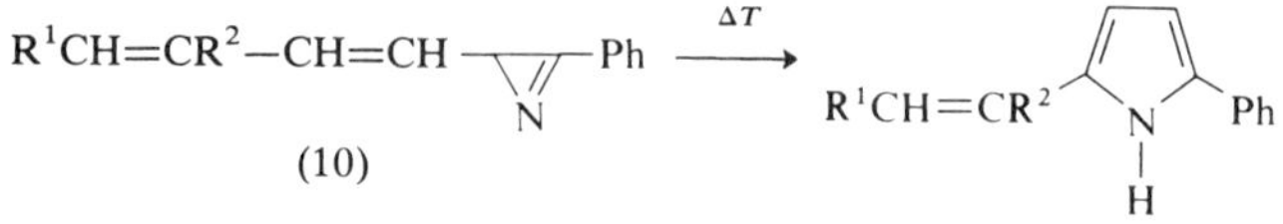

(10)

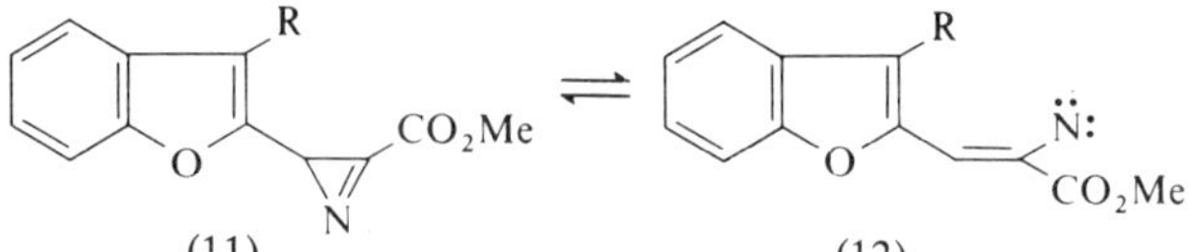

(11) (12)

pyridine. With a phenyl group δ to the nitrene (12; R = Ph) attack on the aromatic ring affords an azepine.

A new method of generating acylnitrenes involves the pyrolysis of *N*-acyl-*S,S*-diphenylsulphimides.[16] The singlet state of acylnitrenes is stabilized by the use of dichloromethane as solvent, leading to enhancement of yields of C—H insertion products in both intermolecular[17] and intramolecular[18] reactions. The latter generally proceed by attack at the γ, δ, or ε C—H bonds and are useful in remote functionalization.[19]

o-Methylbenzenesulphonylnitrenes greatly prefer intermolecular reactions to intramolecular insertion into the side chain C—H bonds, presumably because of the geometry of the sulphonylnitrene. In the case of the sulphonylazide (13), the intramolecular cyclization product (14) results exclusively from attack at the more accessible secondary hydrogen rather than the more reactive tertiary one.[20] The

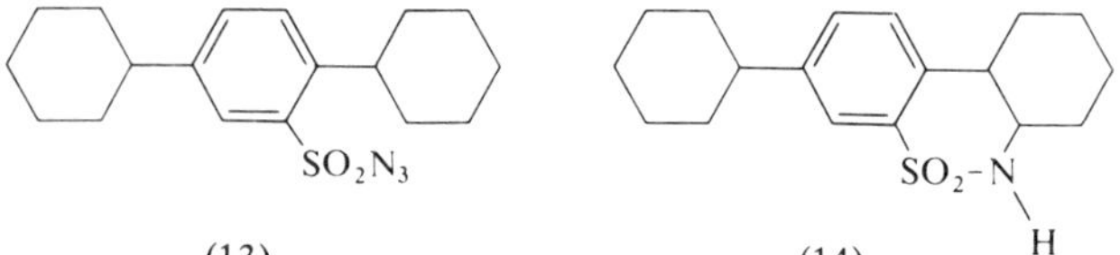

(13) (14)

[16] N. Furakawa, M. Fukumura, T. Nishio, and S. Oae, *J.C.S. Perkin I*, 1977, 96.
[17] H. Takeuchi, N. Murata, Y. Nakagawa, T. Tsuchida, and K. Koyama, *J.C.S. Perkin II*, 1977, 80.
[18] W. Lwowski and S. Linke, *Annalen*, 1977, 8.
[19] P. F. Alewood, M. Benu, J. Wong, and A. J. Jones, *Canad. J. Chem.*, 1977, **55**, 2510; M. R. Czarny, B. W. Benson, and T. A. Spencer, *J. Org. Chem.*, 1977, **42**, 556.
[20] R. A. Abramovitch, R. Chellathurai, W. D. Holcomb, I. T. McMaster, and D. P. Vanderpool, *J. Org. Chem.*, 1977, **42**, 2920.

biaryl ketone derivative (15) cyclizes to a seven-membered ring sultam (16) on thermolysis.[21] There is no evidence for the formation of spiro intermediates or azepines on thermolysis of biarylsulphonylazides,[22] which contrasts with the behaviour of *o*-(arylthio)arylnitrenes.[23]

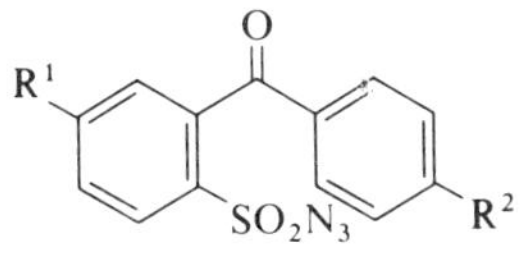

(15)

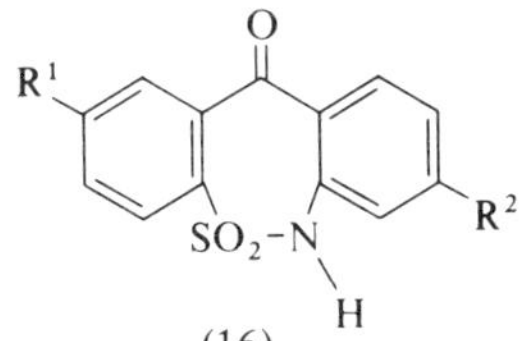

(16)

Evidence in favour of an exclusively singlet pathway for the formation of carbazole from 2-nitrenobiphenyl has been presented.[24]

Nitrenes derived from the amines (4) and (5) will react with allylic sulphides to form ylides, which undergo 2,3-sigmatropic rearrangements (Scheme 5).[25]

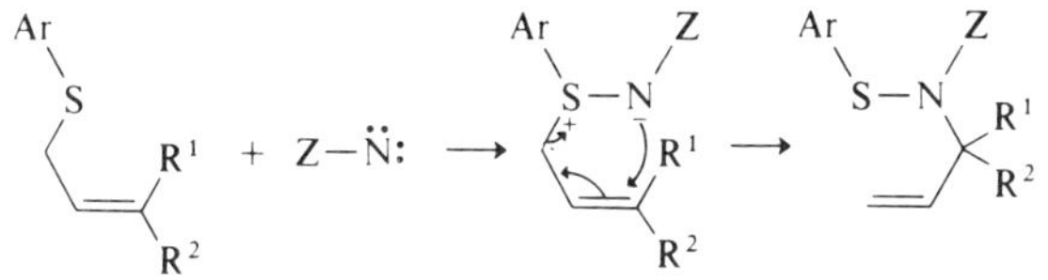

Scheme 5

Acylnitrenes have been reported to abstract oxygen atoms from *N*-nitrosoamines[26] and alkoxycarbonylnitrenes will abstract oxygen from DMSO.[27] In the latter case, the resulting nitrosoformate has been trapped by cycloaddition with thebaine.

Insertion of alkoxycarbonylnitrene into the C—H bonds of *trans*-1,2-dichlorocyclohexane takes place largely at the 4-position, suggesting a controlling interaction of the nitrene with a halogen atom.[28]

The reactions of azides and azirines in the presence of transition metal catalysts appear to involve metal–nitrene complexes.[29] A number of such complexes are now becoming well characterized.[30]

[21] R. A. Abramovitch and D. P. Vanderpool, *J.C.S. Chem. Comm.*, 1977, 18.
[22] R. A. Abramovitch, T. Chellathurai, I. T. McMaster, T. Takaya, C. I. Azogu, and D. P. Vanderpool, *J. Org. Chem.*, 1977, **42**, 2914.
[23] I. M. McRobbie, O. Meth-Cohn, and H. Suschitzky, *J. Chem. Res.* (*S*), 1977, 17.
[24] J. M. Lindley, I. M. McRobbie, and O. Meth-Cohn, *J.C.S. Perkin I*, 1977, 2194.
[25] R. S. Atkinson and S. B. Awad, *J.C.S. Perkin I*, 1977, 346.
[26] K. Nishiyama and J.-P. Anselme, *J. Org. Chem.*, 1977, **42**, 2636.
[27] G. W. Kirby, J. W. M. Mackinnon, and R. P. Sharma, *Tetrahedron Letters*, 1977, 215.
[28] P. A. Tardella and L. Pellacani, *Tetrahedron Letters*, 1977, 4451.
[29] H. Alper, J. E. Prickett, and S. Wollowitz, *J. Amer. Chem. Soc.*, 1977, **99**, 4330; I. Yamamoto, H. Tokanou, H. Uemura, and H. Gotoh, *J.C.S. Perkin I*, 1977, 1241.
[30] O. R. Chambers, M. E. Harman, D. S. Rycroft, D. W. A. Sharp, and J. M. Winfield, *J. Chem. Res.* (*S*), 1977, 150; J. Chatt and J. R. Dilworth, *J. Indian Chem. Soc.*, 1977, **54**, 13; F. Basolo, *ibid.*, p. 7.

3 Carbenes

A number of reviews containing references to carbene chemistry have been published.[31]

Uncertainty still remains concerning the triplet–singlet ($^3B_1$–$^1A_1$) energy separation in methylene. A recent spectroscopic measurement of 19.5 ± 0.7 kcal mol^{-1}, much higher than previous experimental values, has provoked two new state-of-the-art configuration interaction studies.[32] Both place the energy separation in the region of 11 kcal mol^{-1} and it seems unlikely that the true value will turn out to be significantly greater than this.

McBride[33] has emphasized the structural importance of bent bonds in carbenes.

For diphenylcarbene, INDO calculations predict the triplet to be lower than the singlet by 4 kcal mol^{-1}, with the two states interconverting rapidly.[34] In CHBr the triplet is calculated to be only 1 kcal mol^{-1} below the singlet and in other halogenocarbenes (CHCl, CHF, CCl_2, CF_2) the singlet becomes more stable than the triplet by progressively larger amounts.[35]

There is now good evidence[36] for homoallylic conjunction between the carbene and the cyclopropane orbitals in the bridged systems (17) and (18).

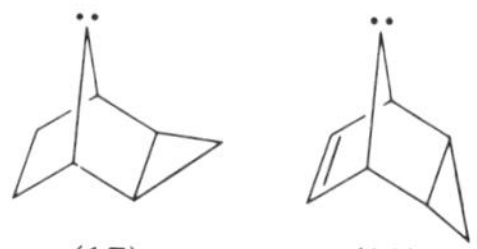

(17) (18)

Generation.—Halogen exchange[37] in the dihalogenocarbenes derived from $CHBrCl_2$ and $CHBr_2Cl$ under phase transfer catalysis (PTC) conditions leads to adducts of all three possible carbenes CBr_2, CBrCl, and CCl_2. This can be avoided[38] by the use of dibenzo[18]crown[6] as the catalyst, which gives exclusively CBrCl adducts from $CHBr_2Cl$.

The use of solid or immobilized phase transfer catalysts for halogenocarbene generation is receiving increasing attention.[39]

When tertiary amines are used as catalysts, selectivity in the product distributions upon addition of dichlorocarbene to polyenes suggests the intermediate formation of nitrogen ylides.[40] In the PTC addition of dihalogenocarbenes to allylic alcohols,

[31] M. D. Roth, *Accounts Chem. Res.*, 1977, **10**, 85; W. M. Jones, *ibid.*, p. 353; S. Braslavsky and J. Heicklen, *Chem. Rev.*, 1977, **77**, 473; R. A. Firestone, *Tetrahedron*, 1977, **33**, 3009.

[32] R. R. Lucchese and H. F. Schaefer, *J. Amer. Chem. Soc.*, 1977, **99**, 6765; B. O. Roos and P. M. Siegbahn, *ibid.*, p. 7716.

[33] J. M. McBride, *J. Amer. Chem. Soc.*, 1977, **99**, 6760.

[34] J. Metcalfe and E. A. Halevi, *J.C.S. Perkin II*, 1977, 634.

[35] C. W. Bauschlicher, jun., H. F. Schaefer, and P. S. Bagus, *J. Amer. Chem. Soc.*, 1977, **99**, 7106.

[36] P. K. Freeman, T. A. Harding, R. S. Raghavan, and D. G. Kuper, *J. Org. Chem.*, 1977, **42**, 3882; K. Okumura and S.-I. Murahashi, *Tetrahedron Letters*, 1977, 3281.

[37] E. V. Dehmlow, M. Lissel, and J. Heider, *Tetrahedron*, 1977, **33**, 363.

[38] M. Fedorynski, *Synthesis*, 1977, 783.

[39] E. Chiellini and R. Solaro, *J.C.S. Chem. Comm.*, 1977, 231; S. L. Regen, *J. Org. Chem.*, 1977, **42**, 875; S. Julia and A. Ginebreda, *Synthesis*, 1977, 682.

[40] Y. Kimura, K. Isagawa, and Y. Otsuji, *Chem. Letters*, 1977, 951. For another example of nitrogen ylide formation see: K. Berg-Nielsen, *Acta Chem. Scand.*, 1977, **B31**, 224.

there is no evidence for any intermediate interaction of the carbene with the OH function.[41]

CFCl and CFBr have been generated by reaction of the corresponding fluorotrihalogenomethanes with butyl-lithium, but attempts to produce monofluorocarbene were not successful.[42]

Carbenes are formed by the metallation of cycloalkane epoxides with lithium dialkylamides and show a highly temperature-dependent behaviour, low temperatures favouring transannular C—H insertion products.[43]

Reactions.—The reactivities of a wide range of electrophilic carbenes in cyclopropenation of olefins have been correlated quantitatively with inductive and conjugative effects of the carbene substituents. An exceptional case is that of ethoxycarbonylcarbene, for which a strong preference for C—H insertion over cyclopropanation leads to deviation between the observed and predicted reactivity.[44] Substituent effects in nucleophilic carbenes have also been discussed.[45]

The ratio of 1,2 to 1,4 addition in the reactions of dihalogenocarbenes with norbornadiene is highly dependent on the carbene substituents. Perturbation theory supports the contention that the dominant electronic effect which augments the relative rate of 1,4 addition is the reduction in electrophilic character of the carbene *p* orbital.[46] The nucleophilic dimethoxycarbene also adds 1,4 to cyclic dienes.[47] Further studies of ketocarbene–oxirene equilibration on the pathway to Wolff rearrangement have been published. Zeller[48] finds 13—16% oxiren participation in the formation of keten on photolysis of the diazo-aldehyde (19). In the case of the isomeric ^{13}C-labelled diazoketones (20) and (21), a marked difference in the extent of oxiren participation from the two precursors suggests a strong displacement of the equilibrium (Scheme 6) in favour of carbene (22).

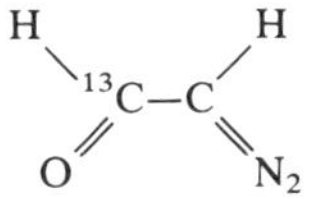

(19)

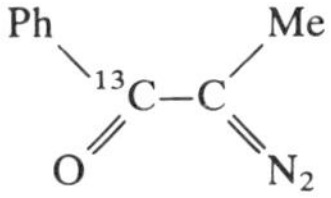

(20)

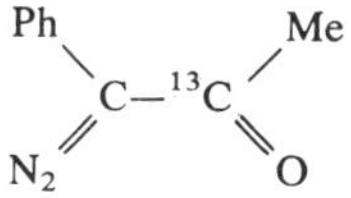

(21)

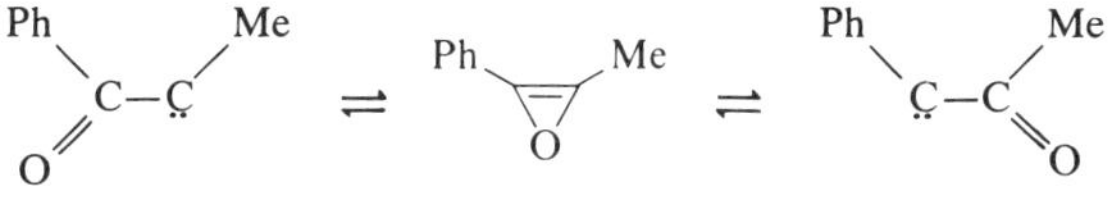

(22)

Scheme 6

[41] K. Kleveland, L. Skattebol, and L. V. Sydnes, *Acta Chem. Scand.*, 1977, **B31**, 463.
[42] D. J. Burton and J. L. Hahnfeld, *J. Org. Chem.*, 1977, **42**, 828.
[43] R. K. Boeckman, jun., *Tetrahedron Letters*, 1977, 4281.
[44] R. A. Moss, C. B. Mallon, and C.-T. Ho, *J. Amer. Chem. Soc.*, 1977, **99**, 4105.
[45] R. W. Hoffmann, B. Hagenbruch, and D. M. Smith, *Chem. Ber.*, 1977, **110**, 23; H. Durr, S. Frolich, and M. Kausch, *Tetrahedron Letters*, 1977, 1767.
[46] Y. Jean, *Tetrahedron Letters*, 1977, 2689.
[47] W. Lilienblum and R. W. Hoffmann, *Chem. Ber.*, 1977, **110**, 3405.
[48] K.-P. Zeller, *Angew. Chem. Internat. Edn.*, 1977, **16**, 781; *Tetrahedron Letters*, 1977, 707.

There has been further confirmation that ring contraction of normal size rings by Wolff rearrangement does not involve oxiren participation.[49]

Thiirens are also attracting considerable attention.[50] A new route to these intermediates involves thermolysis of bis-amine disulphides in the presence of acetylenes, the isomeric thionocarbenes being trapped as 1,3 dipoles by cycloadditions (Scheme 7). In some cases, synthetically useful yields of thiophens are obtained.[51]

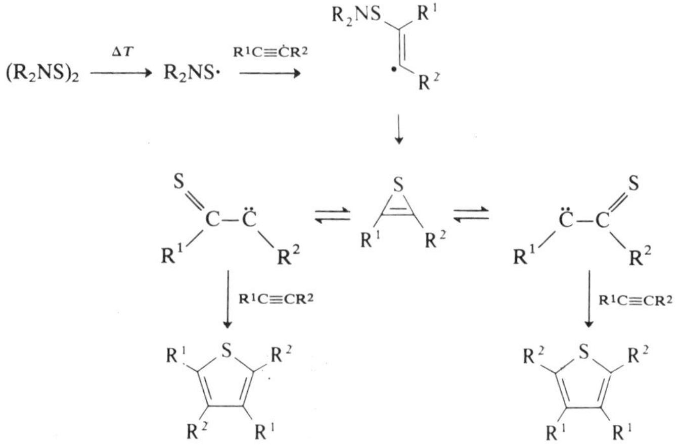

Scheme 7

Cyclopropylidenes may collapse to allenes[52] or insert into neighbouring C—H bonds.[53] The latter reaction is characteristic of tetrasubstituted cyclopropylidenes and generally follows a pattern in which insertion takes place into the α C—H bond of the group geminal to the bulkiest substituent. For example, the carbene (23; $R^1 = Pr^i$, $R^2 = Me$) yields the bicyclobutane (24). However, the aryl derivatives (23; $R^1 = Me$ or Pr^i, $R^2 = Ar$) undergo insertion in the opposite direction,[54] affording the bicyclobutanes (25).

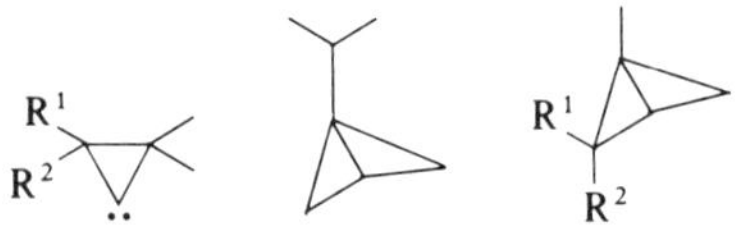

(23) (24) (25)

Two pathways can be envisaged for the rearrangement of vinylcyclopropylidenes to cyclopentadienes (Scheme 8). The results of a labelling study (• = ^{13}C) are consistent with operation of path A, in agreement with earlier work.[55]

[49] U. Timm, K.-P. Zeller, and H. Meier, *Tetrahedron*, 1977, **33**, 453.
[50] L. Benati, P. C. Montevecchi, and G. Zanardi, *J. Org. Chem.*, 1977, **42**, 577; H. Buhl, B. Seitz, and H. Meier, *Tetrahedron*, 1977, **33**, 449; T. Wooldridge and T. D. Roberts, *Tetrahedron Letters*, 1977, 2643.
[51] F. M. Benitez and J. R. Grunwell, *Tetrahedron Letters*, 1977, 3413.
[52] Y. Okude, T. Hiyama, and H. Nozaki, *Tetrahedron Letters*, 1977, 3829.
[53] R. M. Cory and F. R. McLaren, *J.C.S. Chem. Comm.*, 1977, 587.
[54] T. Shono, I. Nishiguchi, T. Komamura, and K. Fujita, *Tetrahedron Letters*, 1977, 4327.
[55] K. H. Holm and L. Skattebol, *Tetrahedron Letters*, 1977, 2347.

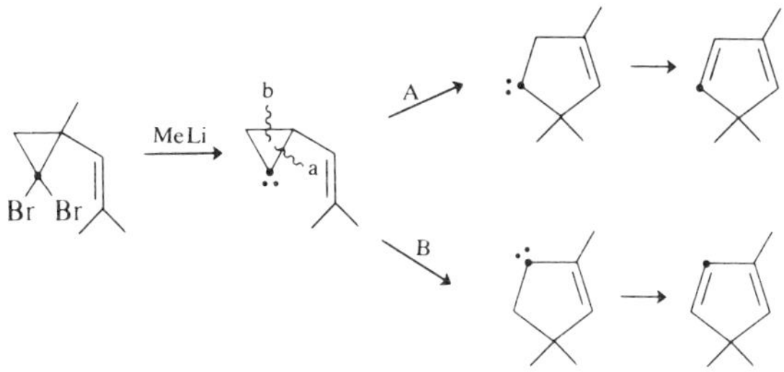

Scheme 8

There is continuing interest in the cyclopropylidene–vinylmethylene rearrangement.[56] Generalized valence bond and MINDO/3 calculations predict[57] a methylene-like triplet ground state for the parent vinylcarbene.

The irradiation of the sulphinyl-3*H*-pyrazole (26) leads to nitrogen extrusion and formation of a sulphinyl cyclopropene which is evidently capable of ring opening to a vinylcarbene at room temperature. Thus, trapping with furan or cyclopentadiene in a dark reaction after photolysis is complete affords the cyclopropane (27; X = O or CH_2) rather than a Diels–Alder adduct of the cyclopropene.[58]

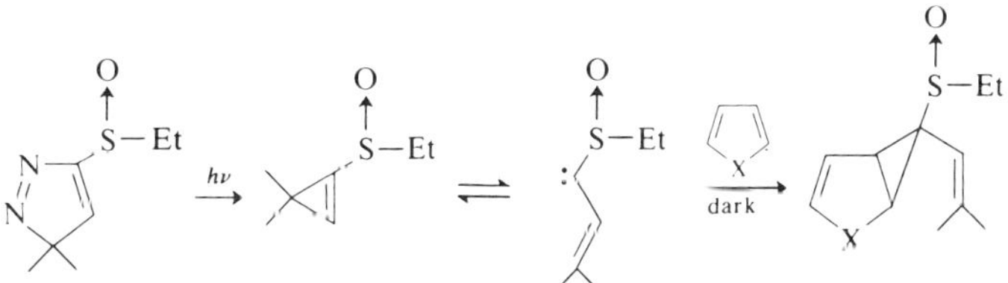

(26) (27)

The vinylcarbene formed by irradiation of the 3-diazo-pyrazole (28) undergoes C—H insertion reactions and also rearrangement *via* the vinylnitrene (29), giving azirine (30) and its decomposition products.[59]

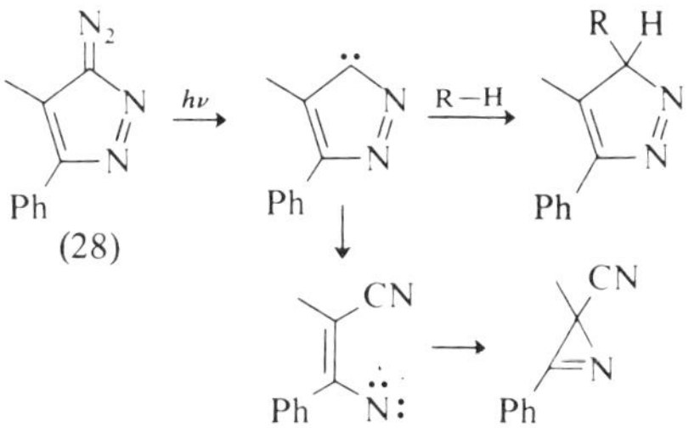

(28) (29) (30)

[56] J. A. Pincock and A. A. Moutsokapas, *Canad. J. Chem.*, 1977, **55**, 979; M. Vincens, A. Dussauge, and M. Vidal, *Tetrahedron*, 1977, **33**, 2937; A. Padwa, R. Loza, and D. Getman, *Tetrahedron Letters*, 1977, 2847.

[57] J. H. Davis, W. A. Goddard, and R. G. Bergman, *J. Amer. Chem. Soc.*, 1977, **99**, 2427; J. A. Pincock and R. J. Boyd, *Canad. J. Chem.*, 1977, **55**, 2482.

[58] M. Franck-Neumann and J.-J. Lohmann, *Angew. Chem. Internat. Edn.*, 1977, **16**, 323.

[59] W. L. Magee and H. Shechter, *J. Amer. Chem. Soc.*, 1977, **99**, 633.

The photochemistry of benzocyclopropenes (31) has been examined[60] and the formation of products accounted for by competition between ring contraction of the vinylcarbene (32) to fulvalene and trapping reactions of the biradical (33).

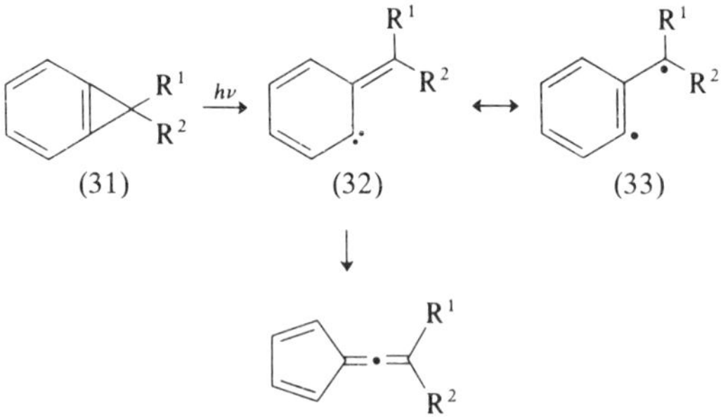

(31) (32) (33)

It has now been shown that thermal reversion of the fulvalene to the phenylcarbene is possible, and consequently the results of all previous attempts to study the mechanism of thermolysis of benzocyclopropenes by labelling studies must be treated with caution.[61]

The intervention of phenylcarbene–cycloheptatrienylidene interconversions results in the formation of indane derivatives (35; M = C or Si) from the *p*-substituted phenylcarbene precursors (34; M = C or Si).[62] The sila-indane (35; M = Si) is also formed[63] in the gas-phase pyrolysis of phenyltrimethylsilyldiazomethane (36) *via* the carbene (37).

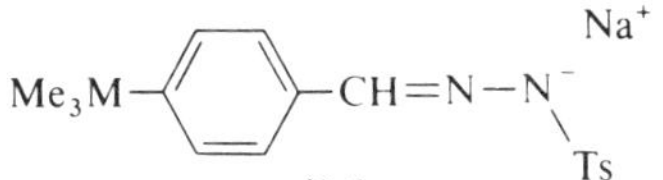

(34)

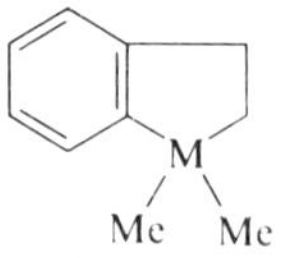

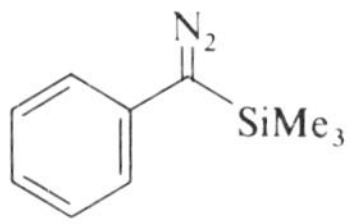

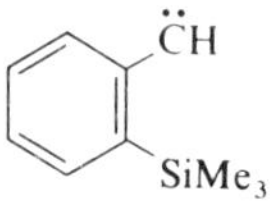

(35) (36) (37)

A new entry into the diphenylcarbene–phenylcycloheptatrienylidene system[64] involves the pyrolysis of the acetate (38) and similarly results in the formation of fluorene.

The participation of bicycloheptatrienes such as (39) in these interconversions offers a further possible entry into the system.[65] Thus, dehydrohalogenation of (40) affords a mixture of the isomeric ethers (41) and (42).

60 H. Durr and A.-J. Ahr, *Tetrahedron Letters*, 1977, 1991.
61 C. Wentrup, E. Wentrup-Byrne, and P. Muller, *J.C.S. Chem. Comm.*, 1977, 211.
62 A. Sekiguchi and W. Ando, *Bull. Chem. Soc. Japan*, 1977, **50**, 3067.
63 W. Ando, A. Sekiguchi, and A. J. Rothschild, *J. Amer. Chem. Soc.*, 1977, **99**, 6995.
64 R. W. Hoffmann, R. Schuttler, and I. H. Loof, *Chem. Ber.*, 1977, **110**, 3410.
65 W. E. Billups and L. E. Reed, *Tetrahedron Letters*, 1977, 2239.

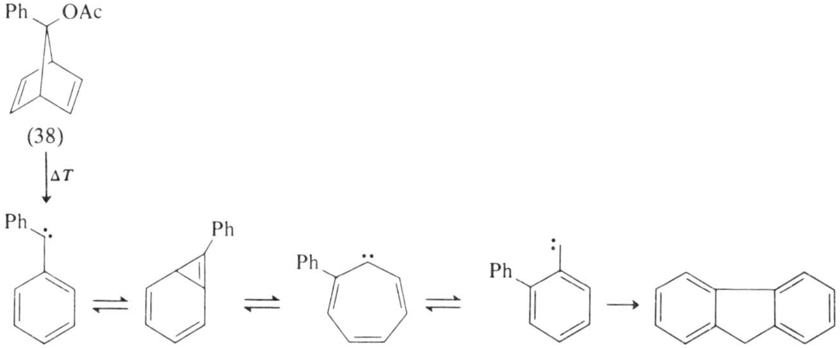

(38) (39)

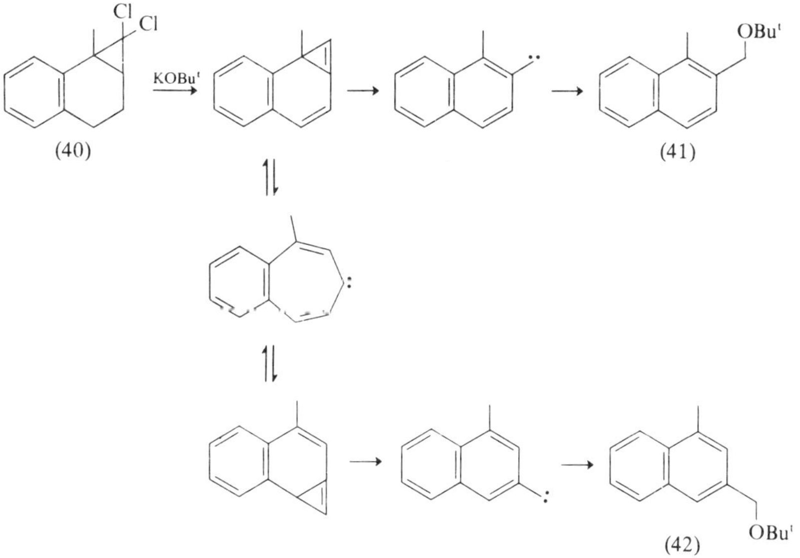

(40) (41) (42)

The first synthesis of a benzothiete (44; X = S) involves photochemical Wolff rearrangement of the diazoketone (43; X = S). Interestingly, the migration ability in (43; X = S, O, or NR) decreases from S to O to nitrogen functions, the latter being unable to rearrange[66] and whereas vapour-phase copyrolysis of (43; X = CO) with alcohols yields 2-carboalkoxybenzocyclobutenones (44; X = CO), irradiation in methanol gives the homophthalate (45).[67]

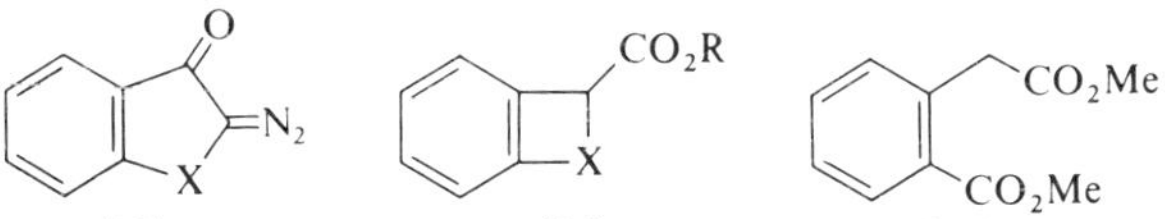

(43) (44) (45)

[66] E. Voigt and H. Meier, *Chem. Ber.*, 1977, **110**, 2242.
[67] R. J. Spangler, J. H. Kim, and M. P. Cava, *J. Org. Chem.*, 1977, **42**, 1697.

Whereas carbenes normally insert into the O—H bonds of alcoholic solvents, the irradiation of aryldiazomethanes in low-temperature alcoholic matrices results in insertion into the C—H bond α to the OH group. This reaction appears to follow an abstraction–recombination mechanism from the triplet carbene under conditions where the mobility of the latter is restricted.[68]

Intramolecular insertion of carbenes into C—H bonds continues to be applied to the synthesis of bridged, polycyclic systems.[69] In the case of (46), dehydrohalogenation results in insertion into the C—H bond α to nitrogen but the initial product undergoes a base-catalysed elimination,[70] yielding the styryl derivative (47).

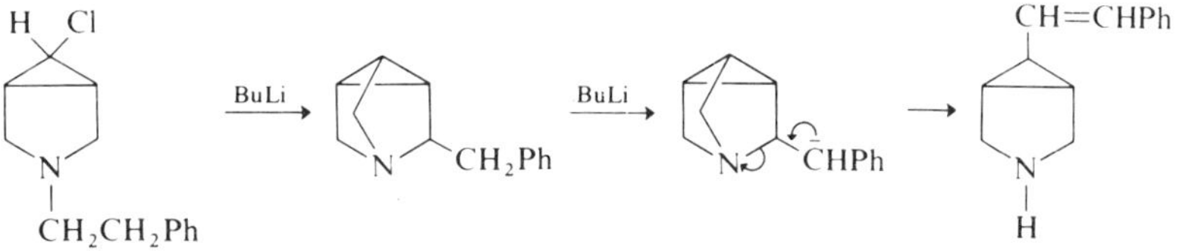

(46) (47)

The directing influences of methyl substituents on the course of C—H insertion reactions of carbenes obtained from 7,7-dibromonorcarane derivatives have been examined in detail.[71] Depending on its relative orientation, a methyl group may have a buttressing effect on the cyclopropylidene ring, which facilitates ring closure, or it may inductively have a direct activating influence on the adjacent C—H bond.

In the carbene (48) the distance for intramolecular C—H insertion to give 2,4-methanoadamantane appears to be too great and only 1,2-shifts to give the olefins (49) and (50) are observed.[72]

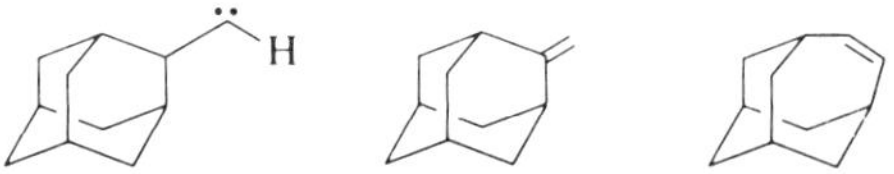

(48) (49) (50)

1,2-H shifts to carbenic centres are known, in general, to show a strong preference for migration of the hydrogen atom into the empty *p* orbital of the carbene[73] and theoretical justification for this has been provided. The poor stereoselectivity in axial versus equatorial H-migration in the cyclohexylidene (51) has therefore

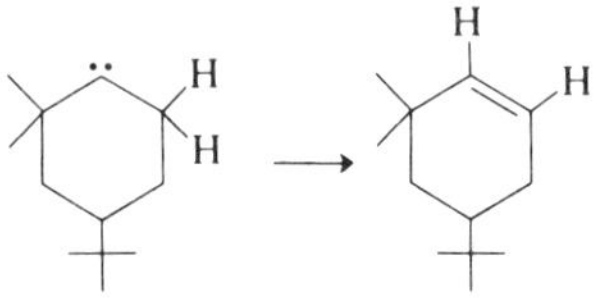

(51)

[68] H. Tomioka and Y. Izawa, *J. Amer. Chem. Soc.*, 1977, **99**, 6128.
[69] D. Farcasiu, H. Bohm, and P. von R. Schleyer, *J. Org. Chem.*, 1977, **42**, 96.
[70] R. F. Boswell and R. G. Bass, *J. Org. Chem.*, 1977, **42**, 2342.
[71] L. A. Paquette and R. T. Taylor, *J. Amer. Chem. Soc.*, 1977, **99**, 5708.
[72] T. Sasaki, S. Eguchi, and Y. Hirako, *J. Org. Chem.*, 1977, **42**, 2981.
[73] E. P. Kyba and C. W. Hudson, *J. Org. Chem.*, 1977, **42**, 1935.

caused some concern. A theoretical reinvestigation by MINDO/3 and MNDO methods supports the argument that migration of *either* hydrogen can proceed through a transition state in which the H atom is aligned with the empty *p* orbital.[74] Further work on this problem is clearly required, but it may well be that stereoselectivities previously observed with bridged systems reflect the influence of torsional interactions rather than the stereoelectronic demand of the migration itself.[75]

Reversal of the reaction by which an olefin is formed by 1,2-migration to a carbene is relatively uncommon. The formation of the carbene (53) from the bridgehead olefin (52) has been authenticated by trapping reactions and is attributed to release of strain.[76]

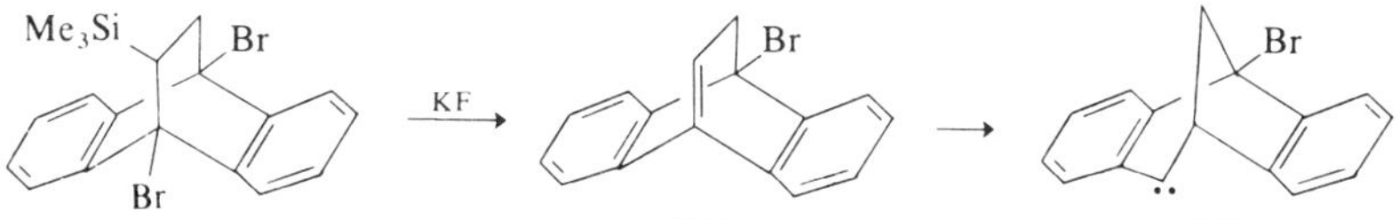

(52) (53)

Direct photolysis of cycloalkenes such as (54) also gives rise to carbene intermediates.[77]

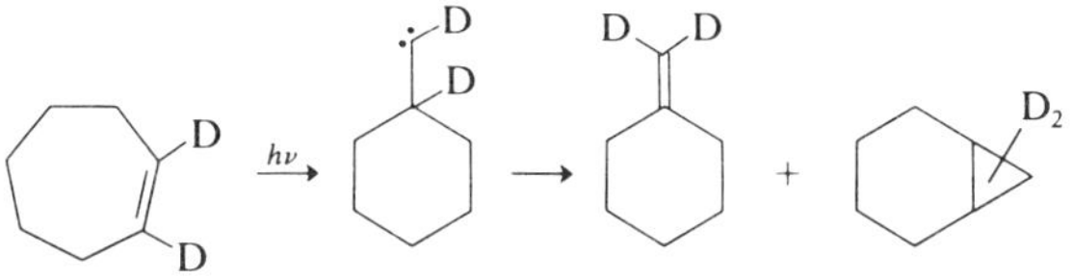

(54)

Ylide formation and rearrangement in the reactions of carbenes with divalent sulphur compounds has been reviewed.[78] An interesting intermolecular example is the rearrangement[79] of the ylide formed on decomposition of the penicillin-derived diazoketone (55).

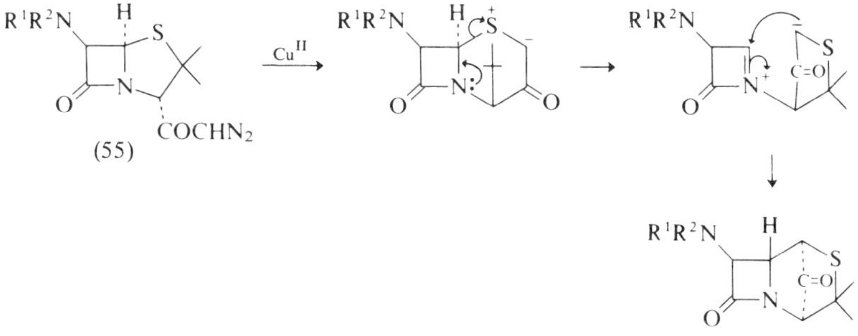

(55)

[74] E. P. Kyba, *J. Amer. Chem. Soc.*, 1977, **99**, 8330.
[75] P. K. Freeman, T. A. Hardy, J. R. Balyeat, and L. D. Wescott jun., *J. Org. Chem.*, 1977, **42**, 3356.
[76] T. H. Chan and D. Massuda, *J. Amer. Chem. Soc.*, 1977, **99**, 936.
[77] Y. Inoue, S. Takamuku, and H. Sakurai, *J.C.S. Perkin II*, 1977, 1635.
[78] W. Ando, *Accounts Chem. Res.*, 1977, **10**, 179. See also: C. Huynh, V. Ratovelomanana, and S. Julia, *Bull. Soc. chim. France*, 1977, 710.
[79] I. Ernest, *Tetrahedron*, 1977, **33**, 547.

Methylenecarbenes can be formed by the treatment of vinyl triflates (56; $R^3 = H$) with strong base.[80] New, considerably milder procedures involve[81] the fluoride-ion initiated decomposition of silylvinyl triflates (56; $R^3 = SiMe_3$) and room temperature thermal decomposition of tosylazo-alkenes (57).

Isopropylidenecarbene, formed by $KOBu^t$-induced elimination from the triflate (56; $R^1 = R^2 = Me$; $R^3 = H$), behaves as an electrophilic singlet in cyclopropanation reactions.[82]

(56) (57)

The ^{14}C-labelled adamantylacetylene (58) shows 25% scrambling of the label to the adjacent alkyne carbon atom on flash vacuum pyrolysis, owing to reversible formation of the methylenecarbene (59) by migration of either the H or adamantyl groups.[83]

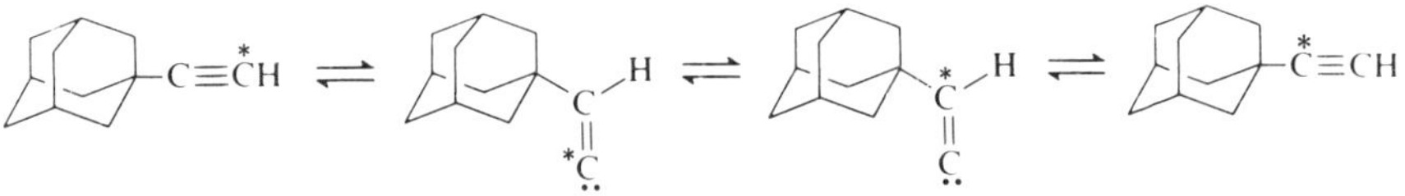

(58) (59a) (59b)

Dimethylvinylidenecarbene, best generated in the free state[84] by phase transfer catalysed elimination from the bromide (60), undergoes mainly addition reactions[85] with the double bonds of unsaturated heterocycles such as 2,5-dihydrofuran.

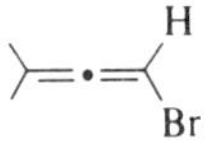

(60)

[80] P. J. Stang and M. G. Mangum, *J. Amer. Chem. Soc.*, 1977, **99**, 2597.
[81] P. J. Stang and D. P. Fox, *J. Org. Chem.*, 1977, **42**, 1667.
[82] P. J. Stang, J. R. Marsden, M. G. Mangum, and D. P. Fox, *J. Org. Chem.*, 1977, **42**, 1802. For a theoretical study see: J. H. Davis, W. A. Goddard, and L. B. Harding, *J. Amer. Chem. Soc.*, 1977, **99**, 2919.
[83] R. F. C. Brown, F. W. Eastwood, and G. P. Jackman, *Austral. J. Chem.*, 1977, **30**, 1757.
[84] T. B. Patrick and D. J. Schmidt, *J. Org. Chem.*, 1977, **42**, 3354.
[85] S. Landor, V. Rogers, and H. R. Sood, *Tetrahedron*, 1977, **33**, 73.

6 Organometallic Chemistry
Part (i) The Transition Elements

By A. STEWART, D. J. THOMPSON, and M. V. TWIGG
I.C.I Corporate Laboratory, P.O. Box 11, The Heath, Runcorn, Cheshire WA7 4QE

1 Introduction

This year we have not included the subsection 'Synthesis of *N*-Heterocyclic Compounds' in order to avoid duplication with the section concerned with heterocyclic chemistry. However, some advances in this area are reported where appropriate.

A number of relevant reviews have been published, including one concerned with transition metal clusters.[1] Others deal with cyclometalation reactions,[2] olefin metathesis,[3] transition metal catalysed cyclizations of acetylenes,[4] carbon monoxide insertion reactions,[5] olefin insertion reactions,[6] and the use of transition metal derivatives in organic synthesis.[7] Amongst new books is the first volume of a series concerned with the application of transition metal organometallic compounds in organic synthesis,[8] and the second volume of the well known book by Wender and Pino.[9]

2 Metal-catalysed Hydrogenation

Continued progress has been made in improving the selectivity of metal-catalysed reductions and more specifically, asymmetric synthesis. The majority of the publications on asymmetric hydrogenation relate to homogeneous rhodium-catalysed reactions and reflect the continuing search for new and more specific chiral ligands. One example for rhodium is the diphosphinite, (+) (1*S*,2*S*)-*trans*-1,2-bis(diphenylphosphinoxy)cyclopentane (1), which is conformationally far more rigid than its cyclohexane analogue (2).[10] In the reduction of α-ethylstyrene, the optical yield, 60%, using (1) was much higher than that found with (2) or (−)-DIOP [(1*S*,2*S*)-2,3-*o*-isopropylidene-2,3-dihydroxy-1,4-bis(diphenylphosphino)-butane], and the highest so far attained for this substrate by homogeneous catalysts.

1 A. K. Smith and J. M. Basset, *J. Mol. Catalysis*, 1977, **2**, 229.
2 M. I. Bruce, *Angew. Chem. Internat. Edn.*, 1977, **16**, 73.
3 J. J. Rooney and A. Stewart in, 'Catalysis', Vol. 1, Specialist Periodical Reports, The Chemical Society, ed. C. Kemball, 1977, Chapter 8.
4 K. P. C. Vollhardt, *Accounts Chem. Res.*, 1977, **10**, 1.
5 F. Calderazzo, *Angew. Chem. Internat. Edn.*, 1977, **16**, 299.
6 G. Henrici-Olivé and S. Olivé, *Topics in Current Chem.*, 1976, **67**, 107.
7 L. Hegedus, *J. Organometallic Chem.*, 1977, **143**, 309.
8 'Transition Metal Organometallics in Organic Synthesis', Vol. 1, ed. H. Alper, Academic Press, 1976.
9 'Organic Synthesis *via* Metal Carbonyls', Vol. 2, by I. Wender and P. Pino, Wiley-Interscience, 1977.
10 T. Hayashi, M. Tanaka, and I. Ogata, *Tetrahedron Letters*, 1977, 295.

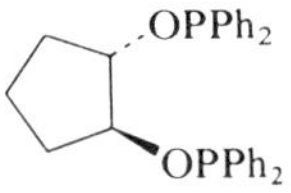

(1)

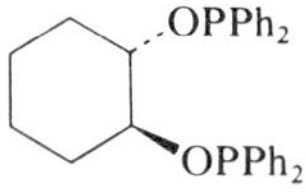

(2)

However, stereoselectivity was lower for unsaturated carboxylic acids but comparable for their esters, indicating that (1) is a ligand unique for the asymmetric hydrogenation of olefinic substrates bearing no polar substituents.

A new class of ligand has been reported whose chirality is the result of atropisomerism, and not to an asymmetric centre on phosphorus or carbon.[11] The described example is (−)-1,1-bi-2-naphthylbis(diphenylphosphinite) (3) and two

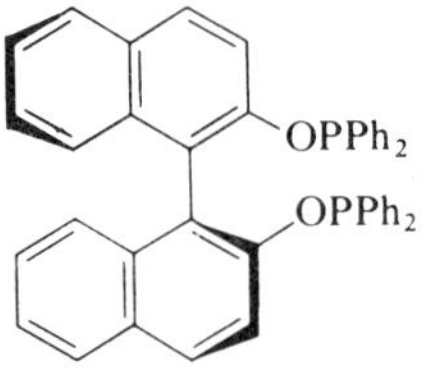

(3)

equivalents of this with one of $[Rh(cyclooctene)_2Cl]_2$ produced a catalyst system which, in the hydrogenation of unsaturated acids and esters, gave better optical yields at lower temperatures. In comparison to catalysts derived from ligands with chirality at phosphorus or carbon, this system was less active but gave similar optical yields.

The preparation of (2*S*,3*S*)-bis(diphenylphosphino)butane, *S*,*S*-chiraphos (4), and its use as a ligand for a Rh^{I} complex have produced remarkably high optical

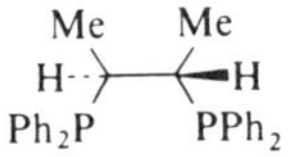

(4)

yields in the hydrogenation of α-*N*-acylaminoacrylic acids to *N*-acylaminoacids at ambient temperature and hydrogen pressure.[12] *X*-ray evidence indicated that (5) is

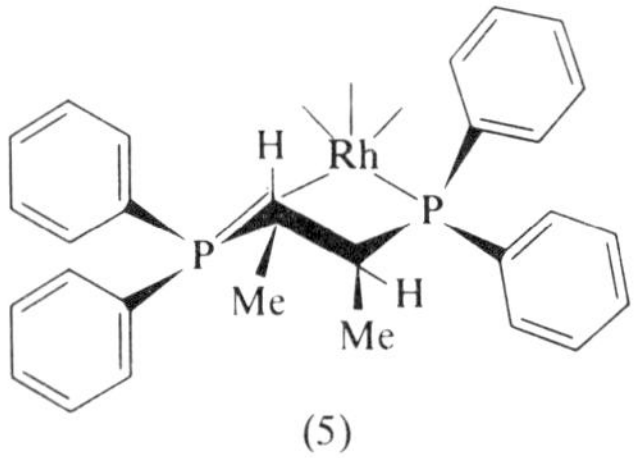

(5)

the preferred conformation of the chelate ring with equatorial methyl groups. With all nine substrates hydrogenated, the (*R*)-amino acid derivatives were produced

[11] R. H. Grubbs and R. A. DeVries, *Tetrahedron Letters*, 1977, 1879.
[12] M. D. Fryzuk and B. Bosnich, *J. Amer. Chem. Soc.*, 1977, **99**, 6262.

and two of these, leucine and phenylalanine, were optically pure. The optical yield was sensitive to both the *N*-acyl and β-vinylic substituents and also to the solvent employed. The authors suggest that the rigidity of the chelated diphosphine with its dissymmetrically orientated phenyl groups is the major source of interaction with the achiral substrates examined and, according to their rationalization in the design of (4), they also prepared (*R*)-1,2-bis(diphenylphosphine)propane, (*R*)-prophos, so that the naturally-occurring amino acids would result. This in fact was the case and optical yields similar to those obtained with (4) were found.

Several ligands (6), derived from L-hydroxylproline, have been investigated in Rh-catalysed reductions of certain achiral substrates.[13] Using neutral Rh^{I} complexes with (6b), esters of pyruvic acid were converted to the corresponding lactates, with higher optical yields than with (−)-DIOP.[13a] In the case of (6b) and propyl pyruvate, an enantiomeric excess of about 76% was found. When dry aprotic solvents (benzene or THF) were employed, optical yields were higher than when methanol was used. Compounds (6b), (6c), and (6d) have been used for synthesis of (*R*)-and (*S*)-*N*-benzyloxycarbonylalanine from their olefinic derivative,[13b] the (*S*)-compound being formed when Et_3N was present in the reaction. The highest optical yields, 59% (*R*) and 21% (*S*), were found with (6d) and it was suggested that the *N*-substituents of PPM (6a) play an important role. In the hydrogenation of substituted cinnamic acids with (6a) and (6b), Et_3N had a remarkable effect on optical yields with (6b) but not with 6a),[13c] indicating some novel interaction which serves to produce 83—91% enantiomeric excess in favourable cases. A new lipophilized bisphosphine–rhodium complex allows greater solubility in aliphatic hydrocarbons and increases lipophilic interaction with substrates.[13d] Here the new ligand (6e), developed from PPM, was applied in the hydrogenation of ethyl and n-butyl pyruvates with optical yields of the (*R*)-configuration of the product as high as 67.3% in the ethyl ester case in cyclohexane solvent. Application of the same catalyst system to the reduction of olefinic double bonds resulted in complete conversions but low optical yields,[13d] certainly lower than systems involving (6b).[13a]

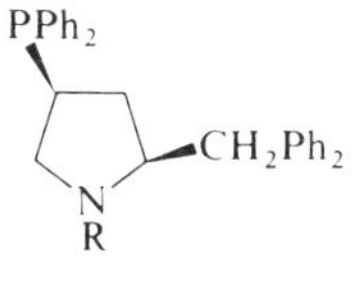

(6) a; (PPM) R = H—
b; (BPPM) R = t-BuO_2C—
c; (APPM) R = MeCO—
d; (PPPM) R = $(Me)_3C{\cdot}CO$—
e; (CPPM) R = cholesteryloxycarbonyl, $C_{27}H_{45}O_2C$—

A set of phosphines chiral at both phosphorus and carbon have been synthesized though only two of them, $(S)_p$– and $(R)_p$–menthylmethylphenylphosphine, were purified sufficiently well for use as ligands in the reduction of olefinic bonds in α,β-unsaturated carboxylic acids.[14] Reduction of (*E*)-β-methyl cinnamic acid gave

[13] (*a*) I. Ojima, T. Kogure, and K. Achiwa, *J.C.S. Chem. Comm.*, 1977, 428; (*b*) K. Achiwa, *Chem. Letters*, 1977, 777; (*c*) K. Achiwa, *J. Amer. Chem. Soc.*, 1976, **98**, 8265; (*d*) K. Achiwa, *Tetrahedron Letters*, 1977, 3735.

[14] C. Fisher and H. S. Mosher, *Tetrahedron Letters*, 1977, 2487.

the (*R*) product in 67.1% optical yield. The opposite configuration of the product was formed with the other phosphine. Further work is in progress to improve the purity of the phosphine ligands.

Discovery of the catalyst precursor $[Rh(COD)L]^+BF_4^-$, where L = (*R*,*R*)-1,2-ethanediylbis(*o*-methoxyphenyl)phenylphosphine, has allowed Knowles and co-workers to operate at higher temperatures and pressures and, as a result, they have found great versatility with their system.[15] Excellent selectivity (about 90% enantiomeric excess) was found in reduction of the α-enol ester, (*Z*)-ethyl-2-acetyloxy-3-phenyl-2-propenoate, the first outstanding result with a nonamide substrate, and optical yields of up to 96% were obtained in the hydrogenation of (*Z*)-α-acylaminocinnamic acids, but only up to 47% for (*E*) isomers at much lower rates. A catalyst picture was presented on the basis of an *X*-ray structure determination and reasons for the discrimination between olefin configurations were discussed.

Bulky (*Z*)-adamantyl or bornyl-α-acetamidocinnamate inhibited the reduction of the (*Z*) methyl ester, catalysed by a neutral Rh-DIOP complex, and underwent (*Z*) to (*E*) isomerization.[16a] Neither in the presence or absence of inhibitors did the methyl ester undergo appreciable isomerization. Further, over a range of α-acylamino unsaturated substrates, with the grouping NHCOR, optical yields of the reaction products decreased as the steric bulk of this group increased from R = Me to R = adamantyl.[16b] The α-formamido analogue (R = H), although having an even smaller amide group than in the corresponding α-acetamide, showed a decrease in optical yield. With a trifluoroacetamido group, a reversal in chirality of the major product was observed.

The use of chiral ruthenium complexes has been extended to asymmetric induction in catalytic hydrogenation of the olefinic bonds in α,β-unsaturated mono- and di-carboxylic acids using precursor cluster complexes.[17] With reaction temperatures of 90—120 °C, $H_4Ru_4(CO)_8[(-)\text{-DIOP}]_2$ and $Ru_8(CO)_{18}[(-)\text{-DIOP}]^3$ gave optical yields which varied with substrate and were improved by the presence of Et_3N, but for (*E*)-α-methylcinnamic acid the former complex produced up to 68% enantiomeric excess, a figure reached by lowering hydrogen pressure.[17a] Lower optical yields were found with the corresponding esters and also with substrates without carboxylic groups, even at lower reaction temperatures. The authors suggested that, since both cluster catalyst precursors gave products with the same predominant configuration and comparable optical purity, both complexes may have led to the same catalytic intermediates and that a free carboxylate group in the substrate is required for a substantial asymmetric bias. Regioselectivity and asymmetric induction in catalytic hydrogenation of α,β-unsaturated dicarboxylic acids, citraconic, and mesaconic acids, using $H_4Ru_4(CO)_8[(-)\text{-DIOP}]_2$ gives, in addition to (−)(*S*)-methylsuccinic acid, a mixture of γ-lactones in ratios which depend on the substrate and the reaction temperature.[17b]

[15] B. D. Vineyard, W. S. Knowles, M. J. Sabacky, G. L. Bachman, and D. J. Weinkauff, *J. Amer. Chem. Soc.*, 1977, **99**, 5946.

[16] (*a*) R. Glaser and J. Blumenfeld, *Tetrahedron Letters*, 1977, 2525; (*b*) R. Glaser and S. Geresh, *ibid.*, p. 2527.

[17] (*a*) C. Botteghi, S. Gladiali, M. Bianchi, U. Matteoli, P. Frediani, P. G. Vergamini, and E. Benedetti, *J. Organometallic Chem.*, 1977, **140**, 221; (*b*) M. Bianchi, F. Piacenti, P. Frediani, U. Matteoli, C. Botteghi, S. Gladiali, and E. Benedetti, *J. Organometallic Chem.*, 1977, **141**, 107.

Two new catalyst systems for reduction of aromatics have appeared. One is the $[Rh(\eta^5\text{-}C_5Me_5)Cl_2]_2$ complex operating under homogeneous conditions with added base (Et_3N) which is required as a cocatalyst.[18a] All *cis* isomers are the chief products and some hydrogenolysis of functional groups is observed. The other catalyst system employs salicylaldehyde complexes of Co, Ni, and Cu, though there are doubts about the homogeneity of these systems.[18b] When lithium aluminium hydride is used as reductant the catalysts are more active.

Rhodium catalysts have been developed for hydrogenation of ketones.[19] The complexes $RhCl(C_8H_{12})PPh_3$ and $Rh_2H_2Cl_2(C_8H_{12})(PPh_3)_2$, in the presence of strong alkali, promoted the reduction of several ketones, although the dirhodium species gave the best results.[19a] Pretreatment with sodium borohydride resulted in higher and more reproducible hydrogenation rates. The authors suggested the existence of a hydroxo complex as an active intermediate. An alkaline medium was again required for the reduction of ketones, at room temperature and at atmospheric pressure, catalysed by complexes of the type $[Rh(2,2'\text{-bipyridine})(diene)]^+PF_6^-$ which also act as hydrogenolysis catalysts for molecular oxygen.[19b] Consequently hydrogenation activity was not destroyed even if molecular oxygen was present in large amounts. Notable points are that selectivity was observed for reduction of carbon—oxygen double bonds even in the presence of olefinic bonds and the system was also active in the selective reduction of dienes and 2-alkynes to alkenes. In non-co-ordinating solvents such as benzene (in the presence of Et_3N) and dichloromethane (with or without Et_3N) the complex $[Rh(COD)(PPh_3)_2]PF_6$ was found to be active in the hydrogenation of ketones, as well as alkenes and alkynes to alkanes.[19c] In contrast the authors found that $[Rh(OCOPh)(COD)(PPh_3)]$ and $[Rh(COD)(PPh_3)_2py]PF_6$ were highly selective in the reduction of 1-alkynes to 1-alkenes if Et_3N and benzoic acid were present.

Ruthenium and iridium complexes have been used as selective catalysts. A report has appeared showing that $RuCl_2(PPh_3)_3$ is a highly selective catalyst for hydrogenation of cyclododeca-1,5,9-triene and other polyenes to monoenes in the presence of Et_3N and for reduction of the olefinic bond in a diene with a terminal and internal double bond.[20a] The same catalyst was found to be effective in the hydrogenation of aldehydes (both aliphatic and aromatic), but not ketones, to alcohols at temperatures of 50—80 °C and 10 atmospheres of hydrogen pressure.[20b] Nitro groups were unaffected. An iridium complex $IrH_3(PPh_3)_3$ in the presence of acetic acid was also selective for the reduction of aldehydes and not ketones.[21]

A new low-valent cobalt complex, $H_3Co[P(O\text{-}Pr^i)_3]_3$, has the notable property of being a soluble selective catalyst for the hydrogenation of α,β-unsaturated ketones and amides to the saturated ketones and amides.[22] Unsaturated aldehydes were not reduced under similar conditions. Catalyst activity was greatly increased

[18] (*a*) M. J. Russell, C. White, and P. M. Maitlis, *J.C.S. Chem. Comm.*, 1977, 427; (*b*) P. Patnaik and S. Sarkar, *Tetrahedron Letters*, 1977, 2531.

[19] (*a*) M. Gargano, P. Giannoccaro, and M. Rossi, *J. Organometallic Chem.*, 1977, **129**, 239; (*b*) G. Zassinovich, G. Mestroni, and A. Camus, *J. Mol. Cat.*, 1977, **2**, 63; G. Mestroni, G. Zassinovich, and A. Camus, *J. Organometallic Chem.*, 1977, **140**, 63; (*c*) R. H. Crabtree, A. Gautier, G. Giordano, and T. Khan, *J. Organometallic Chem.*, 1977, **141**, 113.

[20] (*a*) J. Tsuji and H. Suzuki, *Chem. Letters*, 1977, 1083; (*b*) ibid., p. 1085.

[21] W. Strohmeier and H. Steigerwald, *J. Organometallic Chem.*, 1977, **129**, C43.

[22] M. C. Rakowski and E. L. Muetterties, *J. Amer. Chem. Soc.*, 1977, **99**, 739.

without loss in selectivity by reacting at 70 °C. An example of the use of this catalyst is in the reaction of benzalacetone resulting in a 50% conversion to benzylacetone after 1 day.

Alkynes and conjugated dienes in mixtures with alkenes are selectively hydrogenated to alkenes without significant reduction of alkene to alkane using heterogeneous catalysts with Ni, Pd, or Pt intercalated in graphite.[23]

The comparison of several homogeneous and heterogeneous olefin hydrogenation catalysts has been made for the metals Rh, Co, Ni, and Pd.[24] Phosphine complexes were compared with metal chlorides and bromides supported on phosphine-modified silica carriers and the heterogenized systems were found to be 2 to 4 orders of magnitude more active than their homogeneous counterparts for the reduction of cyclohexene in THF.

Selectivity in the reduction of $\alpha\beta$-unsaturated aldehydes can be influenced to give preferential hydrogenation of the carbon—oxygen double bonds.[25] With rhodium halide catalyst systems, the presence of carbon monoxide and highly basic tertiary amines such as Et_3N and *N*-methylpyrrolidine significantly increased not only catalytic activity but selectivity to cinnamyl alcohol in the reduction of cinnamaldehyde under the oxo reaction conditions, giving no hydroformylation products; up to 85% selectivity to the unsaturated alcohol was achieved with the best system examined, $Rh_2Cl_2(CO)_4$ (7). Prereduction of $RhCl_3$, $3H_2O$ with carbon monoxide suggested that reduction of the trichloride occurred during hydrogenation, because activity after this treatment was such that lower reaction temperatures could be employed. Addition of triphenylphosphine to the systems examined resulted in alteration of the selectivity to one of olefin hydrogenation with no reduction of aldehyde groups. Immobilization of (7) was achieved on a cross-linked chloromethylated polystyrene which was functionalized with pyrrolidine.

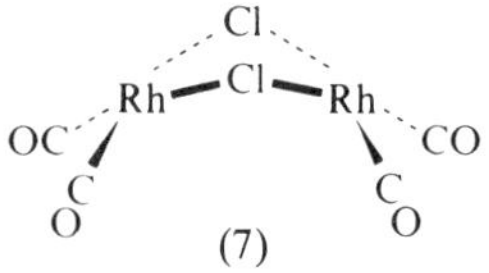

(7)

3 Isomerization

Some interesting cases of transition metal catalysed rearrangements and isomerizations have appeared.

Several reactions of which equations (1) and (2) are examples have been reported where double-bond migrations are catalysed by $RhCl_3$,$3H_2O$, allowing otherwise

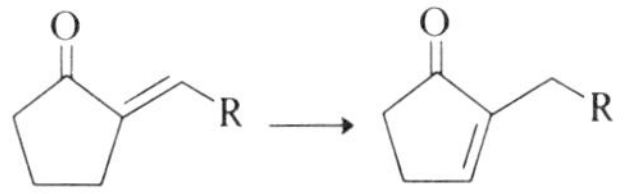

(1)

difficult, or impossible, exocyclic–endocyclic isomerizations to occur under mild conditions and in good yields.[26] Compound (8), readily attainable by isomerization [equation (2)] was previously obtainable only by a three-step synthesis. The iso-

[23] Ventron Corp., Canad. P. 1,000,306, 1976 (*Chem. Abs.*, 1974, **80**, 145 363).
[24] K. Kochloefl, W. Liebelt, and H. Knözinger, *J.C.S. Chem. Comm.*, 1977, 510.
[25] T. Mizoroki, K. Seki, S. Meguro, and A. Ozaki, *Bull. Chem. Soc. Japan*, 1977, **50**, 2148.
[26] J. Andrieux, D. H. R. Barton, and H. Patin, *J.C.S. Perkin I*, 1977, 359.

merization, where the migrating double bond moves into conjugation with the aromatic nucleus [equation (3)] was also found to be catalysed by the same metal

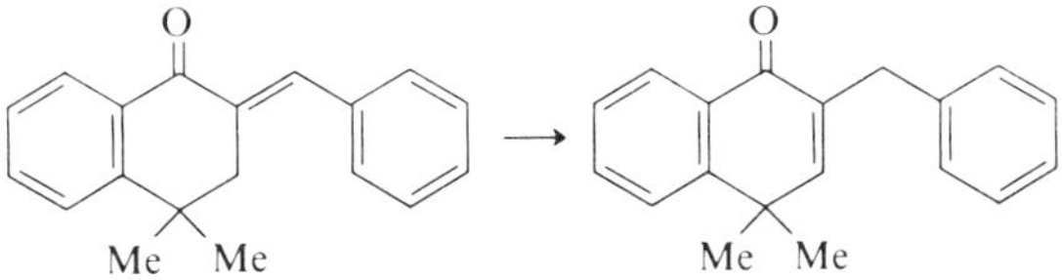

(8) (2)

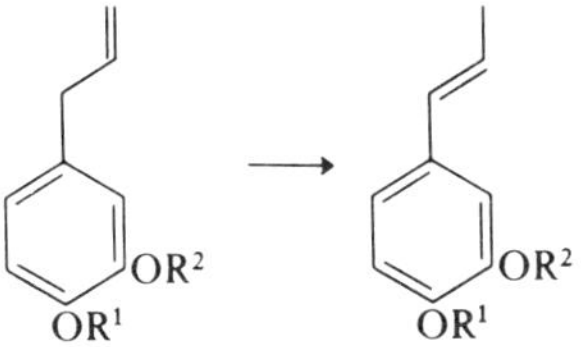

(9) (3)

chloride and the *trans* isomer predominated in the product, in contrast to that found on isomerization of (9) by strong bases. On this evidence even simple transition metal compounds must still have great potential as catalysts in the field of organic synthesis.

The application of an iridium catalyst, $IrCl(CO)(PPh_3)_2$, to migrations of exocyclic double bonds in some cycloalkanones has been described.[27] Isoaromatization and disproportionation reactions were also observed with the substrates examined.

Nickel and rhodium complexes have been shown to catalyse the rearrangement of allyl but-3-enoate to hepta-2,6-dienoic (10) or hepta 3,6 dienoic acid (11) in the presence of phosphine or phosphites.[28] The isomerization can be driven towards either the 2,6- or the 3,6-isomers through selective hydrogen abstraction. With nickel and tri-*o*-tolyl phosphite the ratio of (10) to (11) was highest at 14:1 and, with triethyl phosphite lowest at 0.8:1. In anisole solvent stereoselectivity was observed, the main product being the *trans*-2,6-isomer. With $Rh(PPh_3)_3Cl$ in chloroform the rearrangement gave pre-dominantly the 3,6-isomer. Intermediates such as (12) and (13) were proposed.

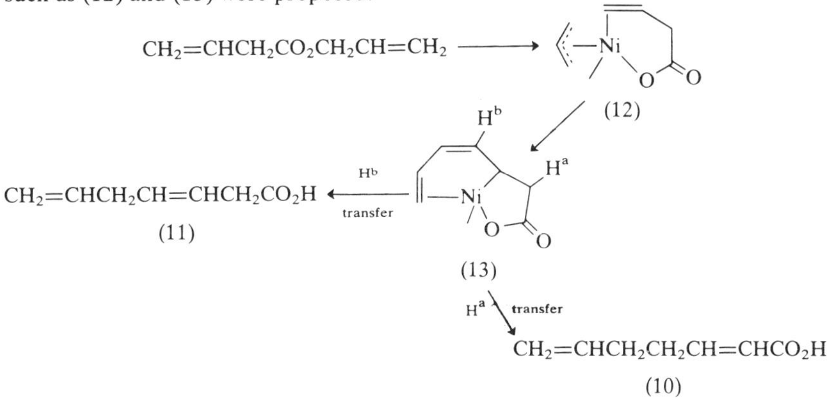

[27] Z. Aizenshtat, M. Hausmann, Y. Pickholtz, D. Tal, and J. Blum, *J. Org. Chem.*, 1977, **42**, 2386.
[28] G. P. Chuisoli, G. Salerno, and F. Dallatomasina, *J.C.S. Chem. Comm.*, 1977, 793.

Palladium(II) catalysis of the formation of 3-phenyl indoles from 2,2-diphenyl-2*H*-azirines has been reported,[29] although as yet the mechanism is not clear. The use of $PdCl_2(PhCN)_2$ at 30 °C and then washing with aqueous ammonia afforded the indole product, the formation of which normally required thermal rearrangement at elevated temperatures.

4 Dimerization, Oligomerization, and Polymerization

Much of the work published in these areas has contributed further to the understanding of the effects of varying ligands and substrates on the selectivities to certain products.[30,31] The palladium-catalysed formation of a 1,4-disilacyclohexa-1,5-diene from a 1-silacyclopropene has been noted[32] and the use of the nitrosyl rhodium complex, $[Rh(NO)(NCMe)_4][BF_4]_2$ as a catalyst for alkene isomerization, oligomerization and polymerization has been reported.[33] Stereospecific polymerization of 1,3-butadiene to *trans*-1,4-polybutadiene occurred with this rhodium system whereas isoprene gave the oligomers from tetramers up to decamers.

Of interest to the organic chemist is the continued work of Vollhardt *et al.* in cobalt-catalysed syntheses of polycyclic ring systems.[34] The elegant synthesis of the steroid nucleus in 71% yield [equation (4)][34a] illustrates the potential of the general reaction of co-oligomerization of $\alpha\omega$-diynes with a monoalkyne, equation (5), and the possibility of a wide variety of final structures. X in the triply unsaturated compound (14) can be varied.[34b] For X = N, the reaction gives annelated pyridines in yields of up to 80%.[34c]

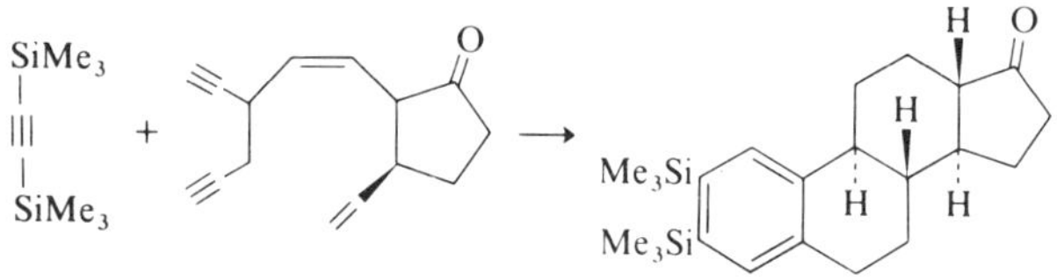

(4)

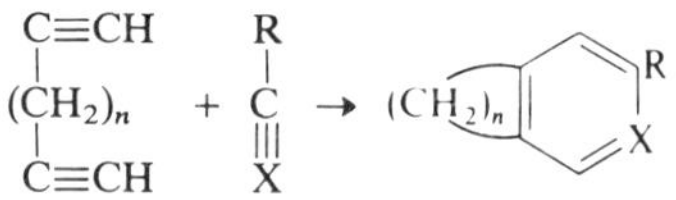

(14)

(5)

$n = 3, 4, 5$; R = $SiMe_3$, CO_2Me, Ph; X = CR, N

[29] K. Isomura, K. Uto, and H. Taniguchi, *J.C.S. Chem. Comm.*, 1977, 664.

[30] See for example, in dimerization and co-oligomerization, (*a*) P. Heimbach, A. Roloff, and H. Schenkluhn, *Angew. Chem. Internat. Edn.*, 1977, **16**, 252; (*b*) P. Heimbach, B. Hugelin, E. F. Nabbefeld, D. Reinehr, A. Roloff, and E. Troxler, *ibid.*, 253; (*c*) K. Kaneda, M. Terasawa, T. Imanaka, and S. Teranishi, *Tetrahedron Letters*, 1977, 2957; (*d*) G. Giacomelli, A. M. Caporusso, and L. Lardicci, *J.C.S. Perkin I*, 1977, 1333; (*e*) H. Suzuki, K. Itoh, Y. Ishii, K. Simon, and J. A. Ibers, *J. Amer. Chem. Soc.*, 1976, **98**, 8494; (*f*) S. Yoshikawa, J. Kiji, and J. Furakawa, *Makromol. Chem.*, 1977, **178**, 1077; (*g*) J. Ficini, J. d'Angelo, and S. Falou, *Tetrahedron Letters*, 1977, 1645; (*h*) G. Henrici-Olivé and S. Olivé, *Transition Met. Chem.*, 1976, **1**, 109; (*i*) H. T. Dieck and H. Bruder, *J.C.S. Chem. Comm.*, 1977, 24.

[31] See for example in polymerization C. Carlini, R. Nocci, and F. Ciardelli, *J. Polym. Sci., Polymer Chem.*, 1977, **15**, 767.

[32] M. Ishikawa, T. Fuchikami, and M. Kumada, *J.C.S. Chem. Comm.*, 1977, 352.

[33] N. G. Connelly, P. T. Draggett, and M. Green, *J. Organometallic Chem.*, 1977, **140**, C10.

[34] (*a*) R. L. Funk and K. P. C. Vollhardt, *J. Amer. Chem. Soc.*, 1977, **99**, 5483; (*b*) R. L. Hillard and K. P. C. Vollhardt, *ibid.*, 4058; (*c*) A. Naiman and K. P. C. Vollhardt, *Angew. Chem. Internat. Edn.*, 1977, **16**, 708.

5 Carbonylation

Several interesting reactions involving cobalt carbonyl intermediates have appeared this year. The reaction of cyclic olefins and ethers with hydrosilanes and carbon monoxide catalysed by $Co_2(CO)_8$ has been reported.[35] Cyclic olefins, which were expected to give product (15) by direct analogy with hydroformylation, in fact gave the enol silyl ethers (16) in good yield (50—74%).[35a] Under similar conditions cyclic ethers reacted to give the silyl protected hydroxyaldehyde (17) in about

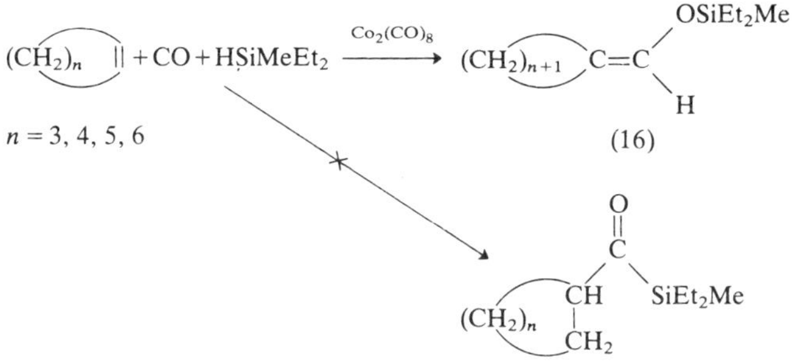

$n = 3, 4, 5, 6$

(16)

(15)

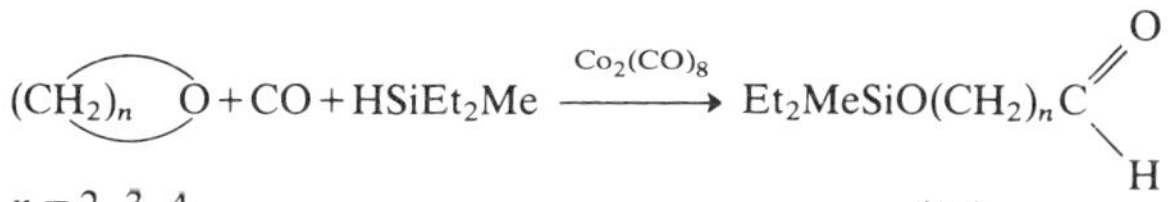

$n = 2, 3, 4$

(17)

50% yield.[35b] The intermediate, $[Co(CO)_4]^-$, which was generated by catalytic disproportionation of $[Co_4(CO)_{12}]$ or $[Co_2(CO)_8]$ using free or complexed halide ions reacted with the dihalide (18) to give 2-indanone (19) in 80% yield.[36]

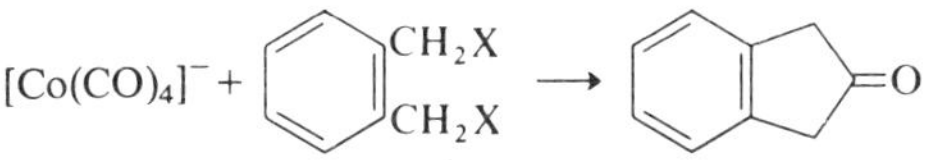

(18) (19)

The use of phase transfer catalysis in carbonylation has been extended to the catalytic carbonylation of benzyl halides using $Co_2(CO)_8$/CO/NaOH in the presence of benzyltriethylammonium chloride as the phase transfer catalyst.[37] Yields appear to be good only for benzyl bromides.

Carbonylation *via* iron carbonyl complexes continues to generate interesting reactions. Two groups have reported the synthesis of lactones from dienes *via* epoxidation followed by carbonylation (Scheme 1).[38] Whereas iron and cobalt catalysts give the α,β-unsaturated lactone (20), rhodium catalysts give the $\beta\gamma$-unsaturated lactone (21).

[35] (*a*) Y. Seki, A. Hidaka, S. Murai, and N. Sonoda, *Angew. Chem. Internat. Edn.*, 1977, **16**, 174; (*b*) Y. Seki, S. Murai, I. Yamamoto, and N. Sonoda, *ibid.*, p. 789.

[36] P. S. Braterman, B. S. Walker, and T. H. Robertson, *J.C.S. Chem. Comm.*, 1977, 651.

[37] (*a*) H. Alper and H. D. Abbayes, *J. Organometallic Chem.*, 1977, **134**, C11; (*b*) L. Cassar and M. Foa, *ibid.*, p. C15.

[38] (*a*) R. Aumann and H. Ring, *Angew. Chem. Internat. Edn.*, 1977, **16**, 50; (*b*) G. D. Annis and S. V. Ley, *J.C.S. Chem. Comm.*, 1977, 581.

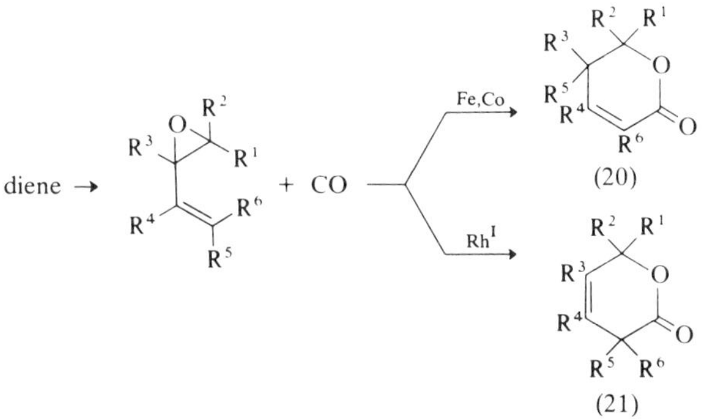

Scheme 1

A new synthesis of α-diketones from aldehydes and alkyl halides using $Fe(CO)_5$ has been reported.[39] The aldehyde, protected as the ethylenedithioacetal, is reacted with butyl lithium and $Fe(CO)_5$ to generate the acyltetracarbonylferrate (22) which then reacts with the alkyl halide to give the α-diketone (23) in overall yield of around 60%.

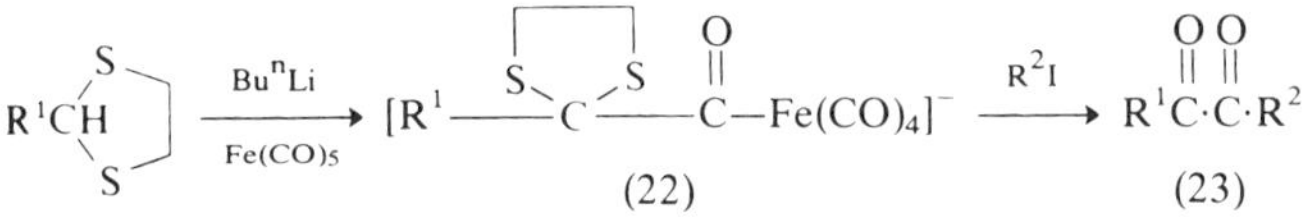

Reaction of the organotetracarbonylferrate (24) with Michael-type acceptors (25) gives the expected product (26) in high yield (*ca.* 90%).[40]

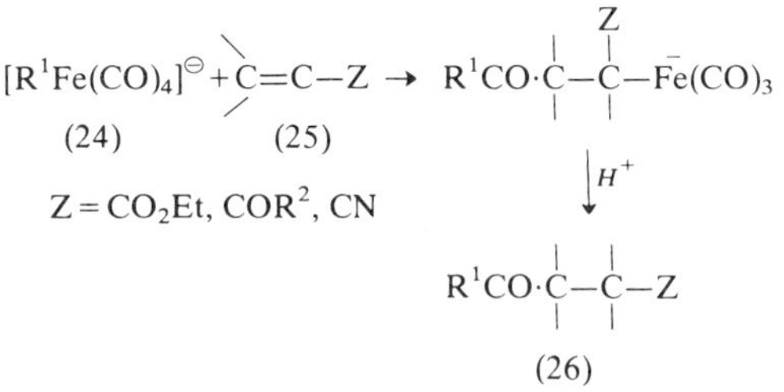

The direct formylation and acylation of pyridine has been achieved using $Fe(CO)_5$/PhLi. Depending on the work-up conditions a variety of products can be obtained (Scheme 2).[41]

6 Reaction of Co-ordinated Ligands

Whereas a lot of work has been done on the reactions of diene tricarbonyl-iron complexes much less work has been done on the olefin tetracarbonyl-iron complexes. A report[42] has appeared this year, however, on the nucleophilic attack

[39] M. Yamashita and R. Suemitsu, *J.C.S. Chem. Comm.*, 1977, 691.
[40] M. P. Cooke, jun., and R. M. Parlman, *J. Amer. Chem. Soc.*, 1977, **99**, 5222.
[41] C. S. Giam and K. Ueno, *J. Amer. Chem. Soc.*, 1977, **99**, 3166.
[42] B. W. Roberts and J. Wong, *J.C.S. Chem. Comm.*, 1977, 20.

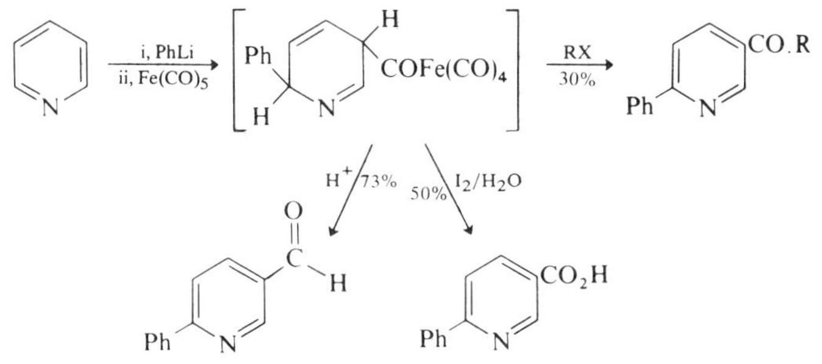

Scheme 2

on tetracarbonyliron complexes which give, after oxidative work up, products of type (27) in good yield.

$$\underset{\displaystyle Fe(CO)_4}{H_2C{=}CHR^1} + [R^2C(CO_2R^3)_2]^- \rightarrow \underset{\displaystyle (27)}{(R^3O_2C)_2CR^2CH_2CH_2R^1}$$

An unusual reaction of a cyclopentadienyl ligand occurs in the reaction of thiobenzophenones (28) with dicarbonylcyclopentadienyliron anion.[43] The two react together to give the fulvene (29) in yields of up to 82%. This cleavage of a cyclopentadienyl ligand followed by desulphurization gives a mild and potentially useful route to fulvenes.

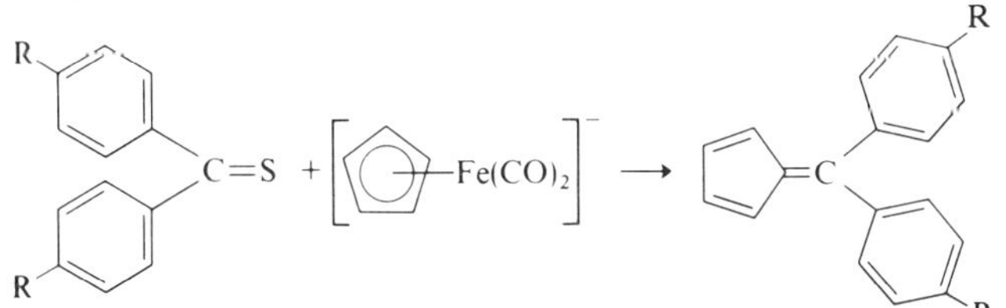

(28) (29)

The use of $[\eta^5\text{-}C_5H_5Fe(CO)_2]$(Fp) complexes in organic chemistry continues this year with the report of a new synthesis of β-lactams.[44] Nucleophilic addition of benzylamine to the complex (30) gives the intermediate (31) which is oxidized at −78 °C with chlorine to give the β-lactam (32) in *ca.* 34% overall yield. The reaction is stereospecific *e.g. trans*-2-butene (30; R^1, R^3 = Me; R^2 = H) gave only *cis*-3,4-dimethylazelidinone (32; R^1, R^3 = Me; R^2 = H). The reaction can also be applied to the synthesis of fused-ring β-lactams.

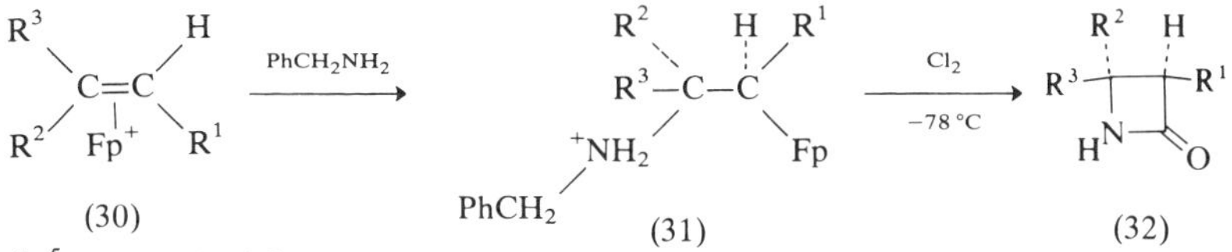

(30) (31) (32)

Fp = $[\eta^5\text{-}C_5H_5Fe(CO)_2]$

[43] H. Alper and H. N. Paik, *J.C.S. Chem. Comm.*, 1977, 126.

[44] P. K. Wong, M. Madhavarao, D. F. Marten, and M. Rosenblum, *J. Amer. Chem. Soc.*, 1977, **99**, 2823.

Carbanion attack on π-anisole and π-toluene chromium tricarbonyl complexes (33) gives, after oxidative work-up, *meta*-substituted aromatics as the major product.[45] With the π-anisole complex the *meta*-substituted product is obtained with greater than 90% selectivity. The π-toluene complex gives mainly the *meta*-substituted product, but with some *ortho*-substituted product.

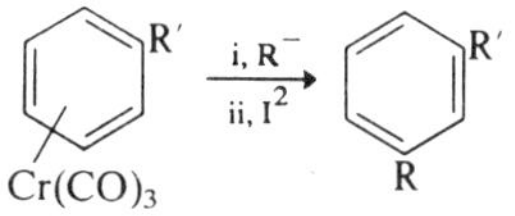

(33)

Intramolecular carbanion attack on arene chromium tricarbonyl complexes, followed by oxidative work up, leads to the formation of the bicyclic product (34) in good yield [(89%) for $n = 3$].[46] By varying the conditions (long reaction time, acid work up) spiro compounds *e.g.* (35) can be obtained in high yield.

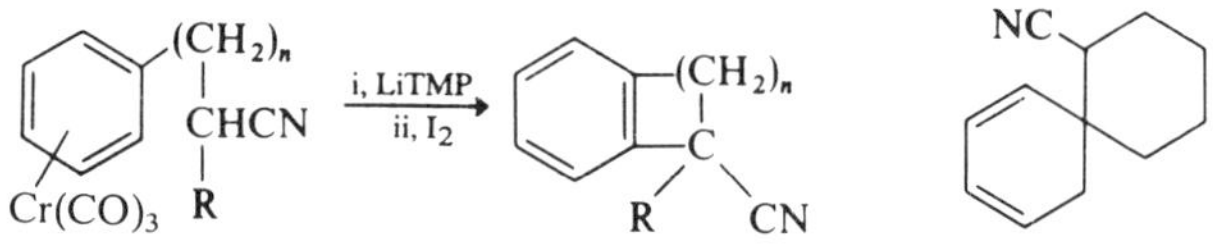

(34) (35)

The use of carbene–chromium complexes in organic synthesis is slowly increasing. Reaction of the carbene complex (36) with alkynes occurs in a stereoselective way to give the substituted α-naphthol chromium tricarbonyl complex (37), which can be readily oxidized to the corresponding 1,4-naphthoquinone.[47]

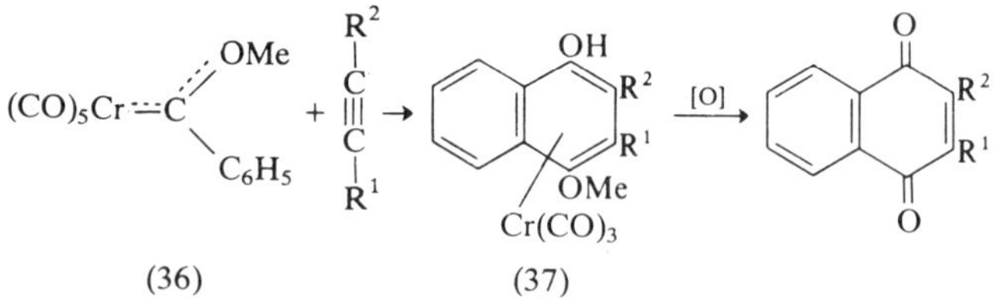

(36) (37)

The lithium enolate of cyclopentanone reacts with the vinyl–carbene complex (38) to give the complex (39).[48] Oxidative work up gives (40) whereas (41) is obtained by treatment of (39) with pyridine.

The cobaltacyclopentanone complex (42) reacts with isocyanates to give 2-oxo-1,2,-dihydropyridines (43) in about 70% yield.[49] With unsymmetrical complexes, *e.g.* (42; R^1, $R^3 = CO_2Me$; R^3, $R^4 = C_6H_5$), reaction proceeds regiospecifically to afford only one product (43; R^1, $R^3 = CO_2Me$; R^3, $R^4 = C_6H_5$).

[45] M. F. Semmelhack and G. Clark, *J. Amer. Chem. Soc.*, 1977, **99**, 1675.
[46] M. F. Semmelhack, Y. Thebtaranonth, and L. Keller, *J. Amer. Chem. Soc.*, 1977, **99**, 959.
[47] K. H. Dotz and R. Dietz, *Chem. Ber.*, 1977, **110**, 1555.
[48] C. P. Casey and W. R. Brunsvold, *Inorg. Chem.*, 1977, **16**, 391.
[49] P. Hong and H. Yamazaki, *Synthesis*, 1977, 50.

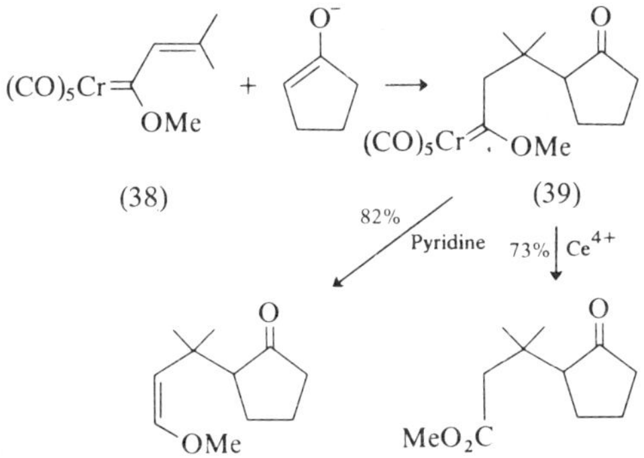

(38) (39) (41) (40)

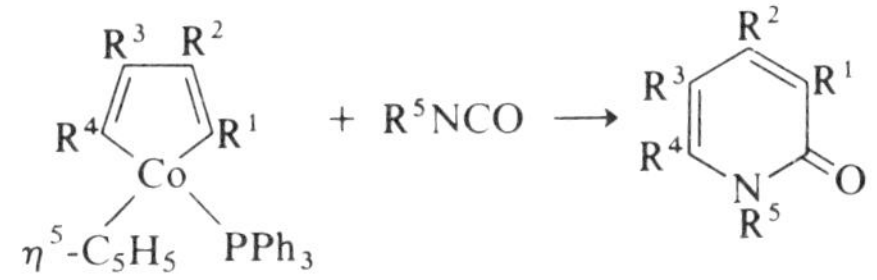

(42) (43)

There have been a number of reports[50] this year on the reactions of the organomanganese complexes (44), which are prepared by reaction of organolithium or organomagnesium compounds with MnI_2. These complexes react with a variety of acid chlorides to give the corresponding ketone (Scheme 3).[50a] This reaction is very selective, no alcoholic bi-products being formed. The ketones formed in this reaction do react further but at a very much slower rate. Aldehydes react much faster than ketones and at low temperatures the organomanganese reagent will selectively attack the aldehyde group even in the presence of an unprotected ketone (Scheme 3).[50b] The manganese complex (44) reacts with ethyl chloroformate to the corresponding alkylated ethyl ester (Scheme 3).[50c]

$$\mathrm{RCO_2Et} \xleftarrow[80\%]{\mathrm{ClCO_2Et}} \underset{(44)}{\mathrm{RMnI}} \xrightarrow[90\%]{\mathrm{R^1CO{\cdot}Cl}} \mathrm{RCO{\cdot}R^1}$$

$$\mathrm{RMnI} \xrightarrow[78\%]{\mathrm{MeCO{\cdot}(CH_2)_3CHO}} \mathrm{MeCO{\cdot}(CH_2)_3CHOH{\cdot}R}$$

Scheme 3

Organozirconium complexes (45), which are produced by hydrozirconation of olefins or acetylenes using $[\eta^5\text{-}C_5H_5Zr(H)Cl]$, react with aluminium chloride to generate the corresponding organoaluminium dichloride (45)[51] *i.e.* a

[50] (*a*) G. Cahiez, D. Bernard, and J. F. Normant, *Synthesis*, 1977, 130; (*b*) G. Cahiez and J. F. Normant, *Tetrahedron Letters*, 1977, 3383; (*c*) G. Cahiez and J. F. Normant, *Bull. Soc. Chim. France*, 1977, 570.

[51] D. B. Carr and J. Schwartz, *J. Amer. Chem. Soc.*, 1977, **99**, 638.

transmetallation reaction from zirconium to aluminium. The reaction proceeds well only for primary, saturated alkyl and alkenyl aluminium dichlorides and the products (46) can be used as mild alkylating agents.

$$[\eta^5\text{-}C_5H_5]_2Zr(R)Cl + AlCl_3 \underset{0\,°C}{\rightleftharpoons} [\eta^5\text{-}C_5H_5]_2ZrCl_2 + [RAlCl_2]_n$$

(45) (46)

The alkenylzirconium complex (47) can be coupled with aryl halides in the presence of a catalytic amount of $Ni(PPh_3)_4$.[52] Yields are generally excellent and the stereochemistry of the product is >98%E. The reaction can tolerate certain functional groups, in particular oxy-functional groups, that are incompatible with hydroaluminations, but it does not work so well for internal alkynes.

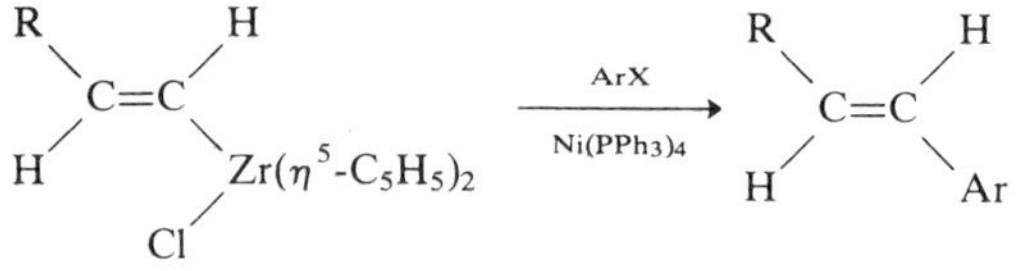

(47)

7 Olefin Metathesis

Interest in olefin metathesis continues unabated, with the number of papers published in 1977 being similar to that during the previous year. Some twenty papers were presented at an 'International Symposium on Metathesis' held in Noordwijkerhout, Holland.[53] Like last year more work was orientated towards mechanistic aspects than to direct application in synthetic organic chemistry. A further extensive review has appeared.[3]

Mechanistic Studies.—Good evidence for the formation of carbenoid species in metathesis systems comes from the formation of methane and ethylene on mixing $[WCl_6]$ with $Sn(CH_3)_4$, or $[Mo(PPh_3)_2(NO)_2Cl_2]$ with $Al_2(CH_3)_3Cl_3$, two typical metathesis catalysts. Production of methane is attributed to metal-carbene formation, and ethylene from its dimerization the latter being a possible termination step in the chain mechanism:

$$L_nM \xrightarrow{CH_3-M'} L_nM{-}CH_3 \xrightarrow{CH_3-M'} L_nM{=}CH_2 + CH_4$$

$$2L_nM{=}CH_2 \rightarrow CH_2{=}CH_2$$

When deca-2,8-diene was added propene was the initial product. This, and products from deuteriated reagents are in accord with the proposed mechanism.[54]

The results of an extensive examination[55] of product ratio and kinetics of metathesis of cyclic and acyclic olefins are also in agreement with the metal carbene chain mechanism, as are results of studies on the stereochemistry of metathesis.

[52] E. Negishi and D. E. Van Horn, *J. Amer. Chem. Soc.*, 1977, **99**, 3168.
[53] For details see: *Rec. Trav. Chim.*, 1977, **96**, M1-M144.
[54] R. H. Grubbs and C. R. Hoppin, *J.C.S. Chem. Comm.*, 1977, 634.
[55] T. J. Katz and J. McGinnis, *J. Amer. Chem. Soc.*, 1977, **99**, 1903.

Stereospecificity of the metathesis of *cis*- and *trans*-pent-2-ene has been considered in terms of steric interactions in the metallocyclobutane intermediate by two groups of workers.[56,57] The use of $[W(CO)_5CPh_2]$ as initiator for the metathesis of *cis*-pent-2-ene leads to much higher yields of *cis*-products than when the more conventional tungsten catalyst systems containing Lewis acids are employed. It is suggested that low stereoselectivity may result from the Lewis acid facilitating carbon–metal bond cleavage in the metallocyclobutane, and formation of a metallopropyl cation which would allow bond rotation.[57]

Gassman and Johnson[58] have been able to generate significant concentrations of 'M=CHR' carbene species in the presence of '$M{=}CH_2$' by conducting the metathesis of a terminal olefin in the presence of ethylcyclopropane. These authors attribute rapid degenerate metathesis of terminal olefins to efficient, highly selective, capture of 'M=CHR' intermediates to which '$\overset{+}{M}{-}\overset{-}{C}HR$' is a major resonance contributor.

The now generally accepted chain mechanism accounts for almost all experimental observations, but Mango[59] has pointed out that, given the known free energies of olefins and cyclopropanes, the absence of the latter as an olefin metathesis product appears to be in conflict with this mechanism. It may be expected that this point will be debated further during 1978.

Applications.—The main reason for the limited application of catalytic olefin metathesis in organic synthesis is due to the strong inhibiting effect of many common functional groups. It is therefore of interest that olefins with an amino-group, that normally do not undergo metathesis, do so when the donor properties of the amine are destroyed by quaternization.[60]

A convenient heterogeneous catalyst system of rhenium heptoxide on alumina promoted by a small amount of tetramethyl tin is a catalyst for metathesis of methyl penta-4-enoate in CCl_4. After an hour the clean conversion of reactant to the diester (48) was 51%.[61] Similar catalysts have also been used[62] for the metathesis

$$2CH_2{=}CH(CH_2)_2CO_2Me \rightleftharpoons MeO_2C(CH_2)_2CH{=}CH(CH_2)_2CO_2Me + CH_2{=}CH_2$$

(48)

of *cis*-1-chloro-octadec-9-ene with *trans*-dec-5-ene, which after five hours afforded a mixture of *cis*- and *trans*-14-chlorotetradec-5-ene in 70% yield. Under similar conditions allylchloride and vinyl chloride failed to react with hex-1-ene.

The metathesis of a series of ω-olefinic esters catalysed by $WCl_6/Sn(CH_3)_4$ gave the thermodynamic mixture of the *cis*- and *trans*-diesters (49). As well as the expected ethylene, the chloroester (50) was formed in low yield by the addition of

$$RO_2C(CH_2)_nCH{=}CH(CH_2)_nCO_2R \qquad CH_3CHCl(CH_2)_nCO_2R$$

(49) (50)

[56] J. L. Bilhou, J. M. Basset, R. Mutin, and W. F. Graydon, *J. Amer. Chem. Soc.*, 1977, **99**, 4083.
[57] T. J. Katz and W. H. Hersh, *Tetrahedron Letters*, 1977, 585.
[58] P. G. Gassman and T. H. Johnson, *J. Amer. Chem. Soc.*, 1977, **99**, 622.
[59] F. G. Mango, *J. Amer. Chem. Soc.*, 1977, **99**, 6117.
[60] J.-P. Laval, A. Lattes, R. Mutin, and J. M. Basset, *J.C.S. Chem. Comm.*, 1977, 502.
[61] E. Verkuijlen, F. Kapteijn, J. C. Mol, and C. Boelhouwer, *J.C.S. Chem. Comm.*, 1977, 198.
[62] R. Nakamura and E. Echigoya, *Chem. Letters*, 1977, 1227.

HCl (derived from the solvent chlorobenzene) across the double bond of the reactant olefin.[63]

It is of considerable interest that complex (51; M = Rh or Ir), which is reminiscent of the proposed metallocyclobutane metathesis intermediate, is a powerful catalyst for 1,1,3,3-tetramethyldisiloxane disproportionation into dimethylsiloxane and higher siloxane oligomers.[64] Here strong Si—O bonds are broken and reformed under unusually mild conditions, and this reaction may provide a route to novel silicone polymers.

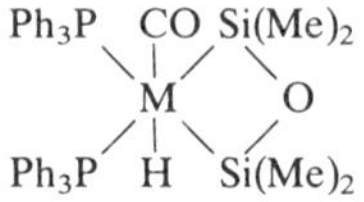

(51)

8 Use of Metal Cluster Complexes in Catalysis

The synthesis and characterization of new metal cluster complexes continues, and their possible application as catalysts for organic reactions is an area of growing interest.[1] A number of studies have been concerned with the catalysis of reactions of small molecules, such as the methanation of carbon monoxide, which are of industrial importance and are at present catalysed by conventional heterogeneous supported metal catalysts. It is possible that mechanistic information for reactions catalysed by metal clusters may lead to an improved understanding of reactions on metal surfaces.

Following their work on the slow homogeneous methanation of carbon monoxide catalysed by $[Os_3(CO)_{12}]$ or $[Ir_4(CO)_{12}]$ in toluene, Muetterties and co-workers report[65] that $[Ir_4(CO)_{12}]$ in molten $NaCl,2AlCl_3$ converts carbon monoxide and hydrogen to methane and ethane with only minor quantities of other hydrocarbons. At 180 °C this relatively fast reaction is homogeneous, but unlike conventional Fischer–Tropsch syntheses has the potential of product selectivity. It is appropriate to note a curious reaction of (52) {formed from $[Zr(C_5H_5)_2Cl_2]$ and AlH^iBu_2},

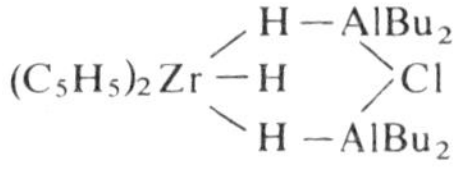

(52)

which, in benzene, absorbs two equivalents of carbon monoxide under ambient conditions to give a complex mixture of aluminium alkyls that does not contain simple aluminium alkoxide. On acid hydrolysis linear alcohols in a ratio comparable to that of a Fischer–Tropsch hydrocarbon synthesis are obtained. The mechanism of this process is thought to be complex involving reduction of carbon monoxide co-ordinated to zirconium, insertion, and transmetallation of organo-group from zirconium to aluminium.[66]

63 R. Baker and M. J. Crimmin, *Tetrahedron Letters*, 1977, 441.
64 J. Greene and M. D. Curtis, *J. Amer. Chem. Soc.*, 1977, **99**, 5176.
65 G. C. Demitras and E. L. Muetterties, *J. Amer. Chem. Soc.*, 1977, **99**, 2796.
66 L. I. Shoer and J. Schwartz, *J. Amer. Chem. Soc.*, 1977, **99**, 5831.

It is not surprising that (53) and (54) are active catalysts for the hydroformylation of pent-1-ene and pent-2-ene.[67] During the reaction, cluster dissociation does not appear to take place, and reaction conditions are milder than those used for the more usual $Co_2(CO)_8$ catalyst. These easily prepared clusters are air stable in the solid and in solution, and offer the advantage that they can be moderately selective for terminal products. Moreover, it is possible to recover the clusters in high yield after reaction.

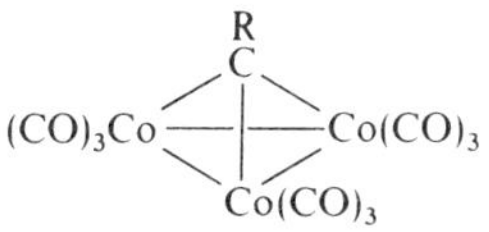

(53)

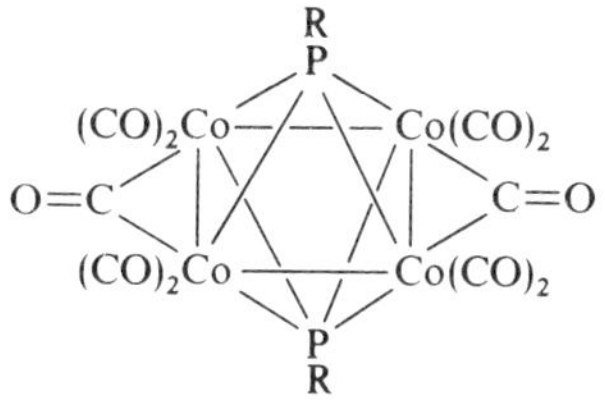

(54)

Carbon monoxide inhibits the isomerization of pent-1-ene and *cis*-pent-2-ene catalysed by $[H_2Ru_4(CO)_{13}]$ and $[H_4Ru_4(CO)_{12}]$, and the results of labelling experiments suggest the presence of σ-alkyl intermediates. With $[HRu_3(CO)_9C_6H_9]$ as catalyst a further isomerization path is operative that appears to involve a π-allyl intermediate.[68]

The cluster $[Ni_4\{CNC(CH_3)_3\}_7]$ is readily prepared from t-butylisocyanide and $[Ni(COD)_2]$ but has the disadvantage of being markedly air-sensitive. It is however an active catalyst for a variety of reactions.[69] Acetylene is smoothly trimerized to benzene, although the corresponding reaction of dialkylacetylenes is slow. Oligomerization of butadiene takes place stereoselectively to cyclo-octa-1,5-diene in the presence of this cluster which maintains high activity over an extended period. While inactive for the hydrogenation of olefins, terminal and internal acetylenes are cleanly hydrogenated to the corresponding *cis*-olefin without alkane formation. However, cluster decomposition products do catalyse olefin hydrogenation.

An investigation[70] of the homogeneous methanation of carbon monoxide is the presence of $[Ti(C_5H_5)_2(CO)_2]$ showed that the reaction is stoicheiometric and not catalytic in titanium. Methane is produced under an atmosphere of pure hydrogen, and a blue solution of the first '$M_6(C_5H_5)_6$' cluster, $[Ti_6(C_5H_5)_6O_8]$, is obtained.

[67] R. C. Ryan, C. U. Pittman, jun., and J. P. O'Connor, *J. Amer. Chem. Soc.*, 1977, **99**, 1986.
[68] G. A. Vaglio, D. Osella, and M. Valle, *Transition Met. Chem.*, 1977, **2**, 94.
[69] M. G. Thomas, W. R. Pretzer, B. F. Beier, F. J. Hirsekorn, and E. L. Muetterties, *J. Amer. Chem. Soc.*, 1977, **99**, 743.
[70] J. C. Huffman, J. G. Stone, W. C. Krusell, and K. G. Caulton, *J. Amer. Chem. Soc.*, 1977, **99**, 5829.

6 Organometallic Chemistry Part (ii) Main-Group Elements

By M. G. HUTCHINGS
I.C.I. Organics Division, Research Department, Blackley, Manchester M9 3DA.

1 Introduction

To maintain continuity with last year's Report[1] and its two predecessors, P, Se, and Te are excluded, as are carbaboranes.

Organometallic species containing Li, Be, and B have been the subjects of various *ab initio* calculations during part of a wider attempt to develop a 'model chemistry' aimed at establishing relationships between structures and energies of simple molecules, rather than merely calculating isolated absolute values. Species studied include phenyl cations,[2a] ethyl and vinyl cations, and their isoelectronic methylboranes,[2b] and small compounds containing Li or Be.[2c]

Considering the number of new organic synthetic routes based on organometallics which have been developed in the past twenty years or so, application in larger synthetic schemes had, until recently, been disappointing. However, the full worth of such reagents is becoming recognized, as evidenced further by the syntheses mentioned below, all dependent on organometallic reagents and procedures which have been recently developed. The enantiomer of the macrocyclic antibiotic dilactone (−)-vermiculin and the closely related fungicide (−)-pyrenophorin, have been synthesized by modifications around lithio dithianes.[3] A 'chemo-, regio-, and stereo-selective' route to the sex peromone (7*E*,9*Z*)-dodecadien-1-yl acetate in 40 % yield from but-1-yne and hept-1-yne has been based largely on organoboron chemistry.[4] Key reagents are $NH_2(CH_2)_3NH^-K^+$, Sia_2BH (Sia = 3-methylbut-2-yl) and EtC≡CLi, I_2, and NaOAc, and finally Sia_2BH followed by HOAc. Syntheses of prostaglandins[5a] and loganin[5b] have included reactions of allylsilanes with electrophiles.

Without wishing to labour the point, it does seem worthwhile reiterating last year's Reporter's trenchant *caveat* concerning overliberal interpretation of product studies when assigning structures to reactive intermediates.[1] Instances of premature inferences based on insufficient data continue to come to light.

[1] K. Smith, *Ann. Reports* (*B*), 1976, **73**, 121.
[2] (*a*) J. D. Dill, P. von R. Schleyer, and J. A. Pople, *J. Amer. Chem. Soc.*, 1977, **99**, 1; (*b*) Y. Apeloig, P. von R. Schleyer, and J. A. Pople, *ibid.*, 1977, **99**, 5901; (*c*) J. D. Dill, P. von R. Schleyer, J. S. Binkley, and J. A. Pople, *ibid.*, 1977, **99**, 6159.
[3] D. Seebach, B. Seuring, H.-O. Kalinowski, W. Lubosch, and B. Renger, *Angew. Chem. Internat. Edn.*, 1977, **16**, 264.
[4] E.-i. Negishi and A. Abramovitch, *Tetrahedron Letters*, 1977, 411.
[5] B.-W. Au-Yeung and I. Fleming, (*a*) *J.C.S. Chem. Comm.*, 1977, 79; (*b*) *ibid.* 1977, 81.

Of a general nature, a review has appeared which deals with the preparation and use in synthesis of reactive metal powders by reduction of their salts with alkali metals.[6] A recently published volume contains articles on organometallic synthesis *via* metal atoms,[7a] π-ligands containing organosilicon groups,[7b] structures of organometallic compounds containing electron-deficient bridge bonds,[7c] and organometallic radical anions.[7d]

Due to an oversight[8] in last year's Report[1] the comments concerning stabilized silicenium ions failed to include reference to a note by Mislow and co-workers.[9] Further evidence relating to the proposed ferrocenyl silicenium ion, $FcSiPh_2^+$, has forced a reappraisal of the data to the effect that this species is now no longer considered to be a necessary discrete intermediate in the observed reaction sequences. Hence, the unequivocal existence of a silicenium ion in solution has yet to be demonstrated.

2 Group I

Reviews have appeared concerning the generation and reactivity of di- and poly-alkali metal derivatives of heterofunctionally substituted organic molecules,[10] the structure and reactivity of alkali metal enolates,[11] and applications of α-metallated isocyanides in organic synthesis.[12]

A useful correlation has been drawn between $^{13}C—^{1}H$ one-bond coupling constants in benzenoid and heterocyclic molecules and the relative reactivity of the corresponding CH moiety toward lithiation.[13] Incorrect predictions caused by heterosubstituents can be related to the propensity for Li complexation and consequent ortho-directing effects. Presumably the technique could be used in reverse to assist in n.m.r. assignments.

Reagents are sometimes favourably modified merely by the change of physical form they undergo when adsorbed on a passive support, and methods have been described for the preparation of Na or K in reactive forms on the high surface area of charcoal, graphite, or alumina.[14a] These reagents, and the intercalation compound C_8K, react with ketones, giving enolates which can be subsequently monoalkylated, preferably in hexane solvent, where competing side reactions are minimized through the reactive species remaining adsorbed on the support. Following a similar procedure, nitriles and esters may be α-metallated heterogeneously with C_8K, and monoalkylated in good yields.[14b]

Lithiation of ketone dimethylhydrazones by more conventional techniques generate useful enolate equivalents. A recent study of the mechanistic stereochemistry of this reaction (Scheme 1)[15] showed that regardless of the hydrazone

6 R. D. Riecke, *Accounts Chem. Res.*, 1977, **10**, 301.
7 (*a*) P. L. Timms and T. W. Turney, *Adv. Organometallic Chem.*, 1977, **15**, 53; (*b*) I. Haiduc and V. Popa, *ibid.*, p. 113; (*c*) J. P. Oliver, *ibid.*, p. 235; (*d*) P. R. Jones, *ibid.*, p. 273.
8 K. Smith, personal communication.
9 P. Bickart, F. M. Llort, and K. Mislow, *J. Organometallic Chem.*, 1976, **116**, C1.
10 E. M. Kaiser, J. D. Petty, and P. L. A. Knutson, *Synthesis*, 1977, 509.
11 L. M. Jackman and B. C. Lange, *Tetrahedron*, 1977, **33**, 2737.
12 U. Schöllkopf, *Angew. Chem. Internat. Edn.*, 1977, **16**, 339.
13 E. B. Pedersen, *J.C.S. Perkin II*, 1977, 473.
14 (*a*) H. Hart, B. Chen, and C. Peng, *Tetrahedron Letters*, 1977, 3121; (*b*) D. Savoia, C. Trombini, and A. Umani-Ronchi, *ibid.*, 1977, 653.
15 M. E. Jung and T. J. Shaw, *Tetrahedron Letters*, 1977, 3305.

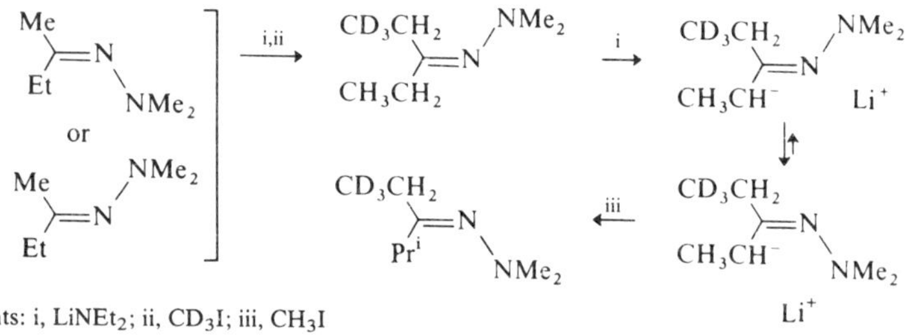

Reagents: i, $LiNEt_2$; ii, CD_3I; iii, CH_3I

Scheme 1

orientation, the primary carbanion is initially generated. The second *anti* deprotonation is kinetically favoured, presumably because of Me_2N hindrance to the alternative, but the *syn* carbanion is thermodynamically favoured, as evidenced by subsequent isomerization. Alkylation of enolates resulting from the addition of lithium organocuprates to cyclopentenones usually gives the 2,3-dialkylated cyclopentanone, with *trans*-orientation of the alkyl substituents. However, the presence of a 3-phenyl group causes the incoming alkylating agent to attack *cis*, in a highly stereoselective fashion.[16a] The reasons for this reversal of stereochemistry are presently unknown, but a possibly related phenomenon concerns the directing effect of a 3-aryl or -benzyl group during the formation of lithium enolates of cyclopentanones.[16b] The thermodynamically favoured 2-enolate is formed preferentially, the 3-substituent possibly acting to co-ordinate, and thereby direct and stabilize the adjacent Li^+ by π-complexation. This model finds support from an independent theoretical study (CNDO/2) which predicts $Li^+—C_6H_6$ to exist as a pyramidal face-metallated species (C_{6v} symmetry), in contrast to $H^+—C_6H_6$ (corner protonated; C_{2v} symmetry).[16c] The preference for this structure can be rationalized by the π-bonding ability of the unoccupied p-orbitals on Li^+, in conjunction with filled π-orbitals of C_6H_6 of suitable symmetry.

An anionic equivalent of the Friedel–Crafts cycloacylation forms the basis of a general ring-forming synthesis (Scheme 2 *e.g.* $Z = CH_2CH_2CH_2$).[17a] For $Z = CH_2$

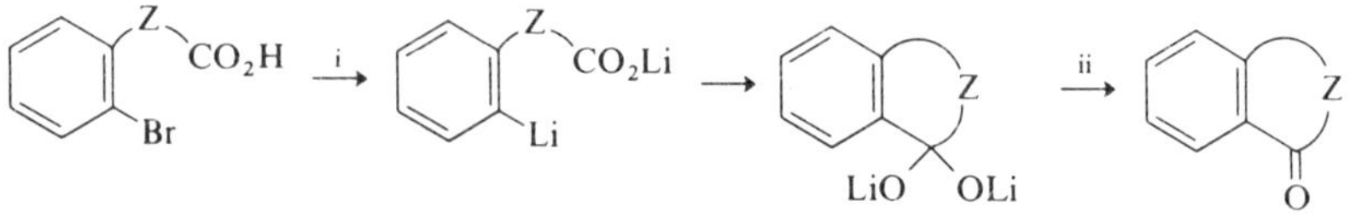

Reagents: i, RLi; ii, H_2O

Scheme 2

or CMe_2 ring formation is not observed, but rather reactions of an alternative type occur, that for $Z = CMe_2$, $E = CO_2H$ being depicted in its most general form in Scheme 3. In principle, Z and E can assume a variety of chemical identities (*e.g.*

[16] G. H. Posner and C. M. Lentz, (*a*) *Tetrahedron Letters*, 1977, 3215; (*b*) *ibid.*, 1977, 3211, (*c*) D. Heidrich and D. Deininger, *ibid.*, 1977, 3751.

[17] (*a*) R. J. Boatman, B. J. Whitlock, and H. W. Whitlock, jun., *J. Amer. Chem. Soc.*, 1977, **99**, 4822; (*b*) D. Hellwinkel, G. Hofmann, and F. Lämmerzahl, *Tetrahedron Letters*, 1977, 3241.

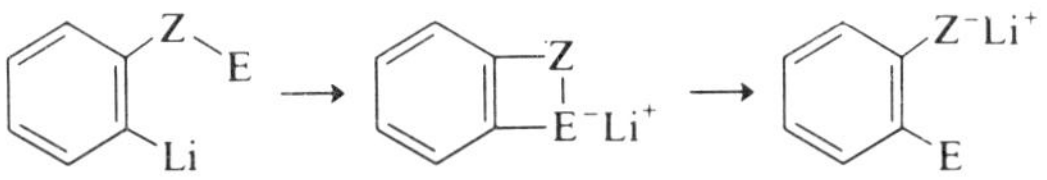

Scheme 3

Z = NR, O; E = COPh), and other examples have been given.[17b] Furthermore, if Z and E themselves are initially part of a ring, the product itself will be bicyclic.

The inclusion of a negatively charged atom α to the carbonyl carbon of an α', β'-unsaturated ketone system suppresses addition of RLi to the carbonyl group, with consequent preference for β'-attack, as exemplified by reactions of the ylid in Scheme 4.[18] An alternative approach to the inhibition of conventional addition to a carbonyl group relies on the effect of extreme steric hindrance to reagent attack.

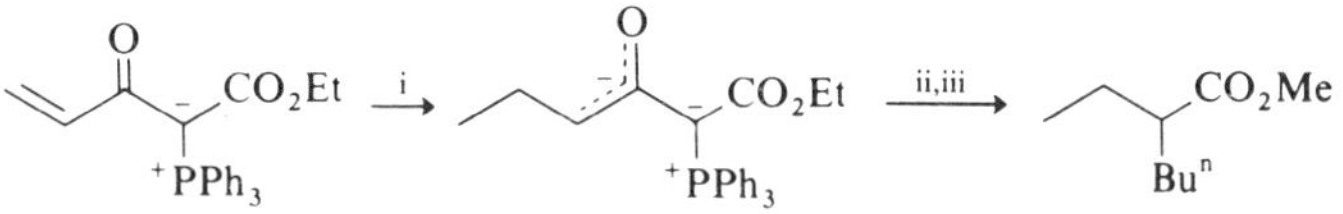

Reagents: i, MeLi; ii, Bu^nI; iii, H^+–MeOH

Scheme 4

Metallations at surprising sites are thereby promoted, as in the generation of (1a) from (1b),[19a] and the metallation of the Y—Me group of (2), where Y = NR, S,[19b] or O.[19c] Reaction of any of these species with electrophiles and suitable work-up leads to products derived from the uncommon synthons $HYCH_2^-$.

By formally adding two electrons to the pentadienone system, electrophilic attack at the terminal 5-position ('E^5 activity') is enhanced at the expense of the conventional nucleophilic attack ('N^5 activity'). In practice, a direct E^5 synthon (3) has been generated by double deprotonation of allylacetophenone and shown to react regio- and stereospecifically with electrophiles.[20]

Studies of polymer-bound organolithium species have given interesting synthetic and mechanistic results, as for instance, in the reaction between Ⓟ—$C_6H_4SCH_2^-Li^+$ and (4a), followed by MeI/NaI.[21a] The product ratio [(4c)]/[(4b)] gives a measure of the extent of site interaction on the polymer; high

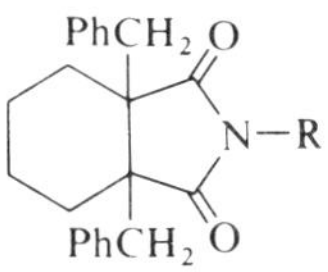

(1) a; R = CH_2Li
b; R = CH_3

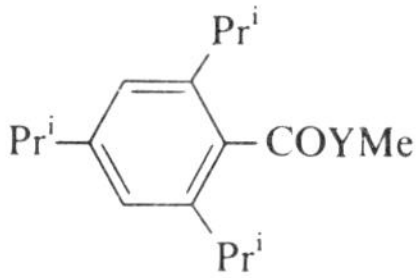

(2)

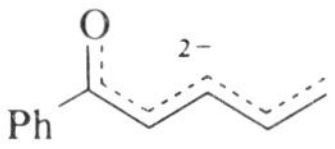

(3)

$I(CH_2)_nI$

(4) a; $n = 4$
b; $n = 5$
c; $n = 6$

[18] M. P. Cooke, jun., and R. Goswami, *J. Amer. Chem. Soc.*, 1977, **99**, 642.

[19] (*a*) R. Schlecker and D. Seebach, *Helv. Chim. Acta*, 1977, **60**, 1459; (*b*) P. Beak, B. G. McKinnie, and D. B. Reitz, *Tetrahedron Letters*, 1977, 1839; (*c*) P. Beak and B. G. McKinnie, *J. Amer. Chem. Soc.*, 1977, **99**, 5213.

[20] M. Pohmakotr and D. Seebach, *Angew. Chem. Internat. Edn.*, 1977, **16**, 320.

[21] (*a*) G. A. Crosby and M. Kato, *J. Amer. Chem. Soc.*, 1977, **99**, 278; (*b*) B. J. Cohen, M. A. Kraus, and A. Patchornik, *ibid.*, 1977, **99**, 4165.

ratios signify ionic clustering, as is in fact observed in this particular system. Polymer-bound trityl-lithium (Ⓟ—$C_6H_4C^-Ph_2Li^+$) has been used to generate the enolate from acetophenone which then reacts with an independent polymer-bound acylating agent (Ⓟ—$C_6H_3NO_2OCOPh$), to give high yields of the diketone ($PhCOCH_2COPh$).[21b] In contrast, the corresponding reaction with wholly soluble reagents gives a much inferior yield. The reaction sequence is capable of extension by use of a third reagent, giving rise to what is termed a 'chemical cascade'.

Among other useful new organolithium reagents are $RCCl_2Li$ (from $RCCl_2H$, including CH_2Cl_2) which reacts with a variety of electrophiles,[22] and *cis*-X—CH=CH—Li (X = OEt, NR_2) which act as $CHOCH_2^-$ equivalents.[23]

In a study of the gas phase acidities of *cis*- and *trans*-but-2-enes by pulsed ion-cyclotron resonance spectroscopy, the *cis*-butenyl anion has been found to be 0.2 kcal mol^{-1} less stable than its *trans*-isomer.[24] This result has relevance to solution-phase studies of butenyl anions where the *cis*-isomer has been found to predominate to a greater extent as the electropositivity of its alkali metal counterion, M^+, increased. Since ion-separation is expected to increase with more electropositive M^+, it had been assumed that the *cis*-preference was to be ascribed to the relative stabilities of the free anions. However, it now appears to arise because of differences in interaction energies among the allylic skeleton, M^+, and the solvent.

Ab initio calculations at minimal basis and split basis levels on the species C_3Li_4, including geometry optimization, have predicted the strange ground-state geometry depicted in (5) (C_{2v} symmetry), where the angle between the two central Li—C bonds is only 93°.[25] The central Li^+ ions bond with the highest occupied nonbonding π-orbitals of the C_3 chain (formally and conveniently considered a C_3^{4-} ion, isoelectronic to CO_2)(6).

(5)

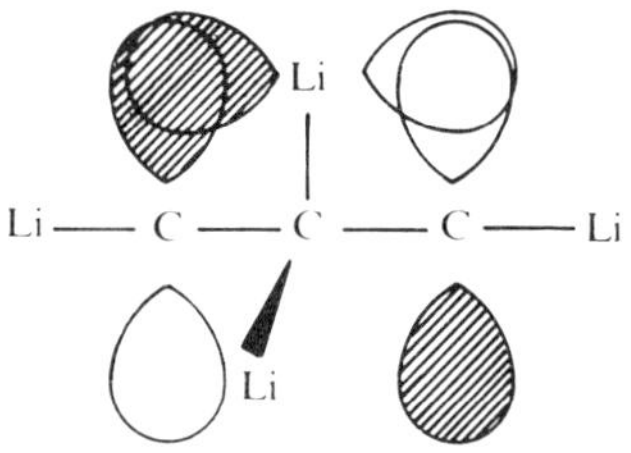

(6)

Attention is drawn to a review of the *umpolung* of the reactivity of carbonyl compounds, with particular reference to lithiated sulphur heterocyclic reagents.[26] Besides this review and work cited elsewhere in this Chapter, Seebach and his prodigious group have published in 1977 a series of papers in *Annalen* and *Chem.*

[22] J. Villiéras, P. Parriot, and J. F. Normant, *Bull. Soc. chim. France*, 1977, 765.

[23] J. Ficini, S. Falou, A.-M. Touzin, and J. d'Angelo, *Tetrahedron Letters*, 1977, 3589; R. H. Wollenberg, K. F. Albizati, and R. Peries, *J. Amer. Chem. Soc.*, 1977, **99**, 7365; L. Duhamel and J.-M. Poirier, *ibid.*, 1977, **99**, 8356.

[24] J. E. Bartmess, W. J. Hehre, R. T. McIver, jun., and L. E. Overman, *J. Amer. Chem. Soc.*, 1977, **99**, 1976.

[25] E. D. Jemmis, D. Poppinger, P. von R. Schleyer, and J. A. Pople, *J. Amer. Chem. Soc.*, 1977, **99**, 5796.

[26] B. T. Gröbel and D. Seebach, *Synthesis*, 1977, 357.

Ber. which give details of previously communicated material (much of which has been highlighted in *Annual Reports* in previous years), but which also include new illuminating insights.

3 Group II

Magnesium.—A useful review has appeared on the one-step reaction of organic halide, Mg, and substrate (Barbier reaction).[27]

Species of general formula HMgR are surprisingly ill-characterized. However, the preparation of a particularly active form of MgH_2 from the reaction of $LiAlH_4$ and Et_2Mg has opened up routes to these species based on 1:1 redistribution reactions with appropriate R_2Mg. Thus, HMgAlkyl may be prepared in THF, in which it is stable at room temperature, contrary to earlier reports.[28a,b] The species is dimeric in dilute solution (bridging hydrogens) and is reported to give the products of both reduction and alkylation on reaction with ketones. Perhaps the most significant result from this series is the observation that HMgCl and HMgBr add to alkenes and alkynes.[28c] No details are yet available, but hopefully we can look forward to a hydrometallation agent analogous to diborane, but which leads to Grignard reagents. Other related species are HMgOAr which are highly stereoselective reducing agents,[28d] and HMgPh, prepared directly by reaction between $LiAlH_4$ and Ph_2Mg.[28e]

An interesting new addition reaction of organomagnesiums is the transition metal catalysed addition of Me_3SiCH_2MgCl to diketene, yielding $Me_3SiCH_2C(:CH_2)CH_2CO_2H$.[29] The reaction provides the first example of vinyl–oxygen bond cleavage in diketene, but conceivably the catalysts, rather than the Grignard itself, could be responsible for the mode of ring opening. Copper-promoted additions of Grignard reagents continue to offer useful synthetic sequences, including the addition to 3-alken-1-ynes predominantly at the 2-position as exemplified in a synthesis of myrcene,[30a] the direct stereospecific replacement of iodide from vinyl iodides,[30b] and regiospecific addition of unsymmetrical allyl Grignards to alkyl halides, used in a simple synthesis of geraniol.[30c]

A purely electrophilic mechanism is proposed for the addition of 2-norbornylmagnesium bromide to formaldehyde, based on similar *exo* : *endo* ratios for product and starting material.[31] In contrast, an electron-transfer mechanism is believed to apply in the rare 'wrong end' addition of RMgX to the carbonyl oxygen in *o*-quinol acetates.[32]

Zinc, Cadmium, and Mercury.—Alkynylzinc reagents couple with alkenyl halides in the presence of Pd catalysts to give high yields of conjugated enynes.[33a] Besides

[27] C. Blomberg and F. A. Hartog, *Synthesis*, 1977, 18.

[28] E. C. Ashby and A. B. Goel, (*a*) *J. Org. Chem.*, 1977, **42**, 3480; (*b*) *J.C.S. Chem. Comm.*, 1977, 169; (*c*) *Inorg. Chem.*, 1977, **16**, 2941; (*d*) E. C. Ashby, A. B. Goel, and J. J. Lin, *Tetrahedron Letters*, 1977, 3133; (*e*) E. C. Ashby and A. B. Goel, *Inorg. Chem.*, 1977, **16**, 1441.

[29] K. Itoh, T. Yogo, and Y. Ishii, *Chem. Letters*, 1977, 103.

[30] (*a*) H. Westmijze, H. Kleijn, J. Meijer, and P. Vermeer, *Tetrahedron Letters*, 1977, 869; (*b*) A. Commerçon, J. F. Normant, and J. Villiéras, *J. Organometallic Chem.*, 1977, **128**, 1; (*c*) F. Derguini-Boumechal, R. Lorne, and G. Linstrumelle, *Tetrahedron Letters*, 1977, 1181.

[31] D. E. Bergbreiter and O. M. Reichert, *J. Organometallic Chem.*, 1977, **125**, 119.

[32] B. Miller, *J. Org. Chem.*, 1977, **42**, 1402, 1408.

[33] (*a*) A. O. King, N. Okukado, and E.-I. Negishi, *J.C.S. Chem. Comm.*, 1977, 683; (*b*) E.-I. Negishi, A. O. King, and N. Okukado, *J. Org. Chem.*, 1977, **42**, 1821.

being >97 % stereospecific, the procedure is compatible with the presence of various electrophilic functionalities in the alkene precursor, and is unique in allowing the direct preparation of terminal enynes, $CH_2{=}CH{-}C{\equiv}C{-}R$, without protection–deprotection sequences. A related reaction involving aryl- or benzylzinc compounds and aryl halides leads to high yields of unsymmetrically coupled products, and is again tolerant of further substitution in the aryl halide.[33b]

The silylated cyclohexenol (7) undergoes synthetically useful spirocyclopropanation to (8) in the presence of excess $Et_2Zn{-}CH_2I_2$ in concentrated benzene solution (Scheme 5).[34a] In contrast, trimethylsiloxymethylenecyclopentane gives (9) under Simmons–Smith conditions (CH_2I_2–Zn–Cu).[34b] In both these reactions, ZnI_2 apparently causes ring opening of the cyclopropane, and either H- or ring C-migration.

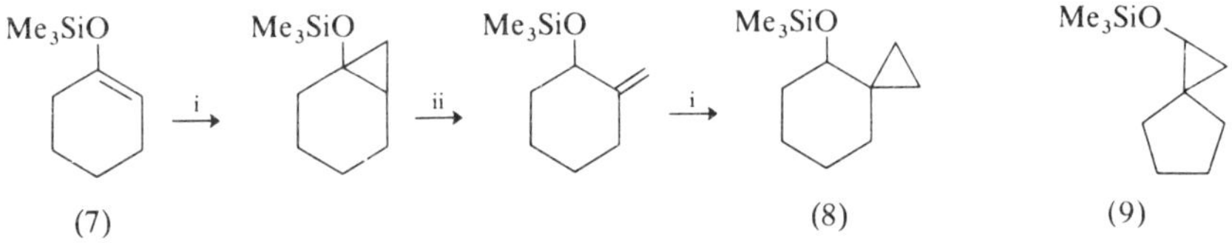

Reagent: i, $Et_2Zn{-}CH_2I_2$; ii, ZnI_2

Scheme 5

The conventional view that substituents preferentially adopt the equatorial orientation in monosubstituted cyclohexanes is no longer tenable in the light of variable temperature ^{13}C and ^{199}Hg n.m.r. studies of various cyclohexylmercury systems.[35] It is shown unambiguously that the mercury substituents preferentially occupy an axial position. Photolysis of α-amino-acids with various aliphatic side chains in the presence of $HgCl_2$ leads to methylmercury chloride, from an apparent fragmentation of the side chain of the amino-acid.[36] It is tempting to infer a connection with the formation of highly neurotoxic methylmercury found in the natural environment.

4 Group III

Boron.—A review has appeared dealing with boraheterocycles prepared *via* hydroboration, their use as reagents and intermediates being particularly emphasized.[37]

By redistribution of Me_2S,BH_3 with BCl_3 and BBr_3 in the appropriate stoicheiometry, Me_2S,BHX_2 and Me_2S,BH_2X (X = Cl, Br) are made readily available.[38] Reagents such as these are particularly important in preparations of unsymmetrical trialkylboranes or variously substituted mono- and di-alkylboranes,

34 (*a*) I. Ryu, S. Murai, and N. Sonoda, *Tetrahedron Letters*, 1977, 4611; (*b*) I. Ryu, S. Murai, S. Otani, and N. Sonoda, *ibid.*, 1977, 1995.
35 P. F. Barron, D. Doddrell, and W. Kitching, *J. Organometallic Chem.*, 1977, **139**, 361.
36 K. Hayashi, S. Kawai, T. Ohno, and Y. Maki, *J.C.S. Chem. Comm.*, 1977, 158.
37 H. C. Brown and E.-I. Negishi, *Tetrahedron*, 1977, **33**, 2331.
38 H. C. Brown and N. Ravindran, (*a*) *J. Amer. Chem. Soc.*, 1977, **99**, 7097; (*b*) *Synthesis*, 1977, 695; (*c*) *J. Org. Chem.*, 1977, **42**, 2533.

all of which have their place in the repertoire of synthetic organic chemistry. Significant observations are that the bromo compounds give appreciable amounts of tertiary alkylboranes with trisubstituted alkenes, and the ease with which $Me_2S,BHBr_2$ effects hydroboration in comparison with $Me_2S,BHCl_2$.[38a] It is relevant that there is evidence for π-interaction between benzene and $Me_2S,BHBr_2$ which is absent from the chloro analogue. Reaction between BBr_3 and $(Me_3Si)_3N$ gives $MeBBr_2$ essentially quantitatively and free from di- or tri-methyl derivatives.[39] Easy routes to methylboranes are always interesting since hydroboration is obviously inapplicable for their preparation. Tertiary alkylboranes are also relatively inaccessible, so the sequence bromination (Br_2–$h\nu$) and alkylation (RM) when applied, *inter alia*, to secondary alkyl boronate esters giving tertiary alkyl boron derivatives is of potential synthetic value.[40] The reagent $Li^+Et_3BH^-$ may be used in a route to other novel boranes and borates, involving *nucleophilic* hydroboration of styrenes under relatively mild conditions.[41] Finally, ω-alkynylalkylboranes, inaccessible by hydroboration, may be prepared by prototropic isomerization [$K^+NH_2(CH_2)_3NH^-$ catalyst] of the available internal alkynylborane, itself prepared by hydroboration of a terminal nonconjugated enyne.[42]

The increasing problems associated with the use of racemic materials as pharmaceuticals (*etc.*) has heightened the need for asymmetric reagents. Therefore, the renewed interest in chiral boranes as reducing agents is particularly timely, and is reflected in several publications dealing with modified isopinocampheylborane systems. The reagent (10) derived from 9-borabicyclononane reduces aldehydes

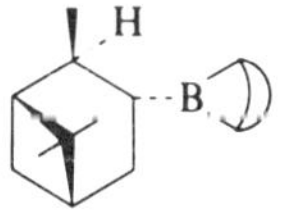

(10)

with concommittant elimination of alkene, and has been used to reduce deuterobenzaldehyde to (S)-benzyl-α-d-alcohol (important for mechanistic biochemical studies) with essentially quantitative asymmetric induction.[43a] Bu^{t}Li donates H^- to the B atom of (10), the new chiral trialkylborohydride reducing even relatively hindered ketones rapidly and quantitatively.[43b] Optical purities of products range only from 3—37 %, but of more importance is the finding that the predominant enantiomers of all alcohols produced to date have the same absolute configuration.

In a reaction analogous to the well known radical addition of R_3B to enones and analogues, B-alkynyl-9-BBN undergoes conjugate addition to those α,β-unsaturated ketones which are capable of assuming a cisoid conformation, but apparently by a nonhomolytic pathway. Further elimination in the reaction with (11) leads to conjugated enynones.[44] Dialkyldialkynylstannanes react with Et_3B or

[39] K. Barlos and H. Nöth, *Chem. Ber.*, 1977, **110**, 3460.
[40] H. C. Brown, N. R. DeLue, Y. Yamamoto, K. Murayama, T. Kasahara, S.-I. Murahashi, and A. Sonoda, *J. Org. Chem.*, 1977, **42**, 4088.
[41] H. C. Brown and S. C. Kim, *J. Org. Chem.*, 1977, **42**, 1482.
[42] C. A. Brown and E.-I. Negishi, *J.C.S. Chem. Comm.*, 1977, 318.
[43] (*a*) M. M. Midland, A. Tramontano, and S. A. Zderic, *J. Amer. Chem. Soc.*, 1977, **99**, 5211; (*b*) S. Krishnamurthy, F. Vogel, and H. C. Brown, *J. Org. Chem.*, 1977, **42**, 2534.
[44] J. A. Sinclair, G. A. Molander, and H. C. Brown, *J. Amer. Chem. Soc.*, 1977, **99**, 954; G. A. Molander and H. C. Brown, *J. Org. Chem.*, 1977, **42**, 3106.

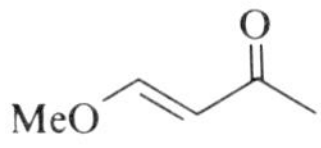

(11)

Bu^n_3B to give the novel heterocyclic dienylboranes (12) which can be used for further synthetic modification.[45] Of relevance to the mechanism of the photochemically induced rearrangement of dialkyl-*trans*-1, 3-dienylboranes is the finding that the independently generated cis-isomers rapidly cyclize, with concurrent alkyl migration, to (13).[46] The rearrangement is formally isoelectronic to the second migration step in the reactions between acylating agents and alkynyl- or cyano-borates.

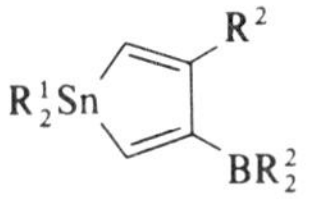

(12)

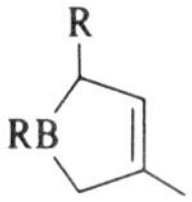

(13)

Reports of new reactions of borate salts have been few in 1977. However, a reaction of great potential value in view of the number of structural features capable of modification, is the Michael addition of alkynylborates to α,β-unsaturated systems, depicted in general form by equation (1).[47] As yet, only nitroethenes have given synthetically useful yields of organic products, although reaction with even these is inhibited by excessive substitution [$R^3 = R^4 = Me$, equation (1)].

$$Li^+R^1_3\bar{B}C{\equiv}CR^2 + R^3R^4C{=}C(X)R^5 \rightarrow R^1_2B\overset{\overset{\displaystyle R^1}{|}}{C}{=}CR^2{-}CR^3R^4{-}\bar{C}R^5XLi^+ \qquad (1)$$

Following on from last year's claim for the generation of 'methylborylene' (MeB:), the analogue 'naphthylboryne' (α-NpB:) has been proposed as a primary product in the photolysis of α-Np_3B.[48] Evidence for this species rests solely on indirect product studies. Firstly, oxidation products observed were cyclohexanol, when cyclohexane was solvent, and cyclohexane-*cis*-1,2-diol for reaction in cyclohexene, the latter *via* the assumed 'stable' species (14); and secondly, Cl^- ions and $COCl_2$ were observed after hydrolysis and oxidation with CCl_4 as solvent. If (14) is

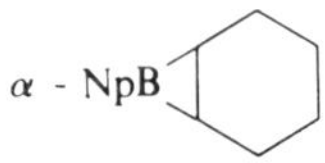

(14)

indeed stable, further data on its characterization are awaited, but in the meantime, readers are invited to make up their own minds on the rightness of the claim for the generation of a boryne.

[45] L. Killian and B. Wrackmeyer, *J. Organometallic Chem.*, 1977, **132**, 213.
[46] G. Zweifel, S. J. Backlund, and T. Leung, *J. Amer. Chem. Soc.*, 1977, **99**, 5192.
[47] A. Pelter and L. Hughes, *J.C.S. Chem. Comm.*, 1977, 913.
[48] B. G. Ramsey and D. M. Anjo, *J. Amer. Chem. Soc.*, 1977, **99**, 3182.

The methylated cyclopentadienylborane (15; X = BCl_2) prepared from (15; X = Me_3Ge) and BCl_3, is fluxional on the ^{1}H n.m.r. time scale down to −80 °C due to rapid BCl_2 sigmatropic shifts.[49a] In contrast (15; X = BI_2) undergoes further reaction with BI_3 yielding the novel salt (16), characterized by a full range of spectroscopic and conductivity techniques.[49b] The bonding of the C_{5v} pyramidal cation in (16) is assumed to be similar to that in the dication $C_6(CH_3)_6^{2+}$. An interesting

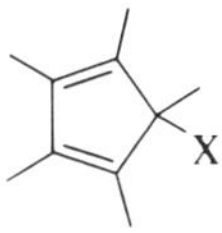

(15)

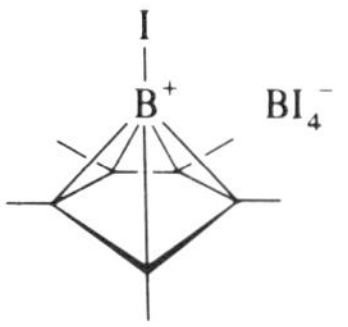

(16)

series of 1,3-shifts has been observed for (17; Scheme 6).[50] Addition of MeLi to (17; R^2 = Me) leads to fragmentation products resulting from migration of R^1 from C to B (Scheme 6). When $R^1 \neq R^3$, R^3BMe_2 is also observed, consistent with the rapid sigmatropic shift of the BMe unit prior to ate-complex formation.

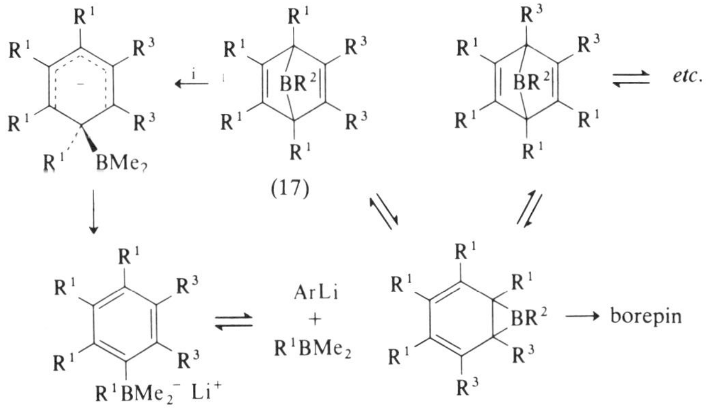

Reagent: i, LiMe (R^2 = Me)

Scheme 6

The recently published MNDO method has been employed to calculate various data for boranes (ΔH_f, geometries, IP, dipole moments, and proton affinities) which agree much more satisfactorily with known experimental data than did the previous MINDO/3 values.[51]

Aluminium, Gallium, and Thallium.—The lack of ready synthetic approaches to alkylalanes has prevented exploitation of their full potential in organic synthesis so approaches reported in 1977 which go some way in removing this deficiency open up important new areas for future research. The alkene- or alkyne-derived metal

[49] P. Jutzi and A. Seufert, (*a*) *Angew. Chem. Internat. Edn.*, 1977, **16**, 41; (*b*) *ibid.*, 1977, **16**, 330.
[50] J. J. Eisch and J. E. Galle, *J. Organometallic Chem.*, 1977, **127**, C 9.
[51] M. J. S. Dewar and M. L. McKee, *J. Amer. Chem. Soc.*, 1977, **99**, 5231.

complexes $Cp_2Zr(Cl)R$ readily transfer an alkyl or alkenyl group, respectively, to $AlCl_3$. The resulting species react rapidly with acyl chlorides at −30 °C to give quantitative yields of ketones, whereas the starting Zr complex reacts, at best, only slowly.[52a] An even greater synthetic potential results from modifications which use the transition metal catalytically. Both AlH_3, and more significantly $LiAlH_4$, in the presence of $TiCl_4$ or $ZrCl_4$,[52b] as well as readily prepared $(Pr^i_2N)_2AlH$ in the presence of Cp_2TiCl_2,[52c] give alkylalanes or alanates on reaction with alkenes. The reactions show analogies to hydroboration with BH_3 in their sensitivity to steric effects (allowing preferential reduction of the terminal alkene bond in nonconjugated dienes), and preference for strongly basic ethers as solvents. By suitable choice of reagents, further reaction generates alkane, alkyl halide, or alcohol (O_2 oxidation) in high yields. The reagents are cheap and the conditions mild, but a critical limitation to the generality of the reaction is the conversion of internal alkenes to terminal alkyl groups. Of course, this may be turned to advantage when mixtures of isomeric alkenes are used as starting material and the terminally substituted product is required.

A further type of novel organoalane, formulated as (18), is also derived by transfer of a group from the more readily formed Zr derivative, $Cp_2Zr(Cl)COR$.[53]

$$(Cl_2Al-\overset{O}{\overset{\|}{C}}-R)_x$$

(18)

While masked carbonyl functionalities have found favour as one source of acyl anion equivalents, the more direct progenitors of this synthon, acylmetallics, have proved singularly disappointing as synthetic intermediates. However, the acyl moiety present in (18) is highly susceptible to electrophilic attack. Thus, (18) is protonated (or deuterated) at the carbonyl carbon by water (or D_2O), and is both *C*- and *O*-acylated with CH_3COCl. The possible existence of a stable Group III acylmetallic species, such as (18), is noteworthy in that no stable acylborane analogue has yet been prepared. Attempts at further structural characterization of (18) would be desirable.

Bu^i_2AlH reacts with Cp_2ZrCl_2 to give (19) which acts as one of the few homogeneous catalysts for the industrially significant Fischer–Tropsch process (reductive polymerization of CO).[54] Straight chain C_1—C_5 alcohols are obtained, remarkably, from a room temperature reaction at atmospheric pressure, all carbon atoms of the products being incorporated from the CO feedstock.

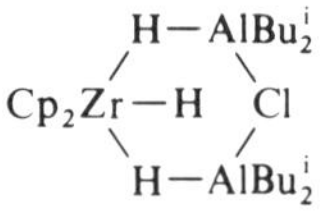

(19)

[52] (*a*) D. B. Carr and J. Schwartz, *J. Amer. Chem. Soc.*, 1977, **99**, 638; (*b*) F. Sato, S. Sato, H. Kodama, and M. Sato, *J. Organometallic Chem.*, 1977, **142**, 71; (*c*) E. C. Ashby and S. A. Noding, *Tetrahedron Letters*, 1977, 4579.

[53] D. B. Carr and J. Schwartz, *J. Organometallic Chem.*, 1977, **139**, C 21.

[54] L. I. Shoer and J. Schwartz, *J. Amer. Chem. Soc.*, 1977, **99**, 5831.

An unusual type of isomerism exists in the oxamide complexes (20) and (21) of gallium.[55]

(20) (21)

Thallium (III) trifluoracetate brings about nonphenolic oxidative coupling of aromatic systems. The reaction is tolerant of nitrogen- and oxygen-containing functionalities, and is applicable to intramolecular couplings as in a synthesis of the aporphine alkaloid (±)-ocoteine.[56] The corresponding thallium (III) acetate effects an unprecedented aromatic acetoxylation during the coupling.

5 Group IV

Silicon.—One of the archetypal components of organic chemistry, the benzene ring, has still to find a Si analogue. However, a report in 1977 claims evidence for the generation and trapping of the first silabenzene (22; Scheme 7).[57] Unfortunately an alternative route to the observed (23) involving carbanionic attack by the lithiated intermediate (24) and subsequent ring closure cannot be excluded on the evidence available. Nor are other anticipated products observed which might be indicative of the intermediacy of (22), such as dimers. In the

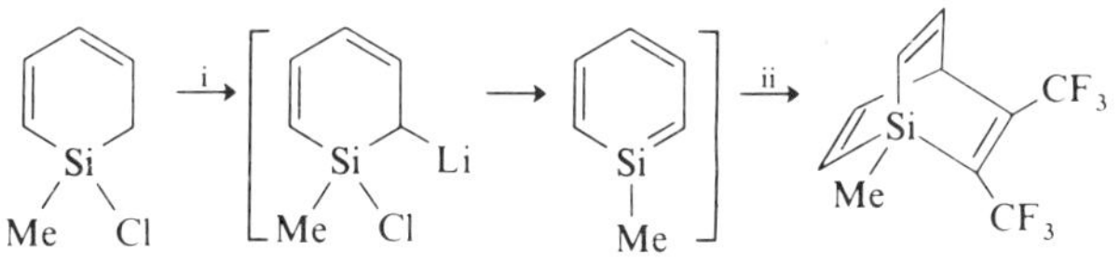

(24) (22) (23)

Reagents: i, $(Me_3Si)_2NLi$, −77 °C; ii, $CF_3C{\equiv}CCF_3$

Scheme 7

absence of more compelling data, the case for a silabenzene must, at best, remain unproven. Two related synthetic approaches have been used to prepare species which undergo reactions now generally ascribed to silaethenes.[58] By varying R (Scheme 8) the silaethene can be generated in solution, and trapped, at temperatures as low as −50 °C (R = Ts) to give products resulting from initial 2 + 2, 2 + 3, and 2 + 4 cycloadditions.[58a] In the absence of trapping agents, dimers are formed.[58] Furthermore, other more stable lithiated intermediates (*e.g.* R = Ph_2PO) can be isolated and subsequently thermolysed *in vacuo*, the silaethene being detected

[55] P. Fischer, R. Gräf, J. J. Stezowski, and J. Weidlein, *J. Amer. Chem. Soc.*, 1977, **99**, 6131.
[56] E. C. Taylor, J. G. Andrade, and A. McKillop, *J.C.S. Chem. Comm.*, 1977, 538.
[57] T. J. Barton and D. S. Banasiak, *J. Amer. Chem. Soc.*, 1977, **99**, 5199.
[58] (*a*) N. Wiberg and G. Preiner, *Angew. Chem. Internat. Edn.*, 1977, **16**, 328; (*b*) P. R. Jones and T. F. O. Lim, *J. Amer. Chem. Soc.*, 1977, **99**, 2013.

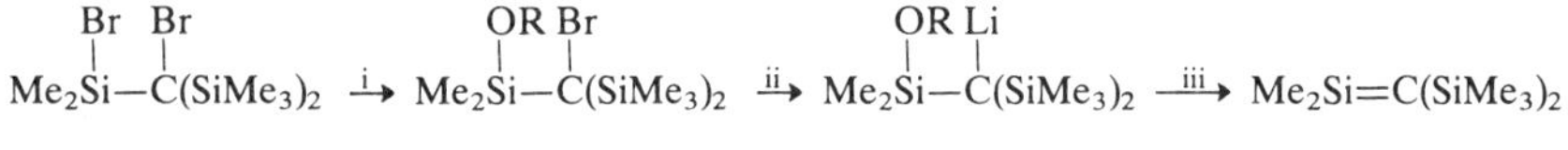

Reagents: i, AgOR; ii, Bu^nLi; iii, Δ

Scheme 8

directly by mass spectrometry.[58a] After last year's report of preparations of stable silacyclopropenes, there have now been several discussions of their reactions with methanol, and with ketones, alkenes, alkynes, dienes, and dimethylsilylene, all leading to insertion and ring expansion.[59a—e] Pd catalyses dimerization to disilacyclohexadienes.[59b,f] Sila-allene (25a) has been proposed as a product of the photolysis of (26) on the basis of isolation of (25b) following initial *in situ* cycloaddition with solvent acetone.[59a] A convenient and efficient synthesis of $Me_2Si{=}O$ results from abstraction of oxygen from DMSO by dimethylsilylene.[60] Intermediacy of the first oxasilacyclopropane (27) has been claimed, but as usual depends solely on product analyses.[61a] Novel intramolecular reactions of the carbene (28) have been reported.[61b]

(25) a; X = Si
b; X = C

(26)

(27)

(28)

In the near absence of experimental data on free sila-alkenes, the latest computational study of $H_2Si{=}CH_2$ offers several useful insights.[62] *Ab initio* calculations, which take into account electron correlation effects, have predicted $H_2Si{=}CH_2$ to be a singlet, planar species in its ground state, 28 kcal mol^{-1} below the first triplet. The difference in energy between the planar and orthogonal geometries, taken as a reflection of the π-bond strength, has the surprisingly high value of 46 kcal mol^{-1}, whereas the barrier to the exothermic dimerization is less than 14 kcal mol^{-1}. The polarity ($\delta-C{=}Si\ \delta+$) and low lying π^* level are responsible for the experimentally observed reactivity of such species. Moreover, the polarization causes the C $2p\pi$ and Si $3p\pi$ orbitals to be enlarged and contracted, respectively, thus vitiating the conventional view that 'mismatching' of these orbitals is responsible for the weakness of the π-bond. A cheerless prospect is that bulky substituents will be insufficient to prevent dimerization, and attempts to alter the polarity will only result in counterproductive weakening of the π-bond. A provocative note concerning *ab initio* calculations on *divalent* Si has predicted that $CH_2{=}Si$ is considerably more stable than $HC{\equiv}SiH$, although further results predict

59 (*a*) M. Ishikawa, T. Fuchikami, and M. Kumada, *J. Amer. Chem. Soc.*, 1977, **99**, 245; (*b*) H. Sakurai, Y. Kamiyama, and Y. Nakadaira, *ibid.*, 1977, **99**, 3879; (*c*) D. Seyferth, S. C. Vick, M. L. Shannon, T. F. O. Lim, and D. P. Duncan, *J. Organometallic Chem.*, 1977, **135**, C 37; (*d*) D. Seyferth and S. C. Vick, *ibid.*, 1977, **125**, C 11; (*e*) M. Ishikawa, K.-I. Nakagawa, and M. Kumada, *ibid.*, 1977, **131**, C 15; (*f*) M. Ishikawa, T. Fuchikami, and M. Kumada, *J.C.S. Chem. Comm.*, 1977, 352.

60 H. S. D. Soysa, H. Okinoshima, and W. P. Weber, *J. Organometallic Chem.*, 1977, **133**, C 17.

61 (*a*) W. Ando, M. Ikeno, and A. Sekiguchi, *J. Amer. Chem. Soc.*, 1977, **99**, 6447; (*b*) W. Ando and A. Sekiguchi, *J. Organometallic Chem.*, 1977, **133**, 219.

62 R. Ahlrichs and R. Heinzmann, *J. Amer. Chem. Soc.*, 1977, **99**, 7452.

$CH_2{=}SiH_2$ to be somewhat more stable than $H_2 + CH_2{=}Si$.[63] Nevertheless, it was suggested that it would be worth searching experimentally for unsaturated divalent organosilicon species.

The burgeoning synthetic chemistry based on vinylsilanes has prompted several new reports of syntheses of these species. Alkynylsilanes react with hydroborating agents leading to α,β-substituted vinylsilanes, useful for conversion to α-silaketones,[64a] carboxylic acids.[64b] and stereospecifically α-alkylated vinylsilanes,[64c] and with Cu-promoted Grignard reagents.[65] An alternative approach relies on the addition of vinyl-lithiums to Me_3SiCl.[66] An important reaction of vinylsilanes is their oxidation to epoxysilanes, themselves precursors to carbonyl compounds[67] and stereochemically defined hetero-substituted alkenes.[68] α-Chloroalkylsilanes may be lithiated, and react with ketones to give epoxysilanes directly. Since these may be cleaved to aldehyde, the overall procedure is one of reductive nucleophilic acylation.[67]

Because of decreased C—Cl polarity, α-chloroalkylsilanes undergo anomalous nucleophilic substitution reactions, both in solution (Scheme 9)[69a] and vapour–solid

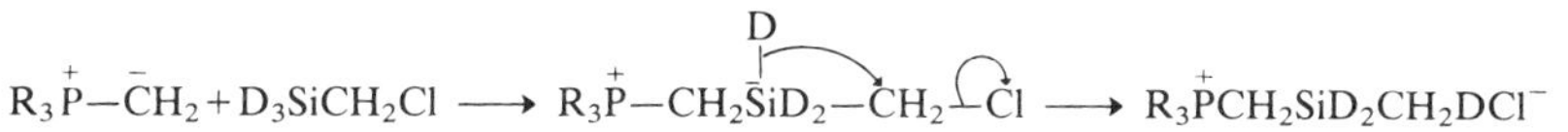

Scheme 9

heterogeneous phases.[69b] Another paper on substitution reactions at Si has pointed out that reagents which favour 1,2-addition to enones (hard bases, charge control) substitute at Si with retention of configuration, whereas others which add 1,4 to enones (soft bases, frontier orbital control) react with inversion at Si. This correlation supports the hypothesis that the electronic character of the nucleophile dictates the stereochemistry of nucleophilic substitution at Si.[70] A range of techniques has shown that both a concerted dyotropic process as well as a 2-step radical process occurs in the thermal rearrangement of (29) to (30),[71a] whereas silylmethyl acetates are believed to rearrange (alkyl migration) *via* an unprecedented inverted Si ylid (31),[71b] reminiscent of species frequently invoked in borate chemistry.

[63] J. N. Murrell, H. W. Kroto, and M. F. Guest, *J.C.S. Chem. Comm.*, 1977, 619.
[64] (*a*) A. Hassner and J. A. Soderquist, *J. Organometallic Chem.*, 1977, **131**, C 1; (*b*) G. Zwiefel and S. J. Backlund, *J. Amer. Chem. Soc.*, 1977, **99**, 3184; (*c*) K. Uchida, K. Utimoto, and H. Nozaki, *Tetrahedron*, 1977, **33**, 2987.
[65] (*a*) H. Westmijze, J. Meijer, and P. Vermeer, *Tetrahedron Letters*, 1977, 1823; (*b*) M. Obayashi, K. Utimoto, and H. Nozaki, *ibid.*, 1977, 1805.
[66] (*a*) R. T. Taylor, C. R. Degenhardt, M. P. Melega, and L. A. Paquette, *Tetrahedron Letters*, 1977, 159; (*b*) D. Seyferth, J. L. Lefferts, and R. L. Lambert, jun., *J. Organometallic Chem.*, 1977, **142**, 39.
[67] (*a*) C. Burford, F. Cooke, E. Ehlinger, and P. Magnus, *J. Amer. Chem. Soc.*, 1977, **99**, 4536; (*b*) F. Cooke and P. Magnus, *J.C.S. Chem. Comm.*, 1977, 513.
[68] P. F. Hudrlik, A. M. Hudrlik, R. J. Rona, R. N. Misra, and G. P. Withers, *J. Amer. Chem. Soc.*, 1977, **99**, 1993.
[69] (*a*) H. Schmidbaur and B. Zimmer-Gasser, *Angew. Chem. Internat. Edn.*, 1977, **16**, 639; (*b*) S. P. Hopper, M. J. Tremelling, R. J. Ginsberg, and P. C. Mendelowitz, *J. Organometallic Chem.*, 1977, **134**, 173.
[70] R. J. P. Corriu and C. Guerin, *J.C.S. Chem. Comm.*, 1977, 74.
[71] (*a*) M. T. Reetz, *Chem. Ber.*, 1977, **110**, 954, 965; (*b*) M. T. Reetz and N. Greif, *Angew. Chem. Internat. Edn.*, 1977, **16**, 712.

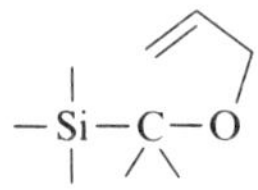
(29)

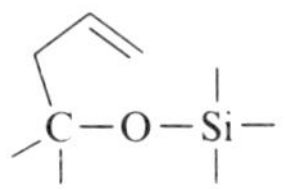
(30)

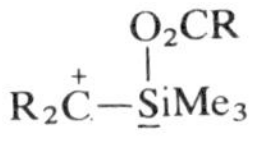
(31)

Further parametrization has permitted extension of empirical force field calculations to polysilanes,[72] and using this technique $Ph_3SiSiPh_3$ is calculated to adopt a D_3 conformation and to be strain free, unlike the unknown Ph_3CCPh_3.[72] $Ph_3SiOSiPh_3$ has a linear Si—O—Si unit in the crystalline phase, and in agreement with predictions of the above calculations, has S_6 symmetry (two staggered antichiral C_3 propellers).[73] Finally, attention is drawn to the remarkable observation that (32) shows Me group inequivalence in its ^1H n.m.r. spectrum up to at least 150 °C, whereas its C and Sn analogues have much lower barriers to substituent rotation.[74] What happens in the Ge and mixed derivatives?

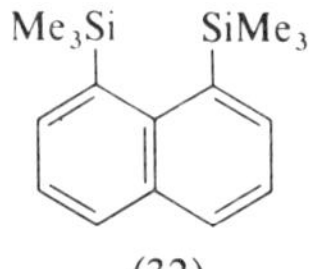
(32)

Germanium, Tin, and Lead.—The rearrangements of benzylideneaminotin (IV) halides (33) to (34) are the first examples of *ortho*-metallation involving Sn.[75] The full scope of the reaction has yet to be demonstrated, but it fails in analogous Si and Ge systems. The cycloperoxidation of bistriflates by means of $R_3SnO_2SnR_3$, has given a route to the bicyclic *endo*peroxide (35) (of relevance in the prostaglandin field) from (36).[76] Me_3SnLi is readily prepared and adds 1,4 to conjugated enones,

(33)

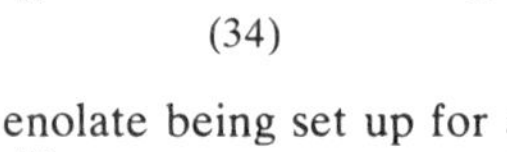
(34)

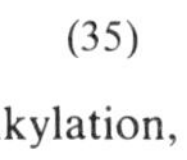
(35)

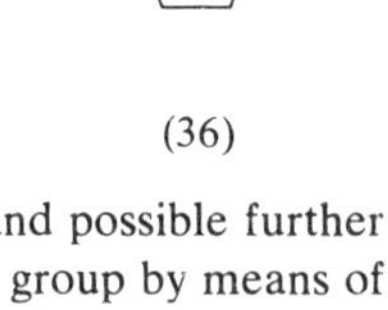
(36)

the resulting stannylated enolate being set up for alkylation, and possible further modification (Scheme 10).[77] The transposition of the carbonyl group by means of oxidative cleavage of the C—Sn bond is particularly noteworthy and demonstrates the superiority of this reagent over Me_3SiLi, which otherwise reacts analogously. The Sn-substituted ylid $Ph_3\overset{+}{P}\overset{-}{C}HCH_2SnMe_3$ undergoes the Wittig reaction with carbonyl compounds to give allyltrimethylstannanes, which are valuable precursors

[72] J. P. Hummel, J. Stackhouse, and K. Mislow, *Tetrahedron*, 1977, **33**, 1925; W. D. Hounshell, D. A. Dougherty, J. P. Hummel, and K. Mislow, *J. Amer. Chem. Soc.*, 1977, **99**, 1916.
[73] C. Glidewell and D. C. Liles, *J.C.S. Chem. Comm.*, 1977, 632.
[74] D. Seyferth and S. C. Vick, *J. Organometallic Chem.*, 1977, **141**, 173.
[75] B. Fitzsimmons, D. G. Othen, H. M. M. Shearer, K. Wade, and G. Whitehead, *J.C.S. Chem. Comm.*, 1977, 215.
[76] M. F. Salomon and R. G. Salomon, (*a*) *J. Amer. Chem. Soc.*, 1977, **99**, 3500; (*b*) *ibid.*, 1977, **99**, 3501.
[77] W. C. Still, *J. Amer. Chem. Soc.*, 1977, **99**, 4836.

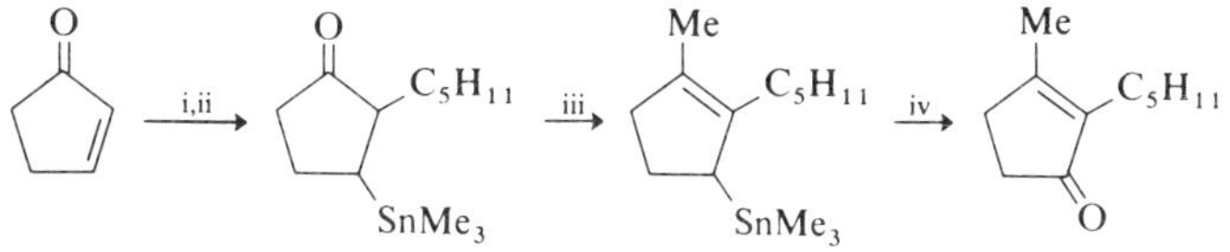

Reagents: i, Me_3SnLi; ii, $C_5H_{11}I$; iii, MeLi; iv, CrO_32py

Scheme 10

to difficultly accessible allyllithiums.[78] Moreover, the Sn compounds themselves are intrinsically useful in that they add to acyl chlorides with Rh catalysis giving good yields of allylketones.[79]

Hyperconjugation by CH_2MMe_3 (M = Si, Ge, Sn, or Pb) has been investigated using various n.m.r. probes in naphthyl[80a] and styryl[80b] systems. The overriding conclusion is that hyperconjugative electron-release in the neutral ground-state follows the order Pb ~ Sn > Ge ~ Si, in contrast to the commonly-accepted order Pb > Sn > Ge > Si based on studies of electron deficient compounds.[80a] Me_3M groups which are bound directly to an aromatic nucleus donate electron density by an inductive/field effect, but act as negative hyperconjugative substituents.[80b]

6 Group V

A synthetically useful series of reactions, based on lithiated alkylarsine oxides, and dependent on the thermal reaction of the halogenated arsorane, RPh_2AsX_2(X = Cl, Br, or I), for the critical cleavage of the R—As bond, is summarized in Scheme 11.[81a,b] Two-carbon homologation of alkyl halides may be effected indirectly, in

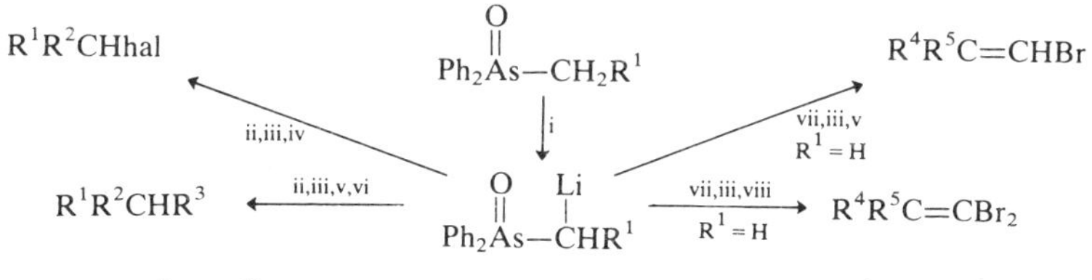

Reagents: i, $LiNPr^i_2$; ii, R^2X; iii, LAH; iv, hal_2 or SO_2Cl_2–Δ; v, Br_2–Δ; vi, R^3(= OH^- or PhS^-); vii, R^4R^5D=O; viii, excess Br_2–Δ.

Scheme 11

good yield, by addition of their alkyllithium derivatives to the vinyl group of diphenylvinylarsane, followed, as above, by bromine or sulphuryl chloride.[81c] The

[78] D. Seyferth, K. R. Wursthorn, and R. E. Mammarella, *J. Org. Chem.*, 1977, **42**, 3104.
[79] M. Kosugi, Y. Shimizu, and T. Migita, *J. Organometallic Chem.*, 1977, **129**, C 36.
[80] (*a*) W. Adcock, D. P. Cox, and W. Kitching, *J. Organometallic Chem.*, 1977, **133**, 393; (*b*) W. F. Reynolds, G. K. Hamer, and A. R. Bassindale, *J.C.S. Perkin II* 1977, 971.
[81] (*a*) T. Kauffmann, H. Fischer, and A. Woltermann, *Angew. Chem. Internat. Edn.*, 1977, **16**, 53; (*b*) T. Kauffmann, R. Joussen, and A. Woltermann, *ibid.* 1977, **16**, 709; (*c*) T. Kauffmann, H. Ahlers, H.-J. Tilhard, and A. Woltermann, *ibid.*, 1977, **16**, 710.

As-containing α-amino-acid analogue $Me_3\overset{+}{As}CH_2CO_2^-, H_2O$ has been isolated from the western rock lobster and fully characterized. Despite its emotive character, the authors say that its existence does not necessarily reflect environmental pollution.[82]

7 Electro-organic Chemistry

By R. LINES

Laboratory of Organic Chemistry, The Norwegian Institute of Technology, The University of Trondheim, N-7034 Trondheim, NTH, Norway

1 Introduction

The current state and growth of electrosynthetic chemistry together with an outline of the most promising reactions of recent years is the subject of a Russian review.[1] Another Russian paper[2] surveys current problems in the anodic fluorination of organic compounds. On the cathodic side, a useful review[3] written from a physical-organic standpoint, has appeared concerning the electroreduction of organic compounds both in aprotic and aqueous media.

2 Anodic Processes

The Anodic Oxidation of Carboxylates.—Alkyl benzenes are reported[4] to be obtained in good yield by the Kolbe electrolysis of 1-alkyl-1,4-dihydrobenzoic acids. Thus decarboxylation of (1) offers a useful synthetic route to the 1-alkyl naphthalenes.

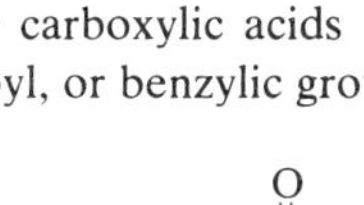

(1)

A systematic study[5] of the migratory aptitudes of the alkyl substituents (R^1 or R^2) in the oxidation of β-hydroxy carboxylic acids (2; $R^3 = H$) revealed preferential migration of olefinic, cyclopropyl, or benzylic groups.

$$\underset{(2)}{R^1R^2\overset{\overset{\text{OH}}{|}}{C}CHR^3CO_2H} \xrightarrow[-CO_2/-H_2O]{-2e} R^2\overset{\overset{\text{O}}{\|}}{C}CHR^1R^3 + R^1\overset{\overset{\text{O}}{\|}}{C}CHR^2R^3$$

Scheme 1

[1] M. Ya. Fioshin, *Soviet Electrochem.*, 1977, **13**, 1.
[2] I. N. Rozhkov, *Russ. Chem. Rev.*, 1977, **45**, 615.
[3] P. J. Elving, *Canad. J. Chem.*, 1977, **55**, 3392.
[4] J. Slobbe, *J.C.S. Chem. Comm.*, 1977, 82.
[5] T. Shono, J. Hayashi, H. Omoto, and Y. Matsumura, *Tetrahedron Letters*, 1977, **31**, 2667.

When $R^3 = Me$ however, the selectivity was much diminished reflecting the importance of the conformation of the intermediate carbenium ion. The results of the study were applied to the synthesis of *dl*-muscone (4) *via* the olefinic acid methyl ester (3) (Scheme 2).

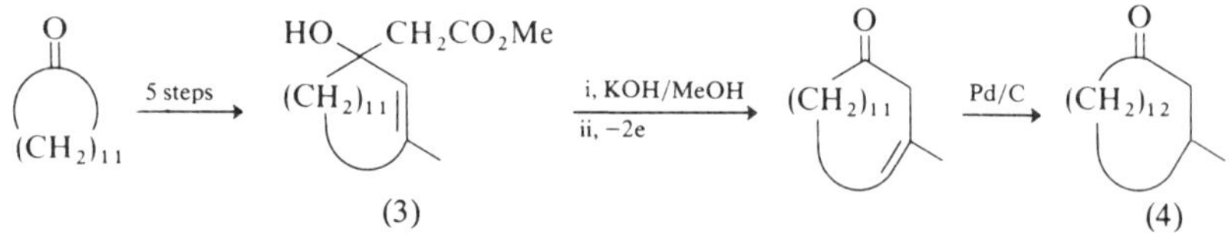

Scheme 2

The electrolysis of acylaminomalonic acid monoesters[6] in acetic acid solutions has been found to produce a good yield of the useful synthetic intermediates 2-acetoxy-2-amino acid derivatives. Similarly, the 3-acetoxy-3-amino acid derivatives were obtained from the oxidation of 3-alkoxycarbonylalanine derivatives in acetic acid–THF media.

Anodic oxidation of 4,4-diphenyl but-3-enoic acid[7] (5; R = H) in methanol results in a readily separable mixture of isomeric diphenyl allyl ethers (6, 7; $R^1 = H$, $R^2 = Me$). Similarly, electrolysis of the Stobbe half-ester (5; $R = CO_2Et$) in ethanol yields the corresponding esters (6, 7; $R^1 = CO_2Et$, $R^2 = Et$).

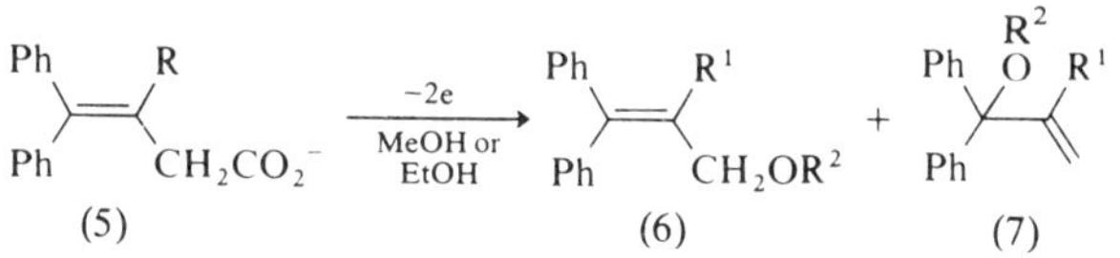

Scheme 3

The relative simplicity of the electrochemical route compared with the alternative chemical syntheses makes this an attractive method for the preparation of these compounds.

The Anodic Oxidation of Neutral Organic Compounds.—The anodic chemistry of the tropane alkaloids has been studied[8] in both acetonitrile and benzonitrile. Controlled potential electrolysis of tropane (8; R = Me) in acetonitrile results in the uptake of 0.79 F mol^{-1} to yield a mixture of (protonated) tropane (57.6%), nortropane (8; R = H, 13.6%) and the *N*-carboxaldehyde (8; R = CHO, 5.3%). Both nortropane and the *N*-carboxaldehyde are believed[8] to be derived from the iminium ion (9). The aldehyde is the major product in the presence of hydroxide.

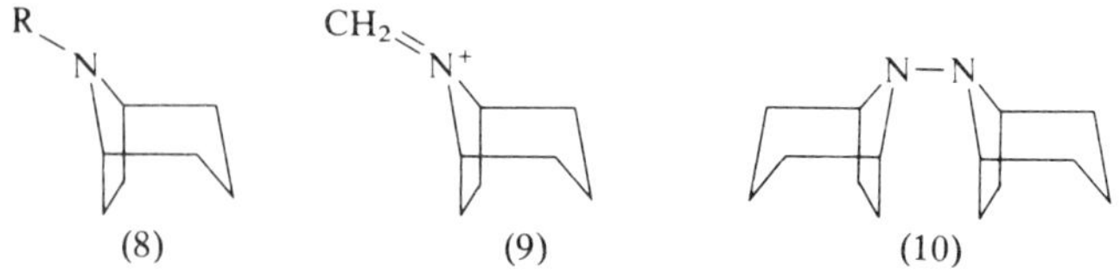

[6] T. Iwasaki, H. Horikawa, K. Matsumoto, and M. Miyoshi, *J. Org. Chem.*, 1977, **42**, 2419.
[7] F. M. Banda and R. Brettle, *J.C.S. Perkin I*, 1977, 1773.
[8] B. L. Laube, M. R. Asirvatham, and C. K. Mann, *J. Org. Chem.*, 1977, **42**, 670.

The oxidation of nortropane results in the formation of both C—N (8; R = CH_2CN) and N—N bonds (10).

The cyanide is thought to arise from generation of solvent radicals formed from the oxidation of hydroxide ions.

$$OH^- \xrightarrow[-e]{CH_3CN} \dot{C}H_2CN + H_2O$$

Scheme 4

This is supported by the observation that binortropane (10) is the major product in benzonitrile where solvent radical formation is prevented. The anodic cyanation of t-aliphatic and heterocyclic amines has been the subject of a systematic study.[9] Preparative electrolysis of the amines at Pt in aqueous methanolic sodium cyanide solution yielded α-cyano amines in reasonable yields (30—60%). The ease of cyanation was found to be in the order $(CH_2)_4 > (CH_2)_5 > Me > Et > Pr^n > Pr^i = 0$. This reactivity series was rationalized[9] by consideration of the geometry of the amine on the electrode surface. Deprotonation, the key step in the reaction, must occur at the α-carbon situated in a position easily accessible to the electrode. Obviously this is not the case for the isopropyl group. Similarly, the configuration of the adsorbed cyclic amines encourages substitution at the α-ring-carbon and for *N*-isopropylpiperidine ring substitution was the sole reaction.

The chemistry of the anodic oxidation of cyclohexene in the presence of cyanide ion turns out to be rather complex.[10] In methanolic sodium cyanide solution at Pt not only cyanation but also methoxylation and isocyanation were observed. The isocyanation reaction was highly potential dependent. It is interesting to note that only tarry products were formed when the electrolyses were performed in acetonitrile.

Anodic cyanation reactions carried out in emulsion systems are claimed to give superior results over the more conventional solvent supporting electrolyte combinations. Eberson and Helgée[11] now have demonstrated the synthetic utility of emulsion electrolysis for the preparation of 4-alkoxy-4-cyano-biphenyls; a class of liquid crystals. The electrochemical method using aqueous sodium cyanide, dichloromethane, and tetra n-butylammonium sulphate gave at least a four-fold yield increase over the alternative five-step chemical synthesis.

A new mode of oxidative coupling of substituted bibenzyls has been reported.[12] Anodic oxidation of (11; Z = O) was expected to yield the aporphine skeleton whereas in acetonitrile the bridged lactone (12) was isolated. The mechanism is thought to involve formation of a cation radical in ring *a* followed by proton loss from the 4-position of the isochromanone ring. Further oxidation of the resulting radical to the cation then serves as an electrophile towards the veratryl ring to give (12). In acidic solution (CH_2Cl–TFA) where proton loss is inhibited, the spirodienone (13; 33%) is formed.

A rather rare example of *ortho*-coupled biphenyls[13] results from the controlled potential electrolysis of methoxylated aryl benzoates. The biphenyls are obtained

[9] T. Chiba and Y. Takata, *J. Org. Chem.*, 1977, **42**, 2973.
[10] K. Yoshida, T. Kanbe, and T. Fueno, *J. Org. Chem.*, 1977, **42**, 2313.
[11] L. Eberson and B. Helgée, *Acta Chem. Scand.*, 1977, **B31**, 813.
[12] I. W. Elliott, jun., *J. Org. Chem.*, 1977, **42**, 1090.
[13] M. Sainsbury and J. Wyatt, *J. C. S. Perkin I*, 1977, 1750.

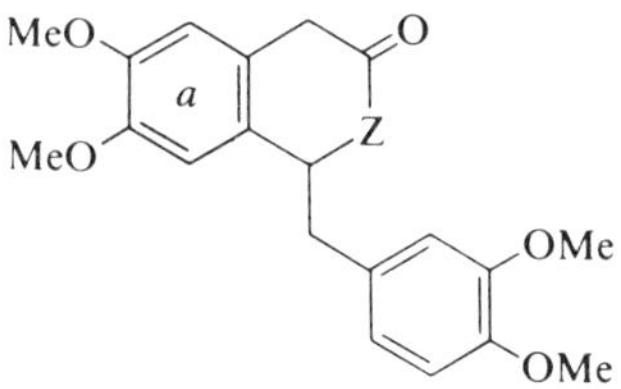

Z = O or NMe

(11)

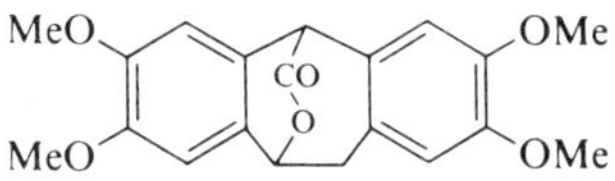

(12)

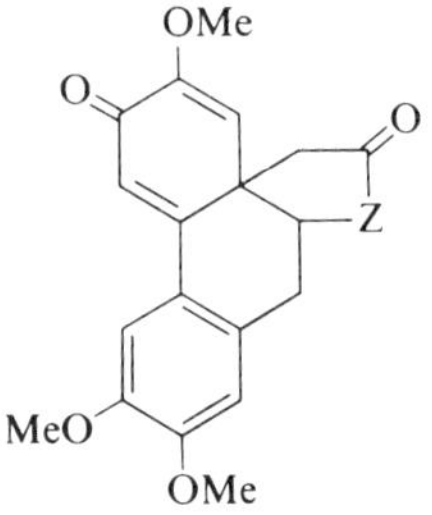

(13)

in high yield after the passage of 1F mol^{-1}. Further oxidation results in *ortho*- and *para*-quinones. An interesting detail here is that the products can be accounted for by a 2F mol^{-1} scheme whereas, experimentally, 3F mol^{-1} is needed for quinone formation.

A useful synthetic route to 5,5′-disubstituted hydruillic acids (15) is reported[14] from the controlled potential oxidation of 5-alkyl substituted barbituric acids (14) at a pyrolitic graphite electrode (Scheme 5).

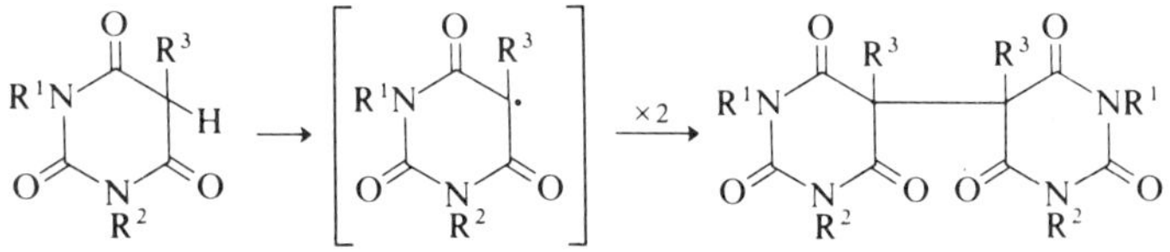

(14) (15)

R^1, R^2 = H or Me
R^3 = Me, Et, or $PhCH_2$

Scheme 5

Examination of the 1-methyl and 1,3-dimethyl barbituric acids over the pH range 0—13 revealed much more complex behaviour.[15] Formation of hydruillic acid (15; R^3 = H) remains the primary process but further oxidation at the 5-position and subsequent reaction with the parent molecule (or the anion at pH > pK_a) yields an open chain trimer which by further oxidation finally gives a cyclic trimer.

The first example of the electrochemical difunctionalization of saturated hydrocarbons has appeared.[16] Anodic oxidation of adamantane in TFA at the potential of the second wave gives, after workup, adamantane-1,3-diol (70%). Similarly, *NN′*-adamantane-1,3-diyl bisacetamide (58%) results from oxidation in acetonitrile. These two nucleophilic solvents are particularly suitable for the high anodic

[14] S. Kato and G. Dryhurst, *J. Electroanalyt. Chem. Interfacial Electrochem.*, 1977, **79**, 391.
[15] S. Kato and G. Dryhurst, *J. Electroanalyt. Chem. Interfacial Electrochem.*, 1977, **80**, 181.
[16] A. Bewick, G. J. Edwards, S. R. Jones, and J. M. Mellor, *J. C. S. Perkin I*, 1977, 1831.

potentials required. Extension of the method to non-bridged hydrocarbons, where concomitant olefin formation becomes important, may require even more efficient nucleophiles to trap the reaction intermediates.

A novel combination of light and electrochemistry has been used to effect the oxidation of organic compounds.[17] A catalytic amount of a quinone is irradiated in the presence of the substrate and a graphite anode at the potential of the quinone–hydroquinone couple. The photochemical reaction produces the oxidized substrate together with the hydroquinone which is electrochemically reconverted into the quinone.

Last year's report [*Ann. Reports* (*B*), 1976, **73**, 143] cited the slow intramolecular coupling of the dication of the [2,2]metacyclophane (16; R = OMe). In a voltammetric study of a series of alkylated [2,2]metacyclophanes, Sato and Kamada[18] suggest the involvement of a transannular cation radical of the 5,13-dimethyl derivative (17).

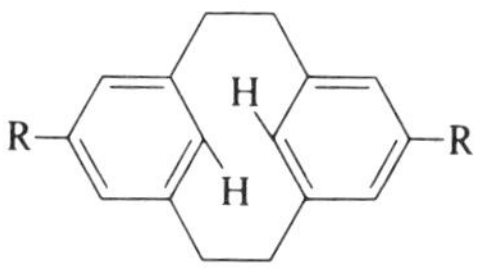

(16)

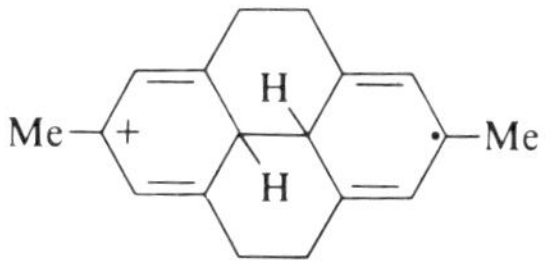

(17)

Preparative electrolyses of (16; R = H or Me) resulted[18] in successive oxidation deprotonation reactions to give ultimately the corresponding pyrenes.

The dications of the tetrathioethylenes (18, $n = 2$ or 3) generated in acetonitrile undergo a novel endocyclic to exocyclic rearrangement[19] (Scheme 6).

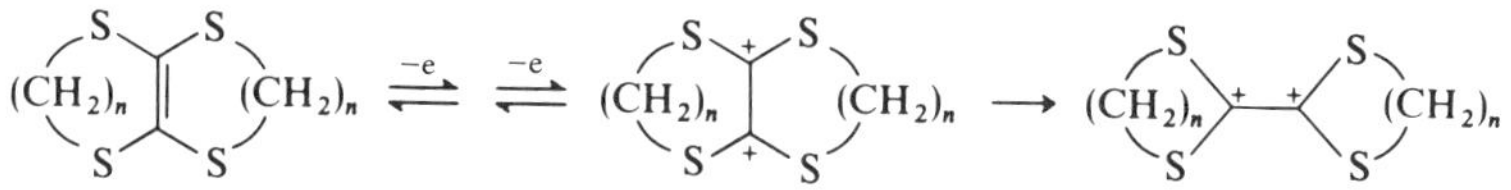

(18)

Scheme 6

Rotation about the central C—C bond of the product provides the driving force of the rearrangement by minimizing the coulombic repulsion between the positive sulphur atoms. The mechanism is suggested[19] to be a simultaneous intramolecular migration of two σ bonds: an example of a dynotropic rearrangement. This same rearrangement reaction characterizes the anodic chemistry of the *ortho*-thio-oxalates[20] of type (19) (Scheme 7). An interesting feature is that (20) is not the primary product of the rearrangement of the cation of (19). Both the pyrolysis and photolysis of (19) give (20) directly which suggests that the electrochemical fragmentation is directed by conformational and/or solvation effects.

[17] J. M. Bobbitt and J. P. Willis, *J. Org. Chem.*, 1977, **42**, 2347.
[18] T. Sato and M. Kamada, *J. C. S. Perkin II*, 1977, 384.
[19] R. M. Harnden, P. R. Moses, and J. Q. Chambers, *J. C. S. Chem. Comm.*, 1977, 11.
[20] P. R. Moses, R. M. Harnden, and J. Q. Chambers, *J. Electroanalyt. Chem. Interfacial Electrochem.*, 1977, **84**, 187.

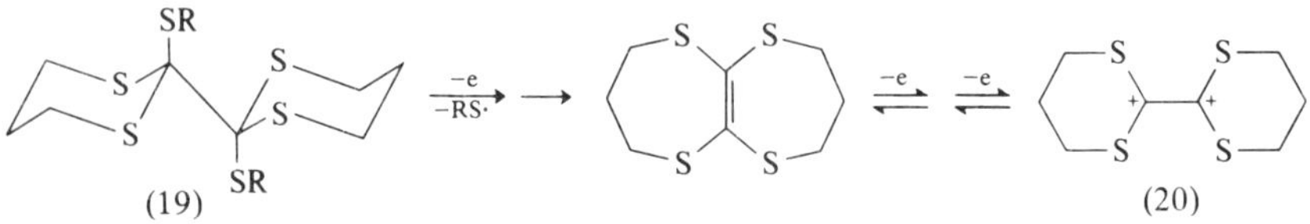

(19) (20)

Scheme 7

3 Cathodic Processes

The Cathodic Reduction of Organic Cations.—A new class of stable cation radicals has been prepared[21] by reduction of the diphosphiacyclohexadiene salts (21; R $\neq$ H). Further reduction yields the highly reactive diphosphabenzenes (22). The cation radicals are unusual in that they decompose by protonation in the presence of strong acids; a property no doubt arising from the strong carbanionic nature of the ring-carbons.

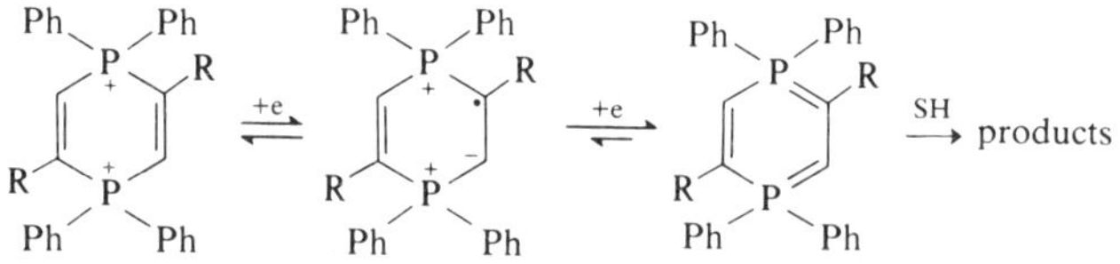

(21) (22)

Scheme 8

The Cathodic Reduction of Neutral Organic Compounds.—Electrogenerated nitrobenzene anion radical in DMF is reported by Wagenknecht[22] to react rapidly with alkyl halides to give high yields of *NO*-dialkylphenyl hydroxylamines. Another useful synthesis involves the electrochemical reductive acylation of electron-deficient olefins.[23] Reduction of ethyl cinnamate at a Hg cathode in DMF containing acetic anhydride gave ethyl 3-phenyl-4-oxopentanoate (75%) and ethyl 2-benzyl-3-oxobutanoate (2%). Anthracene, however, under similar conditions gives the vinyl ester (23) in 66–75% yield.[24]

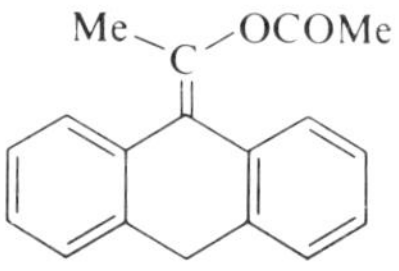

(23)

The reductive coupling of anthracene with 1,2-and 1,3-dihalides[25] has been shown by cyclic voltammetry to proceed catalytically. Coupling of the 1,3-dihalides

[21] R. D. Rieke, R. A. Copenhafer, C. K. White, A. Aguiar, J. C. Williams, jun., and M. S. Chattha, *J. Amer. Chem. Soc.*, 1977, **99**, 6656.
[22] J. H. Wagenknecht, *J. Org. Chem.*, 1977, **42**, 1836.
[23] H. Lund and C. Degrand, *Tetrahedron Letters*, 1977, **40**, 3593.
[24] H. Lund, *Acta Chem. Scand.*, 1977, **B31**, 424.
[25] E. Hobolth and H. Lund, *Acta Chem. Scand.*, 1977, **B31**, 395.

in the 1- or 2- positions of anthracene results in ring-closure to cyclopentanoanthracene derivatives (Scheme 9).

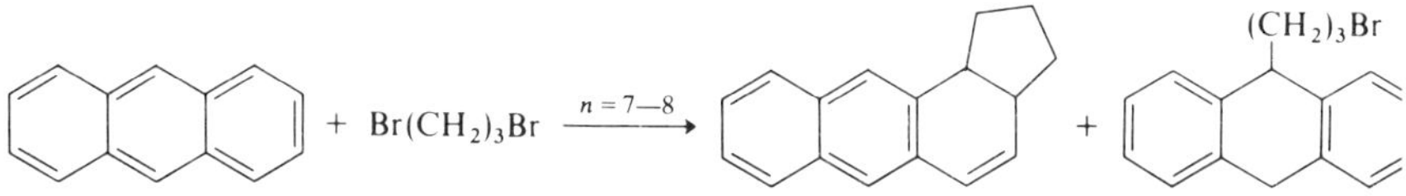

Scheme 9

The corresponding catalytic reductive alkylation of quinolines and isoquinolines with t-butyl halides leads to t-butylated heterocyclic compounds.[26] A large percentage of the alkylation takes place in the carbocyclic ring. Although many isomers are formed and the overall yield of each isomer is low (10—20%) the method has the advantage of experimental simplicity.

Electroreduction of halides continues to provide interesting chemistry. The tetra-bromo compound (24) exhibits two major voltammetric peaks.[27] Electrolysis at the first wave consumed 3F mol^{-1} and yielded 1,2-dibromobenzocyclobutene and polymers. It was deduced[27] that the parent compound (24) is an equilibrium mixture of two conformers and that each product could be ascribed to a single conformer. Reduction of (24) at the second wave gave the tetracyclic product (25), the result of a rearrangement of the Diels–Alder benzocyclobutadiene dimer (Scheme 10).

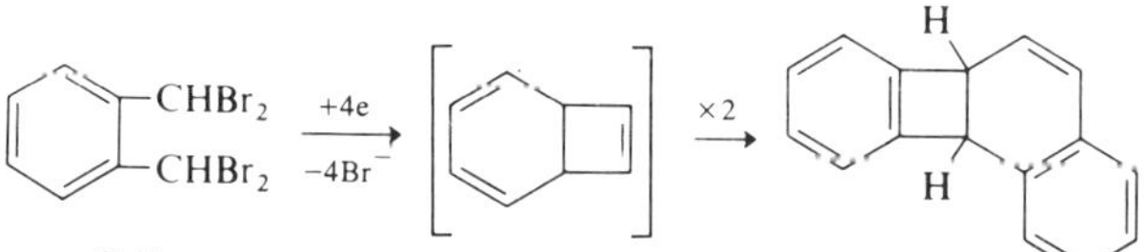

(24) (25)

Scheme 10

The as yet unreported dimethylene cyclopropanone (27) is cited [28] as an intermediate in the electrolysis of the α,α'-dibromoketone (26).

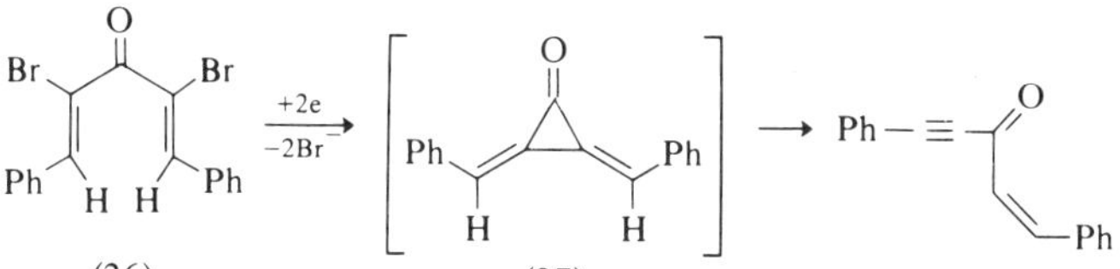

(26) (27)

Scheme 11

Electroreduction of the readily accessible 2,2,2-trichloroethanols[29] offers a direct route to 1,1-dichloro-olefins (Scheme 12). Care must be taken to ensure an acidic solution otherwise the alcohol is the sole product.

[26] C. Degrand and H. Lund, *Acta Chem. Scand.*, 1977, **B31**, 593.
[27] L. Rampazzo, A. Inesi, and R. M. Bettolo, *J. Electroanalyt. Chem. Interfacial Electrochem.*, 1977, **83**, 341.
[28] L. Rampazzo, A. Inesi, and A. Zeppa, *J. Electroanalyt. Chem. Interfacial Electrochem.*, 1977, **76**, 175.
[29] A. Merz, *Angew. Chem. Internat. Edn.*, 1977, **16**, 57.

$$R^1R^2C(OH)CCl_3 \xrightarrow{i} R^1R^2C{=}CCl_2 + R^1R^2C(OH)CHCl_2$$

Reagents: i, Hg cathode, 95% EtOH–Et_3NHCl

Scheme 12

A re-examination[30] of cathodic nucleophilic substitution of *p*-bromobenzophenone [*Ann. Reports* (*B*), 1974, **71**, 226] has led to the conclusion that the process proceeds exclusively *via* a bimolecular radical chain-mechanism. A.c. polarographic measurements as well as careful electrolytic experiments showed that the alternative exchange-current pathway could account for only 2% of the total products. The revised mechanism is shown in Scheme 13.

$$\underset{(28)}{PhCOC_6H_4Br} + e \rightleftharpoons \underset{(29)}{Ph\dot{C}\bar{O}C_6H_4Br} \quad (1)$$

$$Ph\dot{C}\bar{O}C_6H_4Br \rightleftharpoons PhCOC_6H_4^{\cdot} + Br^- \quad (2)$$

$$PhCOC_6H_4^{\cdot} + PhS^- \rightarrow \underset{(30)}{PhCOC_6H_4\dot{S}\bar{P}h} \quad (3)$$

$$(28) + (30) \rightarrow PhCOC_6H_4SPh + (29) \quad (4)$$

$$(30) \xrightarrow[+2H^+]{\times 2} [PhC(OH)C_6H_4SPh]_2 \quad (5)$$

Scheme 13

The same authors[31] have presented an efficient method for the electropinacolization of ketones. The process involves the use of tetra-alkylammonium salts in carefully dried acetonitrile. Under these aprotic conditions the tetra-alkylammonium ions are thought to form tight ion-pairs with the ketyl anion radicals and allow dimerization while inhibiting further charge transfer. It is argued,[31] by consideration of the mechanism given in Scheme 13, that the pinacol is essentially a termination product and therefore pinacolization occurs in bulk solution remote from the electrode surface.

Pinacol formation characterizes the primary step in the reductive coupling of 1,3-diketones.[32] The pinacol from 1,3-diphenyl propane-1,3-dione (31; R = H) is unstable in acidic solution and cyclizes to (32) and a tricyclotrioxanonane (33). Reduction of the corresponding substituted phenyl derivatives (31; R = halogen, Me, or OMe) leads to hexadienediones which can be further reduced to hexenediones. Compound (33) is thought to result from the *dl* form of the intermediate pinacol by a double intramolecular ketalization followed by dehydration.

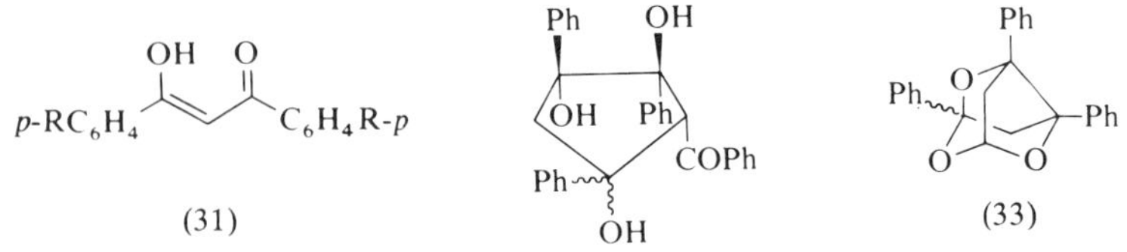

(31) (32) (33)

[30] W. J. M. van Tilborg, C. J. Smit, and J. J. Scheele, *Tetrahedron Letters*, 1977, **24**, 2113.
[31] W. J. M. van Tilborg and C. J. Smit, *Tetrahedron Letters*, 1977, **41**, 3651.
[32] A. J. Klein and D. H. Evans, *J. Org. Chem.*, 1977, **42**, 2560.

A new hydrogen anode[33] [*Ann. Reports* (*B*), 1974, **71**, 228] of composite construction has been evaluated and although more efficient than its predecessors the maximum current density and life-time are still limited and need to be improved for more widespread use.

An interesting example of conformational effects on fragmentation patterns has been noted in the cathodic cleavage of methyl *o*-alkyl substituted aryl sulphones.[34] Phenyl methyl sulphone is known to undergo exclusive methyl–sulphur cleavage but the introduction of *ortho*-substituents promotes aryl–sulphonyl fission and this becomes the major mode with the *ortho*-t-butyl substituted compound. Aryl–sulphonyl cleavage is characteristic of the cyclic aryl sulphones and comparison with molecular models suggested[34] that the conformation of the *ortho*-substituted sulphones approached that of the cyclic compounds.

The electrochemical behaviour of DL-α-lipoic acid[35] (a co-factor for a number of enzyme catalysed reactions) has been studied in order to gain some insight into its biological redox mechanism. The oxidized acid (34) cannot be reduced in water but undergoes direct reduction to the dianion in acetonitrile at -1.92 V (SCE)—an extremely negative potential. The acid is, however, readily reduced by electrochemically generated hydrogen which strongly suggests that reduction of the disulphide linkage occurs by an atom transfer-mechanism.

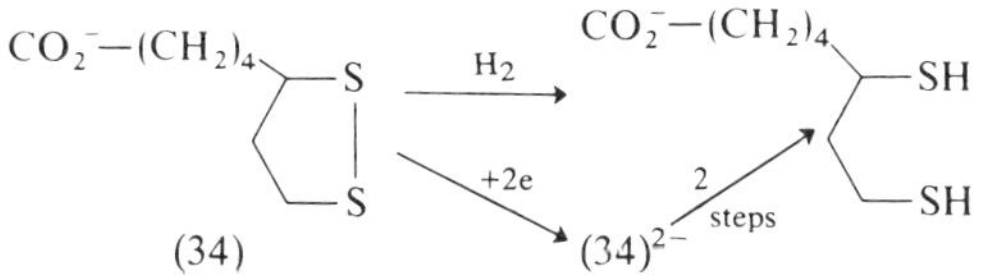

Scheme 14

Martigny and Simonet[36] have devised an autocatalysed cathodic elimination using α,α'-disubstituted 1,2-diphenyl ethanes. Reduction of (35) in DMF gives a product (*trans*-stilbene) which is more easily reduced than the parent. The relatively stable stilbene anion radical then undergoes a homogeneous electron-exchange with the parent molecule. Voltammetric evidence for this process includes a dramatic decay of *Ep* when the sweep rate is decreased. At very slow scans the *Ep* of stilbene is reached.

PhCH(SPh)CH(OAc)Ph

(35)

Labelling experiments often reveal new features of seemingly straightforward reactions. One example is the reduction of phenacyl chloride[37] in DMF–1% D_2O. Recovery of the starting material after short reaction-times reveals not only the reduction product (acetophenone) but also considerable H–D exchange of the methylene hydrogens of the starting material.

[33] C. P. Andrieux, J. M. Dumas-Bouchiai, and J. M. Saveant *J. Electroanalyt. Chem. Interfacial Electrochem.*, 1977, **83**, 355.
[34] B. Lamm and K. Ankner, *Acta Chem. Scand.*, 1977, **B31**, 375.
[35] J. K. Howie, J. J. Houts, and D. T. Sawyer, *J. Amer. Chem. Soc.*, 1977, **99**, 6323.
[36] P. Martigny and J. Simonet, *J. Electroanalyt. Chem. Interfacial Electrochem.*, 1977, **81**, 407.
[37] A. F. Diaz, Y. Y. Cheng, and M. Ochoa, *J. Amer. Chem. Soc.*, 1977, **99**, 6319.

Interestingly, the fraction of phenacyl chloride converted into acetophenone is always less than the fraction of D incorporated into the methylene hydrogens and that the exchange is in excess of the Faraday consumption. A suggestion[37] is that the exchange proceeds (at potentials $\geqslant Ep$ of phenacyl chloride) *via* an anion radical which can undergo several H/D exchanges per electron transfer.

Falsig and Iversen[38] have scaled up the electrolytic reduction of benzotriazole to provide preparative quantities of 2-aminophenyl hydrazine dihydrochloride in 80—85% yields. Treatment of the product with *ortho*-esters represents a good synthetic route to the 3-substituted benzo-1,2,4-triazines.

Evidence has been presented[39] for the intermediacy of a carbene anion radical during the electrolysis of diphenyldiazomethane. The three isolated products diphenyl methane, benzophenoneazine, and diphenylmethylamine are believed to arise by a radical-chain process involving the diphenyl carbene radical anion. The observed properties of this species indicates that it behaves primarily as a radical.

Although the cation radical of chlorophyll-*a* is stable in aprotic solvents, preparation of stable solutions of the corresponding anionic species has proved elusive. A report has now appeared[40] that in DMF which has been treated with active alumina both the anion radical and dianion of chlorophyll-*a* are quite stable. Chemiluminescence has also been observed[40] from the reaction of the radical anion with oxygen.

The electroreduction of phenylglyoxylic acid oxime in the presence of strychnine leads to opically active phenyl glycine,[41] the absolute configuration of which depends on the cathode potential. Thus 17.1% of the *R* isomer is formed at -0.95 V (SCE) whereas at -1.35 V an enantiomeric excess (11.1%) of the *S* isomer is obtained. Although many different parameters are involved the main factor is probably the protonation of the intermediate chiral carbanion (36) or (37). The product arising as a competition between inversion of the carbanion and protonation in the initial configuration at the electrode surface.

$Ph\bar{C}CO_2H$ (with NHOH on the anionic carbon)

(36)

$Ph\bar{C}CO_2H$ (with NH_2 on the anionic carbon)

(37)

4 Miscellaneous

The use of minute working electrodes has made possible voltammetric studies in both benzene and chlorobenzene.[42] Chlorobenzene in particular appears to be an excellent solvent for the study of the redox behaviour of aromatic compounds. In benzene, the very strong ion-pairing effect allows the reversible formation of the trianion radical of 1,2-bis-(9-anthryl) ethane. The potential (*ca.* -2.8 V SCE) required to generate this species is the most negative reversible reduction-potential yet reported for an aromatic system. In contrast, a report has now appeared[43]

[38] M. Falsig and P. E. Iversen, *Acta Chem. Scand.*, 1977, **B31**, 15.
[39] R. N. McDonald, J. R. January, K. J. Borhani, and M. D. Hawley, *J. Amer. Chem. Soc.*, 1977, **99**, 1268.
[40] T. Saji and A. J. Bard, *J. Amer. Chem. Soc.*, 1977, **99**, 2235.
[41] M. Tubault, E. Raoult, J. Armand, and L. Boulares, *J. C. S. Chem. Comm.*, 1977, 250.
[42] R. Lines and V. D. Parker, *Acta Chem. Scand.*, 1977, **B31**, 369.
[43] M. Horner and S. Hünig, *Angew. Chem. Internat. Edn.*, 1977, **16**, 410.

concerning π-system (38) which can exist in five oxidation states separated by four reversible one-electron transfers.

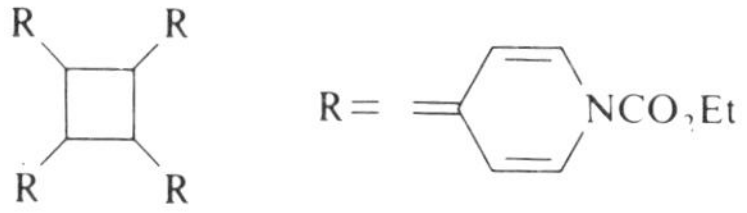

(38)

Other reports have described similar behaviour for two cluster compounds: a tetrameric cyclopentadienyl iron sulphide[44] and a new Pt tetrathiolene cluster.[45]

Oxydipropionitrile (dielectric constant = 60), commonly used as a g.l.c. phase, shows promise as an electrochemical solvent.[46] All the usual supporting electrolytes are soluble and at Pt using $LiClO_4$ the potential range is +2.62 to −3.9 V (SCE). The Ag/Ag^+ reference electrode is stable in this solvent.

An interesting application of cyclic voltammetry to conformational analysis has appeared.[47] The technique requires that the conformers be separated by significant activation barriers as well as oxidizing (or reducing) at substantially different rates. Excellent agreement was obtained between the results of low-temperature cyclic voltammetry of cyclic tetra-alkylhydrazines with the known conformational data of these compounds. The authors[47] point out that for reproducible results a consistent electrode surface is required.

Diaz[48] has noted some differences in stability of some surface bonded pyrazoline derivatives (39) and (40). Both compounds undergo reversible one-electron transfer in acetonitrile (X = OMe) but when bonded to the electrode surface (SnO_2 *via* the silylanyl derivative) the charge transfer process becomes irreversible for (40) but not for (39). The conclusion is that the array of cation radicals bonded to the surface are less stable than in solution but that the structure of the tricyclic derivative (39) inhibits decomposition.

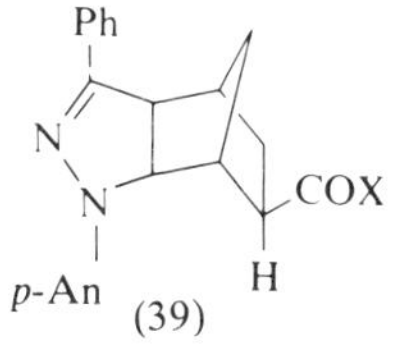

(39)

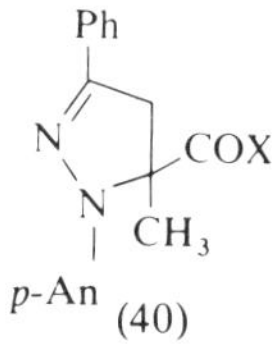

(40)

Strong surface interactions with ions in the electrolyte phase characterize the behaviour of the anisotropic metallic conductor polymeric sulphur nitride, $(SN)_x$, when used as an electrode.[49] The electrodes also exhibit unusual heterogeneous electrode process properties. The authors[49] suggest that such electrodes might be more suitable for modification (chemical or electrochemical) for the fabrication of selective catalysts than, for example, SnO_2 or graphite electrodes.

[44] Trinh-Toan, Boon-Keng Teo, J. A. Ferguson, T. J. Meyer, and L. F. Dahl, *J. Amer. Chem. Soc.*, 1977, **99**, 408.

[45] Boon-Keng Teo, F. Wudl, J. H. Marshall, and A. Kruger, *J. Amer. Chem. Soc.*, 1977, **99**, 2349.

[46] J. Y. Gal and M. Persin, *J. Electroanalyt. Chem. Interfacial Electrochem.*, 1977, **77**, 361.

[47] S. F. Nelsen, L. Echegoyen, E. L. Clennan, D. H. Evans, and D. A. Corrigan, *J. Amer. Chem. Soc.*, 1977, **99**, 1130.

[48] A. Diaz, *J. Amer. Chem. Soc.*, 1977, **99**, 5838.

[49] R. J. Nowak, H. B. Mark, jun., A. G. MacDiarmid, and D. Weber, *J. C. S. Chem. Comm.*, 1977, 9.

8 Photochemistry

By H. A. J. CARLESS

Department of Chemistry, Birkbeck College, Malet Street, London WC1E 7HX

The past year has seen an increased emphasis on the use of photochemical methods for organic synthesis.

The well-known *cis–trans* isomerization of alkenes continues to provide results of interest. Direct irradiation (185 nm) of cyclohexene or cycloheptene in pentane solution produces the highly strained *trans*-isomers, which can be trapped by acidic methanol to form the corresponding methoxycycloalkanes.[1] In this way, *trans*-cycloheptene was found to be a reasonably long-lived species at −10 °C, with a lifetime of 23 minutes. A transient u.v. absorption spectrum produced on irradiation (248 nm) of 1-phenyl-cycloheptene solutions can be assigned to a twisted, *trans*-like isomer of 1-phenylcycloheptene.[2] The photochemical addition of alcohols to cycloalkenes, proceeding *via* the *trans*-intermediate, is generally non-stereospecific. However, irradiation[3] of the benzocycloheptadienone (1) in deuteriated methanol or acetic acid leads to the product (2) of a regio- and stereo-specific *syn*-addition to the *trans*-cycloheptadienone intermediate. It has proved possible to prepare a *trans*-doubly bridged ethylene (3) by xylene-sensitized photo-isomerization of the *cis*-precursor.[4] This photochemical route to (3)

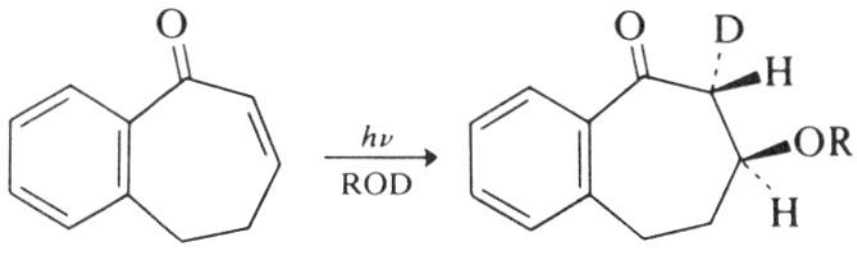

(1) (2) a; R = Me
b; R = MeCO

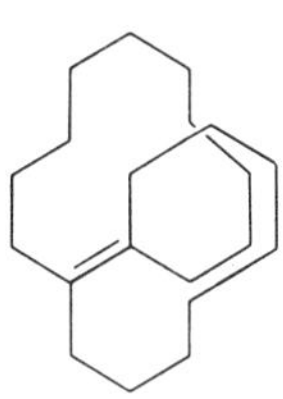

(3)

[1] Y. Inoue, S. Takamuku, and H. Sakurai, *J.C.S. Perkin II*, 1977, 1635.
[2] R. Bonneau, J. Joussot-Dubien, J. Yarwood, and J. Pereyre, *Tetrahedron Letters*, 1977, 235.
[3] E. Dunkelblum and H. Hart, *J. Amer. Chem. Soc.*, 1977, **99**, 644.
[4] M. Nakazaki, K. Yamamoto, and J. Yanagi, *J.C.S. Chem. Comm.*, 1977, 346.

compares very favourably with the non-photochemical approach to such compounds recently reported by Marshall,[5] who would classify compound (3) as a [10,8]betweenanene.

Metathetic reactions of alkenes have recently become topical, and these can now be induced in a route which involves photochemical addition followed by pyrolysis. Two groups have investigated this approach to medium-ring carbocycles.[6,7] For example, [2+2] photocycloaddition of the methyl cyclobutenecarboxylate (4) to cyclopentene gives (5) which can be cleaved on thermolysis to the cyclononadiene ester (7), or under less vigorous conditions to the divinylcyclopentane (6).[6] The reaction of a cyclobutene with a cyclohexenone may

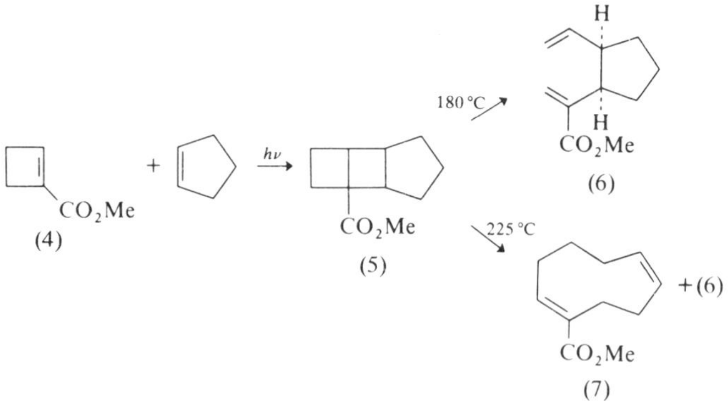

likewise provide a useful route to the ten-membered ring of germacradienes.[6,7] Photochemical cleavage of cyclobutanes is also possible, and Kaupp[8] has summarized the photolysis of substituted cyclobutanes in terms of a '*cis*-effect', in which there is a preference for breaking of the cyclobutane bond linked to the chromophore and also to the substituent which interacts most strongly with the chromophore.

Simultaneous publications by two groups[9,10] discuss the photochemical rearrangement of 3-vinylcyclopropenes to cyclopentadienes in high chemical yield. For example, (8) produces the cyclopentadienes (9) and (10) in the ratio 3:1.[9] The

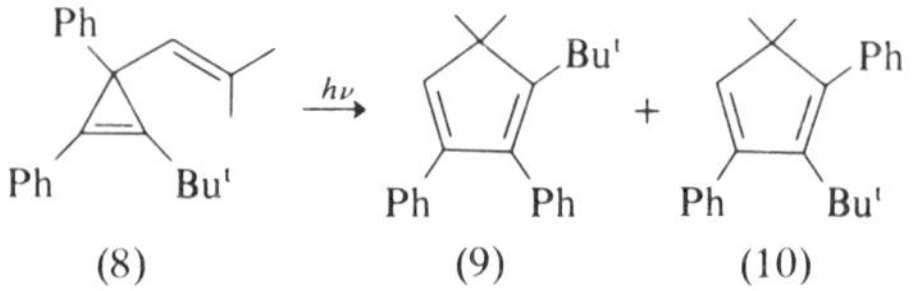

structure of the major product is more in accord with a biradical (11) pathway rather than a carbene (12) route. In confusing contrast, 3-allyl substituted cyclo-

[5] J. A. Marshall and M. Lewellyn, *J. Amer. Chem. Soc.*, 1977, **99**, 3508.
[6] P. A. Wender and J. C. Lechleiter, *J. Amer. Chem. Soc.*, 1977, **99**, 267.
[7] G. L. Lange, M.-A. Huggins, and E. Neidert, *Tetrahedron Letters*, 1976, 4409.
[8] G. Kaupp and M. Stark. *Chem. Ber.*, 1977, **110**, 3084.
[9] H. E. Zimmerman and S. M. Aasen, *J. Amer. Chem. Soc.*, 1977, **99**, 2342.
[10] A. Padwa, T. J. Blacklock, D. Getman, and N. Hatanaka, *J. Amer. Chem. Soc.*, 1977, **99**, 2344.

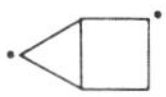

(11)

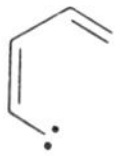

(12)

propenes do seem to undergo photorearrangement *via* a vinyl carbene intermediate.[11] The carbene gives cycloaddition to the allyl double bond, as is also found in the photochemical reactions of allyl azirines (see ref. 33).

Kropp *et al.*[12] have used the technique of photolysis of alkyl iodides at 254 nm as a convenient means for the generation of carbocations, such as the 1-norbornyl cation from 1-iodonorbornane.

There has been a long-standing dispute as to whether the singlet excited state of alkanones possesses reactivity towards hydrogen abstraction comparable with that of the triplet state. Henne and Fischer[13] have determined the yields of various deuteriated pinacols formed in the photoreaction of deuteriated acetone–propan-2-ol mixtures, and the multiplicities of the states which give rise to the pinacols. They conclude that singlet excited acetone has a rate constant for hydrogen abstraction at least as great as that of the acetone triplet. The second excited singlet ($\pi\pi^*$) state of adamantanethione is apparently a very indiscriminate abstractor of hydrogen from alkanes, since it shows only a small kinetic isotope-effect in abstraction and has little preference for abstraction from tertiary rather than primary C—H bonds.[14]

A variety of methods is being used to detect and intercept the 1,4-biradicals believed to be involved in the Norrish Type II photochemical reactions of dialkyl and alkyl aryl ketones. Thus, laser photolysis (347.1 nm) of 5-methylhexan-2-one produces transient absorption spectra (lifetime *ca.* 100 ns) assigned to the 1,4-biradical (13).[15] The paraquat dication (1,1′-dimethyl-4,4′-bipyridylium) is able to trap such biradicals, yielding the easily detected paraquat radical cation whose rate of formation allows the first estimate of the lifetime of the γ-methylvalerophenone biradical (14) ($\tau = 97 \pm 15$ ns) in methanol at room temperature.[16] The spin-trap di-t-butyl selenoketone efficiently intercepts 1,4-biradicals such as (14) which yields $PhCOCH_2CH_2C(Me)_2SeC(Bu^t)_2H$ as the final product.[17]

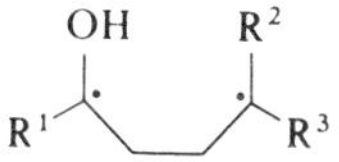

(13) $R^1 = R^2 = R^3 = Me$
(14) $R^1 = Ph, R^2 = R^3 = Me$

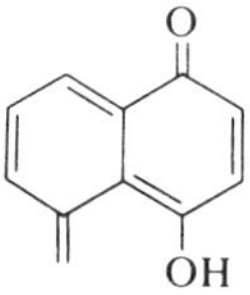

(15)

Irradiation of 5-methyl-1,4-naphthoquinone yields the blue photo-enol (15), which is stable at 77 K, and provides the first unambiguous example of formation of

[11] A. Padwa and T. J. Blacklock, *J. Amer. Chem. Soc.*, 1977, **99**, 2345.
[12] P. J. Kropp, G. S. Poindexter, N. J. Pienta, and D. C. Hamilton, *J. Amer. Chem. Soc.*, 1976, **98**, 8135.
[13] A. Henne and H. Fischer, *J. Amer. Chem. Soc.*, 1977, **99**, 300.
[14] K. Y. Law, P. de Mayo, and S. K. Wong, *J. Amer. Chem. Soc.*, 1977, **99**, 5813.
[15] R. D. Small and J. C. Scaiano, *Chem. Phys. Letters*, 1977, **50**, 431.
[16] R. D. Small and J. C. Scaiano, *J. Phys. Chem.*, 1977, **81**, 828.
[17] J. C. Scaiano, *J. Amer. Chem. Soc.*, 1977, **99**, 1494.

a (Z)-dienol in a photo-enolization reaction.[18] The Norrish Type I reaction of cyclic ketones can produce unsaturated aldehydes, and this route has been used in the synthesis of some prostanoids by photolysis of the appropriately substituted bicyclo[2,2,1]heptan-2-ones,[19] and in production of 5-carbomethoxyhex-4-en-1-al for terpene synthesis from 2-carbomethoxy-2-methylcyclopentanone.[20] A notably short stereoselective synthesis of the insect sex-attractant *exo*-brevicomin (18) relies on the selective excitation (using 1-methylnaphthalene as combined sensitizer and triplet quencher) of the carbonyl group of the dihydropyranyl ketone (16), followed by hydrogenation of the resulting bicyclic compound (17).[21]

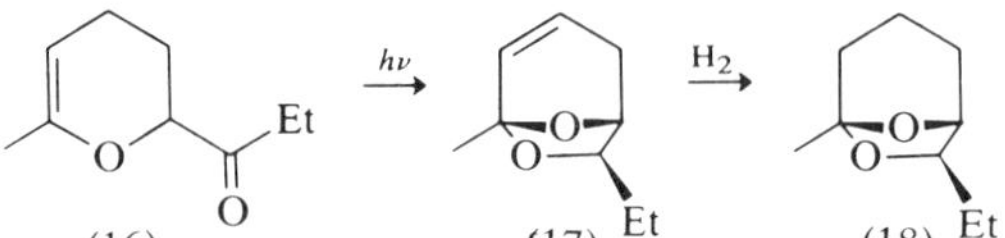

(16) (17) (18)

The absorption of visible light by a dye photosensitizer in the presence of ground-state oxygen ($^3O_2$) is often used to generate singlet oxygen ($^1O_2$), and the importance of $^1O_2$ in synthetic chemistry is increasing rapidly. Singlet oxygen generally gives a 1,4-addition to *cisoid* conjugated dienes, producing *endo*-peroxides which can easily rearrange to *syn*-1,3-diepoxides on heating. The use of this sequence on the annulene (19) can produce an intermediate (20), which on repetition of these steps yields the remarkable naphthalene pentoxide (21).[22] The 1,4-cycloaddition of $^1O_2$ may even occur when one of the diene double bonds forms

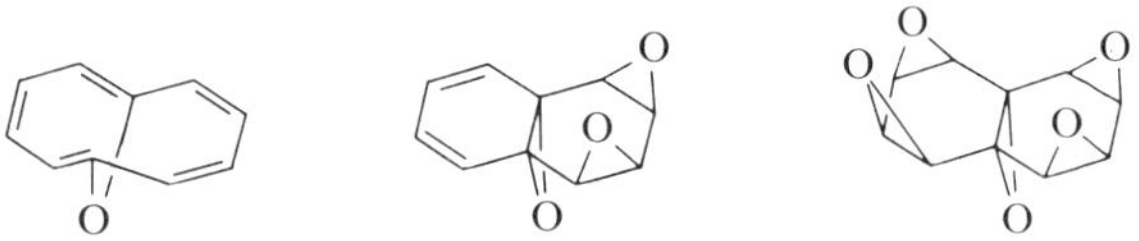

(19) (20) (21)

part of an aromatic ring, as in styrenes. An ingenious synthesis of the anti-tumour agent (±)-crotepoxide (24) relies on photo-oxidation of the styrene derivative (22) to produce a bis(*endo*-peroxide) (23) which can be rearranged and converted into crotepoxide.[23]

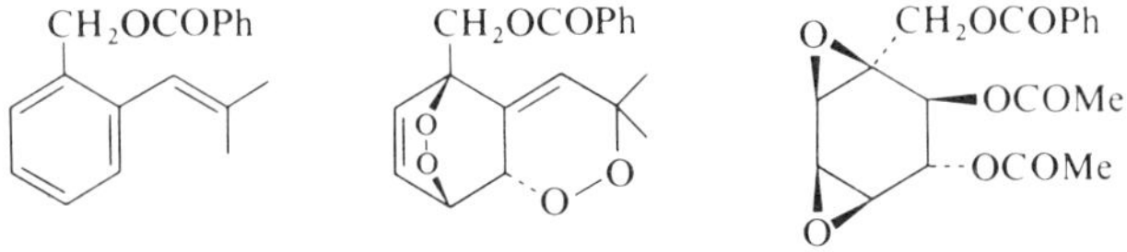

(22) (23) (24)

[18] E. Rommel and J. Wirz, *Helv. Chim. Acta*, 1977, **60**, 38.
[19] N. M. Crossland, S. M. Roberts, and R. F. Newton, *J.C.S. Chem Comm.*, 1977, 886.
[20] G. Bidan, J. Kossanyi, V. Meyer, and J.-P. Morizur, *Tetrahedron*, 1977, **33**, 2193.
[21] P. Chaquin, J.-P. Morizur, and J. Kossanyi, *J. Amer. Chem. Soc.*, 1977, **99**, 903.
[22] E. Vogel, A. Breuer, C.-D. Sommerfeld, R. E. Davis, and L.-K. Liu, *Angew. Chem. Internat. Edn.*, 1977, **16**, 169.
[23] M. Matsumoto, S. Dobashi, and K. Kuroda, *Tetrahedron Letters*, 1977, 3361.

The immediate biological precursors of the prostaglandins are *endo*-peroxides having the 2,3-dioxabicyclo[2,2,1]heptane skeleton. One approach to the synthesis of such peroxides involves the 1,4-cycloaddition of $^1O_2$ to cyclopentadienes, such as 1,4-diphenylcyclopentadiene which yields the peroxide (25).[24] It has now been

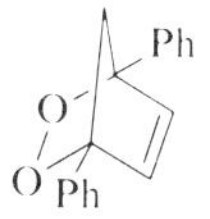

(25)

found that the double bond of (25) can be selectively reduced by di-imide without affecting the peroxidic link, a process which should increase the synthetic uses of cyclic peroxides like (25). Thus, this reaction has made possible a simple synthesis of the previously unknown 2,3-dioxabicyclo[2,2,2]octane from cyclohexa-1,3-diene.[25]

There is a rapidly growing interest in 1,2-dioxetans and their chemistry,[26] and photo-oxidation of hindered alkenes by singlet oxygen presents a good method for their synthesis. In fact, the preparation of the dioxetan (26) by sensitized photo-oxidation of bis(adamantylidene) has become a routine experiment.[27] Thermal

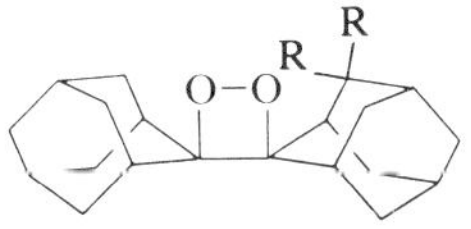

(26) R = R = H

(27) R—R = 

decomposition of such dioxetans has been used because it provides a chemical route to electronically excited species; ring-fission yields excited carbonyl compounds with the subsequent emission of chemiluminescent light. Wynberg and his group[28] have prepared the stable and optically active dioxetan (27) by photo-oxidation of the appropriate optically active alkene, and they[29] have observed that the thermal decomposition of this dioxetan (27) produces chemiluminescence which is circularly polarized.

There has been dispute as to whether dioxetan formation from $^1O_2$ attack on an alkene occurs concertedly, or *via* a perepoxide intermediate as calculated by Dewar *et al.*[30] The findings of McCapra and Beheshti[31] are especially relevant to

[24] D. J. Coughlin and R. G. Salomon, *J. Amer. Chem. Soc.*, 1977, **99**, 655.
[25] W. Adam and H. J. Eggelte, *Angew. Chem. Internat. Edn.*, 1977, **16**, 713.
[26] W. Adam, *Advances in Heterocyclic Chem.*, 1977, **21**, 437.
[27] 'Organic Photochemical Syntheses', Vol. 2, ed. R. Srinivasan, Wiley-Interscience, New York, 1976, p. 10.
[28] H. Wynberg and H. Numan, *J. Amer. Chem. Soc.*, 1977, **99**, 603.
[29] H. Wynberg, H. Numan, and H. P. J. M. Dekkers, *J. Amer. Chem. Soc.*, 1977, **99**, 3870.
[30] M. J. S. Dewar, A. C. Griffin, W. Thiel, and I. J. Turchi, *J. Amer. Chem. Soc.*, 1975, **97**, 4439.
[31] F. McCapra and I. Beheshti, *J.C.S. Chem. Comm.*, 1977, 517.

this question, because $^1O_2$ reacts with camphenylidene-adamantane to give the rearranged 1,2-dioxolan (28) alongside the expected dioxetan. The formation of (28) becomes most easily understood in terms of the rearrangement of a dipolar intermediate, such as the perepoxide (29). Turro, Adam, and co-workers[32] have

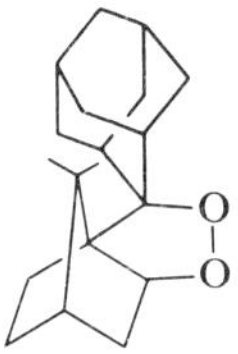
(28)

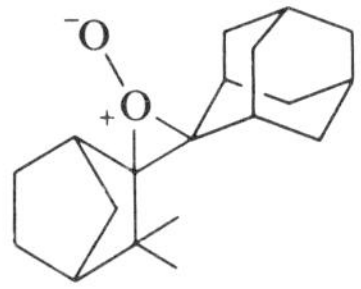
(29)

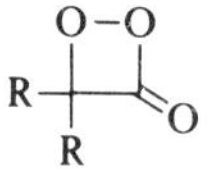
(30) a; R = Me
b; R = Ph
c; R = CF_3

now succeeded in isolating α-peroxylactones such as (30) from reaction of $^1O_2$ with the double bond of ketenes. The dioxetanone (30a) is stable enough to withstand distillation below room temperature, but decarboxylates with intense chemiluminescence on warming.

Padwa has continued his studies on the photochemical generation of nitrile ylides from azirines by examining the intramolecular cycloaddition of the ylide to a double bond. Either 1,1-(carbene-like)[33] or 1,3-cycloaddition[34] have been found, and for some azirines both pathways are competitive, as with (31) yielding (32) and (33) respectively.[35] Padwa and his group believe the predominant photoproduct can be understood in terms of the electronic effect of substituents on the structure of the ylide, as proposed by Houk *et al.*[36]

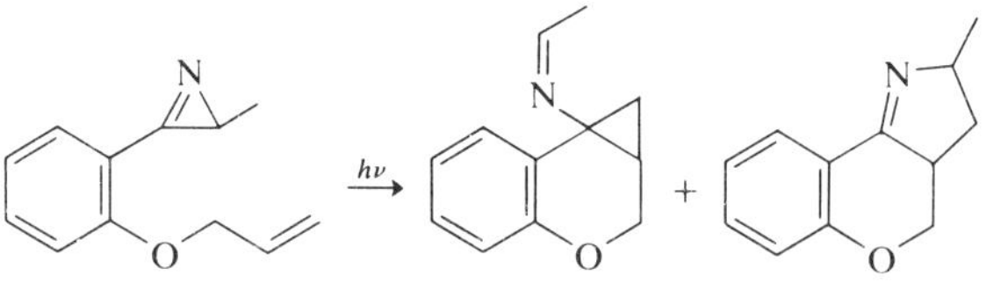
(31) (32) (33)

The first example of a [2+2] photocycloaddition of an alkene to an azo-bond has been noted.[37] Thus, the rigid unsaturated azo compound (34), where a C=C bond and the N=N bond are parallel and close, is claimed to undergo ring-closure to the

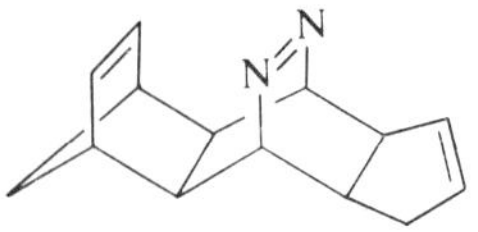
(34)

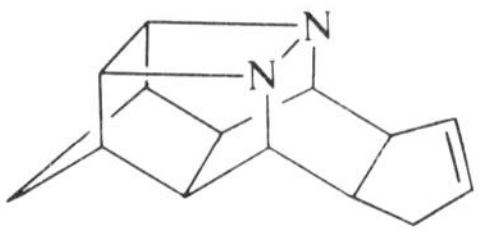
(35)

[32] N. J. Turro, Y. Ito, M.-F. Chow, W. Adam, O. Rodriquez, and F. Yany, *J. Amer. Chem. Soc.*, 1977, **99**, 5836.
[33] A. Padwa and P. H. J. Carlsen, *J. Amer. Chem. Soc.*, 1977, **99**, 1514.
[34] A. Padwa and N. Kamigata, *J. Amer. Chem. Soc.*, 1977, **99**, 1871.
[35] A. Padwa, P. H. J. Carlsen, and A. Ku, *J. Amer. Chem. Soc.*, 1977, **99**, 2798.
[36] P. Caramella and K. N. Houk, *J. Amer. Chem. Soc.*, 1976, **98**, 6397; P. Caramella, R. W. Gandour, J. A. Hall, C. G. Deville, and K. N. Houk, *ibid*, 1977, **99**, 385.
[37] W. Berning and S. Hünig, *Angew. Chem. Internat. Edn.*, 1977, **16**, 777.

remarkable isomer (35) on irradiation in acetonitrile. An unusual photocyclization of *N*-allyliminium salts such as (36) to yield pyrrolidines [*e.g.* (37)] on irradiation in

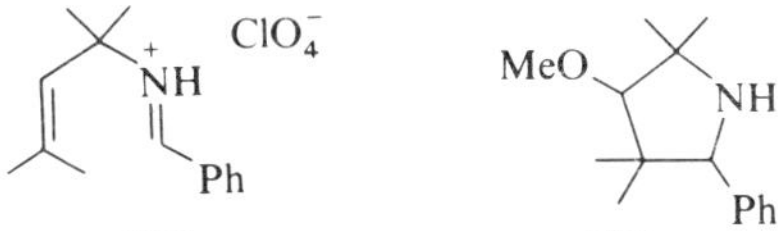

(36) (37)

methanol may well be the result of an intramolecular [2+2] photocycloaddition of the iminium chromophore to the olefinic bond, followed by C—N bond cleavage and capture of a nucleophile.[38]

Photochemical [2+2] cycloaddition reactions of carbonyl compounds to alkenes, yielding oxetans, are well known. Yang and Chiang[39] have now found that the photocycloaddition of aromatic aldehydes (38) to cyclohepta-1,3,5-triene yields a [6+2] adduct (39) as a major product, besides oxetans (40). The reaction is

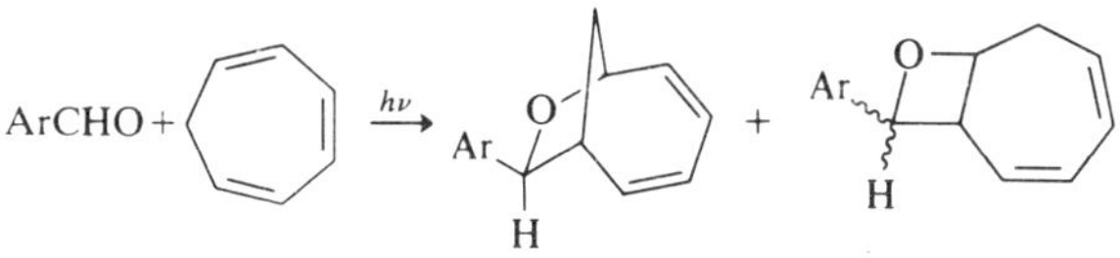

(38) a; Ar = 1-pyrenyl
b; Ar = 2-naphthyl
(39) (40)

believed to occur from the singlet excited state of the aldehyde, and the fact that a [4+2] adduct could not be detected may be evidence for a concerted photochemical mechanism, since the $[_2\pi_s + _2\pi_s]$ and $[_6\pi_s + _2\pi_s]$ additions are allowed on the basis of orbital symmetry rules, whilst the $[_4\pi_s + _2\pi_s]$ is not.

Bryce-Smith, Gilbert, and co-workers[40] have previously published useful guidelines which rationalize the orientation and stereochemistry of cycloaddition of photoexcited benzene to olefins, but the exceptions to these rules which are appearing show that there are still factors to be understood. Thus, the photoaddition of ethylene or propene to benzene (in a u.v. autoclave) gives 1,3-addition to the benzene ring as the major process, rather than the expected 1,2-addition.[41] 1,2-addition to the benzene ring generally gives an *endo*-adduct for donor-substituted alkenes, but an *exo*-adduct for acceptor-substituted alkenes. For example, photoaddition of 2,3-dihydropyran is the most efficient ($\Phi = 0.7$) cycloaddition to benzene so far known, and gives a great preference for the *endo*-adduct (41).[42] In contrast, photoaddition of benzene to 2,2-dimethyl-1,3-dioxolen gives an *exo*-1,2-adduct (42).[43] Photocycloaddition of benzene to methyl acrylate or methacrylate shows surprisingly little selectivity, yielding a mixture of *exo*- and *endo*-1,2-adducts.[44]

38 P. S. Mariano, J. L. Stavinoha, and R. Swanson, *J. Amer. Chem. Soc.*, 1977, **99**, 6781.
39 N. C. Yang and W. Chiang, *J. Amer. Chem. Soc.*, 1977, **99**, 3163.
40 D. Bryce-Smith, A. Gilbert, B. H. Orger, and H. M. Tyrrell, *J.C.S. Chem. Comm.*, 1974, 334.
41 M. F. Mirbach, M. J. Mirbach, and A. Saus, *Tetrahedron Letters*, 1977, 959.
42 A. Gilbert and G. Taylor, *Tetrahedron Letters*, 1977, 469.
43 H.-D. Scharf and J. Mattay, *Tetrahedron Letters*, 1977, 401.
44 R. J. Atkins, G. I. Fray, A. Gilbert, and M. W. bin Samsudin, *Tetrahedron Letters*, 1977, 3597.

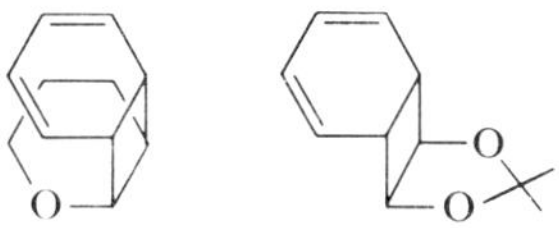

(41) (42)

The photocycloaddition of an acetylene to an aromatic compound usually results in the formation of a cyclo-octatetraene *via* a proposed bicyclo[4,2,0]octatriene intermediate. Two groups[45,46] have presented new evidence on these intermediates. Šket and Zupan[45] have been able to isolate stable bicyclics (43) from the photo-addition of phenyl-substituted acetylenes to hexafluorobenzene, whilst Tinnemans

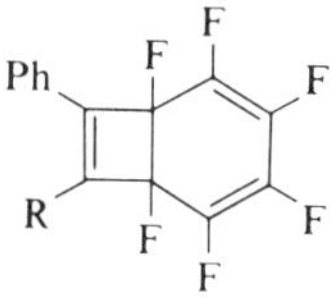

R = Bu^t, Pr^n, or Me

(43)

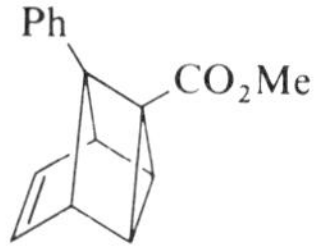

(44)

and Neckers[46] have shown that the bicyclo[4,2,0]octatriene from methyl phenyl-propiolate and benzene can undergo intramolecular photocycloaddition to yield (44).

Five-membered ring heterocycles often undergo ring-atom transpositions on u.v. irradiation [*e.g.* the 2-alkylindazole (45) rearranges to a benzimidazole (46) *via* the detectable tricyclic intermediate (47).[47] Until now, these processes had not been

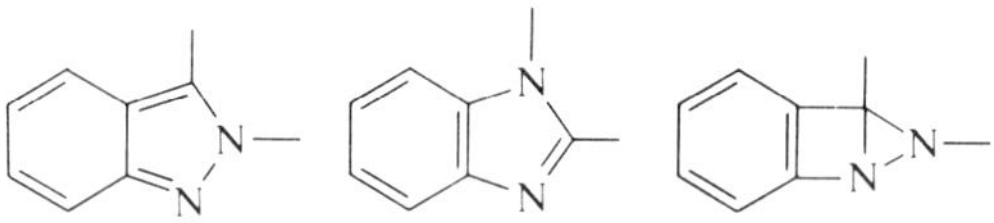

(45) (46) (47)

observed for carbocyclic rings. However, an elegant study of the irradiation of a [$^{13}C_2$]-labelled cyclopentadiene (48) shows that the phototransposition of cyclo-pentadiene [(48) → (50)] occurs, probably by intramolecular [2 + 2] addition to yield bicyclopentene (49), followed by a 1,3-sigmatropic shift.[48] This pathway would account for the fact that, after irradiation, the bicyclopentene is found to have a

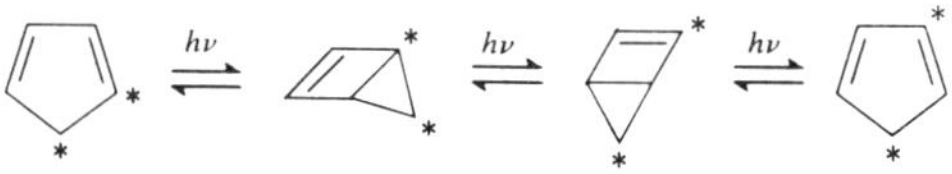

(48) (49) (50)

45 B. Šket and M. Zupan, *J. Amer. Chem. Soc.*, 1977, **99**, 3504.
46 A. H. A. Tinnemans and D. C. Neckers, *J. Amer. Chem. Soc.*, 1977, **99**, 6459.
47 W. Heinzelmann, M. Märky, and P. Gilgen, *Helv. Chim. Acta*, 1976, **59**, 1512.
48 G. D. Andrews and J. E. Baldwin, *J. Amer. Chem. Soc.*, 1977, **99**, 4851.

greater ratio of non-vicinal to vicinal [$^{13}C_2$] label than is found in the recovered cyclopentadiene. Palensky and Morrison[49] have found that a similar skeletal re-arrangement occurs on irradiation (254 nm) of alkyl substituted indenes, *e.g.*, in the interconversion of (51) and (52), and they have independently proposed a similar

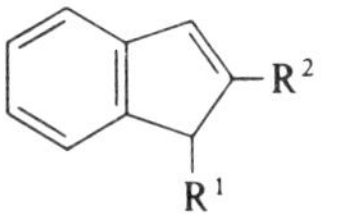

(51) $R^1 = Me, R^2 = H$
(52) $R^1 = H, R^2 = Me$

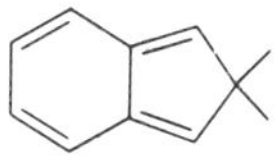

(53)

mechanism. In the case of irradiation of 1,1-dimethylindene, the unstable inter-mediate isoindene (53) becomes the sole product because it cannot undergo a 1,5-hydrogen shift which would normally reform an indene.

Barltrop, Day, and Samuel, who had already shown the intermediacy of a zwitterion [*e.g.* (55)] in the photochemical rearrangement of 4-pyrones to 2-pyrones in trifluoroethanol,[50] have now been able to isolate a further intermediate, the cyclopentadienone epoxide (56), in the conversion of (54) into (57).[51] Barton

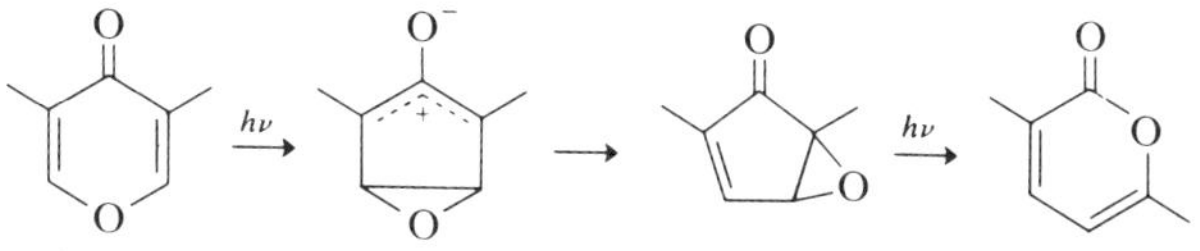

(54) (55) (56) (57)

and Hulshof[52] have used the related photochemical ring contraction of a 4-pyrone (58) to cyclopentenedione (59), followed by *in situ* reduction with sodium cyanoborohydride, in the elegant synthesis of the mould metabolite, terrein

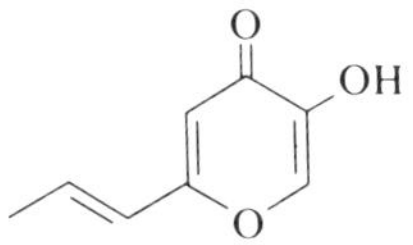

(58)

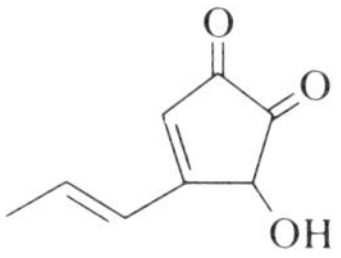

(59)

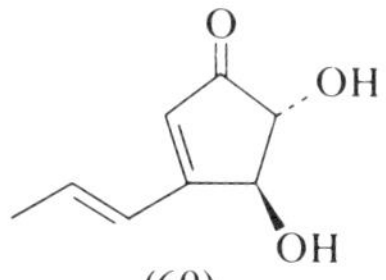

(60)

(60). A similar rearrangement occurs for protonated 4-pyrones (*i.e.* hydroxy-pyrylium cations) on irradiation in sulphuric acid. Thus, the 2,3-dimethyl-4-hydroxypyrylium cation rearranges in part to the corresponding 2-hydroxypyrylium cation; however, in this case Pavlik *et al.*[53] have shown that the major reaction-product is a protonated furyl cation (61). A mechanism for the curious formation of (61) remains to be established.

[49] F. J. Palensky and H. A. Morrison, *J. Amer. Chem. Soc.*, 1977, **99**, 3507.
[50] J. A. Barltrop, A. C. Day, and C. J. Samuel, *J.C.S. Chem. Comm.*, 1976, 822.
[51] J. A. Barltrop, A. C. Day, and C. J. Samuel, *J.C.S. Chem. Comm.*, 1977, 598.
[52] D. H. R. Barton and L. A. Hulshof, *J.C.S. Perkin I*, 1977, 1103.
[53] J. W. Pavlik, D. R. Bolin, K. C. Bradford, and W. G. Anderson, *J. Amer. Chem. Soc.*, 1977, **99**, 2816.

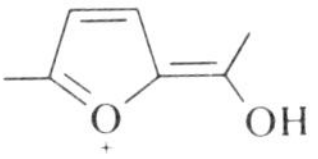

(61)

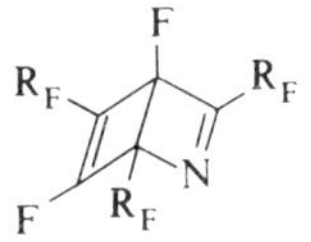

(62) $R_F = CF(CF_3)_2$

Irradiation of perfluorinated pyridines has given the first stable examples of 2-azabicyclic isomers, such as (62).[54] In contrast, a study of the photochemical decomposition of 1,4-[$^{15}N_2$]-*s*-tetrazine to nitrogen and hydrogen cyanide shows that reaction occurs without any preliminary 1,4-nitrogen bonding.[55]

Matrix isolation techniques applied to photochemical reactions have provided some interesting results, such as the first isolation of thiirene (63) by photolysis at 8 K of 1,2,3-thiadiazole,[56] the production of unstable isocyanides [*e.g.* (64)] on

(63)

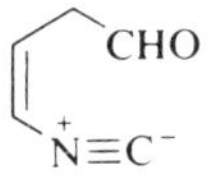

(64)

irradiation of pyridine *N*-oxide at 10 K,[57] and evidence for a process (perhaps involving a ring-opened aldoketene) which scrambles C-2 and C-6 on irradiation of a [^{13}C]-labelled α-pyrone at 8 K.[58]

The use of chiral solvents to induce asymmetric syntheses in photochemical experiments has received some attention this year. Irradiation of 2-styrylbenzo[c]-phenanthrene in chiral solvents gives the optically active hexahelicene in optical yields up to 2%,[59] whilst photochemical rearrangement of nitrones to optically active oxaziridines occurs in optical yields up to 30% on irradiation in a (+)- or (−)-2,2,2-trifluoro-1-phenylethanol and fluorotrichloromethane solvent.[60] The photochemical reduction of carbonyl compounds in the presence of the optically active (*S,S*)-1,4-bis(dimethylamino)-2,3-dimethoxybutane gives both *meso*- and optically active pinacols derived from the carbonyl. For example, acetophenone gives an optical yield of up to 25%, enriched in the (*R,R*)-pinacol enantiomer.[61]

54 R. D. Chambers and R. Middleton, *J.C.S. Chem. Comm.*, 1977, 154.
55 D. S. King, C. T. Denny, R. M. Hochstrasser, and A. B. Smith, *J. Amer. Chem. Soc.*, 1977, **99**, 271.
56 A. Krantz and J. Laureni, *J. Amer. Chem. Soc.*, 1977, **99**, 4842.
57 O. Buchardt, J. J. Christensen, C. Lohse, J. J. Turner, and I. R. Dunkin, *J.C.S. Chem. Comm.*, 1977, 837.
58 B.-S. Huang, R. G. S. Pong, J. Laureni, and A. Krantz, *J. Amer. Chem. Soc.*, 1977, **99**, 4154.
59 W. H. Laarhoven and T. J. H. M. Cuppen, *J.C.S. Chem. Comm.*, 1977, 47.
60 D. R. Boyd and D. C. Neill, *J.C.S. Chem. Comm.*, 1977, 51.
61 D. Seebach, H.-A. Oei, and H. Daum, *Chem. Ber.*, 1977, **110**, 2316.

9 Aliphatic Compounds
Part (i) Hydrocarbons

By D. R. TAYLOR

Department of Chemistry, University of Manchester Institute of Science and Technology, Manchester M60 1QD

1 Acetylenes

The catalytic cyclotrimerization of acetylenes has been reviewed, with special reference to the co-oligomerization of α,ω-diynes with silylacetylenes to produce fused ring systems.[1]

A new general method for the synthesis of 1,3-enynes, which is stereo- and regio-selective, involves the reaction of an alkynylzinc chloride with a vinyl halide in the presence of a palladium(0) complex (Scheme 1).[2] Alkynyl-lithium reagents

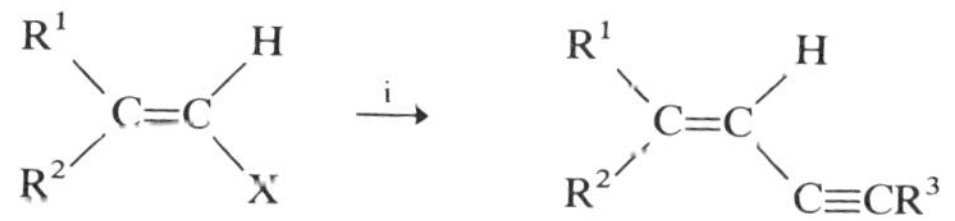

Reagents: i, $R^3C{\equiv}CZnCl$ (R^3 = H or Bu^n), $Pd(PPh_3)_4$–THF

Scheme 1

are unsuitable for the reaction, but are easily converted into the alkynylzinc chlorides, which are stable to disproportionation and do not attack functional groups such as esters. Since enynes are themselves readily converted into dienylmercurials, recently shown to dimerize smoothly in contact with rhodium salts,[3] a new route to conjugated tetraenes has been opened up (Scheme 2).

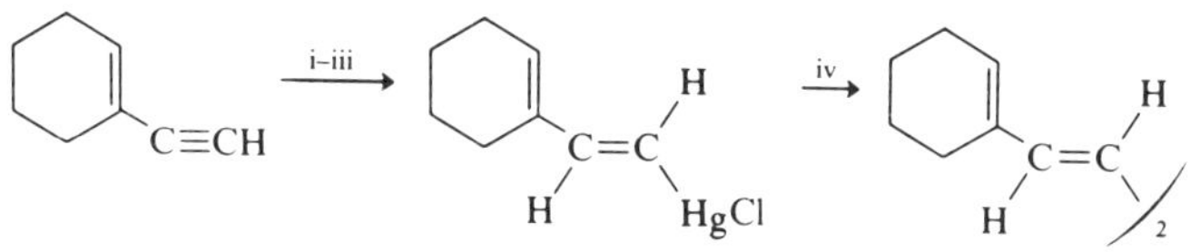

Reagents: i, $(cyclo\text{-}C_6H_{11})_2BH$; ii, $Hg(OAc)_2$; iii, aq. NaCl; iv, $[ClRh(CO)_2]_2$–LiCl–$(Me_2N)_3PO$

Scheme 2

[1] K. P. C. Vollhardt, *Accounts Chem. Res.*, 1977, **10**, 1.
[2] A. O. King, N. O. Okukado, and E.-i. Negishi, *J.C.S. Chem. Comm.* 1977, 683.
[3] R. C. Larock and J. C. Bernhardt, *J. Org. Chem.*, 1977, **42**, 1680.

Enynes bearing remote functionality [*e.g.* (1), required for the synthesis of a sex pheromone of the grape-vine moth] can be obtained by an elegant combination of methods discovered by C. A. and H. C. Brown (Scheme 3).[4] It features the use of the acetylene 'zipper' [$KNH(CH_2)_2NH_2$], which has also been used to effect the migration of the C≡C bond in alkynylborates.[5]

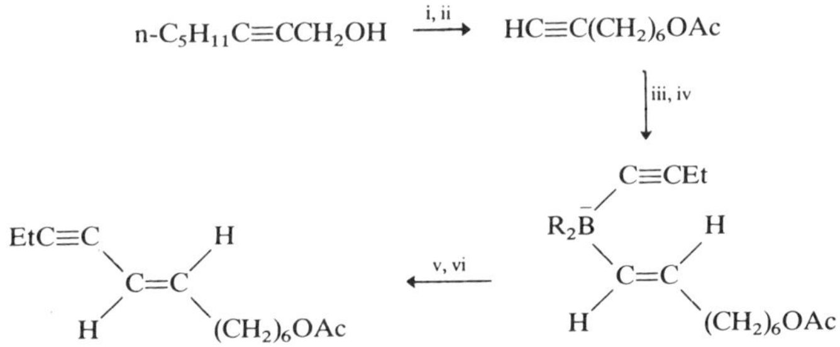

(1)

Reagents: i, $KNH(CH_2)_2NH_2$; ii, Ac_2O; iii, $(Pr^iMeCH)_2BH$; iv, $EtC{\equiv}CLi$; v, I_2; vi, H_2O_2–NaOAc

Scheme 3

The stereospecific *syn*-addition of organometallic compounds (*e.g.* RCu) to the C≡C bond continues to provide fascinating examples of synthetic value. However, treatment of a terminal acetylene (but not C_2H_2) with lithium dialkylcuprates in THF leads to quantitative metal–hydrogen exchange. This undesirable reaction can be avoided (i) by using diethyl ether as solvent or, better, (ii) by adding lithium bromide to the reaction mixture in THF, when deprotonation is completely suppressed (Scheme 4).[6] Organocopper–magnesium halide complexes may also be

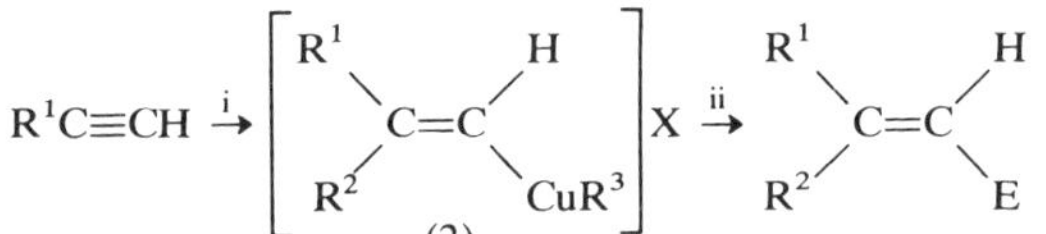

(2)

Reagents: 1, $R^2R^3CuLi–Et_2O$ (X = Li) or R^2R^3CuLi–LiBr–THF (X = Li,LiBr); ii, D^+(E = D) or BrCN–THF (E = Br) or *N*-chlorosuccinimide–$(Me_2N)_3PO$ (E = Cl); (R^1 = alkyl or aryl, R^2, R^3 = alkyl)

Scheme 4

used to prepare vinylcuprates (2; R^3 = Br or R^2, X = MgHal), a reaction which may usefully be followed by copper–halogen exchange using *N*-halogenosuccinimide[7] or cyanogen bromide or iodide,[8] or by copper–nitrile exchange using cyanogen chloride or an arenesulphonyl cyanide.[8] If a silylacetylene is used in this type of reaction sequence, organocopper addition proceeds with the opposite regioselectivity, so providing a route to 1-substituted 1-silylalkenes (Scheme 5).[9,10]

[4] E.-i. Negishi and A. Abramovitch, *Tetrahedron Letters*, 1977, 411.
[5] C. A. Brown and E.-i. Negishi, *J.C.S. Chem. Comm.*, 1977, 318.
[6] H. Westmijze, H. Kleijn, and P. Vermeer, *Tetrahedron Letters*, 1977, 2023.
[7] A. B. Levy, P. Talley, and J. A. Dunford, *Tetrahedron Letters*, 1977, 3545.
[8] H. Westmijze and P. Vermeer, *Synthesis*, 1977, 784.
[9] M. Obayashi, K. Utimoto, and H. Nozaki, *Tetrahedron Letters*, 1977, 1805.
[10] H. Westmijze, J. Meijer, and P. Vermeer, *Tetrahedron Letters*, 1977, 1823.

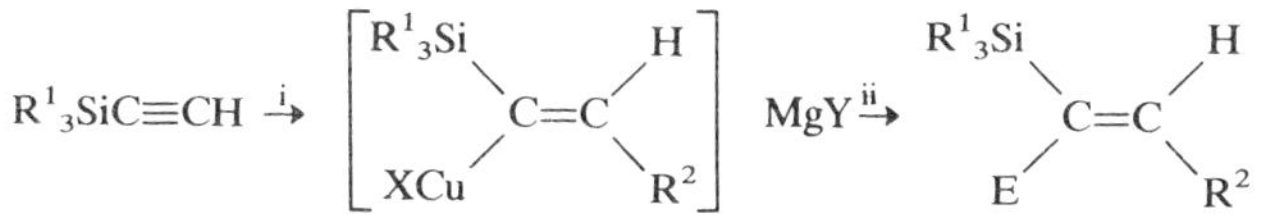

Reagents: i, $R^2CuMgBr_2$ (R^1 = Me, R^2 = Pr^n, n-C_5H_{11}, or n-C_6H_{13}, X = Y = Br) or $R^2{}_2CuMgY$ (R^1 = Ph, Y = Cl or Br, X = R^2 = Et, Pr^i, cyclo-C_6H_{11}, or Bu^t); ii, H_2O (D_2O) (E = H or D), or *N*-halogenosuccinimide (E = halogen), or iodoalkane (E = Me, Pr^n, n-C_5H_{11}, or n-C_6H_{13} *etc.*)

Scheme 5

Vinylcoppers may also be obtained from alkyne-hydrozirconation products by transmetallation;[11] the overall process is both regio- and stereo-specific. Copper(I) arylacetylides react with arylcoppers to yield mixed diarylacetylenes ($Ar^1C{\equiv}CAr^2$) *via* the formation of well-defined and isolable organo-cluster species, $Ar^1{}_4Cu_6(C{\equiv}CAr^2)_2$. The highly selective formation of *mixed* acetylenes was shown to result from the detailed structure of the organo-cluster compounds, in which triangular copper faces are occupied exclusively by one Ar^1 and two $Ar^2C{\equiv}C$ ligands.[12]

The synthesis of acetylenes and allenes *via* polymetallation, usually lithiation, of alkynes continues to receive attention.[13–15] Perlithiopropyne, the structure of which has been determined by *ab initio* calculation,[14] reacts with iodoethane to afford a low yield of an ene-diyne (3), whereas with trimethyl-silyl or -germyl chloride or with diethyl sulphate it yields allenic and/or acetylenic derivatives by straightforward substitution (Scheme 6).[13] But-1-yne and but-2-yne are converted

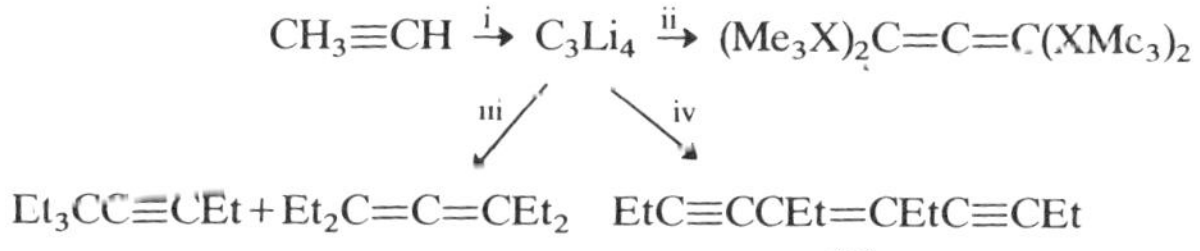

(3)

Reagents: i, Bu^nLi–hexane; ii, Me_3XCl (X = Si or Ge); iii, Et_2SO_4; iv, EtI

Scheme 6

by excess alkyl-lithium into the same trilithiated material, but the 2-yne reacts much more slowly. A trilithiated product is also obtained by lithiation of hexa-2,4-diyne; it is a source of some exotic alkylated, silylated, or germylated allenynes, including the hexasilylbisallene (4).[13]

$$(Me_3Si)_2C{=}C{=}C(SiMe_3)C(SiMe_3){=}C{=}C(SiMe_3)_2$$

(4)

Several publications have appeared reporting investigations into the bromination of acetylenes. Solvent effects on the rate[16] and kinetic product distribution[17] are

[11] M. Yoshifuji, M. J. Loots, and J. Schwartz, *Tetrahedron Letters*, 1977, 1303.
[12] G. van Koten, R. W. M. ten Hoedt, and J. G. Noltes, *J. Org. Chem.*, 1977, **42**, 2705.
[13] W. Priester, R. West, and T. L. Chwang, *J. Amer. Chem. Soc.*, 1976, **98**, 8413; W. Priester and R. West, *ibid.*, pp. 8421, 8426.
[14] E. D. Jemmis, D. Poppinger, P. von R. Schleyer, and J. A. Pople, *J. Amer. Chem. Soc.*, 1977, **99**, 5796.
[15] N. M. Libman, P. A. Brestkin, and S. G. Kuznetsov, *J. Org. Chem. U.S.S.R.*, 1976, **12**, 2480.
[16] M.-F. Ruasse and J.-E. Dubois, *J. Org. Chem.*, 1977, **42**, 2689.
[17] G. D. Mel'nikov and S. P. Mel'nikov, *J. Org. Chem. U.S.S.R.*, 1977, **13**, 625.

considered to be compatible with the formation, possibly *via* a fast, reversibly formed π-complex, of a bromovinyl cation in which the empty *p*-orbital is conjugated with the benzene π-orbitals and therefore coplanar with the β-carbon's σ-bonds (5). The last-mentioned structural element implies hyperconjugation, an

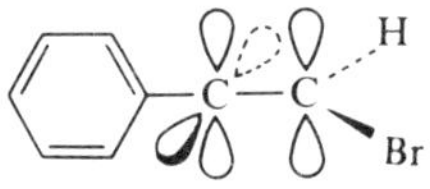

(5)

interaction which would explain the greater difference between bromination rates of similarly substituted olefins and acetylenes[16,18,19] than are found for rates of acid-catalysed hydration:[20] the β-bromine atom is more destabilizing in (5) than in the corresponding intermediate bromocation from styrene. The Ad_EC1 mechanism (a variant of Ad_E2) is thus confirmed. It should be noted that the nature of any substituents attached to their π-bonds greatly affects the relative bromination rates of correspondingly substituted olefins and acetylenes; indeed, some olefinic acids appear to react more slowly than their acetylenic counterparts,[19] though this may be misleading since allowance may have to be made, for example, for their different acidities which may affect ionic strength.

Triethylammonium hydrogen dichloride appears to be the best reagent for achieving the addition of hydrogen chloride to the C≡C bond: no catalyst is required. The alkynes studied, with the exception of phenyl-t-butylacetylene, all reacted stereoselectively to give *syn*-adducts, an observation attributed to the powerfully nucleophilic anion HCl_2^- forcing the adoption of an Ad_E3 mechanism.[21] Hydrogen chloride, also without catalyst, is adequate for hydrochlorination of sulphur- and selenium-substituted acetylenes, an addition which proceeds exclusively in an *anti*-orientation.[22] Mixed acetylenes of the type $R^1SC{\equiv}CSeR^2$ react regiospecifically by protonation next to selenium, presumably because sulphur is better able to stabilize an adjacent cation.

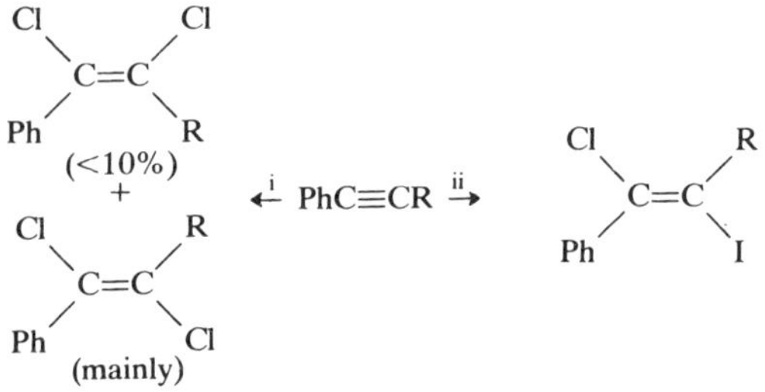

Reagents: i, $CuCl_2$–LiCl (40-fold excess), MeCN (R = H, Me, Pr^n, or Ph); ii, $CuCl_2$–KI or $-I_2$ (five-fold excess), MeCN (R = H, Me, Et, Pr^n, or Pr^i)

Scheme 7

[18] S. De Young, S. Ehrlich, and E. Berliner, *J. Amer. Chem. Soc.*, 1977, **99**, 290.
[19] G. H. Schmid, A. Modro, and K. Yates, *J. Org. Chem.*, 1977, **42**, 2021.
[20] G. Modena, F. Rivetti, G. Scorrano, and U. Tonellato, *J. Amer. Chem. Soc.*, 1977, **99**, 3392.
[21] J. Cousseau and L. Gouin, *J.C.S. Perkin I*, 1977, 1797.
[22] S. I. Radchenko and A. A. Petrov, *J. Org. Chem. U.S.S.R.*, 1977, **13**, 36.

Full details have now appeared of the *anti*-chlorination of acetylenes using copper(II) chloride–lithium chloride in large excess in acetonitrile, and the technique can also be applied to iodochlorination (using $CuCl_2$–I_2 or $CuCl_2$–KI) with good regioselectivity (Scheme 7).[23] Electrophilic attack upon the C≡C bond by both mercurinium and sulphenium cations proceeds almost exclusively to form *anti*-adducts *via* bridged vinyl cations; in the latter addition the intermediate episulphonium ions [*e.g.* (6)] have been isolated as stable antimonates or tetrafluoroborates.[24]

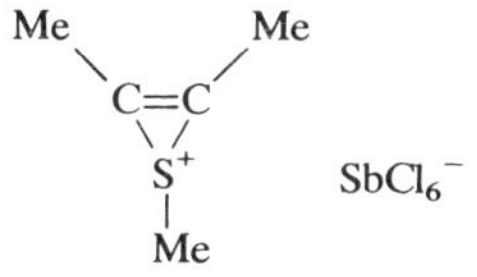

(6)

New reducing agents, termed complex reducing agents (CRA), have been developed by treating sodium hydride with metal salts such as iron(III) chloride (FeCRA) or nickel(II) acetate (NiCRA) in the presence of alkoxide.[25,26] Triple bonds may either be reduced to *Z*-alkenes or fully reduced to alkanes by a judicious choice of reagents and conditions, which are generally mild enough to leave carbonyl groups unaffected.[26]

Conversions of acetylenes into allenes by substitutions which are accompanied by the well-known propargyl rearrangement are often poorly understood in mechanistic terms. The complexities which can arise are exemplified by the reaction of the diphenylbutynol (7; R = Ph_2COH) with lithium aluminium deuteride (Scheme 8). Whereas in dialkyl ethers the main product is the allene (8), none of this product appears to be formed when the solvent is only slightly altered to THF.[27]

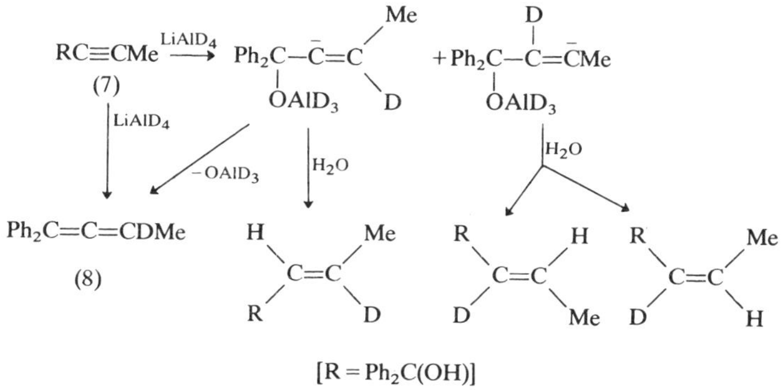

Scheme 8

[23] S. Uemura, H. Okazaki, A. Onoe, and M. Okano, *J.C.S. Perkin I*, 1977, 676.
[24] G. Capozzi, V. Lucchini, G. Modena, and P. Scrimin, *Tetrahedron Letters*, 1977, 911.
[25] J. J. Brunet, L. Mordenti, B. Loubinoux, and P. Caubere, *Tetrahedron Letters*, 1977, 1069.
[26] J. J. Brunet and P. Caubere, *Tetrahedron Letters*, 1977, 3947.
[27] M. P. Hartshorn, R. S. Thompson, and J. Vaughan, *Austral. J. Chem.*, 1977, **30**, 865.

Two ^{13}C n.m.r. studies of acetylenic compounds have recently been reported. One establishes correlations for chemical shifts in $\alpha\beta$-acetylenic ketones,[28] and the other claims to show that chiral recognition is possible across the C≡C bond in molecules such as 2,5-dichlorohex-3-yne.[29]

2 Alkanes

A review has appeared dealing with the conformations of hydrocarbon chains.[30] The synthesis,[31] thermal stability,[31] and thermodynamic properties[32] of some strained alkanes have been reported: ^{13}C n.m.r. spectra can be used to study conformational and rotational barriers in such compounds.[33]

Two new syntheses of alkanes have been developed. The first uses the addition of alkylmercury(II) halides or acetates to electron-deficient olefins to generate the saturated C—C linkage, and is followed by sodium borohydride reduction to cleave the C—Hg bond.[34] The second technique is to convert an alcohol into a chloroformate which may then be conveniently reduced under free-radical conditions by a trialkylsilane.[35]

Further examples have appeared[36] of oxyfunctionalization of alkanes using protonated ozone [see *Ann. Reports* (*B*), 1976, **73**, 172]. As an alternative, hydrogen

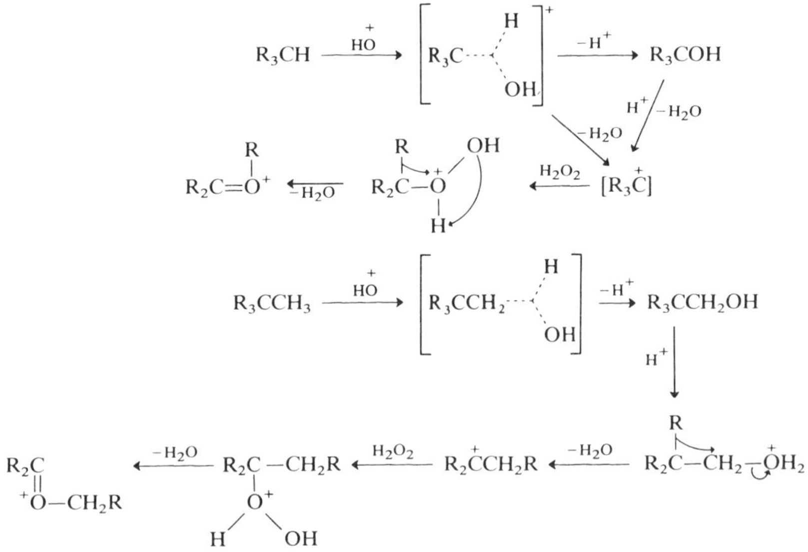

Scheme 9

[28] G. A. Kalabin, A. G. Proidakov, L. D. Gavrilov, and L. I. Vereshchagin, *J. Org. Chem. U.S.S.R.*, 1977, **13**, 449.
[29] A. J. Jones, and P. J. Stiles, *Tetrahedron Letters*, 1977, 1965.
[30] M. A. Winnik, *Accounts Chem. Res.*, 1977, **10**, 173.
[31] H.-D. Beckhaus and C. Rüchardt, *Chem. Ber.*, 1977, **110**, 878.
[32] D. N. J. White and M. J. Bovill, *J.C.S. Perkin II*, 1977, 1610.
[33] S. Brownstein, J. Dunogues, D. Lindsay, and K. U. Ingold, *J. Amer. Chem. Soc.*, 1977, **99**, 2073.
[34] B. Giese and J. Meister, *Chem. Ber.*, 1977, **110**, 2588.
[35] N. C. Billingham, R. A. Jackson, and F. Malek, *J.C.S. Chem. Comm.*, 1977, 344.

peroxide may be used either in superacids[36,37] or in concentrated trifluoroacetic acid.[38] Studies by ^1H n.m.r. in superacid suggest that the reaction proceeds *via* electrophilic attack by HO^+ (or the incipient cation, $HO.\overset{+}{O}H_2$) on the weakest available C—H σ-bond, followed by nucleophilic attack by hydrogen peroxide upon the carbenium ion and subsequent alkyl migration to oxygen (Scheme 9). Predominantly branched alkanes were studied. Reactions between n-alkanes and hydrogen peroxide in trifluoroacetic acid yield alkan-2-ols and -3-ols, with negligible amounts of terminal oxidation.[38]

3 Allenes

An exhaustive review of the synthesis and chemical reactions of the interesting phospha-allene ylides has appeared.[39] These ylides yield allenes when subjected to the Wittig reaction with carbonyl compounds.

An extension of earlier work [see *Ann. Reports (B)*, 1976, **73**, 173] on the synthesis of allenes *via* allenylcoppers has been reported.[40] Electrophilic attack upon the intermediate (9) by iodine yields optically active iodoallenes if optically active propargyl esters are used as starting materials. However, when the electrophile is an acid anhydride or chloromethyl methyl ether (Scheme 10) only

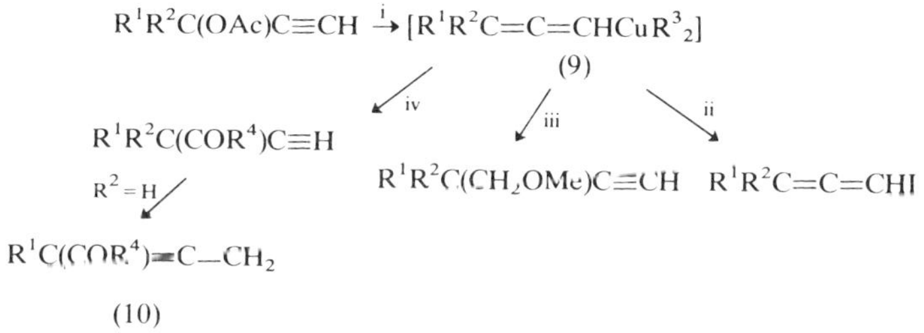

Reagents: i, $R^3{}_2CuLi$ (R^3 = alkyl); ii, I_2, $(MeOCH_2)_2$; iii, $ClCH_2OMe$; iv, $(R^4CO)_2O$ (R^4 = Me, n-C_5H_{11}, or Ph)

Scheme 10

acetylenic products are formed initially, although these may isomerize to, for example, allenones (10). An alternative synthesis of allenes operates upon the same type of starting material [$R_2C(OAc)C{\equiv}CH$], but generates intermediate allenylborates which can be converted into a wider variety of functionalized allenes by protolysis[41] or hydrolysed with a retro-propargyl rearrangement to acetylenes (Scheme 11).

Full details have appeared of the syntheses of higher cumulenes by flash-pyrolytic retro-Diels–Alder reactions[42] and by Skattebøl's method,[43] described in last

36 N. Yoneda and G. A. Olah, *J. Amer. Chem. Soc.* 1977, **99**, 3113.
37 G. A. Olah, N. Yoneda, and D. G. Parker, *J. Amer. Chem. Soc.*, 1977, **99**, 483.
38 N. C. Deno, E. J. Jedziniak, L. A. Messer, M. D. Meyer, S. G. Stroud, and E. S. Tomezsko, *Tetrahedron*, 1977, **33**, 2503.
39 H. J. Bestmann, *Angew. Chem. Internat. Edn.*, 1977, **16**, 349.
40 J.-M. Dollat, J.-L. Luche, and P. Crabbé, *J.C.S. Chem. Comm.*, 1977, 761.
41 M. M. Midland, *J. Org. Chem.*, 1977, **42**, 2650.
42 J. L. Ripoll and A. Thuillier, *Tetrahedron*, 1977, **33**, 1333.
43 G. Karich and J. C. Jochims, *Chem. Ber.*, 1977, **110**, 2680.

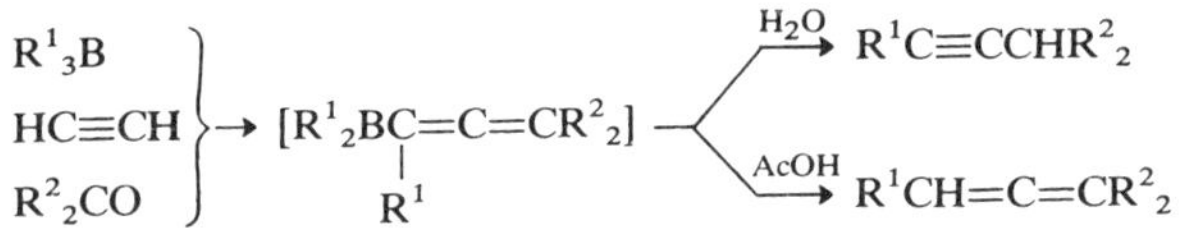

Scheme 11

year's Report (pp. 176–177). As the number of cumulated double bonds increases, the rotational barrier decreases uniformly, and not in an alternating fashion which would have suggested a dependence upon molecular planarity.[44]

A number of allenes of potential commercial interest have been synthesized this year.[45–47] These included 2,4,5-trienamides, regarded as potential insecticides, in which a Wittig reaction was used to introduce the non-allenic C=C bond,[45] and the synthesis of allenic barbiturates from allenylmalonates. The latter derivatives were prepared by an addition–elimination sequence using malonate and 1-halogeno-allenes (Scheme 12), but the correct choice of base to generate the malonate anion must be made: otherwise acetylenic or dienic isomers of the desired allenes may predominate.[46]

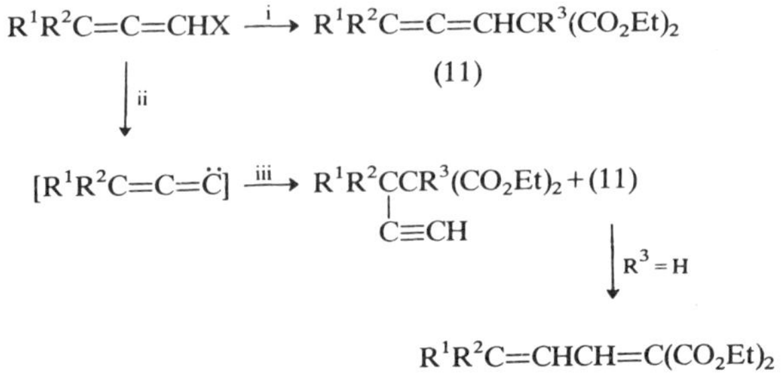

Reagents: i, NaH–C_6H_6–$R^3CH(CO_2Et)_2$ (R^3 = H or Et); ii, NaOEt–EtOH; iii, $R^3CH(CO_2Et)_2$

Scheme 12

Reactions of magnesium or lithium with 1,3-dihalogenopropanes yield cyclo-propanes and not 1,3-propane–dimetal compounds. Hence the importance of the discovery that hydroboration of propadiene followed by mercuration yields propane-1,3-dimercury(II) chloride, which although it cannot be transmetallated directly may nevertheless be converted into propane-1,3-dimagnesium(II) halide in two steps (Scheme 13).[48] This interesting bis-Grignard reagent has so far only been used to generate glutaric esters, propane-1,3-distannyl derivatives, and, more significantly, two silacyclobutanes in good yield. Propadiene has also been used as a precursor in propenyl ketone synthesis; the required C-2 acylation was effected with sodio alkylirontetracarbonyl.[49]

[44] K. Bertsch, G. Karich, and J. C. Jochims, *Chem. Ber.*, 1977, **110**, 3304.
[45] P. D. Landor, S. R. Landor, and O. Odyek, *J.C.S. Perkin I*, 1977, 93.
[46] S. R. Landor, P. D. Landor, and P. F. Whiter, *J.C.S. Perkin I*, 1977, 1710.
[47] L. A. van Dijck, B. Thankwerden, and J. G. C. M. Vermeer, *Rec. Trav. chim.*, 1977, **96**, 200.
[48] L. C. Costa and G. M. Whitesides, *J. Amer. Chem. Soc.*, 1977, **99**, 2390.
[49] A. Guinot, P. Cadiot, and J. L. Roustan, *J. Organometallic Chem.*, 1977, **128**, C35.

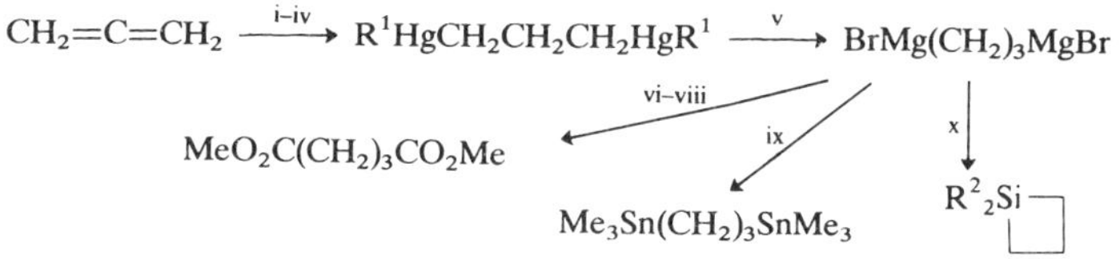

Reagents: i, B_2H_6; ii, $Hg(OAc)_2$; iii, NaCl; iv, $2R^1Li$–THF; v, Mg–$MgBr_2$–THF; vi, CO_2; vii, H^+–MeOH; viii, CH_2N_2; ix, Me_3SnCl; x, $R^2{}_2SiCl_2$ (R^2 = Ph or Me)

Scheme 13

Further additions of resonance-stabilized enolates to allenic sulphonium salts have been investigated.[50] When the addend is a cyano-stabilized enolate a cyano-furan is formed; this and earlier furan syntheses are shown to proceed by a direct substitution mechanism, rather than *via* addition–elimination. Other uni- and bi-dentate nucleophiles were found to yield products of addition–substitution (Scheme 14). When the attacking nucleophile is a malonate anion, the isolable allylic adducts [12; $Z = R^3C(CO_2Et)_2$, R^3 = H or Me] are converted by alkoxide into ylides which undergo [2,3] sigmatropic shifts.[51]

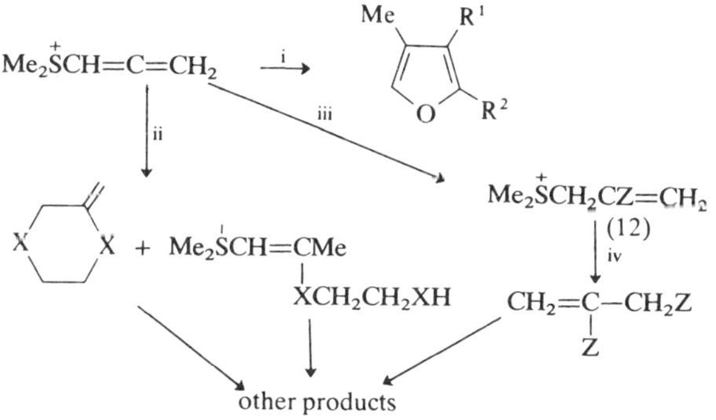

Reagents: i, $R^2CO\bar{C}HR^1$ (R^1 = Ac, $ArSO_2$, CO_2R^3, or CN; R^2 = Ph); ii, $HXCH_2CH_2X^-$ (X = S or SO_2); iii, Z^- (Z = CN or $PhSO_2$); iv, Z^- ($-Me_2S$)

Scheme 14

Further examples of ene reactions of alkylallenes have been reported. The acyclic enophile $EtO_2CN=NCO_2Et$ reacts with allenes, including cyclonona-1,2-diene, to produce the expected 2-bicarbamyl-1,3-dienes in excellent yield,[52] whereas the cyclic enophile *N*-phenyltriazolinedione induces apparent prototropic shifts in some allenes. The ene reactions of such allenes are found to be two to five times faster than those of comparably substituted mono-olefins, presumably as a result of lower steric hindrance and developing resonance interactions in the

[50] B. S. Ellis, G. Griffiths, P. D. Howes, C. J. M. Stirling, and B. R. Fishwick, *J.C.S. Perkin I*, 1977, 286.
[51] G. Griffiths, P. D. Howes, and C. J. M. Stirling, *J.C.S. Perkin I*, 1977, 912.
[52] C. B. Lee and D. R. Taylor, *J.C.S. Perkin I*, 1977, 1463; *J. Chem. Res.* (*S*), 1977, 136; (*M*) 1601.

transition state. Ene adducts were not, however, obtained when cyclonona-1,2-diene was treated with acetylenic enophiles; the 1:1 adducts obtained were simple [2+2] cycloadducts (Scheme 15).

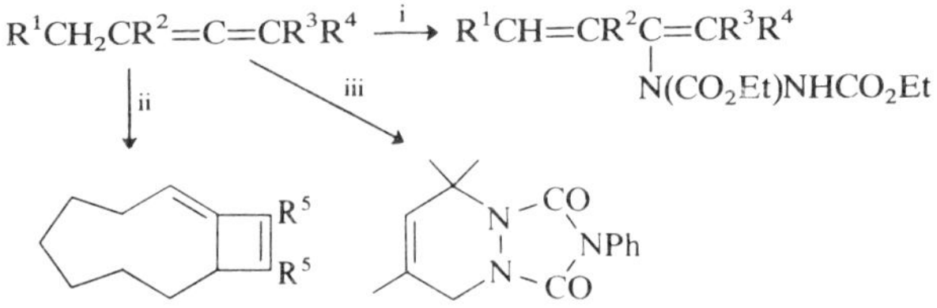

Reagents: i, $EtO_2CN{=}NCO_2Et$ [$R^1 = H$, R^2—$R^4 = Me$, or $R^1 = R^3 = R^4 = H$, $R^2 = Me$, or $R^2 = R^3 = H$, $R^1R^4 = (CH_2)_5$]; ii, $EC{\equiv}CE$ [$E = CF_3$ or CO_2Et, $R^2 = R^3 = H$, $R^1R^4 = (CH_2)_5$]; iii, $\overline{N{=}NCONPhCO}$ ($R^1 = H$, R^2—$R^4 = Me$)

Scheme 15

Further uses of allenes in synthesis have appeared. Allenic nitriles undergo Michael additions when treated with phenylhydrazine: conjugated adducts cyclize thermally to 5-aminopyrazoles, whereas unconjugated adducts give 3*H*-indoles *via* a [3,3]sigmatropic shift.[53] Ketene cycloadducts of cyclohexanylideneallene have also been found useful, since their thermal isomerization leads to annelated products (Scheme 16).[54]

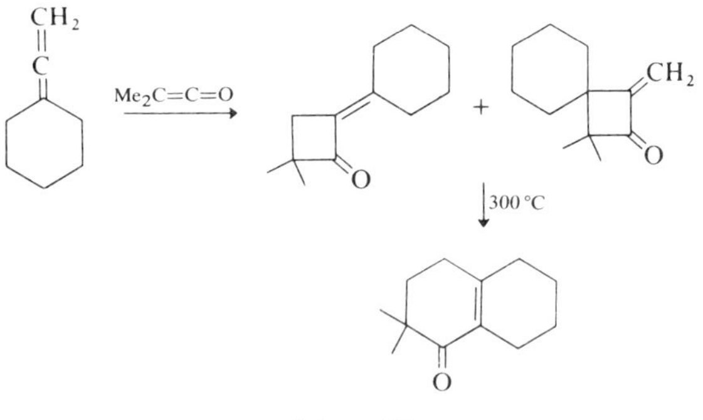

Scheme 16

4 Olefins and Dienes

Advances in the following areas relevant to olefin chemistry have been reviewed recently: stepwise [2+2] cycloadditions of enol ethers,[55] industrial Wittig reactions,[56] structural effects in acid-catalysed hydration,[57] the Prins reaction,[58] and the

[53] S. R. Landor, P. D. Landor, Z. T. Fomum, and G. M. Mpango, *Tetrahedron Letters*, 1977, 3743.
[54] M. Bertrand, G. Gil, A. Junino, and R. Maurin, *Tetrahedron Letters*, 1977, 1779.
[55] R. Huisgen, *Accounts Chem. Res.*, 1977, **10**, 117.
[56] H. Pommer, *Angew. Chem. Internat. Edn.*, 1977 **16**, 423.
[57] V. J. Nowlan and T. T. Tidwell, *Accounts Chem. Res.*, 1977, **10**, 252.
[58] D. R. Adams and S. P. Bhatnagar, *Synthesis*, 1977, 661.

synthesis of insect sex pheromones.[59] The last-mentioned review includes useful discussion of the control of stereochemistry in Wittig and related reactions. Aspects of organometallic olefin chemistry which have also been reviewed include olefin insertions in catalysis[60] and the applications of organosilicon compounds[61] and organopalladium compounds[62] in synthesis. Olefin metathesis reactions continue to receive enthusiastic attention and, besides numerous individual research reports,[63–69] the proceedings of a symposium have been published.[70]

The excellent new olefin synthesis which involves the coupling of two molecules of a carbonyl compound by low oxidation states of titanium[71] ($TiCl_3$–$LiAlH_4$, $TiCl_3$–K or –Li, or $TiCl_4$–Zn) has been used successfully in several laboratories.[72–74] It is especially useful for symmetrical tetrasubstituted olefins, for which a Wittig reaction would be unsuitable. Since the reaction involves a pinacol-type reductive dimerization as the first step (Scheme 17), it is not obvious that unsymmetrical

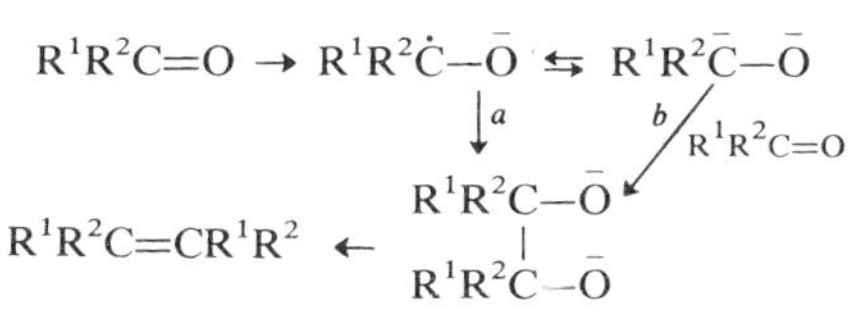

Scheme 17

olefins ($R^1R^2C{=}CR^3R^4$) could be prepared from a mixture of two carbonyl compounds. However, McMurry's group has now shown that the use of an excess of acetone with, for example, a cycloalkanone gives synthetically useful yields of isopropylidene-cycloalkanes. With mixtures of acetone and diaryl ketones, essentially *only* mixed olefins ($Me_2C{=}CAr_2$) are obtained; since the reduction potentials of diaryl ketones are lower than that of acetone, a radical coupling pathway (path *a*) is deemed less likely than the formation of aryldianions ($Ar_2\bar{C}{-}O^-$) which attack acetone preferentially (path *b*).[75] The procedure is exceptionally useful for the preparation of very hindered olefins such as adamantylideneadamantane[71] and led to another first-time achievement, the synthesis and successful resolution by h.p.l.c.

[59] C. A. Henrick, *Tetrahedron*, 1977, **33**, 1845.
[60] G. Henrici-Olivé and S. Olivé, *Topics Current Chem.*, 1976, **67**, 107.
[61] S. S. Washburne, *J. Organometallic Chem.*, 1976, **123**, 1.
[62] B. M. Trost, *Tetrahedron*, 1977, **33**, 2615.
[63] E. Verkuijlen, F. Kapteijn, J. C. Mol, and C. Boelhouwer, *J.C.S. Chem. Comm.*, 1977, 198.
[64] J.-P. Laval, A. Lattes, R. Mutin, and J. M. Basset, *J.C.S. Chem. Comm.*, 1977, 502.
[65] P. G. Gassman and T. H. Johnson, *J. Amer. Chem. Soc.*, 1977, **99**, 622.
[66] T. J. Katz and J. McGinnis, *J. Amer. Chem. Soc.*, 1977, **99**, 1903.
[67] J. L. Bilhou, J. M. Basset, R. Mutin, and W. F. Graydon, *J. Amer. Chem. Soc.*, 1977, **99**, 4083.
[68] R. Baker and M. J. Crimmin, *Tetrahedron Letters*, 1977, 441.
[69] T. J. Katz and W. H. Hersh, *Tetrahedron Letters*, 1977, 585.
[70] Proceedings of the International Symposium on Metathesis, Noordwijkerhout, Sept. 1977, *Rec. Trav. chim.*, 1977, **96**, m1.
[71] J. E. McMurry and M. P. Fleming, *J. Amer. Chem. Soc.*, 1974, **96**, 4708; *J. Org. Chem.* 1976, **41**, 896.
[72] B. Feringa and H. Wynberg, *J. Amer. Chem. Soc.*, 1977, **99**, 602.
[73] D. Lenoir, *Synthesis*, 1977, 553.
[74] G. A. Olah and G. K. S. Prakash, *J. Org. Chem.*, 1977, **42**, 580.
[75] J. E. McMurry and L. R. Krepski, *J. Org. Chem.* 1976, **41**, 3929.

on alumina–(+)-TAPA [2-(2,4,5,7-tetranitro-9-fluorenylideneamino-oxy)propionic acid] of the (±)-*cis*- and (±)-*trans*-bi-hexahydrophenanthrylidenes (13a and b), in which chirality stems solely from torsional distortion.[72]

(13a)

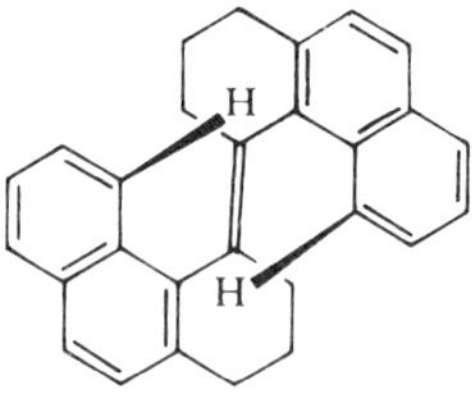

(13b)

Tetrakis(neopentyl)ethylene, synthesized by the Ti^{II} coupling reaction, is unaffected by bromine in carbon tetrachloride and is not even protonated in FSO_3H–SO_2ClF, although the more powerful Magic Acid cleaves the neopentyl groups.[74] Another exceptionally unreactive olefin (14), named [10,10]-betweenane, has been reported; it has a doubly *trans*-cycloalkene π-bond, and so has no accessible face for attack upon the olefinic electrons.[76]

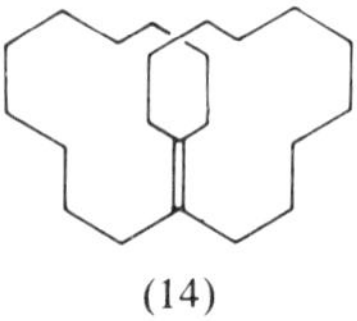

(14)

Another alternative to the Wittig reaction, in this case suitable for preparations of terminal alkenes, is the interaction of a Grignard reagent with Eschenmoser's salt (15) (Scheme 18).[77] The intermediates are amine oxides which decompose at somewhat higher temperatures than the intermediate betaines of the Wittig procedure.

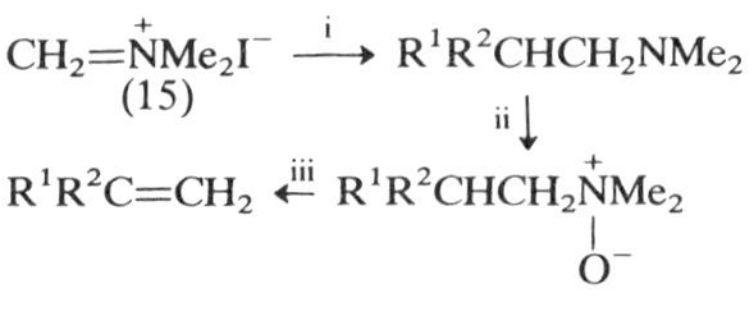

Reagents: 1, R^1R^2CHMgX or R^1R^2CHLi; ii, H_2O_2; iii, 150 °C

Scheme 18

Further papers on the useful olefin synthesis technique based on diphenylphosphinoyl migrations have appeared.[78] The main limitations foreseen for this approach are (i) that the necessary $Ph_2P(O)$ migration no longer occurs when an alternative tertiary migration origin is present, and (ii) that aryl groups at the

[76] J. A. Marshall and M. Lewellyn, *J. Amer. Chem. Soc.*, 1977, **99**, 3508.

[77] J. L. Roberts, P. S. Borromeo, and C. D. Poulter, *Tetrahedron Letters*, 1977, 1299.

[78] A. H. Davidson, I. Fleming, J. I. Grayson, A. Pearce, R. L. Snowden, and S. Warren, *J.C.S. Perkin I*, 1977, 550; A. H. Davidson, C. Earnshaw, J. I. Grayson, and S. Warren, *ibid.*, p. 1452.

migration terminus, or four- or five-membered alicyclic rings at the migration origin, tend to produce substantial amounts of vinyl-, rather than allyl- phosphine oxides (Scheme 19). A Wittig–Horner route to vinyl ethers, and hence ultimately applicable to ketone synthesis, has also appeared.[79]

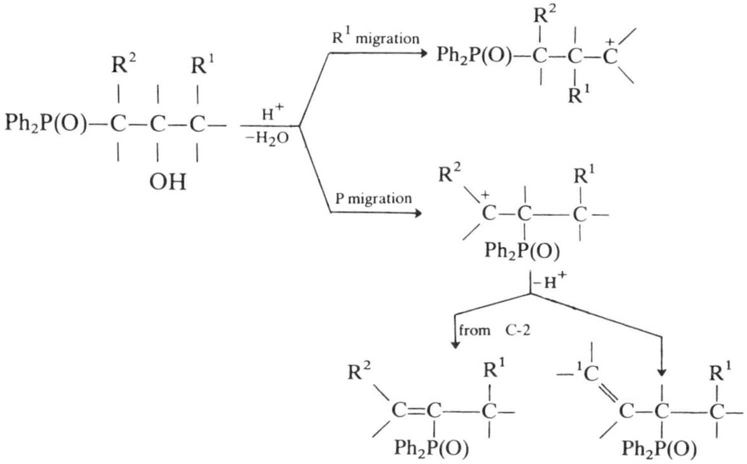

Scheme 19

Olefin syntheses *via* vinylsilanes should prove more easy to undertake now that vinysilanes are accessible from ketones by a two-step procedure (Scheme 20).[80]

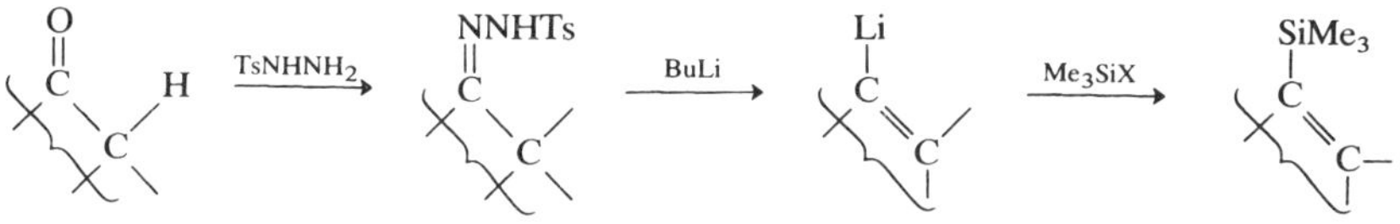

Scheme 20

Trisubstituted vinylsilanes, produced stereospecifically or with a stereochemistry subsequently determined by, for example, ^{13}C n.m.r. spectroscopy, may be converted by a variety of electrophilic reagents into trisubstituted alkenes of known structure (Scheme 21).[81] The hydrolysis of vinylsilanes, which leads to the replacement of Si by H, has been found to occur readily in wet acetonitrile with toluenesulphinic acid as catalyst, but in this reaction the geometry about the double bond is not retained.[82]

79 C. Earnshaw, C. J. Wallis, and S. Warren, *J.C.S. Chem. Comm.*, 1977, 314.

80 R. T. Taylor, C. R. Degenhardt, W. P. Melega, and L. A. Paquette, *Tetrahedron Letters*, 1977, 159.

81 T. H. Chan, P. W. K. Lau, and W. Mychajlowskij, *Tetrahedron Letters*, 1977, 3317; T. H. Chan, W. Mychajlowskij, and R. Amouroux, *ibid.*, p. 1605.

82 G. Büchi and H. Wüest, *Tetrahedron Letters*, 1977, 4305.

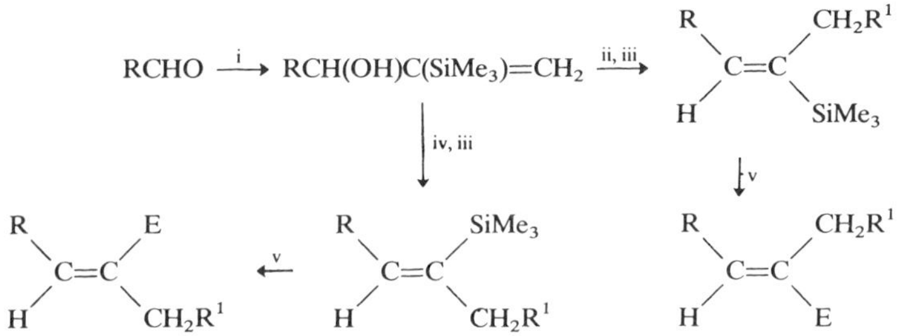

Reagents: 1, $CH_2{=}C(Li)SiMe_3$, $-78\,°C$; ii, Ac_2O; iii, $R^1{}_2CuLi$; iv, $SOCl_2$–Et_2O, 25 °C; v, Br_2 or I_2 or AcCl–$AlCl_3$ or Cl_2CHOMe–$AlCl_3$

Scheme 21

The alkenylboranes derived by *syn*-addition of 9-borabicyclononane (9-BBN) to terminal acetylenes have been found to be much more reactive towards carbonyl groups than saturated organoboranes. Their addition to aldehydes, followed by oxidative hydrolysis in the usual way, provides an ultra-mild pseudo-Grignard synthesis of allylic alcohols of known geometry (Scheme 22).[83] Terminal olefins

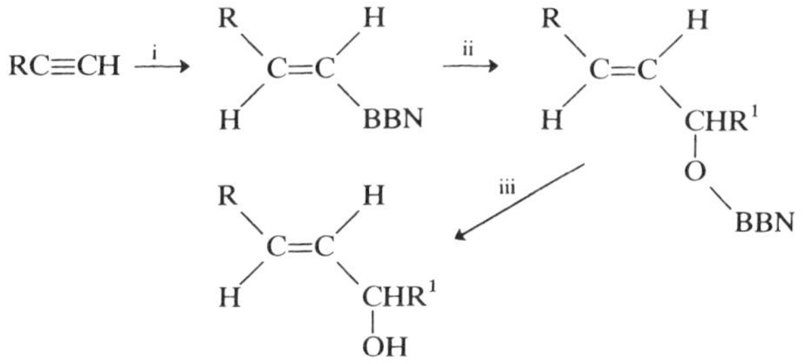

Reagents: i, 9-Borabicyclononane (9-BBN); ii, R^1CHO–THF; iii, NaOH–H_2O_2

Scheme 22

have been prepared by a two-carbon homologation procedure commencing with the reaction of vinylmagnesium bromide with a trialkylborane, yielding a vinylborate(1−) which is cleaved with alkali and iodine.[84] Borate esters, formed quantitatively from a secondary alcohol and boric acid, feature in a mild procedure for olefin formation which is effectively a two-step dehydration of the alcohol. The esters are decomposed by boron trifluoride etherate at 100 °C, and olefin yields are excellent; isomeric mixtures of compositions similar to those obtained by other dehydration methods are formed.[85]

Allyl alcohols can be converted into olefins by replacement of the hydroxy-group by alkyl or aryl residues on treatment with the appropriate alkyl- or aryl-magnesium halide in the presence of nickel(II) complexes, though the formation of

[83] P. Jacob and H. C. Brown, *J. Org. Chem.* 1977, **42**, 579.
[84] K. Utimoto, K. Uchida, M. Yamaya and H. Nozaki, *Tetrahedron*, 1977, **33**, 1945.
[85] M. P. Doyle, S. B. Williams, and C. C. McOsker, *Synthesis*, 1977, 717.

mixtures of isomers indicated that allylic rearrangement accompanied the substitution.[86] A more reliable technique for achieving this rather unusual hydroxyl replacement would seem to be that based on the initial conversion of the alcohol into a lithium allyloxyalkylcuprate, a source of the allyloxy nucleophile for reaction with the phosphonium iodine (16) (Scheme 23).[87]

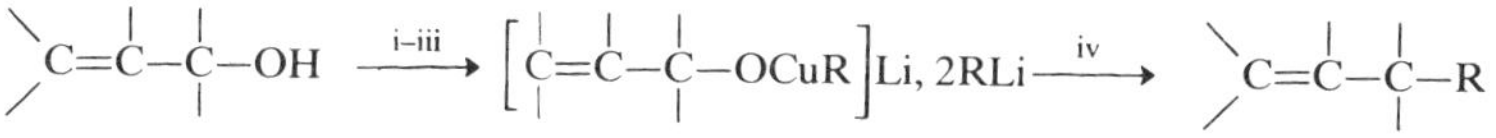

Reagents: i, MeLi; ii, CuI; iii, 3RLi; iv, $Ph_3\overset{+}{P}NMePh\ I^-$ (16), −78 to +25 °C

Scheme 23

Some elegant syntheses of naturally occurring compounds have been reported which are based upon the use of butadiene telomers as starting materials. Examples include the synthesis of *cis*-civetone[88] and queen substance.[89]

Tidwell and co-workers have previously shown that the rates of acid-catalysed hydration of 1,1-disubstituted alkenes can be satisfactorily correlated with the sum of the σ_p^+ constants of the 1-substituents,[57] according to the equation:

$$\log k_2 = -12.3 \sum \sigma_p^+ - 10.1$$

This correlation also applies to the rates of hydration of 2-substituted buta-1,3-dienes, vinyl acetates, and vinylphosphates, and successfully provides a correlation for literature data on hydration rates of 22 substituted styrenes.[90] This finding is taken to confirm the $A d_{SE}2$ (rate-determining *C*-protonation) mechanism. Hydration rates of 1,2-disubstituted olefins require the mathematical model to be slightly modified.[91]

An alternative procedure for effecting the *vic*-addition of halogens to olefins, which avoids the use of elemental bromine or chlorine, has been reported. Positive halogenonium ions are formed from the mixed reagent $HX–H_2O_2$, and the addition of phase-transfer catalyst (PTC) to the two-phase brew minimizes halohydrin formation:[92]

$$R^1CH{=}CHR^2 \xrightarrow[CCl_4\text{–}PTC]{aq.\ HX\text{–}H_2O_2} R^1CHXCHXR^2 + 2H_2O$$

Suitable conditions for vicinal *anti*-addition of $I^+{-}OH^-$, $I^+{-}Cl^-$, and $I^+{-}OCOR^-$ to alkenes[93] and for vicinal alkoxyselenation (addition of RO—SePh) of olefins[94,95] have been reported.

[86] C. Chuit, H. Felkin, C. Frajerman, G. Roussi, and G. Swierczewski, *J. Organometallic Chem.*, 1977, **127**, 371.
[87] Y. Tanigawa, H. Kanamaru, A. Sonoda, and S.-I. Murahashi, *J. Amer. Chem. Soc.*, 1977, **99**, 2361.
[88] J. Tsuji and T. Mandai, *Tetrahedron Letters*, 1977, 3285.
[89] J. Tsuji, K. Masaoka, and T. Takahashi, *Tetrahedron Letters*, 1977, 2267.
[90] W. K. Chwang, P. Knittel, K. M. Koshy, and T. T. Tidwell, *J. Amer. Chem. Soc.*, 1977, **99**, 3395; S. Y. Attia, J. P. Berry, K. M. Koshy, Y.-K. Leung, E. P. Lyznicki, V. J. Nowlan, K. Oyama, and T. T. Tidwell, *ibid.*, p. 3401.
[91] P. Knittel and T. T. Tidwell, *J. Amer. Chem. Soc.*, 1977, **99**, 3408.
[92] T.-L. Ho, B. G. B. Gupta, and G. A. Olah, *Synthesis*, 1977, 676.
[93] R. C. Cambie, D. Chambers, P. S. Rutledge, and P. D. Woodgate, *J.C.S. Perkin I*, 1977, 2231; R. C. Cambie, W. I. Noall, G. J. Potter, P. S. Rutledge, and P. D. Woodgate, *ibid.*, p. 226.
[94] A. Toshimitsu, S. Uemura, and M. Okano, *J.C.S. Chem. Comm.*, 1977, 166.
[95] N. Miyoshi, S. Murai, and N. Sonoda, *Tetrahedron Letters*, 1977, 851.

There have been two useful advances in olefin hydroboration, both stemming from H. C. Brown's group. Firstly, the relatively stable reagent Me_2S,BH_3 is commercially available and makes an excellent starting material for generating borane for addition to hindered olefins or for the preparation of catechylborane.[96] Secondly, it is possible to add 9-BBN selectively to one (usually the less-hindered) double bond of non-conjugated dienes, and hence to introduce functionality at the site of carbon–boron bond formation, a procedure which should find numerous applications in synthesis (Scheme 24).[97]

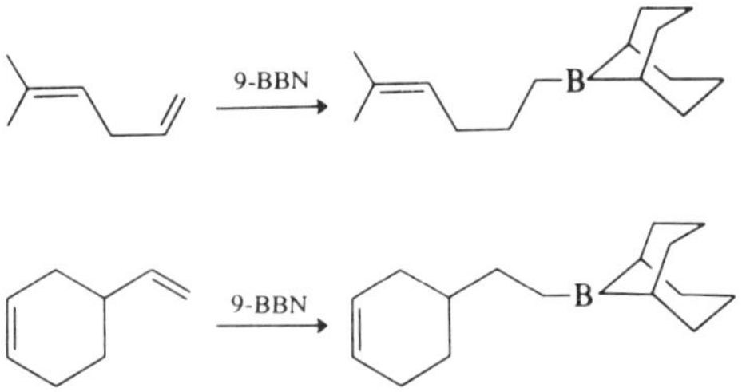

Scheme 24

This year has not been noteworthy for novel developments in olefin cycloadditions, although major papers on the application of the concept of hard and soft acids and bases to the Woodward and Katz model of the Diels–Alder reaction and to [2+2] cycloadditions have appeared.[98] The basis for this treatment of the Diels–Alder reaction is that unsymmetrical reagents will unite to form the two new σ-bonds non-synchronously *via* a non-symmetrical transition state: hence, prediction of the orientation of addition may be made by determining which σ-bond should be formed faster. The assumption is made that this bond will be that linking the *softest* centres, *i.e.* the terminal atoms with greatest overlap stabilization. This leads, for electron-deficient (low LUMO) and electron-rich (high HOMO) reactants, to the rule that atoms with the largest coefficients in these MOs will bond preferentially. Note that all four frontier orbitals must be considered if both reactants have similar electron densities at their terminal atoms.

Applications of Diels–Alder reactions to complex synthetic problems continue to impress this Reporter. Examples which should illustrate this are (i) the use of inter- and intra-molecular Diels–Alder additions of cyanodienes formed *in situ* by electrocyclic ring-opening of benzocyclobutenes (Scheme 25)[99] and (ii) a sequence which may involve the first known example of a bimolecular pericyclic reaction in natural biosynthesis.[100]

Ene reactions have proved a more fruitful area for new developments. A series of conjugated dienes which undergo the ene reaction and cycloaddition simultaneously were treated with a variety of enophiles (maleic anhydride, azo-esters,

96 H. C. Brown, A. K. Mandal, and S. U. Kulkarni, *J. Org. Chem.*, 1977, **42**, 1392.
97 R. Liotta and H. C. Brown, *J. Org. Chem.*, 1977, **42**, 2836.
98 O. Eisenstein, J. M. Lefour, N. T. Anh, and R. F. Hudson, *Tetrahedron*, 1977, **33**, 523; C. Minot and N. T. Anh, *ibid.*, p. 533.
99 T. Kametani, Y. Hirai, F. Satoh, and K. Fukumoto, *J.C.S. Chem. Comm.*, 1977, 16.
100 R. D. Stipanovic, A. A. Bell, D. H. O'Brien, and M. J. Lukefahr, *Tetrahedron Letters*, 1977, 567.

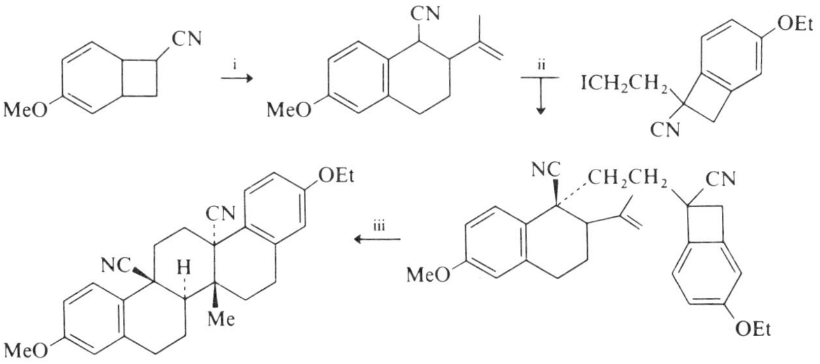

Reagents: i, isoprene, 180 °C, 2h; ii, $NaNH_2$, liq. NH_3; iii, 210–215 °C, toluene, sealed tube

Scheme 25

etc.). The product distributions indicated that steric factors were not of prime importance in the partitioning of a given diene between the two pathways.[101] However, the relative reactivities of different dienes towards a given enophile/dienophile were explicable on a steric basis; for example, (17) is ten times more reactive than (18) towards diethyl azodiformate at 25 °C, because its reactive

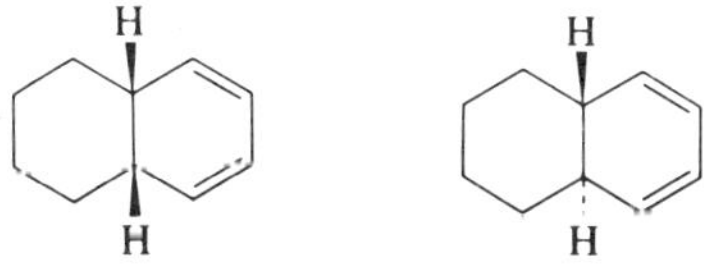

(17) (18)

face is less hindered than either face of (18), in spite of the statistical advantage of (18), which has a suitably placed axial hydrogen on both sides of the ring. A concerted ene reaction seems therefore to be occurring, a conclusion also reached after a kinetic study of ene reactions between maleic anhydride and a series

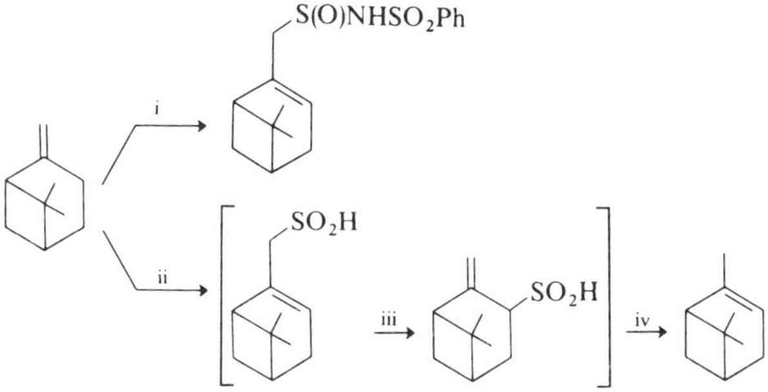

Reagents: i, $PhSO_2N{=}SO$; ii, SO_2; iii, allylic rearrangement; iv, retro-ene, $-SO_2$

Scheme 26

[101] B. M. Jacobson, A. C. Feldstein, and J. I. Smallwood, *J. Org. Chem.*, 1977, **42**, 2849.

of acyclic mono-olefins.[102] Some novel enophiles have been reported: they include *N*-sulphinylbenzenesulphonamide ($PhSO_2N{=}S{=}O$)[103] and sulphur dioxide,[104] the latter causing rapid thermal isomerization of thermodynamically unstable olefins such as β-pinene (Scheme 26). Other novel ene reactions were those induced by the acyl cation (from acetyl hexachloroantimonate), in which a carbocation addition–elimination sequence was precluded by the absence of rearrangement products,[105] and the allyl anion (Scheme 27).[106]

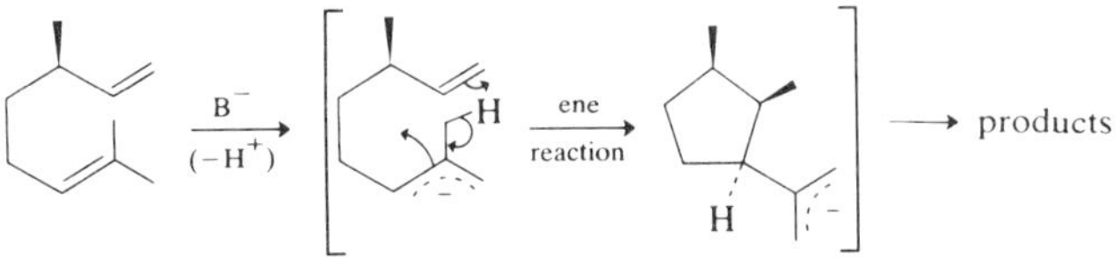

Scheme 27

Catalytic ene reactions of chloral have been the subject of an elegant investigation. Lewis acids which complex with chloral lower the required temperature for ene insertion, which becomes a relatively clean reaction though not always free from by-product formation. Subsequent functional group modification may make this a useful technique for attacking the allylic position in synthesis (Scheme 28).[107]

$Pr^nCH_2CH{=}CH_2 \xrightarrow{i} Pr^nCH{=}CHCH_2CH(OH)CCl_3$

ii, iii ↙ ; iv–vi ↓ ; vii ↘

$Pr^nCH{=}CHCH{=}CHCO_2Et$ (ii, iii)

$Pr^nCH{=}CHCH_2CH(OH)CH_3$ (vii)

$Pr^nCH_2CH_2CH_2CO_2H$ (iv–vi)

Reagents: i, $CCl_3CHO–SnCl_4$ (1–5%), CCl_4; ii, NaH–TsCl; iii, NaOEt–EtOH; iv, Pd/C–H_2; v, CrO_3; vi, aq. NaOH; vii, $Bu^n{}_3SnH$

Scheme 28

With a chiral olefin such as β-pinene, choice of a suitable Lewis acid catalyst enables stereochemical control of the reaction to be achieved.

An improved technique for allylic oxidation by selenium dioxide uses mainly t-butyl hydroperoxide as the oxidant with only 1–2% SeO_2, unless the olefin is very unreactive when higher proportions may have to be used.[108] A detailed study of chromyl chloride oxidation of olefins has shown that all of the three primary products, namely expoxide, chlorohydrin, and vicinal dichloride, arise by *syn*-addition.[109] These observations necessitate a revision of earlier mechanistic pro-

[102] F. R. Benn, J. Dwyer, and I. Chappell, *J.C.S. Perkin II*, 1977, 533.
[103] G. Déléris, J. Kowalski, J. Dunogues, and R. Calas, *Tetrahedron Letters*, 1977, 4211.
[104] M. M. Rogić and D. Masilamani, *J. Amer. Chem. Soc.*, 1977, **99**, 5219.
[105] H. M. R. Hoffmann and T. Tsushima, *J. Amer. Chem. Soc.*, 1977, **99**, 6008.
[106] J. H. Edwards and F. J. McQuillan, *J.C.S. Chem. Comm.* 1977, 838.
[107] G. B. Gill and B. Wallace, *J.C.S. Chem. Comm.*, 1977, 380, 382.
[108] M. A. Umbreit and K. B. Sharpless, *J. Amer. Chem. Soc.*, 1977, **99**, 5526.
[109] K. B. Sharpless, A. Y. Teranishi, and J.-E. Bäckvall, *J. Amer. Chem. Soc.*, 1977, **99**, 3120.

posals based on the intermediate formation of cations, and a scheme involving organochromium(VI) species was advanced (Scheme 29). By contrast, evidence that hindered olefins may react with singlet oxygen *via* a cationic perepoxide pathway was obtained by the detection of substantial amounts of cationic rearrangement products from camphenylideneadamantane.[110]

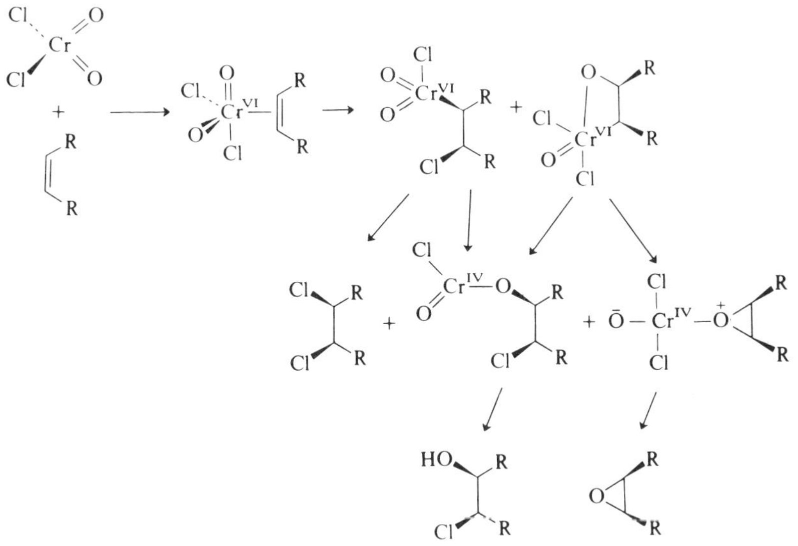

Scheme 29

A new variety of polymer-bound hydrogenation catalyst uses rhodium(III) chloride or other metal halides complexed to a phosphine which is chemically bonded to silica.[111] Photoelectron spectroscopy has shown that ethylene chemisorbed on tungsten has essentially sp^3-hydridized carbon atoms with a carbon–carbon bond length stretched to 1.54 Å,[112] a further manifestation of the diverse applications of this recently developed analytical technique.

[110] F. McCapra and I. Beheshti, *J.C.S. Chem. Comm.*, 1977, 517.
[111] K. Kochloefl, W. Liebelt, and H. Knozinger, *J.C.S. Chem. Comm.*, 1977, 510.
[112] T. V. Vorburger, B. J. Waclawski, and E. W. Plummer, *Chem. Phys. Letters*, 1977, **46**, 42.

9 Aliphatic Compounds
Part (ii) Other Aliphatic Compounds

By R. S. WARD

Department of Chemistry, University College of Swansea, Swansea SA2 8PP

Many recent developments in aliphatic chemistry can be directly attributed either to the continued search for new synthetic methods or to attempts at rationalizing mechanistic or stereochemical principles. Consequently current research reflects a strong interest in the following topics: (a) new and improved asymmetric syntheses of chiral compounds; (b) the development of new methods for preparing lactones; (c) investigations of reversed (*umpolung*) reactivity and equivalent reagents; (d) the generation and reactions of regiospecific enolates; (e) studies of the stereochemistry of reactions in organo-phosphorus compounds; (f) investigations of new reactions of organo-sulphur compounds.

1 Amines and Imines

The polyether antibiotic lasalocid produced by *Streptomyces lasaliensis* has been used to resolve racemic amines by fractional crystallization of their salts.[1] High enantioselectivity is observed (% $R:S \sim 90:10$) in cases where the asymmetric carbon atom is attached directly to the primary amino-group (*e.g.* 1, 2). When the asymmetric centre is separated from the amino-group by a methylene group (*e.g.* 3, 4) the efficiency of the resolution is reduced (% $R:S \sim 25:75$) and the enantiomer in excess has the opposite configuration (*cf.* 1 and 4).

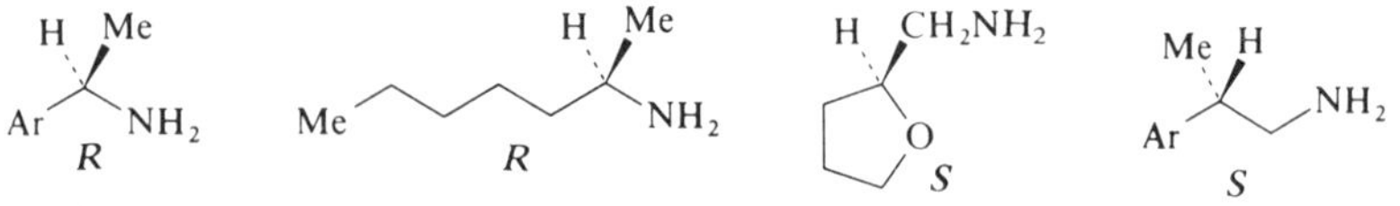

(1) (2) (3) (4)

Many enzyme-catalysed reactions involve the oxidation or reduction of nitrogen containing compounds by NAD(P) or NAD(P)H. Model studies have shown that amines can be oxidized by pyridinium salts.[2] Furthermore, imines can be reduced by 1,4-dihydropyridine derivatives in the presence of metal complexes (Scheme 1).[3]

[1] J. W. Westley, R. H. Evans, and J. F. Blount, *J. Amer. Chem. Soc.*, 1977, **99**, 6057.
[2] Y. Ohnishi, *Tetrahedron Letters*, 1977, 2109.
[3] U. K. Pandit, H. van Dam, and J. B. Steevens, *Tetrahedron Letters*, 1977, 913; see also R. A. Gase and U. K. Pandit, *J.C.S. Chem. Comm.*, 1977, 480.

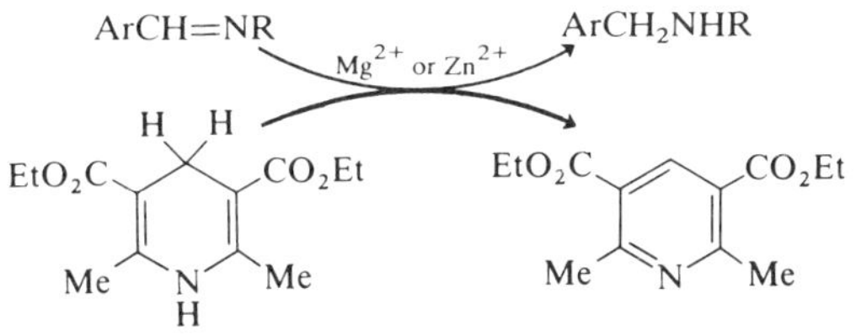

Scheme 1

The metal-ion catalysed synthesis of α-amino-esters from α-keto-esters and amines in the presence of 1,4-dihydropyridines has also been reported.[4] These reactions can be regarded as models for the enzyme-catalysed interconversion of α-keto- and α-amino-acids. Indeed they serve to demonstrate the importance of electrophilic catalysis in NAD(P)H mediated reduction reactions.

The anions derived from dialkylallylamines (5) are useful β-acyl carbanion equivalents.[5] With alkyl halides the γ-substituted products (6) are formed, while with aldehydes and ketones a mixture of the cyclized γ-adduct (7) and the α-adduct (8) is obtained. The lithium reagent (9) gives approximately equal amounts of (7) and (8) while the analogous zinc reagent (10) gives mainly the α-adduct (8).

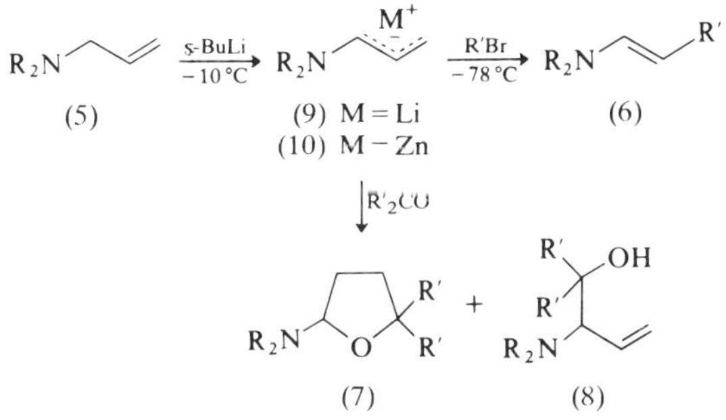

The anions derived from N-nitrosoamines also react with electrophilic reagents (Scheme 2).[6] The reactions are regio- and stereo-selective and can be used to effect

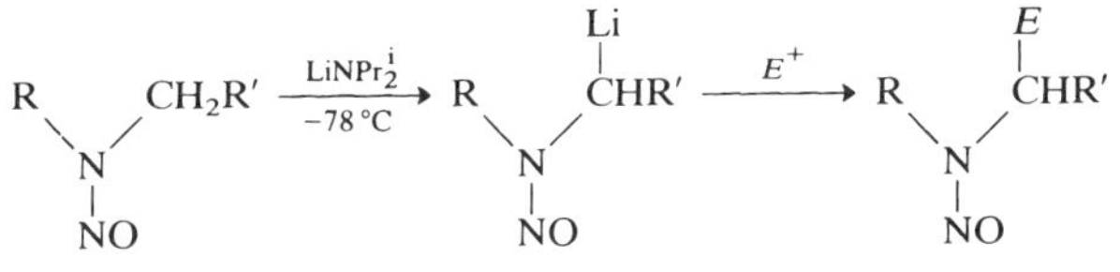

Scheme 2

electrophilic substitution at the α-carbon atom of secondary amines. The anions derived from N-nitrosoallylamines undergo alkylation at the α-position but react

[4] K Nakamura, A. Ohno, and S. Oka, *Tetrahedron Letters*, 1977, 4593.

[5] S. F. Martin and M. T. DuPriest, *Tetrahedron Letters*, 1977, 3925.

[6] D. Seebach, D. Enders, and B. Renger, *Chem. Ber.*, 1977, **110**, 1852; B. Renger, H.-O. Kalinowski, and D. Seebach, *ibid.*, 1866; D. Seebach, D. Enders, R. Dach, and R. Pieter, *ibid.*, 1879; B. Renger and D. Seebach, *ibid.*, 2334.

with aldehydes and ketones reversibly to give a mixture of α- and γ-substituted products. Although reactivity at the α-position is kinetically favoured, the γ-products are formed exclusively under conditions which favour thermodynamic control.

The vinyloxycarbonyl group (VOC) has been shown to be a versatile protecting group for amino-compounds. Amino-acids can be easily converted into their *N*-VOC derivatives by treating them with vinyl chloroformate. The protecting group can be selectively removed by treatment with HCl, HBr, or bromine in dichloromethane, followed by warming in alcohol. Since vinyl chloroformate also attacks tertiary amines it provides a method for selective dealkylation of tertiary amines (11 → 12).[7]

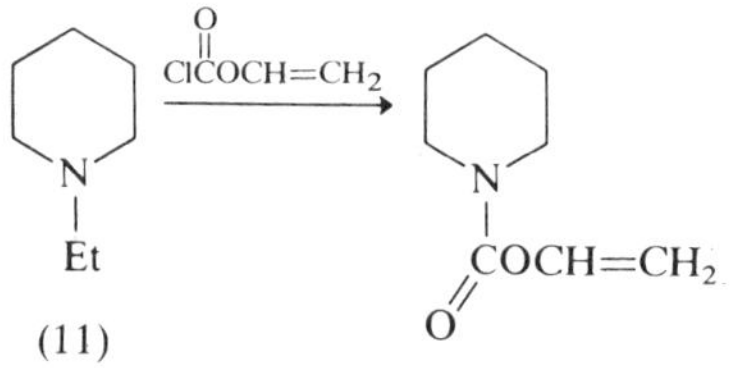

(11) (12)

A new procedure for the *NN*-dimethylation of primary amines involves the preparation and reduction of the bis(alkoxymethyl)amine [$RN(CH_2OCH_3)_2$],[8] while *N*-alkylation of secondary amines by alcohols can be accomplished in high yields (94—98 %) in the presence of aluminium t-butoxide and Raney nickel.[9] The conversion of primary amines into esters (13 → 14) can be carried out by treating

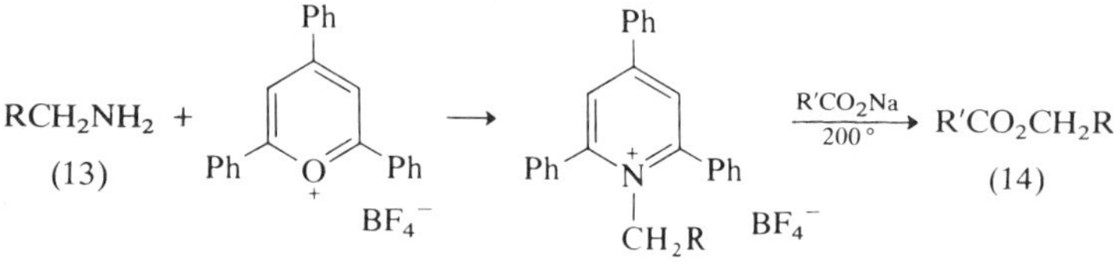

(13) (14)

the amine with 2,4,6-triphenyl-pyrylium tetrafluoroborate to give a pyridinium salt which is then heated *in vacuo* with the sodium salt of a carboxylic acid.[10]

Secondary enamines (15) are thermodynamically unstable since they exist in equilibrium with the tautomeric imines (16). However secondary enamines can be prepared by partial methanolysis of organo-tin or magnesium salts of imines.[11]

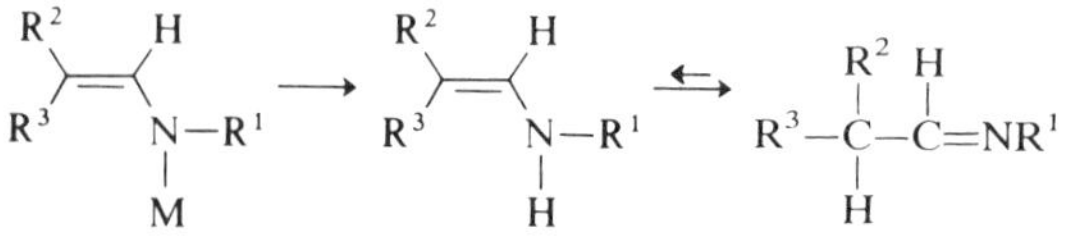

(15) (16)

[7] R. A. Olofson, Y. S. Yamamoto, and D. J. Wancowicz, *Tetrahedron Letters*, 1977, 1563; R. A. Olofson, R. C. Schnur, L. Bunes, and J. P. Pepe, *ibid.*, 1567.
[8] H. Kapnang, G. Charles, B. L. Sondengam, and J. Hentchoya Hemo, *Tetrahedron Letters*, 1977, 3469.
[9] M. Botta, F. De Anngelis, and R. Nicoletti, *Synthesis*, 1977, 722.
[10] U. Gruntz, A. R. Katritzky, D. H. Kenny, M. C. Rezende, and H. Sheikh, *J.C.S. Chem. Comm.*, 1977, 701.
[11] B. De Jeso and J.-C. Pommier, *J.C.S. Chem. Comm.*, 1977, 565.

They are stable at −80 °C but completely isomerize within one hour at room temperature.

Optically active oxaziridines (18) can be prepared by oxidation of imines (17) with *m*-chloroperbenzoic acid in (*S*)-(+)-2,2,2-trifluoro-1-phenylethanol,[12] or by photochemical rearrangement of nitrones (19) in the same chiral solvent.[13] In both cases the enantioselectivity of the reaction is increased by lowering the temperature. The optical yields obtained by both methods compare favourably with those obtained by oxidation of imines with chiral peroxyacids. In the case of an optically active imine (*e.g.* 20) the diastereoselectivity of the reaction is not dependent upon the nature or chirality of the peroxyacid or the solvent. Indeed the configuration of the major product depends only on the configuration of the starting imine. If the substituent has the *R* configuration attack from the *Si* face is preferred.[14]

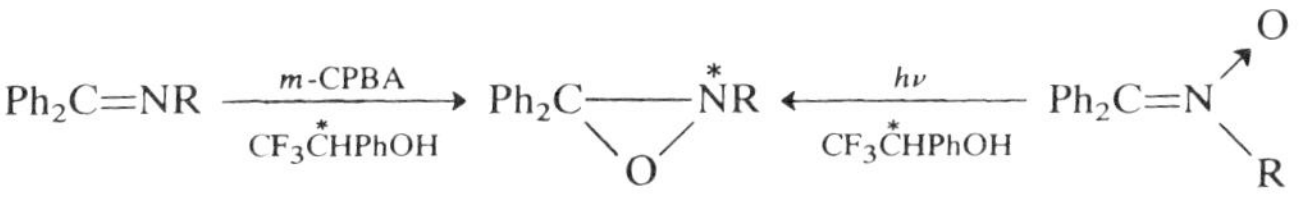

(17) (18) (19)

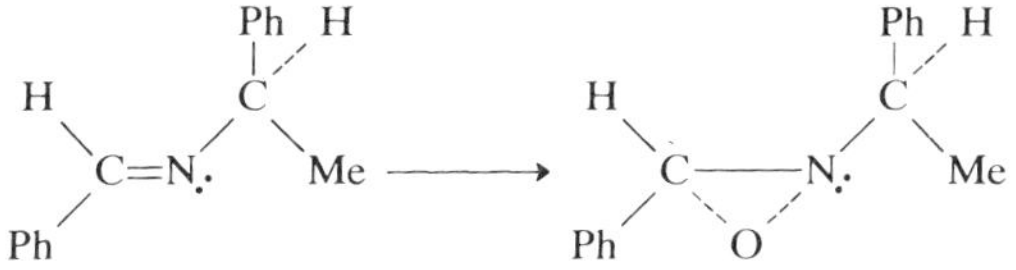

(20) (21)

The enantioselective alkylation of the cyclohexanone imine (22) gives a product of high optical purity (81 %) due to the presence of a suitably situated ether group which reduces the flexibility of the metallated intermediate (23).[15] The alkylation of a chiral enamine of cyclohexanone gives a product of equally high optical purity (83 %).[15]

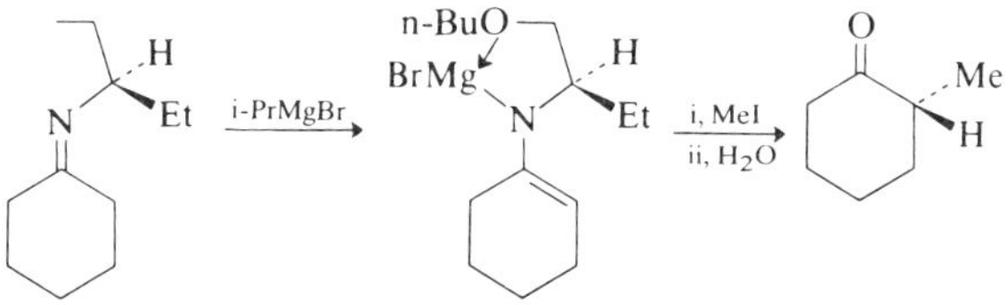

(22) (23)

The cycloaddition of chiral carbodi-imides to achiral ketenes provides an asymmetric synthesis of β-lactams, while 1,3-dipolar cycloaddition of chiral nitrones to achiral alkenes affords optically active isoxazolines.[16]

[12] A. Forni, I. Moretti, and G. Torre, *J.C.S. Chem. Comm.*, 1977, 731; M. Bucciarelli, A. Forni, I. Moretti, and G. Torre, *J.C.S. Perkin II*, 1977, 1339.
[13] D. R. Boyd and D. C. Neill, *J.C.S. Chem. Comm.*, 1977, 51.
[14] D. Mostowicz and C. Belzecki, *J. Org. Chem.*, 1977, **42**, 3917.
[15] J. K. Whitesell and M. A. Whitesell, *J. Org. Chem.*, 1977, **42**, 377; J. K. Whitesell and S. W. Felman, *ibid.*, 1663.
[16] C. Belzecki and Z. Krawczyk, *J.C.S. Chem. Comm.*, 1977, 302; C. Belzecki and I. Panfil, *ibid.*, 1977, 303.

2 Nitriles and Isocyanides

Achiral carbocations generated from racemic alcohols react with chiral nitriles to give a mixture of diastereoisomeric amides which can be hydrolysed to give optically active amines (Scheme 3).[17]

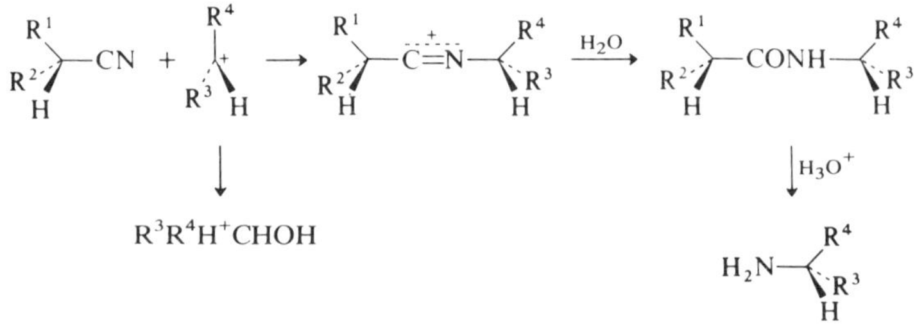

Scheme 3

trans-β-(1-Pyrrolidinyl)acrylonitrile reacts with lithium di-isopropylamide at −105 °C to give the β-metallated derivative, but at higher temperatures the α-metallated derivative is formed (Scheme 4).[18] Both vinyl-lithium reagents react with electrophiles with retention of the configuration of the double bond.

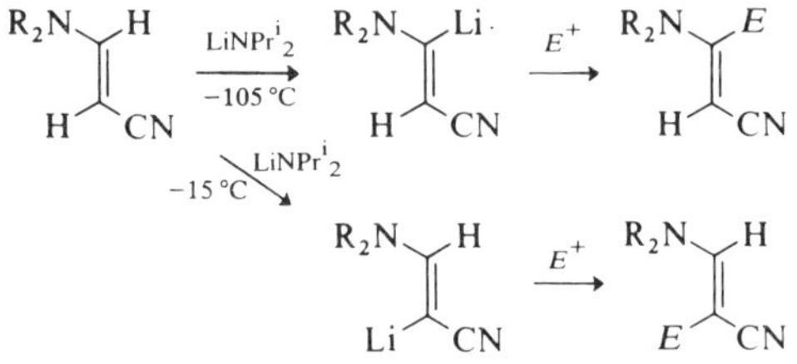

Scheme 4

The thermal rearrangement of nitrile oxides (24) to isocyanates (25) is limited as a preparative method by competition from the dimerization reaction. However the rearrangements can be carried out in excellent yields under mild conditions by using a suitable catalyst, such as sulphur dioxide, which participates in the reaction as a dipolarophile.[19]

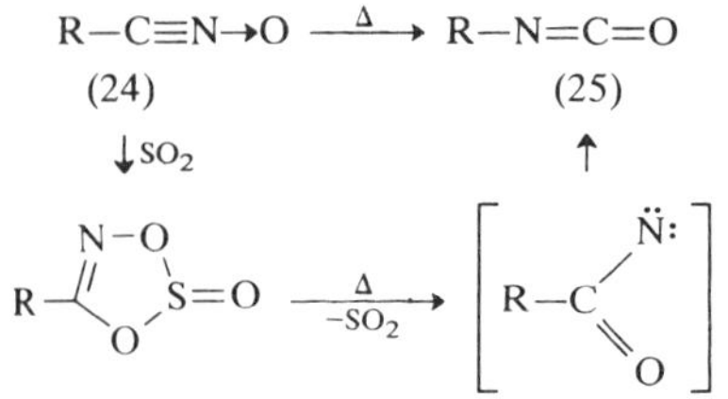

[17] W. Tomasik and C. Belzecki, *J.C.S. Chem. Comm.*, 1977, 86.
[18] R. R. Schmidt and J. Talbiersky, *Angew. Chem. Internat. Edn.*, 1977, **16**, 853.
[19] G. Trickes and H. Meier, *Angew. Chem. Internat. Edn.*, 1977, **16**, 555.

N-Acyl isocyanides (26) can be prepared by treating carboxylic acid iodides with silver cyanide.[20] They react with ynamines to give adducts (27), and undergo cycloaddition to electron-deficient alkynes to give heterocyclic products (28). *N*-Imidoyl-isocyanides have also been prepared from *N*-phenyl-imidoyl bromides.

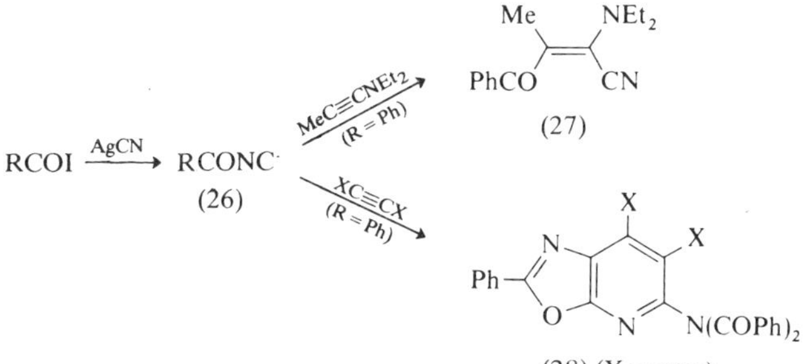

α-Metallated isocyanides undergo cycloaddition to polar double bonds and also act as synthons for α-metallated primary amines (Scheme 5).[21] They contain a

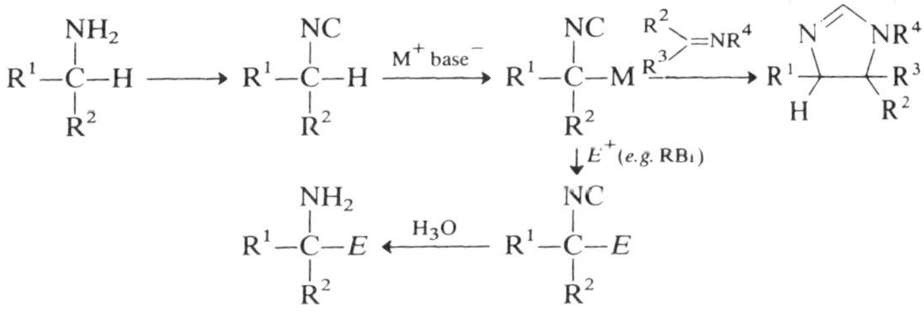

Scheme 5

nucleophilic centre (the metallated carbon atom) and an electrophilic centre (the isocyanide carbon atom) and therefore undergo cycloaddition with imines, *e.g.*, to give 2-imidazolines. The reaction with alkyl halides leads to chain extension of the amine equivalent. The reactions of metallated alkenyl isocyanides with electrophiles and 2-isocyanoacrylic esters with carbanions have also been studied.

3 Nitro-compounds

The powerful nucleophilic toluene *p*-thiolate anion can displace a nitro-group from simple unactivated primary and secondary nitroalkanes yielding alkyl *p*-tolyl sulphides, apparently by a conventional S_N2 displacement process.[22]

Several methods for converting nitroalkanes into carbonyl compounds are available including the acid-catalysed hydrolysis of nitronate salts (Nef reaction).

[20] G. Höfle and B. Lange, *Angew. Chem. Internat. Edn.*, 1977, **16**, 262 and 727.

[21] U. Schöllkopf, *Angew. Chem. Internat. Edn.*, 1977, **16**, 339; U. Schöllkopf, K.-W. Henneke, K. Madawinata, and R. Harms, *Annalen*, 1977, 40; U. Schöllkopf, D. Stafforst, and R. Jentsch, *ibid.*, 1167; U. Schöllkopf and R. Meyer, *ibid.*, 1174; R. Meyer, U. Schöllkopf, and P. Böhme, *ibid.*, 1183.

[22] M. Benn and A. C. M. Meesters, *J.C.S. Chem. Comm.*, 1977, 597.

The oxidative cleavage of nitronate anions under very mild conditions using t-butylhydroperoxide and $VO(acac)_2$ as catalyst affords a simple approach to the prostaglandin intermediate (29).[23] The nitro to carbonyl conversion can also be

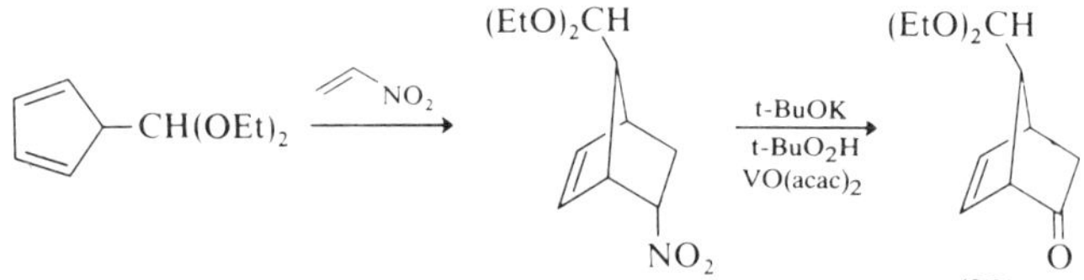

(29)

achieved using basic silica gel. The usefulness of silica, both as a reaction medium and as a reagent, is illustrated by the synthesis of dihydrojasmone (30) from n-heptylamine.[24]

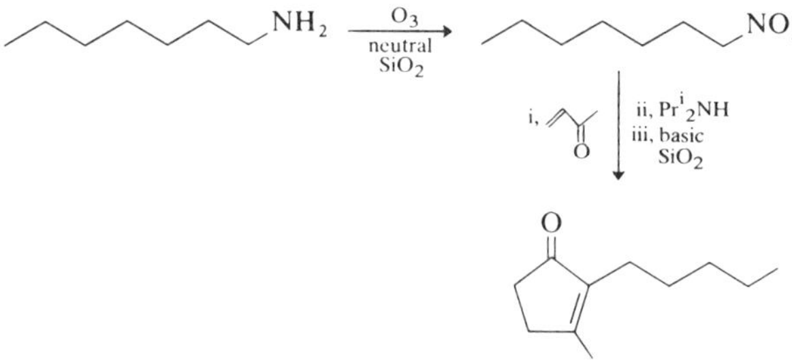

(30)

aci-Nitro-compounds are oxidized to olefinic dimers by silver ions. For example, tetramethyl 1,3,5-hexatriene-1,1,6,6-tetracarboxylate (31) is obtained by heating the silver salt of dimethyl 3-*aci*-nitro-1-propene-1,1-dicarboxylate (32) in acetonitrile.[25]

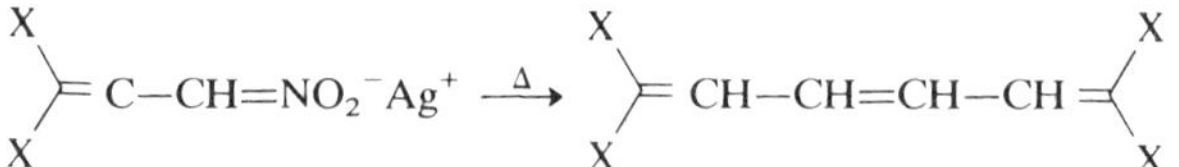

(32) (31)

$(X = CO_2Me)$

A total synthesis of the naturally occurring azoxyalkene, elaiomycin (33) has been dary nitroalkanes have been reported.[26] The dianions are powerful carbon nucleophiles, unlike the simple monoanions which undergo *O*-alkylation.

4 Azo- and Diazo-compounds

A total synthesis of the naturally occurring azoxyalkene, elaiomycin (33) has been reported.[27] The azoxy function is generated by nitrosation of the protected

[23] P. A. Bartlett, F. R. Green, and T. R. Webb, *Tetrahedron Letters*, 1977, 331.
[24] E. Keinan and Y. Mazur, *J. Org. Chem.*, 1977, **42**, 844 and *J. Amer. Chem. Soc.*, 1977, **99**, 3861.
[25] T. Severin, I. Bräutigam, and K.-H. Bräutigam, *Chem. Ber.*, 1977, **110**, 1669.
[26] D. Seebach, R. Henning, F. Lehr, and J. Gonnermann, *Tetrahedron Letters*, 1977, 1161.
[27] R. A. Moss and M. Matsuo, *J. Amer. Chem. Soc.*, 1977, **99**, 1643.

urethane (34), followed by cleavage with potassium t-butoxide, and alkylation of the diazotate so formed.

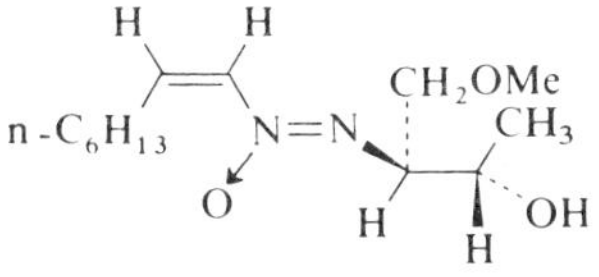

(33)

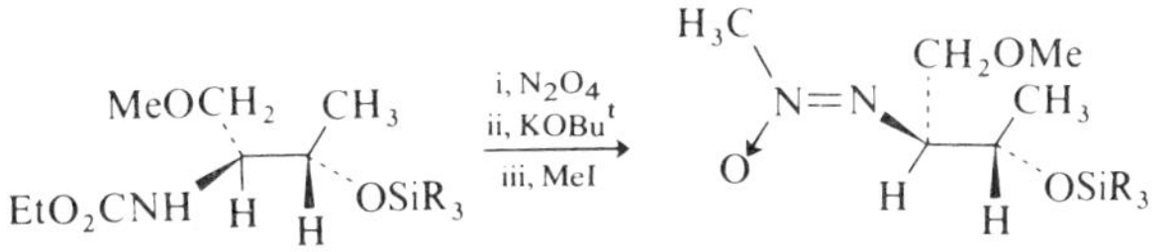

(34)

1-Phenyl-1,2-diaza-allyl-lithium (35) reacts with primary alkyl halides at N-1 to give formaldehyde alkylphenylhydrazones (36) and with aldehydes and ketones at C-3 to give α-hydroxyaldehyde phenylhydrazones (37).[28]

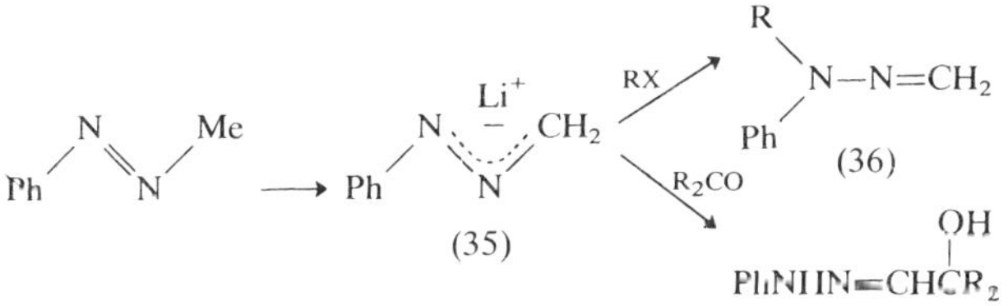

(35) (36) (37)

The copper-catalysed thermolysis of α-diazo-β-dicarbonyl compounds in the presence of enol ethers provides a useful approach to β-methylfurans and their methylenelactone equivalents.[29] The reaction of enol ether (38) with dimethyl diazo-malonate for example, in the presence of trimethoxyphosphine copper(I)iodide, gives the diester (39) which on treatment with sodium hydride followed by lithium aluminium hydride and finally acid yields menthofuran (40).

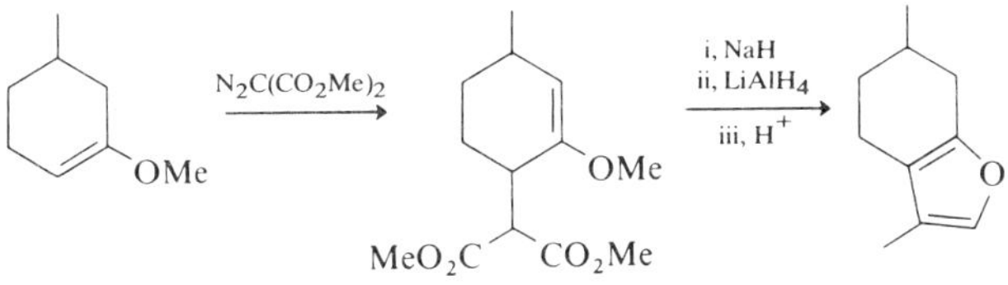

(38) (39) (40)

5 Alcohols and Ethers

A comparison of the acidities of alcohols and thiols in DMSO reveals that replacing the hydroxy group by SH results in a 16—20 kcal/mol increase in the enthalpy of

28 T. Kauffmann, D. Berger, B. Scheerer, and A. Woltermann, *Chem. Ber.*, 1977, **110**, 3034.
29 E. Wenkert, M. E. Alonso, B. L. Buckwalter, and K. J. Chou, *J. Amer. Chem. Soc.*, 1977, **99**, 4778.

deprotonation.[30] However, while the acidity order for alcohols in DMSO is MeOH > EtOH > i-PrOH > t-BuOH the acidity order for thiols shows no apparent trend. The reversal of the acidity order for alcohols, relative to the gas-phase, apparently results primarily from the effect of the alkyl substituent on the heat of solvation of the alkoxide anion. The increased bulk of the alkyl residue hinders solvation of the anion, and destabilizes it relative to the alcohol. In water there is apparently a complete reversal of the acidity order for both alcohols and thiols, relative to the gas-phase. The main factor responsible for the reversed acidity order for thiols is the entropy of ionization in aqueous solution.[31]

Alcohols react with diphenylphosphoryl azide, along with triphenylphosphine and diethyl azodicarboxylate, to give azides in 60—90 % yield.[32] The reaction is stereospecific, involving inversion of configuration, but is sensitive to steric hindrance. Trimethylsilyl iodide converts alcohols into iodides (with inversion of configuration).[33] The latter reagent can also be used to cleave ethers and esters under neutral conditions. The oxidative cleavage of alkyl and cycloalkyl ethers to carbonyl compounds can be accomplished using nitronium tetrafluoroborate or iodine pentafluoride.[34]

The reactions of optically active (R, Z) allylic alcohols with NN-dimethylformamide dimethyl acetal proceeds stereoselectively *via* a [2,3]-sigmatropic rearrangement to give optically active (R, E) amides with 100% transfer of chirality (Scheme 6).[35]

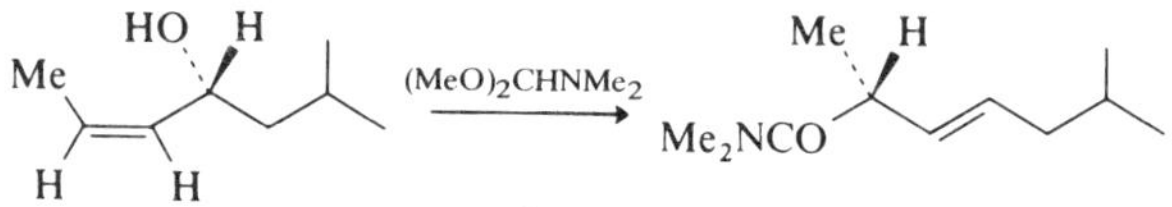

Scheme 6

The reactions of epoxides with the sodium salts of malonate esters afford a novel approach to α-methylene γ-lactones. Bis(α-methylene γ-lactones) can be pre-prepared by treating diketones with 2-(bromomethyl)acrylate.[36]

6 Aldehydes and Ketones

Complete kinetic stereoselectivity can be achieved in aldol condensations of preformed lithium enolates (Scheme 7).[37] The *erythro* : *threo* ratio equals the (Z):(E) ratio under appropriate conditions. By using the lithium enolate derived from a ketone containing a large alkyl group (*e.g.* $R = Bu^t$) one can obtain exclusively the *erythro* aldol, since such ketones yield only the (Z) enolate on deprotonation.

[30] E. M. Arnett and L. E. Small, *J. Amer. Chem. Soc.*, 1977, **99**, 808.
[31] J. E. Bartmess and R. T. McIver, *J. Amer. Chem. Soc.*, 1977, **99**, 4163.
[32] B. Lal, B. N. Pramanik, M. S. Manhas, and A. K. Bose, *Tetrahedron Letters*, 1977, 1977.
[33] M. E. Jung and P. L. Ornstein, *Tetrahedron Letters*, 1977, 2659; M. E. Jung and M. A. Lyster, *J. Amer. Chem. Soc.*, 1977, **99**, 968; see also T.-L.Ho and G. A. Olah, *Synthesis*, 1977, 417 and M. E. Jung and M. A. Lyster, *J. Org. Chem.*, 1977, **42**, 3761.
[34] T.-L.Ho and G. A. Olah, *J. Org. Chem.*, 1977, **42**, 3097; G. A. Olah and J. Welch, *Synthesis* 1977, 419.
[35] K. K. Chan and G. Saucy, *J. Org. Chem.*, 1977, **42**, 3828.
[36] H. Marschall, F. Vogel, and P. Weyerstahl, *Annalen*, 1977, 1557; N. Bensel, K.-D. Klinkmüller, H. Marschall, and P. Weyerstahl, *ibid.*, 1572.
[37] W. A. Kleschick, C. T. Buse, and C. H. Heathcock, *J. Amer. Chem. Soc.*, 1977, **99**, 247; C. T. Buse and C. H. Heathcock, *ibid.*, 8109.

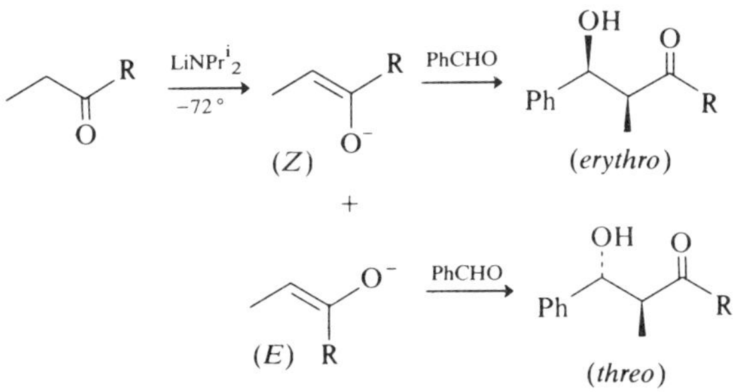

Scheme 7

The stereoselectivity of the reaction can however be completely reversed by using the tetra-alkylammonium enolate derived from the same ketone. The stereoselective condensation of the lithium enolate ($R = Bu^t$) with a chiral aldehyde yields only two of the four possible diastereo-isomeric products.

As a preparation reaction the aldol condensation suffers from the disadvantage that self-condensation products and $\alpha\beta$-unsaturated carbonyl compounds are invariably formed. The aldol condensation between enol silyl ethers and carbonyl compounds (*e.g.* 41 → 42) in the presence of tetrabutyl ammonium fluoride [38] (or titanium tetrachloride[39]) proceeds in a regiospecific manner and gives only minor quantities of the usual by-products. The use of dimethyl(methylene)ammonium trifluoroacetate and the corresponding iodide to quench regiospecifically generated enolates affords a convenient route, *via* the Mannich intermediate (*e.g.* 44), to α-methylene ketones, esters, and lactones.[40]

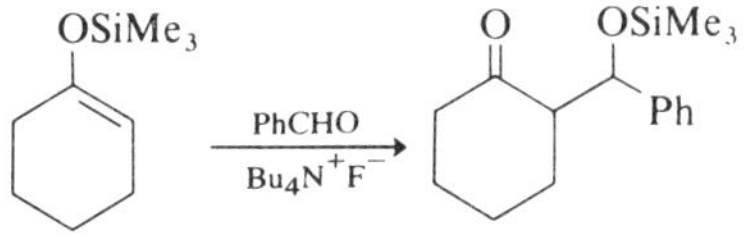

(41) (42)

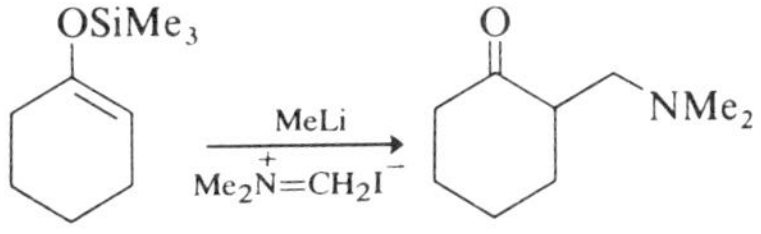

(43) (44)

It has been demonstrated that the intramolecular condensation of 1,5-diketones can be directed to give either of the two possible products (45 and 46), depending upon whether conditions favouring kinetic or thermodynamic control are employed.[41]

38 R. Noyori, K. Yokoyama, J. Sakata, I. Kuwajima, E. Nakamura, and M. Shimizu, *J. Amer. Chem. Soc.*, 1977, **99**, 1265.

39 T. Mukaiyama, *Angew. Chem. Internat. Edn.*, 1977, **16**, 817.

40 N. L. Holy and Y. F. Wang, *J. Amer. Chem. Soc.*, 1977, **99**, 944; J. L. Roberts, P. S. Borromeo and C. D. Poulter, *Tetrahedron Letters*, 1977, 1621.

41 M. Larcheveque, G. Valette, and T. Cuvigny, *Synthesis*, 1977, 424.

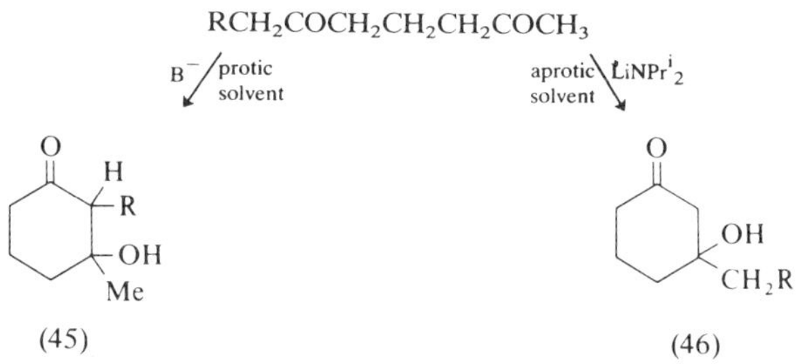

A difficulty which frequently arises in the alkylation of β-dicarbonyl compounds is the concurrent formation of *C*- and *O*-alkylated products. In addition, the attempted monoalkylation of β-dicarbonyl compounds often gives low yields due to competition from dialkylation and other base-catalysed reactions. Several attempts have been made to overcome these difficulties by shielding the oxygen atom by association with a metal cation or a hydrogen-bonding solvent. A new procedure which gives high yields of mono-*C*-alkylated products with no apparent *O*-alkylation or other side reactions involves the use of monosolvates of β-dicarbonyl compounds with tetra-alkylammonium fluorides.[42] The oxygen atom is shielded not only by the large cation but also by the enol hydroxy-fluoride bond. Furthermore the reactivity of the β-dicarbonyl compound is enhanced by transfer of electron density from the fluoride anion *via* the hydrogen bond. The overall result therefore is to shield the oxygen atom, while at the same time increasing the effectiveness of the molecule as a nucleophile.

An n.m.r. study has shown that the anions derived from aldehyde dimethylhydrazones are formed with substantial stereoselectivity and do not equilibrate under the conditions usually employed for alkylation and acylation reactions.[43] However a study of the stereospecificity of deprotonation and alkylation of ketone dimethylhydrazones has revealed a large kinetic preference for formation of the *anti*-anion, followed by isomerization to give the thermodynamically more stable *syn*-anion (Scheme 8).[44] In the case of unsymmetrical ketone dimethylhydrazones

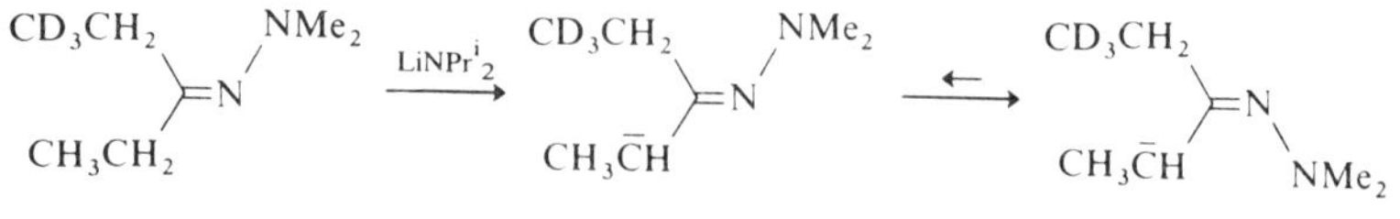

Scheme 8

a primary carbanion is formed in preference to a secondary one irrespective of the hydrazone stereochemistry, but subsequently isomerizes (in the case of an *anti*-anion) to give the *syn*-anion. The enantioselective alkylation of a metallated chiral hydrazone forms the basis of an asymmetric synthesis of chiral aldehydes.[45]

Doubly unsaturated carbonyl compounds react with nucleophiles at C-5. However this reactivity can be reversed by using the dianion formed from the

[42] J. H. Clark and J. M. Miller, *J.C.S. Chem. Comm.*, 1977, 64 and *J.C.S. Perkin I*, 1977, 1743.
[43] M. Newcomb and D. E. Bergbreiter, *J.C.S. Chem. Comm.*, 1977, 486.
[44] M. E. Jung and T. J. Shaw, *Tetrahedron Letters*, 1977, 3305.
[45] D. Enders and H. Eichenauer, *Tetrahedron Letters*, 1977, 191.

analogous $\gamma\delta$-unsaturated carbonyl compound (Scheme 9).[46] The configuration of the product depends upon the nature of the electrophile used.

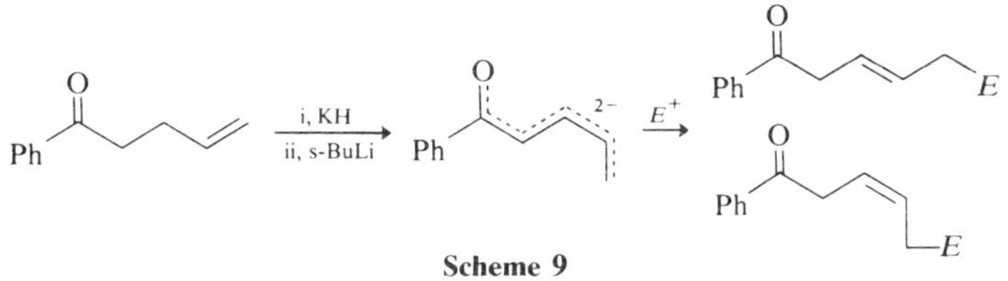

Scheme 9

There is a continued interest in the asymmetric reduction of carbonyl groups. The homogeneous hydrogenation of α-keto-esters using rhodium complexes of chiral diphosphines gives α-hydroxy-esters with an enantiomeric excess of 66—76%.[47] The complex produced by treating lithium aluminium hydride with two equivalents of a chiral amino-alcohol reduces acetylenic ketones to give chiral acetylenic alcohols with an enantiomeric excess of 60—84 %.[48]

7 Carboxylic Acids

Dynamic n.m.r. spectroscopy has been used to study C—S p–$p\pi$-bonding and conformational equilibria (47 $\rightleftharpoons$ 48) of thioacetic acid.[49] The high rotational bar-

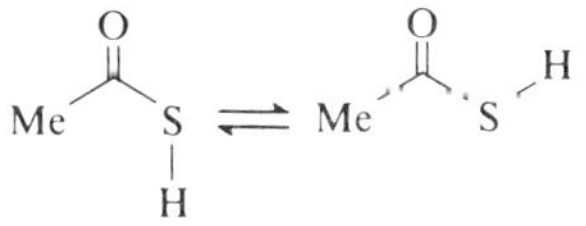

(47) (48)

rier that is observed is taken as evidence for an appreciable contribution from charged resonance forms, and indicates that C—S and C—N p–$p\pi$-bonding are of similar importance in thioacetic acid and acetamide. Contrary to the generally accepted explanation for the activation of thiol esters, the available data suggests that such conjugation is less important in acetic acid itself.

Two new methods for converting carboxylic acids into thiol esters have been reported. The first involves treating the acid with carbonyl di-imidazole (CDI) or carbonyl di-1,2,4-triazole (CDT) to give an activated intermediate which is then treated with the thiol.[50] The second involves treating the carboxylic acid with a tris(alkylthio)borane (Scheme 10).[51]

The kinetics and activation parameters for lactonization of a whole series (n = 3—23) of ω-bromoalkanoic acids $Br(CH_2)_{n-2}CO_2H$ with strong bases in 99 % aqueous dimethyl sulphoxide have now been studied.[52] The reactivities span six

[46] M. Pohmakotr and D. Seebach, *Angew. Chem. Internat. Edn.*, 1977, **16**, 320.
[47] I. Ojima, T. Kogure, and K. Achiwa, *J.C.S. Chem. Comm.*, 1977, 428.
[48] R. S. Brinkmeyer and V. M. Kapoor, *J. Amer. Chem. Soc.*, 1977, **99**, 8339.
[49] E. A. Noe, *J. Amer. Chem. Soc.*, 1977, **99**, 2803.
[50] H.-J. Gais, *Angew. Chem. Internat. Edn.*, 1977, **16**, 244.
[51] A. Pelter, T. E. Levitt, K. Smith, and A. Jones, *J.C.S. Perkin I*, 1977, 1672.
[52] C. Galli, G. Illuminati, L. Mandolini, and P. Tamborra, *J. Amer. Chem. Soc.*, 1977, **99**, 2591.

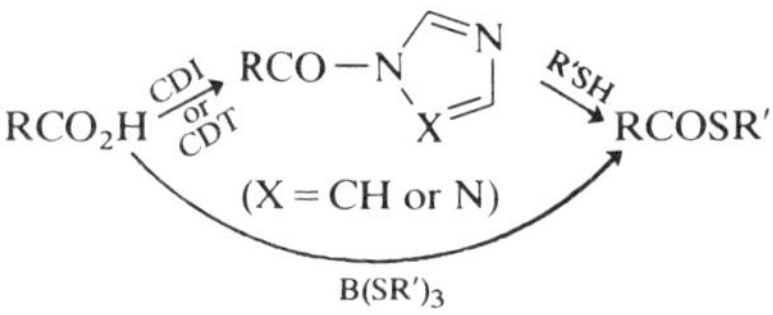

Scheme 10

powers of ten, the most reactive compound being the one for $n = 5$ and the least reactive being the one for $n = 8$. Comparison of the enthalpies of activation for ring-closure with those for the corresponding intermolecular reactions provides a quantitative estimate of the strain energies accompanying ring-closure. The highest strains are experienced for $n = 3$ and 8. The entropies of activation confirm Ruzicka's hypothesis that the probability of mutual approach of the two ends of a bifunctional chain decreases as the length of the chain increases.

The reactions of unsaturated carboxylic acids with phenyl selenenyl halides or benzene sulphenyl halides afford a mild lactonization procedure which has the added advantage that the versatile phenylselenenyl or phenylsulphenyl group is incorporated into the product (*e.g.* 49 → 50).[53] The PhSe or PhS group can then be removed by reduction to give the saturated lactone, or by oxidation followed by elimination to give the unsaturated lactone (51).

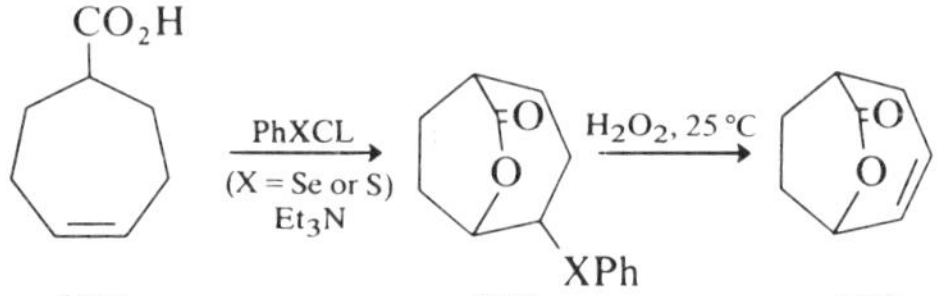

(49) (50) (51)

8 Esters and Lactones

It has been found that the order of introducing alkyl groups (by alkylation using lithium di-isopropylamide) into esters containing highly hindered α-carbon atoms is crucial in determining the overall yield of product obtained.[54] Thus the yields of i-Pr_2EtCCO_2Et obtained by two complementary routes are shown in Scheme 11.

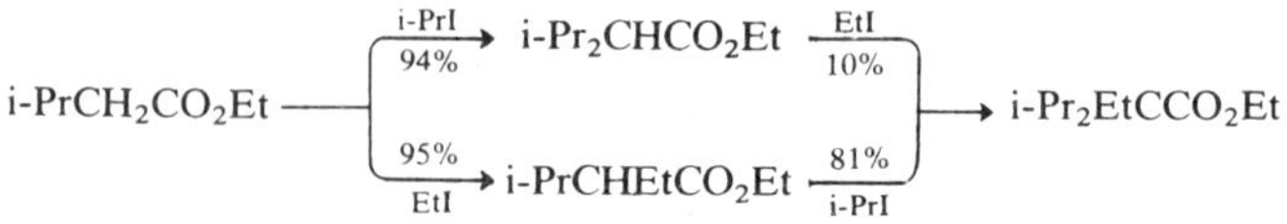

Scheme 11

This, and a number of similar examples demonstrate that a higher overall yield is obtained when the more bulky group is introduced last, which is the reverse of the situation observed in the alkylation of ketones using sodium amide. The low reactivity of hindered esters may be due to the inability of the ester to form an enolate ion, or to the lack of reactivity of the enolate once formed. Even highly

[53] K. C. Nicolaou and Z. Lysenko, *J. Amer. Chem. Soc.*, 1977, **99**, 3185 and *J.C.S. Chem. Comm.*, 1977, 293; D. L. J. Clive and G. Chittattu, *ibid.*, 484.

[54] J. A. MacPhee and J.-E. Dubois, *J.C.S. Perkin I*, 1977, 694.

hindered esters such as i-Pr_3CCO_2Et can be prepared in good yield (68 %) when hexamethylphosphoramide is present during both the enolization and alkylation steps of the reaction.

Various procedures for the mild cleavage of esters and lactones are available (*e.g.* trimethylsilyl iodide[33]). The base-catalysed hydrolysis of esters by anhydrous hydroxide ions ($KOBu^t$: H_2O, 2 : 1) affords a very effective method for hydrolysing hindered esters.[55] The reaction has been shown to involve *O*-acyl cleavage. Alternatively the phenylselenide anion can be used. The latter reagent provides a useful method for converting lactones (*e.g.* 52) into ω-vinyl carboxylic esters (54) *via* the ω-phenylselenenyl derivative (53).[56]

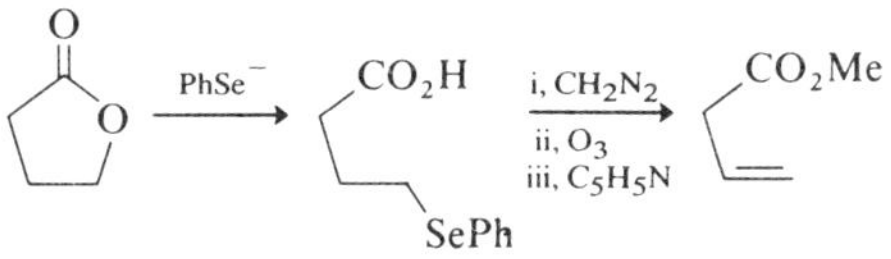

(52) (53) (54)

The rapid developments being made in the area of asymmetric synthesis have increased the need for methods for measuring enantiomeric purities. A useful method for determining the enantiomeric purity of a chiral lactone involves treating the lactone with (−)-(2*R*, 3*R*)-2,3-butanediol to give a mixture of diastereoisomeric *ortho* esters which can be analysed by g.l.c.[57]

The Pd^0 catalysed cyclization of allyl acetates provides a useful method for synthesizing large ring compounds (*e.g.* macrolides) by carbon–carbon bond formation (*e.g.* 55 → 56).[58] The reactions of α-bromo-esters of ω-hydroxy-aldehydes (*e.g.* 57) with activated zinc and diethylaluminium chloride can also be used to prepare β-hydroxy-lactones (58).[59]

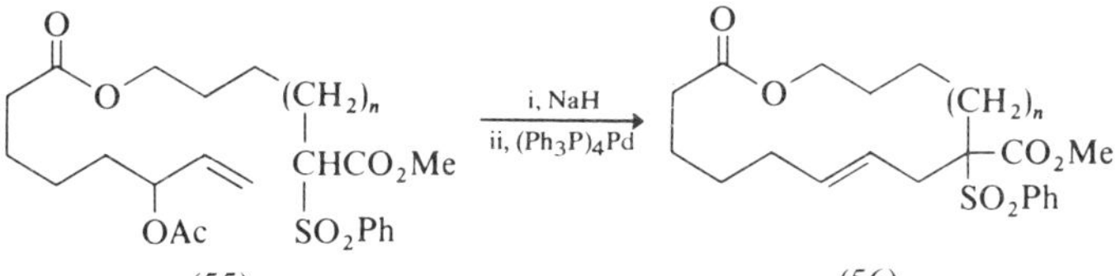

(55) (56)

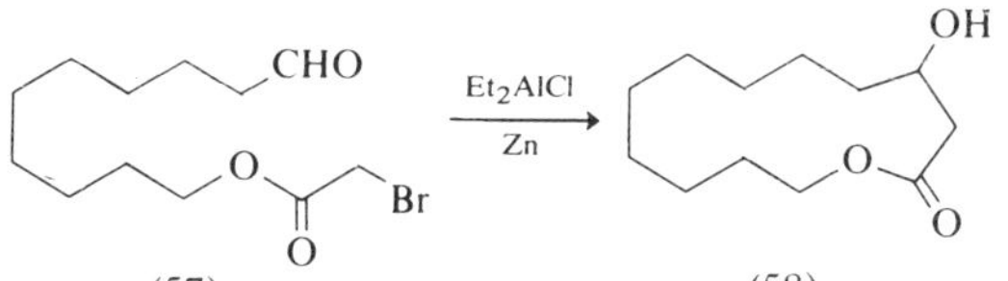

(57) (58)

[55] P. G. Gassman and W. N. Schenk, *J. Org. Chem.*, 1977, **42**, 918.

[56] R. M. Scarborough and A. B. Smith, *Tetrahedron Letters*, 1977, 4361; D. Liotta, W. Markiewicz, and H. Santiesteban, *ibid.*, 4365 and 4369.

[57] G. Saucy, R. Borer, and D. P. Trullinger, *J. Org. Chem.*, 1977, **42**, 3206.

[58] Y. Kitagawa, A. Itoh, S. Hashimoto, H. Yamamoto, and H. Nozaki, *J. Amer. Chem. Soc.*, 1977, **99**, 3864; B. M. Trost and T. R. Verhoeven, *ibid.*, 3867.

[59] K. Maruoka, S. Hashimoto, Y. Kitagawa, H. Yamamoto, and H. Nozaki, *J. Amer. Chem. Soc.*, 1977, **99**, 7705.

9 Amides and Lactams

The barrier to internal rotation about the central C—N bond of *NN*-dialkylthioamides is considerably higher than in *NN*-dialkylamides themselves, presumeably indicating a greater contribution from polar resonance structures in the former case. A linear correlation in fact exists between the activation energy for C—N bond rotation and the ^{15}N chemical shift in *NN*-dimethylamino derivatives.[60] The increased polarization of thioamides also facilitates deprotonation of the NH group in *N*-alkylthioamides and reduces the rate of protonation of thioamides relative to amides.[61]

Relatively few reliable methods are available for the mono-*N*-alkylation of amides. It has however been shown that simple amide methylols (59) can be directly coupled with trialkylaluminium reagents to produce good yields of the *N*-alkyl compounds (60).[62]

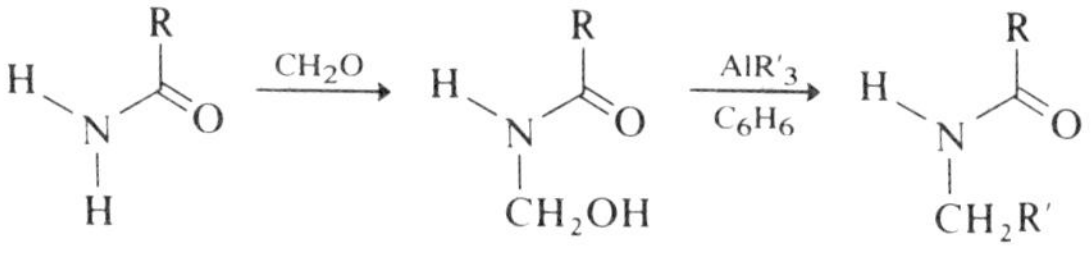

(59) (60)

An efficient method for converting primary amides into nitriles involves treating the amide with trifluoroacetic anhydride and pyridine.[63] Primary thioamides on the other hand can be converted into nitriles by treating them with diethyl azodicarboxylate and triphenylphosphine.[64]

The reactions of various amides with phosgeniminium salts ($Cl_2C{=}\overset{+}{N}Me_2Cl^-$) have been investigated. Dicarboxamides react to give dicyanines which can be hydrolysed to give bis-malonamides, and react with hydrazine to give dipyrazoles.[65] *N*-Alkylamides also react with phosgeniminium salts to give intermediates which on treatment with strong bases yield ynamides.

Diamides (and dinitriles) can be converted into thiono- and dithio-esters of dicarboxylic acids (Scheme 12).[66]

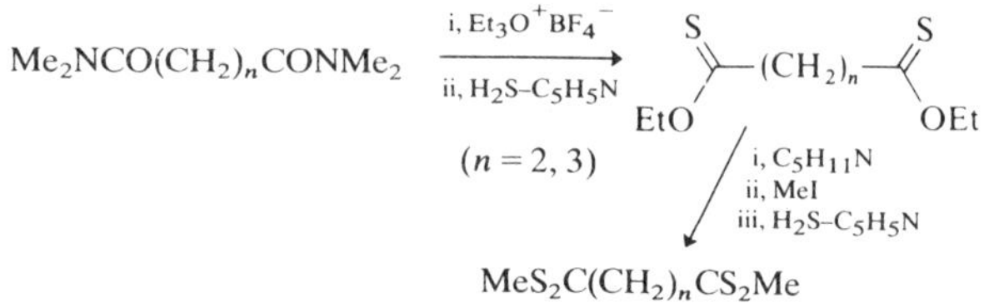

Scheme 12

[60] G. J. Martin, J. P. Gouesnard, J. Dorie, C. Rabiller, and M. L. Martin, *J. Amer. Chem. Soc.*, 1977, **99**, 1381.
[61] B. G. Cox and P. de Maria, *J.C.S. Perkin II*, 1977, 1385.
[62] A. Basha and S. M. Weinreb, *Tetrahedron Letters*, 1977, 1465.
[63] F. Campagna, A. Carotti, and G. Casini, *Tetrahedron Letters*, 1977, 1813.
[64] M. D. Dowle, *J.C.S. Chem. Comm.*, 1977, 220.
[65] M. Huys-Francotte, Z. Janousek, and H. G. Viehe, *J. Chem. Research*, 1977, 100; E. Goffin, Y. Legrand, and H. G. Viehe, *ibid.*, 105; see also H. G. Viehe, *Chem. and Ind.*, 1977, 386.
[66] R. Hoffmann and K. Hartke, *Annalen*, 1977, 1743.

10 Anhydrides and Imides

Bis(thioacyl) sulphides (61) can be prepared by treating aliphatic dithiocarboxylic acids with dicyclohexylcarbodi-imide.[67] They can be used to prepare sterically hindered dithio-esters (62). Acyl-thioacyl sulphides have also been prepared by treating sodium or piperidinium dithiocarboxylates with acetyl chloride.

(61) (62)

Enolate anions derived from acid anhydrides can be prepared by treating the anhydride with a sterically hindered base such as lithium 1,1-bis(trimethylsilyl)-3-methyl-1-butoxide.[68] Such anions react with aldehydes to give β-carbomethoxy γ-butyrolactones (Scheme 13).

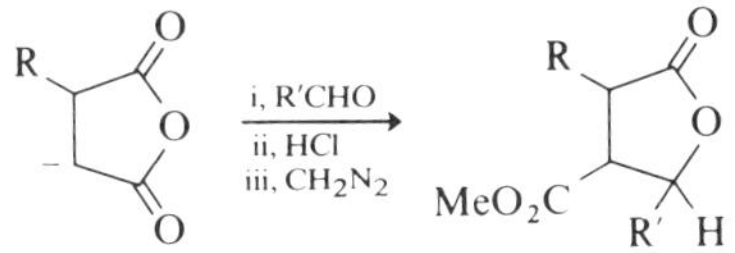

Scheme 13

It is possible to reduce 1,3,3-trisubstituted succinimides regiospecifically at the least hindered carbonyl group by using diborane.[69] This should be contrasted with the reduction of the same compounds by sodium borohydride which attacks the most hindered carbonyl group. The optically active succinimide (63) is reduced by sodium borohydride to give a mixture of the diastereo-isomeric hydroxy-lactams (64) which react with methyl diethylphosphonoacetate and sodium hydride in dimethoxyethane to give a mixture of diastereo-isomeric esters (65).[70] This reaction affords a useful asymmetric synthesis of piperidine derivatives. The nature of the solvent influences the diastereoselectivity of the reaction.

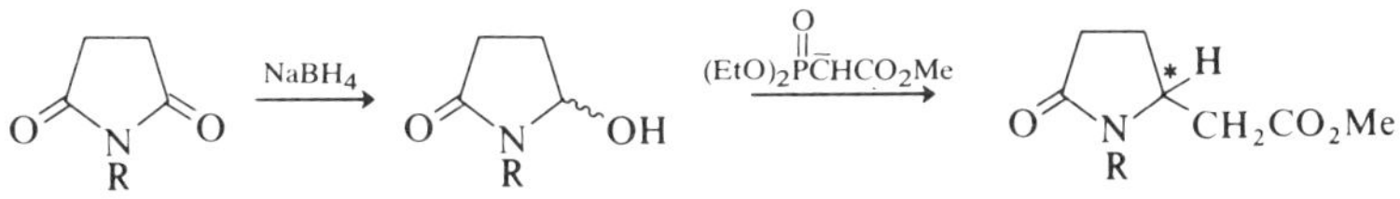

(63) (64) (65)

[R = (*S*)-1-phenylethyl]

11 Phosphorus Compounds

Phosphinothioic methanesulphonic anhydrides react with aluminium chloride to form complexes which can be decomposed by water or methanol to give phos-

[67] S. Kato, T. Takagi, T. Katada, and M. Mizuta, *Angew. Chem. Internat. Edn.*, 1977, **16**, 786; S. Kato, A. Hori, T. Takagi, and M. Mizuta, *ibid.*, 787; S. Kato, K. Sugino, M. Yamada, T. Katada, and M. Mizuta, *ibid.*, 879.

[68] N. Minami and I. Kuwajima, *Tetrahedron Letters*, 1977, 1423.

[69] R. Süess, *Helv. Chim. Acta*, 1977, **60**, 1650.

[70] T. Wakabayashi and M. Saito, *Tetrahedron Letters*, 1977, 93.

phinothioic halides with retention of configuration at the phosphorus atom.[71] The reaction affords a stereoselective synthesis of the phosphinothioic halides. The assignment of configuration to the product is achieved by correlation of its optical rotation with that of the halide obtained by treating the phosphinothioic acid with phosphorus pentachloride, which is known to involve inversion of configuration (Scheme 14).

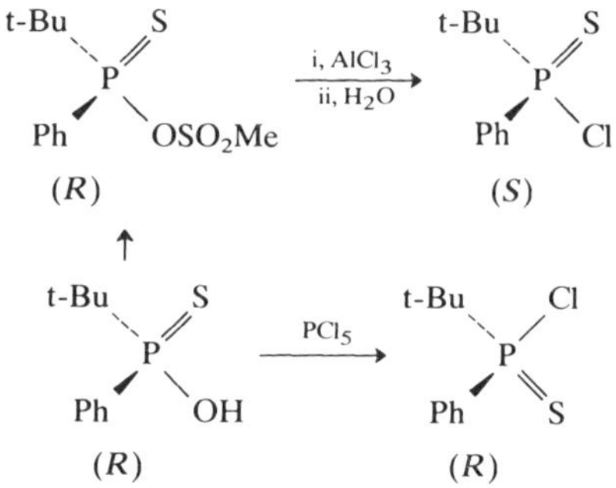

Scheme 14

A new synthesis of imidate esters of phosphorus from amidates has been used to prepare the chiral phosphinimidate (67) from (66).[72]

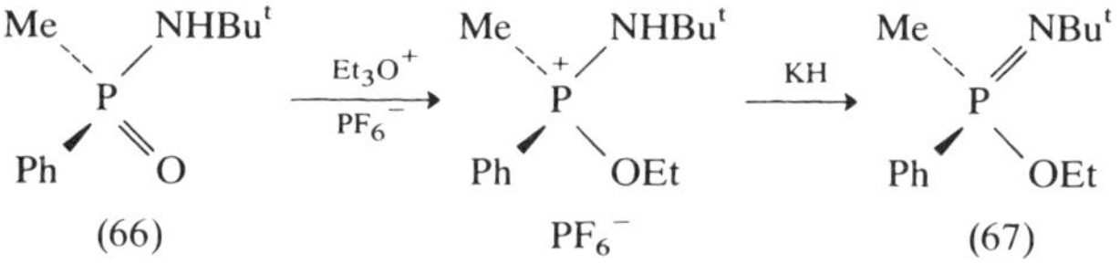

The acid-catalysed alcoholysis of a series of phosphinic amides containing different leaving groups has been studied. At low concentrations of HCl the reactions proceed with inversion of configuration, but at higher concentrations retention becomes increasingly important and eventually exceeds inversion for (68) and (69).[73] At a given acid concentration the proportion of the reaction proceeding with retention of configuration depends on the nucleophilicity of the leaving group.

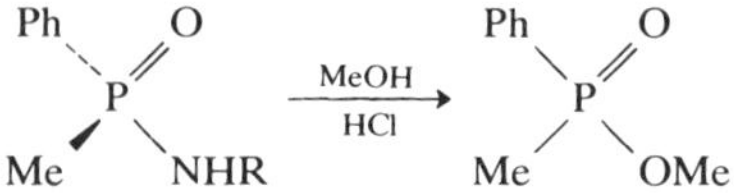

(68) R = Ph
(69) R = p-$O_2NC_6H_4$

A stereospecific route to chiral *OS*-dialkylphosphoramidothioates has been devised and their acid-catalysed alcoholysis studied.[74] Once again two competing reactions are evident, one involving inversion of configuration and favoured by low

[71] J. Michalski and Z. Skrzypczynski, *J.C.S. Chem. Comm.*, 1977, 66.
[72] K. E. DeBruin and L. L. Thomas, *J.C.S. Chem. Comm.*, 1977, 33.
[73] M. J. P. Harger, *J.C.S. Perkin I*, 1977, 2057.
[74] C. R. Hall and T. D. Inch, *Tetrahedron Letters*, 1977, 3761 and 3765.

acid concentrations, the other involving retention of configuration and unaffected by acid strength.

12 Sulphur Compounds

Enolate anions react with allenic sulphonium salts to give a variety of products (*e.g.* 70 and 73).[75] ^{13}C-Labelling experiments have shown that formation of the furan (70) involves attack on the α, rather than the γ-carbon atom, in the cyclization step. The initial adducts (71) formed with monosubstituted malonic esters undergo proton abstraction by sodium ethoxide to form an ylide (72) which then undergoes a [2,3]sigmatropic rearrangement to give (73). In the case of unsubstituted malonic esters the ylide rearrangement is preceded by a [1,3]prototropic shift to give the conjugated tautomer.

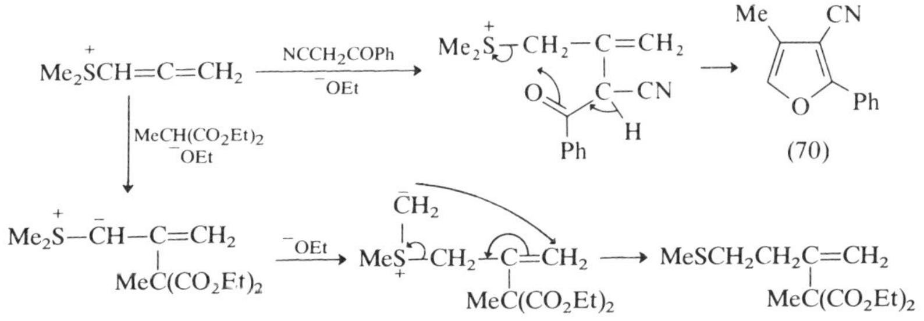

(71) (72) (73)

The anions derived from alkylphenylsulphides react with electrophiles to give α-substituted products.[76] The anions derived from allylphenylsulphides on the other hand react with electrophiles at either the α-position (with alkyl halides) or the γ-position (with carbonyl compounds).[77] Allylphenylsulphides also readily undergo a heat, light, or acid catalysed [1,3]rearrangement of the phenylthio group which further increases the variety of products which can be formed from such compounds. Selenium stabilized carbanions have also acquired a justifiable reputation as useful reagents for carbon–carbon bond formation. Phenylpropargylselenide, for example, reacts with two equivalents of lithium di-isopropylamide to give a dianion which reacts with electrophilic reagents at both the α- and γ-positions (attack occurring first at the α-position) to give a disubstituted product (Scheme 15).[78] Oxidation of the selenide affords the corresponding selenoxide which readily undergoes rearrangement leading to a selenoenone.

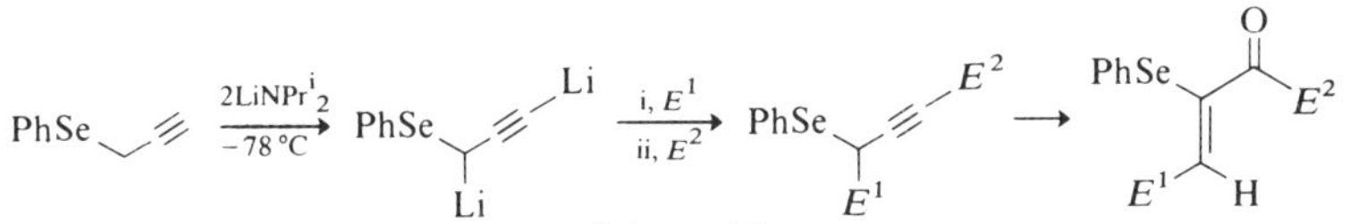

Scheme 15

[75] B. S. Ellis, G. Griffiths, P. D. Howes, C. J. M. Stirling, and B. R. Fishwick, *J.C.S. Perkin I*, 1977, 286; G. Griffiths, P. D. Howes, and C. J. M. Stirling, *ibid.*, 912.
[76] T. M. Dolak and T. A. Bryson, *Tetrahedron Letters*, 1977, 1961.
[77] P. Brownbridge and S. Warren, *J.C.S. Perkin I*, 1977, 1131.
[78] H. J. Reich and S. K. Shah, *J. Amer. Chem. Soc.*, 1977, **99**, 263.

The double deprotonation of allyl thiols gives thiocarbonyl dianions which react with electrophiles preferentially at the γ-position leading to *cis*-vinyl sulphides (Scheme 16).[79]

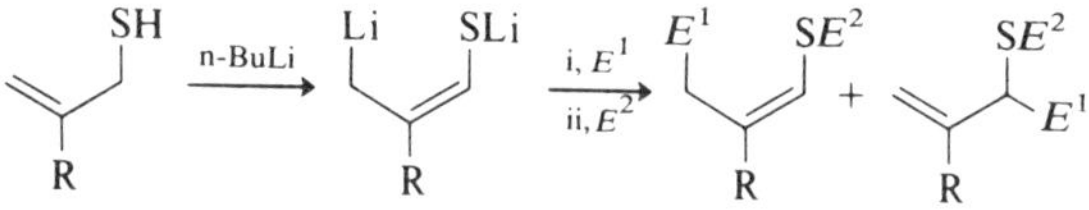

Scheme 16

A new asymmetric synthesis of β-hydroxy-esters (75) involves condensation of the chiral α-sulphinyl ester enolate anion (74) with achiral aldehydes or ketones.[80] In a similar manner, reaction of the anion (76) derived from a chiral sulphoxide with undecanal forms the basis of a stereospecific synthesis of (+)-disparlure (77), the female sex pheremone of the gypsy moth.[81]

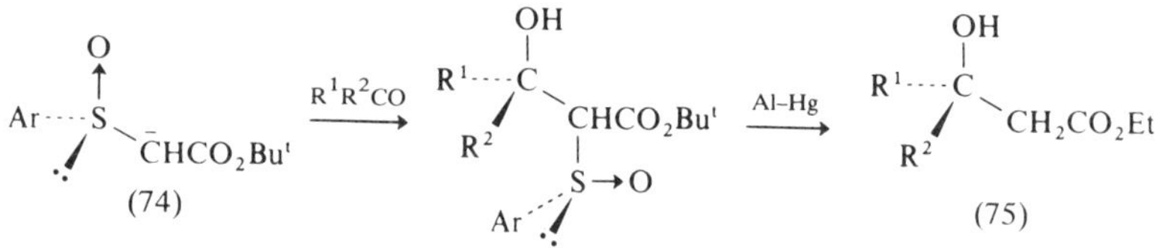

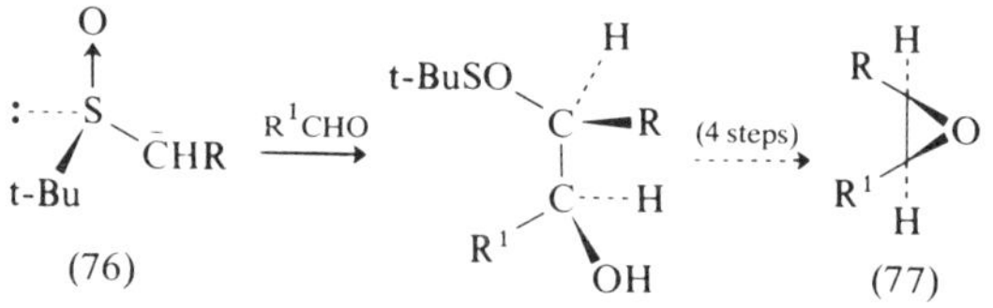

$R = Me_2CH(CH_2)_4$
$R^1 = Me(CH_2)_9$

The synthesis of diselenoacetals (78) affords a convenient route to vinyl selenides (79) which can in turn be converted into α-halogenoalkylselenides (80). The latter yield selenium stabilized carbanions on treatment with n-butyl-lithium.[82]

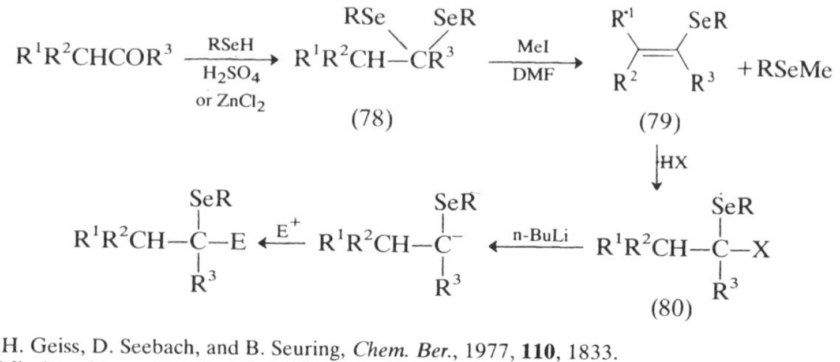

79 K.-H. Geiss, D. Seebach, and B. Seuring, *Chem. Ber.*, 1977, **110**, 1833.
80 C. Mioskowski and G. Solladié, *J.C.S. Chem. Comm.*, 1977, 162.
81 D. G. Farnum, T. Veysoglu, A. M. Cardé, B. Duhl-Emswiler, T. A. Pancoast, T. J. Reitz, and R. T. Cardé, *Tetrahedron Letters*, 1977, 4009.
82 M. Sevrin, W. Dumont, and A. Krief, *Tetrahedron Letters*, 1977, 3835; W. Dumont and A. Krief, *Angew. Chem. Internat. Edn.*, 1977, **16**, 540; W. Dumont, M. Sevrin, and A. Krief, *ibid.*, 541.

The preparation and reactions of a series of *O*-alkyl selenoesters have also been reported.[83]

13 Halogen Compounds

Two useful extensions of the well known Gabriel synthesis have been described which allow primary, secondary, and t-Boc amines to be prepared from alkyl halides. Thus, diphenylphosphinic amide reacts with alkyl halides in the presence of a phase-transfer catalyst to give *N*-monoalkyl derivatives which can be converted into primary amines, or treated with a second alkyl halide molecule (Scheme 17).[84] The potassium salt of t-butyl methyl iminodicarboxylate undergoes smooth

$$Ph_2\overset{O}{\overset{\|}{P}}NH_2 \xrightarrow{RX} Ph_2\overset{O}{\overset{\|}{P}}NHR \xrightarrow{R^1X} Ph_2\overset{O}{\overset{\|}{P}}NRR^1 \longrightarrow RR^1NH$$

Scheme 17

N-alkylation to give derivatives which can be converted by treatment with alkali into t-Boc amines.[85]

The use of an insoluble polymer support as a monoblocking agent to prepare mono-protected derivatives of dicarboxylic acid halides has been reported.[86]

[83] D. H. R. Barton, P.-E. Hansen, and K. Picker, *J.C.S. Perkin I*, 1977, 1723.
[84] A. Zwierzak and I. Podslawczyńska, *Angew. Chem. Internat. Edn.*, 1977, **16**, 702.
[85] J. D. Elliot and J. H. Jones, *J.C.S. Chem. Comm.*, 1977, 758.
[86] C. C. Leznoff and J. M. Goldwasser, *Tetrahedron Letters*, 1977, 1875; see also D. M. Dixit and C. C. Leznoff, *J.C.S. Chem. Comm.*, 1977, 798.

10 Aromatic Compounds

By R. G. COOMBES

Department of Chemistry, The City University, St. John Street, London EC1V 4PB

1 Introduction

The graph-theoretical approach[1a] to conjugation and resonance energies of hydrocarbons involving the concept of conjugated circuits [see *Ann. Reports* (*B*), 1976, **73**, p. 223] has been developed as a new approach to the aromaticity[1b] of polycyclic systems, those having only conjugated circuits with $(4n+2)$ π-electrons, when the Kekulé structures are examined, being defined as aromatic. The approach allows the prediction of the relative aromaticity of some systems containing both $(4n+2)$ and $(4n)$ conjugated circuits. Resonance energies obtained[2a] using Clar's aromatic sextet theory correlate well with the Dewar-type resonance energies from graph theory [see *Ann. Reports* (*B*), 1976, **73**, p. 201],[2b] and the latter approach has been applied to a wide range of systems.[2c] An essentially identical graph-theoretical approach, derived, however, as the result of a different line of argument, has also been reported.[3] The degree of electron delocalization in cyclic systems has been proposed,[4] with theoretical justification, to be a function both of orbital phase continuity requirements and of the disposition of donor and acceptor component systems. The condition under which the normal criterion for delocalization, based on the number of π-electrons, becomes invalid in 'effectively discontinuous systems' has been presented.

An interesting polycyclic system with different possible delocalized structures is the dicyclo-octatetraeno[1,2:4,5]benzene dianion, which has been shown[5] not to sustain a localized 14π-electron diamagnetic ring current. It is probably best described by the planar 20π-electron structure (1), where the electrostatic repulsion is adequate to overcome the strain energy and the antiaromatic character of the ion.

The diamagnetic susceptibility exhaltations of several 7-substituted cycloheptatrienes vary as a function of the size of the substituent, are very large, and are indicative of a substantial ring current.[6] It has been suggested that, in view of the low resonance energies and the bond alternation, these systems should simply be

[1] M. Randić, (*a*) *Tetrahedron*, 1977, **33**, 1905; (*b*) *J. Amer. Chem. Soc.*, 1977, **99**, 444.
[2] J.-I. Aihara, (*a*) *Bull. Chem. Soc. Japan*, 1977, **50**, 2010; (*b*) *ibid.*, p. 3057; (*c*) *J. Amer. Chem. Soc.*, 1977, **99**, 2048.
[3] I. Gutman, M. Milun, and N. Trinajstić, *J. Amer. Chem. Soc.*, 1977, **99**, 1692.
[4] S. Inagaki and Y. Hirabayashi, *J. Amer. Chem. Soc.*, 1977, **99**, 7418.
[5] L. A. Paquette, G. D. Ewing, S. Traynor, and J. M. Gardlik, *J. Amer. Chem. Soc.*, 1977, **99**, 6115.
[6] R. F. Childs and I. Pikulik, *Canad. J. Chem.*, 1977, **55**, 259.

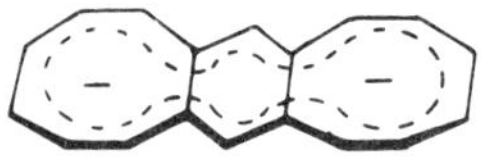

(1)

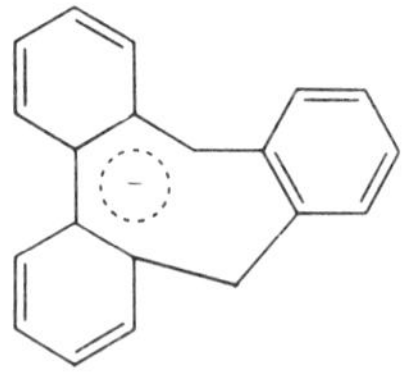

(2)

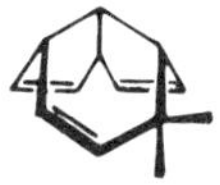

(3) R = $p\text{-}NO_2C_6H_4$

described as diatropic rather than homoaromatic, and the authors question whether diatropicity is a useful or even valid criterion of aromaticity. Two homoaromatic ions have been reported[7] which involve the interruption of the cyclic conjugated system by a three-carbon bridge. The benzannelated homocyclopentadienyl anion (2) is formed by treatment of the parent hydrocarbon with butyl-lithium. Examination of several systems[8] has provided no evidence for trishomoaromatic stabilization. Kinetic and product studies of the hydrolysis of (3), for example, indicate[8b] no participation of the double bonds. A mathematical model has been devised which enables the magnitude of the interaction between homoaromatic centres to be evaluated.[8c]

The sites of protonation of simple substituted benzenes in the gaseous phase have been receiving attention. Ring protonation of phenol has been calculated[9] by *ab initio* MO theory to be 63 kJ mol^{-1} more favourable than *O*-protonation, and a deuterium-exchange experiment gave an approximate value in agreement with this. Measurements of N_{1s} core ionization energy suggest that aniline is protonated on nitrogen in the gaseous phase, but anomalies exhibited by *NN*-dimethylaniline may be due to ring protonation.[10a] Pulsed ion cyclotron resonance spectroscopy confirms[10b] the former result, although *ab initio* MO theory suggests that ring protonation is only 4—13 kJ mol^{-1} less favourable. After comparison of the relative base strengths of some *meta*- and *para*-substituted anilines in the gaseous phase with the values for these compounds in aqueous solutions, it has been suggested that, for *m*-toluidine and *m*-anisidine, protonation occurs on the ring.[10c]

The opposing views concerning the nature of aryl ^{19}F n.m.r. polar field effects have now been largely reconciled. Calculations (CNDO/2) on the 1-substituted 4-*p*-fluorophenylbicyclo[2,2,2]octanes have indicated[11a] the dominance of π-polarization effects, direct field effects accounting for only about 25% of the polar effects on ^{19}F chemical shifts. In this connection, the distinction between π-polarization and the π-inductive effect has been emphasized.[11b] ^{13}C N.m.r. data

[7] A. Dagan, D. Bruck, and M. Rabinovitz, *Tetrahedron Letters*, 1977, 2995.

[8] (*a*) L. A. Paquette, M. J. Kukla, S. V. Ley, and S. G. Traynor, *J. Amer. Chem. Soc.*, 1977, **99**, 4756; L. A. Paquette, H. C. Berk, C. R. Degenhardt, and G. D. Ewing, *ibid.*, p. 4764; (*b*) L. A. Paquette, P. B. Lavrik, and R. H. Summerville, *J. Org. Chem.*, 1977, **42**, 2659; (*c*) L. A. Paquette, T. G. Wallis, T. Kempe, G. G. Christoph, J. P. Springer, and J. Clardy, *J. Amer. Chem. Soc.*, 1977, **99**, 6946.

[9] D. J. DeFrees, R. T. McIver, and W. J. Hehre, *J. Amer. Chem. Soc.*, 1977, **99**, 3853.

[10] (*a*) R. G. Cavell and D. A. Allison, *J. Amer. Chem. Soc.*, 1977, **99**, 4203; (*b*) S. K. Pollack, J. L. Devlin, tert, K. D. Summerhays, R. W. Taft, and W. J. Hehre, *ibid.*, p. 4583; (*c*) K. D. Summerhays, S. K. Pollack, R. W. Taft, and W. J. Hehre, *ibid.*, p. 4585.

[11] (*a*) W. F. Reynolds, *Tetrahedron Letters*, 1977, 675; (*b*) W. Kitching, M. Bullpit, D. Gartshore, W. Adcock, T.-C. Khor, D. Doddrell, and I. D. Rae, *J. Org. Chem.*, 1977, **42**, 2411; (*c*) D. F. Ewing, S. Sotheeswaran, and K. J. Toyne, *Tetrahedron Letters*, 1977, 2041; (*d*) W. Adcock and T.-C. Khor, *ibid.*, p. 3769.

provide unequivocal supporting evidence for π-polarization in the phenyl group of 1-substituted phenylbicyclo[2,2,2]octanes, although there is not quantitative agreement with the above calculations.[11c] The relative importance of the π-polarization and direct field effects has also been estimated by comparison of ^{19}F n.m.r. data on the *para*-fluoro- and *meta*-fluoro-systems, and the result is in good agreement with the estimate mentioned above.[11d] Reversed ^{19}F n.m.r. polar field effects are realized in the *ortho*-fluoro-analogues.

The modified version of the general theory of structure–property relationships, previously proposed,[12a] has been applied[12b] to the problem of the relationships between substituent chemical shifts and substituent reactivity parameters, and some success has been achieved in predicting whether a single parameter or a pair of parameters should be utilized in a particular situation. The non-additive behaviour of ^{13}C chemical shifts in *para*-disubstituted benzenes [see *Ann. Reports* (*B*), 1976, **73**, p. 203] has been reinterpreted.[13] The data for the 1- and 2-positions are in fact predicted additively, and those for the 4-positions may be accurately described by linear proportionality relationships with the appropriate substituent chemical shifts in monosubstituted benzenes. The proportionality relationships were interpreted in terms of localized changes in the excitation term of the paramagnetic shielding expression for ^{13}C shifts, determined largely by the ionization potential of the 4-substituent. Substituent effects on the ^{13}C shifts of the side-chain C_α atoms in *meta*- and *para*-disubstituted benzenes have also been studied and interpreted.[14]

Further reviews illuminate the reactivity–selectivity debate.[15a,b] Considerations concerning change of solvent and temperature have been emphasized, as has the fact that if substituents influence bond cleavage as well as bond formation, a more reactive species can be more selective than a less reactive one in spite of an 'earlier' transition state. An illustrative example is the competition of aniline and 3-chloroaniline for substituted benzenesulphonyl chlorides, where selectivity increases with reactivity. This argument is similar to that concerning desolvation of ions in reactions of charged species [see *Ann. Reports* (*B*), 1976, **73**, p. 203]. This latter effect has, however, been regarded as unimportant.[16]

2 Benzene Isomers, Oxides, and Homobenzenes

The photoelectron spectrum of benzvalene has been reported[17] and this has led to the suggestion that the unusual u.v. absorption (217 nm) may be due to the superimposition of the first member of a low-lying Rydberg series on a valence-shell transition.

Full details of the preparation, separation, and aromatization of *meso*- and *dl*-1,1′-dimethyl-3,3′-bicyclopropenyls have now appeared.[18] The aromatization

12 M. Godfrey, *J.C.S. Perkin II*, (*a*) 1975, 1016; (*b*) 1977, 769.
13 B. M. Lynch, *Canad. J. Chem.*, 1977, **55**, 541.
14 N. Inamoto, S. Masuda, K. Tori, and Y. Yoshimura, *Tetrahedron Letters*, 1977, 737; J. Bromilow, R. T. C. Brownlee, and D. J. Craik, *Austral. J. Chem.*, 1977, **30**, 355.
15 (*a*) B. Giese, *Angew Chem. Internat. Edn.*, 1977, **16**, 125; (*b*) A. Pross, *Adv. Phys. Org. Chem.*, 1977, **14**, 69.
16 C. D. Ritchie and M. Sawada, *J. Amer. Chem. Soc.*, 1977, **99**, 3754.
17 P. J. Harman, J. E. Kent, T. H. Gan, J. B. Peel, and G. D. Willett, *J. Amer. Chem. Soc.*, 1977, **99**, 943.
18 J. H. Davis, K. J. Shea, and R. G. Bergman, *J. Amer. Chem. Soc.*, 1977, **99**, 1499.

passes through an intermediate in which the stereochemical distinction between the diastereoisomers has disappeared. The data are most consistent with Scheme 1 for

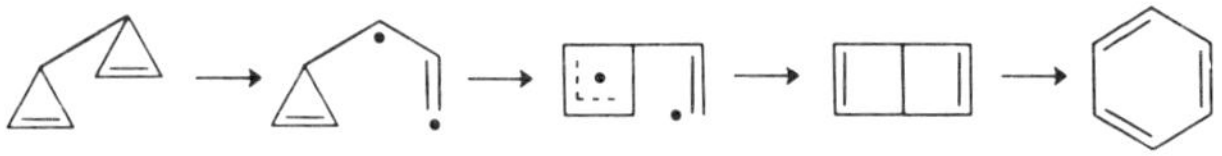

Scheme 1

gas-phase aromatization and probably for reaction in solution also. 1,4-Hexamethylene-Dewar-benzene is formed slowly, and apparently quantitatively, on irradiation of a solution of [6]paracyclophane in cyclohexane; warming the solution to 50—90 °C results in a clean and rapid reversion to the thermodynamically more stable precursor.[19] A polemical reminder of the limitations of the use of correlation diagrams, illustrated by that for the 'allowed' but not spontaneous thermal aromatization of benzvalene, has been given;[20] the words 'allowed' and 'forbidden' are only of use when there is a choice between two stereochemically different modes of reaction. Benzvalene has been shown[21] to undergo a degenerate rearrangement, catalysed by silver(I) ion, in competition with the known catalysed aromatization. Evidence was presented that the reaction of the silver(I) ion with the bicyclobutane moiety is a reversible process involving the intermediacy of (4). Scrambling does not occur with other bicyclobutane derivatives.

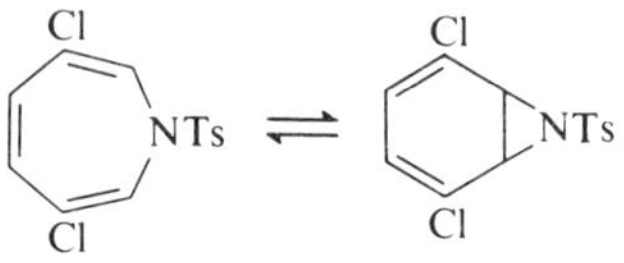

(4) (5) Ts = p-$MeC_6H_4SO_2$ (6)

The synthesis of 3-t-butoxycarbonylbenzene oxide has been accomplished[22] and, in contrast to the 4-isomer, it exists mainly as the benzene oxide valence tautomer. Attempts to effect nucleophilic addition on this molecule failed, although the 4-isomer reacts *via* its oxide tautomer. Benzene oxide itself is converted into (Z,Z)-hexa-2,4-dienedial on oxidation with peroxy-acids,[23] the oxidative ring opening involving the oxepin tautomer. The reaction of $[Rh(CO)_2Cl]_2$ with benzene oxide and arene oxides at room or lower temperatures in chloroform or methanol leads to both deoxygenation and the formation of phenol, by mechanisms involving initial Lewis acid catalysis.[24] The equilibrium between the azepine (5) and its bicyclic valence tautomer (6) has been directly observed[25] in the range −70 to +40 °C [$\Delta G^{\ddagger}$, (5)→(6) = 46 kJ mol^{-1}].

[19] S. L. Kammula, L. D. Iroff, M. Jones, jun., J. W. van Straten, W. H. de Wolf, and F. Bickelhaupt, *J. Amer. Chem. Soc.*, 1977, **99**, 5815.
[20] J. J. C. Mulder, *J. Amer. Chem. Soc.*, 1977, **99**, 5177.
[21] U. Burger and F. Mazenod, *Tetrahedron Letters*, 1977, 1757.
[22] B. A. Chiasson and G. A. Berchtold, *J. Org. Chem.*, 1977, **42**, 2008.
[23] S. G. Davies and G. H. Whitham, *J.C.S. Perkin I*, 1977, 1346.
[24] R. W. Ashworth and G. A. Berchtold, *Tetrahedron Letters*, 1977, 343.
[25] H. Prinzbach, H. Babsch, H. Fritz, and P. Hug, *Tetrahedron Letters*, 1977, 1355.

A new synthetic pathway to substituted *trans*-bishomobenzenes is now available.[26] Thermolysis of dibromotetracyclo[5,1,0,$0^{2,4}$,$0^{3,5}$]octanes, readily available from tricyclo[4,1,0,$0^{2,7}$]hept-3-enes, gave the species (7), which was readily

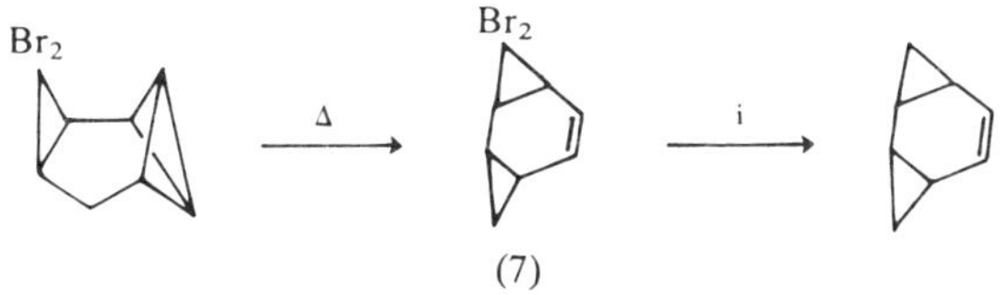

(7)

Reagents: i, Li—Bu^{t}OH in THF

Scheme 2

dehalogenated (Scheme 2). Studies of the flash pyrolysis (485 °C) of $^2H_1$-labelled hexahomobenzene (8) have shown[27] that there is not sufficient interaction energy

(8)

amongst the three constituent rings to allow for bond shifting, a result in agreement with last year's report concerning trioxahexahomobenzenes [see *Ann. Reports* (*B*), 1976, **73**, p. 205].

3 Benzene and its Derivatives

General.—The second part[28] of the review of the organic photochemistry of benzene is concerned with photoreactions with other molecules. The photoaddition of methyl phenylpropiolate to benzene has been shown[29a] to form the tetracyclo[3,3,0,$0^{2,4}$,$0^{3,6}$]oct-7-ene (9) as major product, in addition to the known 1-methoxycarbonyl-8-phenylcyclo-octatetraene, and in proportions which vary with

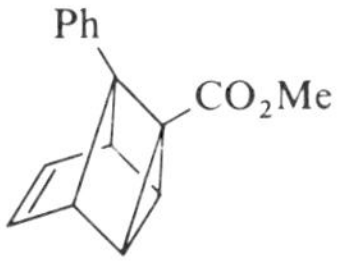

(9)

the wavelength of irradiation. The products are also photochemically interconvertible. The presumed precursor to both products, at least in the photoaddition reaction, is the corresponding bicyclo[4,2,0]octatriene, and species with this structure have now been isolated[29b] from the photoaddition of alkynes to hexafluorobenzene.

[26] R. T. Taylor and L. A. Paquette, *J. Amer. Chem. Soc.*, 1977, **99**, 5824.
[27] M. R. Detty and L. A. Paquette, *Tetrahedron Letters*, 1977, 347.
[28] D. Bryce-Smith and A. Gilbert, *Tetrahedron*, 1977, **33**, 2459.
[29] (*a*) A. H. A. Tinnemans and D. C. Neckers, *J. Amer. Chem. Soc.*, 1977, **99**, 6459; (*b*) B. Šket and M. Zupan, *ibid.*, p. 3504.

The major product of photocycloaddition of 2,3-dihydropyran to benzene is the *endo*-1,2-adduct formed in a process of surprising selectivity having a quantum efficiency greater than those previously observed for any mode of addition of an ethene to benzene.[30a] The 1,3-photocycloadditions of ethyl vinyl ether to toluene, and in particular those to anisole, are promoted by polar solvents, indicating that the 1,3-addition is not the general homopolar process, as was previously believed.[30b] The synthetic utility of the 1,3-addition of alkenes to anisole has been explored as a route to the bicyclo[3,2,1]octen-8-one system.[30c] 1,3-Cycloadducts are the major products of photochemical reactions[30d] between benzene and both

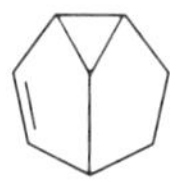

(10)

ethene [to give (10)] and propene, in contrast to the behaviour expected from the ionization potentials of these alkenes.[28] The authors suggest that electron affinities are more relevant for these alkenes, which are electron acceptors relative to benzene.

2,2-Dimethyl-1,3-dioxole undergoes 1,2-photoaddition to benzene, as expected from its ionization potential, but gives, in contrast to normal alkenes, the *exo*-adduct.[30e] Photoreactions of hexafluorobenzene have been the subject of some attention.[29b,31a—c] The [2+2]cycloaddition of cyclopentene in cyclohexane solution proceeds[31a] to give predominantly the *exo*-adduct (85%), and reaction with norbornadiene gives predominantly (80%) the *exo*-*syn*-isomer.[31b] It may be that these results reflect the stereochemistry of precursor complexes formed in the absence of irradiation.[28] Studies of the photoaddition of aliphatic amines to benzene have been reported in full.[31c]

An important review[32a] covers the field of cobalt-catalysed cyclo-oligomerization of alkynes, and full details[32b] of the formation of benzocycloalkenes using this approach, and of their synthetic utility, have appeared. The method has also been applied in an elegant synthesis of the steroid nucleus.[32c] The key step involves the reaction of the precursor (11) (Scheme 3), which is available by convergent series of reactions from hexa-1,5-diyne and 2-methylcyclopent-2-enone. The overall yield is 28%, based on the former. Some attention has been paid to the mechanism(s) of the trimerization. Evidence has been presented[33a] that there exist at least two distinct mechanisms for the production of aromatic compounds from the reaction of alkynes with metallocyclopentadienes, if these are intermediates. One path

[30] (*a*) A. Gilbert and G. Taylor, *Tetrahedron Letters*, 1977, 469; (*b*) *J.C.S. Chem. Comm.*, 1977, 242; (*c*) J. A. Ors and R. Srinivasan, *J. Org. Chem.*, 1977, **42**, 1321; (*d*) M. F. Mirbach, M. J. Mirbach, and A. Saus, *Tetrahedron Letters*, 1977, 959; (*e*) H.-D. Scharf and J. Mattay, *ibid.*, p. 401.

[31] (*a*) B. Šket and M. Zupan, *J.C.S. Chem. Comm.*, 1977, 365; (*b*) *Tetrahedron Letters*, 1977, 2811; (*c*) M. Zupan, B. Šket, and B. Pahor, *ibid.*, p. 4541; (*d*) M. Bellas, D. Bryce-Smith, M. T. Clarke, A. Gilbert, G. Klunkin, S. Krestonosich, C. Manning, and S. Wilson, *J.C.S. Perkin I*, 1977, 2571.

[32] (*a*) K. P. C. Vollhardt, *Accounts Chem. Res.*, 1977, **10**, 1; (*b*) R. L. Hillard, tert., and K. P. C. Vollhardt, *J. Amer. Chem. Soc.*, 1977, **99**, 4058; (*c*) R. L. Funk and K. P. C. Vollhardt, *ibid.*, p. 5483.

[33] (*a*) D. R. McAlister, J. E. Bercaw, and R. G. Bergman, *J. Amer. Chem. Soc.*, 1977, **99**, 1666; (*b*) P. Caddy, M. Green, E. O'Brien, L. E. Smart, and P. Woodward, *Angew. Chem. Internat. Edn.*, 1977, **16**, 648.

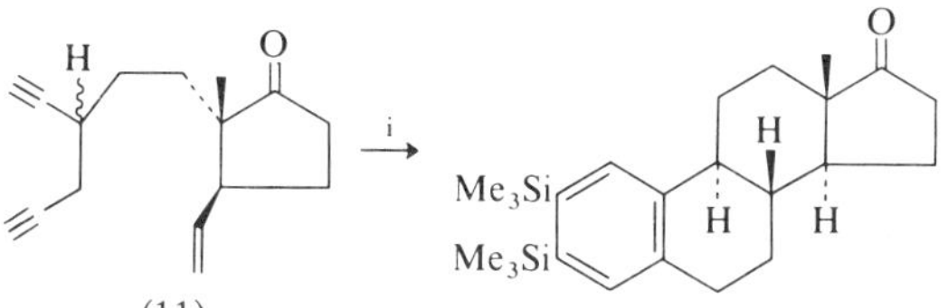

(11)

Reagents: i, bis(trimethylsilyl)acetylene–$CpCo(CO)_2$

Scheme 3

involves initial co-ordination of the alkyne and the other direct pathway may involve Diels–Alder addition. Other studies[33b] of η^5-indenylrhodium(I) complexes have, however, suggested the intermediacy of metallocyclopent-2-enes.

Ynamines ($R^1C{\equiv}CNR^2_2$) undergo cyclotrimerization in moderate yields on warming in acetonitrile solution over Ni^{II} or Ni^0, but not Cu^I, catalysts, and they form the unsymmetrical triaminobenzene (12).[34a] A useful synthesis of hindered

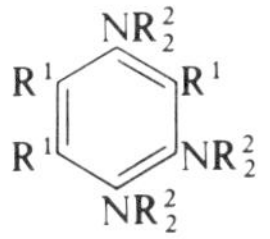

(12)

aromatic amines[34b] also involves ynamines which react with α-pyrone, carbon dioxide being lost spontaneously; methyl 2-(*NN*-diethylamino)-3-methylbenzoate, for example, is formed in 79% yield from 1-diethylaminoprop-1-yne.

Two procedures have been reported for the preparation of *meta*-bisannelated benzenes (13). The first procedure[35a] (Scheme 4) utilizes Corey's modified pinacol

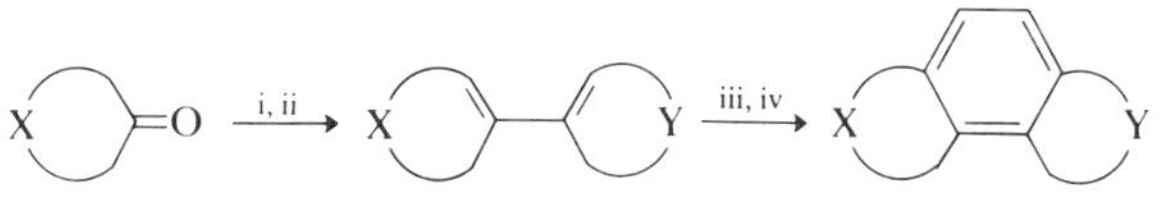

$X = (CH_2)_m$ (13)
$Y = (CH_2)_n$

Reagents: i, Mg(Hg)–$TiCl_4$; ii, $POCl_3$–pyridine; iii, acrylic acid; iv, 10% Pd/C, Δ

Scheme 4

reduction, and the second[35b] involves the initial reaction of 2-cycloalkenylcycloalkanones with malonodinitrile, acid-catalysed cyclization, and efficient removal of the nitrile function from the *ortho*-amino-nitrile. A simpler, albeit low-yield, preparation of benzo[1,2:3,4]dicyclobutane (13; $n = m = 1$) has been described,[36a] and the procedure of the earlier Diels–Alder method has been

[34] (*a*) J. Ficini, J. d'Angelo, and S. Falou, *Tetrahedron Letters*, 1977, 1645; (*b*) T. A. Bryson and D. M. Donelson, *J. Org. Chem.* 1977, **42**, 2930.

[35] (*a*) M. E. Isabelle, D. H. Lake, and R. H. Wightman, *Canad. J. Chem.*, 1977, **55**, 3268; (*b*) J. Sepiol, B. Kawalek, and J. Mirek, *Synthesis*, 1977, 701.

[36] (*a*) E. Giovannini and H. Vuilleumier, *Helv. Chim. Acta*, 1977, **60**, 1452; (*b*) R. P. Thummell and W. Nutakul, *J. Org. Chem.*, 1977, **42**, 300.

applied[36b] to a series of *meta*- and *para*-bisannelated benzenes involving fused 4- and 5-membered rings.

Syntheses of benzocyclobutenes abound. The flash vacuum pyrolysis of tolyl propargyl ethers provides a low-yield source of aromatic-ring-methylated benzocyclopropenes.[37] Full details of the thermal and electron-impact syntheses of benzocyclobutene itself, and examples having oxygenated substituents on the aryl ring, have appeared.[38a] The latter type of compound, however, can indeed be prepared[38b] by the ready and efficient cyclization of the corresponding *ortho*-lithio(2-bromoethyl)benzenes. 3,4-Bis(trimethylsilyl)benzocylobutene has been synthesized for the first time.[39] Interestingly, the trimethylsilyl group at the 4-position is five-hundred times more susceptible to hydrolysis than that at the 3-position. The reaction of $\alpha\alpha\alpha\alpha$-tetrabromo-*o*-xylene with pentacarbonyliron gave[40] a facile synthesis of mainly *trans*-1,2-dibromobenzocyclobutene.

Pyrolysis (600 °C; 0.3 s) of 2-methylbenzoyl chlorides gave the corresponding benzocyclobutenones in good yields.[41] Benzocycloalkenones have been synthesized[42a] in good yield by treatment of *o*-bromophenylalkanoic acids with butyl-lithium at low temperatures, and the procedure allows for the presence of some functionality in the ring being formed. An extension of the oxidative rearrangement of styrene derivatives by thallium(III) nitrate provides[42b] a general method for the ring expansion of benzocycloalkanones of this type. They are first subjected to a Wittig procedure, whereby a methylene carbon, which may be substituted, is inserted selectively between the aromatic ring and the carbonyl group.

The reactions of polyfluoro-aromatic compounds with aryl and alkyl radicals have been reviewed.[43] Some scavenging reactions of phenyl radicals, from the thermal decomposition of phenylazotriphenylmethane, are nearly diffusion-controlled in media of low viscosity, and are so controlled in media of high viscosity, leading to relative rates approaching unity for pairs of scavengers.[44] Competition with slower reactions at higher viscosities has given absolute rate constants. The reactions of the phenyl radicals do not show a clear selectivity–reactivity relationship. Correlations for the rate of hydrogen abstraction from substituted toluenes by isopropyl and t-butyl radicals by a Hammett equation have been recorded.[45] The authors suggest that substituent effects on both the structure of the transition state (consideration of polar resonance structures) and of the initial state (bond dissociation energy) are important and that, contrary to common practice, both should be considered. The ability of aromatic nitro-compounds to increase the yields of biaryls in arylation reactions is now believed[46] to involve

[37] J. M. Riemann and W. S. Trahanovksy, *Tetrahedron Letters*, 1977, 1863.
[38] (*a*) R. J. Spangler, B. G. Beckmann, and J. H. Kim, *J. Org. Chem.*, 1977, **42**, 2989; (*b*) P. D. Brewster, J. Tagat, C. A. Hergrueter, and P. Helquist, *Tetrahedron Letters*, 1977, 4573.
[39] R. L. Hillard, tert., and K. P. C. Vollhardt, *Angew. Chem. Internat. Edn.*, 1977, **16**, 399.
[40] R. Victor, *Transition Metal Chem.* (*Weinheim*), 1977, **2**, 2.
[41] P. Schiess and M. Heitzmann, *Angew. Chem. Internat. Edn.*, 1977, **16**, 469.
[42] (*a*) R. J. Boatman, B. J. Whitlock, and H. W. Whitlock, jun., *J. Amer. Chem. Soc.*, 1977, **99**, 4822; (*b*) E. C. Taylor, C.-S. Chiang, and A. McKillop, *Tetrahedron Letters*, 1977, 1827.
[43] L. S. Kobrina, *Russ. Chem. Rev.*, 1977, **46**, 348.
[44] R. G. Kryger, J. P. Lorand, N. R. Stevens, and N. R. Herron, *J. Amer. Chem. Soc.*, 1977, **99**, 7589.
[45] W. H. Davis, jun., and W. A. Pryor, *J. Amer. Chem. Soc.*, 1977, **99**, 6365.
[46] R. Henriquez and D. C. Nonhebel, *J. Chem. Res.* (*S*), 1977, 253.

oxidation of the intermediate arylcyclohexadienyl radicals by an electron-transfer mechanism, examples having electron-withdrawing substituents being more effective. The authors of the original experimental work on the photobromination of dihalogeno-benzenes have reacted[47] to the suggestion that *ipso*-substitution is involved [see *Ann. Reports* (*B*), 1976, **73**, p. 208], and show that such substitution is neither required nor precluded by their results.

A deamination reaction that uses alkyl nitrite and copper(II) halide has been used in several processes that are of synthetic importance.[48] The reagent rapidly converts arylamines into aryl chlorides and bromides in high yield,[48b] and a Meerwein-type arylation of alkenic substrates, giving α-halogeno-β-aryl compounds, occurs if the reaction is carried out in acetonitrile or acetone solutions containing the substrate.[48c] Good to excellent yields of phenols are formed from aryl bromides under basic conditions on reaction of the derived Grignard reagent with molybdenum peroxide–pyridine–hexamethylphosphoramide.[49] A procedure for the formation of phenols from diazonium ions, which avoids the usual highly acidic conditions and is considered to be the method of choice for new cases, consists in adding copper(I) oxide to a dilute solution of the arenediazonium salt in the presence of a large excess of copper(II) nitrate.[50]

Electrophilic Substitution.—Studies of the gas-phase pyrolysis of some 1-arylethyl acetates have led[51] to the conclusion that the Baker–Nathan order of electron release by alkyl groups is a solvation phenomenon, and also that hyperconjugation from C—C bonds is greater than that from C—H bonds. These studies also suggest that the solvolysis of $\alpha\alpha$-dimethylbenzyl chlorides is exceptionally sensitive to steric hindrance to solvation, and that this limits the suitability of the derived σ^+ constants for the description of electrophilic aromatic substitution. The protonations of toluene and ethylbenzene in super-acid media have been reinvestigated[52] and the presence of the *ortho*-protonated species has been established; the results are now in agreement with gas-phase studies.

Details of the gas-phase reaction of the t-butyl cation with phenol and anisole to give initially predominantly t-butylated oxonium ions have appeared.[53a] In contrast to the t-butyl cation, which exhibits both substrate and positional selectivity, the isopropyl cation does not show substrate selectivity in its reactions with arenes.[53b] It does, however, exhibit a measurable degree of positional selectivity, and its reactions are complicated by secondary dealkylation and isomerization processes. Chlorobenzene is isopropylated at a rate only half that of toluene, and the predominant *ortho*-substitution is again attributed[53c] to initial attack at the substituent. The attack of CH_5^+ and $C_2H_5^+$ ions on halogeno-benzenes causes[54]

[47] P. Gouverneur and J. P. Soumillion, *Bull. Soc. chim. belges*, 1977, **86**, 647.
[48] M. P. Doyle, J. F. Dellaria, jun., B. Siegfried, and S. W. Bishop, *J. Org. Chem.*, 1977, **42**, 3494; M. P. Doyle, B. Siegfried, and J. F. Dellaria, jun., *ibid.*, p. 2426; M. P. Doyle, B. Siegfried, R. C. Elliott, and J. F. Dellaria, jun., *ibid.*, p. 2431.
[49] N. J. Lewis, S. Y. Gabhe, and M. R. DeLaMater, *J. Org. Chem.*, 1977, **42**, 1479.
[50] T. Cohen, A. G. Dietz, jun., and J. R. Miser, *J. Org. Chem.*, 1977, **42**, 2053.
[51] E. Glyde and R. Taylor, *J.C.S. Perkin II*, 1977, 678.
[52] D. Fărcaşiu, M. T. Melchior, and L. Craine, *Angew. Chem. Internat. Edn.*, 1977, **16**, 315.
[53] (*a*) M. Attinà, F. Cacace, G. Ciranni, and P. Giacomello, *J. Amer. Chem. Soc.*, 1977, **99**, 4101, 5022; (*b*) *ibid.*, p. 2611; (*c*) M. Attinà and P. Giacomello, *Tetrahedron Letters*, 1977, 2373.
[54] M. Speranza and F. Cacace, *J. Amer. Chem. Soc.*, 1977, **99**, 3051.

extensive dehalogenation by two competitive routes, which lead to protodehalogenation and methyldehalogenation and are a reflection of the ambident nucleophilic character of the substrates. Mono- and di-substituted benzenes have been brominated by $^{80}Br^+$ in the gas phase.[55] The relative rates led to good σ^+, ρ^+ plots ($\rho^+ = -0.9$) and the data on toluene agreed with the Brown selectivity relationship.

Perrin has asserted[56] that aromatic nitration is not always a conventional electrophilic substitution, but that for some reactive substrates the mechanism involves the transfer of an electron from the aromatic to the nitronium ion, followed by collapse of the radical pair. Experiments quoted in support of this proposal involve the determination of anodic half-wave potentials for nitrogen dioxide and some aromatic compounds, and one involving the generation of the naphthalene radical cation in the presence of nitrogen dioxide, which gives the normal isomer ratio of nitro-naphthalenes. The author appears to confuse and combine the different approaches of others to encounter pairs and π-complexes, and ignores the possibility of catalysis by nitrous acid, which must surely affect the results obtained for some substrates which are used to support the proposed mechanism. Catalysis by nitrous acid is evident in the formation of the *ipso*-substituted σ-complex (14) that is formed on nitration of *NN*-dimethyl-*p*-toluidine in 70—77% sulphuric acid.[57] It slowly rearranges to the *NN*-dimethyl-4-methyl-2-nitroanilinium ion.

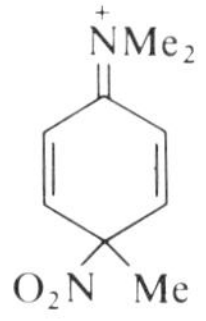

(14)

It seems that the problem of the identity of the electrophile for nitration by nitric acid in acetic anhydride has at last been solved.[58] The observed rates have been shown to be sensitive to the presence of trace impurities at $[HNO_3] < 0.9$ mol dm^{-3}, and a comparison of the true rate constant for the reaction of the electrophile and toluene with that for the reaction in acetic acid indicates that reaction indeed involves the nitronium ion.

A study of some model compounds has led to a reconsideration[59a] of the nitration of acetanilide, and it now seems clear that this molecule is in fact nitrated as the unprotonated species by nitric acid in aqueous sulphuric acid. For anisole in the same medium[59b] the *ortho*:*para* ratio varies considerably, and an explanation has been offered that involves the combined effects of hydrogen bonding and a diffusion-controlled reaction. With *o*- and *p*-methylanisole, *ipso*-substitution at C—Me is of importance, and in the latter case this leads to 4-methyl-2-nitrophenol

[55] E. J. Knust, *J. Amer. Chem. Soc.*, 1977, **99**, 3037.
[56] C. L. Perrin, *J. Amer. Chem. Soc.*, 1977, **99**, 5516.
[57] K. Fujiwara, J. C. Giffney, and J. H. Ridd, *J.C.S. Chem. Comm.*, 1977, 301.
[58] N. C. Marziano, R. Passerini, J. H. Rees, and J. H. Ridd, *J.C.S. Perkin II*, 1977, 1361.
[59] (*a*) R. B. Moodie, P. N. Thomas, and K. Schofield, *J.C.S. Perkin II*, 1977, 1693; (*b*) J. W. Barnett, R. B. Moodie, K. Schofield, J. B. Weston, R. G. Coombes, J. G. Golding, and G. D. Tobin, *J.C.S. Perkin II*, 1977, 248.

as a major product. 4-Methyl-4-nitrocyclohexa-2,5-dienone, a postulated intermediate in this reaction, has been observed[60a] as a product from the nitration of *p*-methylanisole and some other *para*-substituted toluenes in acetic anhydride.

Nitration *ipso* to an ethyl group has been observed[60b] in *p*-diethylbenzene, *p*-ethyltoluene, and almost exclusively in ethylmesitylene. Attempts to activate a biphenyl to nitration *ipso* to a phenyl group have, however, failed. The sensitivity of *ipso*-nitration to substituent effects has been examined[60a] by the competitive study of a series of 5-X-hemimellitene derivatives (X = H, F, Br, OAc, OMe, and NHAc), and a ρ^+ value $(-8 > \rho^+ > -12)$ has been estimated, showing a satisfactory agreement with that for nitrodeprotonation.

Some synthetic aspects of side-reactions attending aromatic nitration have been reviewed.[61] An interesting example of intramolecular trapping of a Wheland intermediate has been observed[62a] in the isolation of (15) from the nitration of methyl β-(2,3,4-trimethylphenyl)isovalerate. Another reaction involving initial *ipso*-substitution is the conversion[62b] of some *p*-xylonitrile derivatives into 1-imino-1,3-dihydroisobenzofurans [*e.g.* (16) from 2,3,4,6-tetramethylbenzonitrile]. Solvolytic studies[63] of the *ipso*-adducts formed on nitration of hemimellitene have reinforced the view [see *Ann. Reports* (*B*), 1976, **73**, p. 210] that, in >50% sulphuric acid, adducts of this type yield the *ipso*-Wheland intermediate, which is then partitioned between trapping by water and rearrangement by, in this case, sequential 1,2-shifts of the nitro-group. Full details of the reactions of nitronium salts with pentamethylbenzene have been given,[64] confirming that the initially detected species is the 1-nitro-1,2,3,4,6-pentamethylbenzenonium ion. The 1,3-dinitro-1,2,4,5,6-pentamethylbenzenonium ion formed from pentamethylnitrobenzene undergoes a degenerate rearrangement of a nitro-group from position 1 to position 5.

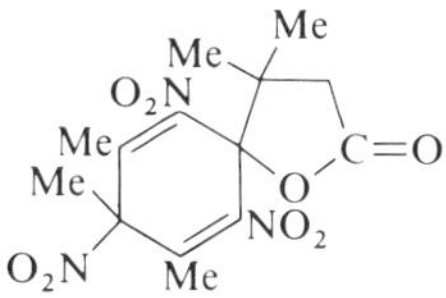

(15)

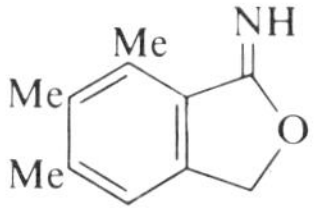

(16)

The mercuric-acetate-catalysed nitration of toluene in acetic acid at 80 °C, which produces 36% *o*-, 12% *m*-, and 52% *p*-nitrotoluenes, has been shown[65] to involve initial mercuration followed by nitrosodemercuration and subsequent oxidation of the nitroso-toluenes.

The sulphonations of some polyethyl- and polyisopropyl-benzenes in concentrated sulphuric acid have been studied.[66] The sulphonic acids from 1,2,3,5- and

[60] A. H. Clemens, M. P. Hartshorn, K. E. Richards, and G. J. Wright, *Austral. J. Chem.*, 1977, **30**, (*a*) p. 113; (*b*) p. 103.
[61] H. Suzuki, *Synthesis*, 1977, 217.
[62] (*a*) M. Shinoda and H. Suzuki, *J.C.S. Chem. Comm.*, 1977, 479; (*b*) H. Suzuki, M. Koge, and T. Hanafusa, *ibid.*, p. 341.
[63] T. Banwell, C. S. Morse, P. C. Myhre, and A. Vollmar, *J. Amer. Chem. Soc.*, 1977, **99**, 3042.
[64] A. N. Detsina, V. I. Mamatyuk, and V. A. Koptyug, *J. Org. Chem.* (*U.S.S.R.*), 1977, **13**, 122.
[65] L. M. Stock and T. L. Wright, *J. Org. Chem.*, 1977, **42**, 2875.
[66] A. Koeberg-Telder and H. Cerfontain, *J.C.S. Perkin II*, 1977, 717; H. Cerfontain, A. Koeberg-Telder, and C. Ris, *ibid.*, p. 720.

1,2,4,5-tetraethylbenzene are unstable, and rearrange into 2,3,4,5-tetraethylbenzenesulphonic acid. With 1,3,5-tri-isopropylbenzene, both rearrangement and dealkylation reactions become of importance. Kinetic and product studies[67a] have shown that the sulphonation of aniline in a large excess of 100—102% sulphuric acid at 25 °C proceeds by direct attack of the electrophile ($H_3S_2O_7^+$) on the aromatic ring of the anilinium ion. Under more aqueous conditions (~97% sulphuric acid), at higher temperatures, some reaction occurs by the same mechanism, as shown by the formation of *m*-aminobenzenesulphonic acid, but reaction involving the intermediacy of phenylsulphamic acid may also occur. The latter compound itself, at 25 °C, yields[67b] only *o*- and *p*-aminobenzenesulphonic acids, by an intermolecular path involving *C*-sulphonation of the minority *O*-protonated species, $PhNHSO_3H$, and subsequent *N*-desulphonation, although at lower acidities solvolysis to the anilinium ion is a competing process.[67c] The sulphonation of *m*-aminobenzenesulphonic acid in fuming sulphuric acid has also been studied.[68a] Studies of methanesulphonanilide ($PhNHSO_2Me$) indicate[68b] that sulphonation occurs on the unprotonated species, which is in the minority above 84% sulphuric acid.

For non-catalytic bromination reactions the process may follow the power-series rate equation, and involve a reaction pathway in which more than one molecule of bromine is concerned in the transition state. The suggestion [see *Ann. Reports* (*B*), 1975, **72**, p. 225] that this will show the same kinetic behaviour as an A-S_E2 process in which the rate of the reverse of the first step becomes significant, and that the rate expressions for these processes are equivalent, is incorrect.[69a] The circumstances under which a distinction may be made have been established and the authors also briefly review the mechanisms available for brominations by molecular bromine. The effect of addition of bromide and perchlorate to the bromination of polymethylbenzenes in 90% aqueous acetic acid shows clearly[69b] that in this case reaction involves a transition state which includes more than one molecule of bromine. Flow ^1H n.m.r. studies have clearly confirmed[70] the generality of the occurrence of cyclohexadienone intermediates in the bromination of substituted phenols in aqueous acetic acid, the lifetime of the intermediates decreasing with the size of alkyl substituents at 2- and 6-positions.

'Free' CT_3^+ cations have been generated from the decay of a radioactive atom in a tritiated methane solution in toluene and benzene.[71] The attack of the cation on the arene is likely to occur in a much shorter time than is necessary for the formation of an organised solvent sphere. The intermolecular selectivity was indicated by an approximate toluene:benzene rate ratio of 2.3:1, and a very low but slightly electrophilic intramolecular selectivity was observed. The vinylation of some aromatic substrates by a variety of vinyl triflates [*e.g.* $PhC(Me){=}C(Ph)OSO_2CF_3$], lacking a β C—H bond, is possible, and it results in

[67] P. K. Maarsen and H. Cerfontain, *J.C.S. Perkin II*, 1977, (*a*) p. 1008; (*b*) p. 921; (*c*) p. 929.
[68] (*a*) P. K. Maarsen, R. Bregman, and H. Cerfontain, *J.C.S. Perkin II*, 1977, 1863; (*b*) P. K. Maarsen and H. Cerfontain, *ibid.*, p. 1003.
[69] (*a*) N. H. Briggs, P. B. D. de la Mare, and D. Hall, *J.C.S. Perkin II*, 1977, 106; (*b*) R. M. Keefer and L. J. Andrews, *J. Amer. Chem. Soc.*, 1977, **99**, 5693.
[70] C. A. Fyfe and L. Van Veen, jun., *J. Amer. Chem. Soc.*, 1977, **99**, 3366.
[71] F. Cacace and P. Giacomello, *J. Amer. Chem. Soc.*, 1977, **99**, 5477.

good yields of styrenes under moderate conditions even in the absence of Friedel–Crafts catalysts.[72]

Nucleophilic Substitution.—Recent developments in the application of linear free-energy relationships between reactivity and physical properties of leaving groups and substrates have been discussed.[73] A comparison of some substitution reactions of both neutral and negative nucleophiles in the 2,4-dinitrohalogeno-benzenes with reactions with organic cations has provided[16] further support for the general applicability of the N_+ nucleophilicity scale. Selectivity again appears not to be a function of reactivity in these reactions.

The photostimulated $S_{RN}1$ reaction between iodobenzene and potassium diethylphosphite in DMSO has been studied,[74a] with particular attention being paid to the initiation and termination stages. The main reaction for the latter is first-order in propagating radical. The initiation, however, may involve initial absorption by iodobenzene or by a molecular complex of iodobenzene with diethylphosphite. An $S_{RN}1$ reaction in aqueous t-butyl alcohol has been identified[74b] in the reaction of halogeno-benzenes with phenoxide ion in the presence of sodium amalgam to form benzene and diphenyl ether. A thermally induced $S_{RN}1$ reaction has been found[74c] in the reaction of bromo- and iodo-benzenes with ketone enolate anions in DMSO at 25 °C in the dark, and in one example this was well separated from the photostimulated process. Large rate accelerations by added oxygen, nitrobenzene, and ferric nitrate were, however, not explained.

The reactions of picryl chloride and bromide with the diethylmalonate anion have been studied[75] by the stopped-flow technique, which enables the stages involving fast formation of σ-complexes followed by slower decomposition to the picryl esters to be identified. In the reaction of 2,4,6-trinitroanisole with n-butylamine to give, in the absence of a large excess of the amine, *N*-(n-butyl)picramide, the ^{1}H n.m.r spectrum of the intermediate σ-complex has been recorded, using a flow technique.[76] The stable intramolecular σ-complex (17), formed under basic conditions from *N*-benzyl-*N*-(2-hydroxyethyl)picramide, cyclizes under the conditions of the Smiles' rearrangement, without prior rearrangement, to form 4-benzyl-5,7-dinitro-2,3-dihydrobenzoxazine.[77] Kinetic and equilibrium data have been reported[78] for the addition of sodium ethoxide to a series of 6-substituted 2,4-dinitrophenetoles to give σ-complexes. The results support the suggestion that interaction with the cation for 1,1-diethoxy-complexes involves the oxygen atoms of the alkoxy-groups attached to C-1 and the *ortho*-substituents.

[72] P. J. Stang and A. G. Anderson, *Tetrahedron Letters*, 1977, 1485.
[73] G. Bartoli and P. E. Todesco, *Accounts Chem. Res.*, 1977, **10**, 125; see also G. Bartoli, P. E. Todesco, and M. Fiorentino, *J. Amer. Chem. Soc.*, 1977, **99**, 6874.
[74] (*a*) S. Hoz and J. F. Bunnett, *J. Amer. Chem. Soc.*, 1977, **99**, 4690; (*b*) S. Rajan and P. Sridaran, *Tetrahedron Letters*, 1977, 2177; (*c*) R. G. Scamehorn and J. F. Bunnett, *J. Org. Chem.*, 1977, **42**, 1449, 1457.
[75] K. T. Leffek and A. E. Matinopoulous-Scordou, *Canad. J. Chem.*, 1977, **55**, 2656, 2664.
[76] C. A. Fyfe, A. Koll, S. W. H. Damji, C. D. Malkiewich, and P. A. Forte, *J.C.S. Chem. Comm.*, 1977, 335.
[77] V. N. Drozd, V. N. Knyazev, and V. M. Minov, *J. Org. Chem.* (*U.S.S.R.*), 1977, **13**, 357.
[78] M. R. Crampton, *J.C.S. Perkin II*, 1977, 1442.

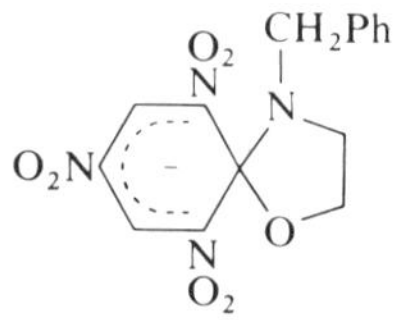

(17)

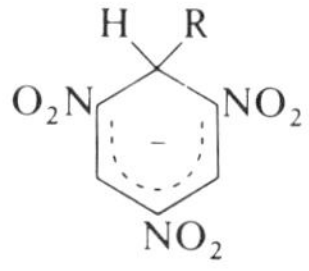

(18)

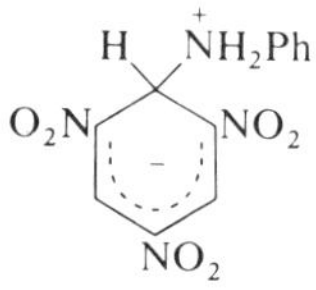

(19)

The formation of the σ-complex (18; R = NHPh) from 1,3,5-trinitrobenzene and aniline, catalysed by a tertiary amine in DMSO, has been shown[79a] to involve rate-limiting deprotonation by the catalyst of the zwitterionic intermediate (19), and it shows only a small primary kinetic isotope effect. The magnitude of the effect should have relevance to the identification of rate-limiting stages in S_NAr reactions (see below). The complex (18; R = NHPh) may also be formed by the reaction of aniline with the complex (18; R = OMe), and the full evidence that in this example a dissociative mechanism involving the intermediacy of trinitrobenzene is involved has been given.[79b] Methoxide ion acts in a similar fashion to the tertiary amine catalyst in the previous reaction, but in this case the formation of the intermediate (19) is rate-limiting. It seems that the unreactivity of aromatic as opposed to aliphatic amines towards nitro-aromatic compounds is due to thermodynamic rather than kinetic factors.

Arguments have been presented[80a] that currently available evidence indicates that the specific base–general acid-catalysed mechanism for substitution in activated aromatic compounds by amines, involving fast deprotonation and rate-limiting catalysed expulsion of the leaving group, is not a significant pathway in protic solvents. Instead the mechanism involves a rate-limiting deprotonation of (20), when the expulsion of the leaving group is relatively fast (*e.g.* X = OPh), or a rapid equilibrium deprotonation of (20) followed by a rate-limiting non-catalysed

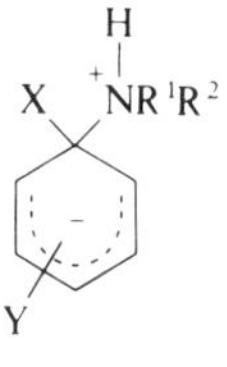

(20)

expulsion of a very poor leaving group (*e.g.* on occasions when X = OMe). Catalysis by base in the reaction of piperidine with 2,4-dinitrophenyl 4-nitrophenyl ether in benzene solution has been discussed[80b] in terms of concerted proton transfer and expulsion of the leaving group, the protic base acting as a bifunctional catalyst. Evidence from the uncatalysed reaction of morpholine with 2,4-dinitrophenyl phenyl ether in dipolar aprotic solvents suggests that here also fast deprotonation and slow expulsion of the leaving group does not occur,[80c] and the authors also

[79] (*a*) E. Buncel, W. Eggimann, and H. W. Leung, *J.C.S. Chem. Comm.*, 1977, 55; E. Buncel and W. Eggimann, *J. Amer. Chem. Soc.*, 1977, **99**, 5958; (*b*) E. Buncel, J. G. K. Webb, and J. F. Wiltshire, *ibid.*, p. 4429.

[80] (*a*) C. F. Bernasconi, R. H. De Rossi, and P. Schmid, *J. Amer. Chem. Soc.*, 1977, **99**, 4090; (*b*) D. Spinelli, G. Consiglio, and R. Noto, *J.C.S. Perkin II*, 1977, 1316; (*c*) D. Ayediran, T. O. Bamkole, J. Hirst, and I. Onyido, *ibid.*, p. 597.

discuss shortcomings of the concerted mechanism for media of low dielectric constant, and suggest a modification involving the formation of aggregates within which bond-making and -breaking is not synchronous.

Biaryls.—An important modification of the nickel–phosphine-complex-catalysed homo-coupling of aryl halides has been reported,[81] and the reaction proceeds catalytically in nickel in the presence of zinc, which reduces the Ni^{II} species formed to the active Ni^0 species. The molar ratio PhBr:Zn:PPh_3:$Ni(PPh_3)_2Cl_2$ of 1:1:0.3:0.5 in DMF solution at 50 °C gave an 89% yield of biphenyl. The presence of iodine greatly accelerated the reaction. Efficient formation of the symmetrical biaryl has been observed[82] when some arylcopper cluster compounds [*e.g.* $(4\text{-}MeC_6H_4)_4Cu_4$] react with equimolar quantities of copper triflate. These reactions involve intermediate complexes which may be isolated in some cases. Treatment of a variety of aromatic substrates with thallium(III) trifluoroacetate in trifluoroacetic acid results in oxidative coupling to give[83] biaryls in good yield in the absence of powerfully electron-withdrawing substituents, probably by a radical cation mechanism; 2,5-dimethylanisole, for example, is converted into 2,2′,5,5′-tetramethyl-4,4′-dimethoxybiphenyl. An improved method[84] has been reported for the conversion of arylmercury (II) chlorides into biaryls, using $[ClRh(CO)_2]_2$ as a catalyst in the presence of lithium chloride. Thermal decomposition of tetra-aryltellurium compounds at 140 °C forms biaryls by a non-radical pathway.[85]

The preparation of a wide variety of mixed biaryls, in improved yields, is possible by the application of phase-transfer catalysis to a Gomberg reaction.[86] Potassium acetate, in the presence of 18-crown-6, apparently reacts with an arenediazonium ion in non-polar solution to form a diazoanhydride, which then decomposes in the presence of various aromatic compounds to form biaryls. The reaction of some arylzinc derivatives with aryl bromides or iodides in the presence of a small amount of a nickel or palladium catalyst [*e.g.* $Ni(PPh_3)_4$] provides a general and highly selective route to unsymmetrical biaryls.[87] Palladium(II) chloride (<0.001 mol equivalent) appears[88] to be a preferred catalyst in Kharasch-type reactions of aryl Grignard reagents with iodobenzenes in THF. Aromatic nitrogen compounds (nitroso-, azo-, azoxy-, and nitro-benzenes and *NN*-dialkylaniline *N*-oxides, in addition to aryl-hydroxylamines) are susceptible to nucleophilic attack by benzene to give biphenyls in varying yields under strongly acidic conditions;[89] 5-(*NN*-dimethylamino)-2-methylbiphenyl, for example, is formed in 46% yield from *NN*-dimethyl-*p*-toluidine *N*-oxide.

A comparison of the rates of solvolysis of some α-(7-X-fluoren-2-yl)benzyl chlorides and α-(4′-X-biphenyl-4-yl)benzyl halides is consistent with regarding

[81] M. Zembayashi, K. Tamao, J.-I. Yoshida, and M. Kumada, *Tetrahedron Letters*, 1977, 4089.
[82] G. van Koten, J. T. B. H. Jastrzebski, and J. G. Noltes, *J.C.S. Chem. Comm.*, 1977, 203; *J. Org. Chem.*, 1977, **42**, 2047.
[83] A. McKillop, A. G. Turrell, and E. C. Taylor, *J. Org. Chem.*, 1977, **42**, 764.
[84] R. C. Larock and J. C. Bernhardt, *J. Org. Chem.*, 1977, **42**, 1680.
[85] D. H. R. Barton, S. A. Glover, and S. V. Ley, *J.C.S. Chem. Comm.*, 1977, 266.
[86] S. H. Korzeniowski, L. Blum, and G. W. Gokel, *Tetrahedron Letters*, 1977, 1871.
[87] E.-I. Negishi. A. O. King, and N. Okukado, *J. Org. Chem.*, 1977, **42**, 1821.
[88] A. Sekiya and N. Ishikawa, *J. Organometallic Chem.*, 1977, **125**, 281.
[89] T. Ohta, K. Shudo, and T. Okamoto, *Tetrahedron Letters*, 1977, 101; K. Shudo, T. Ohta, Y. Endo, and T. Okamoto, *ibid.*, p. 105.

fluorene as a planar biphenyl structure, and yields an interplanar angle for biphenyl in accord with other estimates.[90] Both systems respond to substituent effects in a similar way. The overall reactivity to protiodetritiation of 9,10-dihydrophenanthrene relative to that of fluorene is consistent with the differences in coplanarity between the molecules,[91a] and other results confirm that the low reactivity of the α-position of the latter arises from an increase in strain on going to the corresponding transition state. The reversible rearrangement of 1-fluoro-9*H*-fluoren-9-one into the 3-fluoro-isomer in polyphosphoric acid provides the first direct evidence for the complete reversibility of a Friedel–Crafts acylation.[91b]

Quinones and Related Compounds.—A review of photoinitiated reactions of quinones has appeared.[92] A convenient synthesis[93] of 2-alkyl *p*-benzoquinones in 65—70% yields involves the oxidation of the *p*-alkylphenol with thallium(III) perchlorate in perchloric acid. The reaction of allyl-silanes with various *p*-benzoquinones in the presence of titanium tetrachloride gave[94] allyl-substituted hydroquinones, whilst 2,6-disubstituted quinones gave *p*-allyl-quinols, which were presumed also to be intermediates in the former cases. An interesting base-catalysed rearrangement of *p*-peroxyquinol ethers to give *p*-quinoxyacetic acid derivatives has been observed,[95] and is shown formally in (21). The suggested mechanism involves initiation by homolysis of the O—O bond even at −60 °C.

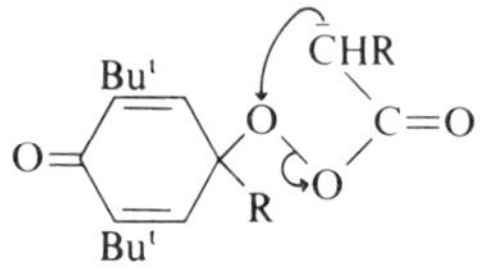

(21)

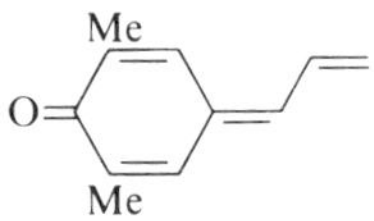

(22)

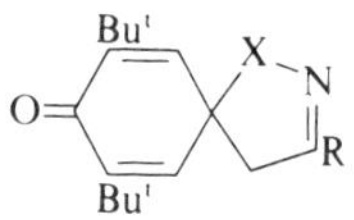

(23)

The quinone methide (22), which has a vinyl group in conjugation with the quinone system, forms, in the solid or in concentrated solution in an aprotic solvent, a dimer involving Diels–Alder addition of the terminal double bond of one molecule to the side-chain diene system of another molecule.[96a] The products of nucleophilic addition to (22) have been studied,[96b] and with water and methanol both 1,6- and 1,8-adducts are observed. Kinetic studies of the related methylene compound have also appeared.[96c] Some reactive 1,3-dipoles ($R-\overset{+}{C}=N-\overset{-}{X}$) add to 2,6-di-t-butyl-4-methylenecyclohexa-2,5-dienone to give the 1:1 cycloadducts (23) in high yield, providing ready access to the hetero-spiro[4,5]decatrienone system.[96d]

[90] R. Bolton and R. E. M. Burley, *J.C.S. Perkin II*, 1977, 426.
[91] (*a*) H. V. Ansell and R. Taylor, *J.C.S. Perkin II*, 1977, 866; (*b*) I. Agranat, Y. Bentor, and Y.-S. Shih, *J. Amer. Chem. Soc.*, 1977, **99**, 7068.
[92] A. V. El'tsov, O. P. Studzinskii, and V. M. Grebenkina, *Russ. Chem. Rev.*, 1977, **46**, 93.
[93] Y. Yamada and K. Hosaka, *Synthesis*, 1977, 53.
[94] A. Hosomi and H. Sakurai, *Tetrahedron Letters*, 1977, 4041.
[95] A. Nishinaga, K. Nakamura, K. Yoshida, and T. Matsuura, *Chem. Letters*, 1977, 303.
[96] (*a*) J. A. Hemingson, *Tetrahedron Letters*, 1977, 2967; (*b*) *J.C.S. Perkin II*, 1977, 616; (*c*) G. Leary, I. J. Miller, W. Thomas, and A. D. Woolhouse, *ibid.*, p. 1737; (*d*) A. D. Woolhouse, *Austral. J. Chem.*, 1977, **30**, 1145.

An electron-diffraction study of *p*-xylylene, which is widely accepted to have a singlet structure, has been made,[97a] and the conclusion reached that the molecule does not simply contain conjugated double and single bonds. The average C=C bond length is longer than typical values for conjugated double bonds, and in good agreement with simple HMO theory. The i.r. and Raman spectra of *o*-xylylene have been observed for the first time.[97b] The molecule was generated in the gaseous phase by the reaction of αα′-dibromo-*o*-xylene with sodium or potassium vapour, which had been excited by a microwave discharge, and it was then collected in an argon matrix (at 8—30 K). 2,2-Dimethyl-2*H*-indene (24) shows more than a

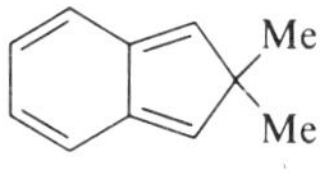

(24)

transient existence.[98a] It was generated and isolated in a matrix at 77 K, but was also generated in a solution at room temperature, where it was sufficiently stable for its ^{1}H n.m.r. spectra to be recorded. Transient isoindenes have been observed and studied in the flash photolysis of a series of 1,1-diaryl-indenes.[98b] Isoindene itself has been implicated[98c] as an intermediate in a photobisdecarbonylation reaction which may be suitable for future matrix-isolation studies. *o*-Xylylenes have also been observed as transients in the flash photolysis of methylated indan-2-ones.[99] The intermediacy of the 2*H*-indenes (26) in the thermal rearrangement of the 1-acyl-1,3-diphenyl- and -dimethyl-indenes (25) to the corresponding 2-acyl-indenes (27) has been established,[100a] and migratory aptitudes have been described. Other studies of the related photochemical rearrangement of 3-substituted 1,1-diphenylindenes have been reported.[100b]

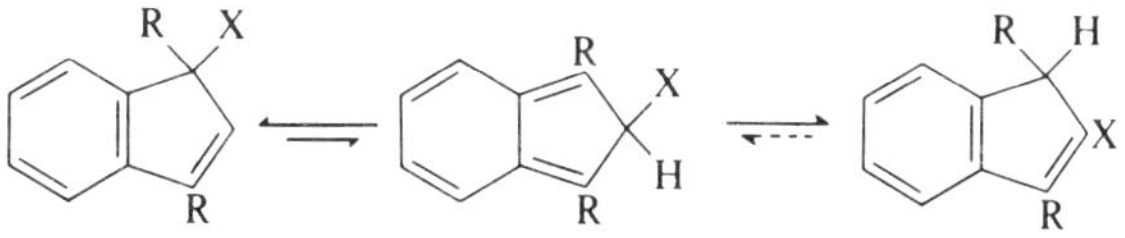

(25) (26) (27)

R = Ph or Me; X = *e.g.* CHO, COMe, COPh, or CO_2Me

Oxidation of 4-aryl-2,6-di-t-butylphenols with oxygen in the presence of a cobalt(II) complex followed by decomposition of the intermediate hydroperoxides with toluene-*p*-sulphonic acid gave the 5-aryl-3-t-butyl-*o*-benzoquinones in good yield.[101] Oxidation of a series of *o*-benzoquinones with lead(IV) acetate in methanol to yield the corresponding dimethyl hexa-2,4-dienedioates in 60—90%

[97] (*a*) P. G. Mahaffy, J. D. Wieser, and L. K. Montgomery, *J. Amer. Chem. Soc.*, 1977, **99**, 4514; (*b*) K. L. Tseng and J. Michl, *ibid.*, pp. 4840, 6154.

[98] (*a*) W. R. Dolbier, jun., K. Matsui, J. Michl, and D. V. Horak, *J. Amer. Chem. Soc.*, 1977, **99**, 3876; (*b*) K. K. de Fonseka, C. Manning, J. J. McCullough, and A. J. Yarwood, *ibid.*, p. 8257; (*c*) R. N. Warrener, R. A. Russell, and T. S. Lee, *Tetrahedron Letters*, 1977, 49.

[99] K. K. de Fonseka, J. J. McCullough, and A. J. Yarwood, *J.C.S. Chem. Comm.*, 1977, 721.

[100] (*a*) D. W. Jones and G. Kneen, *J.C.S. Perkin I*, 1977, 1313; (*b*) W. A. Pettit and J. W. Wilson, *J. Amer. Chem. Soc.*, 1977, **99**, 6372.

[101] A. Nishinaga, K. Nishizawa, H. Tomita, and T. Matsuura, *Synthesis*, 1977, 270.

yield has been reported.[102] A pronounced and unexpected solvent effect has been uncovered[103] in the reaction of some *o*-benzoquinones with butadiene; 4-methoxy-5-methyl-*o*-benzoquinone, for example, gave mainly the abnormal spiro-compound (28) in benzene and (29) in methanol.

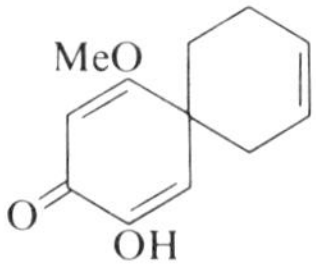

(28)

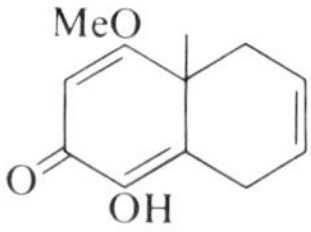

(29)

An unusually stable orange-red crystalline *o*-quinone methide, probably (30), has been isolated[104a] from the oxidation of 2-cinnamyl-4,5-methylenedioxyphenol with silver oxide, and the corresponding methide from 2-(4-methoxybenzyl)-4,5-methylenedioxyphenol rapidly dimerizes in polar solvents. Another stable *o*-quinone methide (m.pt. 284—5 °C) is (31), prepared[104b] by the reaction of 2,4-dimethylphenol with two equivalents of 1,3-benzodithiolylium tetrafluoroborate in

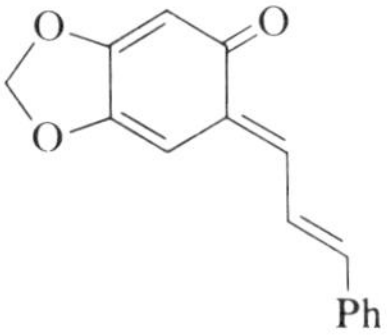

(30)

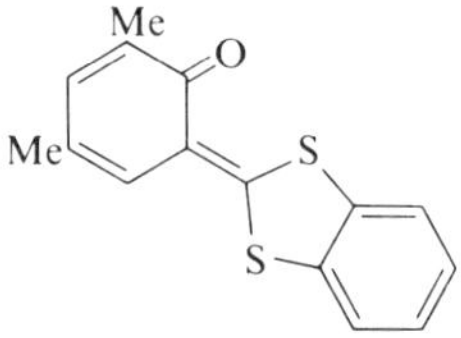

(31)

acetonitrile and subsequent addition of triethylamine. 6-Acetoxycyclohexa-2,4-dien-1-ones react with Grignard and lithium reagents to give, in addition to the expected 1,2- and 1,4-addition, an ether which is the product of the formal addition of a carbanion to the carbonyl oxygen.[105a] Ether formation largely supersedes conjugate addition with tertiary and benzylic organometallics. Substituent effects support an electron-transfer mechanism,[105b] and further evidence[105c] suggests that dialkylmagnesium species are concerned in the case of Grignard reagents.

The metallation reaction of electrochemically derived quinone bisacetals [see *Ann. Reports* (*B*), 1976, **73**, p. 215] has been modified[106] to produce organo-copper-lithium derivatives which will react efficiently with allylic halides to provide a route to some isoprenoid natural products (*e.g.* cymopol).

4 Cyclophanes

A review of the information to be gained concerning steric effects of substituents from conformational studies of cyclophanes has appeared.[107] [5]Metacyclophane,

102 M. Wiessler, *Tetrahedron Letters*, 1977, 233.

103 S. Danishefsky, P. F. Schuda, and W. Caruthers, *J. Org. Chem.*, 1977, **42**, 2179.

104 (*a*) L. Jurd, *Tetrahedron*, 1977, **33**, 163; (*b*) J. Nakayama, K. Yamashita, M. Hoshino, and T. Takemasa, *Chem. Letters*, 1977, 789.

105 (*a*) B. Miller, *J. Org. Chem.*, 1977, **42**, 1402; (*b*) *ibid.*, p. 1408; (*c*) B. Miller, E. R. Matjeka, and J. G. Haggerty, *Tetrahedron Letters*, 1977, 323.

106 P. W. Raynolds, M. J. Manning, and J. S. Swenton, *J.C.S. Chem. Comm.*, 1977, 499.

107 H. Förster and F. Vögtle, *Angew. Chem. Internat. Edn.*, 1977, **16**, 429.

the lowest member of this series to date, is a by-product of a synthesis of 3,3′-hexamethylenebicyclopropenyl.[108] Its spectral properties are indicative of unusually high strain and also of high conformational rigidity of the pentamethylene chain. A novel synthesis[109] of [7]- and [8]-paracyclophanes involves flow pyrolysis of the spiroquinodimethanes (32; $n = 6$ and 7), and a simple synthesis[110] of some [10]paracyclophanes [*e.g.* (33)] involves the reaction of methyl propiolate and aluminium chloride with cyclododecene and thermal dehydrogenation of the resulting 2:1 adducts.

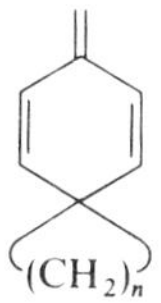

(32)

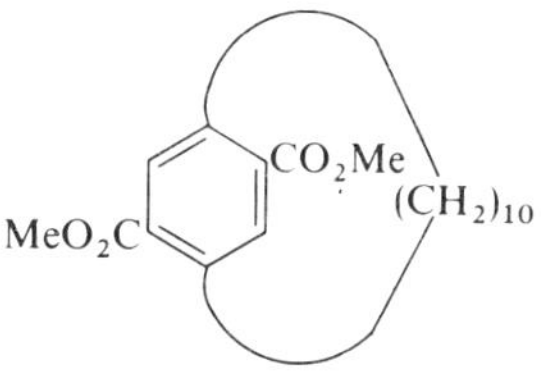

(33)

[3.3]Paracyclophane is now readily available by a route[111a,b] involving the surprisingly efficient (94%) pyrolysis of the disulphone of 2,13-dithia[4.4]paracyclophane. The dithia[4.4]cyclophanes may also yield[111c] the [3.3]cyclophanes on photodesulphurization in triethyl phosphite. Another approach to [3.3]paracyclophanes, also suitable for higher cyclic oligomers, has involved the construction of aliphatic bridges by a reaction between a bromide and a carbanionic species. Charge-transfer interactions in some [3.3]cyclophanes have been studied.[111b,112a] Such interactions may be transmitted through intervening rings in multilayered cyclophanes.[112b] The general usefulness of the sulphone pyrolysis procedure for the synthesis of many-membered hydrocarbon rings has also been illustrated.[113]

Representatives of the highly strained cyclophanes which have two benzene rings held with three or more ethano-bridges have been synthesized. A new approach to [2.2.2](1,2,4)cyclophane involves the formation of the third bridge in a paracyclophane derivative.[114] The chiral [2.2.2](1,2,4)(1,3,5)cyclophane has been synthesized,[115] and it has a marked tendency to resinify at room temperature. Its skew geometry causes an upfield shift of one aromatic proton to τ 4.96. [2.2.2.2](1,2,3,5)Cyclophane, the first to have a consecutive arrangement of three

[108] J. W. van Straten, W. H. de Wolf, and F. Bickelhaupt, *Tetrahedron Letters*, 1977, 4667.
[109] J. W. van Straten, W. H. de Wolf, and F. Bickelhaupt, *Rec. Trav. chim.*, 1977, **96**, 88.
[110] B. B. Snider and N. J. Hrib, *Tetrahedron Letters*, 1977, 1725.
[111] (*a*) D. T. Longone, S. H. Küsefoglu, and J. A. Gladysz, *J. Org. Chem*, 1977, **42**, 2787; (*b*) M. W. Haenel, A. Flatow, V. Taglieber, and H. A. Staab, *Tetrahedron Letters*, 1977, 1733; (*c*) T. Otsubo, M. Kitasawa, and S. Misumi, *Chem. Letters*, 1977, 977; (*d*) T. Synmyozu, K. Kumagae, T. Inazu, and T. Yoshino, *ibid.*, p. 43.
[112] (*a*) T. Shinmyozu, T. Inazu, and T. Yoshino, *Chem. Letters*, 1977, 1347; (*b*) H. A. Staab, U. Zapf, and A. Gurke, *Angew. Chem. Internat. Edn.*, 1977, **16**, 801; see also H. A. Staab and C. P. Herz, *ibid.*, p. 392; C. P. Herz and H. A. Staab, *ibid.*, p. 394.
[113] J. Grütze and F. Vögtle, *Chem. Ber.*, 1977, **110**, 1978; L. Rossa and F. Vögtle, *J. Chem. Res. (S)*, 1977, 264.
[114] S. Trampe, K. Menke, and H. Hopf, *Chem. Ber.*, 1977, **110**, 371.
[115] M. Nakazaki, K. Yamamoto, and Y. Miura, *J.C.S. Chem. Comm.*, 1977, 206.

ethano-bridges, has been synthesized[116a] by a procedure analogous to that applied to the (1,2,4,5)-isomer [see *Ann. Reports* (*B*), 1975, **72**, p. 231]; an *X*-ray structure of the latter has now appeared.[116b]

[2.1.2.1]Paracyclophane has been synthesized in 14% yield by a modified Wurtz reaction, under high dilution, on 4,4′-dichloromethyldiphenylmethane.[117] The application of the Wittig procedure to the synthesis of [2.2.2.2]cyclophanetetraenes [see *Ann. Reports* (*B*), 1975, **72**, p. 231] has been extended,[118a] but an attempt to form [2.2.2]paracyclophanediene from 4,4′-bibenzyldicarbaldehyde and the bisphosphonium salt of 1,4-di(bromoethyl)benzene failed, but gave a small yield of all-*cis*-[2⁶]paracyclophanetetraene, providing an improved route to [2⁶]paracyclophane. The Wittig approach has also led to the synthesis[118b] of [2.0.2.0]metacyclophanediene, which, on photolysis in the presence of iodine, gave bi-4,5-phenanthrylene (34), a non-planar analogue of [8]circulene. The reaction of benzene-1,3,5,-tricarbaldehyde and the bisphosphonium salt of 1,4-di(bromomethyl)benzene has given[118c] a route to the bicyclophane (35), which should exist as a rapidly equilibrating mixture of enantiomeric conformations.

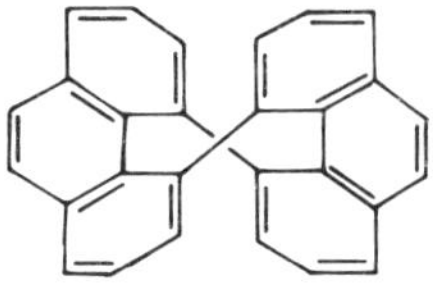

(34)

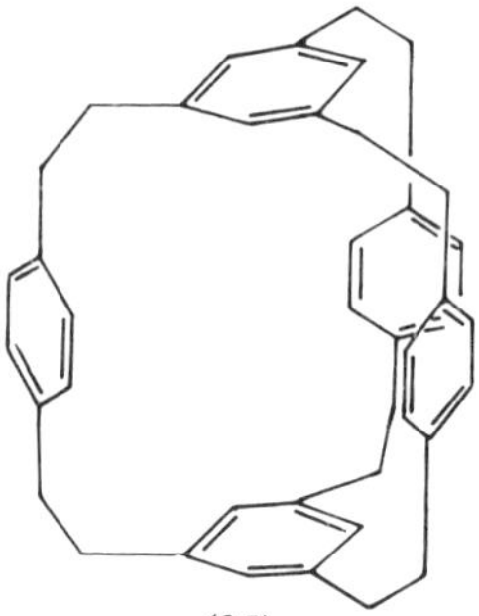

(35)

The 'benzene–Stevens' route [see *Ann. Reports* (*B*), 1975, **72**, p. 231, Scheme 2] has been applied[119] to the syntheses of both *syn*- and *anti*-[2.2](1,4)naphthalenophane-1,13-dienes. [2](1,5)Naphthaleno[2]paracyclophane and its 2,6-isomer, where the bridging is from one benzene ring of the naphthalene moiety to the other, have been synthesized for the first time.[120a] The naphthaleno-paracyclophane (36) has also been prepared[120b] for the first time by two routes, and, in spite

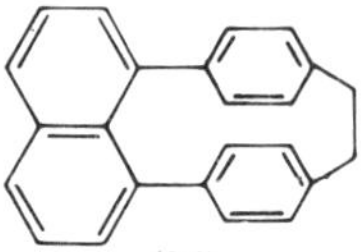

(36)

[116] (*a*) W. Gilb, K. Menke, and H. Hopf, *Angew. Chem. Internat. Edn.*, 1977, **16**, 191; (*b*) A. W. Hanson, *Acta Cryst.*, 1977, **B33**, 2003.

[117] S. Sergheraert, P. Marcinal, and E. Cuingnet, *Tetrahedron Letters*, 1977, 2879.

[118] (*a*) B. Thulin, O. Wennerström, I. Somfai, and B. Chmielarz, *Acta Chem. Scand.* (*B*), 1977, **31**, 135; (*b*) B. Thulin and O. Wennerström, *Tetrahedron Letters*, 1977, 929; (*c*) H. E. Högberg, B. Thulin, and O. Wennerström, *ibid.*, p. 931.

[119] T. Otsubo and V. Boekelheide, *J. Org. Chem.*, 1977, **42**, 1085.

[120] (*a*) M. W. Haenel, *Tetrahedron Letters*, 1977, 4191; (*b*) W. Bieber and F. Vögtle, *Angew. Chem. Internat. Edn.*, 1977, **16**, 175.

of the similarity of its u.v. spectrum to that of a 1,8-diarylnaphthalene, the ^{1}H n.m.r. spectrum shows an enhancement of the anisotropic effect due to the closer approach of the *p*-phenylene rings.

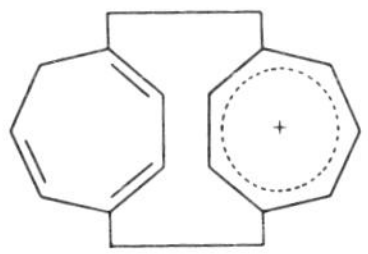

(37)

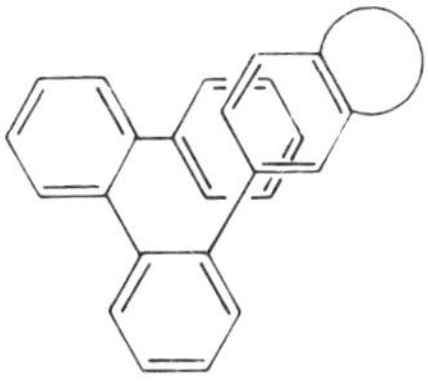

(38)

A further synthesis and an *X*-ray structure of [2.2](2,7)pyrenophane have appeared.[121] [2.2](1,4)tropylio(1,4)cycloheptatrienophane (37) tetrafluoroborate has been formed[122a] from [2.2]paracyclophane by a route involving ring expansions using dibromocarbene. In contrast to its normal behaviour, the cycloheptatriene ring acts as a good intramolecular charge-transfer donor. New cyclophanes having tropylium rings have been reported.[122b]

A new series of compounds which have the combined attributes of helicenes and cyclophanes are the 4,4‴-*o*-quaterphenylophanes (38; $n = 2$ and 6), which have been synthesized[123] by a sulphone-extrusion procedure. Species (38; $n = 2$) exhibits an unexpectedly high conformational flexibility.

5 Molecular Rearrangements

The thermal and acid-catalysed Claisen rearrangements of *N*-allyl-anilines have been investigated.[124] The rearrangement of *N*-(1′,1′-dimethylallyl)-2,6-dimethylaniline into 4-(1′,1′-dimethylallyl)-2,6-dimethylaniline cannot be achieved thermally, but occurs readily in 4 mol dm^{-3} sulphuric acid. A new regioselective synthesis of anilides having carbonyl-functionalized alkyl groups in the *ortho*-position has been accomplished[125a] in 40—70% yields by decarboxylative 1-aza-1′-oxa[3.3]sigmatropic[125b] Claisen rearrangements of enolizable or enolized *N*-aryl-*NO*-diacyl-hydroxylamines, *i.e.* (39)→(40). *N*-Acyl-*N*-aryl-azasulphonium salts

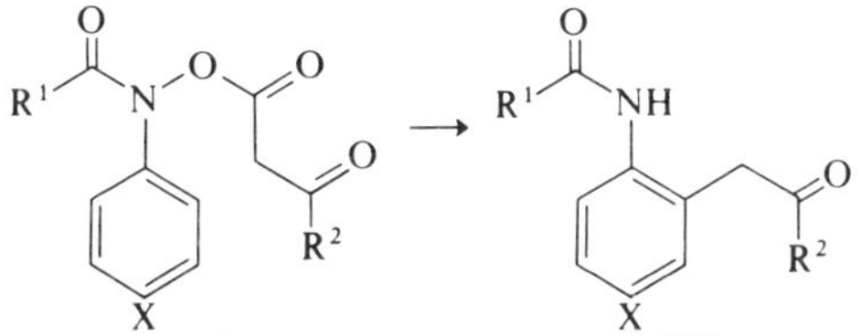

(39) (40)

[121] H. Irngartinger, R. G. H. Kirrstetter, C. Krieger, H. Rodewald, and H. A. Staab, *Tetrahedron Letters*, 1977, 1425.

[122] (*a*) J. G. O'Connor and P. M. Keehn, *Tetrahedron Letters*, 1977, 3711; (*b*) H. Horita, T. Otsubo, and S. Misumi, *Chem. Letters*, 1977, 1309.

[123] F. Vögtle, M. Atzmüller, W. Wehner, and J. Grütze, *Angew. Chem. Internat. Edn.*, 1977, **16**, 325.

[124] S. Jolidon and H.-J. Hansen, *Chimia (Switz.)*, 1977, **31**, 46; *Helv. Chim. Acta*, 1977, **60**, 978.

[125] (*a*) R. M. Coates and I. M. Said, *J. Amer. Chem. Soc.*, 1977, **99**, 2355; (*b*) F. Vögtle and E. Goldschmidt, *Chem. Ber.*, 1976, **109**, 1; (*c*) P. G. Gassman and R. J. Balchunis, *Tetrahedron Letters*, 1977, 2235.

have been prepared for the first time and, when treated with base, they yield ylides (41), which undergo spontaneous [2,3]sigmatropic rearrangement to produce *o*-methylthiomethyl-anilides, and thus provide another route to *ortho*-substituted anilides.[125c] The assumption that the severe conditions necessary for [3,3]sigmatropic rearrangement of *O*-aryl or *O*-vinyl oximes refer to the preliminary tautomerism stage have been confirmed,[126] and some *N*-aryl-*O*-vinyl-hydroxylamines [*e.g.* (42)] and *O*-aryl-*N*-vinyl-hydroxylamines undergo spontaneous rearrangement, demonstrating the ease of cleavage of N—O bonds when a Cope rearrangement path is available. It has been shown[127] by deuterium labelling that 2-vinyl-phenols, at 143 °C, undergo reversible 1,5-hydrogen shifts to give *o*-quinone methides, and the same process leads to rapid (*E,Z*) isomerization of, for example, 2-(prop-1-enyl)-phenols.

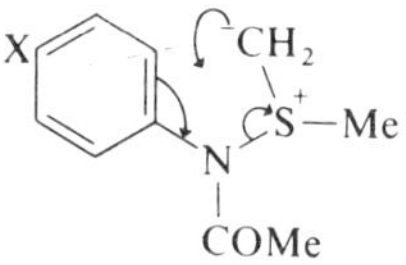

(41)

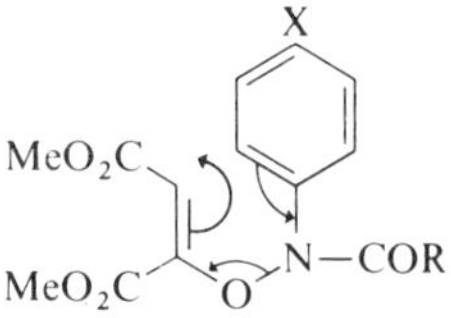

(42)

It has been noted that an autoxidation process to form dimethylbenzylamine and benzoic acid occurs during Stevens rearrangement of benzyldimethylphenacylammonium ylide at low temperatures, and that this predominates under an atmosphere of oxygen.[128] The solvent hexamethylphosphoramide enables the extension of the application of the Smiles' rearrangement of 2-aryloxy-2-methyl-propanamides[129a] into *N*-aryl-2-hydroxy-2-methyl-propanamides to examples which possess deactivated aromatic rings.[129b] This reaction is the basis of a new method for the conversion of phenols into anilines.[129b]

The kinetic nitrogen isotope effect for the benzidine rearrangement of hydrazobenzene is $k_{14}/k_{15} = 1.0203$ (± 0.0007), demonstrating, at last unequivocally, that breaking of the N—N bond in this two-proton process is part of the rate-limiting step. This result[130] rules out the recent proposal involving a second protonation at the C-1 atom, but cannot distinguish between the traditional mechanism having pre-equilibria at the two nitrogen atoms followed by rate-limiting rearrangement, or some concerted process involving N—N bond scission and almost complete second protonation in the rate-limiting stage. The synthetic usefulness of the oxygen analogue of the benzidine rearrangement has been examined.[131] The reactions of salts of *N*-aryl-hydroxamic acids with activated halogeno-benzenes yield 4-amino-4′-hydroxybiphenyls *via* intermediate *NO*-diaryl-hydroxylamines. Biaryls are also formed by a new rearrangement[132] of arylhydrazones of aromatic aldehydes and diaryl ketones, which is stated to be of

[126] T. Sheradsky, E. Nov, S. Segal, and A. Frank, *J.C.S. Perkin I*, 1977, 1827.
[127] H.-J. Hansen, *Helv. Chim. Acta*, 1977, **60**, 2007.
[128] S. H. Pine and E. Fujita, *J. Org. Chem.*, 1977, **42**, 1460.
[129] R. Bayles, M. C. Johnson, R. F. Maisey, and R. W. Turner, *Synthesis*, 1977, (*a*) p. 31; (*b*) p. 33.
[130] H. J. Shine, G. N. Henderson, A. Cu, and P. Schmid, *J. Amer. Chem. Soc.*, 1977, **99**, 3719.
[131] T. Sheradsky and E. Nov, *J.C.S. Perkin I*, 1977, 1296.
[132] F. Fusco and F. Sannicolo, *Tetrahedron Letters*, 1977, 3163.

general applicability, and which is exemplified by the formation of 4-amino-3′-formyl-6′-methoxy-3,5-dimethylbiphenyl (35%) on heating the 2,6-dimethylphenylhydrazone of anisaldehyde in polyphosphoric acid.

The product data from the nitramine rearrangement of *N*-methyl-*N*-nitro-*m*-chloroaniline have been compared[133] with those from the Claisen rearrangement of allyl *m*-chlorophenyl ether, and it is claimed that the 'cartwheel' mechanism for the former is not able to accommodate the findings that isomer formation, except for a small steric effect and the requirement for *ortho,para*-substitution, is random. Photochemical rearrangement of the sulphonyl group has been observed[134] in the conversion of phenyl benzenesulphonate into *o*- and *p*-hydroxyphenyl phenyl sulphones by u.v. irradiation, and a mechanism involving a caged radical pair has been proposed. Further studies[135a] of the Fischer–Hepp rearrangement and concurrent denitrosation have involved methyl-substituted *N*-methyl-*N*-nitroso-anilines, with a view to the evaluation of steric effects; in particular, those on the enigmatic intramolecular rearrangement step. A steric effect was apparent for the 3,5-dimethyl compound. *N*-Nitrosodiphenylamine is denitrosated by various nucleophiles and can also act as a direct ring-nitrosating agent towards anilines.[135b]

6 Condensed Systems

A new route[136] to polymethylated naphthalenes, anthracenes, and phenanthrenes involves the reaction of 3,4-dichloro-1,2,3,4-tetramethylcyclobutene with aromatic hydrocarbons in the presence of aluminium tribromide, and, for example, reaction with naphthalene gave 1,2,3,4-tetramethylphenanthrene (60%). The usefulness of photodehydrocyclization reactions of stilbenes and analogues to give phenanthrenes has been increased, as reaction occurs much more cleanly[137] in the presence of certain π-acceptor molecules (*e.g.* tetracyanoethylene) under anaerobic conditions and, if necessary, at low temperature. The generation of α'-bromo-1,2-naphthoquinodimethane from 1-bromomethyl-2-dibromomethylnaphthalene is involved[138] in an improved one-pot synthesis of phenanthrenes. The species then undergoes Diels–Alder reaction with *N*-phenylmaleimide, elimination of hydrogen bromide, and fortuitous oxidation to give (43). A new phenanthrene synthesis is

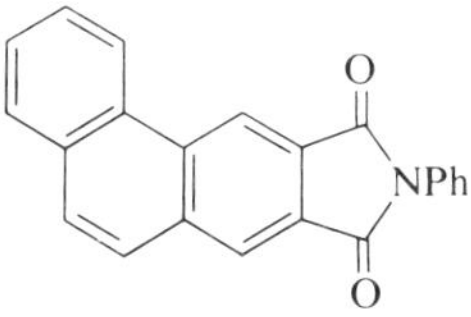

(43)

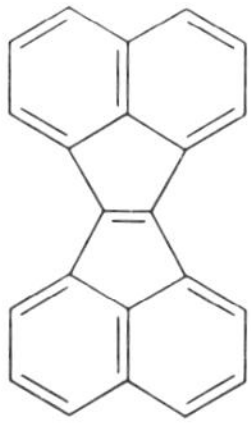

(44)

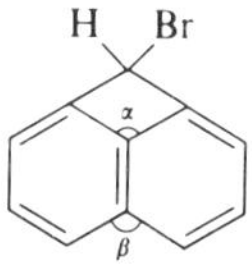

(45)
$\alpha = 99°$
$\beta = 138°$

[133] W. N. White and J. R. Klink, *J. Org. Chem.*, 1977, **42**, 166.
[134] Y. Ogata, K. Takagi, and S. Yamada. *J.C.S. Perkin II*, 1977, 1629.
[135] (*a*) I. D. Biggs and D. L. H. Williams, *J.C.S. Perkin II*, 1977, 44; (*b*) J. T. Thompson and D. L. H. Williams, *ibid.*, p. 1932.
[136] A. P. Krysin, N. V. Bodoev, and V. A. Koptyug, *J. Org. Chem.* (*U.S.S.R.*), 1977, **13**, 1183.
[137] J. Bendig, M. Beyermann, and D. Kreysig, *Tetrahedron Letters*, 1977, 3659.
[138] G. W. Gribble, E. J. Holubowitch, and M. C. Venuti, *Tetrahedron Letters*, 1977, 2857.

involved[139a] in the direct conversion of a 2,2′-di(chloromethyl)-substituted biphenyl into the phenanthrene with sodamide in ammonia, using an iron catalyst. Vanadium trifluoride oxide has been found[139b] to convert a variety of 1,2-diaryl-ethene derivatives into phenanthrenes in high yield.

Full details of the preparation of 1,1-dichlorocyclopropa[*b*]naphthalene from 1,1,1a,7a-tetrachloro-1a,2,7,7a-tetrahydro-2,7-diphenylcyclopropa[*b*]naphthalene have appeared.[140] The *gem*-dibromo-derivative cannot be formed in an analogous way, but the reaction of the dichloro-compound with a small excess of ethyl- or phenyl-magnesium bromide leads to halogen exchange, the isolable *gem*-diethyl compound only being formed with a large excess of the Grignard reagent. The unisolable *gem*-diphenyl compound undergoes ready cleavage of the cyclopropane ring. Acenaphthylene is formed (41% yield) by the elimination of carbon monoxide from 1*H*-phenalen-1-one on flow pyrolysis at 900 °C.[141] This method is of more general utility and, for example, provides the preferred route to indeno[2,1-*a*]indene (dibenzopentalene). Acenaphth[1,2-*a*]acenaphthylene (44) has been synthesized[142] by a new route, and the increasingly strained compounds obtained by introducing ethano-bridges across the *peri*-positions of (44) have been made for the first time. Spectral effects due to strain were noted, and the authors remark on the lack of calculations on these systems. The structure of 1-bromo-1*H*-cyclobuta[*de*]naphthalene (45) has been determined[143a] and, surprisingly, the three rings are essentially coplanar. The strain due to *peri*-bridging is accommodated by adjustments, particularly in bond angles [see (45)], throughout the molecule. The first determination of the crystal structure of a Dewar molecule has been performed[143b] on 1,3,6,8-tetra-t-butylhemi-Dewar-naphthalene. The molecule has a *syn*-structure, as the substituents cause considerable distortion and the benzo-cyclobutene system is not planar.

Decamethylanthracene has been prepared for the first time,[144] utilizing (in part) the removal of a nitrogen bridge in a 1,4-imine by oxidation with peroxide [see *Ann. Reports* (*B*), 1976, **73**, p. 224]. The strain in the structure was illustrated by the ease of photochemical conversion into the 9,10-Dewar-isomer. A convenient method has been found[145] for the introduction of a tertiary carbon carrying an electronegative group into the 9-position of anthracene. The appropriate alkyl radical generated thermally from an azo-compound reacts with anthrone under basic conditions to give the 10-alkyl-anthrone, and the alcohol formed on subsequent reduction could be aromatized to the desired 9-substituted anthracene by phosphorus pentoxide. Dehydrogenation of a series of mono- and di-t-butyldihydroanthracenes has been investigated[146] as a potential synthetic route to t-butyl-anthracenes. 1- and 2-t-Butylanthracene were prepared in this way, although rearrangements complicated other cases. 2,6-Di-t-butylanthracene could be prepared in high yield by direct t-butylation of anthracene.

139 (*a*) M. S. Newman and H. M. Dali, *J. Org. Chem.*, 1977, **42**, 734; (*b*) A. J. Liepa and R. E. Summons, *J.C.S. Chem. Comm.*, 1977, 826.
140 A. R. Browne and B. Halton, *J.C.S. Perkin I*, 1977, 1177.
141 G. Shaden, *Angew. Chem. Internat. Edn.*, 1977, **16**, 50.
142 R. H. Mitchell, T. Fyles, and L. M. Ralph, *Canad. J. Chem.*, 1977, **55**, 1480.
143 (*a*) M. Gessner, P. Card, H. Shechter, and G. G. Christoph, *J. Amer. Chem. Soc.*, 1977, **99**, 2371; (*b*) R. W. Franck, R. Gruska, and J. G. White, *Tetrahedron Letters*, 1977, 509.
144 H. Hart and B. Ruge, *Tetrahedron Letters*, 1977, 3143.
145 T. Mitsuhashi, S. Otsuka, and M. Ōki, *Tetrahedron Letters*, 1977, 2441.
146 P. P. Fu and R. G. Harvey, *J. Org. Chem.*, 1977, **42**, 2407.

Some unusual structures in the anthracene series have been stabilized. 2,4,9-Trichloro-1,10-anthraquinone is a stable crystalline solid prepared[147] by refluxing 9-chloro-10-hydroxy-1,4-anthraquinone with thionyl chloride and triethylamine, and some stable 9,10-quinodimethanes [*e.g.* (46), m.pt. 105 °C] have been prepared[148a] by the initial reaction of the anthraquinone with methylmagnesium iodide followed by treatment with phosphoryl chloride in pyridine. Related compounds (47) have been formed[148b] from the appropriate anthrone. Compounds which had at least two *peri*-methyl–methyl interactions were formed with the alkylidenedihydroanthracene structures shown, whereas simpler compounds were formed as the anthracene tautomer. Other examples have been reported[144,148c] of the ease of loss of aromaticity of the central rings of anthracenes caused by double *peri* interactions.

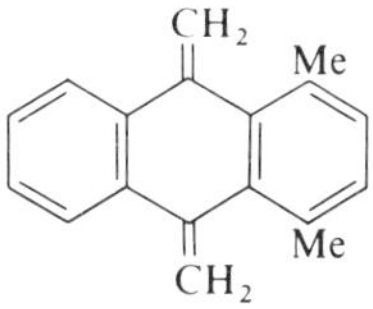

(46)

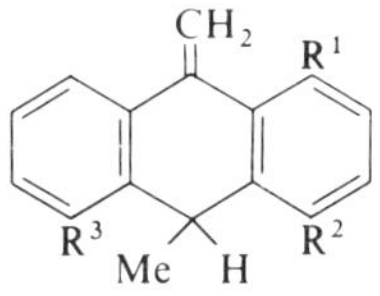

(47) R^1, R^2, R^3 = Me, H, Me; Me, Me, H; H, Me, Me; or Me, Me, Me

A ^{13}C-labelling experiment has revealed[149a] the scrambling of α- and β- (but not angular) carbon atoms in naphthalene at 1035 °C ('automerization'), and a reasonable (but unproven) hypothesis involves the intermediacy of azulene. Mass spectrometric studies indicate[149b] that complete scrambling of the carbon atoms of naphthalene occurs in an energy range between the ionization potential and 18.8 eV, and prior to loss of C_2H_2. The oxidation of naphthalene with *m*-chloroperbenzoic acid under carefully controlled conditions has led[150a] to a synthesis of *anti*-naphthalene 1,2:3,4-dioxide, which complements that of the *syn*-isomer reported last year [see *Ann. Reports* (*B*), 1976, **73**, p. 225]. A naphthalene pentaoxide has been synthesized[150b] from 1,6-epoxy[10]annulene by two successive sequences of addition of singlet oxygen and thermolysis, and it has been shown to have the *syn,syn,syn,syn*-1,2;3,4;5,6;7,8-*anti*-9,10-structure.

The mechanism proposed for the photocatalysed nitrosation of 2-naphthol appeared unlikely[151a] [see *Ann. Reports* (*B*), 1975, **72**, p. 223], but new evidence for acid-catalytic action of singlet excited 2-naphthol on a ground-state reaction has been presented for the production of acetals from the irradiation of alkanals in alcohols.[151b] Although 1-naphthol, in contrast to phenol, does not undergo photodimerization, it has now been reported[152] that irradiation of the 1-naphthoxide ion

[147] M. V. Gorelik, S. P. Titova, and V. A. Trdatyan, *J. Org. Chem.* (*U.S.S.R.*), 1977, **13**, 424.

[148] (*a*) B. F. Bowden and D. W. Cameron, *Tetrahedron Letters*, 1977, 383; (*b*) *J.C.S. Chem. Comm.*, 1977, 78; (*c*) H. Hart and H. Wachi, *ibid.*, p. 409.

[149] (*a*) L. T. Scott and G. K. Agopian, *J. Amer. Chem. Soc.*, 1977, **99**, 4506; (*b*) H. Budzikiewicz and R. Stolze, *Monatsh.*, 1977, **108**, 869.

[150] (*a*) K. Ishikawa and G. W. Griffin, *Angew. Chem. Internat. Edn.*, 1977, **16**, 171; see also K. Ishikawa, H. C. Charles, and G. W. Griffin, *Tetrahedron Letters*, 1977, 427; (*b*) E. Vogel, A. Brewer, C.-D. Sommerfeld, R. E. Davis, and L.-K. Liu, *Angew. Chem. Internat. Edn.*, 1977, **16**, 169.

[151] (*a*) E. A. Chandross, *J. Amer. Chem. Soc.*, 1976, **98**, 1053; (*b*) M. Hisaoka and K. Tokumaru, *Chem. Letters*, 1977, 533.

[152] T. Kitamura, T. Imagawa, and M. Kawanisi, *J.C.S. Chem. Comm.*, 1977, 81.

remarkably gives a dimer (48), probably *via* a photoinduced nucleophilic addition reaction. In a seemingly unprecedented nucleophilic substitution reaction, the formation of 1-piperidinomethyl-4-methyl-3-nitronaphthalene has been recorded[153] from the reaction of 1,4-dimethyl-2,3-dinitronaphthalene with piperidine, and the authors suggest redefining the term '*tele*-substitution' to cover reactions where the entering group occupies a position one or two atoms away from that vacated by the leaving group.

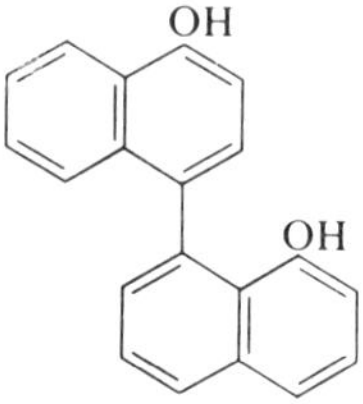

(48)

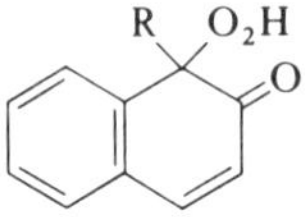

(49)

Heating potassium 2-naphthoxide in a primary alcohol (benzyl, or containing more than three carbon atoms) has been found[154] to give a novel route to the 1-alkyl-2-naphthol in good yield. The syntheses of 1-t-butyl- and 1-t-pentyl-2-naphthol have been reported,[155a] and, although these molecules are considerably strained, any ketone tautomer present was undetectable. The bulk of the alkyl group does lead, however, to an extremely rapid autoxidation to the hydroperoxynaphthalenone (49). It has been suggested[155b] that the unreactivity to oxygen of some 1-aryl- and 1-aralkyl-2-naphthols is due to intramolecular hydrogen bonding of the hydroxy-group to the aryl group of the substituent.

The first quantitative determination of the reactivity towards electrophiles of anthracene results from studies of protiodetritiation.[156] The 1- and 2- positions are more reactive than the corresponding positions in naphthalene, but are closer in reactivity than are those in naphthalene.

Irradiation of 2-styrylbenzo[*c*]phenanthrene in chiral solvents gives[157] the optically active hexahelicene, albeit with low optical yields (0.2—2%). 3,15-Ethano- and 3,15-(2-oxapropano)-[7]helicene have been synthesized,[158] and their conformations

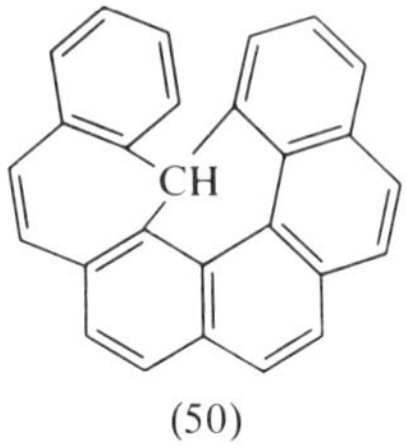

(50)

[153] G. Guanti, S. Thea, M. Novi, and C. Dell'Erba, *Tetrahedron Letters*, 1977, 1429.
[154] T. Kito and K. Ota, *J. Org. Chem.*, 1977, **42**, 2020.
[155] (*a*) J. Carnduff and P. A. Brady, *J. Chem. Res.* (*S*), 1977, 235; (*b*) P. A. Brady, J. Carnduff, and F. Monaghan, *Tetrahedron Letters*, 1977, 3295.
[156] H. V. Ansell, M. M. Hirschler, and R. Taylor, *J.C.S. Perkin II*, 1977, 353.
[157] W. H. Laarhoven and Th. J. H. M. Cuppen, *J.C.S. Chem. Comm.*, 1977, 47.
[158] M. Joly, N. Defay, R. H. Martin, J. P. Declerq, and G. Cermain, *Helv. Chim. Acta*, 1977, **60**, 537.

and ^{1}H n.m.r. spectra compared with those of [7]helicene and 3,15-dimethyl[7] helicene. Thermolysis of 1-formyl[6]helicene tosylhydrazone in benzene in the presence of sodium hydride gave[159] the carbene-insertion product (50), which was resolved, using h.p.l.c. [see *Ann. Reports* (*B*), 1976, **73**, p. 226].

7 Non-benzene Systems

Three- and Four-membered Rings.—The 11-methyl-11-tricyclo[4,4,1,0^{1,6}]undecyl cation has been formed,[160] below −60 °C, and its ^{13}C n.m.r. spectrum suggests that it has significant 2π-homoaromatic character, leading to an unsymmetrical structure (51). The planar 1-(2,3-diphenylcycloprop-2-enylidenemethyl)-2,3-diphenylcyclopropenium ion has been synthesized[161] by treatment of bis(2,3-diphenylcycloprop-2-enyl)methane with trityl perchlorate. Spectral measurements indicate a significant contribution of (52) to its structure. Tetra(*NN*-dialkylamino)triafulvalene dications [*e.g.* (53)] have been synthesized[162] for the first time.

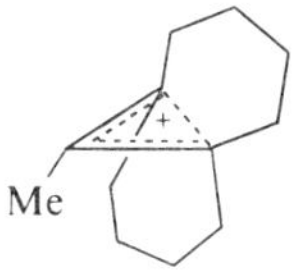

(51)

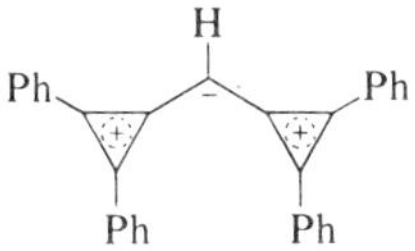

(52)

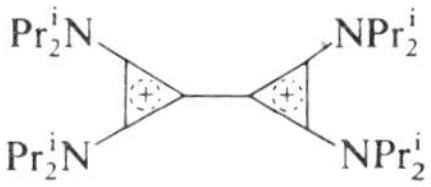

(53)

The mono-, bis-, and tris-(*p*-*NN*-di-isopropylamino)triphenylcyclopropenium ions have been formed[163a] by the reaction of *NN*-di-isopropylaniline with the appropriate chlorocyclopropenium ion; the fully substituted ion is one of the most stable cyclopropenium ions yet reported. Studies of these ions allow an interesting comparison to be made with analogous triphenylmethyl cations, showing, for example, the reduced importance of the immonium canonical form in the former cases. The reactions of 1,2-bisdialkylamino-3-halogenocyclopropenium ions and aryl Grignard reagents have been studied,[163b] and reactions of an intermediate cyclopropenium Grignard reagent identified.

Two electrochemical studies[164] have yielded values for 'antiaromatic' destabilization upon the formation of a cyclobutadiene ring. The value for cyclobutenedione upon reduction to the radical anion was at least 59 kJ mol^{-1}.[164b] Dewar has reinforced[165a] his suggestion that the cyclobutadiene formed by the matrix-isolation technique is a metastable triplet [see *Ann. Reports* (*B*), 1975, **72**, p. 240], and he has used MINDO/3 theory to calculate (with apparent success) the i.r.

159 J. Jespers, N. Defay, and R. H. Martin, *Tetrahedron*, 1977, **33**, 2141.
160 G. A. Olah, G. Liang, D. B. Ledlie, and M. G. Costopoulos, *J. Amer. Chem. Soc.*, 1977, **99**, 4196.
161 K. Komatsu, K. Masumoto, and K. Okamoto, *J.C.S. Chem. Comm.*, 1977, 232.
162 Z.-I. Yoshida, H. Konishi, S.-I. Sawada, and H. Ogoshi, *J.C.S. Chem. Comm.*, 1977, 850.
163 (*a*) K. Komatsu, R. West, and D. Stanislawski, *J. Amer. Chem. Soc.*, 1977, **99**, 6286; (*b*) Z.-I. Yoshida, H. Konishi, Y. Miura, and H. Ogoshi, *Tetrahedron Letters*, 1977, 4319.
164 (*a*) M. Horner and S. Hünig, *Angew. Chem. Internat. Edn.*, 1977, **16**, 410; (*b*) R. D. Rieke, C. K. White, L. D. Rhyne, M. S. Gordon, J. F. W. McOmie, and N. P. Hacker, *J. Amer. Chem. Soc.*, 1977, **99**, 5387.
165 (*a*) M. J. S. Dewar and A. Komornicki, *J. Amer. Chem. Soc.*, 1977, **99**, 6174; (*b*) H. Kollmar and V. Staemmler, *ibid.*, p. 3583.

spectrum of such a species. The paper seems, however, to have been overtaken in publication by the doubts raised concerning the experimental values [see *Ann. Reports* (*B*), 1976, **73**, p. 228]. In contrast to Dewar's calculations, and with relevance to his arguments concerning metastability, another treatment suggests[165b] that, even for a square geometry, the singlet is lower in energy than the triplet, which is in violation of Hund's rule, due to spin polarization effects. Both observed and calculated electron densities for a rectangular cyclobutadiene, for which the structure is known, have shown[166a] that the density maxima for the four-membered ring lie significantly off the lines joining the carbon atoms, indicating that the bonds are 'bent'. The symmetrical and apparently uncomplexed tetrakis(trifluoromethyl)cyclobutadiene has been fully characterized[166b] spectrally in solution and matrix at low temperatures. It has a singlet ground state and appears to be rectangular. Tri-t-butylcyclobutadiene, for which a rectangular structure is to be assumed, has now been shown[166c] to undergo reactions typical of a triplet diradical species; for example, it undergoes addition with carbon tetrachloride. It was suggested that the two structures may be in rapid equilibrium or that the reaction is initiated by transfer of an electron from the cyclobutadiene within a donor–acceptor complex.

Labelling experiments[167] have confirmed the intermediacy of butalene (54) in the reaction of 1-chloro[2,2,0]bicyclohexa-2,5-diene with lithium dialkylamides to give dialkyl-anilines, but this pathway only accounts for about 50% of the product.

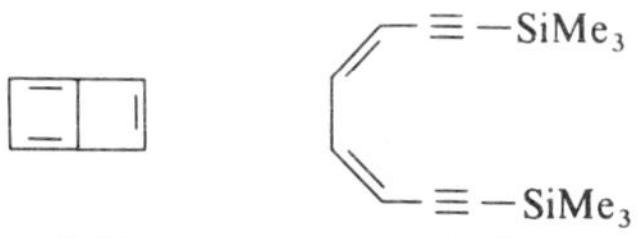

(54) (55)

1,2-Bis(trimethylsilyl)benzocyclobutadiene has been prepared[168] by the vacuum pyrolysis of the *cis,cis*-dienediyne (55) as a thermally stable air-sensitive orange-red oil. The ^{1}H n.m.r. spectrum provides convincing evidence of strong paramagnetic ring-current contributions to the induced ring-current. The pathways of oxidation of benzocyclobutadienes have been clarified[169a] by the observation that in solution the tetraphenylbenzo[1,2:4,5]dicyclobutadiene (56) is oxidized initially to the oxide (57). In the solid, however, the stable *o*-quinone methide (58) is formed.

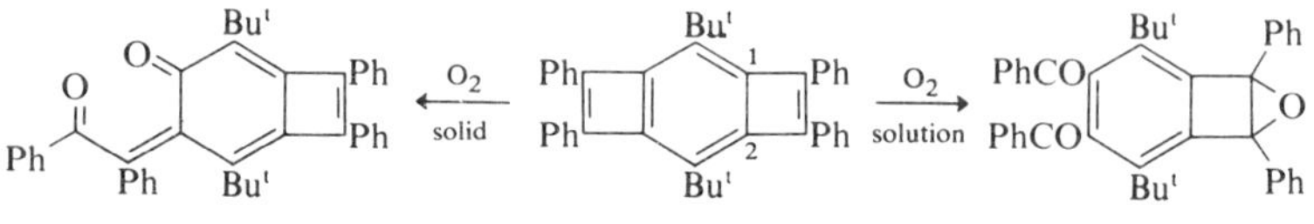

(58) (56) (57)

[166] (*a*) H. Irngartinger, H.-L. Hase, K.-W. Schulte, and A. Schweig, *Angew. Chem. Internat. Edn.*, 1977, **16**, 187; (*b*) S. Masamune, T. Machiguchi, and M. Aratani, *J. Amer. Chem. Soc.*, 1977, **99**, 3524; (*c*) G. Maier and W. Sauer, *Angew. Chem. Internat. Edn.*, 1977, **16**, 51.

[167] R. Breslow and P. L. Khanna, *Tetrahedron Letters*, 1977, 3429.

[168] K. P. C. Vollhardt and L. S. Yee, *J. Amer. Chem. Soc.*, 1977, **99**, 2010.

[169] (*a*) F. Toda, N. Dan, K. Tanaka, and Y. Takehira, *J. Amer. Chem. Soc.*, 1977, **99**, 4529; (*b*) F. Toda and T. Yoshioka, *Chem. Letters*, 1977, 561.

Cycloadditions of tetracyanoethylene to substituted analogues of (56) occur in a [2 + 2] manner across the bond indicated (1,2) to give highly strained propellanes.[169b]

Five- and Seven-membered Rings.—A fulvene synthesis has resulted[170] from a study of the reactions of thiobenzophenones with the dicarbonylcyclopentadienyl-iron anion. Yields are improved by the inclusion of *para*-electron-releasing substituents in the aryl group. The synthesis of fulvenes from the reaction of 1-chloroalkyl acetates and sodium pentadienide, followed by base-catalysed elimination, has been extended[171a] to some 6-substituted fulvenes, and 1,2-benzofulvene, 1,2,3,4-dibenzofulvene, and the corresponding 6-methyl- and 6-phenyl-derivatives have been prepared by an analogous procedure[171b] from sodium indenide or sodium fluorenide. 1,2-Benzofulvene has also been formed[172] by the flash vacuum pyrolysis of the 11-diphenylmethylene derivative (59), which was prepared from the benzyne adduct of 6,6-diphenylfulvene by selective hydrogenation.

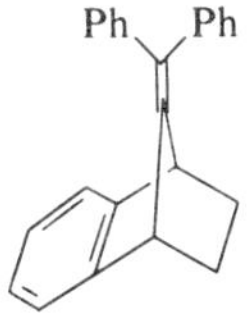

(59)

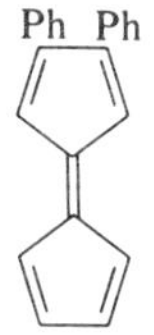

(60)

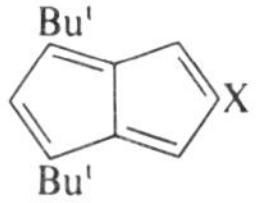

(61) X = CO_2Me, CHO, or CN

A series of 6,6-disubstituted fulvenes has been protonated under super-acid conditions at low temperatures.[173] Protonation takes place exclusively at the C-2 position of the fulvene ring in accord with calculated electron densities. The crystalline isolable 1,2,3,4-tetraphenyl- and 1,2:3,4-dibenzofulvalenes, and 2,3-diphenylfulvalene (60), which is only stable in solution, have been synthesized.[174] Species (60) readily dimerizes by [4 + 2]addition. The scope of the facile pentalene synthesis from 1,3-di-t-butyl-6-dimethylaminofulvene [see *Ann. Reports (B)*, 1976, **73**, p. 230] has been extended,[175a] by using other alkynes, to give the pentalenes (61). These pentalenes are formed as dimers, with which the monomers are in rapid equilibrium and which form the monomers in organic solvents. The dimerization reactions and other dimerization products that are formed after longer times have been studied,[175a] as have the reactivities of the pentalene monomers to acids, bases, dienophiles, and dienes.[175b]

3-Amino-tropones have been formed[176] by rapid pyrolysis of a corresponding enamine; 3-amino-4-methyltropone, for example, being formed from (62). 3-Ethoxy- and 3-phenoxy-tropones have been prepared[177] by the reaction of 3-tosyloxytropone with sodium ethoxide and lithium phenoxide respectively. A

[170] H. Alper and H.-N. Paik, *J.C.S. Chem. Comm.*, 1977, 126.
[171] (*a*) M. Neuenschwander and R. Iseli, *Helv. Chim. Acta*, 1977, **60**, 1061; (*b*) M. Neuenschwander, R. Vögeli, H.-P. Fahrni, H. Lehmann, and J.-P. Ruder, *ibid.*, p. 1073.
[172] R. N. Warrener, K. I. Gell, and M. N. Paddon-Row, *Tetrahedron Letters*, 1977, 53.
[172] G. A. Olah, G. K. S. Prakash, and G. Liang, *J. Org. Chem.*, 1977, **42**, 661.
[174] H. Prinzbach, H. Sauter, H.-G. Hörster, H.-H. Limbach, and L. Knothe, *Annalen*, 1977, 869.
[175] M. Suda and K. Hafner, *Tetrahedron Letters*, 1977, (*a*) p. 2449; (*b*) p. 2453.
[176] J. Ficini and A. Duréault, *Tetrahedron Letters*, 1977, 809.
[177] M. Cavazza, R. Cabrino, and F. Pietra, *Synthesis*, 1977, 298.

reported preparation[178a] of 3-hydroxytropone derivatives appears, however, to be incorrect,[178b] the products being 2-hydroxyacetophenones.

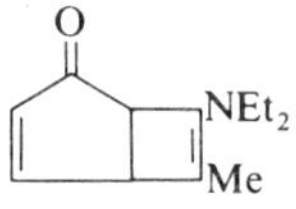

(62)

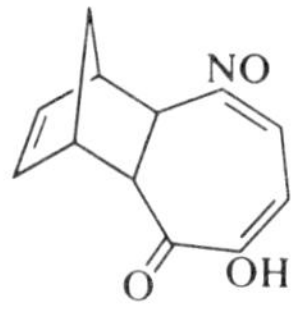

(63)

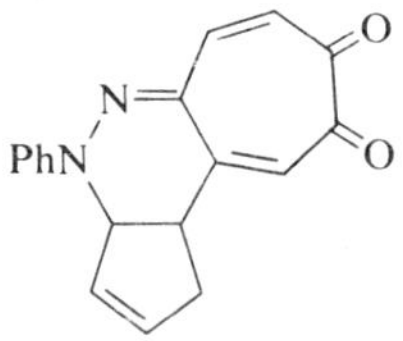

(64)

Electrophilic substitution of 5-methyltropolone by arenediazonium ions occurs first at the 3- and then at the 7-position.[179a] 5-Arylazotropolones undergo arylazo exchange (61—92%) on reaction with a different arenediazonium ion.[179b] The reaction of 5-nitrosotropolone with cyclopentadiene gives (63), and provides an example of a stable tropolone derivative reacting as a 2π compound in a [4+2]cycloaddition.[179c] 5-Phenylazotropolone, however, which is known to react as its diketo-form, adds to cyclopentadiene, utilizing its N=N—C=C linkage to give (64).[179d] Further reactions with cyclopropene have established the *endo*-stereochemistry required for the latter reaction.[179e] An analogous side-chain reaction has also been established[179f] as a minor pathway in one case, in a homodiene system, on reaction of 2-methoxy-6-styryltropone with maleic anhydride.

The first benzo[3,4]cyclobuta[1,2-*c*]tropolone, the 9-chloro-derivative, has been synthesized[180] and appears to exist predominantly or exclusively as the tautomer (65). The geometrically isomeric 2,3:6,7-dioxa-bis-σ-homotropones [*e.g.* (66)] have been formed[181] by the direct oxidation of tropone with alkaline hydrogen peroxide. The unreactivity of the remaining double bond, however, limits their synthetic utility. A series of methoxy-substituted heptalene-1,8- and -3,8-diones have been formed[182a] by oxidation of 8*H*-cyclohepta[*d*]tropolone.[182b] In solution in fluorosulphonic acid, these compounds are diprotonated, and provide the first examples of the 10π-heptalenium dication structure [*e.g.* (67)].

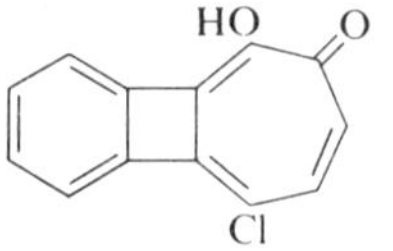

(65)

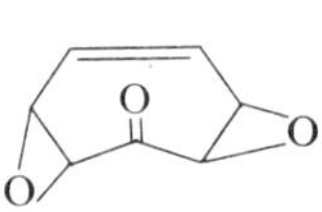

(66)

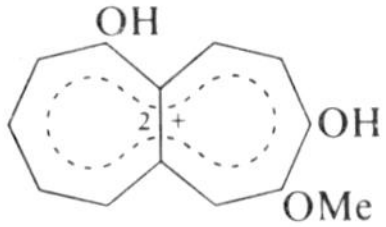

(67)

The MINDO/3 method has been applied to the prediction of the rearrangement of benzyl and substituted benzyl cations to tropylium ions and related processes and

[178] (*a*) J. H. Clark and J. M. Miller, *Tetrahedron Letters*, 1977, 139; (*b*) H. Takeshita, *ibid.*, p. 1657.
[179] (*a*) T. Ide, K. Imafuku, and H. Matsumura, *Chem. Letters*, 1977, 717; (*b*) *ibid.*, p. 511; (*c*) I. Saito, K. Sakan, and S. Itô, *ibid.*, p. 253; (*d*) S. Itô and I. Saito, *Tetrahedron Letters*, 1977, 1203; (*e*) Y. Fujise, M. Sakaino, and S. Itô, *ibid.*, p. 2663; (*f*) I. Saito, Y. Watanabe, and S. Itô, *ibid.*, p. 3049.
[180] M. Sato, S. Ebiñe, and J. Tsunetsugu, *Tetrahedron Letters*, 1977, 855.
[181] H. Prinzbach, W. Seppelt, and H. Fritz, *Angew. Chem. Internat. Edn.*, 1977, **16**, 198.
[182] S. Kuroda and T. Asao, *Tetrahedron Letters*, 1977, (*a*) p. 289; (*b*) p. 285.

of the properties of these ions.[183] A pure heptamethyltropylium salt has been prepared[184] for the first time by the use of phosphorus pentachloride as a reagent for hydride abstraction from 1,2,3,4,5,6,7-heptamethyltropilidene. The 9,10-dihydro-9-10-(1,2-tropylio)anthracene ion (68) has been synthesized,[185] and the u.v. spectrum is indicative of strong intermolecular charge-transfer interaction despite the small overlap of the orbital systems involved.

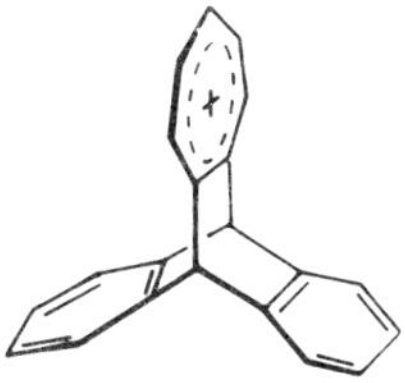

(68)

8-Phenylheptafulvene has been prepared[186] by the reaction of the benzyltropylium ion with an equivalent of triethylamine in dichloromethane. It is stable for several days in solution at room temperature under nitrogen. U.v. spectroscopy suggests even less π-electron delocalization than in heptafulvene itself, and only a small contribution of the dipolar canonical form for this system. 8-Amino- and 8-methoxy- 8-(triethylsiloxy)heptafulvene have also been synthesized; they are hygroscopic and thermolabile.[187] A facile synthesis of 8-azaheptafulvenes involves the reaction of tropylium borofluoride with anilines to give 8-aryl-8-azaheptafulvenium salts, which are deprotonated with aqueous sodium carbonate.[188a] The cycloaddition reactions of these compounds with, *inter alia*, isocyanates and ketens, to give (usually) [8+2]cycloadducts, have also been studied.[188b]

Two independent reports[189] have been given of an improved route to azulenes by [6+4]cycloaddition of 6-*NN*-dimethylaminofulvene with thiophen 1,1-dioxides; 3,4-dichlorothiophen, for example, yielding the hitherto unknown 5,6-dichloroazulene in 60%[189a] or 46%[189b] yield. 2,6-Dihydroxyazulene has been prepared[190a] from diethyl 2,6-dihydroxyazulene-1,3-dicarboxylate[190b] and undergoes a heavily solvent-dependent keto–enol tautomerism. In chloroform the species exists mainly as 1,3-dihydroazulene-2,6-dione. 1,3-Dicarbethoxyazulene-2,6-dione has also been synthesized,[190b] but was isolated only in the form of a dimer. 1-Alkyl-azulenes are converted into 1-acyl-azulenes in 59—98% yields by a novel oxidation, using DDQ.[190c]

[183] C. Cone, M. J. S. Dewar, and D. Landman, *J. Amer. Chem. Soc.*, 1977, **99**, 372; M. J. S. Dewar and D. Landman, *ibid.*, pp. 2446, 4633, 7439.
[184] K.-I. Takeuchi, Y. Yokomichi, and K. Okamoto, *Chem. Letters*, 1977, 1177.
[185] T. Nakazawa and I. Murata, *J. Amer. Chem. Soc.*, 1977, **99**, 1996.
[186] K. Komatsu, M. Fujimori, and K. Okamoto, *Tetrahedron*, 1977, 2791.
[187] K. M. Rapp and J. Daub, *Tetrahedron Letters*, 1977, 227.
[188] (*a*) K.-I. Sanechika, S. Kajigaeshi, and S. Kanemasa, *Synthesis*, 1977, 202; (*b*) K. Yamamoto, S. Kajigaeshi, and S. Kanemasa, *Chem. Letters*, 1977, 85, 91.
[189] (*a*) S. E. Reiter, L. C. Dunn, and K. N. Houk, *J. Amer. Chem. Soc.*, 1977, **99**, 4199; (*b*) D. Copland, D. Leaver, and W. B. Menzies, *Tetrahedron Letters*, 1977, 639.
[190] (*a*) T. Morita, H. Kanzawa, and K. Takase, *Chem. Letters*, 1977, 753; (*b*) T. Morita and K. Takase, *ibid.*, p. 513; (*c*) T. Amemiya, M. Yasunami, and K. Takase, *ibid.*, p. 587.

Annulenes.—A series of substituted cyclo-octatetraene dications has now been prepared[191] by oxidation of the parent hydrocarbon. Most are stable only at low temperatures, but the *sym*-dibenzocyclo-octatetraene dication is stable even at 0 °C, and spectral evidence suggests that delocalization of all fourteen π-electrons about the sixteen-carbon-atom periphery is occurring. Oxidation of cyclo-octa[*def*]fluorene by a different method gave[192] the protonated cyclo-octa[*def*]fluorenone dication (69), a 14π-electron aromatic system which is stable at room temperature. The 9,10-diphenylbicyclo[6,2,0]decapentaenes (70) have been synthesized[193] by a route involving initial photoaddition of diphenylacetylene to bicyclo[4,2,0]octa-3,7-diene, and other stages, one of these involving the thermal isomerization of bicyclo[4,2,0]octatriene to cyclo-octatetraene. The preliminary evidence suggests that π-electron delocalization in (70; R = H), but not in (70; R = Me), overcomes the strain involved in attaining (near) planarity.

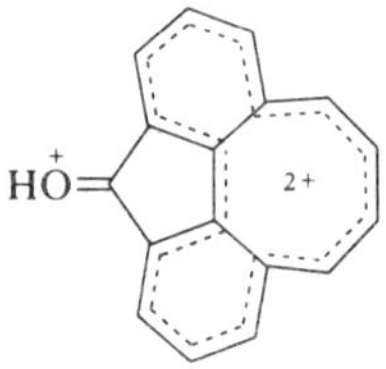

(69)

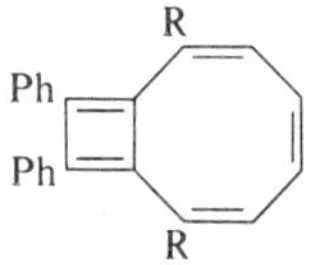

(70) R = H, D, or Me

PMO theory has demonstrated[194] that the small variations in the perimeter bond lengths of 1,6-methano[10]annulene are due mainly to transannular 1,6-homoaromatic interaction. This type of interaction is smaller, but still important, for the 1,6-methano[11]annulenium cation [*cf. Ann. Reports* (*B*), 1976, **73**, p. 234]. 1,6-Methano[10]annulene and some hetero-bridged analogues undergo Diels–Alder addition with 4-substituted-1,2,4-triazoline-3,5-diones, giving both mono- and bis-adducts, formed by *anti*-attack.[195a] The methylene bridge also hinders sterically the approach of *NNN'N'*-tetramethyl-*p*-phenylenediamine to 2,7-dinitro-1,6-methano[10]annulene, leading to the formation of only a 1:1 donor–acceptor complex.[195b] Unsubstituted 1,5-methano[10]annulene has been synthesized, and appears to maintain a diamagnetic ring current.[196] In contrast to its 1,6-analogue, however, its electronic spectrum extends into the visible region, and it is an orange crystalline solid.

Two sets of correlations[197] have been reported between experimentally determined properties of annulenes and variously calculated values of 'resonance energies per π-electron' (REPE). In the first case[197a] the correlation involved the rates of formation of some annulenes determined by Sondheimer *et al.* in connection

[191] G. A. Olah, J. S. Staral, G. Liang, L. A. Paquette, W. P. Melega, and M. J. Carmody, *J. Amer. Chem. Soc.*, 1977, **99**, 3349.
[192] I. Willner, A. L. Gutman, and M. Rabinovitz, *J. Amer. Chem. Soc.*, 1977, **99**, 4167.
[193] M. Oda, H. Oikawa, N. Fukazawa, and Y. Kitahara, *Tetrahedron Letters*, 1977, 4409.
[194] R. C. Haddon, *J. Org. Chem.*, 1977, **42**, 2017.
[195] (*a*) P. Ashkenazi, D. Ginsburg, and E. Vogel, *Tetrahedron*, 1977, **33**, 1169; (*b*) J. A. Chudek, R. Foster, and E. Vogel, *J.C.S. Perkin II*, 1977, 994.
[196] S. Masamune and D. W. Brooks, *Tetrahedron Letters*, 1977, 3239.
[197] (*a*) B. A. Hess, jun., and L. J. Schaad, *J.C.S. Chem. Comm.*, 1977, 243; (*b*) B. A. Hess, jun., L. J. Schaad, and M. Nakagawa, *J. Org. Chem.*, 1977, **42**, 1669.

with their 'reactivity criterion' of aromaticity [see *Ann. Reports* (*B*), 1976, **73**, p. 237]. In the second case,[197b] the differences in ^{1}H n.m.r. chemical shifts of the inner (i) and outer (o) protons of some dehydroannulenes (71; $m = p = 0$, 1, 2, or 3) gave a good linear correlation with REPE, suggesting that chemical-shift data, at least if used in this way, can give quantitative information concerning aromaticity.

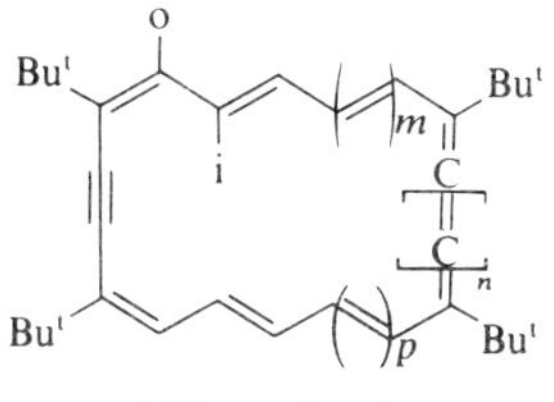

(71) $n = 1$ or 3

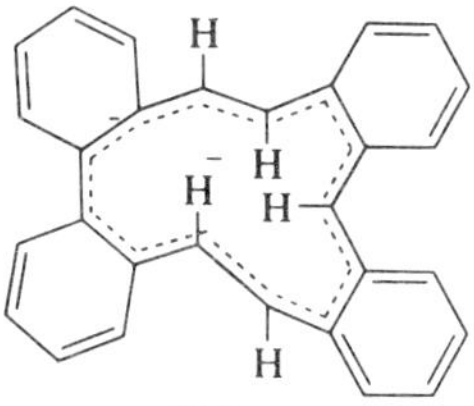

(72)

Studies of some newly synthesized model compounds have led to the conclusion that the [12]annulene rings in some previously studied di- and tri-benzannelated bisdehydro[12]annulenes are paratropic.[198a] A necessarily planar benzo-naphtho-analogue was also synthesized,[198b] and the ^{1}H n.m.r. shifts were interpreted in terms of paratropicity of the [12]annulene ring. The 2,3:6,7:8,9:12,13-tetra-benzo[13]annulenyl anion (72), the first such ion to contain only formal double and single bonds, has been prepared[199a] from the corresponding hydrocarbon.[199b] The ^{1}H n.m.r. spectrum is indicative of the reduced diatropicity caused by benzannelation, and, in addition, of the considerable change in ring conformation from that of the parent hydrocarbon.

Full details[200a] of the synthesis of 3,7,10,14-tetrasubstituted 1,8-bisdehydro[14]annulenes by the method involving cyclic dimerization of 2,3-dien-4-yne ketones, and the details[200b] of the greatly improved synthetic route to bisdehydrobenzannulenes, have appeared. In the latter publication, the syntheses of bisdehydrobenz-[14]-, -[16]-, -[18]-, and -[20]-annulenes, as well as of some methyl derivatives, are described. It seems that the structural features which allow one to draw equivalent Kekulé structures for some dehydroannulenes [*e.g.* (73)] have little effect on their properties.[201] Diatropicity differences between (73) and

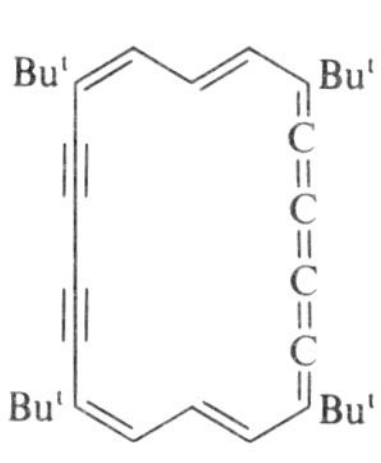

(73)

[198] (*a*) H. A. Staab and P. Günthert, *Chem. Ber.*, 1977, **110**, 619; (*b*) H. A. Staab and H. J. Shin, *ibid.*, p. 631.

[199] I. Willner, A. Gamliel, and M. Rabinovitz, (*a*) *Chem. Letters*, 1977, 1273; (*b*) *Synthesis*, 1977, 410.

[200] (*a*) K. Fukui, T. Nomoto, S. Nakatsuji, S. Akiyama, and M. Nakagawa, *Bull. Chem. Soc. Japan*, 1977, **50**, 2758; (*b*) N. Darby, T. M. Cresp, and F. Sondheimer, *J. Org. Chem.*, 1977, **42**, 1960.

[201] (*a*) M. Osuka, S. Akiyama, and M. Nakagawa, *Tetrahedron Letters*, 1977, 1649; (*b*) Y. Yoshikawa, S. Nakatsuji, F. Iwatani, S. Akiyama, and M. Nakagawa, *ibid.*, p. 1737.

(71); $m = 0$, $n = 3$, $p = 1$), for example, appear to be insignificant,[201a] and similar conclusions were also drawn from studies of a paratropic bisdehydro[16]annulene system.[201b] It seems that energy differences between non-equivalent canonical forms [*e.g.* in (71; $m = 1$, $n = 3$, $p = 1$)] are not enough to depress significantly the π-electron delocalization.

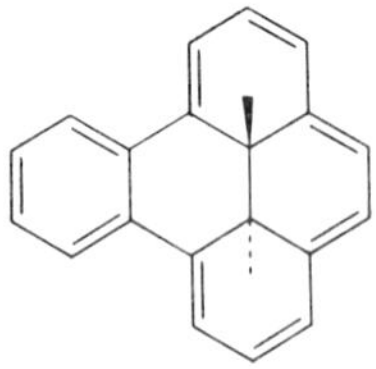

(74)

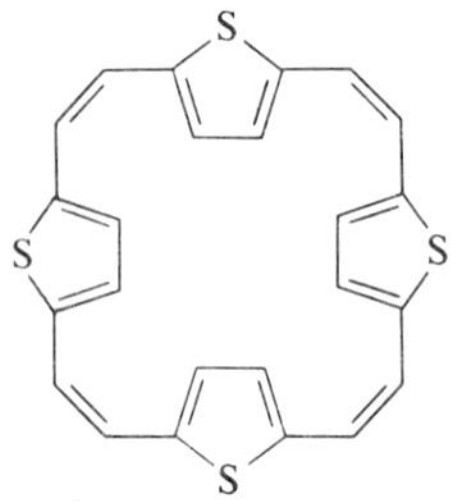

(75)

Benzo[18]annulene has been synthesized,[202] in low yield, by a route involving prototropic isomerization of an intermediate 13,14,15,16-bisdehydro-derivative. Again, annelation of the benzene ring reduces the diatropicity of the macrocyclic ring. Two further annulenes have been synthesized[203] where extensive benzannelation has removed the diatropicity of $(4n+2)\pi$-electron macrocyclic rings, and, for example, tribenzo[*a,g,m*]-15,17-bisdehydro[18]annulene is atropic. The dimethyldihydrobenzo[*e*]pyrene (74) sustains[204] about 55% of the ring-current of the parent non-benzannelated dimethyldihydropyrene. This amount is much greater than the value for other benzannelated [14]annulenes, and reflects the greater aromaticity of the rigid dihydropyrene nucleus. All-*cis*-[2^4](2,5)thiophenophanetetraene (75) has been prepared[205] in a one-step reaction from thiophen-2,5-dicarbaldehyde and the Wittig reagent from 2,5-bis(chloromethyl)thiophen. The compound shows a small paramagnetic ring current over the 24π-electron periphery; this is enhanced on cooling, when planar conformations become of more importance. The presence of the thiophen nuclei does not seem to interfere with the paratropicity of the molecule.

The syntheses of the dibenzannulenones (76; $n = 1$, 2, and 3) have been completed[206a] and the results for (76; $n = 1$) and (76; $n = 2$) have now been reported[206b] in full. The paratropicity exhibited by (76; $n = 1$) and (76; $n = 2$), which are, however, less paratropic than monobenzannelated analogues, disappears in (76; $n = 3$), and comparison with an acyclic analogue indicates that the latter is atropic. The first monocyclic large-ring annulenones (77; R = H) and (77; R = Me) have been synthesized.[206c] The introduction of the additional methyl group (77; R = Me) causes a change in conformation at the other *trans* double bond. Both species exhibit paratropicity, and this is enhanced on dissolution in

[202] U. E. Meissner, A. Gensler, and H. A. Staab, *Tetrahedron Letters*, 1977, 3.

[203] J. Ojima, M. Enkaku, and C. Uwai, *Bull. Chem. Soc. Japan*, 1977, **50**, 933.

[204] S. Içli, V. J. Nowlan, P. M. Rahimi, C. Thankachan, and T. T. Tidwell, *Canad. J. Chem.*, 1977, **55**, 3347.

[205] A. Strand, B. Thulin, and O. Wennerström, *Acta Chem. Scand.* (*B*), 1977, **31**, 521.

[206] (*a*) J. Ojima, M. Enkaku, and M. Ishiyama, *J.C.S. Perkin I*, 1977, 1548; (*b*) J. Ojima, Y. Yokoyama, and M. Enkaku, *Bull. Chem. Soc. Japan*, 1977, **50**, 1522; J. Ojima, M. Ishiyama, and A. Kimura, *ibid.*, p. 1584; (*c*) T. M. Cresp, J. Ojima, and F. Sondheimer, *J. Org. Chem.*, 1977, **42**, 2130.

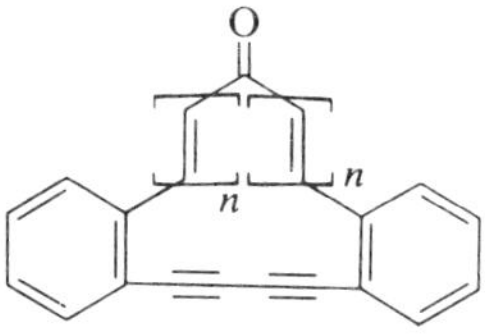

(76)

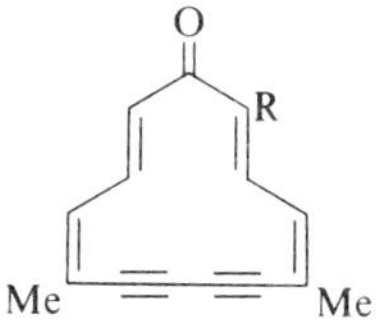

(77)

(78)

deuteriotrifluoroacetic acid. *syn*-4,13;6,11-Dimethano[15]annulenone (78) has been synthesized as an air-stable red crystalline solid, and an *X*-ray crystallographic analysis has been performed. The molecule is protonated in trifluoroacetic acid to give a delocalized annulenium ion.[207]

The syntheses of the first bridged [22]annulenes, *i.e. syn*- and *anti*-(79), have been accomplished,[208] in one step, by the dimerization of 6,11-methano[11]annulenylidene, followed by ring closure and loss of hydrogen. ^{1}H N.m.r. measurements show no diatropicity, and suggest that there is localization of bonds as shown in (79). The first derivative to be formed that is derived from the hitherto unknown nonalene skeleton is the planar delocalized diatropic dibenzo[*gh*, *op*]nonalenide dianion (80), formed[209] from the reaction of 6,14-dihydrodibenzo[*gh*, *op*]nonalene with butyl-lithium.

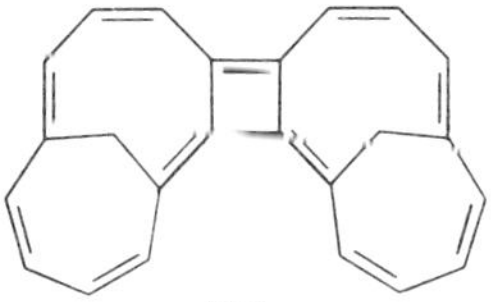

(79)

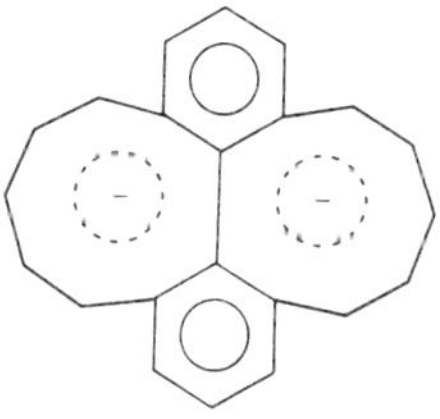

(80)

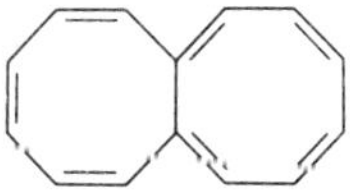

(81)

Further bicyclic compounds containing two fused macrocyclic conjugated π-systems have been reported.[210] Octalene (81) has been prepared[210a] from 1,4,5,8-tetrahydronaphthalene as a lemon-yellow air-sensitive compound. Structures involving a central double bond, or aromatic structures, were ruled out. Peripheral C=C bond shifting occurs above 80 °C. *ortho*-Fused [14]annuleno-[16]- and -[18]-annulenes analogous to the original [14]annulene [see *Ann. Reports* (*B*), 1975, **72**, p. 248] have been synthesized.[210b] The diatropicity of the 14-membered ring is reduced as the size of the fused annulene is decreased, irrespective of whether this latter annulene is $(4n+2)$- or $4n$-membered. Other annuleno-annulenes are (82; $m=1$)[210c] and (82; m = 2),[210d] which are both strongly diatropic and quite stable, in

207 W. Wagemann, K. Müllen, E. Vogel, T. Pilati, and M. Simonetta, *Angew. Chem. Internat. Edn.*, 1977, **16**, 170.

208 U. H. Brinker, R. W. King, and W. M. Jones, *J. Amer. Chem. Soc.*, 1977, **99**, 3175.

209 I. Willner and M. Rabinovitz, *J. Amer. Chem. Soc.*, 1977, **99**, 4507.

210 (*a*) E. Vogel, H.-V. Runzheimer, F. Hogrefe, B. Baasner, and J. Lex, *Angew. Chem. Internat. Edn.*, 1977, **16**, 871; J. F. M. Oth, K. Müllen, H.-V. Runzheimer, P. Mues, and E. Vogel, *ibid.*, p. 872; (*b*) T. M. Cresp and F. Sondheimer, *J. Amer. Chem. Soc.*, 1977, **99**, 194; (*c*) S. Nakatsuji, S. Akiyama, and M. Nakagawa, *Tetrahedron Letters*, 1977, 3723; (*d*) M. Osuka, Y. Yoshikawa, S. Akiyama, and M. Nakagawa, *ibid.*, p. 3719.

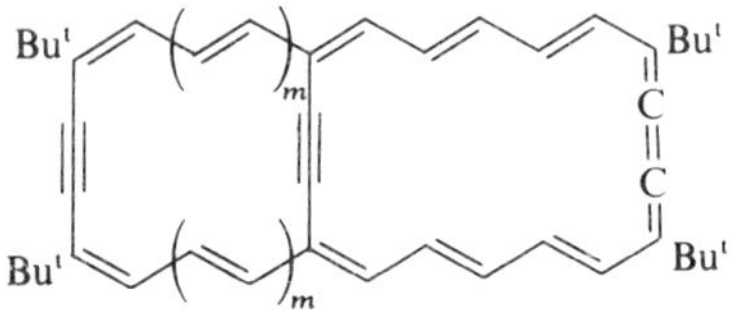

(82)

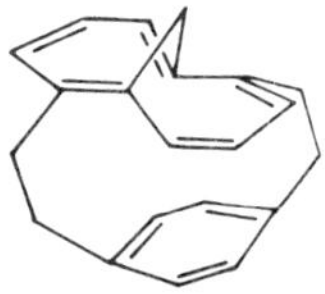

(83)

contrast to the appropriate [26]- and [30]-annulenes. In the ^1H n.m.r. spectra the inner protons of the 14-membered ring appear at a much higher field than those of the 18-membered ring, and the latter and all outer protons shift to higher field as the distance from the central bridge increases.

Finally, the first examples of annulenophanes, *e.g.* (83), have been reported.[211] The benzenoid protons of (83) show the expected upfield shift due to the ring current in the [10]annulene ring. The indications are, however, that in this example transannular interaction of π-electrons is small.

[211] M. Matsumoto, T. Otsubo, Y. Sakata, and S. Misumi, *Tetrahedron Letters*, 1977, 4425.

11 Heterocyclic Chemistry

By A. J. BOULTON

School of Chemical Sciences, University of East Anglia, Norwich NR4 7TJ

1 Heterocycles in Functional Group Transformations

Mukaiyama has continued to exploit quaternary salts containing reactive halogen atoms, as a means of converting stable organic functions into good leaving groups. The conversions of trialkylammonium dithiocarbamates into isothiocyanates,[1] of thioureas into carbodi-imides[2] (using reagent 1), of alcohols into chlorides,[3] formamides into isocyanides,[4] and epoxides into 1,2-dichlorides (using 2),[5] are briefly reported. More preliminary notes describe the further work of Katritzky's group on the reaction of pyrylium salts (3) with primary amino-functions, to give

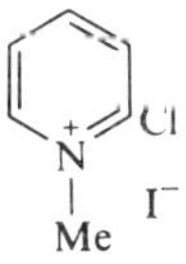

(1)

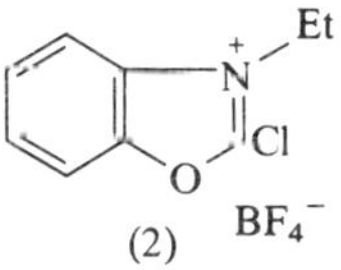

(2)

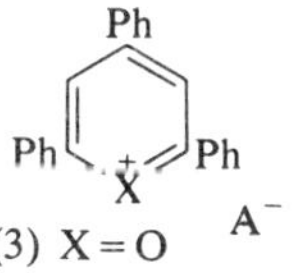

(3) X = O
(4) X = NCH_2R

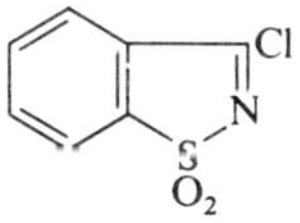

(5)

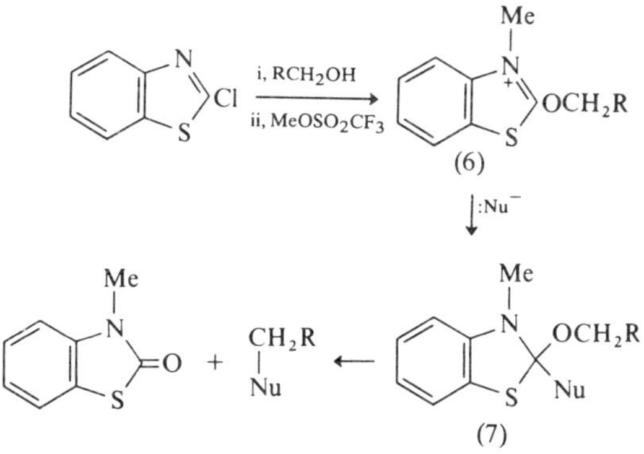

Scheme 1

[1] T. Shibanuma, M. Shiono, and T. Mukaiyama, *Chem. Letters*, 1977, 573.
[2] T. Shibanuma, M. Shiono, and T. Mukaiyama, *Chem. Letters*, 1977, 575.
[3] T. Mukaiyama, S. Shoda, and Y. Watanabe, *Chem. Letters*, 1977, 383.
[4] Y. Echigo, Y. Watanabe, and T. Mukaiyama, *Chem. Letters*, 1977, 697.
[5] Y. Echigo, Y. Watanabe, and T. Mukaiyama, *Chem. Letters*, 1977, 1013.

pyridinium salts (4), which then lead to iodides[6] (using 3, A = I) and esters[7] (using 3, A = BF_4) from primary amines RCH_2NH_2, and by rearrangement, carbodi-imides ArN:C:NAr′ from amidrazones ArNHCAr′:NNH_2[8] (using 3, A = ClO_4). ‘Saccharin chloride’, the benzisothiazole (5), is suggested as a useful reagent for the production of β-lactams from acids and imines.[9] Another condensing technique uses 2-chlorobenzothiazole, which activates a primary alcohol as outlined in Scheme 1. The quaternary salt (6) may alkylate another nucleophile (the method works with a wide variety of these) possibly with the pseudobase (7) as an intermediate.[10]

2 General Heterocyclic Synthesis

The salt (8) is a versatile reagent for the synthesis of many heterocycles (with five-, six-, and seven-membered rings) with the structural element (A);[11] the isomeric arrangement (B) had earlier been shown to be derived similarly from (9).[12] Both starting-materials are made from dimethyl phosgenimmonium chloride (Scheme 2). *N*-Ethoxycarbonylthioamides (10) provide smaller pieces for building into heterocycles: three-atom and one-atom units are incorporated, on reaction (a) with 1,2- and 1,3-binucleophiles, and (b) with 1,4- and larger binucleophiles, respectively.[13]

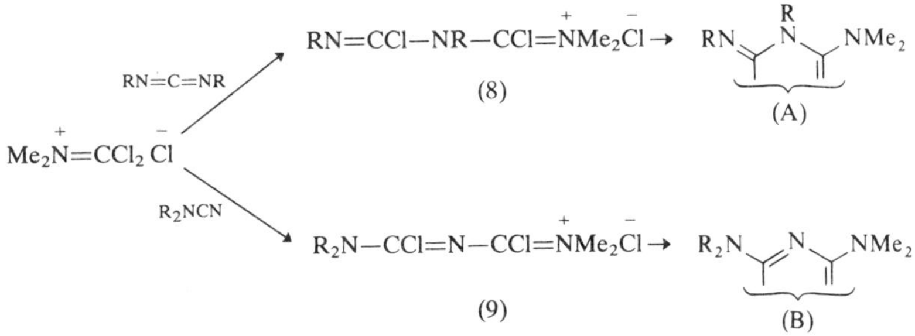

Scheme 2

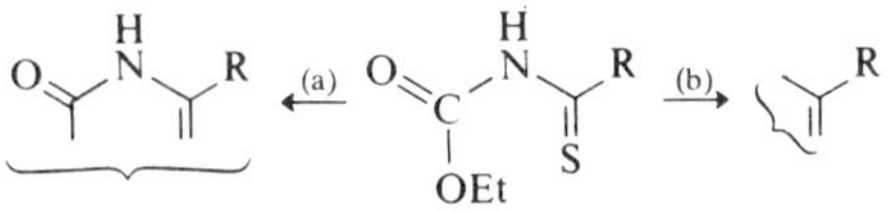

(10)

3 Three-membered Rings

Attempts to prepare oxirene by heating epoxy-barrelenes have been unsuccessful. At 300–400 °C cycloheptatriene aldehyde derivatives were obtained from (11)

[6] N. F. Eweiss, A. R. Katritzky, P. L. Nie, and C. A. Ramsden, *Synthesis*, 1977, 634.
[7] U. Gruntz, A. R. Katritzky, D. H. Kenny, M. C. Rezende, and H. Sheikh, *J.C.S. Chem. Comm.*, 1977, 701.
[8] A. R. Katritzky, P. L. Nie, A. Dondoni, and D. Tassi, *Synthetic Comm.*, 1977, **7**, 387.
[9] M. S. Manhas, H. P. S. Chawla, S. G. Amin, and A. K. Bose, *Synthesis*, 1977, 407.
[10] F. A. Souto-Bachiller and S. Masamune, *Tetrahedron Letters*, 1977, 1881.
[11] A. Elgavi and H. G. Viehe, *Angew. Chem. Internat. Edn.*, 1977, **16**, 181.
[12] Z. Janousek and H. G. Viehe, *Angew. Chem. Internat. Edn.*, 1973, **12**, 74.
[13] B. George and E. P. Papadopoulos, *J. Org. Chem.*, 1976, **41**, 3233; *ibid.*, 1977, **42**, 441, 2530.

(acidic alumina brought about the same transformation), but at 600–800 °C flash thermolysis did produce some ketene, the expected rearrangement product of oxirene.[14] The bridged anthracenes (12; R = H and Me) rearranged, on heating, rather than eliminating the bridge.[15] Developments in the study of the Wolff rearrangement, and the involvement of oxirenes and their interconversion with α-ketocarbenes, are treated in Chapter 5 of this Volume.

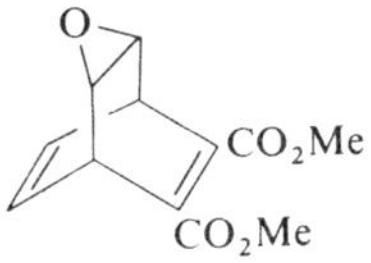

(11)

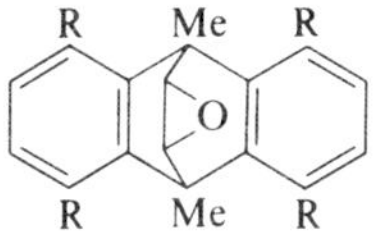

(12)

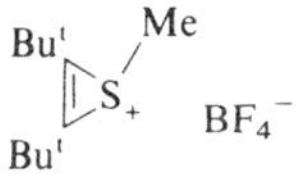

(13)

Thiirene seems to be slightly less elusive than oxirene: its matrix-isolation has been reported, along with an i.r. study, and data were also obtained on mono- and di-deuterio- and methyl derivatives, and on selenirene.[16] Benzothiirene appears to be an intermediate in the pyrolytic formation of two isomeric thianthrenes from a monosubstituted 1,2,3-benzothiadiazole,[17] although this view is challenged by another group.[18] Thiirenium salts appear to be more stable still,[19,20] and one of these (13) has been studied by *X*-ray crystallography. The ring C—S bonds are unusually long (1.82 Å; *cf.* 1.80 Å for the S—Me bond).[19] Thiirenium salts are prepared by methods analogous to those used for the thiiranium, seleniranium, and selenirenium salts reported earlier.[21]

Oxidation of thioketen monomers, *e.g.* (14), with a pyrroline-1-oxide forms thiiran-2-ones (thiol-α-lactones), *e.g.* (15), the structures of which were established by i.r. ($\nu_{C=O}$ *ca.* 1800 cm^{-1}), and the X-ray crystallography of (15).[22]

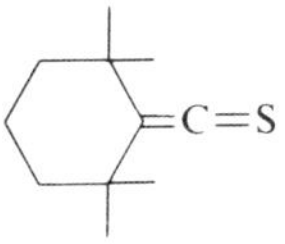

(14)

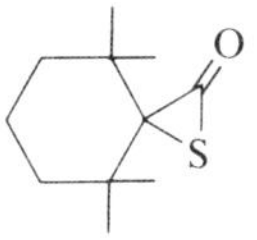

(15)

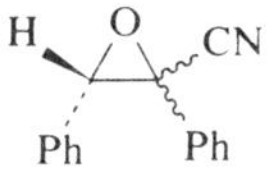

(16)

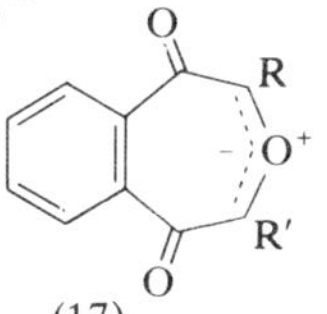

(17)

Flash photolysis of the *cis* and *trans* isomers of α-cyanostilbene oxide (16) forms carbonyl ylids ($PhCH{=}\overset{+}{O}{-}\overset{-}{C}PhCN$), which are highly coloured, and the rates of their reactions, and their interconversion, could be studied. Upper and lower limits are suggested for the depths of the energy troughs in which the ylids lie.[23] Irradiation of naphthoquinone oxides in the presence of norbornene and other dipolarophiles forms bridged benzoxepin cycloadducts derived from the valence tautomer (17).[24]

14 E. Lewars and G. Morrison, *Tetrahedron Letters*, 1977, 501.
15 H. Hart, J. B. C. Jiang, and M. Sasaoka, *J. Org. Chem.*, 1977, **42**, 3840.
16 A. Krantz and J. Laureni, *J. Amer. Chem. Soc.*, 1977, **99**, 4842.
17 T. Wooldridge and T. D. Roberts, *Tetrahedron Letters*, 1977, 2643.
18 L. Benati, P. C. Montevecchi, and G. Zanardi, *J. Org. Chem.*, 1977, **42**, 575.
19 R. Destro, T. Pilati, and M. Simonetta, *J.C.S. Chem. Comm.*, 1977, 576.
20 G. Capozzi, V. Lucchini, G. Modena, and P. Scrimin, *Tetrahedron Letters*, 1977, 911.
21 *cf. Ann. Reports* (*B*), 1975, **72**, 253.
22 E. Schaumann and U. Behrens, *Angew. Chem. Internat. Edn.*, 1977, **16**, 722.
23 R. Huisgen, V. Markowski, and H. Hermann, *Heterocycles*, 1977, **7**, 61.
24 S. Arakawa, *J. Org. Chem.*, 1977, **42**, 3800.

cis-2-Vinyl-3-ethynyloxirane (18) has been shown to rearrange cleanly, *via* two Cope rearrangements, to the cyclopropane (19). With the corresponding *N*-t-butylaziridine (20), however, the presumed intermediate (21; X = NBu^t) is side-tracked by an intramolecular proton shift, and the product is 1-t-butylazepine.[25]

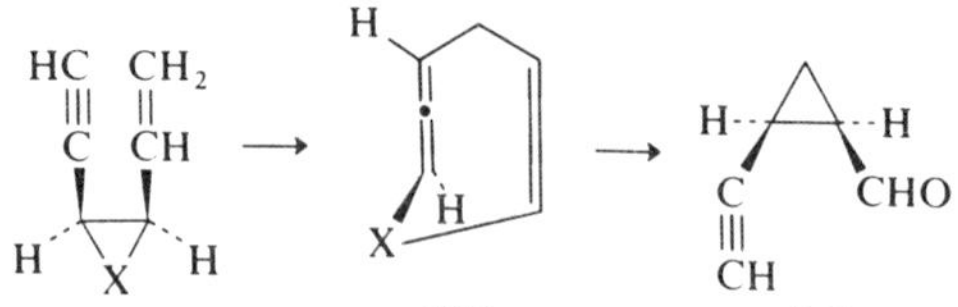

(18) X = O
(20) X = NBu-t
(21)
(19)

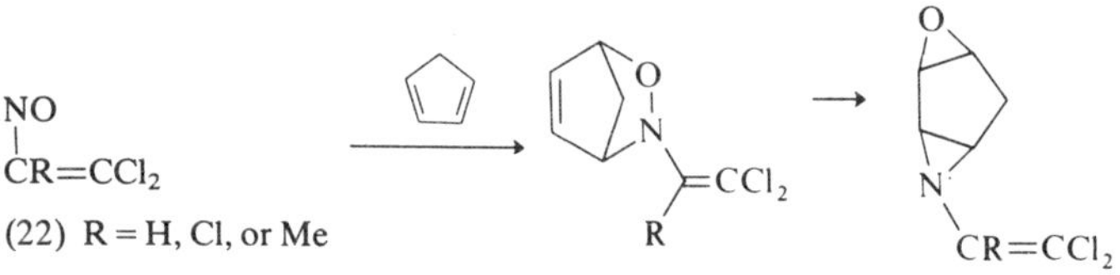

(22) R = H, Cl, or Me
(24)
(23)

Viehe *et al.*[26] have discovered an interesting analogy to the rearrangement of singlet oxygen–diene adducts (Δ^2-1,4-peroxides) to bis-epoxides: the nitroso-olefins (22) with cyclopentadiene form the diene mono-oxide–mono-imines (23), *via* (4 + 2) cycloadducts (24), which are detectable intermediates at −60 °C.

The addition of nitronic esters (25) to acetylenes (26) does not form the expected *N*-alkoxyisoxazolines (27); these are probable intermediates, but they immediately rearrange to 2-acyl-1-alkoxyaziridines. The stereochemistry of this reaction has been studied, and it was found that the products (28, 29) are formed stereo-specifically only with respect to the relative configurations of the substituents on the nitrone.[27]

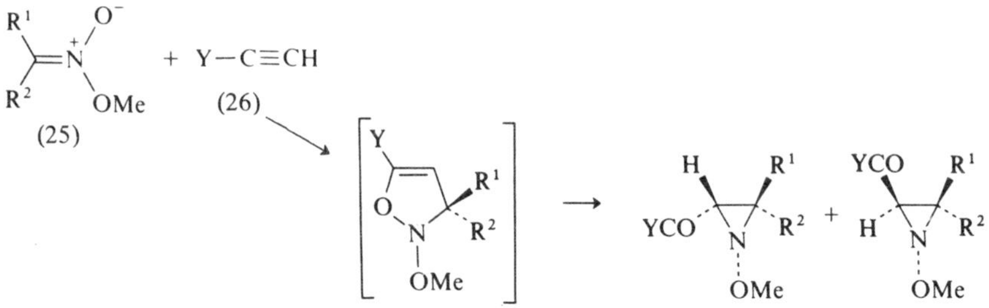

(25) (26)
(27) (28) (29)

R^1 = CN or CO_2Me, R^2 = H, or *vice versa*
Y = COPh, COMe, or CO_2Me

[25] N. Manisse and J. Chuche, *J. Amer. Chem. Soc.*, 1977, **99**, 1272.
[26] H. G. Viehe, R. Merenyi, E. Francotte, M. Van Meerssche, G. Germain, J. P. Declerq, and J. Bodart-Gilmont, *J. Amer. Chem. Soc.*, 1977, **99**, 2340.
[27] R. Greé and R. Carrié, *J. Amer. Chem. Soc.*, 1977, **99**, 6667; *J.C.S. Chem. Comm.*, 1975, 112.

The automerization rates of perfluorotetramethyl Dewar thiophene (30) and its *S*-oxide have been compared.[28] The former exchanges slowly ($\Delta G^{\ddagger}$ 93.8 ± 0.5 kJ mol^{-1} at 157 °C), the latter extremely rapidly ($\Delta G^{\ddagger}$ 28 ± 0.5 kJ mol^{-1} at −136 °C), but the authors suggest that the mechanism is probably similar.

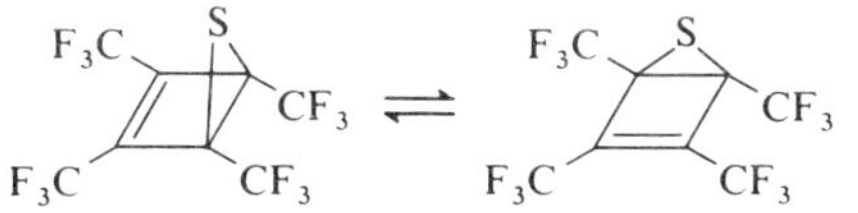

(30)

(31) R = subst. phenyl
(32) R = NMe_2

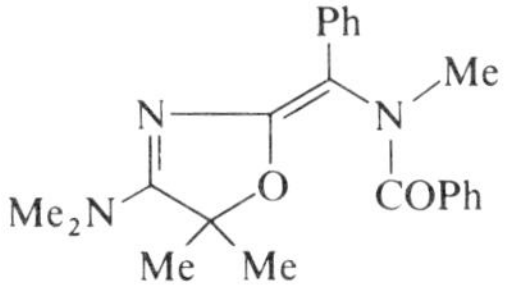

(33)

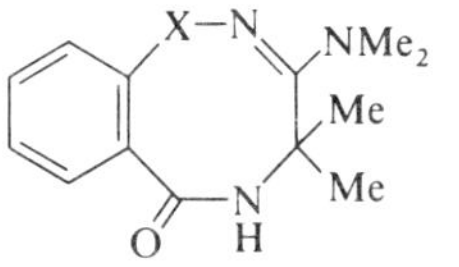

(34)

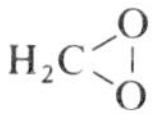

(37)

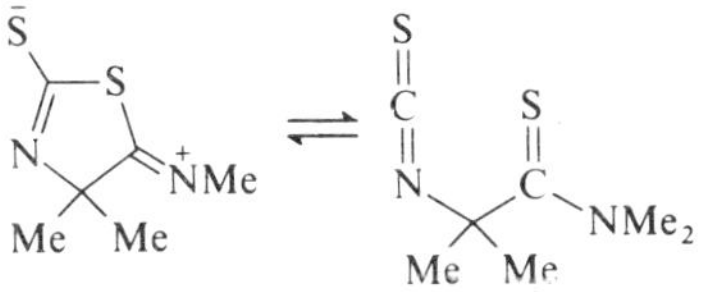

(35) (36)

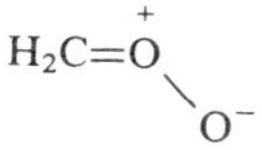

(38)

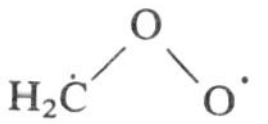

(39)

1977 saw the publication of a number of papers from the Zurich group on the chemistry of 2*H*-azirines. The u.v. spectra, and also the regiospecificity of cycloaddition to various dipolarophiles, of the benzonitrile isopropylides formed by photochemical ring-opening of the azirines (31), are investigated.[29] 3-Dimethylamino-2,2-dimethylazirine (32) is a fertile source of interesting reactions. With a munchnone (1,3-oxazolium-5-olate) it appears to add to the acylamidoketen valence tautomer (r.t., in MeCN); subsequent rearrangement and recyclization leads to the amino-oxazoline (33).[30] Azirine (32) forms eight-membered heterocycles (34; X = CO, SO_2) by insertion into the heterocyclic rings of phthalimide and saccharin.[31] The adduct with CS_2 exists as the betaine (35) in the solid phase, but mainly as the open-chain structure (36) in non-polar solvents.[32] Isothiocyanates form adducts of structure similar to (35), amongst other things.[33] With substituents other than methyl at C-2, other types of product are sometimes formed.[34]

[28] C. H. Bushweller, J. A. Ross, and D. M. Lemal, *J. Amer. Chem. Soc.*, 1977, **99**, 629.
[29] U. Gerber, H. Heimgartner, H. Schmid, and W. Heinzelmann, *Helv. Chim. Acta*, 1977, **60**, 687.
[30] J. Lukáč, J. H. Bieri, and H. Heimgartner, *Helv. Chim. Acta*, 1977, **60**, 1657.
[31] S. Chaloupka, P. Vittorelli, H. Heimgartner, H. Schmid, H. Link, K. Bernauer, and W. E. Oberhänsli, *Helv. Chim. Acta*, 1977, **60**, 2476.
[32] S. Chaloupka, H. Heimgartner, H. Schmid, H. Link, P. Schönholzer, and K. Bernauer, *Helv. Chim. Acta*, 1976, **59**, 2566.
[33] E. Schaumann, E. Kausch, and W. Valter, *Chem. Ber.*, 1977, **110**, 820.
[34] E. Schaumann, S. Grabley, K. D. Seidel, and E. Kausch, *Tetrahedron Letters*, 1977, 1351.

The detection of dioxirane (37), by mass[35] and microwave[36] spectrometry, as a transient intermediate in the gas-phase ozonolysis of ethylene, has been claimed. Calculations[37] had earlier suggested that it should compare favourably in stability with the zwitterionic (38) and diradical (39) open-chain forms, and that its decomposition might proceed by O—O cleavage; some kinetic evidence appears to support this view.[38]

L'abbé *et al.*[39] have described the preparation of alkylidenediaziridines (40). They do not isomerize to aziridinimines on heating, but rather form methacrylic amidines (41). With phenyl isocyanate the adduct (42) is obtained. The thermal rearrangement of aziridinimines (*e.g.* 43⇌44) has been studied kinetically. Unfortunately, they also decompose very easily, and these reactions compete with the rearrangement.[40]

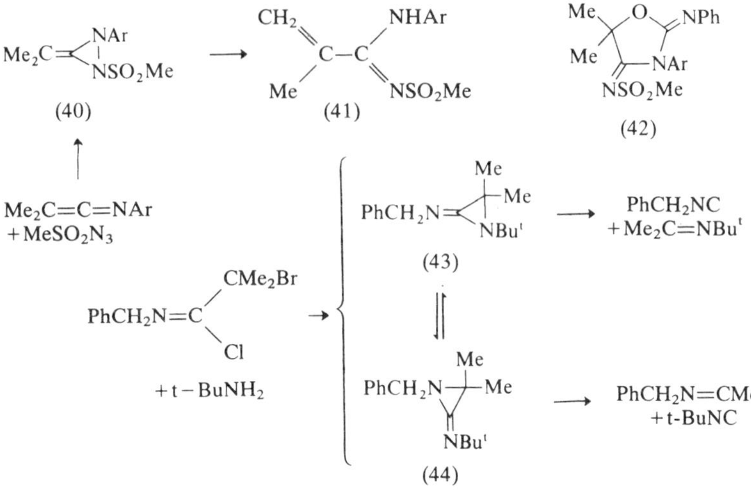

Quast's group have made further contributions to this general theme ('hetero-trimethylenemethanes'). Phenyl isocyanate adds to a diaziridinimine (45; $R^1 = R^2 = R^3 = Pr^i$) to form a 1,2,4-triazolidinonimine (46); deuterium-labelling [$(CD_3)_2CH$] demonstrated: (*a*), that the diaziridinimine did *not* scramble the label between the N atoms, and (*b*), that addition occurred across the exocyclic and one of the ring N atoms, as expected.[41] The addition of (45) to dimethyl acetylenedicarboxylate eventually yields a 1:3 adduct of spiro-construction (47).[42] The possibility of thiaziridinimine–diaziridinethione involvement in some reactions of

[35] R. I. Martinez, R. E. Huie, and J. T. Herron, *Chem. Phys. Letters*, 1977, **51**, 457.
[36] F. J. Lovas and R. D. Suenram, *Chem. Phys. Letters*, 1977, **51**, 453.
[37] W. R. Wadt and W. A. Goddard, *J. Amer. Chem. Soc.*, 1975, **97**, 3004.
[38] J. T. Herron and R. E. Huie, *J. Amer. Chem. Soc.*, 1977, **99**, 5430.
[39] G. L'abbé, C. C. Yu, and S. Toppet, *Angew. Chem. Internat. Edn.*, 1977, **16**, 475.
[40] H. Quast and P. Schäfer, *Tetrahedron Letters*, 1977, 1057.
[41] H. Quast and E. Spiegel, *Angew. Chem. Internat. Edn.*, 1977, **16**, 109.
[42] H. Quast, K. H. Ross, E. Spiegel, K. Peters, and H. G. von Schnering, *Angew. Chem. Internat. Edn.*, 1977, **16**, 177.

thiatriazoles and tetrazolinethiones has been raised by L'abbé,[43,44] but this seems to be speculative, at present.

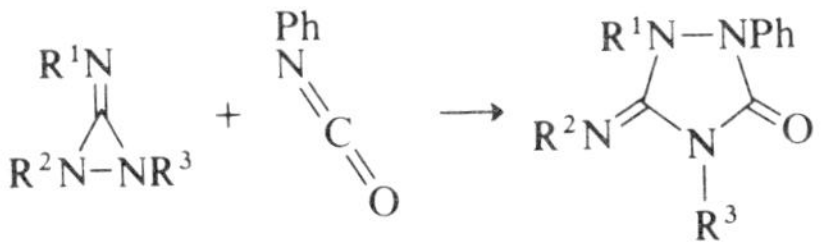

(45) (46)

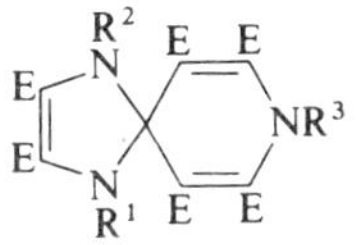

(47) $E = CO_2Me$

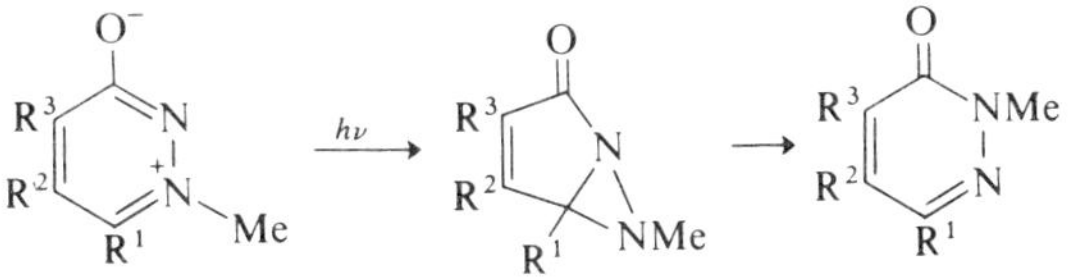

(48) $R^1 = Me, R^2 = R^3 = H$
(49) $R^1 = H, R^2R^3 = C_4H_4$

(50) (51)

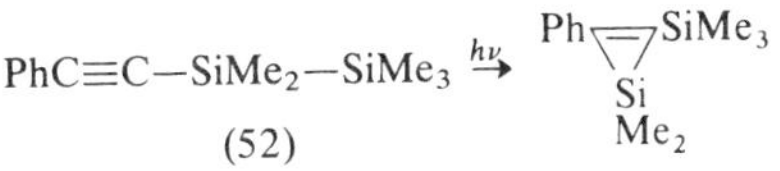

(52) (53)

The irradiation of 3-oxidopyridazinium betanes (48) and the corresponding phthalazines (49) in acetonitrile forms bicyclic valence isomers (50). These compounds show little or no amide resonance ($\nu_{C=O}$ *ca.* 1750 cm^{-1}), and in protic solvents the five-membered ring is opened; recyclization and aromatization produces the rearranged pyridazinones (51).[45]

A further development on the silirene front[46] is the reported photochemical conversion of the acetylenic disilane (52) into the cyclic isomer (53). In several reactions of (53) which were investigated, external groups were inserted into the Si—C(Ph) bond.[47]

4 Four-membered Rings

By a careful choice of temperature, wavelength of illumination (270 nm), and matrix material, it has proved possible to convert the oxazinone (54; $R^1 = R^2 = H$) into its bicyclic valence isomer (55). Shorter wavelengths (254 nm) produced the keten (56); both valence isomerizations were reversed thermally. Substituted oxazinones behaved similarly. Prolonged irradiation of the lactone (55; $R^1 = Me$, $R^2 = Bu^t$) gave the dimer(s) of the azete (57), when conducted at −70 °C in 2-methyltetrahydrofuran, but at 7K (Ar matrix) fission products (CO_2, acetylenes, and cyanides) were formed, which were consistent with the formation of (57), rather than with cleavage of (55) to a nitrile and oxetone.[48] (Scheme 3)

43 G. L'abbé, G. Verhelst, and S. Toppet, *J. Org. Chem.*, 1977, **42**, 1159.
44 G. L'abbé, G. Verhelst, and G. Vermeulen, *Angew. Chem. Internat. Edn.*, 1977, **16**, 403.
45 Y. Maki, M. Kawamura, H. Okamoto, M. Suzuki, and K. Kaji, *Chem. Letters*, 1977, 1005.
46 *Ann. Reports* (*B*), 1976, **73**, 243.
47 H. Sakurai, Y. Kamiyama, and Y. Nakadaira, *J. Amer. Chem. Soc.*, 1977, **99**, 3879.
48 G. Maier and U. Schäfer, *Tetrahedron Letters*, 1977, 1053. *cf.* also A. Krantz and B. Hoppe, *J. Amer. Chem. Soc.*, 1975, **97**, 6590.

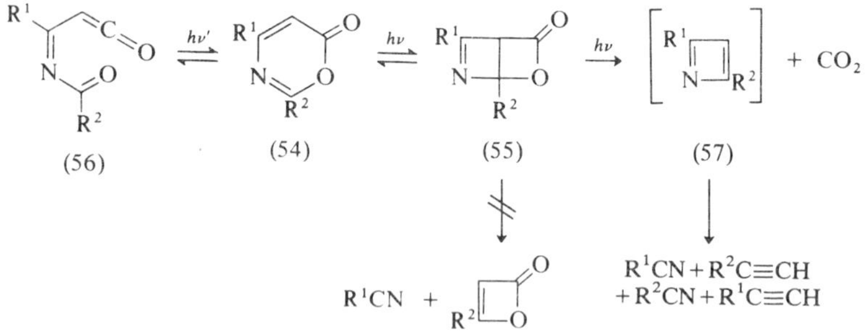

(56) (54) (55) (57)

Scheme 3

(59) (58) (60) (61)

R = Tosyl or CO_2Me

Further examples[49] of azetines have been reported. The Δ^1-azetin-3-one acetal (58) is formed by thermal decomposition of the photoadduct (59) of 6-methyl-2-phenyl-1,3-oxazin-4-one and 1,1-dimethoxyethene,[50] while photolysis of the tetracyclic compounds (60) gave the Δ^2-azetine derivatives (61).[51]

Adam *et al.* have given details for their synthesis of 1,2-dioxetanones (62), by carbodi-imide cyclization of α-hydroperoxy-carboxylic acids,[52] for which a convenient preparation has been reported.[53] An optically active 1,2-dioxetane has been prepared[54] and its crystal structure was determined;[55] it was found to emit circularly polarized chemiluminescence when it decomposed.[56] Singlet oxygen adds to the dihydrodioxin (63) at −78 °C to form the (4+2) cycloadduct, which rearranges to the dioxetane (64) (silica gel, r.t., in xylene); the luminescent decomposition of (64) occurs at 80 °C.[57]

Benzothiet (65) has been prepared, by flash vacuum pyrolysis of benzo[*b*]thiophene 1,1-dioxide.[58]

Thiete dioxides (66), on heating, form vinyl sulphenes which undergo (4+2) cycloaddition to olefins. Yields are poor, however, unless conditions are favourable (as, for instance, when the olefin is strained).[59] The remarkable diphosphete (67) is reported by Appel *et al.*[60] as a stable white solid.

[49] *cf. Ann. Reports* (*B*), 1975, **72**, 254; *ibid.*, 1976, **73**, 243.
[50] T. H. Koch, R. H. Higgins, and H. F. Schuster, *Tetrahedron Letters*, 1977, 431.
[51] R. N. Warrener, G. Kretschmer, and M. N. Paddon-Row, *J.C.S. Chem. Comm.*, 1977, 806.
[52] W. Adam, A. Alzerreca, J. C. Liu, and F. Yany, *J. Amer. Chem. Soc.*, 1977, **99**, 5768.
[53] W. Adam and O. Cueto, *J. Org. Chem.*, 1977, **42**, 38.
[54] H. Wynberg and H. Numan, *J. Amer. Chem. Soc.*, 1977, **99**, 603.
[55] H. Numan, J. H. Wieringa, H. Wynberg, J. Hess, and A. Vos, *J.C.S. Chem. Comm.*, 1977, 591.
[56] H. Wynberg, H. Numan, and H. P. J. M. Dekkers, *J. Amer. Chem. Soc.*, 1977, **99**, 3870.
[57] A. P. Schaap, P. A. Burns, and K. A. Zaklika, *J. Amer. Chem. Soc.*, 1977, **99**, 1270.
[58] W. J. M. van Tilborg and R. Plomp, *J.C.S. Chem. Comm.*, 1977, 130.
[59] D. C. Dittmer, J. E. McCaskie, J. E. Babiarz, and M. V. Ruggeri, *J. Org. Chem.*, 1977, **42**, 1910.
[60] R. Appel, F. Knoll, and H. D. Wihler, *Angew. Chem. Internat. Edn.*, 1977, **16**, 402.

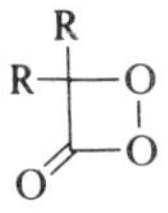
(62)

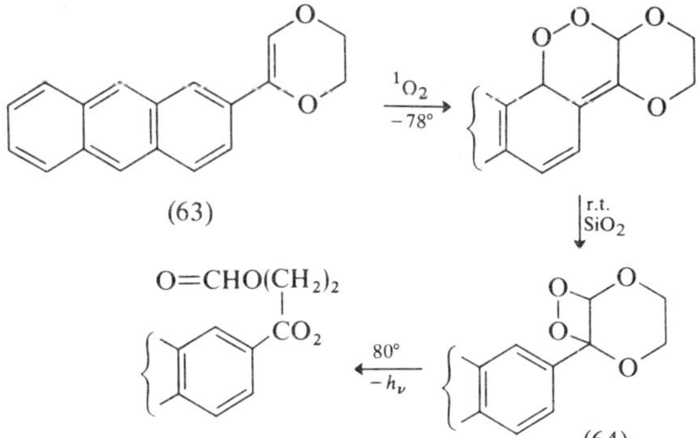
(63) (64)

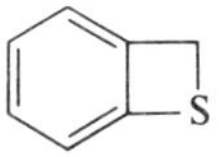
(65)

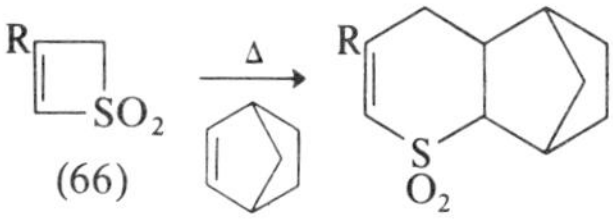
(66)

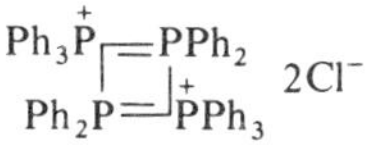
(67)

5 Five-membered Rings

Simple furans can be elaborated into a variety of useful 'synthons', according to several research groups. The acid-catalysed rearrangement of 2-furylcarbinols (68) to 5-hydroxycyclopenten-3-ones has been used in prostaglandin-intermediate synthesis.[61] Rather than undergo simple acyloin reduction, furoic esters form the

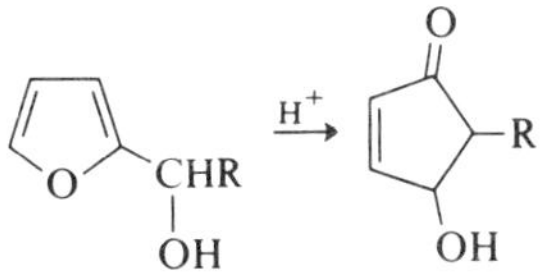
(68)

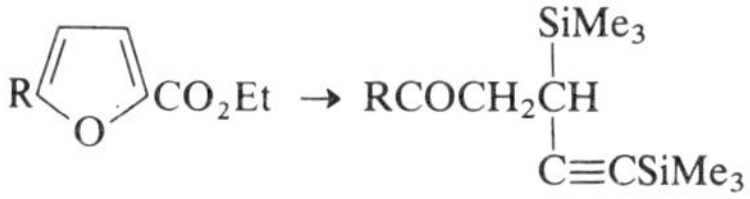
(69)

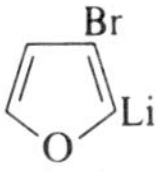
(70)

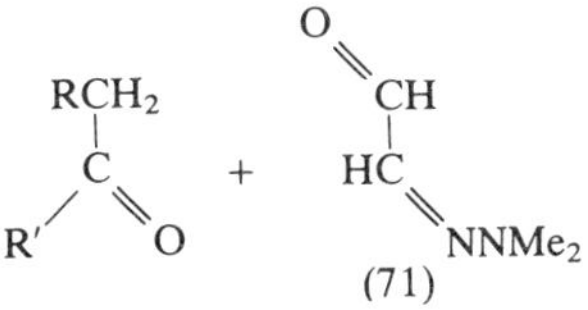
(71)

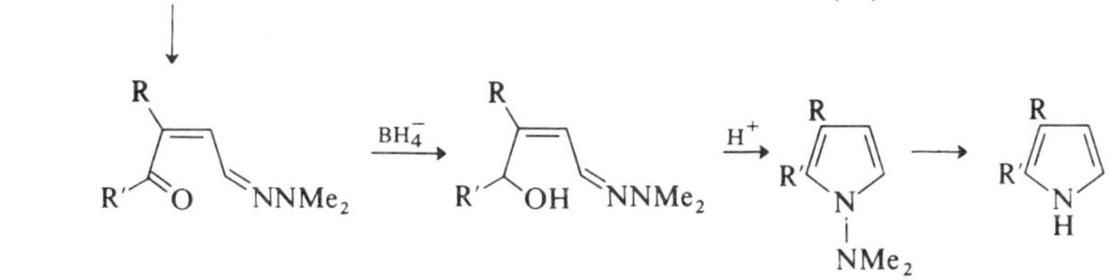

Scheme 4

[61] G. Piancatelli, A. Scettri, and S. Barbadoro, *Tetrahedron Letters*, 1976, 3555; G. Piancatelli and A. Scettri, *ibid.*, 1977, 1131.

acetylenic derivatives (69) by ring cleavage, with sodium and trimethylsilyl chloride in tetrahydrofuran followed by acid workup.[62] Further synthetic possibilities of this product are explored.[63] 2,5-Dilithiated furan, thiophene, and pyrroles have been investigated,[64] and the lithiated bromofuran (70) has provided another route to 'rose furan'.[65]

Glyoxal monohydrazones (*e.g.* 71) provide a route to 2,3-disubstituted pyrroles and their 1-amino-substituted derivatives, as outlined in Scheme 4.[66] A convenient synthesis of azulenes is claimed from the (6+4) cycloaddition of 6-dialkylamino-fulvenes to thiophene dioxide.[67,68]

Indoles (72) are formed by intramolecular cyclization of α-lithiated *o*-tolyl isocyanides (a favourable '5-*Endo-Dig*' process, in Baldwin's scheme[69]). The lithio derivatives (73; R = H) can be alkylated at −70 °C, and the product again metallated (Pr^i_2NLi) to (73; R = alkyl); cyclization occurs on heating to room temperature.[70]

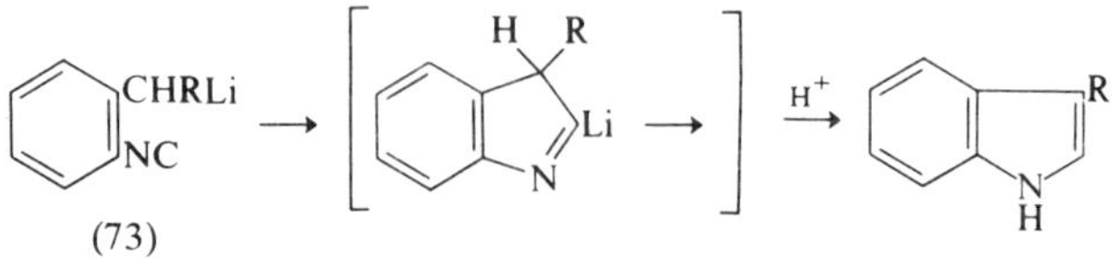

(73)

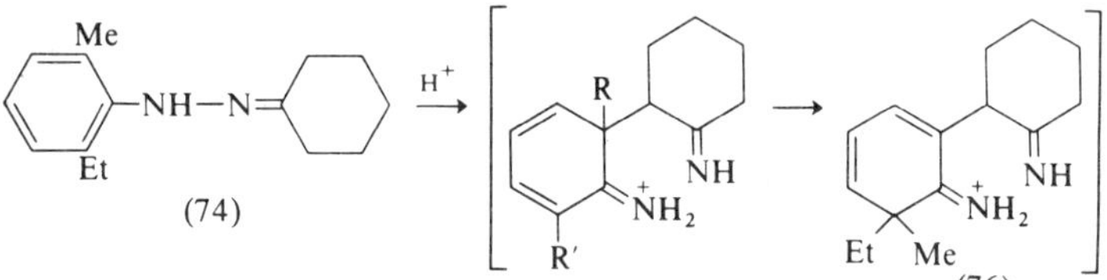

(74) (77) (76)

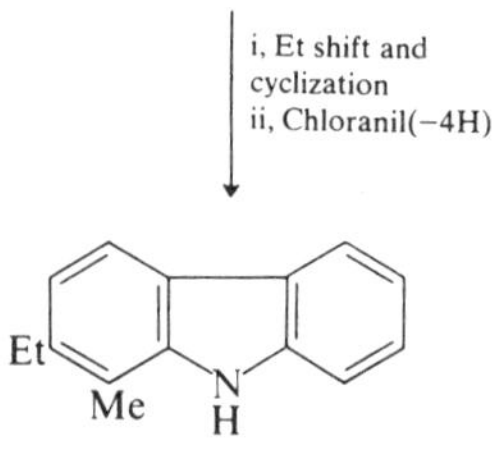

(75)

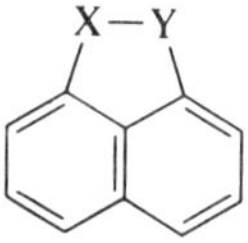

(78)

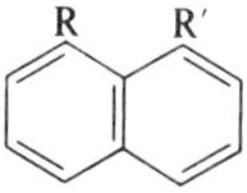

(79)

Mechanisms of alkyl group migration in the Fischer indolization reaction of the hydrazone (74) have been examined.[71] The major product (75) is suggested to be

[62] I. Kuwajima, K. Atsumi, and I. Azegami, *J.C.S. Chem. Comm.*, 1977, 76.
[63] K. Atsumi and I. Kuwajima, *Tetrahedron Letters*, 1977, 2001.
[64] D. J. Chadwick and C. Willbe, *J.C.S. Perkin I*, 1977, 887.
[65] N. D. Ly and M. Schlosser, *Helv. Chim. Acta*, 1977, **60**, 2085.
[66] T. Severin and H. Poehlmann, *Chem. Ber.*, 1977, **110**, 491.
[67] D. Copland, D. Leaver, and W. B. Menzies, *Tetrahedron Letters*, 1977, 639.
[68] S. E. Reiter, L. C. Dunn, and K. N. Houk, *J. Amer. Chem. Soc.*, 1977, **99**, 4199.
[69] J. E. Baldwin, *J.C.S. Chem. Comm.*, 1976, 734.
[70] Y. Ito, K. Kobayashi, and T. Saegusa, *J. Amer. Chem. Soc.*, 1977, **99**, 3532.
[71] B. Miller and E. R. Matjeka, *Tetrahedron Letters*, 1977, 131.

formed *via* (76), which arises from two 1,2-alkyl shifts (or a 1,5-shift?) from the primary rearrangement ions (77; R = Me, R′ = Et, and *vice versa*). In the final migration step, the ethyl group moves in preference to the methyl. Minor products are produced by 1,2-alkyl shifts in (77), leading to 1,4-dialkyl-5,6,7,8-tetrahydrocarbazoles.

Peri-bridged naphthalene systems (78) continue to receive attention. The reactions of methyl-lithium, followed by methyl iodide, on the three congeners (78; X = Y = S, Se, and Te) contrast, in that the dithio-compound forms 1,8-bis(methylthio)naphthalene (79; R = R′ = SMe), the diseleno- gives a 4:1 mixture of the bis(methylseleno)naphthalene and the mono-*C*-alkylated compound (79; R = SeMe, R′ = Me), while the ditelluride loses the tellurium altogether, giving 1,8-dimethylnaphthalene.[72] The synthesis of analogues with mixed group six elements (78; X = S, Y = Se and Te, and X = Se, Y = Te) has been reported.[73] The dithiole 1,1-dioxide (78; X = S, Y = SO_2) forms a mixture with the monoxide and trioxide (78; X = SO, Y = S and SO_2, respectively) when treated with NaOH, followed by HCl.[74,75]

An extensive series of interesting papers describes the work of J. C. Martin's group on the heterocyclic sulphuranes. The optically active chlorosulphurane (80) was found to racemize slowly,[76] and the two isomers of the spiro-compound (81) interconverted, also slowly ($\Delta G^{\ddagger}$ *ca.* 125 kJ mol^{-1}).[77] The mechanism of these reactions is uncertain. The X-ray crystal structure of the asymmetric compound (82) shows that the S—O bond lengths are significantly different (*a* 1.955 Å, *b* 1.713 Å),[78] a state of affairs reminiscent of that obtaining with the S—S distances found in unsymmetrically substituted thiathiophthenes.

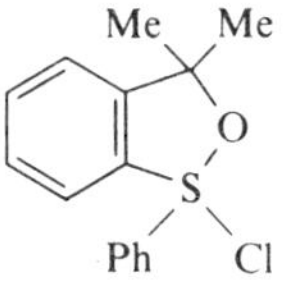

(80)

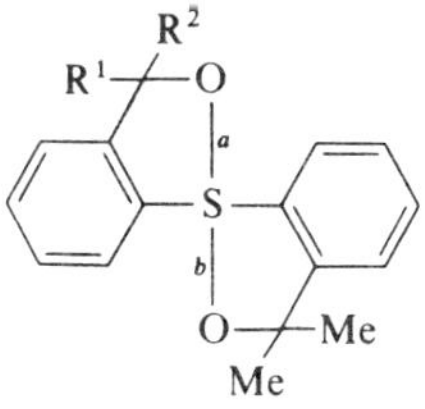

(81) R^1 = Me, R^2 = Et
(82) R^1 = R^2 = CF_3

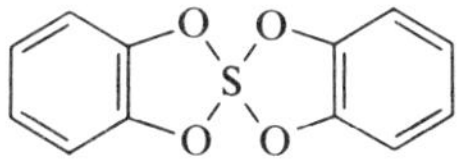

(83)

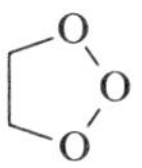

(84)

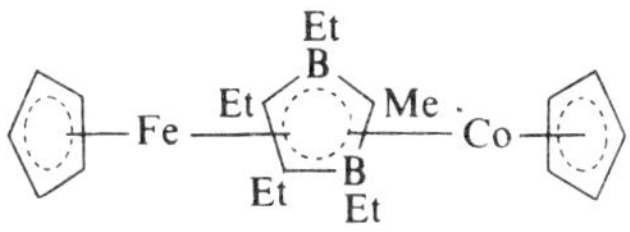

(85)

(86)

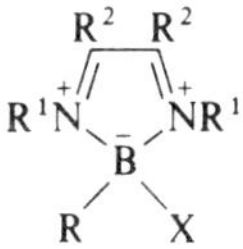

(87)

[72] J. Meinwald, D. Dauplaise, F. Wudl, and J. J. Hauser, *J. Amer. Chem. Soc.*, 1977, **99**, 255.
[73] J. Meinwald, D. Dauplaise, and J. Clardy, *J. Amer. Chem. Soc.*, 1977, **99**, 7743.
[74] B. I. Stepanov, V. Ya. Rodionov, M. A. Andreeva, V. P. Perevalov, and S. E. Voinova, *Zhur. obshchei Khim.*, 1977, **47**, 234. These authors claim the trioxide as a new compound; see, however, ref. 75.
[75] J. L. Kice and H. Margolis, *J. Org. Chem.*, 1975, **40**, 3623; M. M. Chau and J. L. Kice, *J. Org. Chem.*, 1977, **42**, 3265.
[76] J. C. Martin and T. M. Balthazor, *J. Amer. Chem. Soc.*, 1977, **99**, 152.
[77] L. J. Adzima and J. C. Martin, *J. Org. Chem.*, 1977, **42**, 4006.
[78] L. J. Adzima, E. N. Duesler, and J. C. Martin, *J. Org. Chem.*, 1977, **42**, 4001.

Phenylene orthosulphite (83) has been reported as showing an AA′BB′ 220 MHz ^{1}H n.m.r. spectrum, indicating a four-fold (D_{2d}) symmetry, at least when time-averaged. It is very unstable to moisture.[79]

The reaction between ethylene and ozone has been studied by i.r. in CO_2 and CCl_4 matrices, at temperatures down to 65 K. The secondary ozonide (1,2,4-trioxolan) was the major product observed, but a less volatile minor compound remained after that, along with ethylene, ozone, and CO_2, had been pumped off at −120 °C, and this was assumed to be the primary ozonide, 1,2,3-trioxolan (84).[80] For the gas-phase ozonolysis of ethylene, see Section 3.

A number of appetizing sandwich compounds containing boron heterocycles, including some of the triple-decker type, were reported during 1977. Both sides of the 1,3-diborole ring are complexed in (85),[81] and both sides of a 1,2,5-thiadiborolene (86) accept cobalt[82] and manganese[83] tricarbonyl bonding units. 1,3,2-Diazaborolenium salts (87) are formed by the reaction of α-di-imines with dihaloboranes.[84] They can be reduced to the very unstable 1,3,2-diazaborolenes,[84] which may be stabilized in the form of $Cr(CO)_3$ complexes.[85] Some water-stable iron tricarbonyl compounds of boroles have been prepared.[86]

4-Oxazolin-2-ones (88; R = H or Ph, X = NH or NAc) with dienes form Diels–Alder adducts, which can be hydrolysed to provide a synthesis of β-amino-alcohols, in the same way that 1,2-diols can be made using vinylene carbonate (88; R = H, X = O).[87] Another illustration of the synthetic use of oxazoles is in the heating of 3-sulphoximino-2-oxazolidinones (89); these decompose as indicated, to leave an olefin, and they have been used in attempts to prepare the (*Z*) (successfully) and (*E*) isomers of bicyclo[3,3,1]non-1-ene.[88]

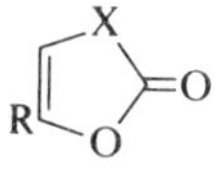

(88)

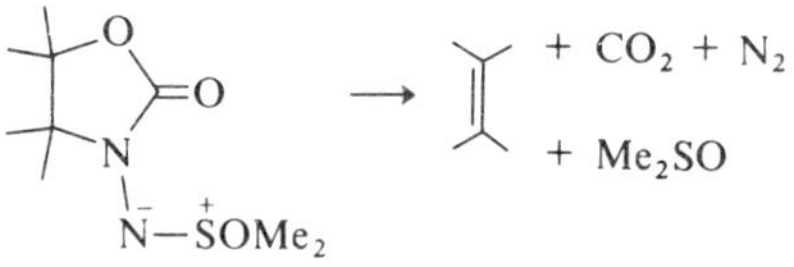

(89)

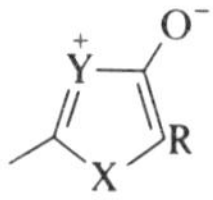

(90)

Potts has extended his work on the mesoionic heterocycles of general structure (90; X, Y = O, S, Se, or NR′). These compounds are conveniently prepared by reaction of a 1,2-bielectrophile RCHBrCOCl (R = Ph or CO_2Et) with a wide variety of 1,3-binucleophiles.[89]

[79] G. E. Wilson and B. A. Belkind, *J. Org. Chem.*, 1977, **42**, 765.
[80] B. Nelander and L. Nord, *Tetrahedron Letters*, 1977, 2821.
[81] W. Siebert and M. Bochmann, *Angew. Chem. Internat. Edn.*, 1977, **16**, 857.
[82] W. Siebert and W. Rothermel, *Angew. Chem. Internat. Edn.*, 1977, **16**, 373.
[83] W. Siebert and K. Kinberger, *Angew. Chem. Internat. Edn.*, 1976, **15**, 434.
[84] G. Schmid and J. Schulze, *Chem. Ber.*, 1977, **110**, 2744.
[85] G. Schmid and J. Schulze, *Angew. Chem. Internat. Edn.*, 1977, **16**, 249.
[86] G. E. Herberich, J. Hengesbach, O. Kölle, and W. Oschmann, *Angew. Chem. Internat. Edn.*, 1977, **16**, 42.
[87] J. A. Deyrup and H. L. Gingrich, *Tetrahedron Letters*, 1977, 3115.
[88] M. G. Kim and J. D. White, *J. Amer. Chem. Soc.*, 1977, **99**, 1172; *cf. Ann. Reports (B)*, 1975, **72**, 312.
[89] K. T. Potts, S. J. Chen, J. Kane, and J. L. Marshall, *J. Org. Chem.*, 1977, **42**, 1633; K. T. Potts and S. J. Chen, *ibid.*, p. 1639; K. T. Potts, F. Huang, and R. K. Khattak, *ibid.*, p. 1644.

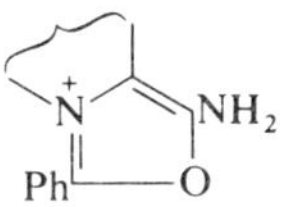

(91)

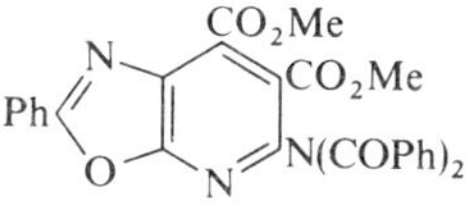

(92)

The structures of Reissert salts have been reformulated as fused 5-amino-oxazolium systems (91), rather than the 4*H*-imino-structures previously assigned.[90] The fused oxazolo[5,4-*b*]pyridine (92) is formed as a 3:1 addition product of benzoyl isocyanide and dimethyl acetylenedicarboxylate.[91]

The thermal rearrangement of 5-hydrazinoisoxazoles has been studied in some detail.[92] The products are 1- and 4-aminopyrazolin-5-ones, and tetrahydro-1,2,4-triazin-6-ones, in varying proportions, depending on the solvent and substituents. Another interesting rearrangement product was observed when the photochemistry of a variety of *N*-alkyl-2,1-benzisoxazolium salts was investigated.[93] In protic solvents *N*-substituted *o*-amino-phenyl ketones are formed, with further substitution in the benzene ring (*e.g.* 93→94); this is perhaps unexceptional, but in acetonitrile the 1-adamantyl-3-phenyl compound gives a red salt for which the structure of the cation is suggested to be (95) or (96). The former is supported by

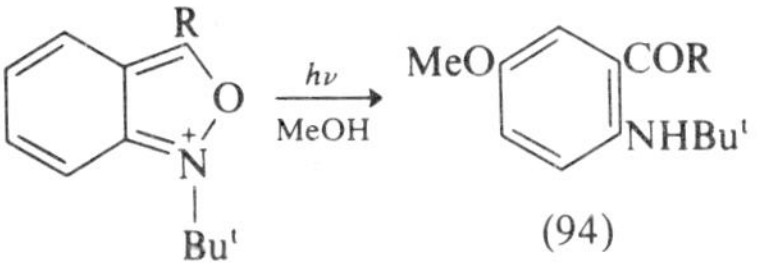

(93) (94)

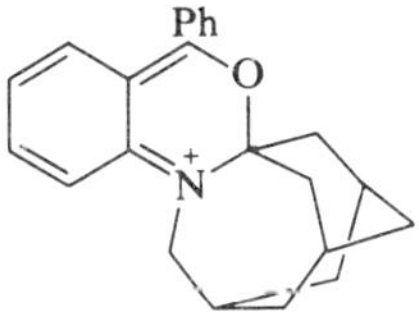

(95)

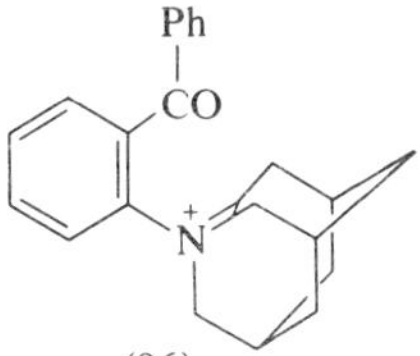

(96)

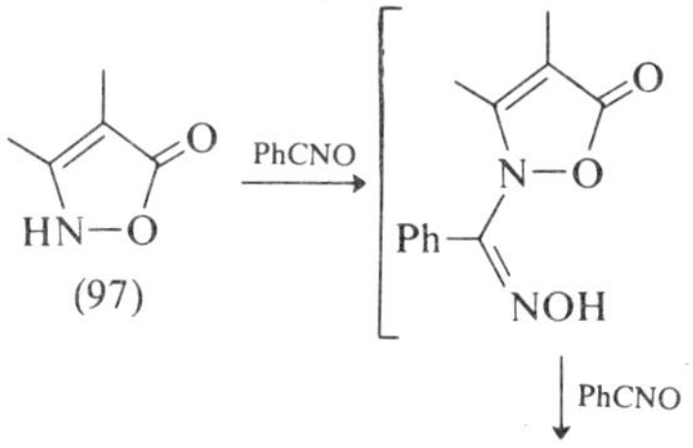

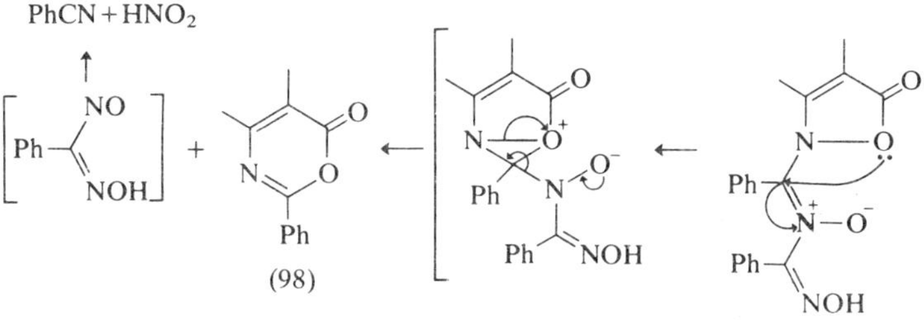

[90] M. J. Cook, A. R. Katritzky, and A. D. Page, *J. Amer. Chem. Soc.*, 1977, **99**, 165.
[91] G. Höfle and B. Lange, *Angew. Chem. Internat. Edn.*, 1977, **16**, 262.
[92] G. Adembri, A. Camparini, F. Ponticelli, and P. Tedeschi, *J.C.S. Perkin I*, 1977, 971.
[93] N. F. Haley, *J. Org. Chem.*, 1977, **42**, 3929.

^{13}C n.m.r., the latter by its i.r. spectrum. Another interesting observation in the isoxazole series is the curious insertion of the PhC element of benzonitrile oxide into some isoxazolin-5-ones (97; substituents unspecified), which forms 1,3-oxazin-6-ones (98); the route suggested is shown.[94]

Benzothiazoles with a nitro-group in the benzene ring react with alkylmagnesium bromides, followed by addition of boron trifluoride etherate to the complex, to produce alkyl-nitroso-benzothiazoles (*e.g.* 99→100) in fair to good yields.[95,96] It remains to be seen what role, if any, the thiazole ring plays in this rather unexpected reaction.

The full paper has appeared from Meth-Cohn's group on the preparation of benzimidazole-1,3-dioxides (101) from benzofuroxan and secondary nitroparaffins, and of 1-hydroxybenzimidazole-3-oxides similarly from primary nitroparaffins. The products undergo some curious and fascinating reactions with a variety of reagents [$(NC)_2C{=}C(CN)_2$, $MeO_2CC{\equiv}CCO_2Me$, malononitrile]; with the acetylenic ester the structure (102) is formed.[97] 2*H*-Benzimidazoles (*e.g.* 103) suffer nucleophilic attack by amines and sulphides; the products are oxidized by the starting compounds to the correspondingly substituted derivatives (104).[98]

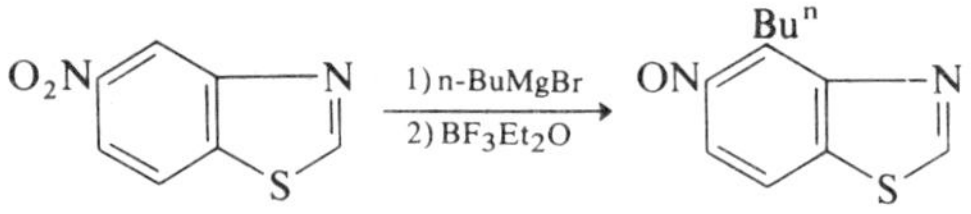

(99) (100)

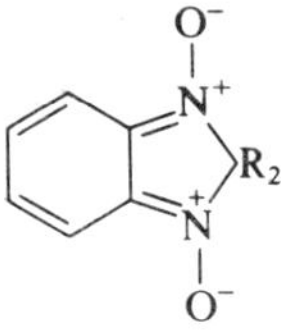

(101)

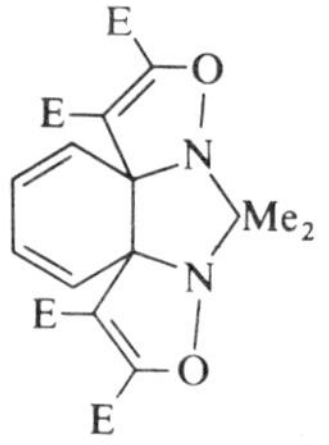

(102) $E = CO_2Me$

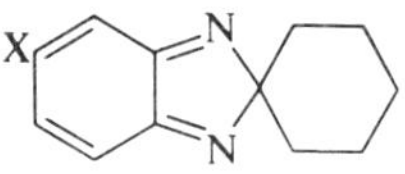

(103) X = H
(104) X = Nu

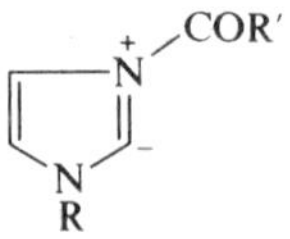

(105)

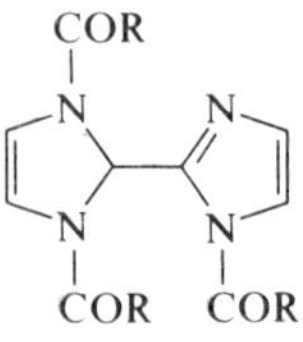

(106)

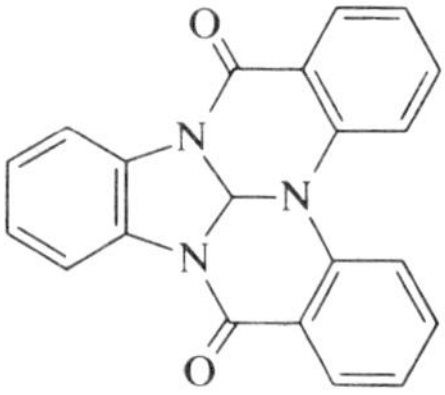

(107)

1-Alkyl and 1-arylimidazoles react with acyl chlorides in the presence of triethylamine in polar solvents (*e.g.* MeCN) to form the 2-acyl derivatives. Possibly the zwitterion (105) is involved in this reaction.[99] Imidazole itself gives the dimeric

[94] G. Lo Vecchio, F. Foti, G. Grassi, and F. Risitano, *Tetrahedron Letters*, 1977, 2119.
[95] G. Bartoli and G. Rosini, *Synthesis*, 1976, 270.
[96] G. Bartoli, R. Leardini, M. Lelli, and G. Rosini, *J.C.S. Perkin I*, 1977, 884,
[97] D. W. S. Latham, O. Meth-Cohn, H. Suschitzky, and J. A. L. Herbert, *J.C.S. Perkin I*, 1977, 470.
[98] A. M. Jefferson and H. Suschitzky, *J.C.S. Chem. Comm.*, 1977, 189.
[99] E. Regel and K. H. Büchel, *Annalen*, 1977, 145.

compound (106),[100] under the same conditions, while with isocyanates in hot solution 2-carboxamides are formed.[101] Benzimidazole suffers a three-pronged attack by 2,2′-dichlorocarbonyldiphenylamine; the structure of the product (107) has been confirmed by X-ray crystallography.[102]

Prolonged irradiation of the pyridoindazole (108) in tetrahydrofuran forms δ-carboline (109), and a spiro-intermediate is proposed, (110), being one of many canonical forms by which it can be represented.[103] However, the thermal rearrangement of the 3-imidazolyl and 3-pyrazolylanthranils (111) and (112) leads to a variety of products which are reasonably suggested to arise by rearrangement of spiro-intermediates (113).[104] The Authors argue that, in the case of the rearrangement of the 3-(2-pyridyl)anthranil, which cleanly forms pyrido[1,2-*b*]-cinnolin-11-one (114),[105] a spiro-2-pyridine intermediate of type (113) is improbable.

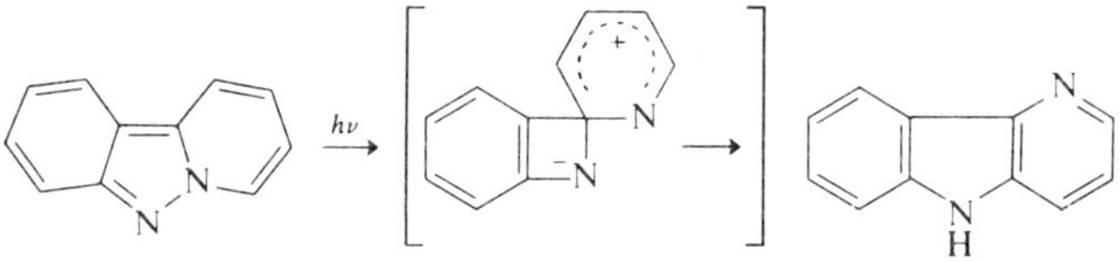

(108) (110) (109)

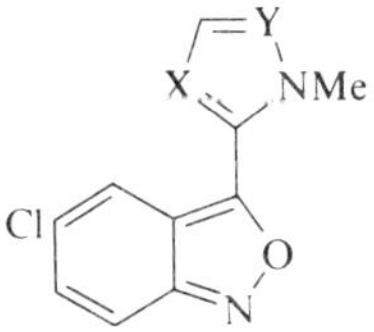

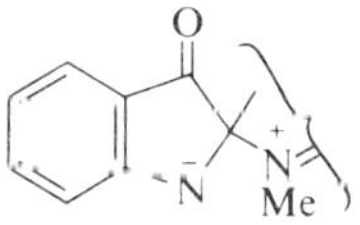

(113)

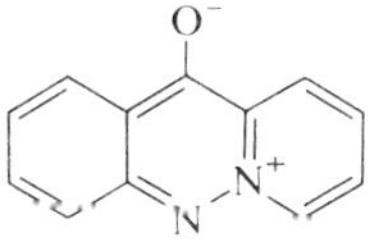

(114)

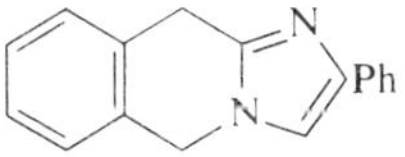

(115)

(111) X = N, Y = CH
(112) X = CH, Y = N

The structure (115) of the product of reaction of 2-phenacylisoquinolinium bromide and ammonium acetate[106] has been confirmed by *X*-ray crystallography.[107]

Two pyrazole syntheses from azines were noted in the year under review. α′-Oxo-αβ-unsaturated azines (116), for which a variety of synthetic methods are available, give pyrazolyl ketones and esters (117; R^1 = alkyl, aryl, or alkoxy) on heating.[108] This is an extension of a reaction reported some ten years ago,[109] and appears to be efficient and versatile. The other synthesis involves cyclization of simple ketazines with lithium di-isopropylamide in hexamethylphosphoramide, and

[100] E. Regel, *Annalen*, 1977, 159.
[101] E. P. Papadopoulos, *J. Org. Chem.*, 1977, **42**, 3925.
[102] A. Banerji, J. C. Cass, and A. R. Katritzky, *J.C.S. Perkin I*, 1977, 1162.
[103] J. H. Boyer and C. C. Lai, *J.C.S. Perkin I*, 1977, 74.
[104] R. Y. Ning, J. F. Blount, P. B. Madan, and R. I. Fryer, *J. Org. Chem.*, 1977, **42**, 1791.
[105] R. Y. Ning, W. Y. Chen, and L. H. Sternbach, *J. Heterocyclic Chem.*, 1974, **11**, 125.
[106] R. F. Cookson, D. P. Nowotnik, R. T. Parfitt, J. E. Airey, and A. S. Kande, *J.C.S. Perkin I*, 1976, 201.
[107] H. J. Lindner and B. Kitschke, *Tetrahedron Letters*, 1977, 2091.
[108] T. A. Albright, S. Evans, C. S. Kim, C. S. Labaw, A. B. Russiello, and E. E. Schweizer, *J. Org. Chem.*, 1977, **42**, 3691.
[109] R. L. Stern and J. G. Krause, *J. Org. Chem.*, 1968, **33**, 213; *J. Heterocyclic Chem.*, 1968, **5**, 263.

is exemplified by pinacolone azine (118). The product (119) contains an extra carbon atom; this is evidently C-5 of the ring, and is derived from the solvent.[110] The propensity for azines to undergo 'criss-cross' electrocyclic reactions is also illustrated in some work of Bürger on what he terms 'dipole metathesis'. The azomethineimine (120) adds to electrophilic olefins, and with acetylenes it undergoes a series of somersaulting additions to provide a variety of 1,5- and 1,7-dipolar systems.[111] Other work from Delaware[112] is relevant to this area.

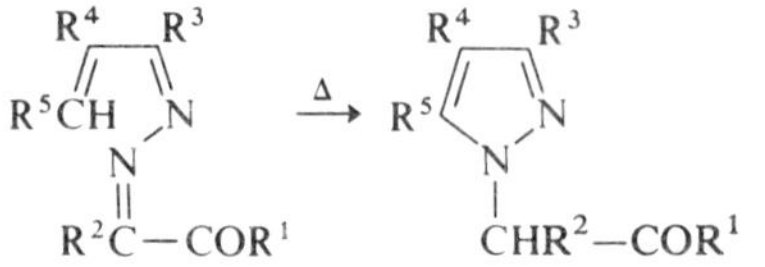

(116) (117)

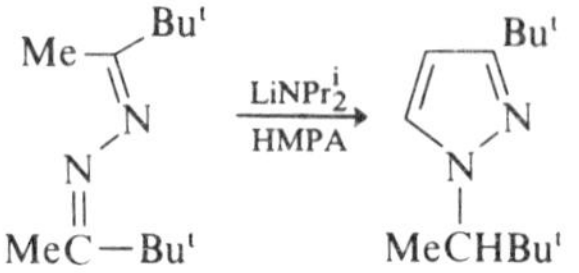

(118) (119)

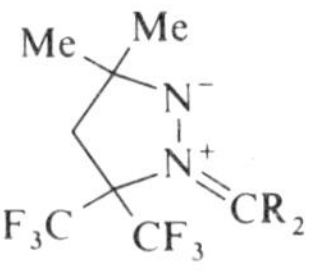

(120)

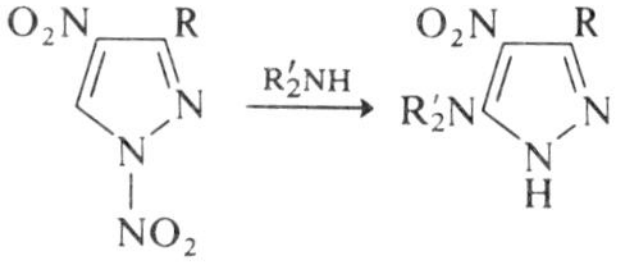

(121) R = H or Me

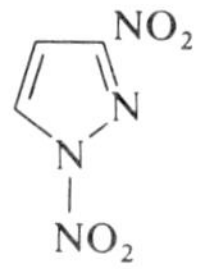

(122)

1,4-Dinitropyrazoles (121) with secondary amines undergo interesting *cine*-substitution, with departure of the leaving group from the nitrogen. With the isomeric 1,3-dinitro compound (122), the amine simply removes the 1-nitro-group.[113]

A potentially useful synthesis of 1-substituted pyrazole-2-oxides (123) is the reaction of nitrile oxides with imidoyl-substituted oxosulphonium ylids (124).[114] Freeman's group has published further results in the 4*H*-series. The chlorination of pyrazoles, 1-hydroxypyrazoles, and their 2-oxides forms 4-chloro-derivatives;[115,116] these with methoxide introduce a methoxy substituent into a

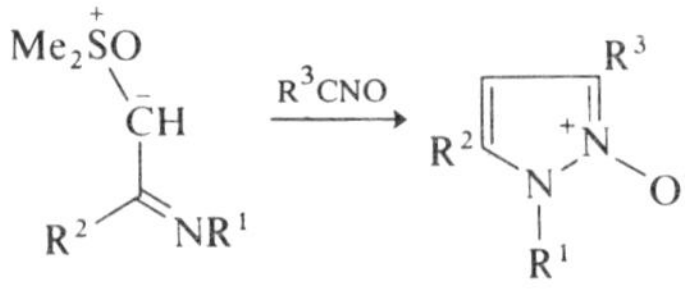

(124) (123)

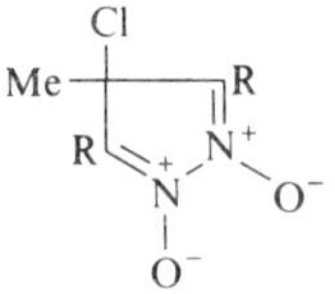

(125)

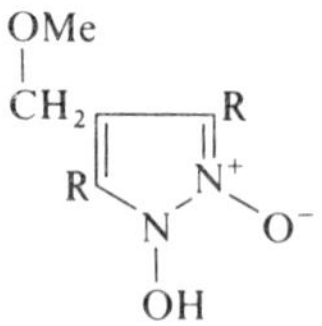

(126)

methyl group, either in the 4- or 3-position, probably *via* diazafulvene derivatives, *e.g.* (125)→(126).[116] Silver acetate converts 4-chloro-4*H*-pyrazole-1,2-dioxides into the 4-acetoxy-dioxides, but the 4-chloromonoxides (127) give 3-acetoxy

[110] Y. Tamaru, T. Harada, and Z. Yoshida, *Tetrahedron Letters*, 1977, 4323.
[111] K. Burger, H. Schickaneder, and C. Zettl, *Angew. Chem. Internat. Edn.*, 1977, **16**, 54, 55.
[112] S. Evans, R. C. Gearhart, L. J. Guggenberger, and E. E. Schweizer, *J. Org. Chem.*, 1977, **42**, 452.
[113] C. L. Habraken and E. K. Poels, *J. Org. Chem.*, 1977, **42**, 2893.
[114] R. Faragher and T. L. Gilchrist, *J.C.S. Perkin I*, 1977, 1196.
[115] J. P. Freeman and E. R. Janiga, *J. Org. Chem.*, 1974, **39**, 2663.
[116] J. P. Freeman and J. F. Lorenc, *J. Org. Chem.*, 1977, **42**, 177.

derivatives (128). With silver nitrate the monoxides (127) decompose to acetylenes, probably through the 3-hydroxy-3*H* compounds (129), but 3-acetoxy compounds in part revert to 4-hydroxy derivatives, besides decomposing to acetylenes (not invariably the expected ones, carrying the 4- and 5-substituents).[117]

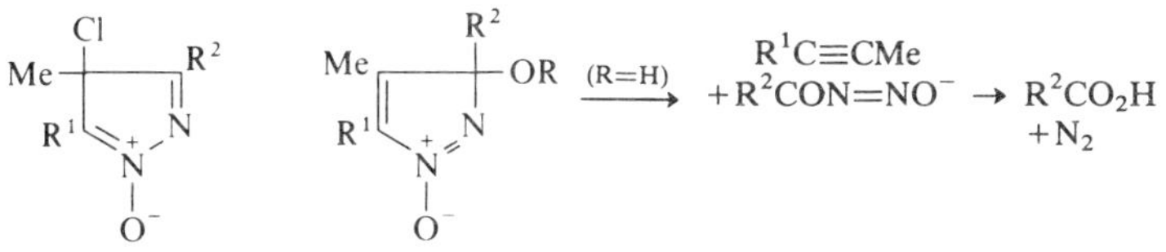

(127) (128) R = Ac (129) R = H

3*H*-Pyrazoles readily lose nitrogen on irradiation. The sulphur-containing derivatives (130; *n* = 0, 1, and 2) form carbenes (131), which can be intercepted with olefins, and cyclopropenes (132), which can be trapped by 1,3-dipolar cycloaddition of dimethyldiazomethane, and in the case of the sulphone (132; *n* = 2) by

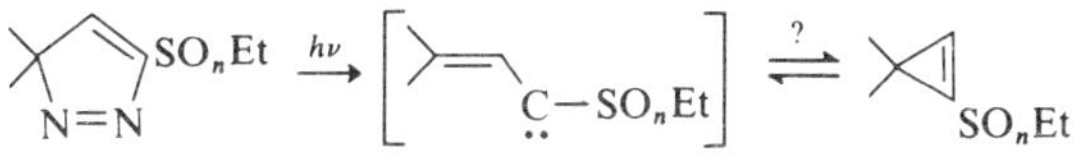

(130) (131) (132)

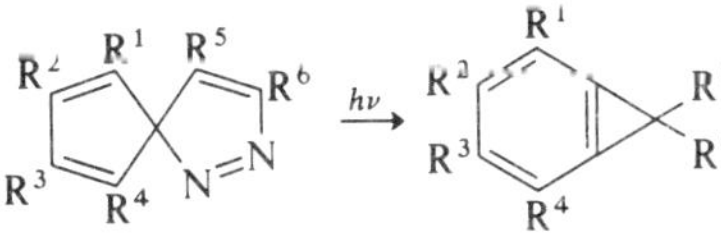

(133) (134)

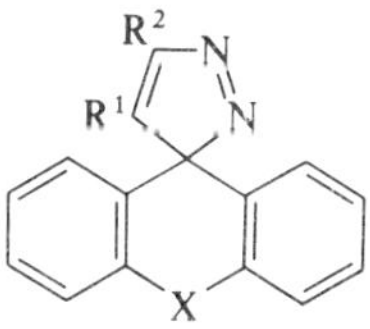

(135) X = S, O, CO

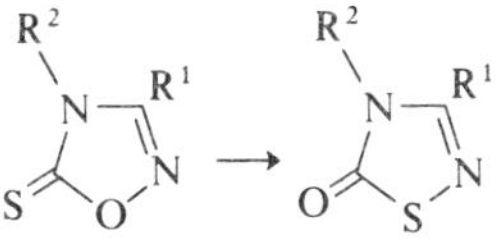

(138)

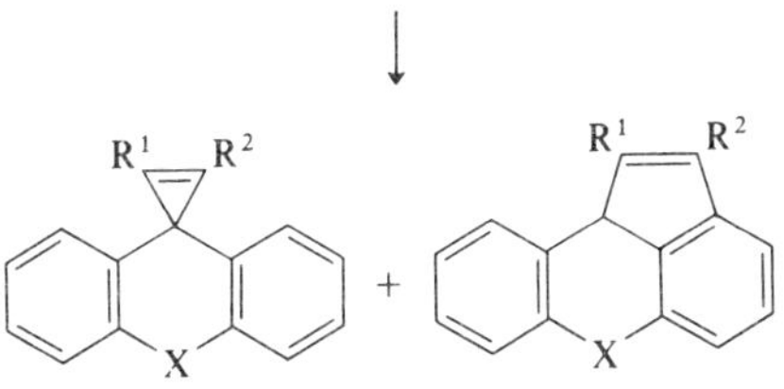

(136) (137)

Diels–Alder addition to furan. The question whether or not an equilibrium exists between the carbenes (131) and the cyclopropenes (132) was not settled, but products of the carbene mode of addition are produced even after the irradiation has ceased.[118] An account of a careful study of the photodecomposition of spiropyrazoles (133), using monochromatic radiation, has been given.[119] The benzocyclopropenes (134), which are the first identifiable products, undergo further

[117] J. P. Freeman, E. R. Janiga, and J. F. Lorenc, *J. Org. Chem.*, 1977, **42**, 3721.
[118] M. Franck-Neumann and J. J. Lohmann, *Angew. Chem. Internat. Edn.*, 1977, **16**, 323.
[119] E. Lüddecke, H. Rau, H. Dürr, and H. Schmitz, *Tetrahedron*, 1977, **33**, 2677.

photoreactions, which depend on the nature of the substituents and the illumination. Spiropyrazoles (135) on irradiation form spirocyclopropenes (136), and, in some cases, fused indenes (*e.g.* 137). It is not yet clear whether (137) is formed *via* (136).[120] Pyrolysis of 1,2,3-thiadiazole cleanly forms thioketene, which was characterized by its photoelectron spectrum.[121] For the pyrolysis of 1,2,3-benzothiadiazoles see Section 3.

The rearrangement of the oxadiazolinethiones (138) to thiadiazolinones has been studied for a wide range of alkyl and aryl substituents. The reaction proceeds well photochemically, and also thermally when catalysed by copper; without the copper only the 3-aryl compounds give thiadiazolinones. These results are in striking contrast to the very ready thermal rearrangement of open-chain oxime thionocarbamates.[122]

An extension of an old-fashioned rearrangement forms 1,2,4-thiadiazoles (139; X, Y = N, CH, CMe, or CPh) from 3-amino-1,2-oxazoles (140). The intermediate thioamides can sometimes be isolated.[123] The reaction of an aroyl chloride in pyridine with 5-amino-1,2,3,4-thiatriazole (141) eventually provides the dioxathiadiazapentalene system (142), with loss of nitrogen.[44,124] Imidoyl chlorides behave similarly, to give thiatetra-azapentalenes (143),[44] but with acetyl chloride a more profound reorganization produces the thiadiazole (144).[124] Further examples (145; X = O or S) of the 'extended thiathiophthene' system have been prepared.[125]

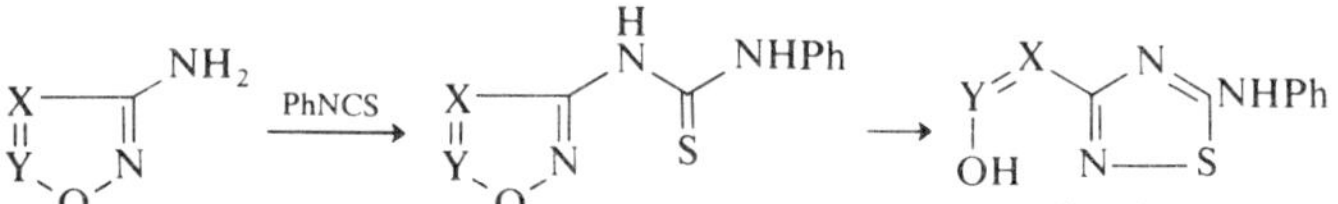

(140) (139)

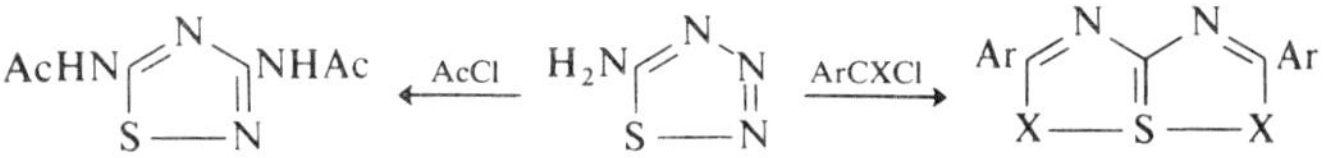

(144) (141) (142) X = O

(143) X = NAr′

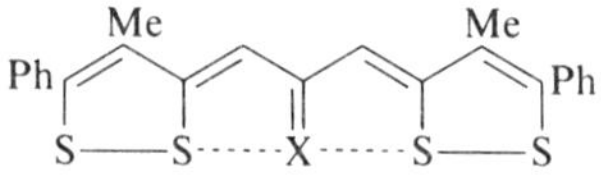

(145)

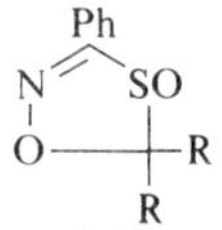

(148)

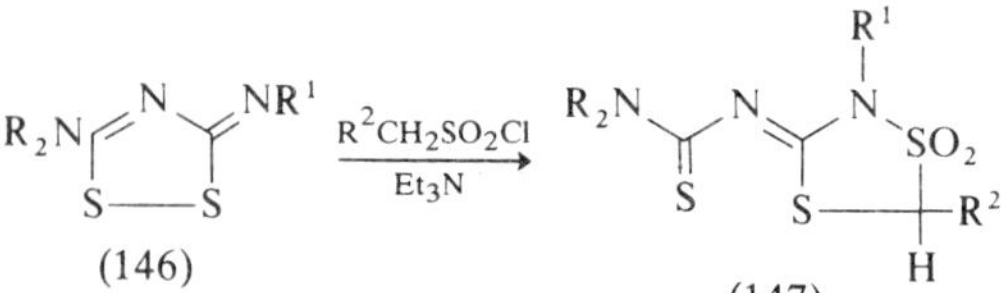

(146) (147)

[120] H. Dürr, S. Fröhlich, B. Schley, and H. Weisgerber, *J.C.S. Chem. Comm.*, 1977, 843.
[121] H. Bock, B. Solonki, G. Best, and P. Rosmus, *J. Amer. Chem. Soc.*, 1977, **99**, 1663.
[122] A. Pelter and D. Sumengen, *Tetrahedron Letters*, 1977, 1945.
[123] N. Vivona, G. Cusmano, and G. Macaluso, *J.C.S. Perkin I*, 1977, 1616.
[124] R. J. S. Beer and I. Hart, *J.C.S. Chem. Comm.*, 1977, 143.
[125] E. G. Frandsen, *J.C.S. Chem. Comm.*, 1977, 851.

The addition of a sulphene to the dithiazolinimines (146) is reported to form the rearranged system (147).[126] Addition of benzonitrile oxide to sulphines generally produces 1,4,2-oxathiazole-4-oxides (148), but in the case of fluorenethione-*S*-oxide the isomeric 1,5,2-oxathiazole-5-oxide is also formed.[127]

A very interesting reaction occurs between 2,5-dimethylfuran and nitrosocarbonylbenzene, PhCONO. The 1,3,4-dioxazole (149) is formed, possibly by rearrangement of the primary Diels–Alder adduct as indicated: and it is reversible! The product (149) on heating regenerates the two original Diels–Alder participants.[128]

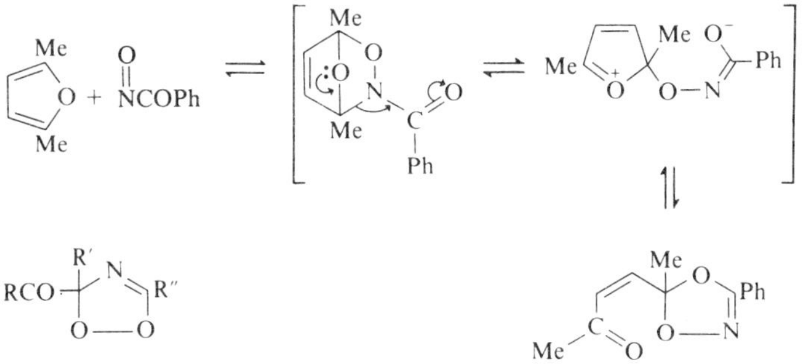

(150) (149)

1,2,4-Dioxazoles (150) are produced by the reaction of singlet oxygen with oxazoles.[129] Their decomposition gives a variety of products, including diacylamines.[130]

The reaction of 1,3,4-thiadiazolium pseudobases (151) with diethyl acetylenedicarboxylate in boiling benzene involves an ylid intermediate, and an unusual *p*-nitrophenyl group migration. The products (152) undergo further addition of an alkynamine.[131] (Scheme 5).

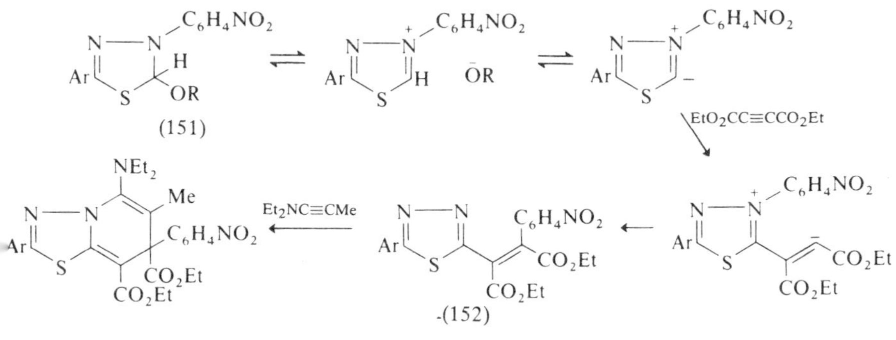

Scheme 5

[126] H. W. Linden and J. Goerdeler, *Tetrahedron Letters*, 1977, 1729.
[127] B. F. Bonini, G. Maccagnani, G. Mazzanti, L. Thijs, P. M. M. Ambrosius, and B. Zwanenburg, *J.C.S. Perkin I*, 1977, 1468.
[128] C. J. B. Dobbin, D. Mackay, M. R. Penney, and L. H. Dao, *J.C.S. Chem. Comm.*, 1977, 703.
[129] M. L. Graziano, M. R. Iesce, A. Carotenuto, and R. Scarpati, *J. Heterocyclic Chem.*, 1977, **14**, 261; *Synthesis*, 1977, 572.
[130] M. L. Graziano, A. Carotenuto, M. R. Iesce, and R. Scarpati, *Tetrahedron Letters*, 1977, 447.
[131] G. Scherowsky, K. Dünnbier, and G. Höfle, *Tetrahedron Letters*, 1977, 2095.

Dehydrodithizone (153) and its oxygen analogue (154), both 'Type B mesoionic compounds' in the classification of Ollis and Ramsden,[132] are thermally rearranged to the isomeric 'Type A' heterocycles (155).[133] Dehydrodithizone reacts with tetracyclone to give an interesting product (156) containing an open-chain azimine grouping, rather than the isomeric fused dihydrotetrazole, which seems to be an elusive system.[134] The structure of (156) was proved by *X*-ray crystallography, and the same technique was applied to solve the structure of the reaction product (157) of glyoxal and the bis-azo compound (158).[135]

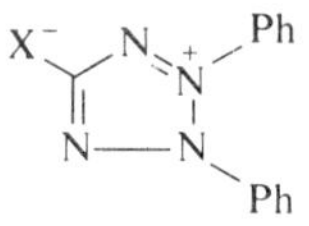

(153) X = S
(154) X = O

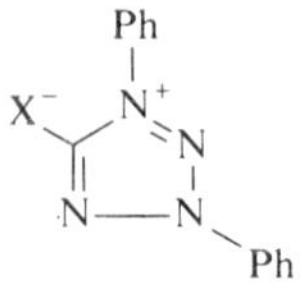

(155)

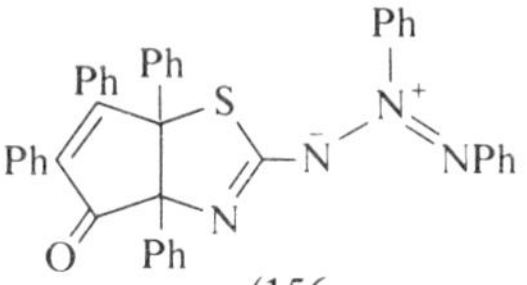

(156

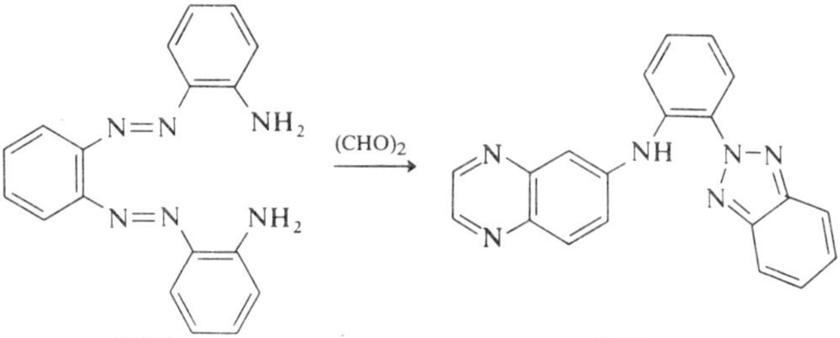

(158) (157)

The 1,2,3,5-dithiadiazolium system (159; R = Bu^t, CCl_3, or Ph) has been prepared, by the reaction of the nitriles RCN with trithiazyl chloride, $S_3N_3Cl_3$.[136] The reactions of this and other thiazyl chlorides with a variety of olefinic substrates have been studied by Barton and Bubb;[137] isothiazoles and thiadiazoles are among the products observed.

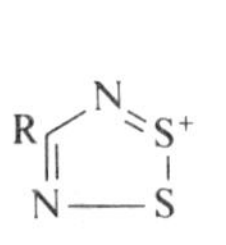

(159)

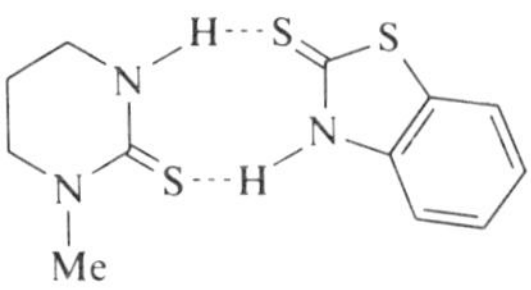

(160) (161)

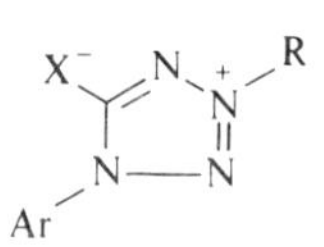

(162)

An interesting study of thioamide base-pairing through hydrogen bonds has been reported.[138] In CCl_4 the tetrahydropyrimidine-2-thione (160) (weakly associated

[132] D. W. Ollis and C. A. Ramsden, *Adv. Heterocyclic Chem.*, 1976, **19**, 1.
[133] P. N. Preston and K. Turnbull, *J.C.S. Perkin I*, 1977, 1229.
[134] G. V. Boyd, T. Norris, and P. F. Lindley, *J.C.S. Perkin I*, 1976, 1673; *ibid.*, 1977, 965.
[135] P. Skrabal and M. Hohl-Blumer, *Helv. Chim. Acta*, 1976, **59**, 2906; P. Luger, J. Malkowski, and P. Skrabal, *ibid.*, 1977, **60**, 1545.
[136] G. G. Alange, A. J. Banister, B. Bell, and P. W. Millen, *Inorg. Nuclear Chem. Letters*, 1977, **13**, 143.
[137] D. H. R. Barton and W. A. Bubb, *J.C.S. Perkin I*, 1977, 916.
[138] E. Gentric, J. Lauransan, C. Roussel, and J. Metzger, *Tetrahedron Letters*, 1977, 251.

by itself because of the weak hydrogen bonding tendency on the part of its NH group), when mixed with an equal proportion of the benzothiazolinethione (161) (which is fairly strongly dimerized, on account of the very acidic NH group, and in spite of its weakly basic thione S-atom), forms a hydrogen-bonded dimer pair as shown, with an association constant almost twice the value of that for the homo-associated pair $(161)_2$. I.r. spectroscopy was used to determine the equilibrium constants, and the results agreed well with the predictions based on calculation.

The structures (162; R = alkyl, X = S or Se) are assigned, on the basis of ^{13}C n.m.r. evidence, to the products of alkylation, followed by demethylation, of 1-aryl-5-methylthio- and -methylseleno-tetrazoles. The same technique is used to distinguish between tautomeric thiol and thione structures, and the corresponding selenol–selenones, in a series of azoles.[139]

An interesting case of tautomerism arises with the bicyclic phosphorus compound (163). The exchange is sufficiently rapid at room temperature that the n.m.r. spectra indicate overall two-fold symmetry of the molecule; the signals separate at low temperature (−60 °C; ^{13}C n.m.r. at 22.63 MHz). Compound (163) arises by disproportionation of the diazaphospholane (164), as shown.[140]

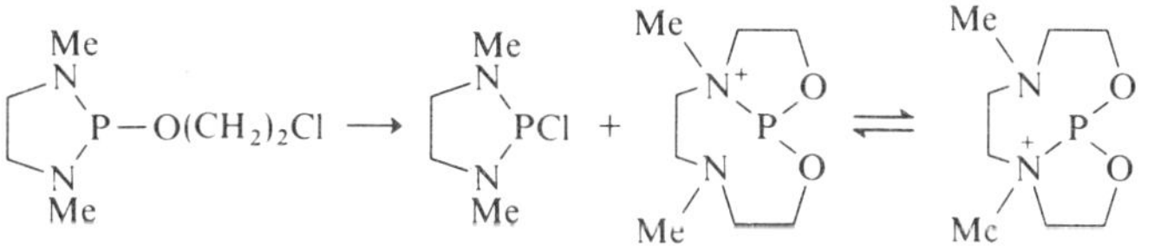

(164) (163)

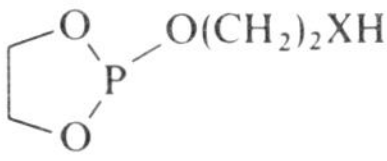

(165)

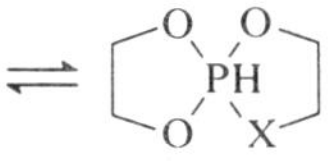

(166)

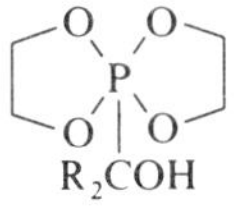

(167)

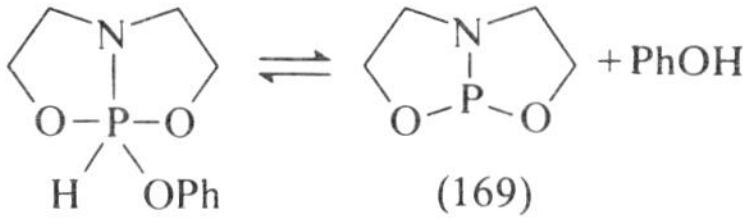

(168) (169)

Several 1977 papers illustrate the propensity for P^{III} to expand its valency to P^{V}, particularly when the atom is contained in a five-membered ring. Equilibria have been studied for a variety of systems of the type (165)⇌(166) (X = O or NR), and the results were rationalized using Pearson's HSAB treatment.[141] The *P*-H phosphoranes (166; X = O) react with aldehydes and ketones to form products (167),[142] which have been studied by i.r. and n.m.r. spectroscopy.[143] Bicyclic phosphoranes

139 J. R. Bartels-Keith, M. T. Burgess, and J. M. Stevenson, *J. Org. Chem.*, 1977, **42**, 3725.
140 H. Sliwa and J. P. Picavet, *Tetrahedron Letters*, 1977, 1583, and personal communication.
141 A. Munoz, *Bull. Soc. chim. France*, 1977, 728.
142 H. Germa and R. Burgada, *Bull. Soc. chim. France*, 1975, 2607.
143 H. Germa, M. Bon, and F. Mathis, *Bull. Soc. chim. France*, 1977, 508.

containing nitrogen (*e.g.* 168) are reversibly formed from the P^{III} compounds (169).[144]

6 Six-membered Rings

The relationship between odd alternant hydrocarbon anions and six-membered mesomeric betaines has been used by Ramsden[145] to formulate general rules for predicting whether 1,3-dipolar reactivity can be expected. The statement is made, that 'neutral heterosystems isoelectronic with irreducible odd alternant hydrocarbon anions cannot be represented by non-polar structures if the lone-pair originates at an unstarred position.' This arises from simple parity arguments. The treatment places, for example, the pyrazinium derivative (170) into the class of fully conjugated mesomeric betaines, entirely distinct from the pyrimidine (171), which is a cross-conjugated dipole. In fact, the former behaves as a 1,3-dipole, the latter as a 1,4-dipole. Sammes's group exploited the 1,4-dipolar reactivity of compounds of type (171) earlier,[146] and they have now presented further examples (172→173), in which intramolecular cyclization of olefinic, acetylenic, or cyano substituents to pyrimidine rings – not necessarily dipolar – leads to useful syntheses of annulated pyridines and pyrimidines.[147]

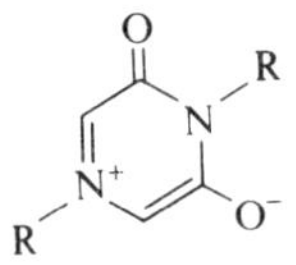

(170)

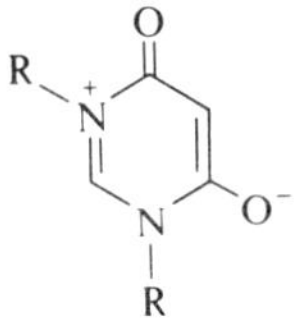

(171)

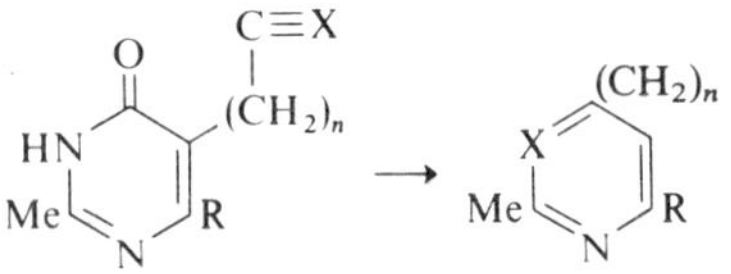

(172) X = CH or N
R = OH, Me or Ph

(173)

Considerable interest is still shown in conformational analysis, and several papers have illustrated the great value of the information provided by ^{13}C n.m.r. Some years ago, 1-chloropiperidine was studied by ^{1}H n.m.r., and a barrier to ring inversion of 56.5 kJ mol^{-1} was found.[148] This technique gave no information on the N-inversion process, except that it had to be faster. The ^{13}C n.m.r. has now shown that, at −32 °C, the Cl-axial form is *ca.* 4% abundant ($\Delta G°$ *ca.* 6 kJ mol^{-1}), with $\Delta G^{\ddagger}$ 43 kJ mol^{-1} for the N-inversion process.[149] For such heavily-biased equilibria the dynamic ^{13}C n.m.r. method is particularly useful, as illustrated in the case of the perhydro-1,2,4-triazines and -1,3,4-thiadiazines.[150] The conformations of 1-methylpiperidinium salts in water provided evidence for solvation of the NH^{+} group as a factor in influencing equilibria.[151]

[144] D. Houlla, F. H. Osman, M. Sanchez, and R. Wolf, *Tetrahedron Letters*, 1977, 3041.
[145] C. A. Ramsden, *J.C.S. Chem. Comm.*, 1977, 109.
[146] *Ann. Reports* (*B*), 1975, **72**, 312.
[147] L. B. Davies, P. G. Sammes, and R. A. Watt, *J.C.S. Chem. Comm.*, 1977, 663.
[148] J. B. Lambert, W. L. Olivers, and B. S. Packard, *J. Amer. Chem. Soc.*, 1971, **93**, 933.
[149] F. A. L. Anet and I. Yavari, *Tetrahedron Letters*, 1977, 3207.
[150] A. R. Katritzky, R. C. Patel, and D. M. Read, *Tetrahedron Letters*, 1977, 3803.
[151] E. L. Eliel, C. Y. Yen, and G. Zuniga Juaristi, *Tetrahedron Letters*, 1977, 2931.

Further use of ion cyclotron resonance to determine gas-phase basicities, and hence to obtain estimates of tautomeric equilibrium constants, has been made.[152] 1*H*-2- and 4-Pyridinethione structures, strongly favoured in polar solvents, are of minor importance in the gas phase. 3-Acetyl-2-ethoxy-3,4,5,6-tetrahydropyridine (174a) exists practically completely as such, in $CDCl_3$, but the similarly constructed furopyridine (175) strongly prefers the NH form (b). It is suggested that a destabilizing interaction exists between the lone pairs of the heteroatoms in (175a); this is avoided in (174a) by rotation of the ethoxy group.[153]

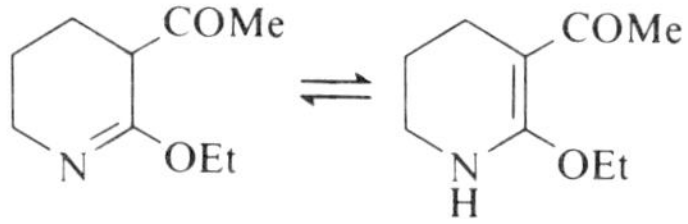

(174a) (174b)

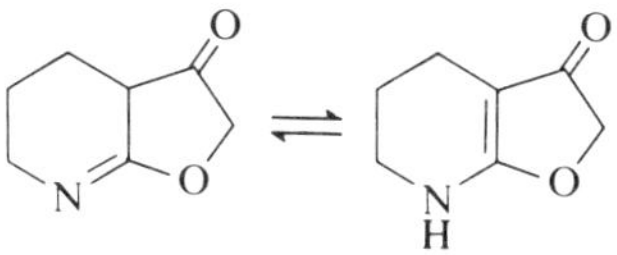

(175a) (175b)

A number of papers featured the thermal and photochemical destruction of heteroaromatic six-membered rings. The irradiation of [1,4-$^{15}N_2$]-1,2,4,5-tetrazine under a variety of conditions (exciting both singlet and triplet states) produced N_2 and HCN, but no C_2H_2 or doubly-labelled N_2, suggesting that 1,4-bonding plays no significant role in the photodecomposition of tetrazine.[154] The fused 1,2,3-triazine (176) loses nitrogen to form the nitrile (177).[155] Pyridine-1-oxide, when irradiated at 10 K in an argon matrix forms an unstable isocyanide, the anion (178) of which is much more stable, being formed if the irradiation is carried out at room temperature in aqueous base.[156] Pyridazine-1,2-dioxide (179) has been found (after a false start[157]) to be isomerized to the dihydroisoxazoloisoxazole (180).[158] It is tempting to ascribe this to cleavage of (179) to 1,4-dinitrosobutadiene, which then undergoes serpentine bicyclization in a manner for which several precedents

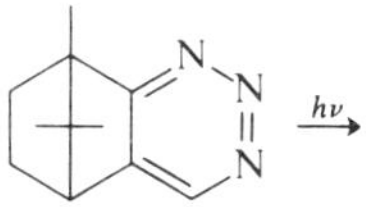

(176)

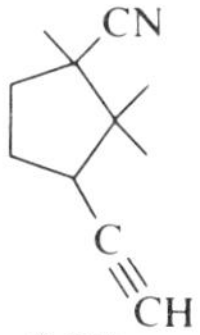

(177)

$^{-}O{-}CH{=}CH{-}CH{=}CH{-}NC$

(178)

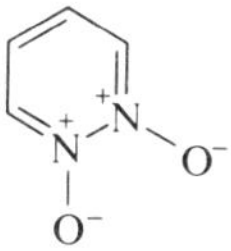

(179)

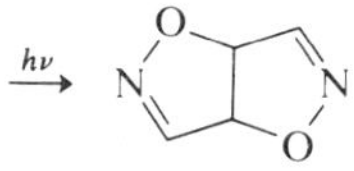

(180)

[152] C. B. Theissling, N. M. M. Nibbering, M. J. Cook, S. El-Abbady, and A. R. Katritzky, *Tetrahedron Letters*, 1977, 1777. See also M. J. Cook, A. R. Katritzky, M. Taagepera, T. D. Singh, and R. W. Taft, *J. Amer. Chem. Soc.*, 1976, **98**, 6048.

[153] H. Sliwa and L. Delaunay, *Tetrahedron Letters*, 1977, 2793.

[154] D. S. King, C. T. Denny, R. M. Hochstrasser, and A. B. Smith, *J. Amer. Chem. Soc.*, 1977, **99**, 271.

[155] I. Ito, N. Oda, S. I. Nagai, and Y. Kudo, *Heterocycles*, 1977, **8**, 319.

[156] O. Buchardt, J. J. Christensen, C. Lohse, J. J. Turner, and I. R. Dunkin, *J.C.S. Chem. Comm.*, 1977, 837.

[157] H. Arai, A. Ohsawa, K. Saiki, and H. Igeta, *J.C.S. Chem. Comm.*, 1977, 133.

[158] H. Arai, A. Ohsawa, K. Saiki, H. Igeta, A. Tsuji, T. Akimoto, and Y. Iitaka, *J.C.S. Chem. Comm.*, 1977, 856.

exist.[159] Vacuum pyrolysis of benzo[*c*]cinnoline derivatives has provided further examples of aza-biphenylenes.[160]

An attempted halogen–lithium exchange in the acetal (181) led, after hydrolytic work-up, to the pyridone (182). The proposed explanation, which was consistent with deuterium-labelling experiments, involved metallation at the pyridine 3-position indicated, followed by cleavage to the nitrile (183), and recyclization.[161] In the reaction of the pyridinium salt (184) with a secondary amine, it is the pyridine 2-position which is deprotonated; the ylid cyclizes to the ester group, and the pyridine ring is then opened, leaving a 4-isoxazolinone (185).[162]

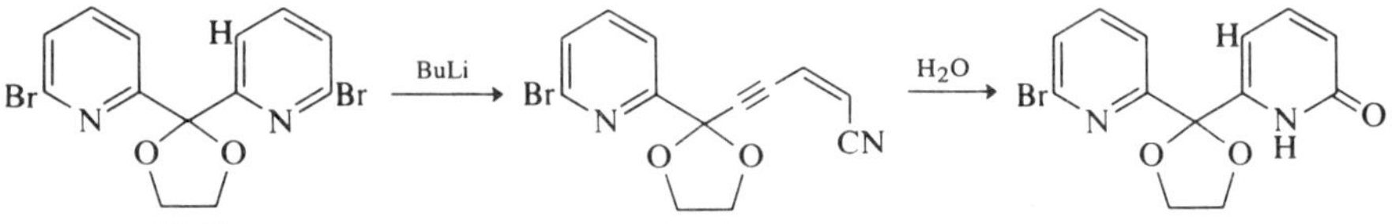

(181) (183) (182)

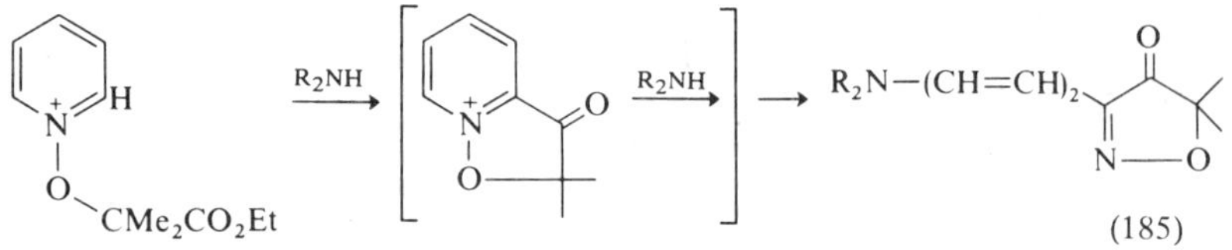

(184) (185)

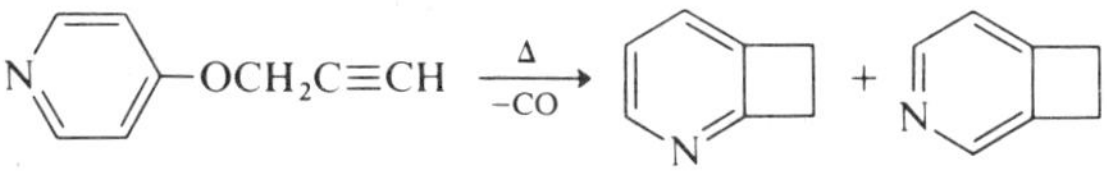

(186) (187)

The pyrolysis of aryl propargyl ethers provides a goldmine of interesting chemistry. The two isomeric cyclobutapyridines (186) and (187) (ratio 2 : 1, 50% overall yield) are formed in a curious reaction (the mechanism of which requires fuller investigation) from 4-propargyloxypyridine at 550 °C.[163]

Abramovitch *et al.*[164] have reported the preparation and some reactions of 1-aryloxypyridinium salts and 1-aryloxy-2-pyridones. The salts, prepared from the *N*-oxides and diaryliodonium fluoborates, are unstable, and rearrange, usually under the influence of bases. With phosphorus oxychloride the pyridones form pyrido[2,3-*b*]benzofurans.

159 see, *e.g.* G. Maier and M. Wiessler, *Tetrahedron Letters*, 1969, 4987. A recent dihydro-pyrrolopyrrole (R. M. Carr, R. O. C. Norman, and J. M. Vernon, *J.C.S. Chem. Comm.*, 1977, 854) also conforms to the same structure type, and may be formed analogously.

160 S. Kanoktanaporn and J. A. H. MacBride, *Tetrahedron Letters*, 1977, 1817.

161 G. R. Newkome, J. D. Sauer, and S. K. Staires, *J. Org. Chem.*, 1977, **42**, 3524.

162 H. Sliwa and A. Tartar, *Tetrahedron Letters*, 1977, 311.

163 J. M. Riemann and W. S. Trahanovsky, *Tetrahedron Letters*, 1977, 1867.

164 R. A. Abramovitch and M. N. Inbasekaran, *Tetrahedron Letters*, 1977, 1109; R. A. Abramovitch, G. Alvernhe, and M. N. Inbasekaran, *ibid.*, p. 1113.

The full paper[165] has appeared on the preparation and reactions of 1,1′-bipyridinium salts, which have been mentioned in these Reports earlier.[166]

The pyridine-*N*-imine derivatives (188) undergo a variety of thermal and photochemical cyclization reactions, an outline of which is presented in Scheme 6.[167]

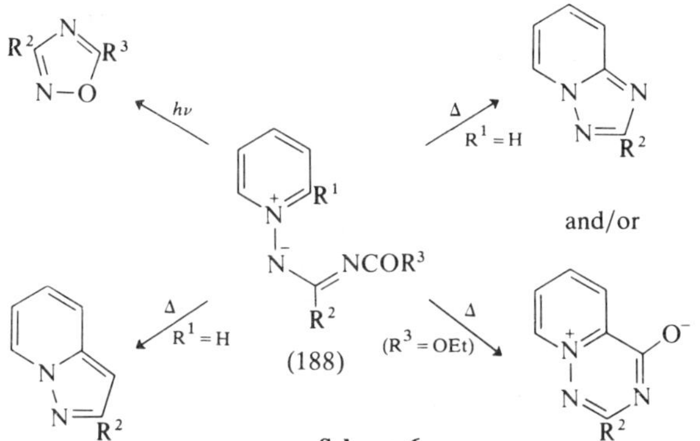

Scheme 6

[1,2,4]-Triazino[1,6,5-*ji*]quinolines (189) can be prepared by *N*-amination, followed by cyclization, of 8-acylaminoquinolines (190). The red products behave as 1,3- (or 1,11-) dipoles, (189; R = H) giving the cyclazine derivative (191) with dimethyl acetylene dicarboxylate.[168]

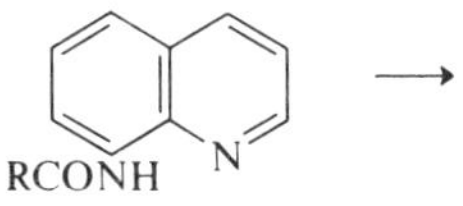

(190)

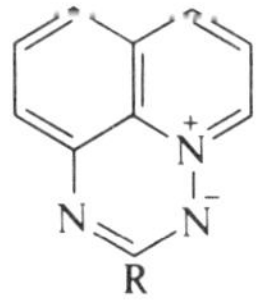

(189)

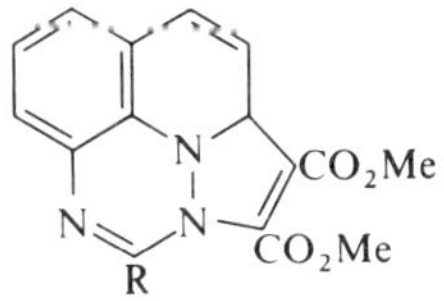

(191)

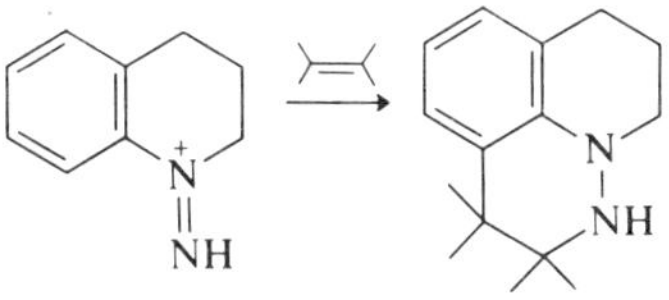

(192)

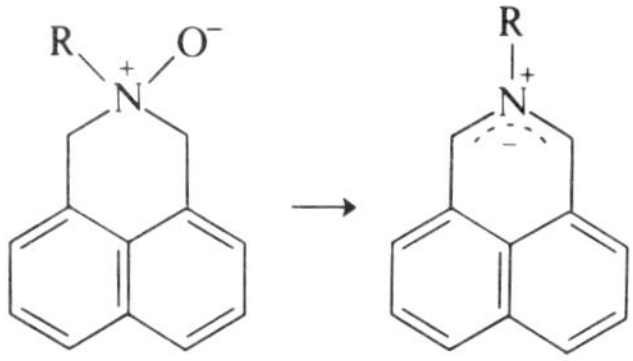

(194) (193)

The diazenium salt (192) can be obtained by electrochemical oxidation of 1-amino-1,2,3,4-tetrahydroquinoline. In acid, it adds to olefins; in base it forms a 2-tetrazene.[169] Cycloaddition reactions also trap the azomethine ylid (193), generated by acetic anhydride–triethylamine dehydration of the *N*-oxide (194).[170]

[165] M. P. Sammes, H. K. Wah, and A. R. Katritzky, *J.C.S. Perkin I*, 1977, 327.
[166] *Ann. Reports (B)*, 1975, **72**, 269; *ibid.*, 1976, **73**, 429.
[167] A. Kakehi, S. Ito, K. Uchikama, Y. Konno, and K. Kondo, *J. Org. Chem.*, 1977, **42**, 443.
[168] Y. Tamura, Y. Miki, H. Hayashi, Y. Sumida, and M. Ikeda, *Heterocycles*, 1977, **6**, 281.
[169] G. Cauquis, B. Chabaud, and Y. Gohee, *Tetrahedron Letters*, 1977, 2583.
[170] M. Ikeda, Y. Miki, S. Kaita, Y. Nishikawa, and Y. Tamura, *J.C.S. Perkin I*, 1977, 44.

The preparation[171] and stability[172] of *N*-ethoxycarbonyl-1-pyrindine (195) is discussed by Anastassiou *et al.* They claim that the ethoxycarbonyl grouping robs the heterocyclic system of none of its aromaticity, in contrast to what is observed with 1-ethoxycarbonylazonine.

Dihydropyridines are featured in a number of noteworthy papers. 1-Carbomethoxy-1,2-dihydropyridine was the starting-point for a synthesis of the dihydroazabarrelene (196), which is a remarkably unstable compound, decomposing to benzene in a few hours.[173] Surprisingly, organomagnesium and organocadmium reagents attack 1-acylpyridinium salts at the 2-position so quickly that competition from attack at the carbonyl group is not serious. The acid chloride can even be added to a mixture of the organometallic reagent and the pyridine: competition from the acid chloride, under these conditions, is also negligible.[174] 1,2-Dihydropyridines have been used in syntheses of di- and tetra-hydroazocines, and of pyrrolizidines.[175] A new technique for preparing Reissert compounds, using trimethylsilyl cyanide and aluminium chloride, allows their formation in water-free solvents, and so permits use of acid chlorides which are sensitive to hydrolysis.[176] Spiro-1,4-dihydropyridines (197) are formed by addition of dimethyl acetylenedicarboxylate to a wide variety of heterocyclic imines.[177]

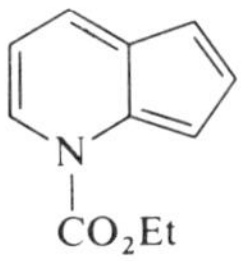

(195)

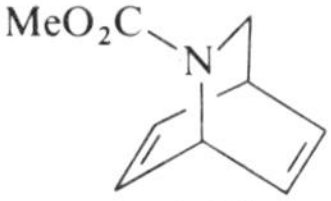

(196)

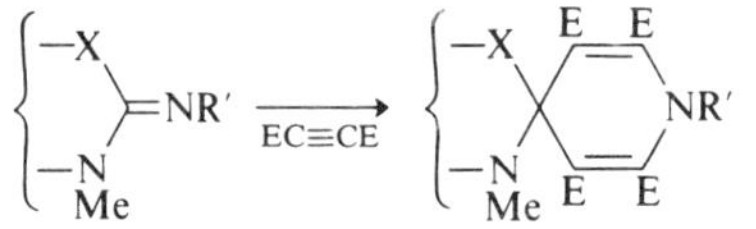

(197)

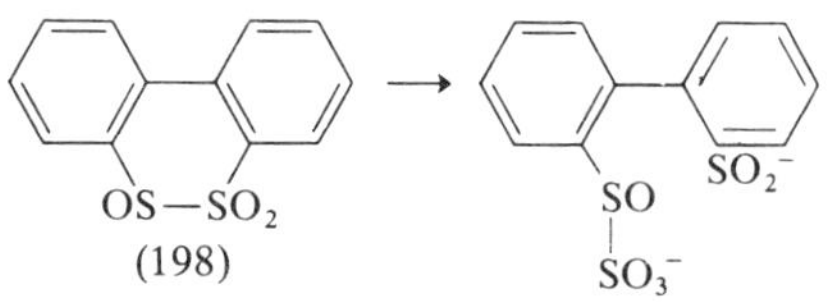

(198) (199)

X = S or NMe, E = CO_2Me

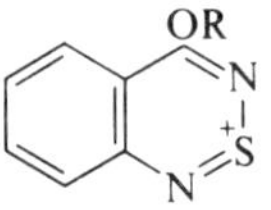

(200)

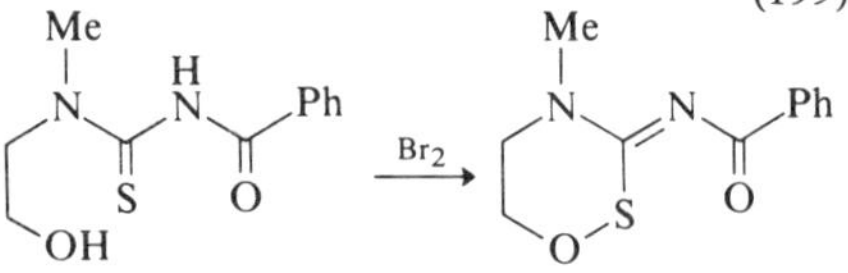

(202) (201)

Chau and Kice have found that the dibenzodithiin trioxide (198) with sulphite ion produces the Bunte salt *S*-oxide (199), an unusual and unstable type of

[171] A. G. Anastassiou, S. J. Girgenti, R. C. Griffith, and E. Reichmanis, *J. Org. Chem.*, 1977, **42**, 2651.
[172] A. G. Anastassiou and E. Reichmanis, *J. Amer. Chem. Soc.*, 1977, **99**, 7392.
[173] H. Sliwa and Y. le Bot, *Tetrahedron Letters*, 1977, 4129.
[174] R. E. Lyle, J. L. Marshall, and D. L. Comins, *Tetrahedron Letters*, 1977, 1015.
[175] P. S. Mariano, M. E. Osborn, D. Dunaway-Mariano, B. C. Gunn, and R. C. Pettersen, *J. Org. Chem.*, 1977, **42**, 2903.
[176] S. Ruchiwarat, N. Phadungkul, M. Chuankamnerdkarn, and C. Thebtaranonth, *Heterocycles*, 1977, **6**, 43.
[177] H. Quast and E. Spiegel, *Tetrahedron Letters*, 1977, 2705.

structure.[178] Interesting sulphur-containing six-membered rings that were noted were the 2,1,3-benzothiadiazinium system (200),[179] and the 1,2,4-oxathiazine (201),[180] the latter arising by bromine oxidation of the thiourea (202). The synthetic utility of 2-chloro-1,3-dithian has been investigated.[181]

The rearrangement of optically active 10-thia-anthracenes (203) occurs with predominant racemization, but some retention of optical activity, in the product (204).[182]

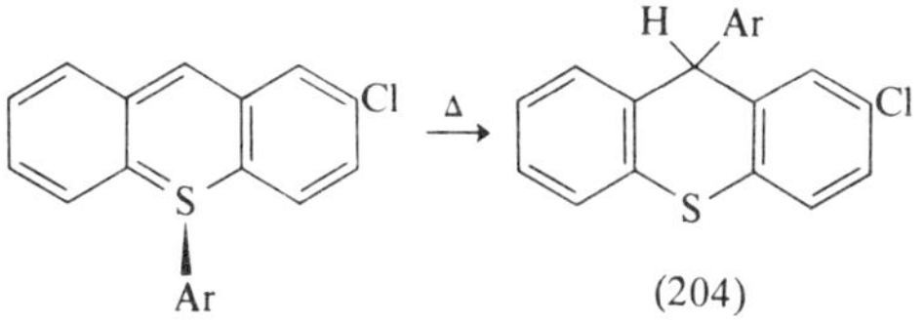

(203) (204)

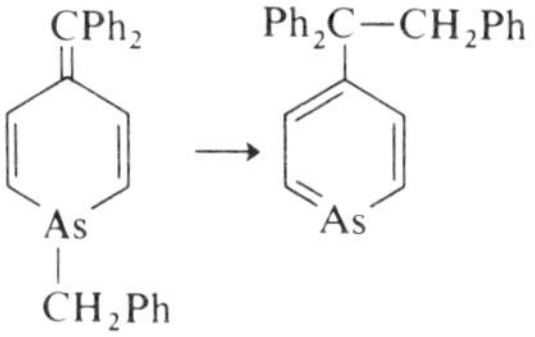

(205)

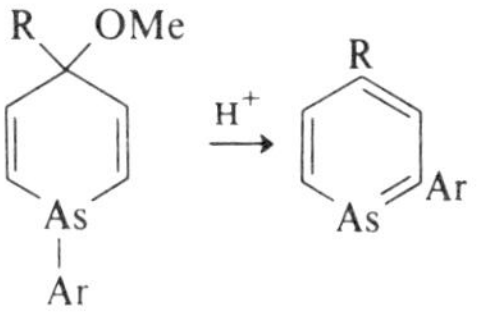

(206)

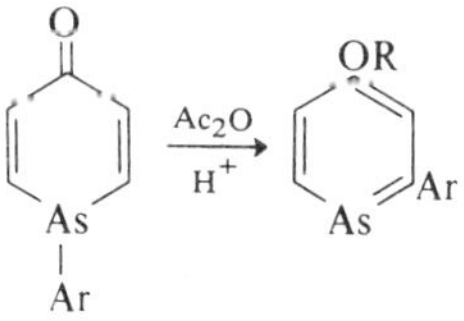

(207) (208) R = Ac
(209) R = H

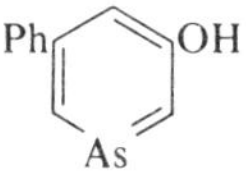

(210)

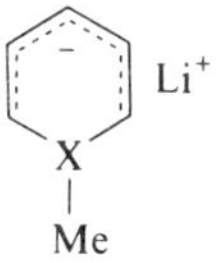

(211) X = P, As, Sb
(212) X = CMe

Research on six-membered rings with more unusual heteroatoms has been the subject of much activity, and here too several rearrangements have been discovered. The benzyl group migrates, on heating, from the arsenic atom to the 4-substituent in (205); this has been known for some time in the phosphorin series.[183] Aryl groups undergo 1–2 migration with acid catalysis in (206)[184] and (207);[185] hydrolysis of the acetoxy-derivative (208) forms the hydroxy-compound (209), which appears to exist as such,[185] although alkylation of 4-hydroxyarsenin

[178] M. M. Chau and J. L. Kice, *J. Org. Chem.*, 1977, **42**, 3103, 3265.
[179] W. Kosbahn and H. Schäfer, *Angew. Chem. Internat. Edn.*, 1977, **16**, 780.
[180] S. Solyom, P. Sohar, L. Toldy, A. Kalman, and L. Parkanyi, *Tetrahedron Letters*, 1977, 4245.
[181] C. Kruse, N. Broekhof, A. Wijsman, and A. van der Gen, *Tetrahedron Letters*, 1977, 885; E. C. Taylor and J. L. LaMattina, *ibid.*, 1977, 2077.
[182] C. A. Maryanoff, K. S. Hayes, and K. Mislow, *J. Amer. Chem. Soc.*, 1977, **99**, 4412.
[183] G. Märkl and J. B. Rampal, *Tetrahedron Letters*, 1977, 2569.
[184] G. Märkl and R. Liebl, *Angew. Chem. Internat. Edn.*, 1977, **16**, 637.
[185] G. Märkl and J. B. Rampal, *Tetrahedron Letters*, 1977, 3449.

gives the 1-alkyl-4-one.[183] The 'phosphaphenol' (210), too, appears to be preferred over a ketonic tautomer.[186]

Diels–Alder addition and retro-addition of dimethyl acetylenedicarboxylate and methyl propiolate to arsabenzene and its 2-methyl derivative have been studied. Some dimethyl phthalate is produced, indicating that 1-arsabarrelene derivatives are not exclusively formed, but that also some of the 2-arsa compounds arise; these can lose HC≡As, to produce the phthalate.[187] Methyl-lithium adds the alkyl group to the heteroatoms of phospha-, arsa-, and stiba-benzene, in contrast to pyridine, to give the anions (211). The chemical shifts of the protons on the pentadienyl fragments are very similar to those in (212), suggesting that little or no stabilizing interaction is derived from the presence of the heteroatom in (211).[188]

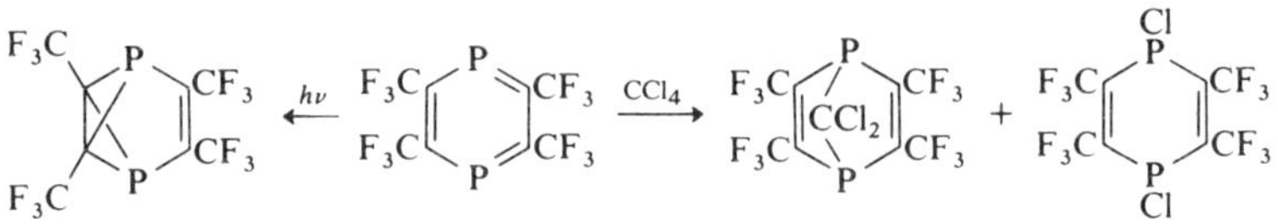

(214) (213) (215) (216)

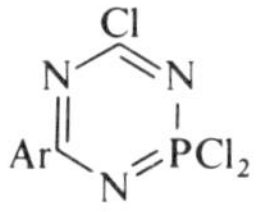

(217)

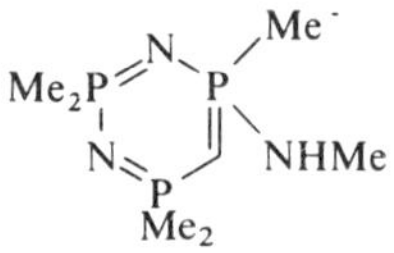

(218)

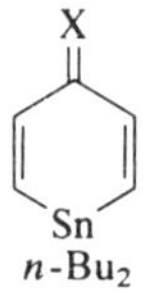

(219) X = O, CR_2

Irradiation of the 1,4-diphosphorin (213) is reported to form the diphosphabenzvalene (214), a remarkably stable compound,[189] while with carbon tetrachloride it reacts in an odd way, giving the 1,4-bridged compound (215), and the unstable dichloro-derivative (216), which was not isolated.[190] Representatives of C_2N_3P (217)[191] and CN_2P_3 (218)[192] ring systems were noted. With group IV elements, the stannins (219), with sp^2-hybridization of all the ring carbon atoms, were prepared,[193] and a report on the generation of a silabenzene appeared: this is mentioned in Chapter 6.

7 Seven-membered and Larger Rings

Like the thiirenes mentioned earlier, so 1-benzothiepins are stabilized appreciably by quaternization at sulphur. The benzothiepinium salt (220) has a half-life of 69 hours at 80 °C (CD_3NO_2), while (221) is nearly half gone in three hours (80 °C, CCl_4). The sulphoxide (222) is much less stable than either, the sulphone (223) much more.[194]

186 G. Märkl, G. Adolin, F. Kees, and G. Zander, *Tetrahedron Letters*, 1977, 3445.
187 A. J. Ashe and H. S. Friedman, *Tetrahedron Letters*, 1977, 1283.
188 A. J. Ashe and T. W. Smith, *Tetrahedron Letters*, 1977, 407.
189 Y. Kobayashi, S. Fujino, H. Hamana, I. Kumadaki, and Y. Hanzawa, *J. Amer. Chem. Soc.*, 1977, **99**, 8511.
190 Y. Kobayashi, I. Kumadaki, H. Hamana, and S. Fujino, *Tetrahedron Letters*, 1977, 3057.
191 P. P. Kornuta, N. V. Kolotilo, and L. N. Markovskii, *Zhur. obshch. Khim.*, 1977, **47**, 342.
192 R. T. Oakley and N. L. Paddock, *Canad. J. Chem.*, 1977, **55**, 3651.
193 G. Märkl and J. B. Rampal, *Tetrahedron Letters*, 1977, 2325.
194 H. Hofmann and A. Molnar, *Tetrahedron Letters*, 1977, 1985.

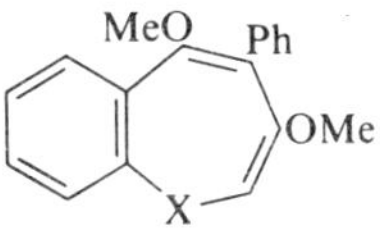

(220) X = $\overset{+}{S}$Me
(221) X = S
(222) X = SO
(223) X = SO_2

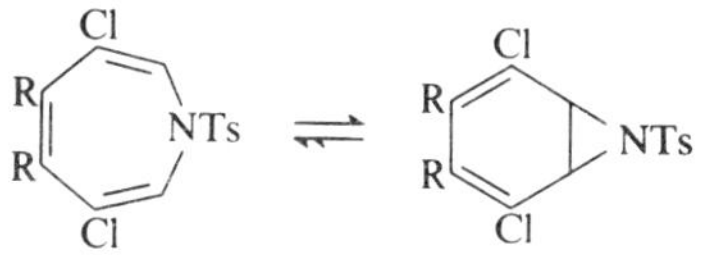

(224) (a) R = H
(b) R = CO_2Me

(225)

The equilibrium between 1-tosylazepines (224) and the corresponding benzene *N*-tosylimines (225) has been investigated by ^{13}C and ^{1}H n.m.r. spectroscopy. Although the signals from the imine (225a) were not directly observed, line-broadening phenomena of the azepine bands revealed its presence (*ca.* 1%), and gave $\Delta G^{\ddagger}$ *ca.* 46 kJ mol^{-1} for the activation barrier. The imine (225b) was more abundant at equilibrium (*ca.* 10% at −70 °C, in [$^2H_6$]-acetone).[195]

A full paper has appeared on Becker and Gustafsson's work on the action of amines on the 'spiro-benzoxete' oxidation dimers of 2,4-di-t-alkylphenols.[196] The azepinone product (226) is rapidly oxidized ($NaBiO_3$) to the tricyclic compound (227); this forms a tetracyclic photo-isomer which can be oxidized to the spiro-benzofuranone (228).[197] Some further work has been reported on the interesting rearrangement of 3,4-diazanorcaradienes to 6*H*-1,4-diazepines, which was mentioned in these Reports last year,[198] and some ambiguities in the details of the various steps have been clarified.[199] Details of the preparation of 1*H*- and 3*H*-1,2-benzodiazepines have been published.[200]

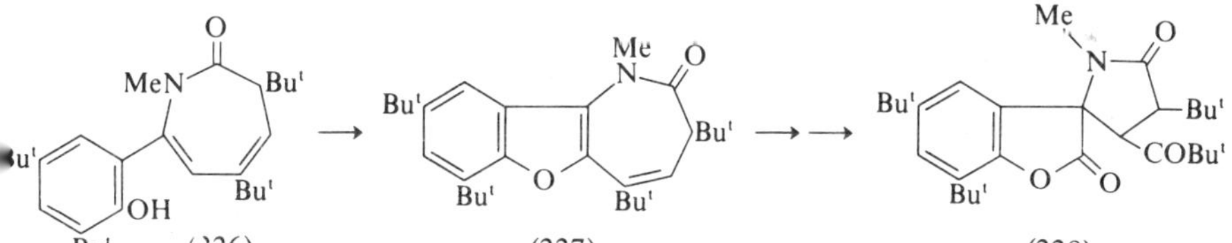

(226) (227) (228)

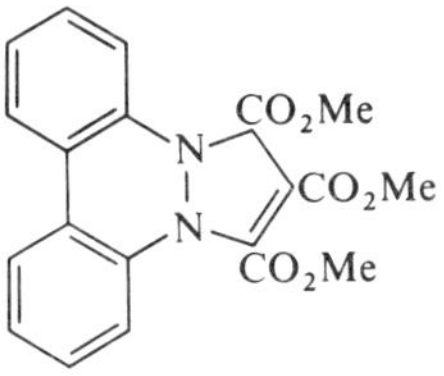

(229)

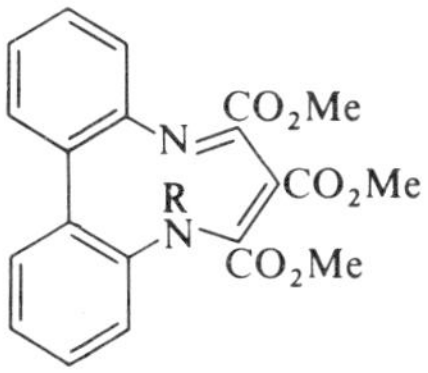

(230) R = Li
(231) R = H
(232) R = Me

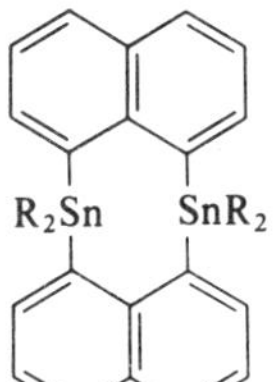

(233) R = Me, Ph

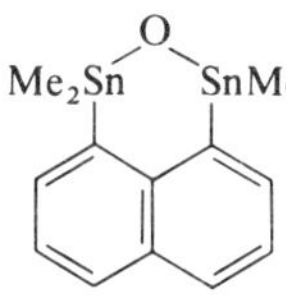

(234)

[195] H. Prinzbach, H. Babsch, H. Fritz, and P. Hug, *Tetrahedron Letters*, 1977, 1355.

[196] *cf. Ann. Reports* (*B*), 1976, **73**, 271. The dimers have recently been shown to possess oxepino[2,3-*b*]-benzofuran structures: H. Meier, H. P. Schneider, A. Ricker, and P. B. Hitchcock, *Angew. Chem. Internat. Edn.*, 1978, **17**, 121.

[197] H. D. Becker and K. Gustafsson, *J. Org. Chem.*, 1977, **42**, 2966.

[198] *Ann. Reports* (*B*), 1976, **73**, 271.

[199] G. Reissenweber and J. Sauer, *Tetrahedron Letters*, 1977, 4389; H. D. Fühlhuber and J. Sauer, *ibid.*, 1977, 4393.

[200] T. Tsuchiya, J. Kurita, and V. Snieckus, *J. Org. Chem.*, 1977, **42**, 1856; *cf. Ann. Reports* (*B*), 1974, **71**, 354.

A neat procedure for preparing the dizenzo[*f,h*][1,5]diazonine system is reported,[201] in the base-promoted cleavage of a transannular bond in the dihydropyrazole derivative (229). The lithium salt (230) can be protonated and alkylated, to give the neutral species (231) and (232).

1,8-Dilithionaphthalene, on reaction with dichlorostannanes, gave the *peri*-naphtho-fused distannocins (233). *X*-ray crystallography showed the central ring to be strongly distorted from planarity, but the molecule (233; $R = Me$) was conformationally mobile: down to −100 °C the methyl protons appeared as a singlet in the n.m.r. spectrum. The oxadistannepin (234) was also prepared.[202]

The dimerization of 3-(1-imidazolyl and 1-benzimidazolyl)quinolines, by, *e.g.* metalation (Pr^i_2NLi) followed by oxidation, yields, *e.g.* interesting fused diazocines (235).[203] Other ring assemblies of note include the circulenes and helicenes produced by oligomerization by acid of quinones, which have been studied by *X*-ray crystallography. The 'closed tetramer' (236) was found to be planar.[204] The preparation of the 'phenanthrolinophane' (237)[205] parallels that of the corresponding hexa-aza compound (238), reported earlier.[206]

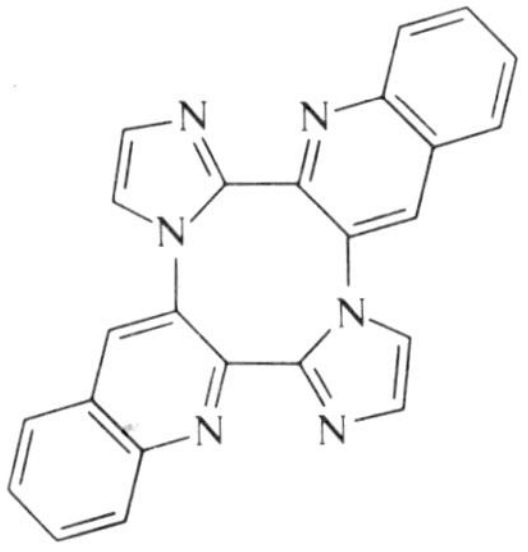

(235)

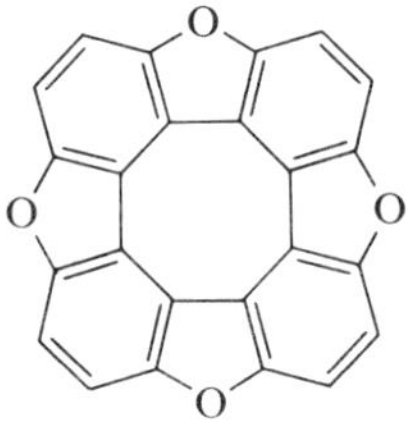

(236)

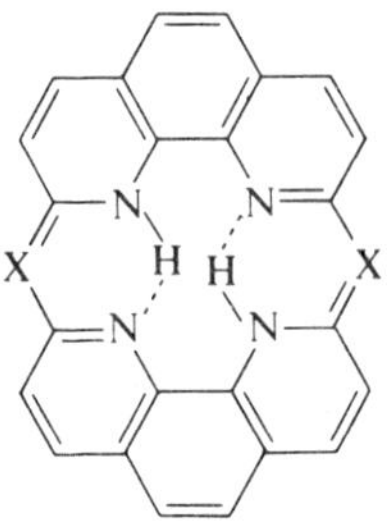

(237) X = CH
(238) X = N

A method for the repetitive ring-expansion of cyclic allylsulphonium allylides is described from two laboratories.[207,208] For example (Scheme 7), 2-vinyltetra-

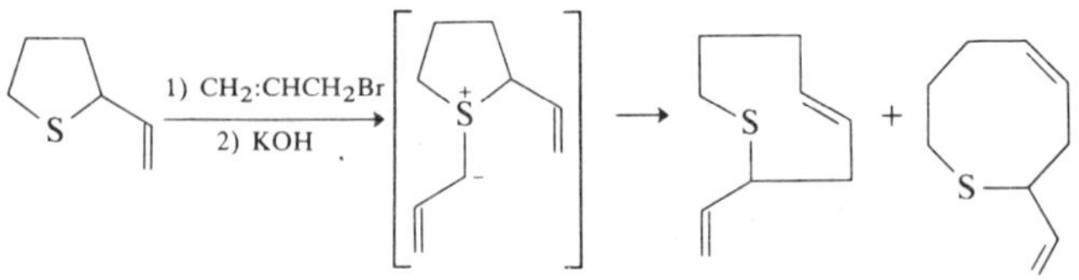

(239)

Scheme 7

[201] D. G. Farnum and K. Rasheed, *J. Org. Chem.*, 1977, **42**, 573. The ring system (230) is incorrectly named and numbered, in this paper.
[202] J. Meinwald, S. Knapp, and T. Tatsuoka, *Tetrahedron Letters*, 1977, 2247.
[203] T. Kaufmann, D. Tigler, and A. Woltermann, *Tetrahedron Letters*, 1977, 741.
[204] J. E. Bug, H. Erdtman, H. E. Högberg, B. Karlsson, A. M. Pilotti, and A. C. Söderholm, *Tetrahedron Letters*, 1977, 1831.
[205] S. Ogawa, *J.C.S. Perkin I*, 1977, 214.
[206] S. Ogawa, T. Yamaguchi, and N. Gotoh, *J.C.S. Perkin I*, 1974, 976.
[207] R. Schmid and H. Schmid, *Helv. Chim. Acta*, 1977, **60**, 1361.
[208] E. Vedejs, M. J. Mullins, J. M. Renga, and S. P. Singer, *Tetrahedron Letters*, 1978, 519.

hydrothiophen can be converted into a 2-vinylthiocin (239), and the process could be repeated to produce 11-, 14-, and 17-membered monothia-cycloalkenes.

Although 1,4-diazabicyclo[2,2,2]octane is a well-known compound, its homologues with bridges of other lengths are difficult to obtain. Alder *et al.*[209] have now described a procedure, starting from bridged hydrazines, which promises to be quite versatile. The formation of 1,4-diazabicyclo[3,3,3]undecane (240) is shown in Scheme 8.

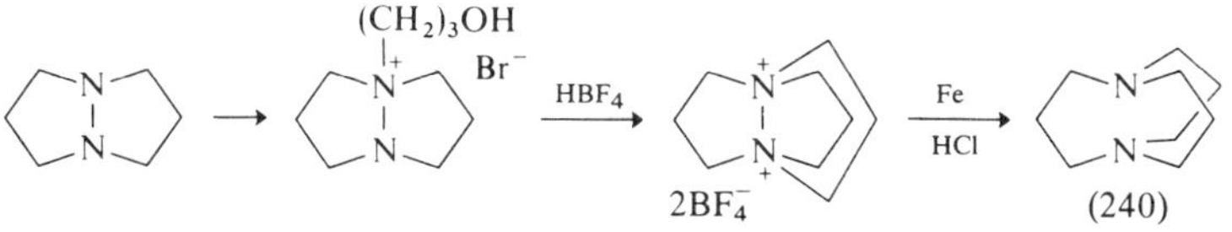

(240)

Scheme 8

Polydentate ligands containing oxygen, nitrogen, and sulphur are an active area of research, and the preliminary notes of previous years are being followed by detailed papers. Cram's series of six[210] are particularly noteworthy, as are Stoddart's reports on chiral crown ethers[211] and on macro-bicyclic polyethers with bridgehead carbon atoms,[212] in which the question of 'in-out' isomerism is discussed. An 'in-out – out-in' isomerization process has been observed in this type of compound: the chiral 'cryptand' (241) shows two bridgehead carbon signals at −70 °C, but only one at +112 °C, and the achiral homologue (242) also isomerized, apparently much more rapidly.[213]

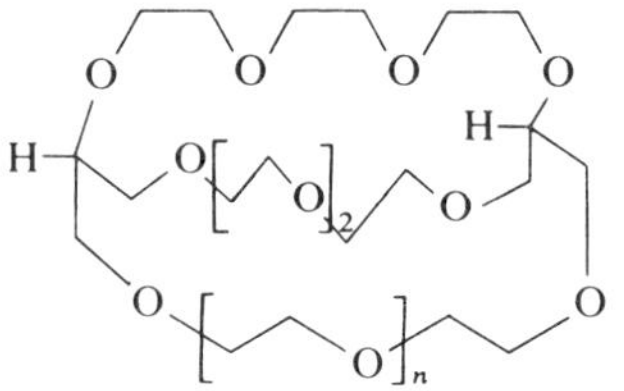

(241) $n = 1$
(242) $n = 2$

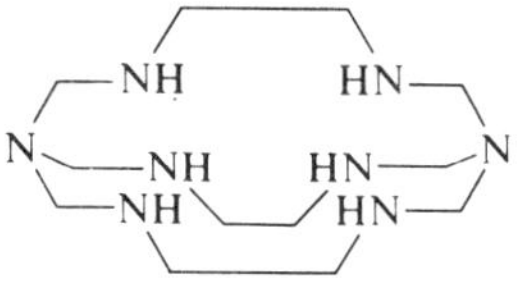

(243)

An octa-aza cage compound (243) (for which the trivial name 'sepulchrate' was coined, in keeping with the somewhat morbid nomenclature for this type of compound) has been assembled by reaction of the tris(ethylenediamine)cobalt(III) ion with formaldehyde and ammonia.[214] The nitrogen atoms in the bridges preserve their (chiral) octahedral arrangement about the metal atom, disinterment of which was not accomplished.

[209] R. W. Alder, R. B. Sessions, J. M. Mellor, and M. F. Rawlins, *J.C.S. Chem. Comm.*, 1977, 747.
[210] Part 6: R. C. Helgeson, T. L. Tarnowski, J. M. Timko, and D. J. Cram, *J. Amer. Chem. Soc.*, 1977, **99**, 6411.
[211] I. J. Burden, A. C. Coxon, J. F. Stoddart, and C. M. Wheatley, *J.C.S. Perkin I*, 1977, 220; W. D. Curtis, D. A. Laidler, J. F. Stoddart, and G. H. Jones, *ibid.*, p. 1756.
[212] A. C. Coxon and J. F. Stoddart, *J.C.S. Perkin I*, 1977, 767.
[213] B. H. Gregory, A. H. Haines, and P. Karntiang, *J.C.S. Chem. Comm.*, 1977, 918.
[214] I. I. Creaser, J. M. Harrowfield, A. J. Herlt, A. M. Sargeson, J. Springborg, R. J. Geue, and M. R. Snow, *J. Amer. Chem. Soc.*, 1977, **99**, 3181.

8 Reviews

A two-volume monograph on heterocyclic stereochemistry has been published.[215] Other aspects of general heterocyclic chemistry which have been reviewed include (2+2)-cycloaddition and cycloreversion reactions,[216] reactions involving cycloelimination of nitrogen,[217] the thermochemistry of the alkylation of tautomeric heterocycles,[218] the gas-phase thermal and photochemical decomposition of heterocycles containing O, N, or S,[219] and the rearrangements of *N*-oxides.[220] Macrocycles containing 'subheterocyclic rings' – heterocyclophanes, for the most part – have been collected.[221]

Among reviews of specific ring systems or groups of systems are two of particular interest to phosphorus chemists: 'Compounds of Two-co-ordinated Phosphorus'[222] include the λ^3-phosphorins, and the diaza- and triaza-phospholes, and 'Polycyclic Carbon–Phosphorus Heterocycles'[223] encompass a wide variety of structural types. Other topics reviewed are pyrrolodiazines with bridgehead nitrogen,[224] boraheterocycles,[225] 6,8-dioxabicyclo[3,2,1]octanes,[226] the heterocyclic analogues of prostaglandins,[227] thiiranium and thiirenium ions,[228] tetrazoles,[229] tellurophenes, [230] thienopyridines,[231] and oxazolones,[232] isoxazolidines,[233] 1,2-dioxetans,[234] and the mesomeric betaine derivatives of heteropentalenes.[235] Reviews on the electrocyclic ring-opening reactions of ethylene oxides,[236] the action of hydroxylamines and hydrazines on γ-pyrone derivatives,[237] and the photochemistry of 2*H*-azirines[238] deal with special aspects of specific heterocyclic rings. A monograph devoted to pyrroles has appeared.[239] Chemical Society publications of particular heterocyclic interest include the regular specialist reports on

215 W. L. F. Armarego, 'Stereochemistry of Heterocyclic Compounds', Parts I and II, Wiley-Interscience, New York, 1977.
216 D. N. Reinhoudt, *Adv. Heterocyclic Chem.*, 1977, **21**, 253.
217 H. Meier and K. P. Zeller, *Angew. Chem. Internat. Edn.*, 1977, **16**, 835.
218 P. Beak, *Accounts Chem. Research*, 1977, **10**, 186.
219 S. Braslavsky and J. Heicklen, *Chem. Rev.*, 1977, **77**, 473.
220 S. Oae and K. Ogino, *Heterocycles*, 1977, **6**, 583.
221 G. R. Newkome, J. D. Sauer, J. M. Roper, and D. C. Hager, *Chem. Rev.*, 1977, **77**, 513.
222 N. I. Shvetsov-Shilovskii, R. G. Bobkova, N. P. Ignatova, and N. N. Mel'nikov, *Uspekhi Khim.*, 1977, **46**, 967; *Russ. Chem. Rev.*, 1977, **46**, 514.
223 S. D. Venkataramu, G. D. MacDonell, W. R. Purdum, M. El-Deek, and K. D. Berlin, *Chem. Rev.*, 1977, **77**, 121.
224 D. E. Kuhla and J. G. Lombardino, *Adv. Heterocyclic Chem.*, 1977, **21**, 1.
225 H. C. Brown and E. Negishi, *Tetrahedron*, 1977, **33**, 2331.
226 P. P. Mundy, K. B. Lipkowitz, and G. W. Dirks, *Heterocycles*, 1977, **6**, 51.
227 E. I. Levkoeva and L. N. Yakhontov, *Uspekhi Khim.*, 1977, **46**, 1074; *Russ. Chem. Rev.*, 1977, **46**, 565.
228 G. H. Schmid, in 'Topics in Sulfur Chemistry' ed. A. Senning and P. S. Magee, Thieme, Stuttgart, 1977, vol. 3.
229 R. N. Butler, *Adv. Heterocyclic Chem.*, 1977, **21**, 223.
230 F. Fringuelli, G. Marino, and A. Taticchi, *Adv. Heterocyclic Chem.*, 1977, **21**, 119.
231 J. M. Barker, *Adv. Heterocyclic Chem.*, 1977, **21**, 65.
232 R. Filler and Y. S. Rao, *Adv. Heterocyclic Chem.*, 1977, **21**, 175.
233 Y. Takeuchi and F. Furusaki, *Adv. Heterocyclic Chem.*, 1977, **21**, 207.
234 W. Adam, *Adv. Heterocyclic Chem.*, 1977, **21**, 437.
235 C. A. Ramsden, *Tetrahedron*, 1977, **33**, 3203.
236 R. Huisgen, *Angew. Chem. Internat. Edn.*, 1977, **16**, 572.
237 C. Morin and R. Beugelmans, *Tetrahedron*, 1977, **33**, 3183.
238 P. Gilgen, H. Heimgartner, H. Schmid, and H. J. Hansen, *Heterocycles*, 1977, **6**, 143.
239 R. A. Jones and G. P. Bean, 'The Chemistry of Pyrroles' (Organic Chemistry monographs), Academic Press Inc., London, 1977, vol. 34.

alkaloids,[240] aromatic compounds,[241] and Group VI elements,[242] and a volume on the β-lactam antibiotics,[243] which should prove extremely valuable to the many active research workers in this field.

[240] 'The Alkaloids', ed. M. F. Grundon (Specialist Periodical Reports) The Chemical Society, London, 1977, vol. 7.
[241] 'Aromatic and Heteroaromatic Chemistry', ed. C. W. Bird and G. W. H. Cheeseman (Specialist Periodical Reports) The Chemical Society, London, 1977, vol. 5.
[242] 'Organic Compounds of Sulphur, Selenium, and Tellurium', ed. D. R. Hogg, (Specialist Periodical Reports) The Chemical Society, London, 1977, vol. 4.
[243] 'Recent Advances in the Chemistry of β-Lactam Antibiotics', ed. J. Elks, (Special Publication no. 28) The Chemical Society, London, 1977.

12 Alicyclic Chemistry

By A. COX
Department of Chemistry and Molecular Sciences, University of Warwick, Coventry CV4 7AL

1 Introduction

This year has seen the publication of a further volume of 'Organic Syntheses'[1] and a survey of the 1975 literature on mechanisms has appeared.[2] Reviews have been published on the chemistry of rings of various sizes,[3] intramolecular [4+2] and [3+2] cycloadditions,[4] pericyclic synthesis,[5] the Robinson Annelation[6] and strained organic molecules.[7] Finally, further work is reported on the rules for ring closure[8] and a new mechanistic criterion has appeared for early and late transition states.[9]

2 Synthesis

Three- and Four-membered Rings.—A high-yield route to cyclopropyl and cyclopropenyl ethers has appeared[10] which involves trapping of an alkoxycarbenoid as the key step. For example, treatment of a mixture of $EtOCH_2Cl$ and cyclohexene with lithium 2,2,6,6-tetramethylpiperidine (LiTMP) leads to 7-ethoxynorcarane (55%). In a related paper, an improved synthesis of cyclopropanols is reported.[11] Use is again made of LiTMP to generate simple alkoxycarbenoids which are then allowed to react with an alkene, but in this case they are formed by α-elimination of HCl from β-chloroethyl ethers. The product ethers are subsequently cleaved using n-butyl-lithium; use of acyloxycarbenes generated similarly is also reported.[12] A method has been described[13] for the conversion of $\alpha\beta$-unsaturated aldehydes and esters into cyclopropanes. The transformation (Scheme 1) involves conversion of the carbonyl oxygen into a leaving group while the β-carbon is rendered nucleophilic by Birch reduction of the thiophenyl moiety. A communication describing an

1 'Organic Syntheses', ed. G. H. Büchi, Wiley, New York, 1977, Vol. 56.
2 'Organic Reaction Mechanisms', ed. A. R. Butler and M. J. Perkins, Interscience, London, 1977.
3 'Alicyclic Chemistry', ed. W. Parker, (Specialist Periodical Reports), The Chemical Society, London, 1977, Vol. 5.
4 W. Oppolzer, *Angew. Chem. Internat. Edn.*, 1977, **16**, 10.
5 A. G. Anastassiou, *Accounts Chem. Res.*, 1976, **9**, 453.
6 R. E. Gawley, *Synthesis*, 1976, 777.
7 J. F. Liebman and A. Greenberg, *Chem. Rev.*, 1976, **76**, 311.
8 J. E. Baldwin and L. I. Kruse, *J.C.S. Chem. Comm.*, 1977, 233.
9 R. Huisgen and R. Schug, *J. Amer. Chem. Soc.*, 1976, **98**, 7819.
10 R. A. Olofson, K. D. Lotts, and G. N. Barber, *Tetrahedron Letters*, 1976, 3779.
11 G. N. Barber and R. A. Olofson, *Tetrahedron Letters*, 1976, 3783.
12 R. A. Olofson, K. D. Lotts, and G. N. Barber, *Tetrahedron Letters*, 1976, 3381.
13 Y.-H. Chang, D. E. Campbell, and H. W. Pinnick, *Tetrahedron Letters*, 1977, 3337.

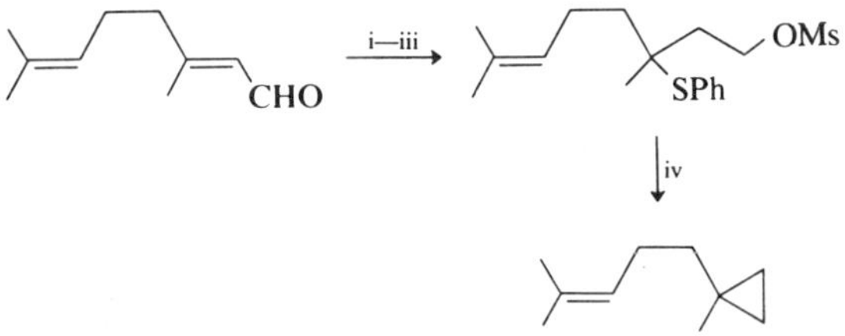

Reagents: i, PhSH–Na; ii, $LiAlH_4$; iii, $MeSO_2Cl$–py; iv, Li–NH_3

Scheme 1

efficient and operationally straightforward route to divinylcyclopropanes has appeared.[14] Conversion of *trans*-1-bromo-2-vinylcyclopropane into *trans*-lithiovinylcyclopropane is accomplished using t-butyl-lithium in ether–pentane at −78 °C and, following addition of 3-ethoxy-2-methylcyclopent-2-enone and subsequent hydrolysis (Scheme 2), the *trans*-divinylcylopropane product can be isolated in high yield.

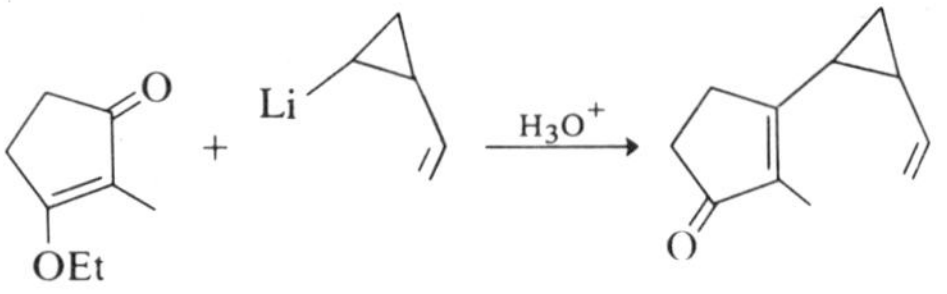

Scheme 2

The generation and trapping of the interesting hydrocarbon methylenecyclopropane has been reported (Scheme 3).[15] Reaction of (1) with Bu^tOK in

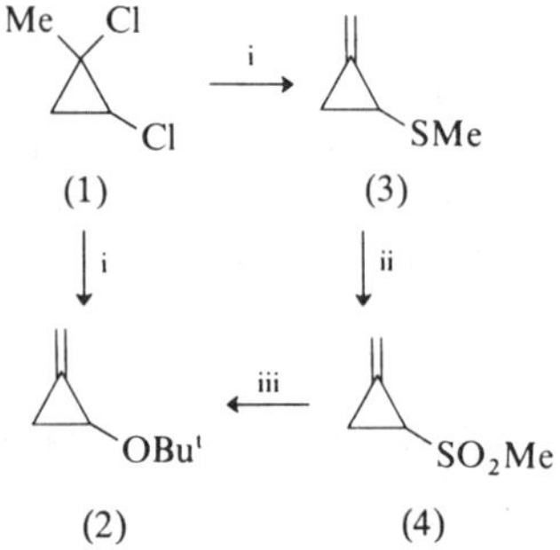

Reagents: i, Bu^tOK–MeS^-; ii, H_2O_2–$MeCO_2H$; iii, Bu^tOK–Me_2SO

Scheme 3

Me_2SO in the presence of MeS^- yields (2) together with the sulphide (3). Oxidation of (3) gives (4), which with Bu^tOK leads to (2) as the only volatile product. Since

14 P. A. Wender and M. P. Filosa, *J. Org. Chem.*, 1976, **41**, 3490.
15 W. E. Billups, A. J. Blakeney, and W. T. Chamberlain, *J. Org. Chem.*, 1976, **41**, 3771.

sulphones are well known to give alkenes *via* β-elimination, this provides compelling evidence for the intermediacy of methylenecyclopropane. Also reported[16] is the synthesis of methylenecyclopropenes lacking a stabilizing polar substituent on the exocyclic double bond (Scheme 4). It has been found that treatment of (5) with

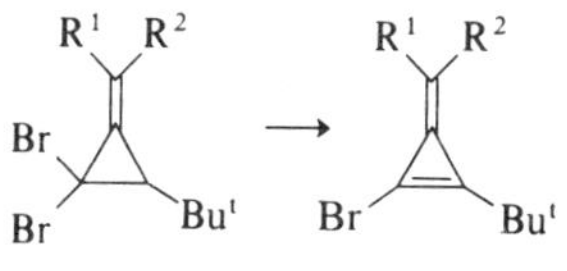

(5) $R^1 = H$ or Bu^t
$R^2 = Bu^t$ or H

(6)

Scheme 4

Bu^tOK in dry THF at low temperatures leads to (6), a structure observed to be extremely reactive towards water. A synthesis of four-membered rings has appeared[17] which allows cyclobutane, cyclobutene, and cyclobutanone to be generated from both saturated and αβ-unsaturated aldehydes and ketones *via* the intermediacy of the adducts of 1-lithiocyclopropyl phenyl sulphide to the carbonyl partner. In another communication, further syntheses of cyclobutanones, potentially useful in the terpene field, are reported.[18] The reaction involves acid rearrangement of a variety of alcohols having the 1-oxycyclopropyl structure and prepared from α-alkoxy-αβ-unsaturated ketones and esters by way of reduction followed by Simmons–Smith reaction of the resultant α-alkoxyallyl alcohols.

Five- and Six-membered Rings.—A new synthesis of five-membered rings according to a ($C_3 + C_2$) pattern has been announced[19] and provides a unique route to derivatives having the functionality shown in two readily executed steps (Scheme 5). Slow addition of solutions of the lithium enolates of the ketones (7) to a solution

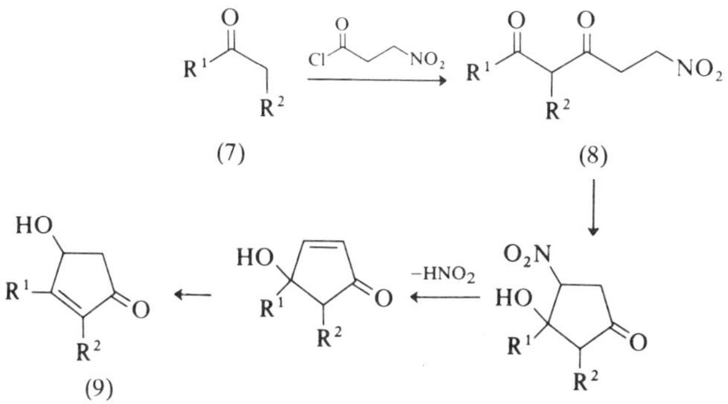

Scheme 5

[16] W. E. Billups and A. J. Blakeney, *J. Amer. Chem. Soc.*, 1976, **98**, 7817.
[17] B. M. Trost, D. E. Keeley, H. C. Arndt, and M. J. Bogdanowicz, *J. Amer. Chem. Soc.*, 1977, **99**, 3088.
[18] E. Wenkert, N. F. Golob, R. P. Hatch, D. Wenkert, and R. Pellicciari, *Helv. Chim. Acta*, 1977, **60**, 1.
[19] D. Seebach, M. S. Hoekstra, and G. Protschuk, *Angew. Chem. Internat. Edn.*, 1977, **16**, 321.

of β-nitropropionyl chloride in THF at −80 °C furnishes the diketone (8). Cyclization can be effected by a buffer solution of pH 8.5 to give hydroxycyclopentenone which rearranged to (9) on prolonged treatment. In another method based on the (C_3+C_2) pattern, and one of considerable generality, the very highly reactive dimethoxytetrachlorocyclopentadiene is heated in the presence of olefins.[20] This leads to the Diels–Alder adducts which on reaction with sodium in liquid ammonia–ethanol removes all the chlorine atoms. Oxidation of the norbornene-type double bond is accomplished using potassium permanganate in a pH 7 buffer system (Na_2HPO_4,–KH_2PO_4) with t-butyl alcohol as co-solvent. Yields of between 22 and 40% overall have been achieved for the four steps involved. The construction of five-membered rings of potential use in prostaglandin synthesis continues to attract interest, and one such example is a newly reported[21] ene route to 3-hydroxycyclopentanone derivatives. The vinylogous aldol (10), protected as its t-butyldimethylsilyl derivative, is submitted to the kinetic aldol reaction with propynal to give (11). Reaction of (11) with trimethylsilyl cyanide promotes cyanohydrin formation and simultaneously protects the two free hydroxyls as their trimethylsilyl ethers. By heating in toluene at 250 °C, thermal cyclization of (12) is readily effected, giving (13) (Scheme 6). An interesting report has appeared[22] of the

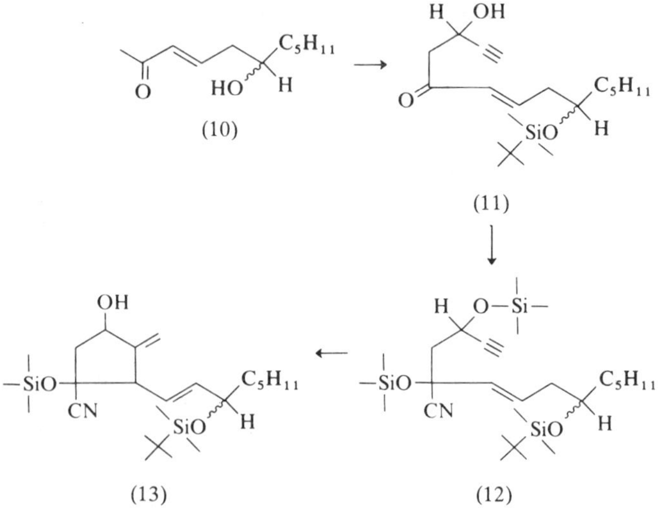

Scheme 6

application in 11-deoxyprostaglandin E_1 synthesis of thiophenol behaving as a nucleophile in stereospecific ring-opening reactions. The key transformation involves treatment of methyl *exo*-6-(*trans*-hept-1-enyl)-2-oxobicyclo[3,1,0] hexane-1-carboxylate (14) with potassium thiophenolate in Bu^{t}OH at room temperature to produce the 2,3-disubstituted cyclopentanone (Scheme 7) in 89%

[20] M. E. Jung and J. P. Hudspeth, *J. Amer. Chem. Soc.*, 1977, **99**, 5508.
[21] G. Stork and G. Kraus, *J. Amer. Chem. Soc.*, 1976, **98**, 6747.
[22] K. Kondo, T. Umemoto, Y. Takahatake, and D. Tunemoto, *Tetrahedron Letters*, 1977, 113.

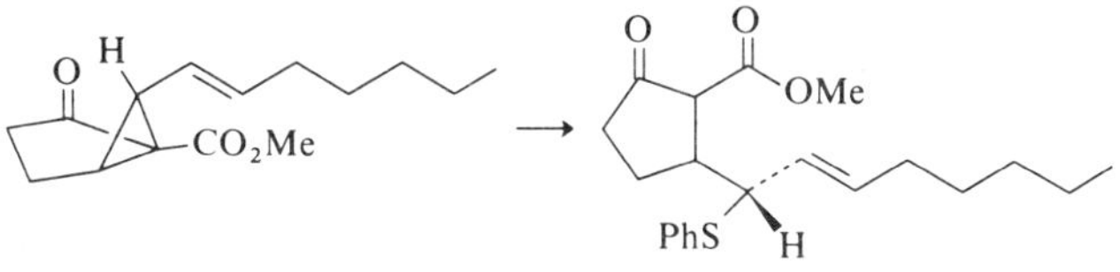

(14)

Scheme 7

yield. The desired 11-deoxyprostaglandin E_1 was then reached in four steps by standard means. A synthesis of PGA_2 using the same procedure has also been reported.[23]

The reaction between some cyclic olefins (C_5—C_8 inclusive) and diethyl(methyl)silane and CO in the presence of $Co_2(CO)_8$ has been studied.[24] In the particular case of cyclopentene, a 48% yield of the enol silyl ether of cyclohexanone has been achieved and much better yields are reported for the higher homologues. The reaction affords direct access to enol silyl ethers from olefins. A report has appeared[25] of the preparation of a versatile new diene 2-methoxy-3-phenylthiobutadiene, which can serve as an annelating agent to introduce a masked β-ketosulphide moiety as an integral part of an annelation; this masking allows structural modification elsewhere. A second point of importance is that the regiochemistry complements that normally obtained with 2-oxygenated dienes, and the ease with which sulphur can be removed from organic molecules suggests this as a possible general approach to reversing the normal orientation of Diels–Alder reactions studied. However, although this diene reacts highly successfully with maleic anhydride at room temperature, yields of adducts derived from the easily polymerizable methacrylyl dienophiles are low. An investigation of the iodocyclization of homologous dienes reveals[27] that, in the case of hepta-1,6-diene, introduction of dialkyl substitutents at the 4-position completely changes the reaction path from that of addition to cyclization (Scheme 8). The scope of the iodocyclization of several other dienes has also been investigated.

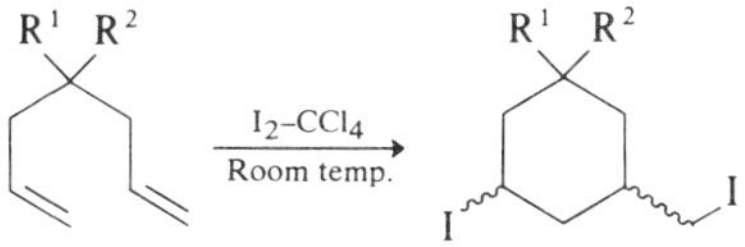

Scheme 8

Seven- and Eight-membered Rings.—A description has been given[28] for the construction of functionalized cycloheptadienes which is based on the use of 2-vinylcyclopropylcopper reagents. For example, treatment of the homocuprate (15) with

[23] D. F. Taber, *J. Amer. Chem. Soc.*, 1977, **99**, 3513.
[24] Y. Seki, A. Hidaka, S. Murai, and N. Sonoda, *Angew. Chem. Internat. Edn.*, 1977, **16**, 174.
[25] B. M. Trost and A. J. Bridges, *J. Amer. Chem. Soc.*, 1976, **98**, 5017.
[26] S. Danishefsky, C. F. Yan, and P. M. McCurry, jun., *J. Org. Chem.*, 1977, **42**, 1819.
[27] H. J. Günther, V. Jäger, and P. S. Skell, *Tetrahedron Letters*, 1977, 2539.
[28] J. P. Marino and L. J. Browne, *Tetrahedron Letters*, 1976, 3245.

methyl propiolate at −78 °C gives the *trans*-divinylcyclopropene (16) (80%) and the rearranged cycloheptadiene (17) (15%) as shown in Scheme 9. Subsequent rearrangement of (16) at 140 °C yields (17) quantitatively. A new general route[29] to 3-substituted cyclohepta-1,4-dienes involves [4+2] cycloaddition of the appropriate 7-substituted cycloheptatriene to 4-phenyl-1,2,4-triazoline-3,5-dione.

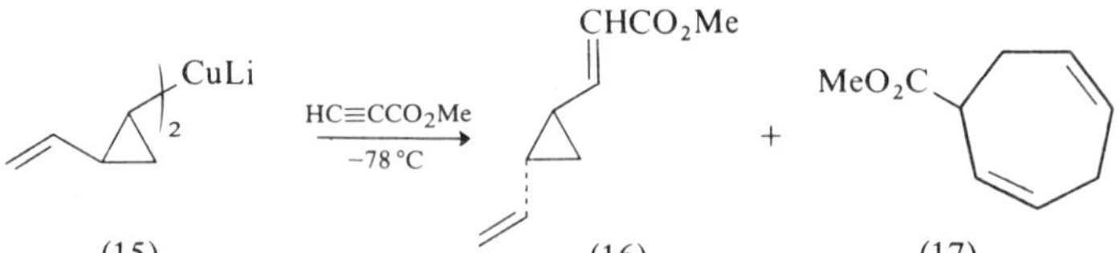

(15) (16) (17)

Scheme 9

Following catalytic reduction, basic hydrolysis of the heterocyclic ring of these products, and Cu^{2+} oxidation of the resulting hydrazo-compounds, nitrogen is readily lost to give 3-substituted cyclohepta-1,4-dienes.

An improved one-step method for preparing *trans*-cyclo-octene has been published[30] involving direct photoisomerization of the *cis*-isomer in pentane. The practical procedure consists of direct irradiation at 185 nm of the 0.12 or 0.6 molar solution of *cis*-cyclo-octene and subsequent extraction of the product with aqueous silver nitrate. The same authors also report[31] a detailed investigation of the direct and sensitized *cis–trans* photoisomerizations of cyclo-octene in the liquid phase along with photosensitization in the vapour phase. A discussion is presented of the effects of spin multiplicity and vibrational activation of the excited states on the photostationary state, and potential curves accounting for the results are proposed. A convenient synthesis of some substituted *cis,trans*-cyclo-octadienes (18) has been reported[32] which can be either isolated or trapped depending on their reactivity.

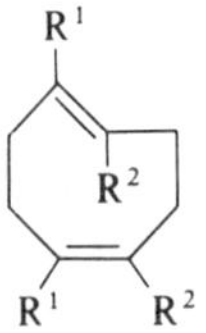

(18) $R^1 = CO_2Me$, $R^2 = H$
$R^1 = CN$, $R^2 = H$
$R^1 = R^2 = CO_2Me$

Annulenes.—The first report has appeared of the 1,5-methano[10]annulene (bicyclo[5,3,1]undeca-1,3,5,7,9-pentaene) system[33] and several of its derivatives.[34] The

29 I. Pikulik and R. F. Childs, *Canad. J. Chem.*, 1977, **55**, 251.
30 Y. Inoue, S. Takamuku, and H. Sakurai, *Synthesis*, 1977, 111.
31 Y. Inoue, S. Takamuku, and H. Sakurai, *J. Phys. Chem.*, 1977, **81**, 7.
32 H.-D. Martin, M. Hekman, G. Rist, H. Sauter, and D. Belluš, *Angew. Chem. Internat. Edn.*, 1977, **16**, 406.
33 S. Masamune and D. W. Brooks, *Tetrahedron Letters*, 1977, 3239.
34 S. Masamune, D. W. Brooks, K. Morio, and R. L. Sobczak, *J. Amer. Chem. Soc.*, 1976, **98**, 8277.

^{13}C n.m.r. spectrum of (19 $R^1 = R^2 = H$) exhibits only seven signals even at $-80\,°C$ and the molecule, therefore, possesses two-fold symmetry. Induction of a diamagnetic ring current in this monocyclic 10π-electron system is suggested by the high-field H signals due to the C-11 protons and the appearance of signals in the aromatic region. In contrast to the 1,6-methano counterpart, the u.v. spectrum of

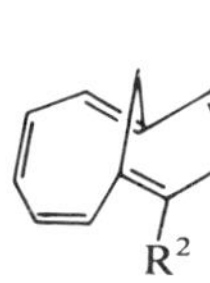

(19)

(19) extends deeply into the visible region. Sondheimer has described[35] a new and improved synthesis of derivatives of [14]annuleno[14]annulene, [14]annuleno[16]annulene, and [14]annuleno[18]annulene. These vinylogues of naphthalene are the first examples of bicyclic compounds consisting of two fused macrocyclic conjugated π-systems, and the effect of one ring upon another has been investigated by ^{1}H n.m.r. spectroscopy. The synthesis and properties of a third example of a fused non-benzenoid aromatic hydrocarbon (20), incorporating two 14π-electron systems has been reported.[36] Analysis of the 100 MHz ^{1}H n.m.r. spectrum reveals that (20) is strongly diatropic, showing the inner proton signals at δ -2.85 and those of the outer protons at δ 10.16. The electronic spectrum of (20) shows a considerable hypsochromic shift as compared with that of 3,11,14,22-tetra-t-butyl-1,12-bisdehydro[22]annulene. Both of these observations suggest that the bicyclic annulene (20) is a higher analogue of napthalene and may be represented by the delocalized formula (21). The synthesis of the tosylhydrazone sodium salt (22)

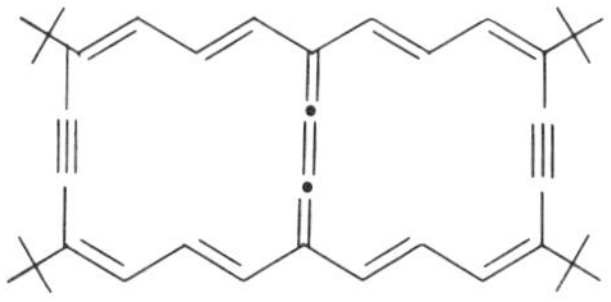

(20)

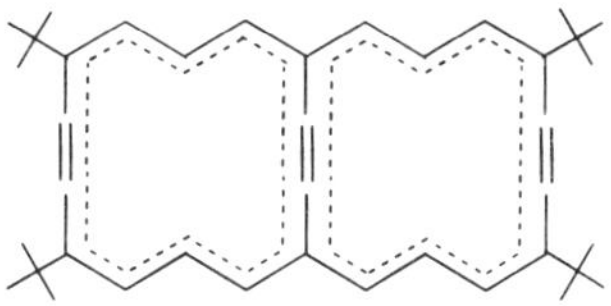

(21)

has been reported[37] and pyrolysis of the salt in refluxing diglyme gives, contrary to anticipation, not the [4,4,2]propellane (24) via a carbene–carbene rearrangement, but rather only products believed to originate from carbene (or allene) dimers. The probable intermediates (23) are thought to lose hydrogen spontaneously to give the unsaturated hydrocarbon (25) with the double bonds localized as shown in Scheme 10.

Polycyclic Systems.—A new reversible formation of bicyclobutanes involving a 1,3-bismethylenecyclobutane has been described.[38] Specifically, solutions of (26) in ethanolic lithium perchlorate react rapidly with iodine to give the bicyclobutanes

[35] T. M. Cresp and F. Sondheimer, *J. Amer. Chem. Soc.*, 1977, **99**, 194.
[36] S. Akiyama, M. Iyoda, and M. Nakagawa, *J. Amer. Chem. Soc.*, 1976, **98**, 6410.
[37] U. H. Brinker, R. W. King, and W. M. Jones, *J. Amer. Chem. Soc.*, 1977, **99**, 3175.
[38] M. Horner and S. Hünig, *J. Amer. Chem. Soc.*, 1977, **99**. 6120.

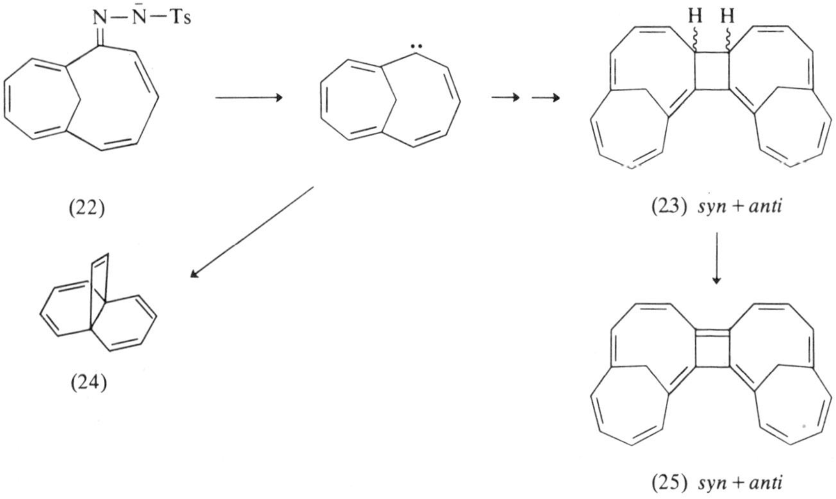

Scheme 10

(27) in high yield (Scheme 11). It is also observed that the bridging bond in (27) is readily cleaved by reduction, *e.g.* Na–Hg in acetonitrile, to regenerate the original

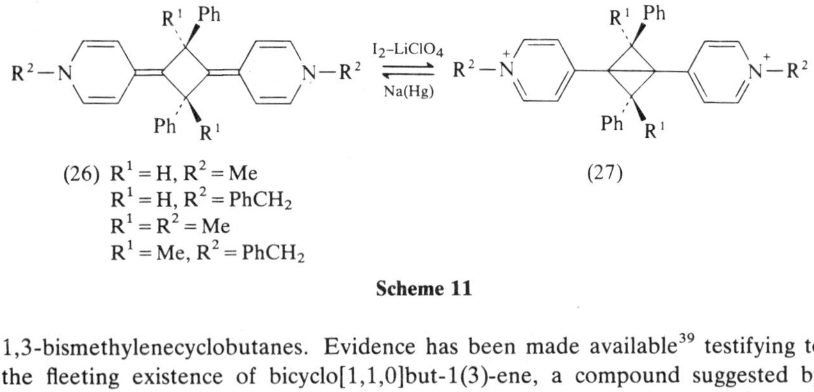

(26) $R^1 = H, R^2 = Me$
$R^1 = H, R^2 = PhCH_2$
$R^1 = R^2 = Me$
$R^1 = Me, R^2 = PhCH_2$

Scheme 11

1,3-bismethylenecyclobutanes. Evidence has been made available[39] testifying to the fleeting existence of bicyclo[1,1,0]but-1(3)-ene, a compound suggested by calculations to exist on a local minimum on the C_4H_4 potential energy hypersurface. It has been observed that reaction of (28) with n-butyl-lithium leads to an 87% yield of 1-n-butyltricyclo[4,1,0,$0^{2,7}$]heptane in a very rapid reaction. Of the three mechanistic possibilities considered for this conversion, elimination of hydrogen chloride from (28) by the organometallic reagent with the formation of a bicyclo[1,1,0]but-1(3)-ene derivative and addition of the organolithium compound to the strained double bond is considered the most likely. The first member has been reported[40] of a class of hitherto unknown fused bicylic *trans*-cycloalkenes in which two rings share a common double bond. This class of chiral olefin is

[39] G. Szeimies, J. Harnisch, and O. Baumgärtel, *J. Amer. Chem. Soc.*, 1977, **99**, 5183.
[40] J. A. Marshall and M. Lewellyn, *J. Amer. Chem. Soc.*, 1977, **99**, 3508.

designated [*a*,*b*]betweenanenes (29) and the one reported is [10,10]betweenanene (30). No report is available of the chemical and physical properties, which could be

(28)

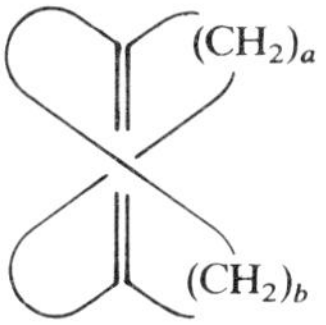

(29)

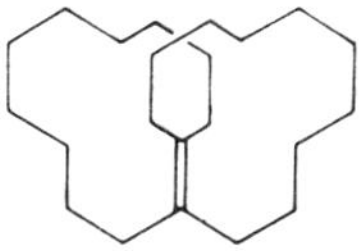

(30)

unusual as a consequence of the buried nature of the double bond. Semibenzilic acid ring contraction of 6-halogenotricyclo[3,2,1,$0^{3,6}$]octan-7-one (31) has been observed to give[41] tricyclo[3,1,1,$0^{3,6}$]heptane-6-carboxylic acid (32), a compound

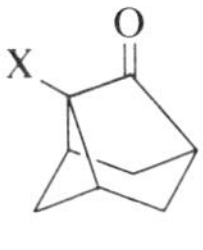

(31) X = Cl or Br

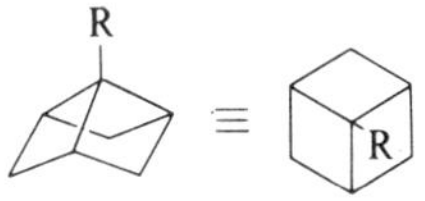

(32) R = CO_2H

containing a new ring system characterized by three cyclobutane rings fused to a common carbon atom. The structure of this acid was determined unequivocally by examination of the ^1H and ^{13}C n.m.r. spectra of the corresponding methyl ester. Studies are reported [42] of the synthesis of the novel bridgehead olefin tricyclo[3,3,3,$0^{2,6}$] undec-2(6)-ene (33). In Wiseman's analysis of bridgehead olefins this compound may be viewed as a bisethano-derivative of *trans*-cycloheptene and, although the olefin carbons in such molecules may be somewhat pyramidalized, overlap between the atomic orbitals forming the π-part of the double bond is principally diminished by torsion. Although (33) has not been isolated, decomposition of (34) in refluxing triethyl phosphite, containing diphenylisobenzofuran as trapping agent, gives the Diels–Alder adduct of (33) in 85% isolated yield. Detailed

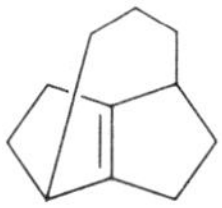

(33)

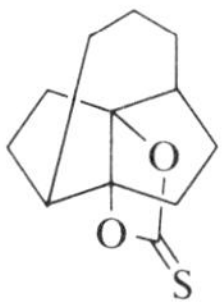

(34)

reactivity studies are reported[43] of substrates which can serve as precursors to the 'anchored' trishomoaromatic ions (35)—(37). Particular reference is made to gaining access to (35) and a number of remarkable interconversions on the $(CH)_{12}$ energy surface are described. For example, reduction of tricyclo[5,4,1,$0^{4,12}$]dodeca-2,5,8,10-tetraene (38) using potassium metal in ammonia gives a mixture of four hydrocarbons of which the major product (39) has been

[41] S. A. Monti and J. M. Harless, *J. Amer. Chem. Soc.*, 1977, **99**, 2690.
[42] R. Greenhouse, T. Ravindranathan, and W. T. Borden, *J. Amer. Chem Soc.*, 1976, **98**, 6738.
[43] L. A. Paquette, M. J. Kukla, S. V. Ley, and S. G. Traynor, *J. Amer. Chem. Soc.*, 1977, **99**, 4756.

defined spectroscopically. On the basis of these measurements it is not possible to distinguish between direct two-electron reduction of (38) to dianion (35) or a stepwise reduction sequence mediated by a radical anion. In a related paper,[44] consideration is given to (37). However, all attempts to deprotonate nortriquinacene (40) failed and this lack of acidity is interpreted to mean that the

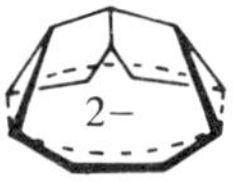

(35)

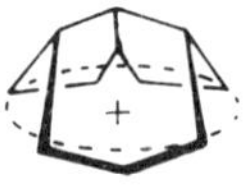

(36)

(37)

(38)

(39)

(40)

transition state for *exo*-hydrogen abstraction does not experience significant stabilizing interaction with the proximate double bonds. Metal-reduction studies also failed to provide evidence for formation of a stabilized carbanion. Further studies are reported[45] on attempts to synthesize dodecahedrane (41) and of the successful synthesis of the pentasecododecahedrane derivative (42), the most highly condensed precursor of (41) known. Elaboration of dodecahedrane from (42) appears to require development of a dehydrative retro-Baeyer–Villiger sequence which will not trigger transannular reactions, and ultimate closure by threefold C—C bond formation.

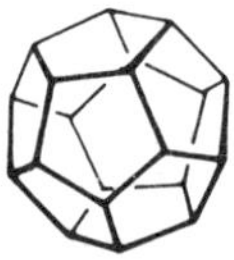

(41)

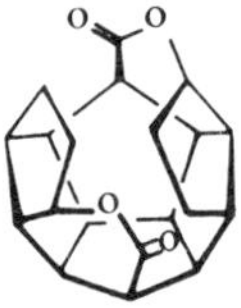

(42)

The first synthesis of two vinylidenedicyclopropanes doubly halogenated in their allylic positions has been reported[46] together with their intramolecular coupling to give the dispiroheptanes (43) and (44). By cyclopropanation using the Simmons–Smith reagent, (43) can be quantitatively transformed into [3]rotane, and the higher rotanes (45) can all be prepared by way of (44). In an accompanying paper[47]

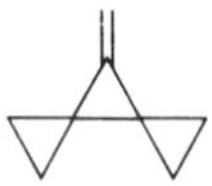

(43)

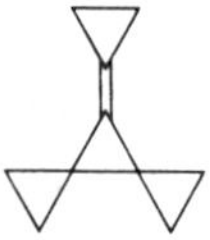

(44)

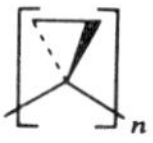

(45) $n = 4, 5,$ or 6

[44] L. A. Paquette, H. C. Berk, C. R. Degenhardt, and G. D. Ewing, *J. Amer. Chem Soc.*, 1977, **99**, 4764.
[45] L. A. Paquette, M. J. Wyvratt, O. Schallner, D. F. Schneider, W. J. Begley, and R. M. Blankenship, *J. Amer. Chem. Soc.*, 1976, **98**, 6744.
[46] L. Fitjer, *Angew. Chem. Internat. Edn.*, 1976, **15**, 762.
[47] L. Fitjer, *Angew. Chem. Internat. Edn.*, 1976, **15**, 763.

by the same author, a method permitting synthesis of each of the higher rotanes (45) is reported which is based upon a closed homologation sequence and which can, in principle, be repeated without limit.

3 Stereochemistry

The ^{1}H n.m.r. spectrum of a solution of phenylcyclopropane has been measured[48] in a CS_2 solution and the spin–spin coupling constant between the *para*-proton of the phenyl group and the α-proton in the cyclopropyl moiety found to be -0.23 ± 0.02 Hz. The authors conclude that this indicates a two-fold barrier to internal rotation of 8.4 ± 1.3 kJ mol^{-1} and implies that the low-energy conformation is such that the C—H$_\alpha$ bond prefers the plane of the phenyl group. This result is consistent with other theoretical and experimental data. Silver-ion assisted methanolysis of the epimeric tricyclic derivatives (46) and (47) has been studied[49] in an attempt to provide insight into the stereochemistry of the cyclopropyl cation, predicted by calculation to be planar. The results suggest that a planar, encumbered ion is not formed and also exclude extensive participation of unsaturation in the transition state during the solvolysis of (47). A study of the effect of increasing pressure on

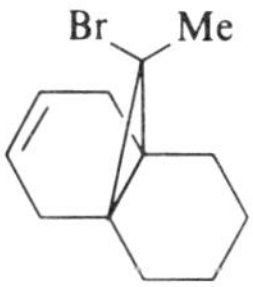

(46)

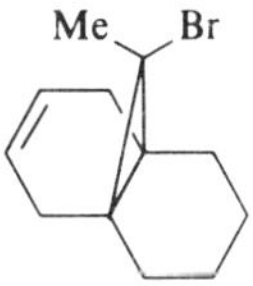

(47)

the ring inversion of cyclohexane reports[50] that pressure rises promote an increase in inversion rate. In qualitative terms, this can be explained by assuming that the twist or boat conformations, adopted by the molecule during ring inversion, have slightly smaller molar volumes than the energetically more favourable chair forms. Results are reported[51] of *ab initio* calculations involving *cis*-1,4-di-t-butylcyclohexane made in an attempt to decide whether a chair or boat form of this molecule is the more stable. By employing standard single-determinant molecular orbital theory on the STO-3G minimal basis set level, the total energies have been calculated and a boat emerges as the most stable conformer. Similar calculations show that in *trans*-1,2-di-t-butylcyclohexane both the diaxial chair and diaxial boat are of importance in its ground state. SCF-INDO-FPT calculations have been performed for various geometries of the flat cyclohexa-1,4-diene molecule and the values of $^5J_{cis}/^5J_{trans}$ found[52] to remain within the relatively narrow range 1.22—1.29 Hz. This result is claimed to preclude the need to postulate a rapidly inverting cyclohexa-1,4-diene ring. Similar studies[53] on [2,3,4,5,6-$^2H_5$,*carboxy*-^{13}C]-1,4-dihydrobenzoic acid (48), [2,3,4,5,6,7,8-$^2H_7$,*carboxy*-^{13}C]-1,4-dihydro-1-

[48] W. J. E. Parr and T. Schaefer, *J. Amer. Chem. Soc.*, 1977, **99**, 1033.
[49] D. B. Ledlie, W. Barber, and F. Switzer, *Tetrahedron Letters*, 1977, 607.
[50] H.-D. Lüdemann, R. Rauchschwalbe, and E. Lang, *Angew. Chem. Internat. Edn.*, 1977, **16**, 331.
[51] M. Askari, D. L. Merrifield, and L. Schäfer, *Tetrahedron Letters*, 1976, 3497.
[52] P. W. Rabideau, J. W. Paschal, and J. L. Marshall, *J.C.S. Perkin II*, 1977, 842.
[53] J. L. Marshall, L. G. Faehl, C. R. McDaniel, jun., and N. D. Ledford, *J. Amer. Chem. Soc.*, 1977, **99**, 321.

naphthoic acid (49), and [*carboxy*-^{13}C]-9,10-dihydro-9-anthroic acid (50) show that the ratio J_{eq-eq}/J_{ax-eq} (for J_{HH}) decreases and the ratio J_{ax-ax}/J_{ax-eq} (for J_{CH}) increases in the series (48)—(50), indicating that the puckering in (49) is intermediate between that of (48) and (50). This leads to the identification of three

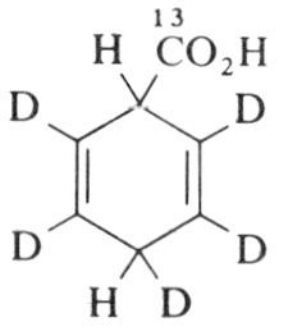

(48)

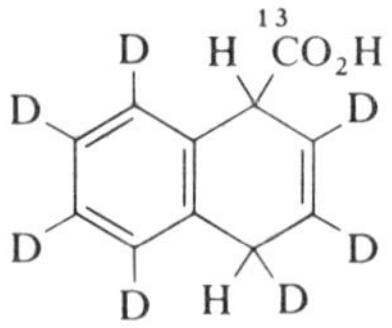

(49)

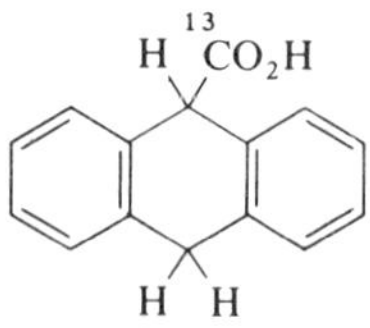

(50)

categories, 'flat', 'flattened boat', and 'true boat'. The results of calculations based on a potential function derived using vibrational data[54] to determine conformations of cycloheptane, oxepan, and 1,3-dioxepan are compared[55] with n.m.r. results and with rotational Raman spectra. This leads to a consistent picture of the seven-membered rings, with twist-chairs as the lowest-energy conformations. The conformation of 1-dimethylphosphono-1-hydroxycycloheptane has been determined[56] in the solid state by *X*-ray analysis and it has been found that both the twist-chair and chair conformers occur in the crystal structure. The relative abundances as determined by a refinement of occupancy factor are 0.93 and 0.07. Values of energy barriers associated with conformational ring inversion processes occurring in certain *trans*-cycloalkenes have been obtained[57] by analysis of their temperature-dependent ^{13}C n.m.r. spectra. In *cis, trans*-cyclo-octa-1,5-diene (51) and the derived compounds (52) these barriers are estimated to be 146.4, 173.6, 188.3,

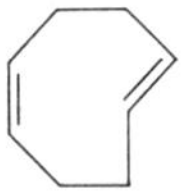

(51)

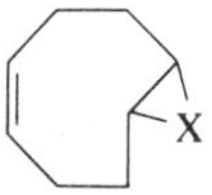

(52) X = O, CH_2, or NPh

and 186.2 kJ mol^{-1} respectively. An investigation of the 251 MHz ^{1}H and 63.1 MHz ^{13}C n.m.r. of *cis, cis*-cyclonona-1,5-diene extending over the temperature range −5 to −170 °C has been carried out[58] and the data interpreted in terms of a pair of chiral chair-like conformations with C_2 symmetry. A ^{13}C n.m.r. study has been reported[59] in which the thermodynamic parameters for equilibration of the chair–chair conformational enantimorphs of *cis*-10-methyldecalin-9-ol and *cis*-10-methyldecal-1-one have been determined. N.m.r. spectroscopy has also been used to determine the conformational preferences of a number of mobile ring systems including a series of 9-R-9,10-dihydrophenanthrenes (R = Me, CN, Bu^t, COMe, CO_2Me, CO_2Et, OH, or $SiMe_3$) as well as two related 5,6-dihydrochrysenes, with special regard to the location of substituents in pseudoaxial or

[54] D. F. Bocian and H. L. Strauss, *J. Amer. Chem. Soc.*, 1977, **99**, 2866.
[55] D. F. Bocian and H. L. Strauss, *J. Amer. Chem. Soc.*, 1977, **99**, 2876.
[56] G. I. Birnbaum, G. W. Buchanan, and F. G. Morin, *J. Amer. Chem. Soc.*, 1977, **99**, 6652.
[57] S. G. Davies, P. F. Newton, and G. H. Whitham, *J.C.S. Perkin II*, 1977, 1371.
[58] F. A. L. Anet and I. Yavari, *J. Amer. Chem. Soc.*, 1977, **99**, 6496.
[59] J. W. Blunt, J. M. Coxon, N. B. Lindley, and G. A. Lane, *Austral. J. Chem.*, 1976, **29**, 967.

pseudoequatorial positions. With the exception of CN, all the substituents are found[60] to adopt the pseudoaxial conformation preferentially. *Ab initio* energies have been reported[61] for several minimum-energy conformations using standard single-determinant MO theory with an STO-3G minimal basis set for a number of compounds including *trans*-decal-2-one, 10-methyl-*trans*-decal-2-one, 1α-methyl-*trans*-decal-2-one, 3α-methyl-*trans*-decal-2-one, and 1,1,10-trimethyl-*trans*-decal-2-one. The calculations predict a significant population of flexible states for the latter two compounds at moderate temperatures. The results of calculations carried out using a 1973 force field on the 25 isomers of dodecahydrophenanthrene are presented[62] in a report which lists the values of ΔH_f^0 for each conformation together with the values of ΔH for each compound (conformational mixture) and the relative entropies.

4 Structural Properties and Orbital Interactions

In order to account for the fact that substituents on cyclopropane are known to alter the bond lengths in the ring, *ab initio* wavefunctions have been obtained[63] for cyclopropane and a number of relevant related molecules. Comparison of orbital composition, energies, and overlap populations of the parent with substituted molecules has been utilized to ascertain the origin of these substituent and ring-closure effects. The analysis shows that the substituent induced electron withdrawal and charge redistribution in the cyclopropane orbitals so that the adjacent bonds were shortened while the opposite bond was lengthened. Although no cases are known in which simple cyclopropyl cations have been unambiguously demonstrated as intermediates in solvolysis reactions, by use of ion cyclotron double-resonance techniques it has been found[64] that cyclopropene forms a $C_3H_5^+$ ion by proton transfer from dimethyloxonium ion. Such experiments suggest that the $C_3H_5^+$ ion produced has a heat of formation of 996 kJ mol^{-1}. This value is not consistent with the heat of formation of the allyl cation and suggests that a cyclopropyl cation may have been formed. The surfaces of the ground state and first excited singlet and triplet states for the conversion of cyclopropene into vinylcarbene have been calculated[65] using the MINDO/3 semi-empirical MO method. In general, the results appear to be in agreement with the known thermal and photochemical reactivities of cyclopropenes and the results of some *ab initio* calculations on these species. Generalized valence-bond calculations on cyclopropene and vinylmethylene have been carried out.[66] These predict that (*a*) the allyl-type π-system has a methylene-like triplet but a 1,3-biradical-like singlet state, (*b*) the lowest singlet state of vinylmethylene is of the form of a singlet methylene, and (*c*) the ground state of vinylmethylene should exist in two geometrical forms. These results suggest that the ring opening of cyclopropene proceeds directly to a biradical planar intermediate. The reaction of tri-t-butyl-

[60] R. G. Harvey, P. P. Fu, and P. W. Rabideau, *J. Org. Chem.*, 1976, **41**, 3722.
[61] M. Askari, N. S. Ostlund, and L. Schäfer, *J. Amer. Chem. Soc.*, 1977, **99**, 5246.
[62] H. Hönig and N. L. Allinger, *J. Org. Chem.*, 1977, **42**, 2330.
[63] C. A. Deakyne, L. C. Allen, and V. W. Laurie, *J. Amer. Chem. Soc.*, 1977, **99**, 1343.
[64] D. H. Aue, W. R. Davidson, and M. T. Bowers, *J. Amer. Chem. Soc.*, 1976, **98**, 6700.
[65] J. A. Pincock and R. J. Boyd, *Canad. J. Chem.*, 1977, **55**, 2482.
[66] J. H. Davis, W. A. Goddard, and R. G. Bergman, *J. Amer. Chem. Soc.*, 1977, **99**, 2427.

cyclopropene with photochemically generated t-butoxyl radicals has been monitored[67] using e.s.r. It is found that the magnitude of the ^{13}C coupling constants is inconsistent with (53), involving an aromatic π-radical, and is in agreement with (54), involving equilibrium of energetically equivalent σ-radicals. The 1-isocyano-

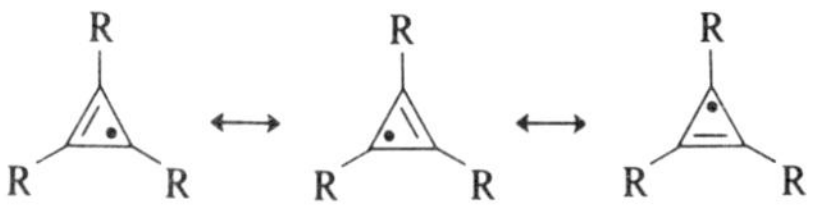

(53)

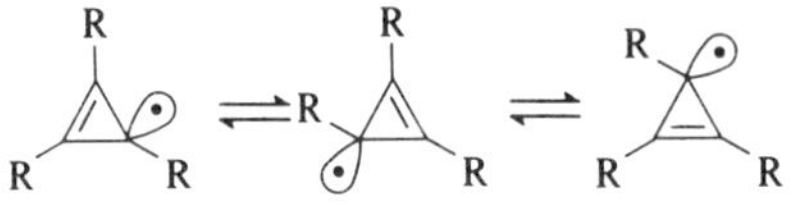

(54)

2,2-diphenylcyclopropyl anion has been found[68] to be configurationally stable at −72 °C, in contrast to the corresponding 1-cyano-derivative. This stability is not affected by changing the cation from Li to Na to K nor is it affected by change in solvent or by the addition of co-ordinating agents such as crown ethers, triglyme, or HMPA. A series of 1-methylcyclopropylcarbinyl cations has been prepared and the ^{1}H and ^{13}C n.m.r. spectra investigated[69] under stable-ion conditions. The parent ion has been found to be best represented in terms of a dynamic equilibrium involving significant contributions from both the 1-methylcyclobutyl cation and the set of three equivalent δ-delocalized 1-methylcyclopropyl cations, a conclusion supported by *ab initio* MO calculations at the split-valence-shell 4-31G level.

Ab initio calculations of the equilibrium structure of cyclobutane have been reported[70] in which use is made of a single-determinant restricted Hartree–Fock theory involving an extended basis set of Gaussian orbitals augmented by polarization functions. The molecule is found to be non-planar with a degree of puckering of 0.23 Å and a barrier to planarity of 3.8 kJ mol^{-1}; tilting of the methylene groups proves to be essential for relieving steric strain. Semi-empirical SCFMO calculations (CNDO/I, CNDO/II, and MINDO/3) have been used[71] to estimate the contribution of 1,3-carbon–carbon interactions to the strain energy of cyclobutane, and perturbation energies are found to lie between 85.4 and 140.2 kJ mol^{-1}, depending upon the particular calculation and molecular geometry. The analyses suggest that cognisance of 1,3-carbon–carbon interactions is obligatory in the cyclobutane series. Cyclobutadiene continues to attract interest. The energy relationships between the singlet and triplet states of cyclobutadiene have been discussed[72] in terms of MINDO/3 calculation and *ab initio* SCF calculations using the 4-31G basis set; calculations have also been made of molecular vibration frequencies. The results support the suggestion that the $(CH)_4$ species obtained by pho-

[67] K. Schriener and A. Berndt, *Angew. Chem. Internat. Edn.*, 1976, **15**, 698.
[68] M. P. Periasamy and H. M. Walborsky, *J. Amer. Chem. Soc.*, 1977, **99**, 2631.
[69] G. A. Olah, R. J. Spear, P. C. Hiberty, and W. J. Hehre, *J. Amer. Chem. Soc.*, 1976, **98**, 7470.
[70] D. Cremer, *J. Amer. Chem. Soc.*, 1977, **99**, 1307.
[71] N. L. Bauld and J. Cessac, *J. Amer. Chem. Soc.*, 1977, **99**, 942.
[72] M. J. S. Dewar and A. Komornicki, *J. Amer. Chem. Soc.*, 1977, **99**, 6174.

tolyses in matrices at low temperatures is a metastable triplet. The fluxional isomerism of cyclobutadiene has been analysed[73] at the π-electronic level in terms of an orbital correspondence analysis in maximum symmetry, which shows that the process is allowed under a skew-planar distortion. Analysis of the ground-state eigenvector indicates that the most significant structure is the singlet biradical which is consistent with cycloaddition reactions of 1,3-di-t-butylcyclobutadiene. A full spectral characterization of tetrakis(trifluoromethyl)cyclobutadiene has been reported.[74] The compound has a singlet ground state and if the assignment of the i.r. band at 1700 cm^{-1} to the C=C stretching vibration proves to be correct then its geometry is not square but rectangular. This study also shows that, at low temperature, direct irradiation does not lead to the formation of the tetrahedrane system at least as a major course of reaction. In another communication,[75] tri-t-butylcyclobutadiene (55), a molecule for which a rectangular structure is assumed, has been shown to undergo reactions typical of a triplet molecule. Thus, if (55), prepared by matrix isolation of 2,3,4-tri-t-butylcyclopentadienone, is allowed to react after thawing with an excess of CCl_4, the product is found to contain no dimer of (55) but exclusively the adduct (56). A possible mechanism for the formation of (56) is shown (Scheme 12). A number of substituted cyclobutadiene dications have

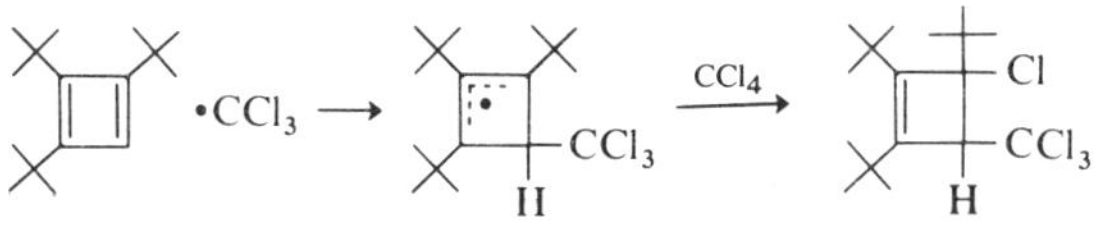

(55) (56)

Scheme 12

been prepared under stable-ion conditions and studied[76] by ^{1}H, ^{19}F, and ^{13}C n.m.r. spectroscopy. The ^{13}C n.m.r. spectra of the phenylated cyclobutadiene dications indicate that the aromatic 2π-electron cyclobutenediylium ring interacts mesomerically with phenyl rings and accepts a significant amount of π-electron density from them. The question as to whether the n.m.r. spectroscopic observations of cyclobutadiene dications correspond to several species in dynamic equilibrium or to static, non-equilibrating dications has been systematically investigated and unequivocal evidence for the latter presented. The photoelectron spectrum of bicyclo[4,1,1] octa-2,4-diene has been recorded.[77] Based upon assignment of the first bands in the spectrum, it is concluded that the interaction between the Walsh orbitals of the cyclobutane ring and the olefinic moiety can be described by a resonance integral $\beta = -183$ kJ mol^{-1}.

The e.s.r. spectra of the radical anions of 1,4- and 1,5-dimethylcyclo-octatetraene have been measured[78] and it is found that in the former compound the π-spin density is spread almost equally over the eight ring carbons whereas in the

[73] E. A. Halevi, F. A. Matsen, and T. L. Welsher, *J. Amer. Chem. Soc.*, 1976. **98**, 7088.
[74] S. Masamune, T. Machiguchi, and T. Aratani, *J. Amer. Chem. Soc.*, 1977, **99**, 3524.
[75] G. Maier and W. Sauer, *Angew. Chem. Internat. Edn.*, 1977, **16**, 51.
[76] G. A. Olah and J. S. Staral, *J. Amer. Chem. Soc.*, 1976, **98**, 6290.
[77] R. Gleiter, P. Bischof, W. E. Volz, and L. A. Paquette, *J. Amer. Chem. Soc.*, 1977, **99**, 8.
[78] J. H. Hammons, C. T. Kresge, and L. A. Paquette, *J. Amer. Chem. Soc.*, 1976, **98**, 8172.

latter compound it is localized on the four odd-numbered ring carbons. It is argued that these results provide support for symmetry-orbital models of substituted COT radical anions. The structures of the 1,4-dimethyl-, 1,3,5,7-tetramethyl-, 1,3,5,7-tetraphenyl-, and *sym*-dibenzo-cyclo-octatetraene dications, prepared by oxidation of suitably substituted cyclo-octatetraenes using SbF_5 in SO_2ClF solution, have been investigated[79] by ^1H and ^{13}C n.m.r. spectroscopy. The observations reported indicate that these ions possess those characteristics expected of a C_8 6π aromatic system. Calculations using MINDO/3 have been reported[80] for a number of potential homoaromatic molecules, including (57) and (58), and the homocon-

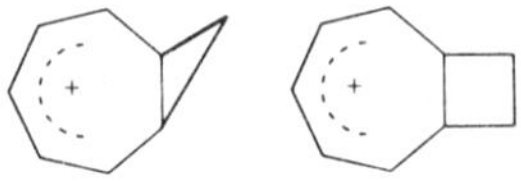

(57) (58)

jugate interactions in the cyclobutyl-fused compounds studied were shown to be negligible, whereas (57) appears to be homoaromatic. Judging the compounds investigated as a whole, the conclusion is drawn that the energetic impact of monohomoaromaticity is only profound for the cyclobutenyl and homotropylium cations. Metallation routes have been announced[81] to the trilithium salts of the cycloheptatrienyl trianion, its n-butyl derivative, and its linearly conjugated analogue, the heptatrienyl trianion. Among the evidence given is the appearance of the e.s.r. signal expected for cycloheptatrienyl dianion radical dilithium (a_H = 3.52 and a_{Li} = 0.83 G) and a singlet in the ^1H n.m.r. and also in the ^{13}C n.m.r. spectra. By use of the newly synthesized bicyclobutanes (27) in which the bridgehead carbons are connected to pyridinium rings and positions 2 and 4 to substituents of different size, quantitative data have been provided[82] concerning the effect of π-substituents and geometry on the LUMO energies in bicyclobutanes. Electrochemical oxidation transforms the 1,3-bismethylenecyclobutane (26) into (27), which in turn is reduced back to (26), and the evidence gained from cyclic voltammograms indicates a large decrease in the energy of the LUMO in (27) into which the two electrons are added as compared with that required for reduction of normal bicyclobutanes. The u.v. spectrum of bicyclo[1,1,0]butane has been examined[83] and the band occurring near 50 000 cm^{-1} assigned to a transition between bonding and antibonding 'π-like' central C—C bond orbitals. The lowest electronic state ($\tilde{A}$) has been examined[84] with a resolution of 0.3 cm^{-1}, permitting the rotational band contours for the vibronic components to be observed. Analysis of the band contours has shown the first excited state to have A_2 symmetry. A number of spiroconjugated hydrocarbons have been studied[85] theoretically, and by use of a modification within the MINDO framework an orbital splitting is obtained which is within 5% of experiment in the case of spiro[4,4] nonatetraene. The

79 G. A. Olah, J. S. Starel, G. Liang, L. A. Paquette, W. P. Melega, and M. J. Carmody, *J. Amer. Chem. Soc.*, 1977, **99**, 3349.
80 W. L. Jorgensen, *J. Amer. Chem. Soc.*, 1976, **98**, 6784.
81 J. J. Bahl, R. B. Bates, W. A. Beavers, and C. R. Launer, *J. Amer. Chem. Soc.*, 1977, **99**, 6126.
82 M. Horner and S. Hünig, *J. Amer. Chem. Soc.*, 1977, **99**, 6122.
83 K. B. Wiberg, G. B. Ellison, and K. S. Peters, *J. Amer. Chem. Soc.*, 1977, **99**, 3941.
84 K. B. Wiberg, K. S. Peters, G. B. Ellison, and F. Alberti, *J. Amer. Chem. Soc.*, 1977, **99**, 3946.
85 M. D. Gordon, T. Fukunaga, and H. E. Simmons, *J. Amer. Chem. Soc.*, 1976, **98**, 8401.

conclusion is also drawn that the MINDO technique using properly corrected overlap can yield realistic predictions on the magnitude of 'through space' interactions. The orbital energy differences for a number of spiro-compounds of type (59) and for the analogues in which the spiro-centre is replaced by a four-membered ring (60) have been calculated.[86] This shows that the interaction between two mutually perpendicular π-systems linked by a spiro-centre becomes considerably stronger on replacing the spiro-centre by a four-membered ring, and this is considered to occur by the Walsh orbitals of the four-membered ring acting as a 'relay'.

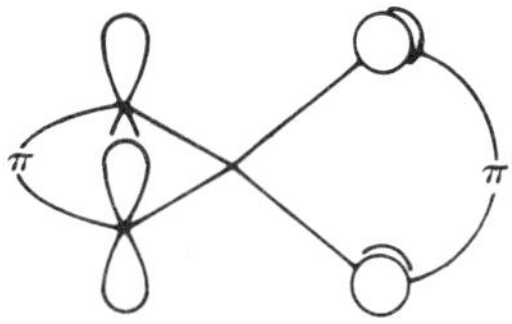

(59)

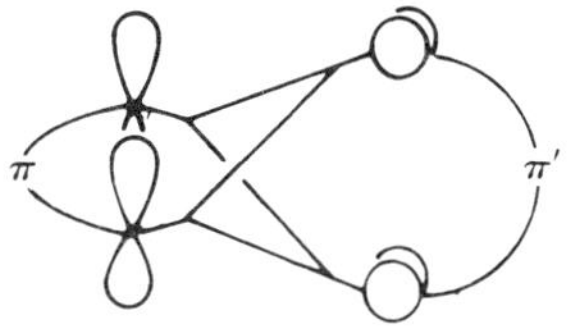

(60)

5 Reactions

Metal-promoted Reactions.—Methylenecyclopropane has been catalytically trimerized using Ni^0 compounds modified by trialkylphosphines to give a mixture of three open-chained and three cyclic $C_{12}H_{18}$ isomers.[87] The composition of the mixture depends on the nature of the alkyl groups attached to the phosphorus and a suggested route to one of the products (61) is shown in Scheme 13. Exclusive

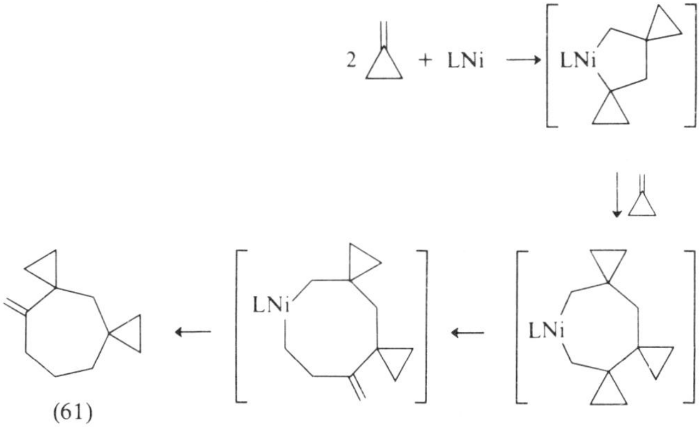

Scheme 13

$[2\sigma + 2\pi]$ cycloaddition of methylenecyclopropane with alkenes has been found[88] on Pd^0 catalysts, where, in contrast to the Ni^0-catalysed reactions, the three-membered ring is opened between C-2 and C-3. The catalytic cycloadditions

[86] P. Bischof, R. Gleiter, and R. Haider, *Angew. Chem. Internat. Edn.*, 1977, **16**, 110.
[87] P. Binger, A. Brinkmann, and J. McMeeking, *Annalen*, 1977, 1065.
[88] P. Binger and U. Schuchardt, *Angew. Chem. Internat. Edn.*, 1977, **16**, 249.

described might proceed *via* palladacyclobutanes of type (62) and palladacyclohexanes of type (63). Vinylcyclopropanes have been found[89] to undergo a novel

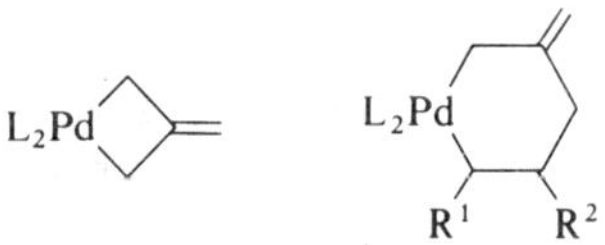

(62) (63)

epimerization in the presence of dicarbonylrhodium(I) chloride dimer and this has been shown to be coupled with *trans–cis* isomerization about the C—C π-bond. In related work, kinetic studies involving rearrangement of 1,1-dimethyl-2-vinylcyclopropane and other compounds provide evidence in support of a mechanism for ring cleavage involving initial co-ordination of the vinyl group with Rh^{I} and stereospecific *cis*-β-hydride elimination from a subsequent metallocyclic allylrhodium(III) alkyl intermediate. Methylenecyclobutane and its 3-substituted derivatives are reported[90] to undergo a Pd^{II}-catalysed ring expansion to the corresponding cyclopentanones in good yield. Analogous but somewhat less efficient reactions are observed for camphene, methylenecamphor, and methylenecyclopentane. The transformation has been found to be solvent sensitive, yields being highest in benzene but falling off with increasing solvent polarity. An examination has been made[91] of the silver(I) perchlorate-catalysed rearrangement of [1-^{2}H]benzvalene to [$^{2}H_{1}$]benzene in which rupture of two diagonally opposed bonds of the bicyclobutane moiety occurs. By following the reaction with the ^{2}H-{^{1}H} FT n.m.r. technique, it is found that the benzvalene deuterium label is scrambled by a catalysed degenerate rearrangement in competition with the aromatization. Together with labelling studies, the observations provide unequivocal evidence that the reaction of the Ag^{I} ion with the cyclobutane moiety is a reversible process involving rupture and then reclosure of a C—C bond. A study has been made[92] of the heats of metal-catalysed rearrangements of some small-ring hydrocarbons in an attempt to rectify the lack of experimental thermochemical data. In particular, the isomerizations of quadricyclane to norbornadiene and of tricyclo[4,1,0,0^{2,7}]heptane to 3-methylenecyclohexene have been studied. The synthesis of *anti*-tetramantane (64), a $C_{22}H_{28}$ hydrocarbon having the regular topology of the diamondoid lattice, has been achieved,[93] in which the key step is a ring expansion–cyclization reaction of a polycyclic diene (65) in hydrogen in the gas phase on a platinum–silica catalyst. The mode of expansion of the four-membered ring in the [2+2] dimers of adamantane in the gas phase on platinum has also been investigated[94] and the results support the view that loss of a single bridgehead hydrogen atom from the alkane is sufficient to initiate rearrangements. This is to say that essentially covalently bonded alkyl groups rear-

89 R. G. Salomon, M. F. Salomon, and J. L. C. Kachinski, *J. Amer. Chem. Soc.*, 1977, **99**, 1043.
90 P. Boontanonda and R. Grigg, *J.C.S. Chem. Comm.*, 1977, 583.
91 U. Burger and F. Mazenod, *Tetrahedron Letters*, 1977, 1757.
92 K. B. Wiberg and H. A. Connon, *J. Amer. Chem. Soc.*, 1976, **98**, 5411.
93 W. Burns, T. R. B. Mitchell, M. A. McKervey, J. J. Rooney, G. Ferguson, and P. Roberts, *J.C.S. Chem. Comm.*, 1976, 893.
94 W. Burns, M. A. McKervey, J. J. Rooney, N. G. Samman, J. Collins, P. von R. Schleyer, and E. Osawa, *J.C.S. Chem. Comm.*, 1977, 95.

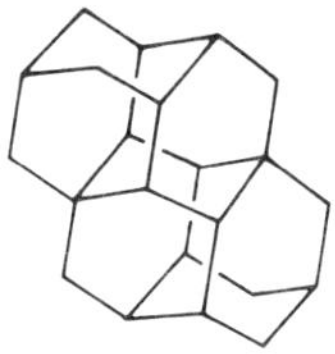

(64)

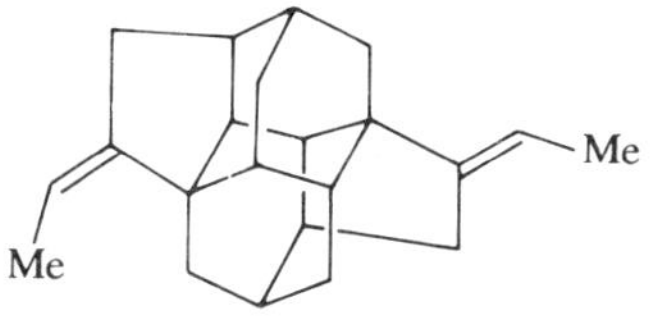

(65)

range on a platinum surface in a manner commonly observed with carbonium ions. This represents an alternative pathway to the formation and further reaction of $\alpha\alpha\lambda$-triadsorbed intermediate (66) and is presumably followed as the $\alpha\alpha\lambda$

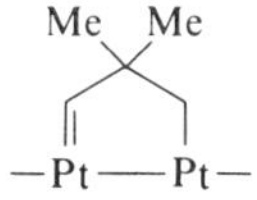

(66)

mechanism is ruled out on simple geometric grounds. It has been reported[95] that (diphenylcarbene)pentacarbonyltungsten, without added co-catalysts, is an effective initiator of metatheses of numerous *cis*-cycloalkenes, converting them into polyalkenamers, 90% or more of whose double bonds are *cis*. Cyclobutene, cyclopentene, cyclohexene, cycloheptene, cyclo-octene, and norbornene are all transferred in this way. In an accompanying paper,[96] (phenylmethoxy-carbene)pentacarbonyltungsten is also reported as being able to bring about metatheses of cyclobutene and norbornadiene to the corresponding polyalkenamers. The importance of this discovery is that, although ways have yet to be found to replace the stabilizing substituents in the metal carbenes such as (phenylmethoxycarbene)pentacarbonyltungsten in order to transform them into initiators for olefin metathesis, it emerges that this carbene is itself an effective catalyst. The photochemical decomposition of a number of substituted metallocyclobutane derivatives of the general form $[(\eta^5\text{-}C_5H_5)_2\overline{WCH_2CHR^2CHR^1}]$ have been studied[97] and the major olefin formed on decomposition is shown to have one carbon less than the appropriate carbon chain of the initial metallocyclobutane. It is suggested that the initial photochemical reaction causes a $\eta^5\text{-}C_5H_5$ to-$\eta^3\text{-}C_5H_5$ ring shift and that this 16-electron intermediate allows the rearrangement of the metallocyclobutane ring to occur. Equilibria of the type shown in Scheme 14 have been postulated as the central step in the olefin dismutation reaction and these results provide strong support for this mechanism. It has been found[98] that 1-methyl-*trans*-cyclo-octene undergoes metathesis to yield a polymer that, within the limits of the detection method, is perfectly alternating. Thus, to the extent that this reaction is a valid measure, the selectivity for equation (1) is >50 times that for equation (2) (Scheme 15). The experiments also show the stereochemistry of trisubstituted olefin metathesis (E-olefins yield mainly E-products) and indicate

95 T. J. Katz, S. J. Lee, and N. Acton, *Tetrahedron Letters*, 1976, 4247.
96 T. J. Katz and N. Acton, *Tetrahedron Letters*, 1976, 4251.
97 M. Ephritikhine and M. L. H. Green, *J.C.S. Chem. Comm.*, 1976, 926.
98 S. J. Lee, J. McGinnis, and T. J. Katz, *J. Amer. Chem. Soc.*, 1976, **98**, 7818.

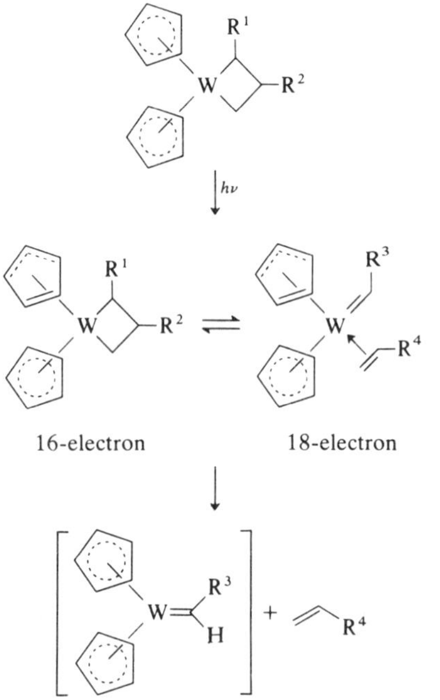

$R^1 = R^2 = R^3 = R^4 = H$; $R^1 = R^3 = H$, $R^2 = R^4 = Me$;
or $R^1 = Me$, $R^2 = H$, $R^3 = Me$ or H, $R^4 = H$ or Me

Scheme 14

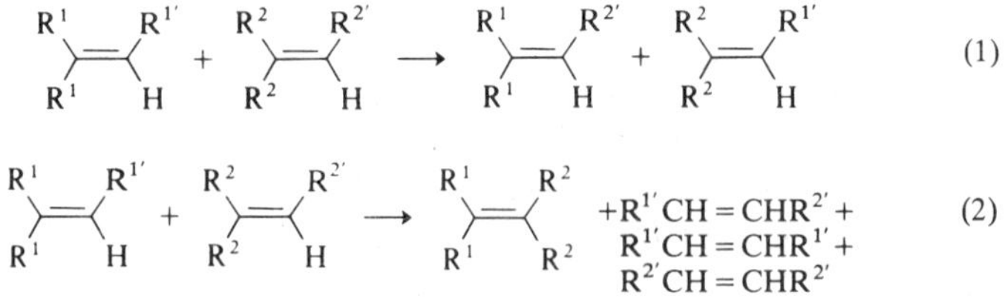

Scheme 15

the presence of the metathesis initiator at the ends of the polymer chains. A model has been proposed[99] to account for both the observed metathesis and the *cis-trans* isomerization, as measured by the relative amounts of the various olefins produced at low conversion from the reaction of *cis*-pent-2-ene on a tungsten catalyst. A stereochemical model is proposed based on the belief that an olefin, co-ordinated to a metallo-carbene unit, passes through a four-centre transition state to form a new olefin and a new metallo-carbene.

[99] J. L. Bilou, J. M. Basset, R. Mutin, and W. F. Graydon, *J.C.S. Chem. Comm.*, 1976, 970.

Thermally Induced Reactions.—A laminar flow reactor has been described[100] which is useful for gas-phase kinetic measurements on small amounts of compounds; the vinylcyclopropane rearrangement and the retro-Diels–Alder reaction of dicyclobutadiene were the reactions used in test runs to demonstrate the reliability of the results. Light has been cast on the stereochemistry of the cyclopentene-forming reaction by reports[101] of a study of the thermal isomerization of (+)-(1*S*,2*S*)-*trans,trans*-2-methyl-1-propenylcyclopropane for which the figures shown in Scheme 16 have been obtained. It is suggested that there is competition among four concerted processes, two allowed and two forbidden, with the possible intervention of a planar π-vinylcyclopropane intermediate having an orbital structure appropriate for disrotatory ring closure to racemic (67). A study has been

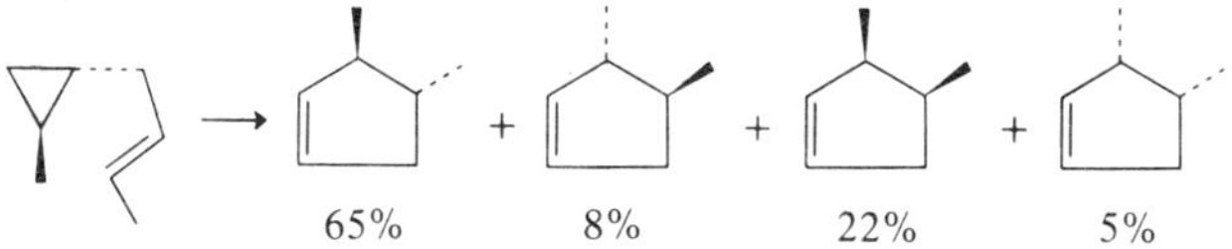

65% 8% 22% 5%

(67)

Scheme 16

made[102] of the pyrolysis of 6-*exo*-vinylbicyclo[3,1,0]hex-2-ene in order to determine whether isomerization to the 6-*endo*-isomer occurs through a one-centre epimerization at C-1 and C-5. The results show that the one-centre epimerization pathway is followed and for the first time this is demonstrated in a cyclopropane for which the alternative two-centre option is geometrically possible. Investigation of the thermal decomposition of bicyclic divinyl-Δ^1-pyrazolines and divinylcyclopropanes has shown[103] that identical product decomposition patterns arise from (68) and (69) (Scheme 17). Comparison of the kinetic parameters for decomposition of (68) with those for *trans*-3,5-dimethyl-Δ^1-pyrazoline and 3-vinyl-Δ^1-pyrazoline suggests a contribution towards a lowering of the activation enthalpy

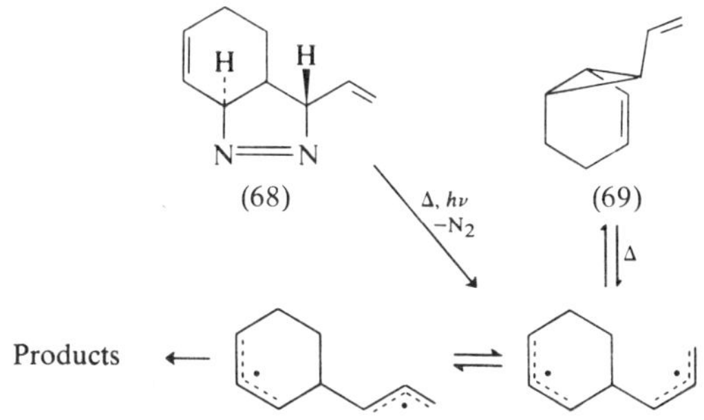

Scheme 17

[100] L.-U. Meyer and A. de Meijere, *Chem. Ber.*, 1977, **110**, 2545.
[101] G. D. Andrews and J. E. Baldwin, *J. Amer. Chem. Soc.*, 1976, **98**, 6705.
[102] J. E. Baldwin and K. E. Gilbert, *J. Amer. Chem. Soc.*, 1976, **98**, 8283.
[103] M. P. Schneider and B. Csacsko, *J.C.S. Chem. Comm.*, 1977, 330.

consistent with formation of a diallylic 1,3-biradical and this is taken as supporting evidence that the thermal rearrangement of *trans*-divinylcyclopropane proceeds through a biradical intermediate. In order to attempt to define the mechanism for thermal geometrical isomerizations of cyclopropanes the rates have been measured[104] for three distinct degenerate processes shown by the cyclopropyl moiety of (*Z*)-1-methyl-2-deuteriospiro[2,4]hepta-4,6-diene. The results show that even in this favourable case, the first cyclopropane constituted to enhance prospects for one-centre epimerization through a planar intermediate, epimerization at C-3 is definitely not prominent and may well be absent. Studies[105] of the thermal rearrangement of methylenecyclobutane indicate that electronic factors are least likely to be the cause of the observed stereospecificity in this reaction, and they also reveal a similarity between the reaction of cyclopropane and cyclobutane with that of methylenecyclobutane, in that all pass through a biradical transition state (or intermediate) which is not bound by large rotational barriers. Experiments have been reported[106] which are claimed to demonstrate unambiguously 'hot molecule' effects at an atmosphere of nitrogen and exclusive operation of the C-1—C-4 bond-cleavage mechanism in the transformation of bicyclo[2,1,0]pent-2-ene to cyclopentadiene. It is suggested that the assumption that collisional deactivation in solution always pre-empts thermal reactions of vibrationally excited molecules can no longer be credited. The first experimental evidence has been reported[107] which demonstrates that stabilization of norcaradiene relative to cycloheptatriene can be achieved by replacement of hydrogen at C-7 with a π-donor. The equilibria (70) ⇄ (71) (Scheme 18) have been established and the time-averaged values of $\delta_{H_{1,6}}$ in (70a) ⇄ (71b) (4.45) and (70b) ⇄ (71b) (4.88) indicate that in contrast to (71c) (5.74) substantial quantities of the norcaradiene form are present in both mixtures. From n.m.r. data can be obtained $K_{25\,°C}$[(70a) ⇄ (71b)] ~ 0.7 [41% (71a)] and $K_{25\,°C}$[(70b) ⇄ (71b)] ~ 0.3 [24% (71b)]. This establishes that a piperidino-group stabilizes the norcaradiene form relative to the cycloheptatriene

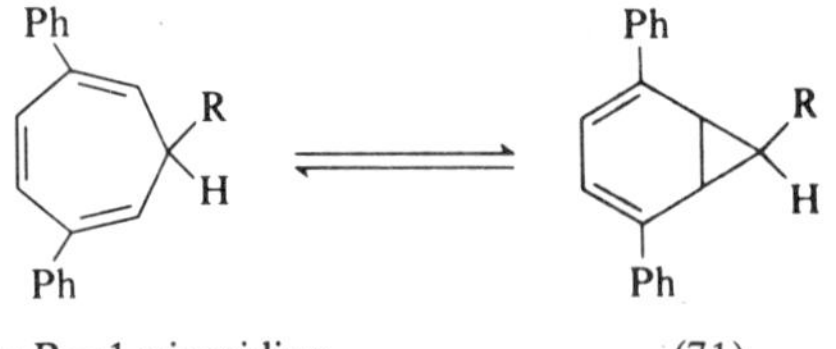

(70) a; R = 1-piperidino
b; R = cyclohexyl
c; R = H

(71)

Scheme 18

form slightly more than does the sterically similar cyclohexyl group. It is suggested that HOMO–LUMO interactions between π-electron donors and a cyclopropyl ring whether in a perpendicular or a bisected conformation must cause stabiliza-

[104] K. E. Gilbert and J. E. Baldwin, *J. Amer. Chem. Soc.*, 1976, **98**, 1593.
[105] W. W. Schoeller, *J. Amer. Chem. Soc.*, 1977, **99**, 5919.
[106] G. D. Andrews and J. E. Baldwin, *J. Amer. Chem. Soc.*, 1977, **99**, 4853.
[107] S. W. Staley, M. A. Fox, and A. Cairncross, *J. Amer. Chem. Soc.*, 1977, **99**, 4524.

tion, not destabilization. An investigation has been undertaken[108] of the thermal rearrangements of spiro[4,4]nonatetraene (72) and the related less unsaturated analogues spiro[4,4]nona-1,3,6-triene (73), spiro[4,4]nona-1,3,7-triene (74), and spiro[4,4]nona-1,3-diene (75). Low activation barriers are observed for (72) and (73) relative to (74) and (75) and a transition state is favoured which implicates the π-system at the migrating carbon for (72) and (73). The transition state (76) for a

(72) (73) (74) (75)

simple sigmatropic shift of a carbon unit attached to a cyclopentadiene ring is appropriate for reactions of (74) and (75), but may be modified in the cases of (72) and (73) owing to interactions of a *p*-orbital on the migrating carbon (LUMO of an olefin or 1,3-diene unit with the cyclopentadienyl π-system HOMO). This HOMO–LUMO interaction (77) would stabilize the transition states, leading to the

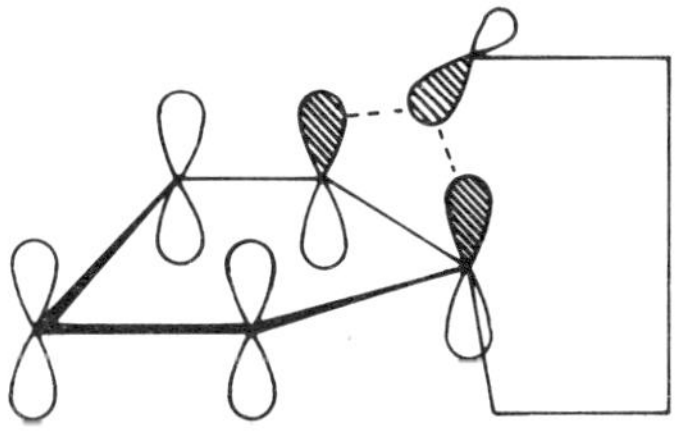

(76) Orbital interaction during 1,5-alkyl shift

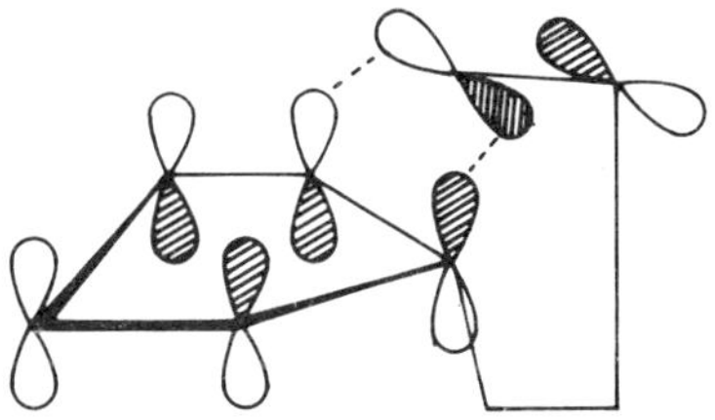

(77) Secondary π-orbital interaction during 1,5-vinyl shift

low barriers to rearrangement for (72) and (73) as compared with (74) and (75). Thermal isomerization[109] of cyclopropylallene at 380 °C using a contact time of *ca.* 2 minutes leads to a mixture of five components of which two, 1,4-dimethylcyclohexa-1,3-diene and *p*-xylene, cannot be explained without invoking a facile 1,7-hydrogen migration. The thermolysis of bicyclo[2,2,0]hex-2-ene has been approached[110] by generating a complete list of mechanistic alternatives, both plausible and otherwise and then by experiment rigorously excluding as many as possible. By this method there is found no pericyclic alternative to the symmetry-forbidden least-motion mechanism shown in Scheme 19. The kinetic distribution of isomeric xylenes formed on thermal aromatization of *dl*- and *meso*-1,1-dimethyl-3,3′-bicyclopropenyls and

Scheme 19

[108] M. F. Semmelhack, H. N. Weller, and J. S. Foos, *J. Amer. Chem. Soc.*, 1977, **99**, 292.
[109] D. E. Minter and G. J. Fonken, *Tetrahedron Letters*, 1977, 1717.
[110] M. J. Goldstein, R. S. Leight, and M. S. Lipton, *J. Amer. Chem. Soc.*, 1976, **98**, 5717.

of 1,1′-dimethyl-3,3′-bicyclopropenyl has been determined.[111] The data are most consistent with a mechanism involving initial cleavage of one of the cyclopropane rings, followed by expansion of the other ring, closure to a Dewar-benzene, and finally opening of the Dewar intermediate to form aromatic products. A report has appeared[112] of a study of the kinetic and thermochemistry of the valence isomerization of benzvalene to benzene. The transformation is found to be essentially non-chemiluminescent and the valence isomerization may be viewed as a symmetry-forbidden $[_{\sigma}2+_{\sigma}2]$ process and as such it may be supposed that a biradical mechanism is involved.

[111] J. H. Davis, K. J. Shea, and R. G. Bergman, *J. Amer. Chem. Soc.*, 1977, **99**, 1499.
[112] N. J. Turro, C. A. Renner, and T. J. Katz, *Tetrahedron Letters*, 1976, 4133.

13 Synthetic Methods

By G. PATTENDEN
Department of Chemistry, The University, Nottingham, NG7 2RD

1 Introduction

The first volume of a new Specialist Periodical Report entitled 'General and Synthetic Methods' has been published by the Chemical Society.[1]

2 Alkanes

A number of useful procedures for the reduction of secondary alcohols, based on free-radical intermediates, have emerged in the past few years. Billingham *et al.*[2] have now demonstrated that primary and secondary alcohols are converted into the corresponding hydrocarbon by radical-initiated reduction of their chloroformates, using tri-n-propylsilane in the presence of t-butyl peroxide at 140 °C (Scheme 1).

$$\mathrm{ROH} \xrightarrow{\mathrm{i}} \mathrm{ROCOCl} \xrightarrow{\mathrm{ii}} [\mathrm{R{-}O\dot{C}O} \rightarrow \dot{\mathrm{R}}] \rightarrow \mathrm{RH}$$

Reagents: i, $COCl_2$; ii, Pr_3SiH–$(Bu^tO)_2$

Scheme 1

Quantitative yields of hydrocarbons are realized when primary halides are reduced with $LiAlH_4$ in the presence of first-row transition-metal halides (*e.g.* $CoCl_2$,$NiCl_2$).[3] The same modified reducing agents also effect reductions of alkenes to alkanes, and of disubstituted alkynes to *cis*-alkenes, in high yields.

A novel method for the direct substitution of a hydroxy-group in an alcohol by an alkyl group proceeds *via* aminophosphonium salts [*viz.* (1)] produced by the reaction of the lithium alkoxyalkylcuprate with *NN*-methylphenylamino-triphenylphosphonium iodide (Scheme 2).[4]

$$\mathrm{ROH} \xrightarrow{\mathrm{i-iii}} \mathrm{R{-}O{-}CuBu_3Li_3} \xrightarrow{\mathrm{iv}} \underset{(1)}{[\mathrm{R{-}O{-}PPh_3}]^{+}[\mathrm{Bu_3CuNMe_2PhLi_2}]^{-}} \rightarrow \mathrm{R{-}Bu}$$

Reagents: i, MeLi; ii, CuI; iii, BuLi; iv, $Ph_3\overset{+}{P}N(Me)Ph\ I^-$

Scheme 2

[1] 'General and Synthetic Methods', ed. G. Pattenden, (Specialist Periodical Reports), The Chemical Society, London, 1978, Vol. 1.
[2] R. C. Billingham, R. A. Jackson, and F. Malek, *J.C.S. Chem. Comm.*, 1977, 344.
[3] E. C. Ashby and J. J. Lin, *Tetrahedron Letters*, 1977, 4481.
[4] Y. Tanigawa, H. Kanamaru, A. Sonoda, and S. I. Murahashi, *J. Amer. Chem. Soc.*, 1977, **99**, 2361.

Vedejs and Stolle[5] have shown that alkyl-lithium reagents add readily to the C—N linkage in aldehyde hydrazones, producing adducts [*viz.* (2)] which fragment, leading to the corresponding hydrocarbon, 'reductive alkylation', products (Scheme 3).

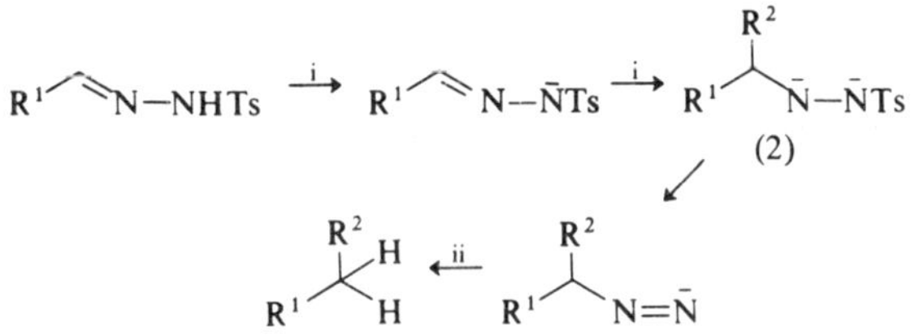

Reagents: i, R^2Li; ii, H_2O ($-N_2$)

Scheme 3

3 Alkenes

In no other area of natural product synthesis has the Wittig reaction had such a profound effect as in carotenoid synthesis. It is timely therefore that in an article based on an address delivered on the occasion of Professor Wittig's 80th birthday, H. Pommer should provide an account of the importance of the reaction in industrial practice generally, and in the synthesis of carotenoids and Vitamin A in particular.[6]

Leznoff and his co-workers[7] have extended their work on the use of insoluble polymer supports in synthesis, and have shown that the approach can be used to synthesize unsymmetrical carotenoids by the Wittig reaction. Thus the reaction between the di-aldehyde (7) and the cross-linked divinylbenzene–styrene copolymer (3) containing vicinal diol groups led to the mono-blocked polymer-bonded aldehyde (4), which was condensed with phosphoranylides to give the polymer-bonded Wittig reaction product (5), as shown in Scheme 4; hydrolysis with acid then led to (6). The synthesis of the strained methylene-bridged bicyclic alkenes (8), (9), and (10) *via* intramolecular olefinations using phosphoranylides attests further to the synthetic versatility of the Wittig reaction.[8]

The Wadsworth–Emmons modification of the Wittig reaction, using phosphonate carbanions, is well known to lead to predominantly (~95%) *trans*-olefins in condensation reactions with aldehydes. Thus condensation between iso-butyraldehyde and the phosphonate (11) produces a sample of the alkene (12) that contains less than 7% of the *cis*-isomer (14). Observations of this type are usually rationalized on the basis of reversible formation and interconversion of the *erythro* and *threo* 'betaine' intermediates, producing ultimately the more stable *threo*-form, which then undergoes elimination to the *trans*-olefin. It is known that phosphorus compounds that have five-membered rings react faster than their acyclic counterparts, and the observation has now been exploited in a synthetically useful

[5] E. Vedejs and W. T. Stolle, *Tetrahedron Letters*, 1977, 135.
[6] H. Pommer, *Angew. Chem. Internat. Edn.*, 1976, **15**, 161.
[7] C. C. Leznoff and W. Symanyk, *J. Org. Chem.*, 1977, **42**, 3203.
[8] K. B. Becker, *Helv. Chim. Acta*, 1977, **60**, 69, 81.

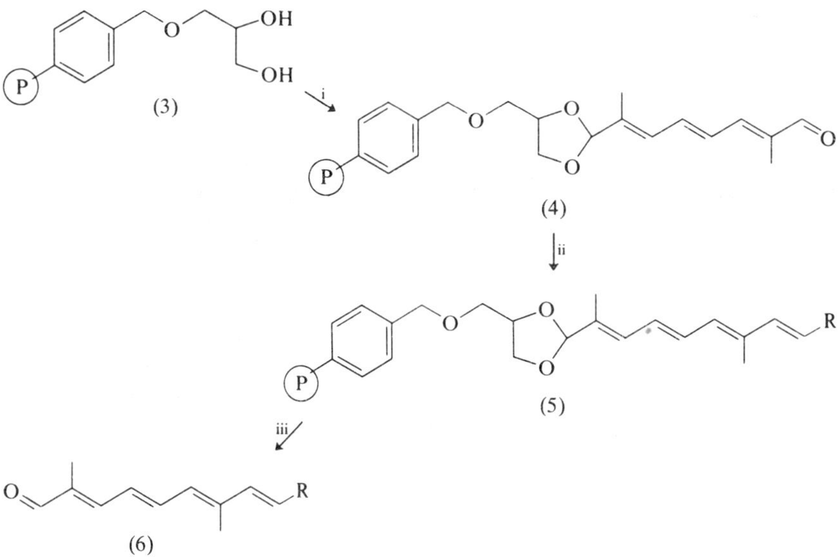

Reagents: i, 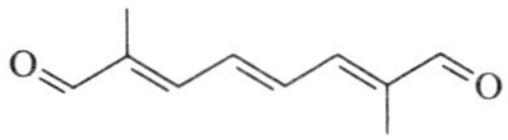 ; ii, $R\bar{C}H{-}\overset{+}{P}{<}$; iii, HCl–THF

(7)

Scheme 4

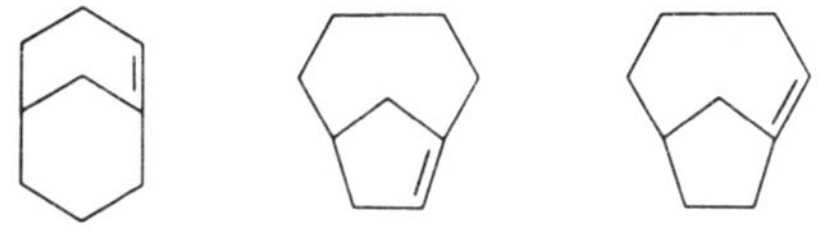

(8) (9) (10)

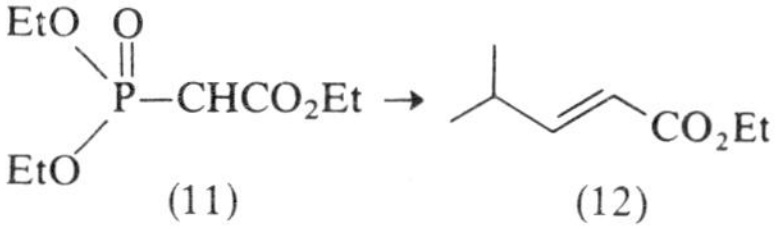

(11) (12)

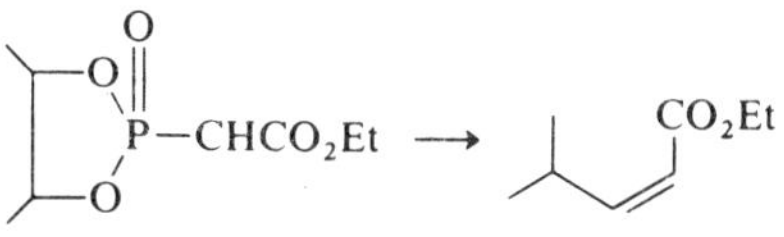

(13) (14)

approach to *cis*-alkenes by the Wadsworth–Emmons reaction, using the cyclic phosphonate (13); this is found to react with isobutyraldehyde, for example, leading to largely *cis*-alkene (14) (~70%).[9]

[9] E. Breuer and D. M. Banet, *Tetrahedron Letters*, 1977, 1141; *cf.* B. Deschamps, J. P. Lampin, F. Mathey, and J. Segden-Penne, *ibid.*, 1977, 1137 and I. F. Wilson and J. C. Tebby, *J. Chem. Soc.*, (*C*), 1972, 2713.

Allyl-tin and allyl-silicon reagents are increasingly used as reagents in synthesis, and both reagents are easily synthesized *via* Wittig reactions involving β-trimethylstannyl and β-trimethylsilyl phosphoranylides.[10]

The scope provided by carbonyl olefination with trimethylsilylmethyl-lithium and its derivatives (*i.e.* the Peterson reaction), and the importance of the reaction as an alternative to the Wittig reaction, has been stressed in earlier Reports [see *Ann. Reports* (*B*), 1975, **72**, p. 311; 1976, **73**, p. 304]. Now Kauffman *et al.*[11] have shown that the analogous reaction with triphenyl*stannyl*methyl lithium could offer even more advantages as an alternative olefination sequence. Vinylsilanes are valuable 'starters' for the synthesis of the β-hydroxy-silane intermediates in the Peterson reaction, and a number of alternative approaches to these molecules have been developed in the past few years. A new approach starts from a carbonyl compound, which is first converted into the corresponding tosylhydrazone; treatment with BuLi in tetramethylethylenediamine, followed by trimethylsilyl chloride, then leads to the required vinylsilane (Scheme 5).[12]

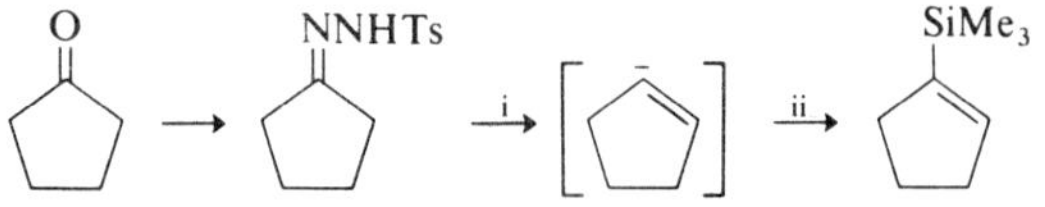

Reagents: i, BuLi–$Me_2NCH_2CH_2NMe_2$ (TMEDA); ii, Me_3SiCl

Scheme 5

Details of the conversion of 1,2-diols into alkenes *via* reductive elimination of cyclic phosphoric amide derivatives with dissolving metals have been presented,[13] and Barton and his co-workers[14] have shown that the conversion is just as readily accomplished by treatment of the corresponding bis-dithiocarbonates with Bu^n_3SnH. In related studies, Lythgoe and Waterhouse[15] demonstrate that treatment of β-hydroxy-sulphides, β-hydroxy-sulphones, and β-chloro-sulphides with Bu^n_3SnH also provides an expeditious synthesis of alkenes.

The decarboxylative dehydration of β-hydroxy-acids can be accomplished literally within seconds at 0 °C, using the adduct of triphenylphosphine and ethyl azodicarboxylate,[16] and in one of the first illustrations of the use of organotellurium(I) reagents in synthesis, Clive and Menchen have shown that epoxides can be converted into the corresponding alkenes by treatment with *OO*-diethyl phosphorotelluroate.[17]

The use of di-imide as a reducing agent for the stereospecific hydrogenation of alkynes to *cis*-alkenes is well documented. In a procedure which seems to have several practical advantages over existing methods, Kondo *et al.*[18] have now shown

[10] D. Seyferth, K. R. Wursthorn, and R. E. Mammarella, *J. Org. Chem.*, 1977, **42**, 3104.
[11] T. Kauffman, R. Kriegesmann, and A. Wothermann, *Angew. Chem. Internat. Edn.*, 1977, **16**, 862.
[12] R. F. Taylor, C. R. Degenhardt, W. P. Melega, and L. A. Paquette, *Tetrahedron Letters*, 1977, 159.
[13] J. A. Marshall and M. E. Lewellyn, *J. Org. Chem.*, 1977, **42**, 1311.
[14] A. G. M. Barrett, D. H. R. Barton, R. Bielski, and S. W. McCombie, *J.C.S. Chem. Comm.*, 1977, 866.
[15] B. Lythgoe and I. Waterhouse, *Tetrahedron Letters*, 1977, 4223.
[16] J. Mulzer and G. Brüntrup, *Angew. Chem. Internat Edn.*, 1977, **16**, 255.
[17] D. L. J. Clive and S. M. Menchen, *J.C.S. Chem. Comm.*, 1977, 658.
[18] K. Kondo, S. Murai, and N. Sonoda, *Tetrahedron Letters*, 1977, 3727.

that elemental selenium provides a useful alternative oxidizing agent for the generation of di-imide from hydrazine. Disubstituted alkynes are also converted into the corresponding *cis*-alkenes with the organo-copper reagent prepared from CuI and two equivalents of a primary Grignard reagent;[19] this reaction presumably proceeds *via* copper hydride species.

Vinyl organometallic reagents, which are easily available from the corresponding alkynes, have featured in several useful routes to functionalized alkenes[20—22] and 1,3-dienes[23,24] published this year (Scheme 6).

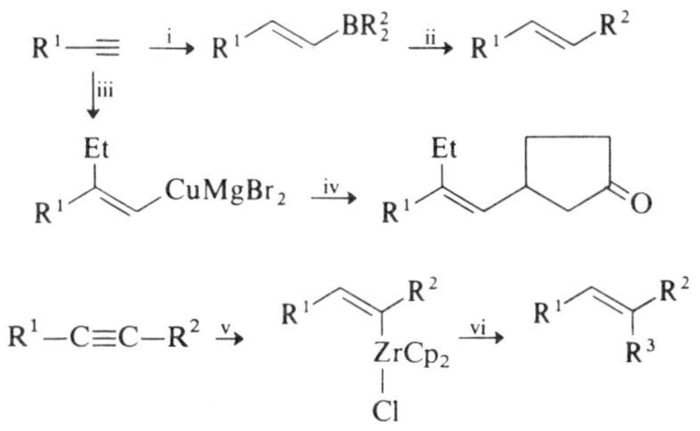

Reagents: 1, $R_2{}^2BH$; ii, $Pd(OAc)_2$–Et_3N; iii, $EtCuMgBr_2$; iv, 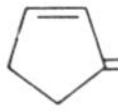=O; v, $Cl(H)ZrCp_2$; vi, R^3X–$Ni(PPh_3)_4$

Scheme 6

The boron-trifluoride-catalysed elimination of borate esters of secondary alcohols provides a useful synthesis of disubstituted alkenes,[25] and the combination of nucleophilic addition to vinyl sulphones, followed by *in situ* Ramberg–Backlund rearrangement, provides an interesting approach to olefins of type (15).[26]

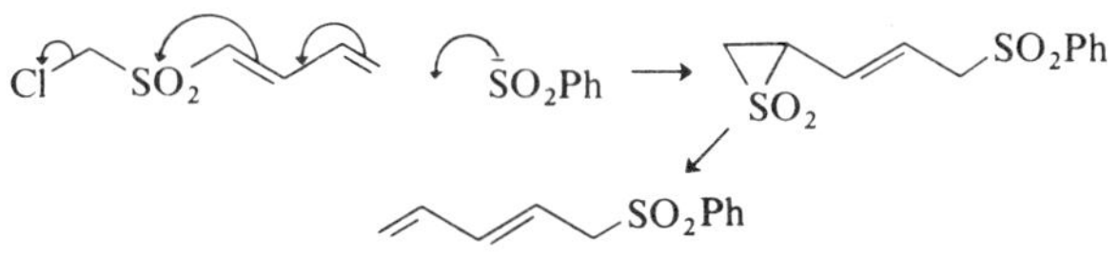

(15)

In a procedure which may have scope for the synthesis of more difficultly accessible terminal alkenes, Poulter and his co-workers[27] have shown that this type of alkene is produced by the addition of Grignard reagents to Eschenmoser's salt (16) followed by oxidation and elimination (Scheme 7).

[19] J. K. Crandall and F. Collonges, *J. Org. Chem.*, 1977, **41**, 4089.
[20] H. Yatagai, Y. Yamamoto, K. Maruyama, A. Sonoda, and S. I. Murahashi, *J.C.S. Chem. Comm.*, 1977, 852.
[21] A. Marfat, P. R. McGuirk, R. Kramer, and P. Helquist, *J. Amer. Chem. Soc.*, 1977, **99**, 253.
[22] E.-I. Negishi and D. E. Van Horn, *J. Amer. Chem. Soc.*, 1977, **99**, 3168.
[23] M. Yoshifuji, M. J. Loots, and J. Schwartz, *Tetrahedron Letters*, 1977, 1303.
[24] H. Westmijze, H. Kleijn, J. Meijer, and P. Vermeer, *Tetrahedron Letters*, 1977, 869.
[25] M. P. Doyle, S. B. Williams, and C. C. McOsker, *Synthesis*, 1977, 717.
[26] T. B. R. A. Chen, J. J. Burger, and E. R. de Waard, *Tetrahedron Letters*, 1977, 4527.
[27] J. L. Roberts, P. S. Borromeo, and C. D. Poulter, *Tetrahedron Letters*, 1977, 1299.

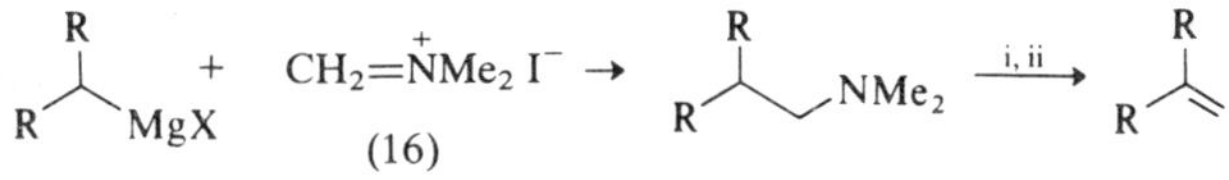

Reagents: i, H_2O_2; ii, Δ, 150 °C

Scheme 7

Although these Reports have commented frequently on the use of the 'ene' reaction in the elaboration of complex ring systems, in general the synthetic potential of the reaction is diminished somewhat by the forcing thermal conditions required to make it go. Gill and Wallace[28] have now shown that the ene reaction of chloral with olefins is greatly accelerated in the presence of a Lewis acid catalyst, and furthermore that these catalysed reactions have high regio- and stereo-specificities [*e.g.* (17)→(18)].

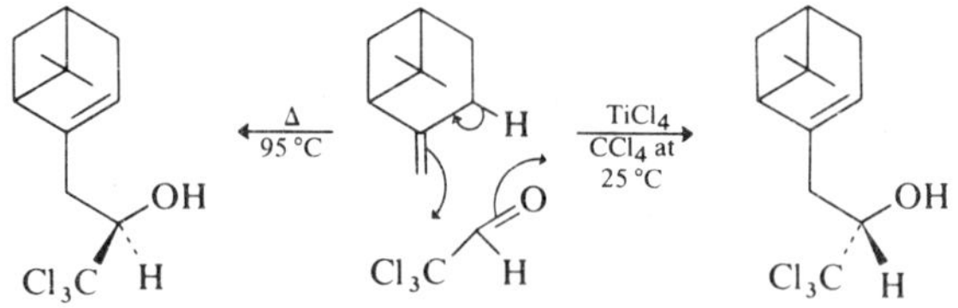

(83%) (17) (100%) (18)

In a remarkably short synthesis of the steroid nucleus, Funk and Vollhardt[29] have shown that the pentacyclic ketone (22) is produced when the 1,5-diyne (19) is added to bis(trimethylsilyl)acetylene. This synthesis, which avoids the isolation of any of the intermediates, proceeds through the benzocyclobutane (20) and *via* intramolecular Diels–Alder reaction of the *ortho*-xylylene (21). The intermediate (23), which is suitable for pentacyclic triterpene synthesis, has also been synthesized by an intramolecular Diels–Alder reaction.[30]

In a novel approach towards the polyene dimethylcrocetin (28), Quinkert *et al.*[31] have demonstrated that the sunlight of Frankfurt is an effective enough 'reagent' for the photochemical cleavage of the cyclohexadienones, (24) and (25) affording the seco-isomeric dienyl ketones (26) and (27) respectively, which were used subsequently to synthesize (28).

4 Alkynes, Enynes, and Allenes

1,1-Dichloroalkenes are useful intermediates in the synthesis of a wide range of mono- and di-substituted alkynes [see *Ann. Reports* (*B*), 1975, **72**, p. 316). A useful alternative olefination procedure for the preparation of these compounds[32]

[28] G. B. Gill and B. Wallace, *J.C.S. Chem. Comm.*, 1977, 380, 382.
[29] R. L. Funk and K. P. C. Vollhardt, *J. Amer. Chem. Soc.*, 1977, **99**, 5483.
[30] T. Kametani, Y. Hirai, F. Satoh, and K. Fukumoto, *J.C.S. Chem. Comm.*, 1977, 16.
[31] G. Quinkert, K. R. Schmieder, G. Durner, K. Hache, A. Stegk, and D. H. R. Barton, *Chem. Ber.*, 1977, **110**, 3582.
[32] P. Coutrot, C. Laurenco, and J. F. Normant, *Synthesis*, 1977, 615.

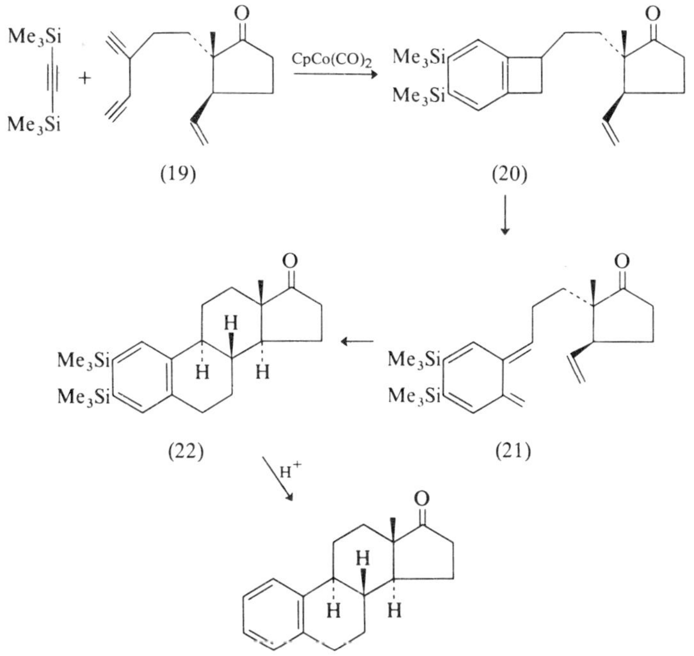
Me3Si
Me3Si
O
CpCo(CO)2
(19)
(20)
H
H
H
(22)
(21)
H+

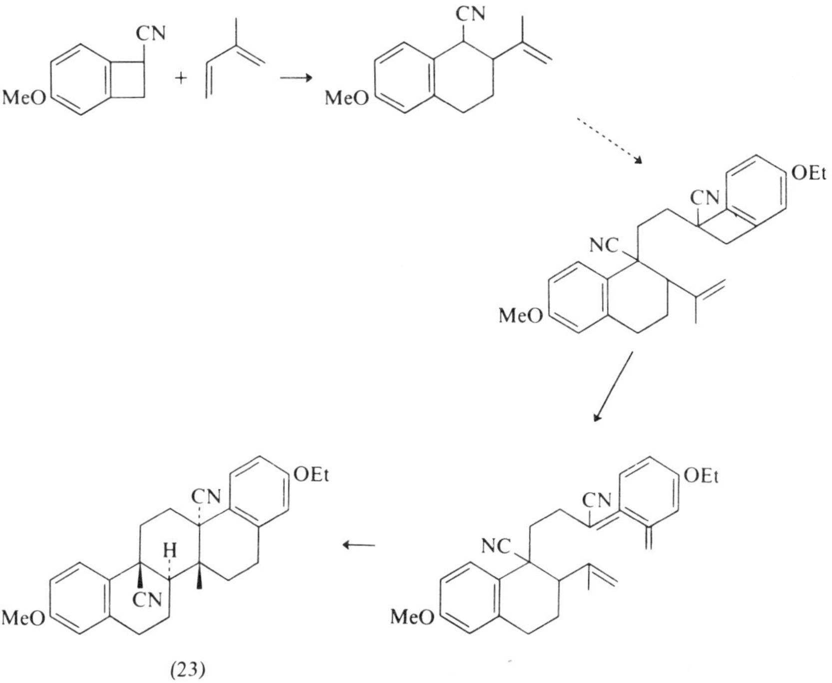
CN
MeO
OEt
NC
H
(23)

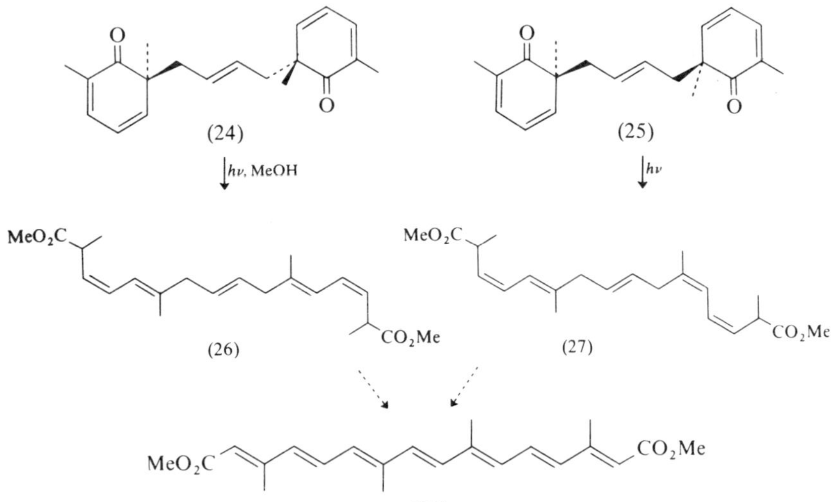

(28)

uses the phosphorane (29) prepared from bromotrichloromethane and hexamethylphosphortriamide (Scheme 8).[33]

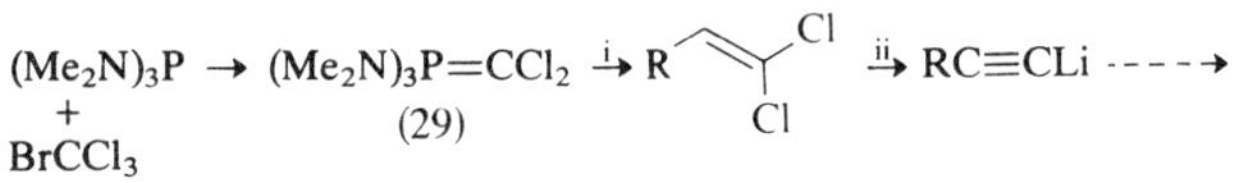

Reagents: i, RCHO; ii, BuLi

Scheme 8

Brown and Negishi[34] have shown that the isomerization of the triple bond in alkynyl-organoboranes can be effected in high yield, using the 'acetylene zipper' potassium 3-aminopropylamide (KAPA). The resulting ω-alkynyl-organoborane can then be used in typical organoborane reactions (Scheme 9). Full details have

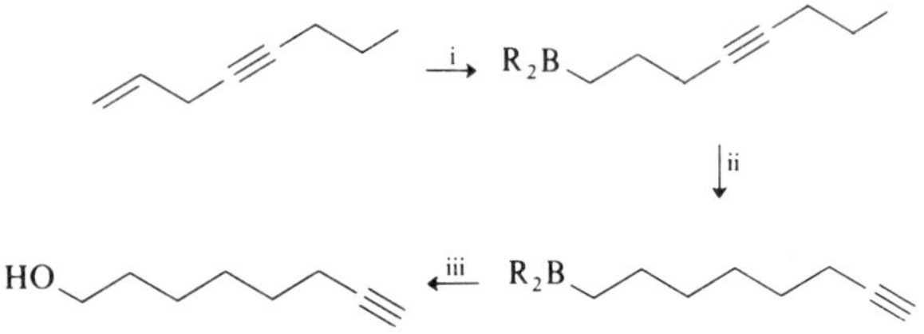

Reagents: 1, R_2BH; ii, $KHN(CH_2)_3NH_2$ (KAPA); iii, H_2O_2–NaOH

Scheme 9

[33] W. G. Salmond, *Tetrahedron Letters*, 1977, 1239.
[34] C. A. Brown and E.-I. Negishi, *J.C.S. Chem. Comm.*, 1977, 318.

appeared of Coke's work on the synthesis of acetylene ketones, which uses a variant of the Eschenmoser fragmentation [*e.g.* (30 → (31)].[35]

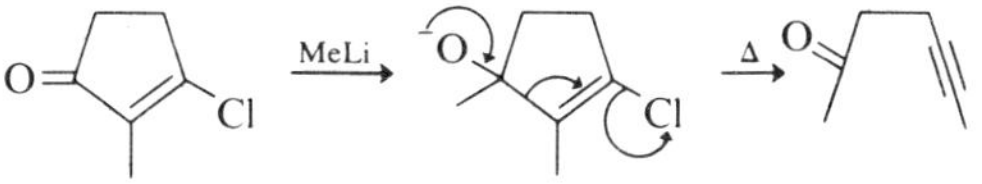

(30) (31)

The coupling of alkynyl-zinc reagents with (*Z*)- and (*E*)-alkenyl halides in the presence of a Pd catalyst provides a refreshingly new and efficient method for the regio- and stereo-selective synthesis of conjugated enynes (Scheme 10).[36] Symmetrically substituted *trans*-enynes are produced by the hydroalumination of conjugated diynes followed by protonation (Scheme 10),[37] and in their total synthesis of laurencin (32) from marine algae, Murai *et al.* used the familiar Wittig reaction approach to introduce the sensitive terminal enyne unit.[38]

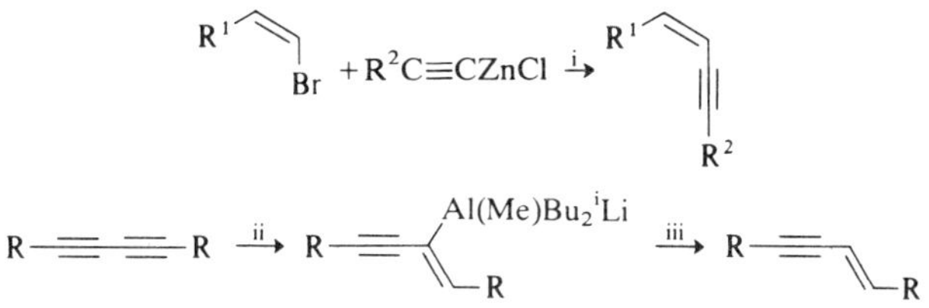

Reagents: i, Pd complex; ii, MeLi–Bu^i_2AlH; iii, H^+

Scheme 10

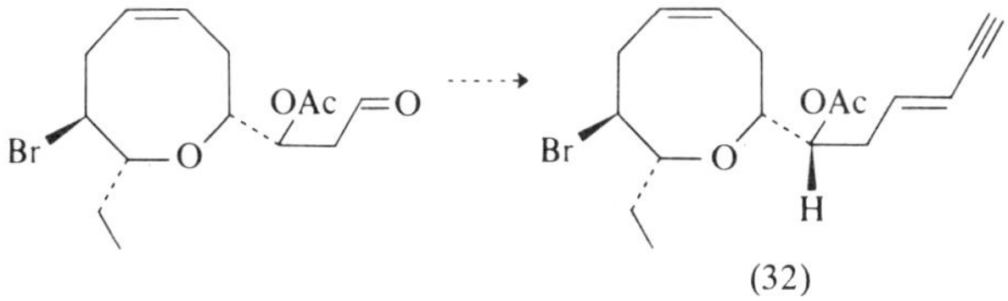

(32)

The Wadsworth–Emmons olefination procedure has proved useful in the synthesis of allenic amides of type (34), using the phosphonate amide (33),[39] and allenic boranes feature in a useful synthesis of substituted allenes of the type (35).[40]

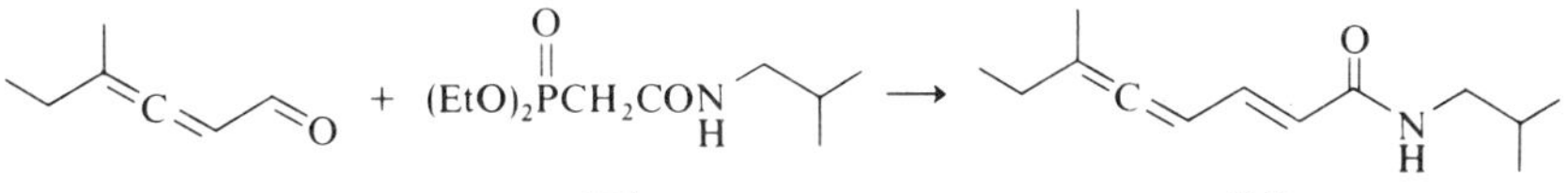

(33) (34)

[35] J. L. Coke, H. J. Williams, and S. Natarajan, *J. Org. Chem.*, 1977, **42**, 2380.
[36] A. O. King, N. Okukado, and E.-I. Negishi, *J.C.S. Chem. Comm.*, 1977, 683.
[37] G. Zweifel, R. A. Lynd, and R. E. Murray, *Synthesis*, 1977, 52.
[38] A. Murai, H. Murase, H. Matsue, and T. Masamune, *Tetrahedron Letters*, 1977, 2507.
[39] P. D. Landor, S. R. Landor, and O. Odyek, *J.C.S. Perkin I*, 1977, 93.
[40] M. M. Midland, *J. Org. Chem.*, 1977, **42**, 2650.

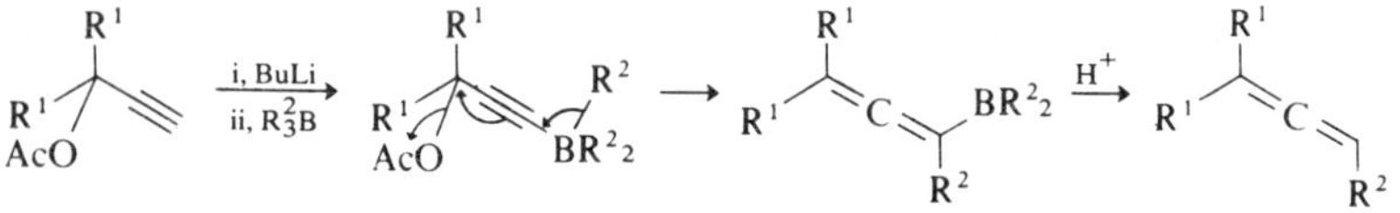

(35)

In a demonstration of the application of 1,4-addition of organocuprates to allenic esters, Bertrand *et al.* have developed a useful synthetic approach to the irregular terpenoid carbon skeleton present in lavandulol (36) (Scheme 11).[41]

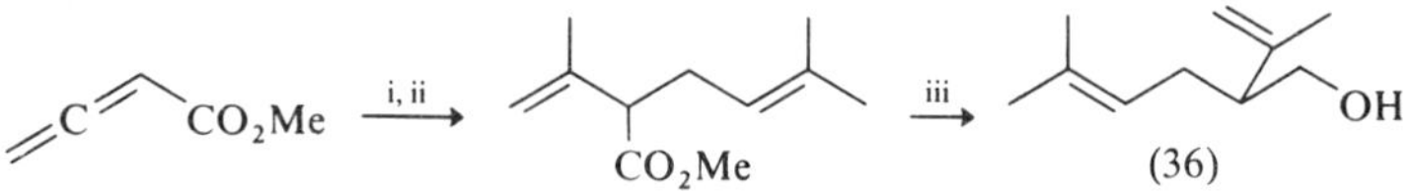

(36)

Reagents: i, Me_3CuLi; ii, $Me_2C{=}CHCH_2Br$; iii, $LiAlH_4$

Scheme 11

5 Halogeno-compounds

A very simple and efficient one-step method for the conversion of primary, secondary, and tertiary alcohols into the corresponding iodides is *via* treatment with trimethylsilyl iodide.[42] Primary alkyl chlorides are quantitatively converted into their corresponding bromides in the presence of ethyl bromide, *N*-methyl-2-pyrrolidinone, and a catalytic amount of metal bromide.[43] A new synthetic method for the transformation of alcohols into alkyl chlorides uses 2-chlorobenzoxazolium salts in the presence of tetraethylammonium chloride.[44]

The tertiary phosphine–carbon tetrachloride reagent has proved extremely useful for chlorinations of alcohols. Magid *et al.*[45] have now shown that the corresponding 'reagent' from hexachloroacetone and PPh_3 offers several practical advantages over CCl_4–PPh_3, particularly for the stereo- and regio-selective conversions of allyl alcohols into allyl chlorides [*e.g.* (37) → (38)]. Allylic bromides are produced by the 'anti-Markovnikov' addition of phenylselenyl bromide to terminal alkenes followed by oxidation with H_2O_2 [*e.g.* (39) → (40)].[46]

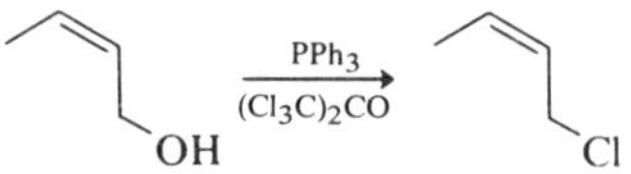

(37) (38)

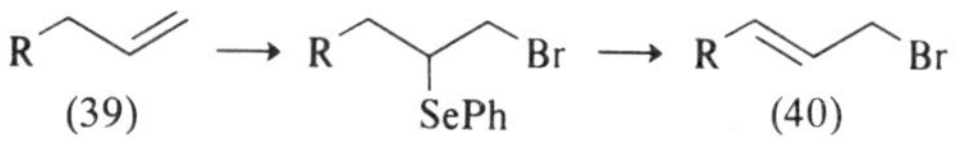

(39) (40)

Vinylic halides feature prominently in a number of useful synthetic transformations. Two general approaches to vinyl bromides published this year employ

41 M. Bertrand, G. Gil, and J. Viala, *Tetrahedron Letters*, 1977, 1785.
42 M. E. Jung and P. L. Ornstein, *Tetrahedron Letters*, 1977, 2659.
43 W. E. Willy, D. R. McKean, and B. A. Garcia, *Bull. Chem. Soc., Japan*, 1976, **49**, 1989.
44 T. Mukaiyama, S.-I. Shoda, and Y. Watanabe, *Chem. Letters*, 1977, 383.
45 R. M. Mazid, O. S. Fruehey, and W. L. Johnson, *Tetrahedron Letters*, 1977, 2999.
46 S. Raucher, *Tetrahedron Letters*, 1977, 3909.

vinyl-silanes[47] and $\alpha\beta$-epoxy-silanes[48] as starting materials (Scheme 12), whilst a third method proceeds *via* vinylcopper reagents (Scheme 13).[49]

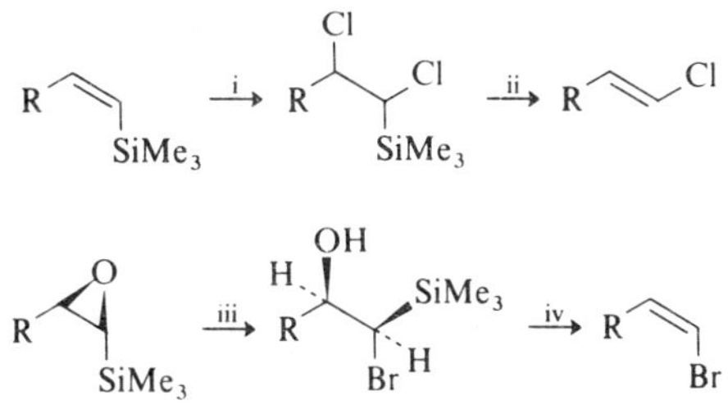

Reagents: i, Cl_2; ii, NaOMe; iii, HBr; iv, BF_3

Scheme 12

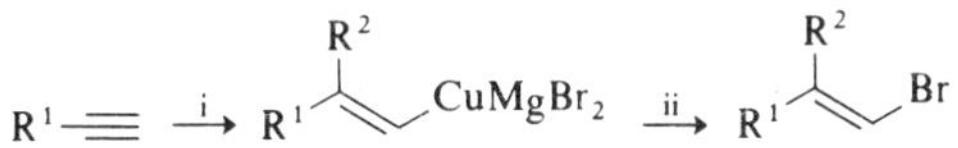

Reagents: i, $R^2CuMgBr_2$; ii, NBS

Scheme 13

6 Alcohols

Posner and his co-workers[50] have reported on the scopes and limitations of the two 'reagents' isopropyl alcohol and di-isopropyl alcohol on dried Woelm alumina, for the controlled reduction of the carbonyl function to alcohols. These reagents are weaker reducing agents than sodium cyanoborohydride and 9-borabicyclononane (9-BBN), and permit the selective reduction of aldehydes in the presence of ketones, and also the selective reductions of $\alpha\beta$-unsaturated aldehydes to the corresponding allylic alcohols; other functional groups, *e.g.* nitro, ester, and nitrile, are unaffected by the reagents. Two alternative methods which permit the specific reductions of aldehydes in the presence of ketones use either $NaBH_4$ in the presence of thiol[51] or organomagnesium iodides.[52]

Gibson and Bailey[53] have described the preparation of a series of polymer-supported reducing agents containing units of structure (41), from anion-exchange

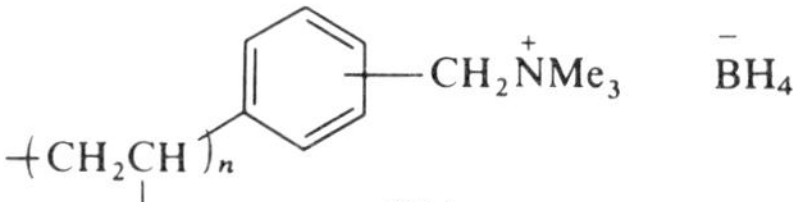

(41)

[47] R. B. Miller and G. McGarvey, *Synth. Comm.*, 1977, 475.
[48] P. F. Hudrlik, A. M. Hudrlik, R. J. Jona, R. N. Misra, and G. P. Withers, *J. Amer. Chem. Soc.*, 1977, **99**, 1993.
[49] A. B. Levy, P. Talley, and J. A. Dunford, *Tetrahedron Letters*, 1977, 3545.
[50] G. H. Posner, A. W. Runquist, and M. J. Chapdelaine, *J. Org. Chem.*, 1977, **42**, 1202.
[51] Y. Maki and K. Kikuchi, *Tetrahedron Letters*, 1977, 263.
[52] G. Cahiez and J. F. Normant, *Tetrahedron Letters*, 1977, 3383.
[53] H. W. Gibson and F. C. Bailey, *J.C.S. Chem. Comm.*, 1977, 815.

resins of the quaternary ammonium type, and aqueous $NaBH_4$. The method clearly offers scope for easy work-up of reduction products by simple filtration or by the use of columns.

Lithium β-isopinocamphenyl-9-borabicyclo[3,3,1]nonyl hydride is a newly described reagent for the asymmetric reduction of ketones,[54] and Crumbie *et al.*[55] have reported the selective reduction of *one* enantiomer of racemic ketones by actively fermenting yeast; the latter method yields optically active secondary alcohols, and leaves the ketone in optically pure form.

Although selenium dioxide is one of the most useful reagents for the direct introduction of oxygen into activated C—H bonds, the method suffers from the drawback that the reduced forms of selenium that are produced concurrently are often exceedingly difficult to remove. Umbreit and Sharpless[56] have now made the interesting observation that the SeO_2 can be employed in a *catalytic* sense in the presence of t-butyl hydroperoxide, which rapidly re-oxidizes the selenium species to SeO_2 as soon as it is consumed in the oxidation. Using this method, yields are comparable with those obtained by SeO_2 alone, and the messy work-up operations normally associated with oxidation by SeO_2 are avoided.

7 Ethers

Ethers are produced directly from alcohols and halides in the presence of bis(acetylacetonato)nickel as catalyst;[57] by contrast, phenols do not give the corresponding ethers under these conditions. A frequent limitation to the use of alkyl ethers as protecting groups for aliphatic alcohols has been the difficulty associated with their removal. Jung and Lyster[58] have now shown that the interesting electrophile trimethylsilyl iodide is an exceptionally efficient reagent for this purpose.

Olefins undergo smooth expoxidations when one uses benzene–peroxyseleninic acid, which is readily generated *in situ* from benzeneseleninic acid and 50% H_2O_2.[59] Epoxides of the Darzen type [*viz.* (42)] can be prepared by oxidation of β-hydroxy-ester dianions with iodine, as shown in Scheme 14.[60] A feature of this method is that it leads to the formation of the least hindered compound, whereas the classical Darzens procedure generally leads to diastereoisomeric mixtures.

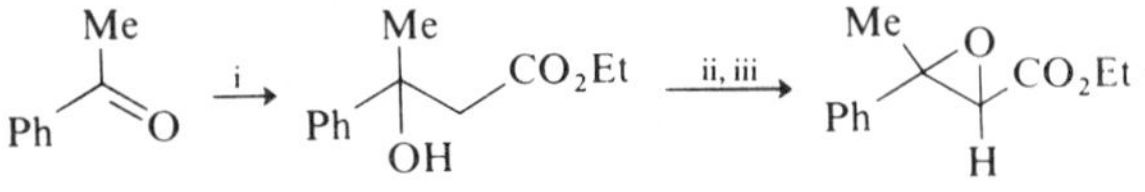

(42)

Reagents: i, $LiCH_2CO_2Et$; ii, $LiNPr^i_2$; iii, I_2

Scheme 14

[54] S. Krishnamurthy, F. Vogel, and H. C. Brown, *J. Org. Chem.*, 1977, **42**, 2534.
[55] R. L. Crumbie, D. D. Ridley, and G. W. Simpson, *J.C.S. Chem. Comm.*, 1977, 315.
[56] M. A. Umbreit and K. B. Sharpless, *J. Amer. Chem. Soc.*, 1977, **99**, 5526.
[57] M. Yamashita and Y. Takegami, *Synthesis*, 1977, 803.
[58] M. E. Jung and M. A. Lyster, *J. Org. Chem.*, 1977, **42**, 3761.
[59] P. A. Grieco, Y. Yokoyama, S. Gilman, and M. Nishizawa, *J. Org. Chem.*, 1977, **42**, 2034.
[60] G. A. Kraus and M. J. Taschner, *Tetrahedron Letters*, 1977, 4575.

The antifungal antibiotic 'cerulenin' (45), containing an interesting epoxide moiety, has attracted the attentions of the synthetic chemist, and two syntheses of the antibiotic have been published during the period covered by this Report. In each synthesis the epoxide moiety was introduced at an early stage, in the form of the butanolide (43) and the anhydride (44).[61,62]

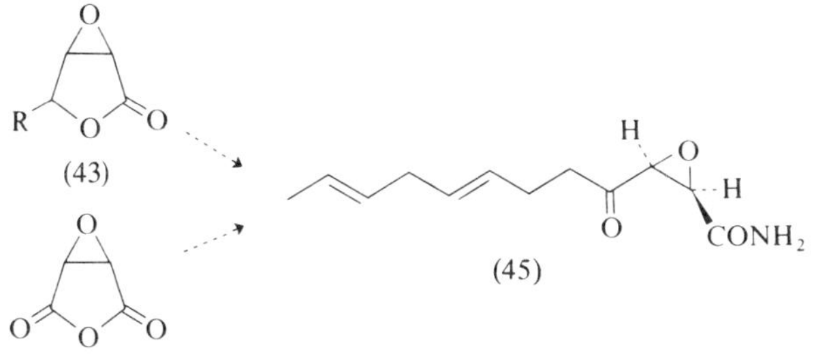

(43) (45) (44)

Both the Wadsworth–Emmons[63] and the Horner[64] variants of the Wittig reaction offer considerable advantages over the Wittig method for the synthesis of vinyl ethers from α-alkoxyalky phosphorus ylides (Scheme 15).

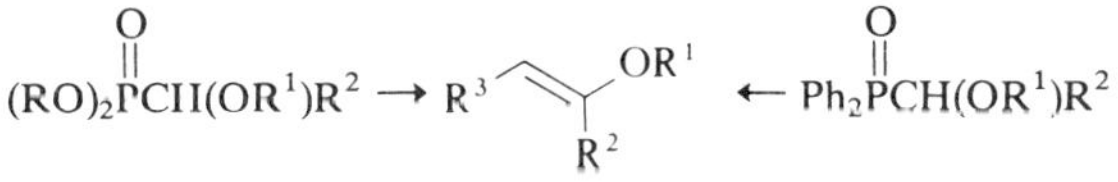

Scheme 15

8 Amines

Phase-transfer catalysis features in a useful synthesis of primary amines from alkyl halides *via* the corresponding diethoxyphosphine oxide [*viz*. (46)].[65] By the same general procedure, secondary amines are produced from primary amines by the alkylation, under phase-transfer conditions, of the amide (46), followed by treatment with gaseous HCl in THF at 25 °C, as shown in Scheme 16.

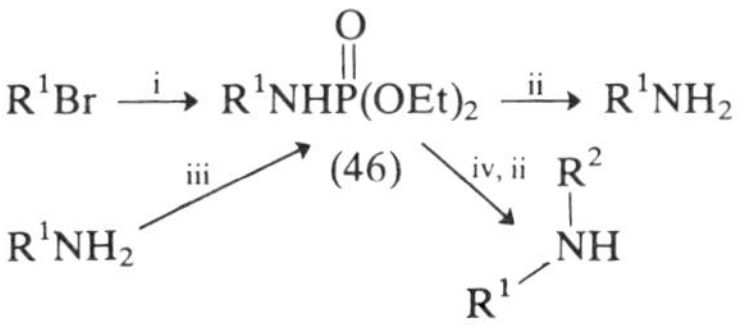

Reagents: i, $(EtO)_2PONH_2$; ii, HCl, iii, $(EtO)_2POH$, iv, R^2I

Scheme 16

[61] A. A. Jakabouski, F. S. Guzier, and M. Tishler, *Tetrahedron Letters*, 1977, 2399.
[62] E. J. Corey and D. R. Williams, *Tetrahedron Letters*, 1977, 3847.
[63] E. Schaumann and F. F. Grabley, *Annalen*, 1977, 88.
[64] C. Earnshaw, C. J. Wallis, and S. Warren, *J.C.S. Chem. Comm.*, 1977, 314.
[65] A. Zwierzak and J. Brylikowska-Piotrowicz, *Angew. Chem. Internat. Edn.*, 1977, **16**, 107; A. Zwierzak and I. Podstawczynska, *ibid.*, p. 702.

The merits of 1,2-diphenylmaleyl as an alternative to phthaloyl as a group for the protection of amino-functions have been discussed,[66] and a new method for the protection of amine functions is based on their conversion into 9-anthrylmethyl carbamates (47). In spite of the fact that carbamates are resistant to the action of

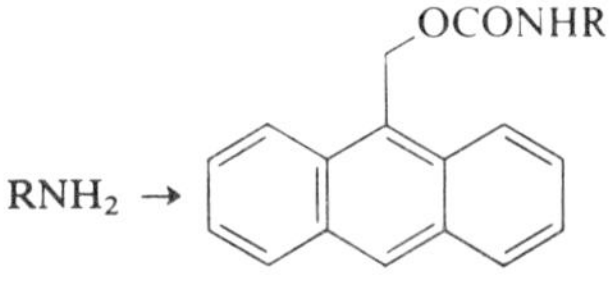

(47)

various bases and acids, the removal of these protecting groups is achieved simply by treating them with the sodium salt of methanethiol (4 min at 25 °C!).[67]

9 Aldehydes and Ketones

In the general pursuit of more regioselective oxidizing agents for poly-alcohol substrates, Jung and Speltz[68] have shown that secondary alcohols are oxidized selectively in the presence of primary alcohols, by treatment of the trimethylsilyl or t-butyl ethers of the alcohols with triphenylcarbonium salts [*e.g.* (48) → (49)]; these oxidations proceed *via* initial hydride abstraction, followed by loss of the group on oxygen.

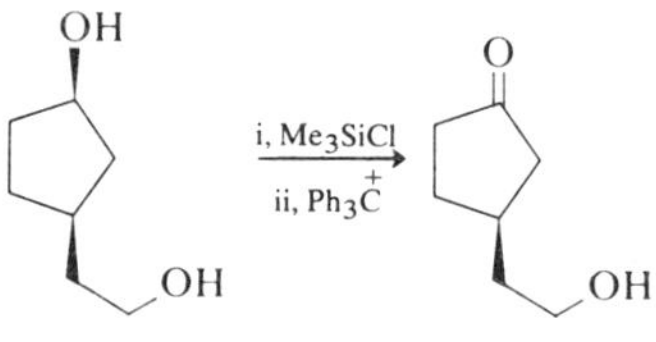

(48) (49)

Chromyl chloride adsorbed on silica-alumina is an effective reagent for the oxidation of alcohols to carbonyl compounds when the use of neutral, non-aqueous conditions is important.[69] The method also offers the manipulative convenience provided by the recently described method based on the use of chromic acid on anion-exchange resins [see *Ann. Reports* (*B*), 1976, **73**, p. 320]. Yet another oxidizing agent is pyridine oxodiperoxychromium(VI), one of the family of modified chromic acids, which is prepared by adding hydrogen peroxide to a mixture of pyridine and CrO_3 in water.[70]

The homogeneous oxidation of secondary alcohols to ketones by molecular oxygen can be effected at ambient temperature in the presence of $PdCl_2$–NaOAc,[71] and the technique of 'dry ozonization' has proved particularly useful for the

66 U. Zehavi, *J. Org. Chem.*, 1977, **42**, 2819.
67 N. Kornblum and A. Scott, *J. Org. Chem.*, 1977, **42**, 399.
68 M. E. Jung and L. M. Speltz, *J. Amer. Chem. Soc.*, 1976, **98**, 7882.
69 J. S. Filippo and C.-I. Chern, *J. Org. Chem.*, 1977, **42**, 2182.
70 G. W. J. Fleet and W. Little, *Tetrahedron Letters*, 1977, 3749.
71 T. F. Blackburn and J. Schwartz, *J.C.S. Chem. Comm.*, 1977, 157.

oxidation of cyclopropanes in the α-position, leading to cyclopropyl ketones [*e.g.* (50) → (51)].[72]

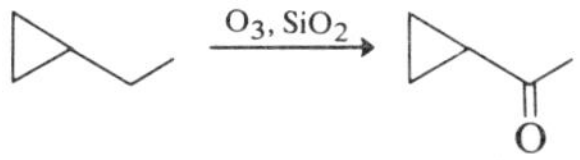

(50) (51)

Shono and his co-workers[73] have elaborated further on the potential provided by the synthesis of carbonyl compounds by anodic cleavage of glycols and their alkyl ethers (Scheme 17), and Ho and Olah[74] have shown that the oxidative cleavage of alkyl methyl ethers to carbonyl compounds can be smoothly accomplished with nitronium tetrafluoroborate.

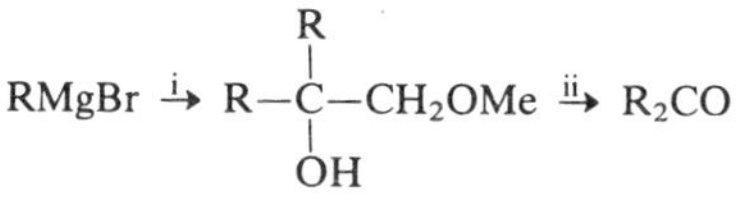

Reagents: i, $MeOCH_2CO_2Et$; ii, −e

Scheme 17

The reduction of acid chlorides to aldehydes can be effected in excellent yields, using the hydridotetracarbonylferrate anion (52) in aprotic solvents;[75] an acyl-hydride complex (53) is implicated in the reaction. An alternative practical

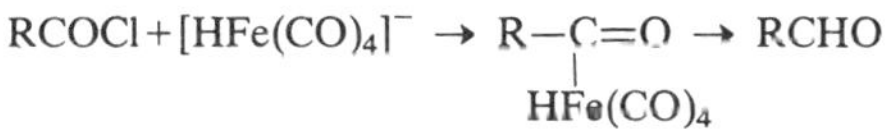

(52) (53)

synthesis of aldehydes from carboxylic acids is *via* reduction of the corresponding *sym*-triazolium salts with $NaBH_4$, followed by hydrolysis of the resulting *sym*-triazolines.[76]

Highly substituted ketones are produced by the acylation of α-anions of carboxylic acid salts with acid chlorides (Scheme 18),[77] and Pittman and Hanes[78] have shown that the polymer-bound bis(triphenylphosphine)chlorocarbonylrhodium compound (54) is an excellent regeneratable reagent for the synthesis of ketones from acid chlorides and organolithium reagents (Scheme 19).

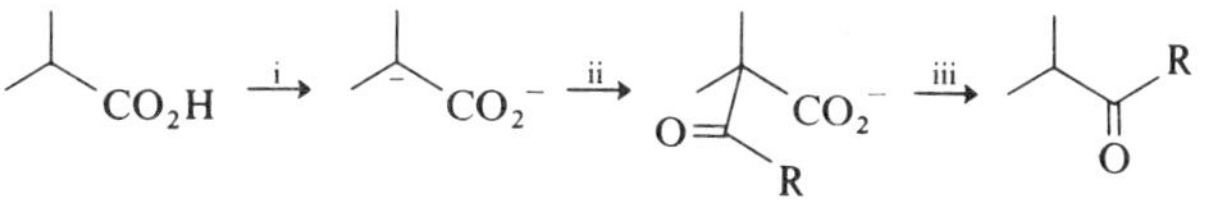

Reagents: i, $LiNPr_2^i$; ii, RCOCl; iii, Δ

Scheme 18

[72] E. Prokseh and A. de Meijere, *Angew. Chem. Internat. Edn.*, 1976, **15**, 761.
[73] T. Shono, H. Hamaguchi, Y. Matsumura, and K. Yoshida, *Tetrahedron Letters*, 1977, 3625.
[74] T.-L. Ho and G. A. Ohal, *J. Org. Chem.*, 1977, **42**, 3097.
[75] T. E. Cole and R. Pettit, *Tetrahedron Letters*, 1977, 781.
[76] G. Doleschall, *Tetrahedron*, 1977, **32**, 2549.
[77] A. P. Krapcho, D. S. Kashdan, E. G. E. Jahngen, and A. J. Lovey, *J. Org. Chem.*, 1977, **42**, 1189.
[78] C. U. Pittman and R. M. Hanes, *J. Org. Chem.*, 1977, **42**, 1194.

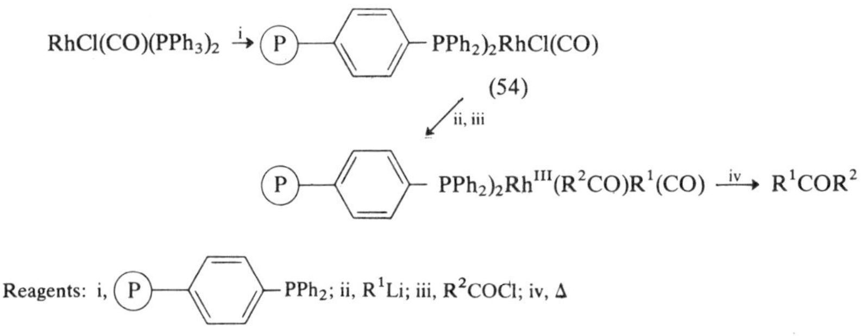

Scheme 19

The conversion of nitro-groups into carbonyl groups is well known, and of considerable value in synthesis. An extraordinarily simple method for effecting this conversion, which is based on the Nef reaction, involves oxidation by 'dry oxidation', whereby the nitro-compound is simply embedded into activated basic silica gel for 2 days; the carbonyl compound is then simply eluted from the silica gel.[79] t-Butyl hydroperoxide, in the presence of [VO(acac)$_2$] as catalyst, can be used in an alternative mild oxidative procedure for the nitro → carbonyl conversion.[80]

In what amounts to a unique combination of electrolytic decarboxylation and the Moffatt oxidation, Mandell *et al.*[81] have shown that carboxylic acids can be reduced to the lower homologous aldehyde by simple electrolysis in DMSO solution (Scheme 20). A useful method for the homologation of aldehydes (and ketones) is by direct conversion into the corresponding silyl-epoxide, using α-chloro-α-trimethylsilyl carbanion, followed by hydrolysis (Scheme 21).[82,83]

Scheme 20

Acid chlorides are smoothly coupled in the presence of [$Fe_2(CO)_9$], producing symmetrical ketones,[84] and Olah and his co-workers have shown that the simple Friedel–Crafts acylation of tetra-alkyl-silanes provides a useful route to unsymmetrical dialkyl ketones (Scheme 21).[85]

Organo-tin reagents have featured in a number of synthetically useful transformations in the past few years. In a further demonstration of their unique reactivity, Still[86] has shown that the ease with which alkyl-stannanes undergo oxidation provides a useful entry to carbonyl compounds of type (55) (Scheme 23).

[79] E. Keinan and Y. Mazur, *J. Amer. Chem. Soc.*, 1977, **99**, 3861.
[80] P. A. Bartlett, F. R. Green, tert., and T. R. Webb, *Tetrahedron Letters*, 1977, 331.
[81] L. Mandell, R. F. Daly, and R. A. Day, *J. Org. Chem.*, 1977, **42**, 1461.
[82] C. Burford, F. Cooke, E. Ehlinger, and P. Magnus, *J. Amer. Chem. Soc.*, 1977, **99**, 4536.
[83] F. Cooke and P. Magnus, *J.C.S. Chem. Comm.*, 1977, 513.
[84] T. C. Flood and A. Sarhangi, *Tetrahedron Letters*, 1977, 3861.
[85] G. A. Olah, T.-L. Ho, G. K. S. Prakash, and B. G. B. Gupta, *Synthesis*, 1977, 677.
[86] W. C. Still, *J. Amer. Chem. Soc.*, 1977, **99**, 4836.

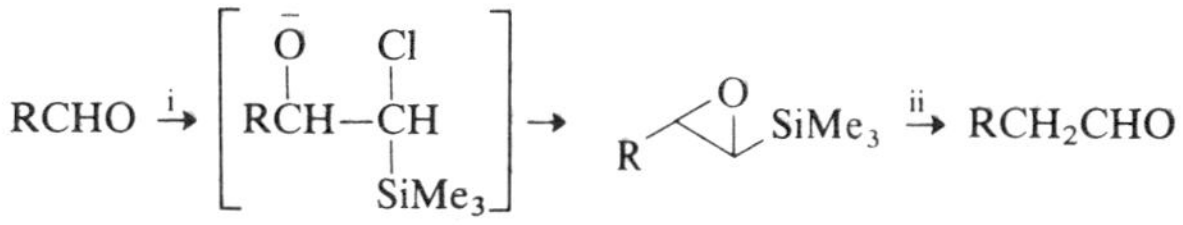

Reagents: i, $Me_3SiCH(Li)Cl$; ii, H_2SO_4–MeOH–H_2O

Scheme 21

$$RCOCl + Fe_2(CO)_9 \rightarrow RCOR$$

$$R^1COCl + R_4{}^2Si \xrightarrow{AlCl_3} R^1COR^2$$

Scheme 22

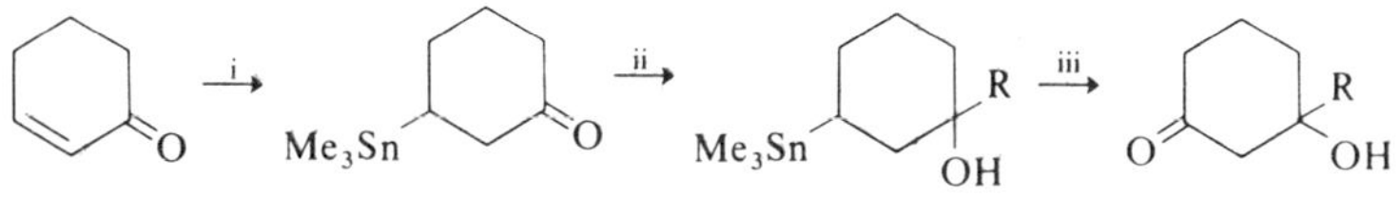

(55)

Reagents: i, Me_3SnLi; ii, RLi; iii, $CrO_3,2C_5H_5N$

Scheme 23

Thioketals enjoy widespread use for the protection of carbonyl compounds, and are also used as a means of inverting the normal polarity of the $\geq$C=O group (*i.e.* carbonyl umpolung). A useful new method for their formation is by reaction with thiosilanes in the presence of various acid catalysts.[87] Benzeneseleninic anhydride holds promise as a useful alternative to existing reagents for the regeneration of carbonyl compounds from their 1,3-dithiolan derivatives.[88] The same reagent is also useful for the hydrolysis of oximes and semicarbazones to carbonyl compounds.[89]

A simple alternative to conventional methods for the conversion of thioacetals into the parent carbonyl compound is to irradiate them in the presence of oxygen.[90] The interesting electrophilic reagent Me_3SiI is also highly recommended for the regeneration of carbonyl compounds from ketals under non-aqueous conditions.[91]

Functionalized Aldehydes and Ketones.—Ketone enolates play pivotal roles in a wide range of synthetic transformations, yet their *regioselective formation* is by no means an easy task. In an excellent review on this subject d'Angelo has summarized the methods available for accomplishing this end, and also outlined the major uses of the enolates in synthesis.[92] Contemporaneously, Jackman and Large[93] have reviewed the 'Structure and Reactivity of Alkali Metal Enolates', and Rasmussen[94] has summarized the recent developments in the preparation and

[87] D. A. Evans, L. K. Truesdale, K. G. Grimm, and S. L. Nesbitt, *J. Amer. Chem. Soc.*, 1977, **99**, 5009.
[88] D. H. R. Barton, N. J. Cussans, and S. V. Ley, *J.C.S. Chem. Comm.*, 1977, 751.
[89] D. H. R. Barton, D. J. Lester, and S. V. Ley, *J.C.S. Chem. Comm.*, 1977, 445.
[90] T. T. Takahashi, C. Y. Nakamura, and J. Y. Satoh, *J.C.S. Chem. Comm.*, 1977, 680.
[91] M. E. Jung, W. A. Andrus, and P. L. Ornstein, *Tetrahedron Letters*, 1977, 4175.
[92] J. d'Angela, *Tetrahedron*, 1976, **32**, 2979.
[93] L. M. Jackman and B. C. Lange, *Tetrahedron*, 1977, **33**, 2737.
[94] J. K. Rasmussen, *Synthesis*, 1977, 91.

reactions of silyl enol ethers derived from aldehydes and ketones. In the light of these reviews it is interesting that treatment of α-halogeno-ketones with zinc, trimethylsilyl chloride, and tetramethylethylenediamine (TMEDA) provides a regioselective method for the preparation of silyl enol ethers of the *more substituted type* [*viz.* (56)],[95] and that the reaction between n-butyl-lithium and THF at 25 °C

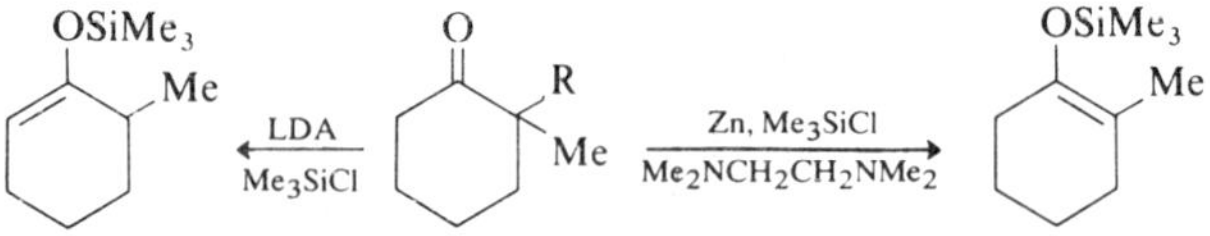

(R = H) (56) (R = Cl)

affords an exceptionally mild method for the preparation of the lithium enolate of acetaldehyde in essentially quantitative yield [(57) → (58)].[96]

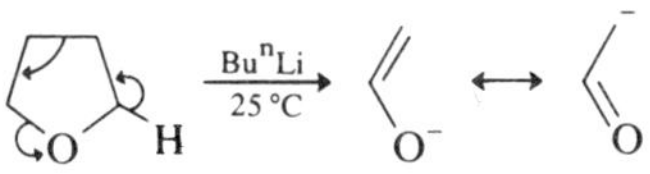

(57) (58)

Developments in 'methods' for controlling the directionality of the familiar aldol condensation continue. Heathcock and his co-workers[97] have observed complete regioselectivity in a number of simple aldol condensations where there is a bulky group R present in the enolate, *i.e.* (Z)-enolate (59) leads to the *erythro*-product while (E)-enolate (60) leads to the *threo*-product.

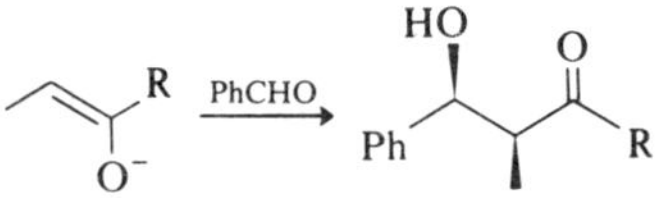

(Z)-(59)

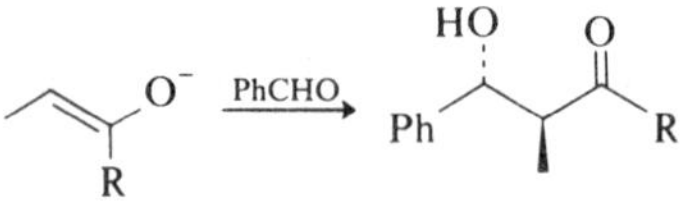

(E)-(60)

Efficient crossed aldol condensations between enol silyl ethers and carbonyl compounds, under the influence of a catalytic amount of tetrabutylammonium fluoride,[98] or in the presence of $TiCl_4$,[99] have been described. The interesting boron enolates (61) and (63) feature in directed aldol condensations between two carbonyl compounds, the di-n-butyl derivative (61) producing aldol (62), and the 9-BBN derivative (63) leading to the opposite regioselectivity [*viz.* (64)].[100]

[95] G. M. Rubottom, R. C. Mott, and D. S. Krueger, *Synth. Comm.*, 1977, **7**, 327.
[96] M. E. Jung and R. B. Blum, *Tetrahedron Letters*, 1977, 3791.
[97] W. A. Kleschick, C. T. Buse, and C. H. Heathcock, *J. Amer. Chem. Soc.*, 1977, **99**, 247.
[98] R. Noyori, K. Yokoyama, J. Sakata, I. Kuwajima, E. Nakamura, and M. Shimizu, *J. Amer. Chem. Soc.*, 1977, **99**, 1265.
[99] K. Banno, *Bull. Chem. Soc. Japan*, 1976, **49**, 2286.
[100] I. Inoue, T. Uchimaru, and T. Mukaiyama, *Chem. Letters*, 1977, 153.

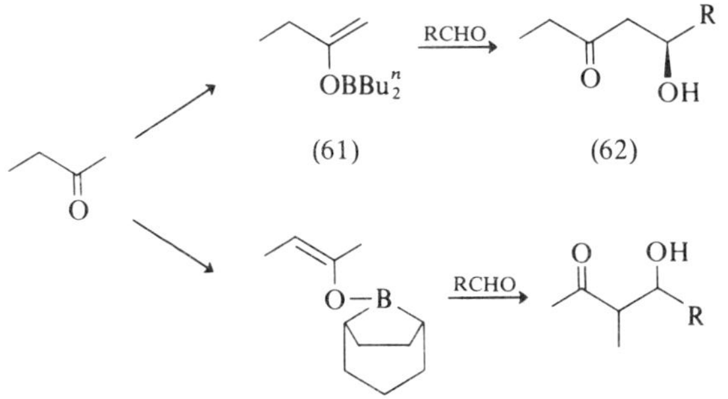

(61) (62)

(63) (64)

Two apparently useful methods for controlling carbonyl group condensations, and which ultimately lead to practical syntheses of $\alpha\beta$-unsaturated carbonyl systems, have been developed (Scheme 24). In the first of these, a keto-ester (65) is employed as a ketone-specific enol equivalent in reactions with aldehydes,[101] and in the second method use is made of an enol-silane of an acyl-silane as a specific enol of an aldehyde in condensation with an acetal, leading to (66).[102]

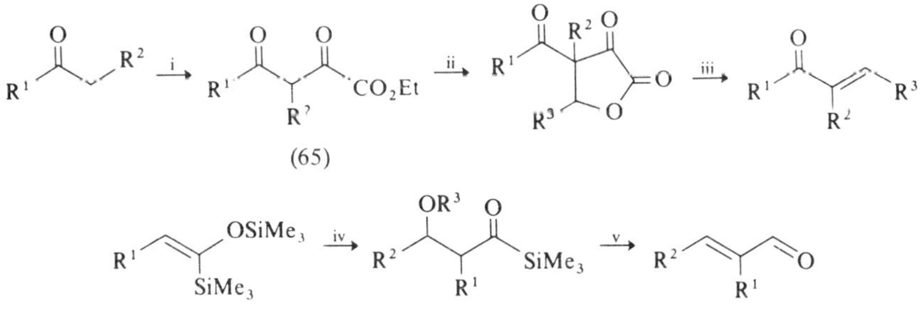

(65)

(66)

Reagents: i, NaOEt–$(EtO_2C)_2$; ii, R^3CHO; iii, $KHCO_3$; iv, BF_3–$R^2CH(OR^3)_2$; v, $Bu_4\overset{+}{N}\overset{-}{X}$

Scheme 24

A number of alternative syntheses of $\alpha\beta$-unsaturated aldehydes and ketones have been reported during the period covered by this Report, and these are collected in Scheme 25.[103—106] Methods for the synthesis of $\alpha\beta$-unsaturated aldehydes from carbonyl compounds by two- and three-carbon homologation procedures have also been reported (Scheme 26).[107,108]

[101] G. M. Ksander, J. E. McMurry, and M. Johnson, *J. Org. Chem.*, 1977, **42**, 1180.
[102] T. Sato, M. Arai, and I. Kuwajima, *J. Amer. Chem. Soc.*, 1977, **99**, 5827.
[103] K. Yakakoto, O. Nunokawa, and J. Tsuji, *Synthesis*, 1977, 721.
[104] A. A. Schegoler, W. A. Smit, S. A. Khurshudyan, V. A. Chertkov, and V. F. Kucherov, *Synthesis*, 1977, 324.
[105] V. Reutrakul and W. Kanghae, *Tetrahedron Letters*, 1977, 1377.
[106] E. Friedrich and W. Lutz, *Angew. Chem. Internat. Edn.*, 1977, **16**, 413.
[107] T. Nakai, T. Mimura, and A. Ari-Izumi, *Tetrahedron Letters*, 1977, 2425.
[108] S. F. Martin and P. J. Garrison, *Tetrahedron Letters*, 1977, 3875.

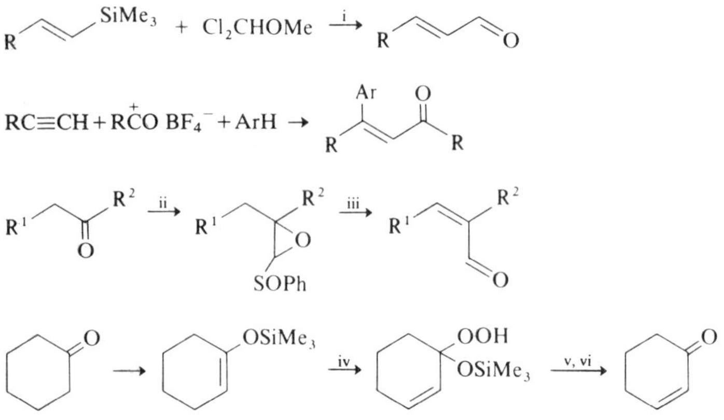

Ref. 103

Ref. 104

Ref. 105

Ref. 106

Reagents: i, $TiCl_4$; ii, $PhSOCH_2Cl$; iii, Δ; iv, $^1O_2$; v, PPh_3; vi, MeOH

Scheme 25

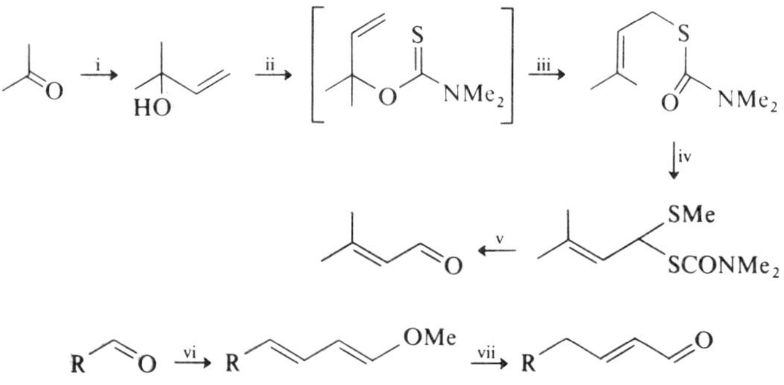

Reagents: i, MgBr; ii, NaH–ClC(=S)NMe_2; iii, Δ; iv, $LiNPr^i_2$–MeS_2Me; v, Hg^{II};
vi, $Ph_3\overset{+}{P}$ OMe; vii, H^+

Scheme 26

The Claisen rearrangement of 1-siloxy-3-oxahexa-1,5-dienes provides a useful new approach to $\beta\gamma$-unsaturated ketones (Scheme 27),[109] and Reuter and Salomon[110] have shown that tris(triphenylphosphine)ruthenium is an efficient catalyst for the thermal rearrangement of diallyl ethers to $\gamma\delta$-unsaturated aldehydes [(67) → (68)]; in the latter sequence, the catalyst first effects regiospecific isomerization to an allyl vinyl ether, which then undergoes the conventional Claisen rearrangement. The acylation of olefins with acetyl hexachloroantimonate at low temperatures in the presence of a hindered tertiary amine also affords an interes-

[109] J. L. C. Kachinski and R. G. Salomon, *Tetrahedron Letters*, 1977, 3235.
[110] J. M. Reuter and R. G. Salomon, *J. Org. Chem.*, 1977, **42**, 3360.

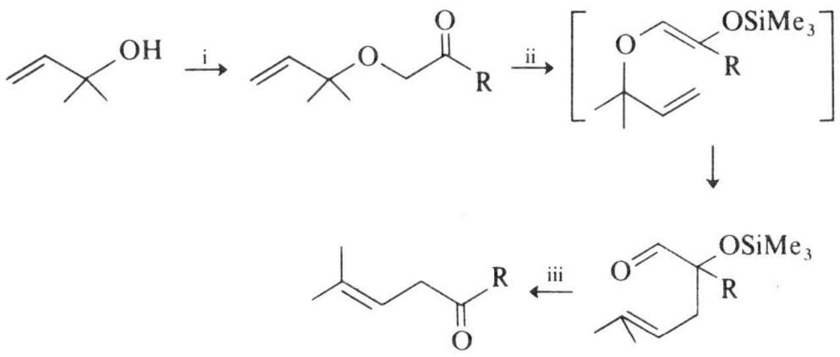

Reagents: i, $RCOCHN_2$; ii, Me_3SiCl; iii, HIO_4–MeOH

Scheme 27

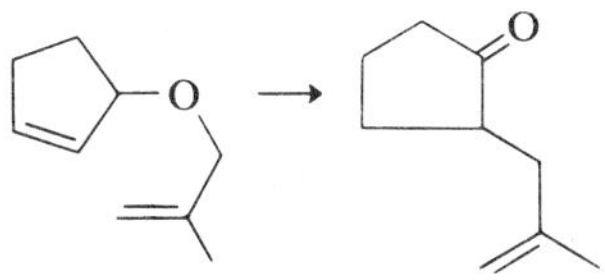

(67) (68)

ting approach to $\beta\gamma$-unsaturated ketones [*e.g.* (69)→(70)];[111] this sequence probably involves an 'ene' reaction.

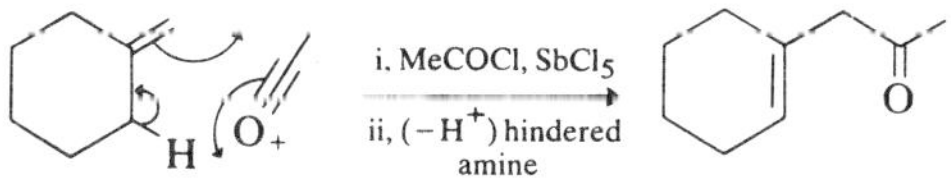

(69) (70)

The oxidation of a methylene group α- to a ketone function to form an α-diketone is usually effected by reagents like SeO_2 or alkyl nitrites. Wasserman and Ives[112] have now shown that the same transformation can be smoothly accomplished *via* reaction between the corresponding enamine of the ketone and singlet oxygen. Aldehydes can be converted into α-diketones *via* their dithian anions, using pentacarbonyliron as a carbonylating agent (Scheme 28).[113] In another

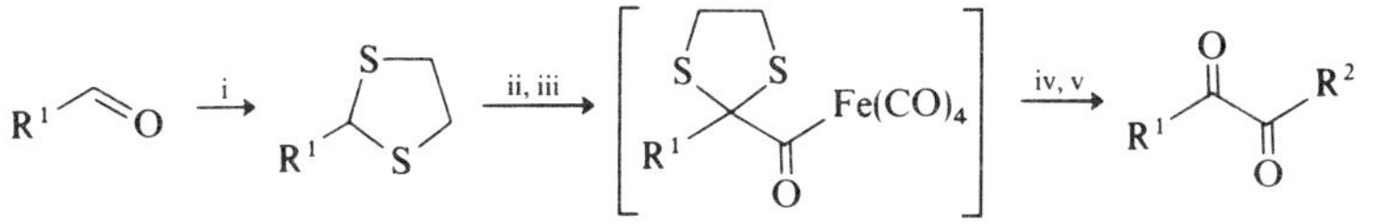

Reagents: i, $(HSCH_2)_2$; ii, Bu^nLi; iii, $Fe(CO)_5$; iv, R^2I; v, H_2O

Scheme 28

interesting new approach to α-diketones, use is made of tosylmethyl isocyanide as a masked formaldehyde (Scheme 29).[114]

[111] H. M. R. Hoffmann and T. Tsue Rima, *J. Amer. Chem. Soc.*, 1977, **99**, 6008.
[112] H. H. Wasserman and J. L. Ives, *J. Amer. Chem. Soc.*, 1976, **98**, 7868.
[113] M. Yamashita and R. Suemitsu, *J.C.S. Chem. Comm.*, 1977, 691.
[114] D. van Leusen and A. M. van Leusen, *Tetrahedron Letters*, 1977, 4233.

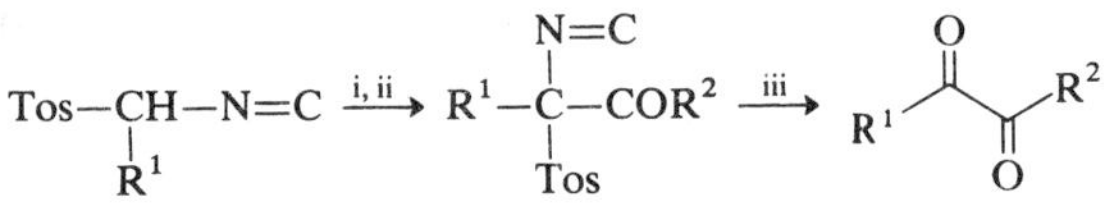

Reagents: 1, BuLi; ii, R^2COCl; iii, H^+

Scheme 29

Several new, useful syntheses of 1,4-diones have been described. Lithium enolates derived from ketones can be coupled with copper(I) halides or with $Cu(OTf)_2$ to give symmetrical 1,4-diones (Scheme 30),[115,116] and Cooke and Parlman[117] have shown that organo-tetracarbonylferrates undergo Michael addition with enones, leading to unsymmetrical 1,4-diones of type (71) (Scheme 30).

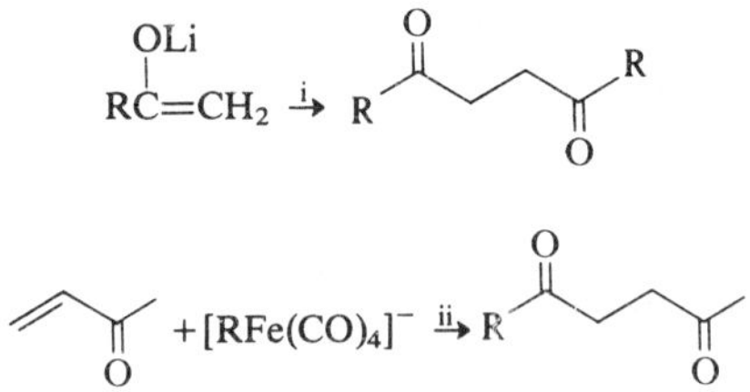

(71)

Reagents: i, $Cu(OTf)_2$ or $CuCl_2$; ii, H^+

Scheme 30

2-Hydroxycyclobutanes feature in an alternative route to 1,4-diones,[118] and the acylation of butenolides, followed by acid work-up, provides another attractive route to this functionality (Scheme 31).[119]

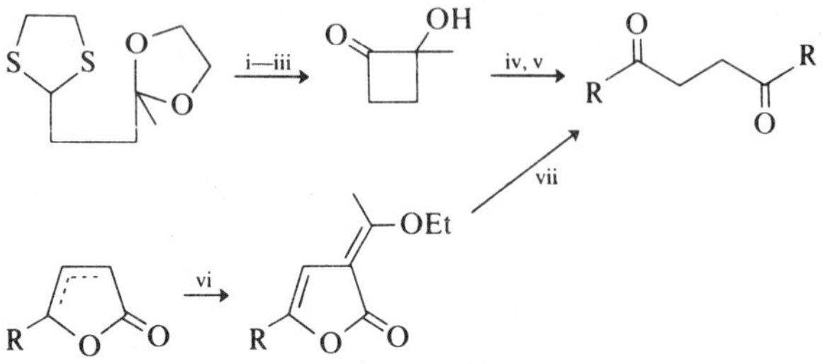

Reagents: i, BuLi; ii, DHP–H^+; iii, $Hg^{II}Cl$; iv, RMgX; v, HIO_4; vi, Ac_2O–$MeC(OEt)_3$; vii, AcOH–H_2O

Scheme 31

[115] Y. Ito, T. Konoike, T. Harada, and T. Saegusa, *J. Amer. Chem. Soc.*, 1977, **99**, 1487.
[116] Y. Kobayashi, T. Taguchi, and E. Tokuno, *Tetrahedron Letters*, 1977, 3741.
[117] M. P. Cooke and R. M. Parlman, *J. Amer. Chem. Soc.*, 1977, **99**, 5222.
[118] A. Murai, M. Ono, and T. Masamune, *J.C.S. Chem. Comm.*, 1977, 573.
[119] M. Asaoka, N. Sugimura, and H. Takei, *Chem. Letters*, 1977, 171.

Gröbel and Seebach[120] have provided an excellent account of the umpolung of carbonyl reactivity through sulphur-containing reagents. They stress in their review that the term *masked* functionality should be reserved for those cases in which the normal polarity of a group is *inverted*, and that *latent* functionality should be used in cases where the reagent exhibits the normal polarity of the hidden (or 'masked') group. It is interesting that Dolak and Bryson[121] have demonstrated that two sulphur atoms, as in dithians, are not actually necessary for the stabilization of adjacent carbanion centres. These workers have shown that when *tertiary*-butyllithium is used, anions from simple alkyl phenyl sulphides can be prepared. In some related studies, Fuji *et al.*[122] have described the use of 1,3-oxathians as equivalents of acyl anions; since these are cleaved under milder conditions than the corresponding dithians, they may ultimately have wider utility as equivalents of acyl anions.

Anions derived from cyclic vinyl ethers constitute a useful series of substituted acyl anions [*e.g.* (72) → (73)],[123] and Magnus and his co-workers[124] have outlined

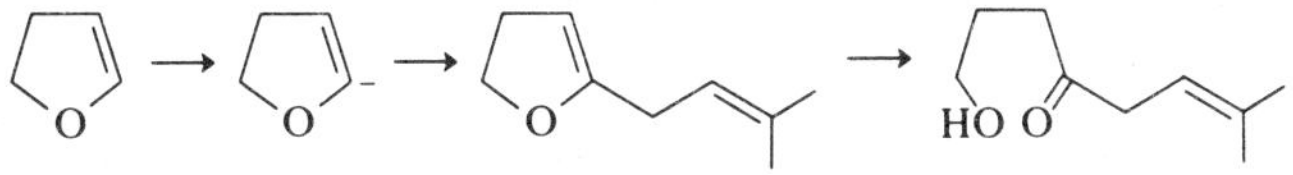

(72) (73)

the use of the allyl anion derived from allyltrimethylsilane as an effective β-acylcarbanion equivalent (Scheme 32).

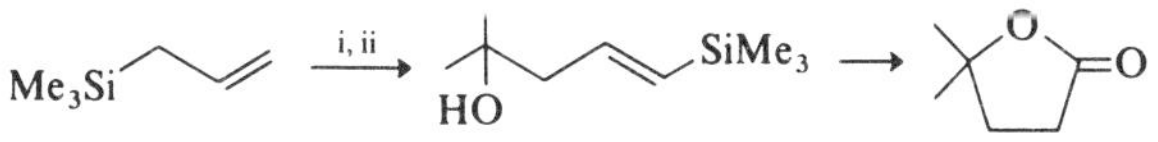

Reagents: i, BuLi; ii, (CH$_3$)$_2$C=O ; iii, *m*-ClC$_6$H$_4$CO$_3$H; iv, H$^+$; v, Jones oxidation

Scheme 32

N-Nitromethylphthalimide has been employed in synthesis as an equivalent of formyl anion,[125] and 2-lithio-3,3-diethoxypropene (74) is a useful α-acrolein carbanion equivalent.[126]

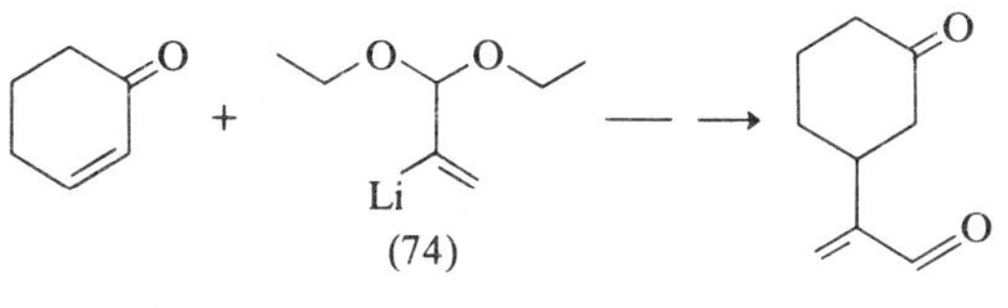

(74)

In spite of all the current interest in dithian anions as equivalents of carbonyl-group anions, the corresponding formyl cation is just as inaccessible as its anion. In

[120] B.-T. Gröbel and D. Seebach, *Synthesis*, 1977, 357.
[121] T. M. Dolak and T. A. Bryson, *Tetrahedron Letters*, 1977, 1961.
[122] K. Fuji, M. Ueda, and E. Fujita, *J.C.S. Chem. Comm.*, 1977, 814.
[123] R. K. Boeckman and K. J. Bruza, *Tetrahedron Letters*, 1977, 4187.
[124] D. Ayalon-Chass, E. Ehlinger, and P. Magnus, *J.C.S. Chem. Comm.*, 1977, 772.
[125] F. G. Cowherd, M.-C. Doria, E. Galeazzi, and J. M. Muchowski, *Canad. J. Chem.*, 1977, **55**, 2515.
[126] R. K. Boeckman and M. Ramaiah, *J. Org. Chem.*, 1977, **42**, 1581.

this case Taylor and La Mattina[127] have now shown that the chloride (75) is a useful equivalent.

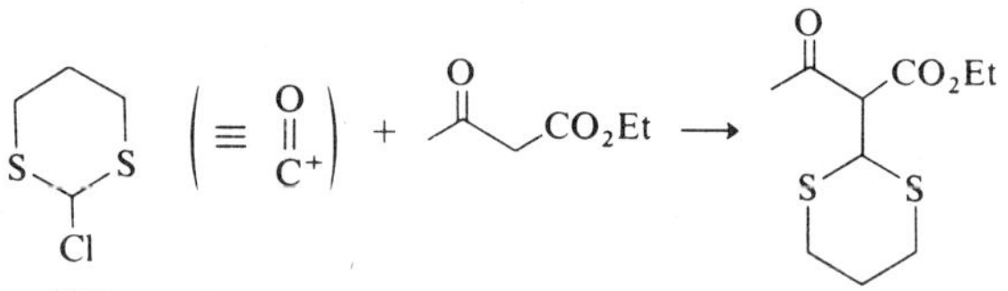

(75)

10 Carboxylic Acids and Anhydrides

Trimethylsilyl iodide has emerged as an exceptionally efficient reagent for the cleavage of esters to carboxylic acids under neutral conditions.[128,129] Hindered esters undergo hydrolysis at room temperature, using so-called 'anhydrous hydroxide', generated *via* the reaction of two equivalents of potassium t-butoxide with one equivalent of water.[130]

The monohydroboration of 1-alkynyl-silanes followed by oxidation of the resulting 1-boro-1-silyl-alkene (76) with H_2O_2–NaOH provides an attractive synthesis of carboxylic acids from terminal acetylenes (Scheme 33);[131] a feature of

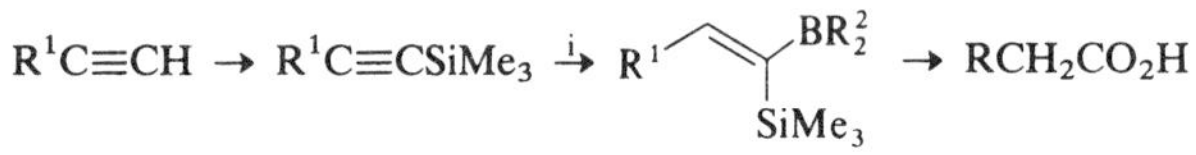

(76)

Reagents: i, $R_2{}^2BH$; ii, NaOH–H_2O_2

Scheme 33

the method is that it can be applied to the synthesis of $\beta\gamma$-unsaturated acids starting with conjugated enynes.

Optically pure α-hydroxy-acids are produced from β-keto-acids by reduction with an actively fermenting mutant of *Saccharomyces cerevisiae.*[132] They are also obtained from $\alpha\beta$-unsaturated carboxylic acids *via* the corresponding amide with *S*-proline, according to Scheme 34.[133]

The novel fragmentation of tetramethylsilyl ethers derived from bicyclic cylopropanols, initiated by oxidation with $Pb(OAc)_4$, has provided an expeditious synthesis of ω-unsaturated acids [*viz.* (77)→(78)].[134] ω-Unsaturated acids are also produced from the nucleophilic ring opening of lactones with phenylselenide ion, followed by elimination of the selenium residue (Scheme 35).[135]

[127] E. C. Taylor and J. L. La Mattina, *Tetrahedron Letters*, 1977, 2077.
[128] M. E. Jung and M. A. Lyster, *J. Amer. Chem. Soc.*, 1977, **99**, 968.
[129] T.-L. Ho and G. A. Olah, *Angew. Chem. Internat. Edn.*, 1977, **15**, 774.
[130] P. G. Gassman and W. N. Schenk, *J. Org. Chem.*, 1977, **42**, 918.
[131] G. Zweifel and S. J. Backlund, *J. Amer. Chem. Soc.*, 1977, **99**, 3184.
[132] B. S. Deol, D. D. Ridley, and G. W. Simpson, *Austral. J. Chem.*, 1976, **29**, 2459.
[133] S. Terashima and S.-S. Jew, *Tetrahedron Letters*, 1977, 1005.
[134] G. M. Rubottom, R. Marrero, D. S. Krueger, and J. L. Schreiner, *Tetrahedron Letters*, 1977, 4013.
[135] D. Liotta and H. Santiesteban, *Tetrahedron Letters*, 1977, 4369.

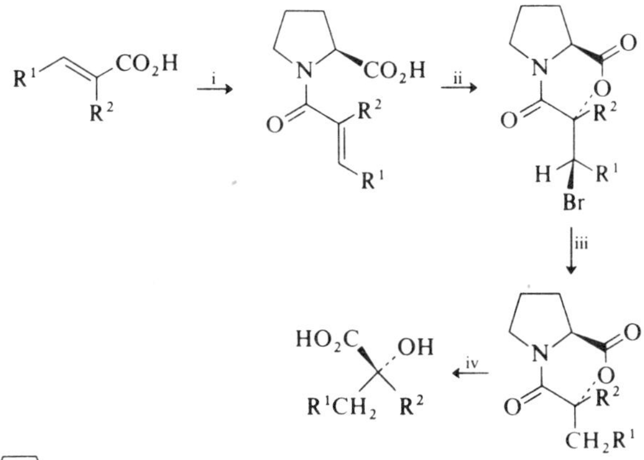

Reagents: i, 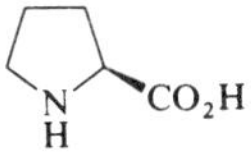 ; ii, NBS; iii, Bu_3SnH; iv, H^+

Scheme 34

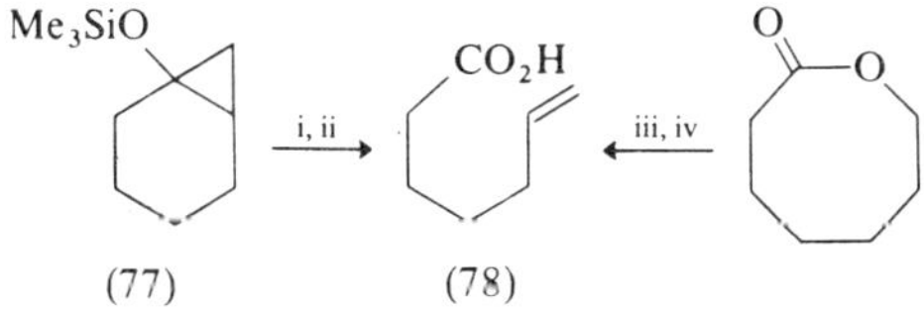

Reagents: i, $Pb(OAc)_4$—HOAc; ii, H_2O; iii, $PhSe^-$; iv, H_2O_2.

Scheme 35

Minami and Kuwajima have found that the hindered base (79) can be used to produce the relatively inaccessible enolate anions of saturated anhydrides; the anions can then be used in condensation reactions with aldehydes, for example; this provides a useful synthesis of lactones (Scheme 36).[136] Nucleophilic addition of

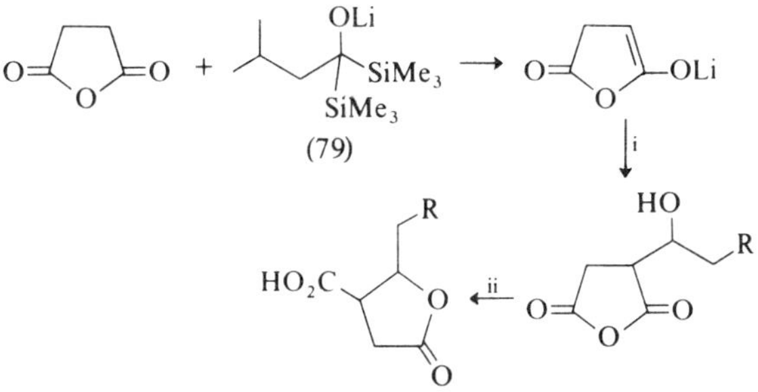

Reagents: i, RCH_2CHO; ii, $NaHCO_3$

Scheme 36

[136] N. Minami and I. Kuwajima, *Tetrahedron Letters*, 1977, 1423.

methoxycarbonylmethylenetriphenylphosphorane to the unsymmetrical anhydride (80) has led to a facile synthesis of the ylidenetetronic acid derivatives (81) and (82); the former has been isolated from *Penicillium multicolor.*[137]

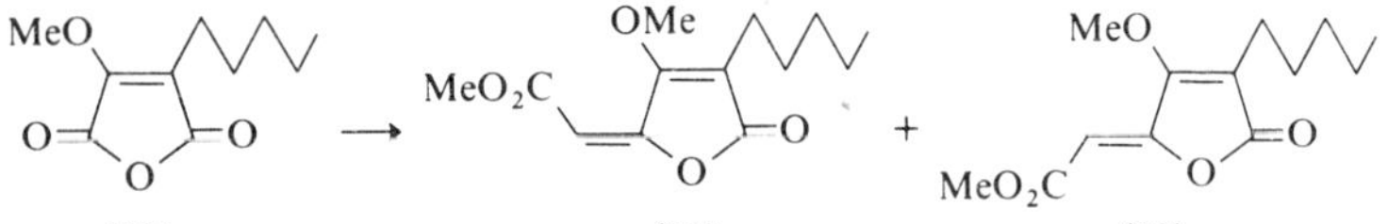

(80) (81) (82)

11 Esters and Functionalized Esters

Nowhere is the synthetic application of catalysis by crown-ethers better illustrated this year than in the synthesis of esters by the traditional malonic ester synthesis. Hunter and Perry[138] have shown that in the presence of 18-crown-6 ether, the whole malonic ester synthesis can be performed essentially in 'one pot', where the crown ether (*a*) accelerates the hydrolysis of the ester in stage one, by potassium hydroxide, and (*b*) allows decarboxylation of the potassium carboxylate under mild conditions in stage two (Scheme 37).

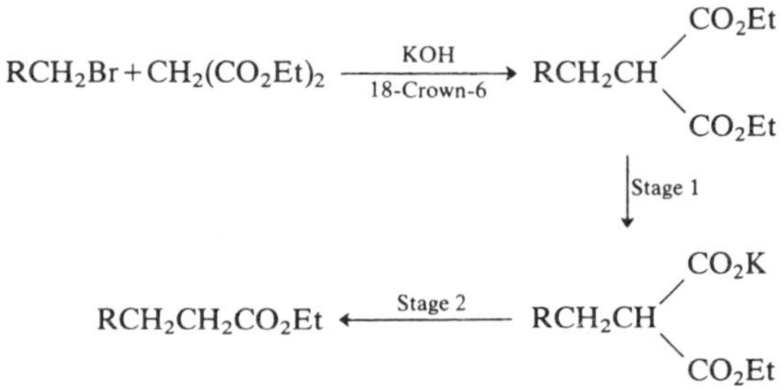

Scheme 37

Primary amines can be converted into esters in a two-step process involving reaction with a pyrylium salt, followed by pyrolysis of the resulting pyridinium salt with the corresponding sodium carboxylate (Scheme 38).[139]

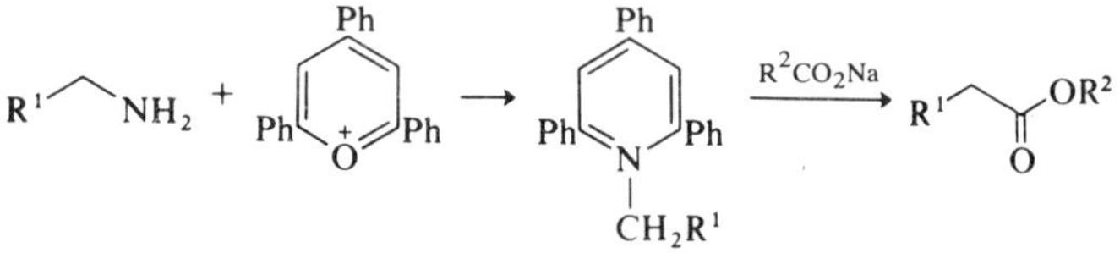

Scheme 38

[137] M. J. Begley, D. R. Gedge, and G. Pattenden, *J.C.S. Chem. Comm.*, 1978, 60.
[138] D. H. Hunter and R. A. Perry, *Synthesis*, 1977, 37.
[139] U. Gruntz, A. R. Katritzky, D. H. Kenny, M. C. Rezende, and H. Sheikh, *J.C.S. Chem. Comm.*, 1977, 701.

Two vinylation methods provide useful routes to $\alpha\beta$- and $\beta\gamma$-unsaturated esters. In one method, Miyaura *et al.*[140] have found that the reaction between Cu^{I} methyltrialkylborates and β-bromo-acrylates proceeds in a stereospecific manner, leading to (*Z*)- and (*E*)-$\alpha\beta$-unsaturated esters, and in the second method Millard and Rathke[141] show that $\beta\gamma$-unsaturated esters are produced by vinylations of lithium ester enolates in the presence of a catalyst produced from the addition of $Bu^{n}Li$ to Ni^{II} bromide (Scheme 39). $\beta\gamma$-Unsaturated esters are also obtained from β-keto-

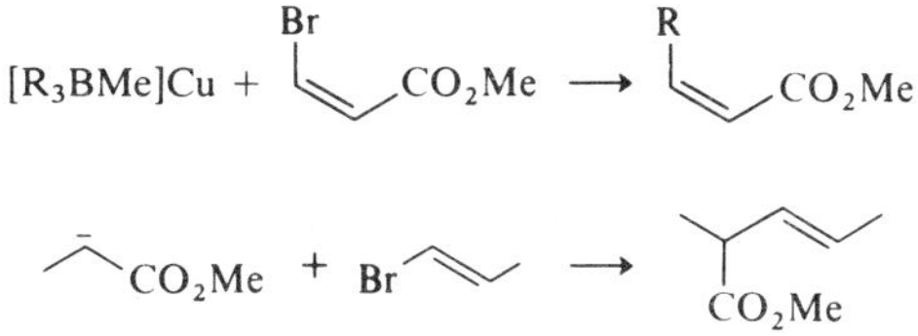

Scheme 39

esters *via* the Shapiro reaction, involving conversion into the corresponding tosylhydrazone, metallation, alkylation, and quenching with ammonium chloride (Scheme 40).[142]

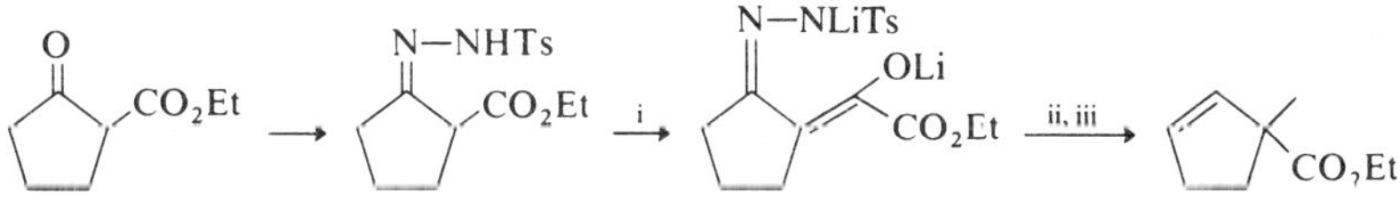

Reagents: i, $LiNPr^{i}_{2}$; ii, MeI; iii, H^{+}

Scheme 40

Lactones.—Benzeneperoxyseleninic acid, which can be generated *in situ* from benzeneseleninic acid and hydrogen peroxide, holds promise as a useful alternative reagent in the Baeyer–Villiger synthesis of lactones from ring ketones.[143]

1,3-Dienes are smoothly transformed into lactones by conversion into the mono-epoxide followed by complexation with $[Fe_2(CO)_5]$ and oxidation [*e.g.* (83) → (84)],[144] and two groups of workers have shown that certain unsaturated carboxylic acids react with benzeneselenyl chloride to afford phenylselenyl lactones [*e.g.* (85) → (86)].[145]

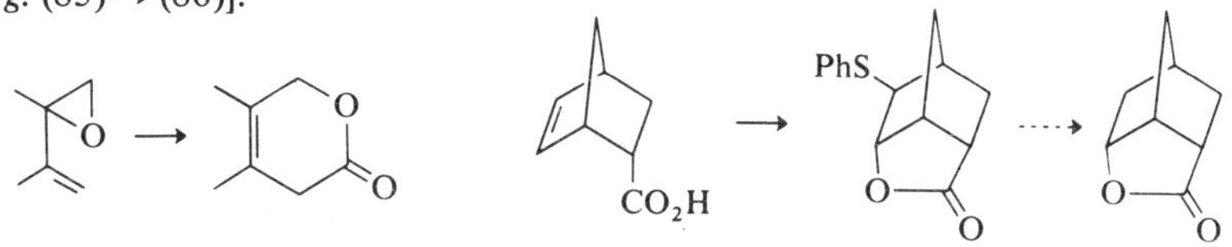

(83) (84) (85) (86)

[140] N. Miyaura, N. Sasaki, M. Itoh, and A. Suzuki, *Tetrahedron Letters*, 1977, 3369.
[141] A. A. Millard and M. W. Rathke, *J. Amer. Chem. Soc.*, 1977, **99**, 4833.
[142] C. A. Bunnell and P. L. Fuchs, *J. Amer. Chem. Soc.*, 1977, **99**, 5184.
[143] P. A. Grieco, Y. Yokoyama, S. Gilman, and Y. Ohfune, *J.C.S. Chem. Comm.*, 1977, 870.
[144] G. D. Annis and S. V. Ley, *J.C.S. Chem. Comm.*, 1977, 581.
[145] K. Nicolaou and Z. Lysenko, *J.C.S. Chem. Comm.*, 1977, 293; D. L. J. Clive and G. Chittattu, *ibid.*, p. 484.

3-Substituted furanoterpenes occur widely in nature, and can be obtained by reduction of the corresponding but-2-enolides with di-isobutylaluminium hydride. Two approaches to substituted but-2-enolides proceed *via* the corresponding β-ylidenelactone (87),[146] and by direct prenylation of a but-2-enolide (Scheme 41).[147]

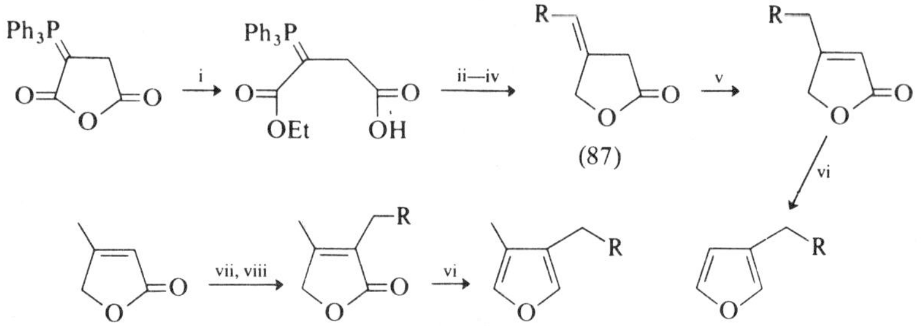

Reagents: i, EtOH; ii, RCHO; iii, $NaAlEt_2H_2$; iv, *p*-$MeC_6H_4SO_3H$; v, Al_2O_3; vi, $Bu^i{}_2AlH$; vii, $LiNPr^i{}_2$–HMPA; viii, RCH_2Br

Scheme 41

Macrolides.—Most of the procedures employed in the past few years for the synthesis of macrolides from ω-hydroxy-carboxylic acids have adopted a strategy whereby the carboxy-group is activated; it is then attacked by the hydroxy-group, acting as nucleophile.[148—150] The 'reverse-activation' technique is best applied using azodicarboxylate–triphenylphosphine, and in certain instances this strategy offers considerable advantages over the more conventional method, *e.g.* vermiculine (89) from (88)[151,152] and pyrenophorin (91) from (90).[152,153] In some related studies, Vorbruggen and Krolikiewicz[154] have shown that *NN*-dimethylformamide di-neopentylacetal functions in the same manner as azodicarboxylate–triphenylphosphine. In spite of these developments, the carboxyl activation method is much valued in macrolide synthesis, and has been used recently; for example, in the synthesis of (*a*) enterobactin (92), a macrolide iron-transporting agent,[155] (*b*) brefeldin (93), a fungal metabolite,[156,157] and (*c*) diploidialide A (94), a steroid-hydroxylation inhibitor.[158]

In an alternative approach to the macrolides, Trost and his co-workers[159] have used anions of sulphones in the presence of 2—6% tetrakis(triphenyl-

[146] J. E. McMurry and S. F. Donovan, *Tetrahedron Letters*, 1977, 2869.
[147] D. R. Gedge and G. Pattenden, *Tetrahedron Letters*, 1977, 4443.
[148] T. G. Back, *Tetrahedron*, 1977, **33**, 3041.
[149] K. C. Nicolaou, *Tetrahedron*, 1977, **33**, 683.
[150] S. Masamune, G. S. Bates, and J. W. Corcoran, *Angew. Chem. Internat. Edn.*, 1977, **16**, 585.
[151] Y. Fukuyama, C. L. Kinkemo, and J. D. White, *J. Amer. Chem. Soc.*, 1977, **99**, 646.
[152] D. Seebach, B. Seuring, H.-O. Kalinowski, W. Lubosch, and B. Renger, *Angew. Chem. Internat. Edn.*, 1977, **16**, 264.
[153] H. Gerlach, K. Oertle, and A. Thalmann, *Helv. Chim. Acta*, 1977, **60**, 2860.
[154] H. Vorbruggen and K. Krolikiewicz, *Angew. Chem. Internat. Edn.*, 1977, **16**, 876.
[155] E. J. Corey and S. Bhattacharyya, *Tetrahedron Letters*, 1977, 3919.
[156] E. J. Corey, R. H. Wollenberg, and D. R. Williams, *Tetrahedron Letters*, 1977, 2243.
[157] R. Baudoug, P. Crabbé, A. E. Greene, C. Le Drian, and A. F. Orr, *Tetrahedron Letters*, 1977, 2973.
[158] T. Ishida and K. Wada, *J.C.S. Chem. Comm.*, 1977, 337.
[159] B. M. Trost and T. R. Verhoeven, *J. Amer. Chem. Soc.*, 1977, **99**, 3867.

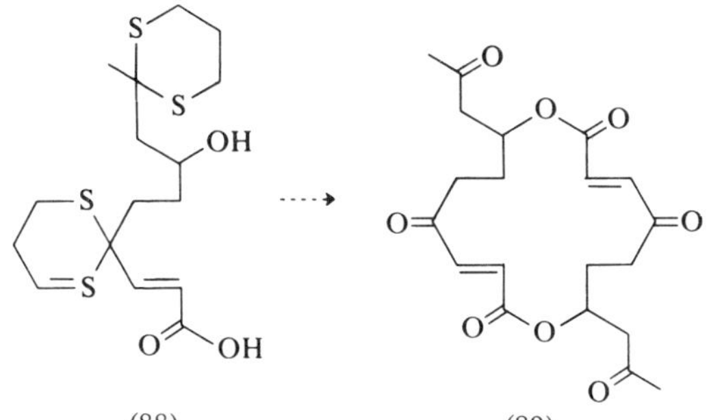

(88) (89)

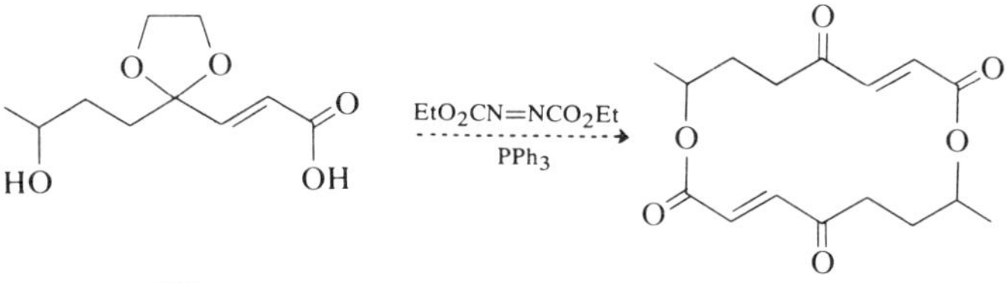

(90) (91)

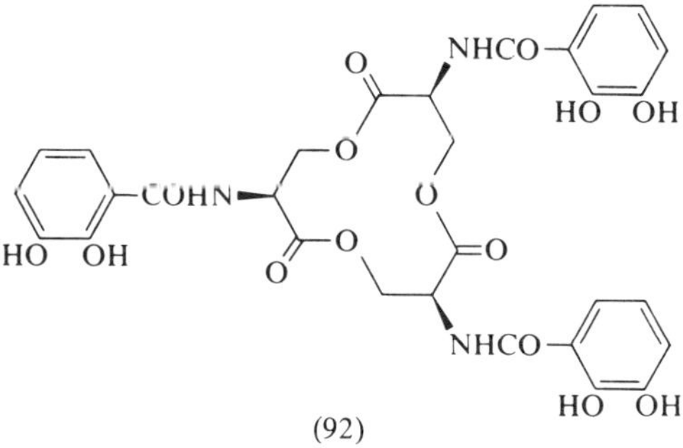

(92)

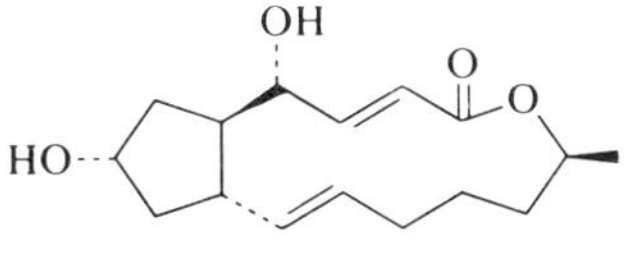

(93)

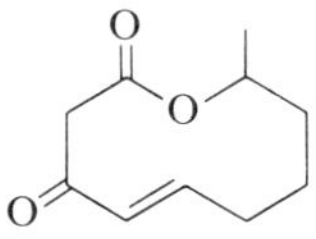

(94)

phosphine)palladium as a strategem for the synthesis of the appropriate macrocycles [*e.g.* (95) → (96)].

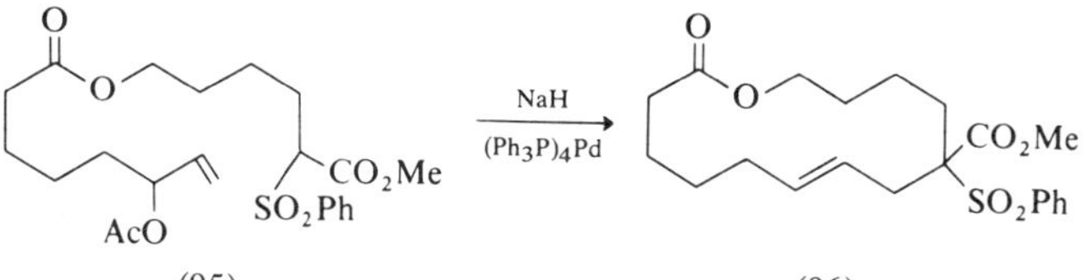

(95) (96)

12 Nitriles

Ziegler and Wender[160] have provided further details of the method of converting carbonyl compounds into nitriles containing an additional carbon atom, based on the decomposition of methyl dialkylcyanodiazenecarboxylates (Scheme 42); a modification of this approach uses the corresponding benzenesulphonates.[161]

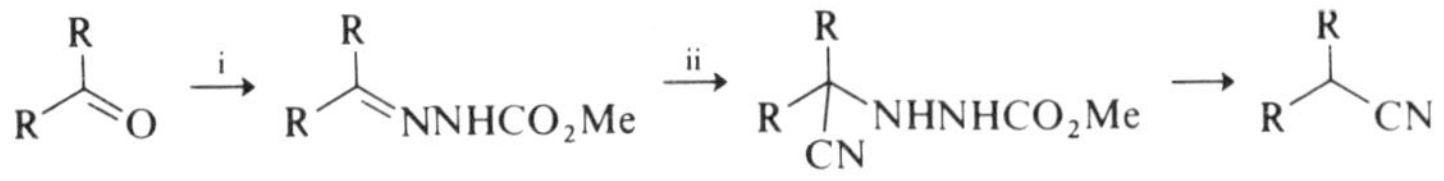

Reagents: i, H_2NNHCO_2Me; ii, HCN–MeOH; iii, NaOMe

Scheme 42

Vinyl nitriles are conveniently produced from vinyl halides by treatment with KCN in the presence of a catalytic amount of Pd^O complex and crown ethers,[162] and the Vilsmeir formylation approach, outlined in Scheme 43, provides a useful

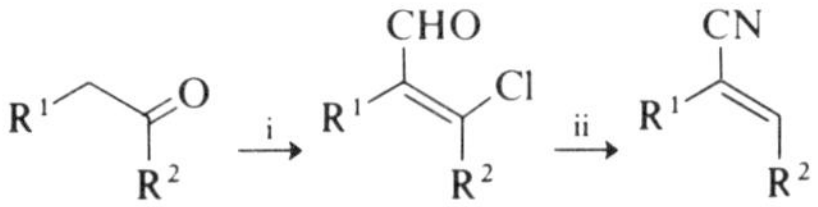

Reagents: i, $POCl_3$–DMF; ii, NH_2OH

Scheme 43

synthesis of vinyl nitriles from carbonyl compounds.[163] Aldehydes are smoothly converted into vinyl nitriles of the type (97), following cyanoselenylation and oxidative removal of the selenium residue (Scheme 44).[164]

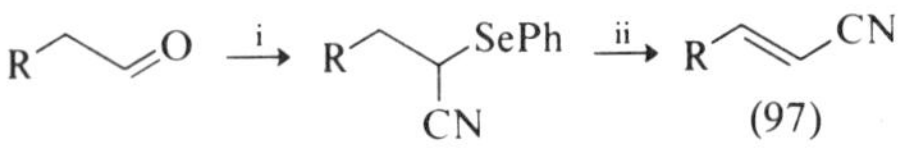

Reagents: i, $PhSeCN–Bu_3P$; ii, H_2O_2

Scheme 44

13 Alkylation

In an extension of their earlier studies, Enders and Eichenauer[165] have demonstrated that optically pure α-alkyl-substituted aldehydes can be produced by metallation and alkylation of chiral hydrazones. In related studies, Whitesell and his co-workers[166,167] have observed enantioselective alkylation (optical purities

160 F. E. Ziegler and P. A. Wender, *J. Org. Chem.*, 1977, **42**, 2001.
161 D. M. Orere and C. B. Reese, *J.C.S. Chem. Comm.*, 1977, 280.
162 K. Yamamura and S.-I. Murahashi, *Tetrahedron Letters*, 1977, 4429.
163 P. C. Traas, H. J. Takken, and H. Boelens, *Tetrahedron Letters*, 1977, 2027.
164 P. A. Grieco and Y. Yokoyama, *J. Amer. Chem. Soc.*, 1977, **99**, 5210.
165 D. Enders and H. Eichenauer, *Tetrahedron Letters*, 1977, 191.
166 J. K. Whitesell and S. W. Felman, *J. Org. Chem.*, 1977, **42**, 1663.
167 J. Whitesell and M. A. Whitesell, *J. Org. Chem.*, 1977, **42**, 377.

80—90%) of the cyclohexanone enamine (98) prepared from (+)-*trans*-2,5-dimethylpyrrolidine, and also of the imine (99) containing *C*-2 *O*-butyl substitution.

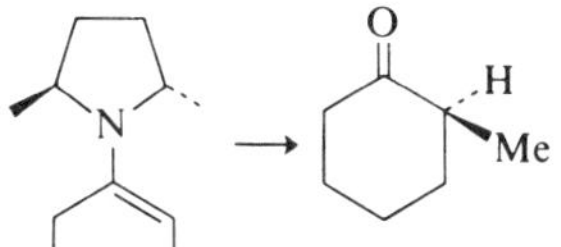

(*S*; 83%)

(98)

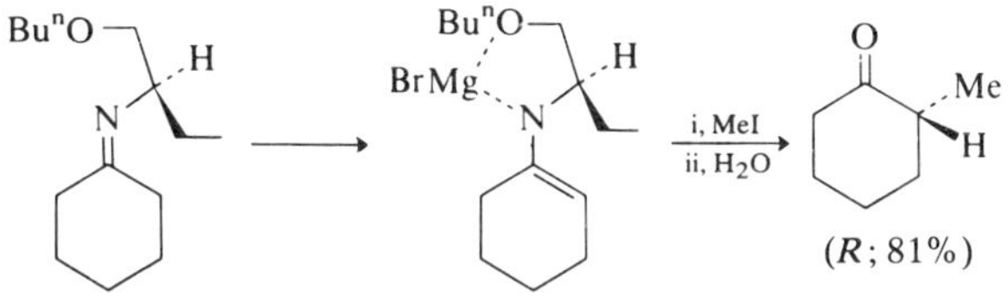

(*R*; 81%)

(99)

A perennial problem that arises with the alkylation of β-dicarbonyl compounds is the concurrent formation of both *C*- and *O*-alkylated products. Clark and Miller[168] have now found that exclusive *C*-alkylation of β-dicarbonyl compounds can be accomplished, using the corresponding H-bonded solvates formed with tetraethylammonium fluoride; the method is a useful alternative to that based on the utilization of thallium(I) enolates. *C*-2-Substituted cyclohexane-1,3-diones are most conveniently produced from the bis-enol ether (100) according to Scheme 45.[169]

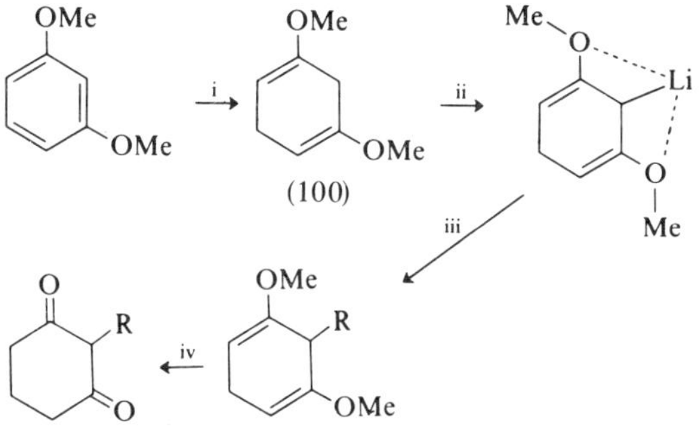

(100)

Reagents: i, Na–NH_3; ii, Bu^tLi; iii, RX; iv, H^+

Scheme 45

Hájek and Malek[170] have made the interesting observation that alkylation of carbanions of the doubly activated type [*viz.* (101)] with alk-1-enes can be initiated by certain metal oxides, and it has been shown that angular allylation of fused cyclic

168 J. H. Clark and J. M. Miller, *J.C.S. Chem. Comm.*, 1977, 64.
169 E. Piers and J. R. Grierson, *J. Org. Chem.*, 1977, **42**, 3755.
170 J. Hájek and J. Malek, *Synthesis*, 1977, 454.

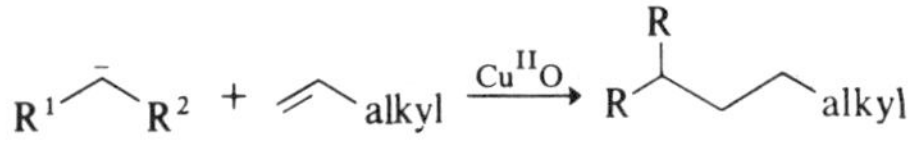

(101) $R^1 = R^2 = CO_2Et$, COMe, or CN

$\alpha\beta$-enones can be readily accomplished, using allyl-silanes in the presence of $TiCl_4$.[171]

Although copper derivatives are well known to add in a conjugate sense to $\alpha\beta$-enones, most reactive carbon nucleophiles undergo direct attack at the carbonyl group. Cooke and Goswami show that the carbonyl group in the phosphorane (102) has so little reactivity that even alkyl-lithiums add at the conjugate posi-

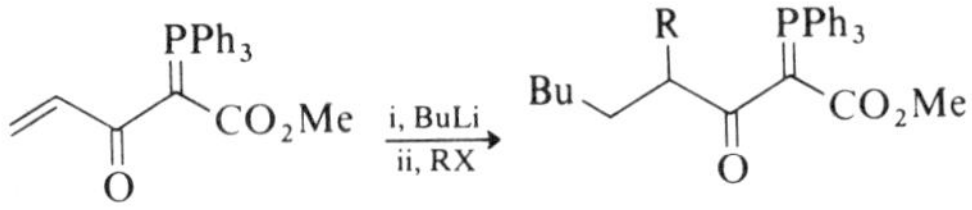

(102)

tion.[172] In a remarkable demonstration of the directing effect of the phenyl group on the stereochemistry of enolate alkylation, it is found that the vicinal bis-alkylation of cyclopentenone with lithium di*phenyl*cuprate, followed by alkylation leads to *cis*-2-alkyl-3-phenylcyclopentanone (103); all other vicinal substitutions have been shown to lead to the corresponding *trans*-isomers.[173]

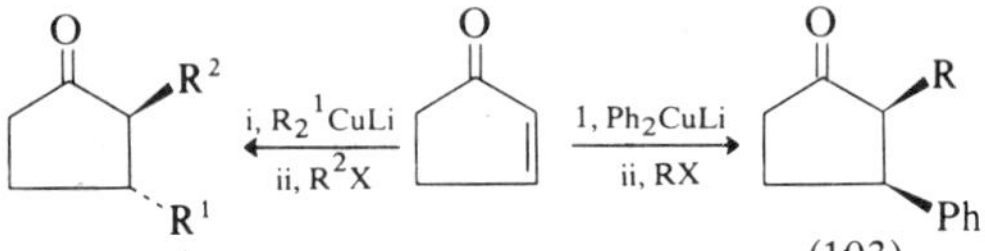

(103)

14 Ring Synthesis

Further experimental support for some of Baldwin's Rules for ring closure, announced last year [see *Ann. Reports* (*B*), 1976, **73**, p. 335], has been obtained by Baldwin and his colleagues.[174,175] They have shown, for example, that the enolate (104) cyclizes to the enol ether (105) (*i.e.* 5-*endo-trig* is disfavoured), whereas the corresponding enolate (106) undergoes ring closure, with alkylation on carbon, to give the ring ketone (107) (*i.e.* 6-*endo-trig* is favoured).

The intramolecular coupling of dicarbonyl compounds to cycloalkenes induced by low-valent titanium shows promise as a useful general route to medium- and

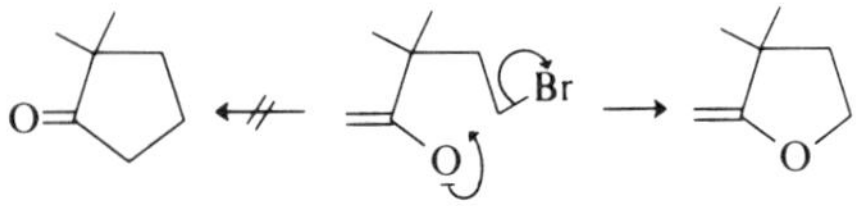

(104) (105)

[171] A. Hosomi and H. Sakurai, *J. Amer. Chem. Soc.*, 1977, **99**, 1673.
[172] M. P. Cooke and R. Goswami, *J. Amer. Chem. Soc.*, 1977, **99**, 642.
[173] G. H. Posner and C. M. Lentz, *Tetrahedron Letters*, 1977, 3215.
[174] J. E. Baldwin, R. C. Thompson, L. I. Kruse, and L. Silberman, *J. Org. Chem.*, 1977, **42**, 3856.
[175] J. E. Baldwin and L. I. Kruse, *J.C.S. Chem. Comm.*, 1977, 233.

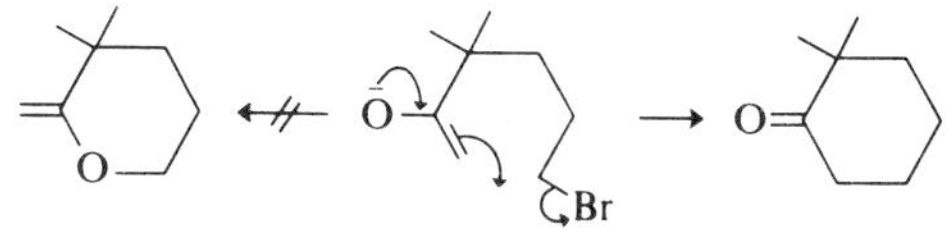

(106) (107)

large-ring carbocycles; McMurry and Kees[176] show that the method can be applied in the synthesis of anything between four- and sixteen-membered rings in yields between 70 and 90%. Treatment of propane-1,3-diols under the same conditions leads to cyclopropanes.[177]

In an interesting synthesis of the tetracyclic sesquiterpene ishwarane (109) Corey and McLaren have applied a 'double insertion' reaction of dibromocarbene into the olefinic and the C—H bond (asterisked) in the cyclohexene (108).[178]

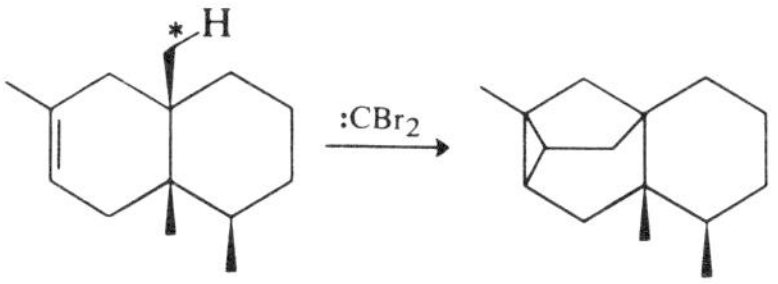

(108) (109)

Photochemistry has featured in a number of approaches to interesting naturally occurring cyclopentanes published this year. Roberts and his co-workers,[179] for example, have shown that photolysis of the bicyclo[2,2,1]heptan-2-one (110) provides a facile entry to the important prostaglandin intermediate (111), and Barton

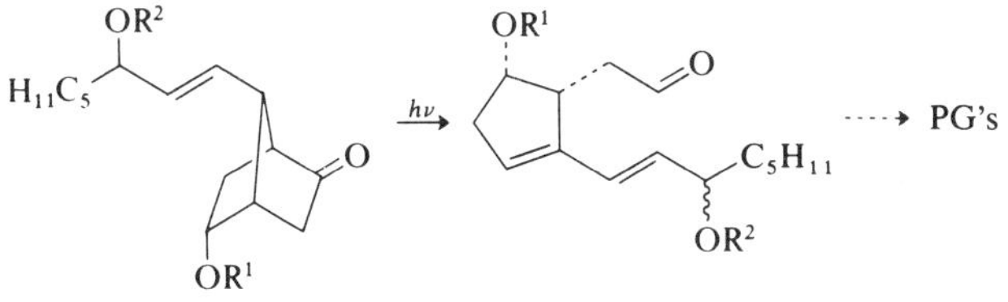

(110) (111)

and Hulshof[180] have found that photochemical ring contraction of the pyrone (112) in the presence of sodium cyanoborohydride leads to the mould metabolite terrein (113). Intramolecular [2+2] cycloaddition of dicyclopent-1-enylmethanes (114),

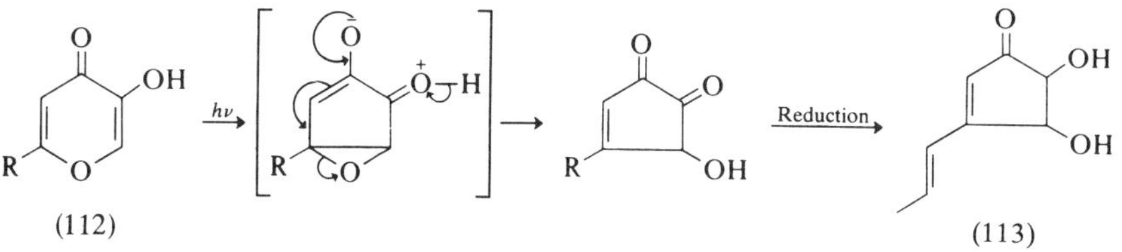

(112) (113)

[176] J. E. McMurry and K. L. Kees, *J. Org. Chem.*, 1977, **42**, 2655.
[177] A. L. Baumstark, C. J. McCloskey, T. L. Tolson, and G. T. Syriopoulus, *Tetrahedron Letters*, 1977, 3003.
[178] R. M. Cory and F. R. McLaren, *J.C.S. Chem. Comm.*, 1977, 587.
[179] N. M. Crossland, S. M. Roberts, and R. F. Newton, *J.C.S. Chem. Comm.*, 1977, 866.
[180] D. H. R. Barton and L. A. Hulshof, *J.C.S. Perkin I*, 1977, 1103.

followed by *in situ* addition of methanol to the strained three-membered ring in the presumed intermediate, has provided a new entry to fused [5,5,5]-ring systems [*viz.* (115)] found in hirsutic acid and related natural sesquiterpenes.[181]

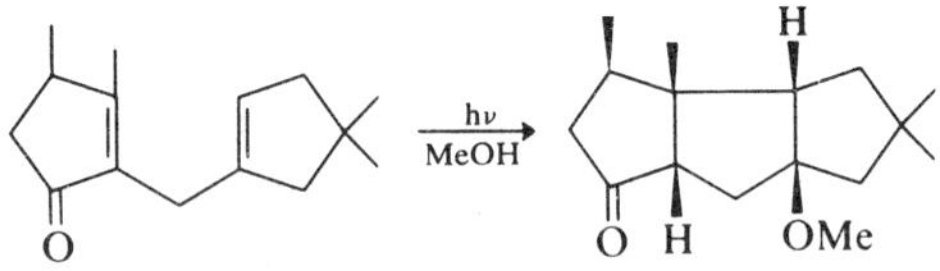

(114) (115)

The synthesis of biologically important macrocycles continues to be an active area. Yamamoto and his co-workers[182] have described a stereoselective approach to the humulene molecule, which has an eleven-membered ring, based on the intramolecular cyclization of the (π-allyl)palladium complex (117) derived from (116), and Kato *et al.*[183] have used an intramolecular Friedel–Crafts reaction in a synthesis of several cembrenoid (fourteen-membered) carbon skeletons [*e.g.* (118)$\rightarrow$(119)].

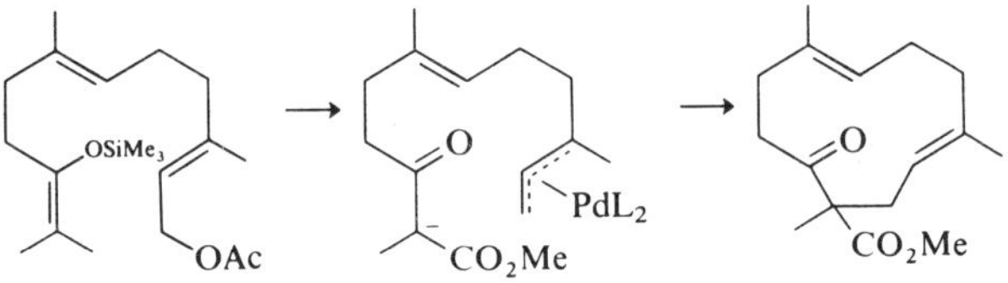

(116) (117)

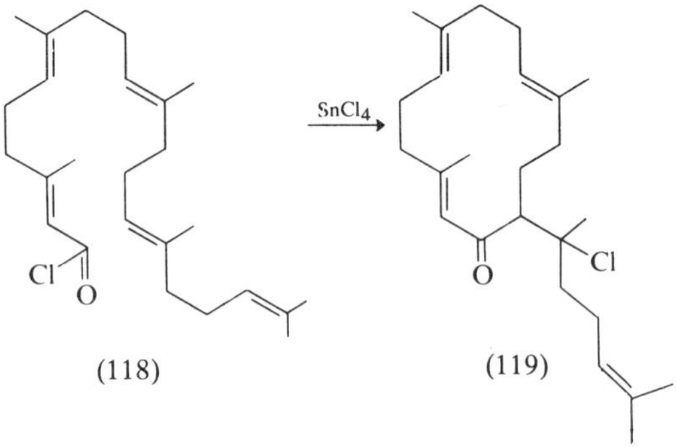

(118) (119)

[181] J. S. H. Kueh, M. Mellor, and G. Pattenden, *J.C.S. Chem. Comm.*, 1978, 5.

[182] Y. Kitagawa, A. Itoh, S. Hashimoto, H. Yamamoto, and H. Nozaki, *J. Amer. Chem. Soc.*, 1977, **99**, 3864.

[183] T. Kato, M. Suzuki, Y. Nakazima, K. Shimizu, and Y. Kitahara, *Chem. Letters*, 1977, 705.

14 Biological Chemistry
Part (i) Monosaccharides

By P. M. COLLINS
Department of Chemistry, Birkbeck College
Malet Street, London, WC1E 7HX

1 Introduction

It is four years since monosaccharides were last reviewed in *Annual Reports* and for a comprehensive survey of the work published during that period the reader is referred to the *Specialist Periodical Reports* on carbohydrate chemistry, the 10th volume of which appears this year. The present report is necessarily highly selective.

2 Glycosides

The search for efficient stereospecific syntheses of *O*-glycosides continues. Angyal's group[1,2] have shown that calcium or strontium chloride can affect the outcome of the Fischer glycosidation of sugars in which the two hydroxy-groups adjacent to the anomeric centre have the *erythro*-configuration. Complex formation occurs with the anomer which has three oxygen atoms in an axial–equatorial–axial sequence on six-membered rings or in a *cis–cis* sequence on five-membered rings. In this way good yields of methyl aldosides[1] and ketosides[2] were obtained which are not otherwise readily available.

Glycosidations utilizing halide displacements have undergone many refinements over the years and are now used with much success. Lemieux's group proposed three conditions for the best yields of α-glycosides by halide-ion-catalysed glycosidations from glycosyl halides with non-participating groups at C-2. Firstly the concentration of halide ion should be such that anomerization of the glycosyl halide is faster than the displacement of the less reactive α-halide by the alcohol, secondly the glycosyl halide should react with the alcohol without assistance from metal ions or highly polar solvents, and finally the reaction conditions should minimize the formation of glycosidic products other than the desired α-glycoside.[3] Based upon these considerations several α-linked disaccharides have been synthesized in good yields and in a highly stereoselective manner by the reaction of perbenzylated glycopyranosyl bromides with suitably protected sugar derivatives in the presence of

[1] S. J. Angyal, C. L. Bodkin, and F. W. Parrish, *Austral. J. Chem.*, 1975, **28**, 1541.
[2] S. J. Angyal, C. L. Bodkin, J. A. Mills, and P. M. Pojer, *Austral. J. Chem.*, 1977, **30**, 1259.
[3] R. U. Lemieux, K. B. Hendriks, R. V. Stick, and K. James, *J. Amer. Chem. Soc.*, 1975, **97**, 4056.

tetraethylammonium bromide. The Le[a] blood-group antigenic determinant, *O*-(α-L-fucopyranosyl)-(1 → 4)-[*O*-(β-D-galactopyranosyl)-(1 → 3)]-2-acetamido-2-deoxy-D-glucose, was synthesized using similar principles,[4] as was *O*-(α-L-fucopyranosyl)-(1 → 2)-[*O*-(α-D-galactopyranosyl)-(1 → 3)]-D-galactose, which forms part of the antigenic determinant of blood-group B substance.[5] One of the key steps in the synthesis of this trisaccharide is the α-D-galactosylation of disaccharide (1) with (2). The D-galactopyranoside derivative (3) is an important intermediate in the synthesis of disaccharide (1) since after fucosylation, the 3,4-*O*-isopropylidene group was replaced by a 3,4-*O*-(ethoxyethylidene) group, which was regiospecifically opened on partial hydrolysis to give the corresponding axial 4-acetate. β-Glycosides of *O*-(α-L-fucopyranosyl)-(1 → 4)-[*O*-(β-D-galactopyranosyl)-(1 → 3)]-2-acetamido-2-deoxy-D-glucose have also been prepared using either 8-ethoxycarbonyl- or 8-methoxycarbonyl-octanol.[6] Semisynthetic antigens were then prepared from these esters by conversion into an 8-azidocarbonyloctyl derivative, which reacts with the free amino-groups in bovine serum albumin. Antibodies raised with the trisaccharide antigen precipitated blood-group Le[a] substance and agglutinated Le[a] red-blood cells.

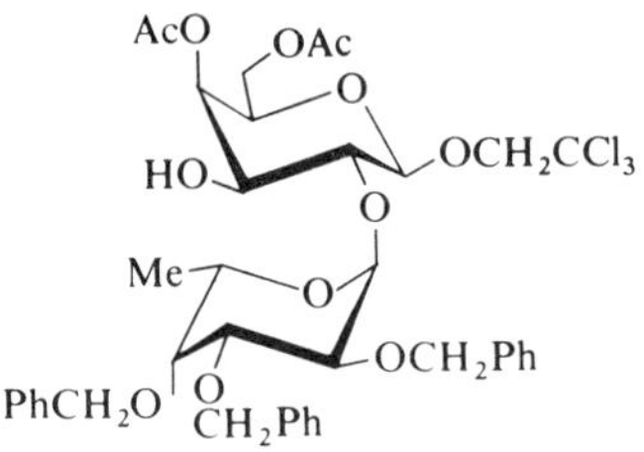

(1)

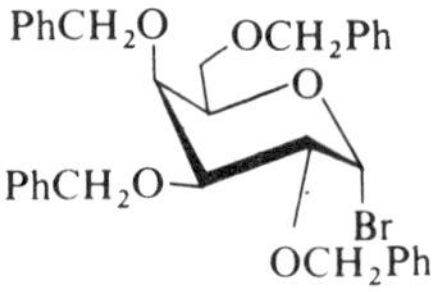

(2)

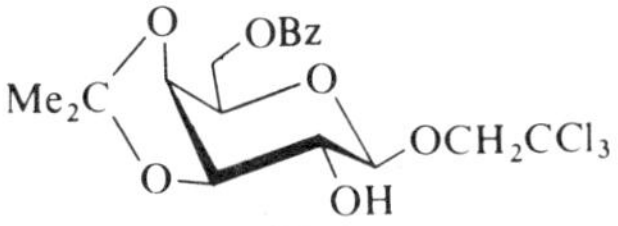

(3)

The 2-*O*-benzyl group is the most popular non-participating group in glycosylations, although others have been used. For example, 6-*O*-acetyl-2-azido-3,4-di-*O*-benzyl-β-D-glucopyranosyl chloride reacted with benzyl 2-azido-3,4-*O*-benzyl-β-D-glycopyranoside under silver-perchlorate-catalysed conditions to give an α-linked (1 → 6) disaccharide without participation from the azido-group.[7]

Syntheses of 1,2-*trans*-linked disaccharides from glycopyranosyl bromides possessing participating groups at C-2 have been reported[8] which use silver trifluoromethanesulphonate as the catalyst and 1,1,3,3-tetramethylurea as the proton acceptor. Thus glycosylations at hydroxy-groups situated individually at C-2, C-3, C-4, and C-6 in suitably protected sugar derivatives have been achieved by this method with 2,3,4,6-tetra-*O*-acetyl-α-D-glucopyranosyl bromide. The resulting (1→2)-, (1→3)-, (1→4)-, and (1→6)-β-linked disaccharide derivatives were

[4] R. U. Lemieux and H. Driguez, *J. Amer. Chem. Soc.*, 1975, **97**, 4063.
[5] R. U. Lemieux and H. Driguez, *J. Amer. Chem. Soc.*, 1975, **97**, 4069.
[6] R. U. Lemieux, D. R. Bundle, and D. A. Barker, *J. Amer. Chem. Soc.*, 1975, **97**, 4076.
[7] H. Paulsen and W. Stenzel, *Angew. Chem. Internat. Edn.*, 1975, **14**, 558.
[8] S. Hanessian and J. Banoub, *Carbohydrate Res.*, 1977, **53**, C13.

respectively formed in a high state of anomeric purity in preparatively significant yields and with much manipulative simplification compared with previously used methods.

Glycosylating agents other than glycosyl halides have been studied in the past few years. The well-established orthoester procedure continues to attract attention, and it has been shown that the basicity of the alcohol used influences the ratio of α- and β-glycosides formed.[9] Thus the ethyl derivative (4) gave only the β-glycoside, whereas the chloroethyl derivatives (5), (6), and (7) yielded mixtures containing 16, 50, and 67%, respectively, of the corresponding α-glycoside.

Disaccharide derivatives are formed when 1-*O*-aryloxycarbonyl-pyranose derivatives which possess non-participating groups at C-2 are fused with partially blocked sugar derivatives, as illustrated for example by the condensation of the 1-*O*-aryloxycarbonyl mannose derivative (11) with 1,2,3,4-tetra-*O*-acetyl-β-D-glucopyranose to give the α (1→6)-linked disaccharide derivative (15).[10] When participating groups are present at C-2 then the reaction takes a different course and orthoesters are produced in good yields.

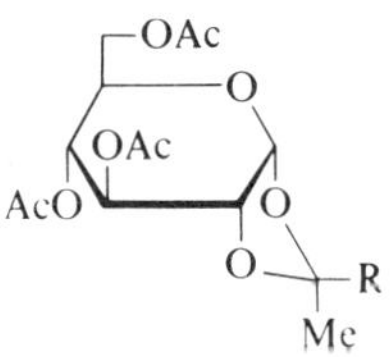

(4) R = OEt
(5) R = OCH_2CH_2Cl
(6) R = OCH_2CHCl_2
(7) R = OCH_2CCl_3
(8) R = SEt
(9) R = $\overset{+}{S}EtCPh_3$
(10) R = 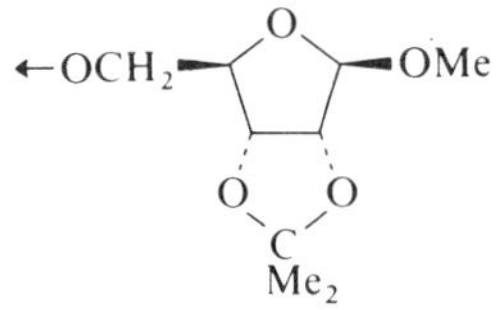

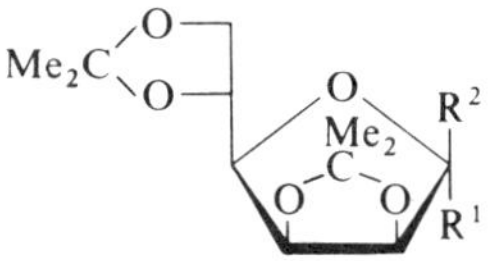

(11) $R^1 = OCO_2Ph$, $R^2 = H$
(12) $R^1 = OH$, $R^2 = H$
(13) $R^1 = Cl$, $R^2 = H$
(14) $R^1 = H$, $R^2 = C\equiv CH$
(15) $R^2 = H$, $R^1 =$ 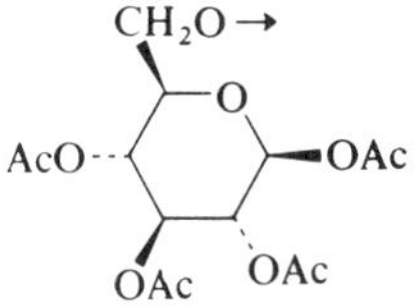

Sinaÿ and his co-workers[11] have introduced what appears to be an excellent α-glycosylating reagent. They have shown that 2,3,4,6-tetra-*O*-benzyl-α-D-glycopyranosyl chloride reacts with *N*-methylacetamide to give 1-*O*-(*N*-methyl)acetimidyl-2,3,4,6-tetra-*O*-benzyl-β-D-glucopyranose in 88% yield. This compound readily glucosylates free hydroxy-groups in partially protected sugars [see compound (16)]. For example, with methyl 2,3,6-tri-*O*-benzyl-α-D-glucopyranoside in anhydrous benzene containing toluene-*p*-sulphonic acid it gave methyl hepta-*O*-benzyl-α-D-maltoside in 85% yield, which is remarkably good for

9 P. J. Garegg and I. Kvarnström, *Acta Chem. Scand.*, 1976, **B30**, 655.
10 Y. Ishido, S. Inaba, H. Komura, and A. Matsuno, *J.C.S. Chem. Comm.*, 1977, 90.
11 J.-R. Pougny, J.-C. Jacquinet, M. Nassr, D. Duchet, M.-L. Milat, and P. Sinaÿ, *J. Amer. Chem. Soc.*, 1977, **99**, 6762.

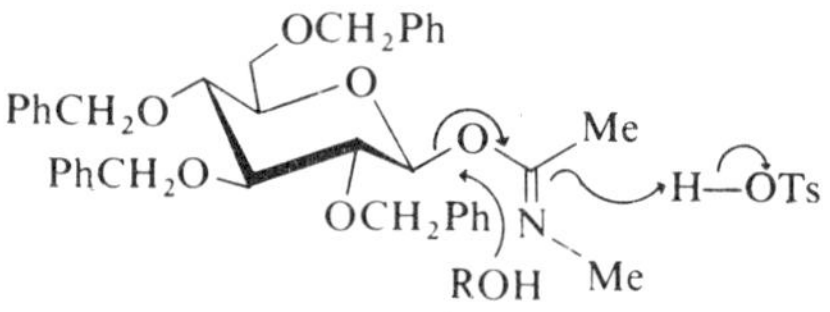

(16)

the glycosylation of this 4-OH group. Eleven glycosides were prepared in 70—92% yield by this method and it was also extended to include α-L-fucosylation, with obvious application in the synthesis of natural products of biological significance.

Schuerch[12] has extended the use of glycosyl sulphonates to the synthesis of disaccharides. Reactions related to this have been developed in which a free anomeric hydroxy-group, in an otherwise fully protected sugar derivative, is sulphonated and the product is used *in situ* to glycosylate the aglycone hydroxy-group. For example, Perlin and co-workers[13] activated the anomeric hydroxy-group of 2,3,4,6-tetra-*O*-benzyl-D-glucopyranose with trifluoromethanesulphonic anhydride and thence formed an α-D-(1 → 6)-linked disaccharide in 60% yield by the addition of 1,2,3,4-tetra-*O*-acetyl-β-D-glucopyranose.

In a different approach[14] to the activation of the anomeric position, the hydroxy-group at C-1 in 2,3 : 5,6-di-*O*-isopropylidenemannofuranose (12) was treated with dichlorocarbene, which was generated from chloroform and aqueous sodium hydroxide, in benzene, in the presence of the phase-transfer catalyst benzyl-triethylammonium chloride. This gave the α-furanosyl chloride (13), which could be isolated in reasonable yield or treated directly with simple alcohols or partially protected sugars to give glycosides or disaccharides, respectively, in yields in the range 65–85%. Presumably the dichlorocarbene inserts into the O—H bond and the dichloromethyl aglycone so formed (or a hydrolysed form of it) undergoes nucleophilic displacement by chloride ion attack at the anomeric centre.

Several groups have tried for improvements in glycoside synthesis by activation of the alcohol from which the aglycone is derived. Cyclic amide acetals derived from simple vicinal diols or carbohydrates have been used as a source of aglycones in glycoside and disaccharide syntheses.[15] For example, condensation of 1-*O*-acetyl-2,3,5-tri-*O*-benzoyl-β-D-ribofuranose with the 3,4-cyclic amide derived from 2-*O*-mesyl-β-D-arabinopyranoside (17) in the presence of stannic chloride gave the intermediate (18), which after hydrolysis yielded the formyl ester of the disaccharide (19). The elegance of this method stems from the regiospecific opening of the cyclic amide ring in which the disaccharide linkage is formed at the equatorially oriented hydroxy-group at C-3.

Trialkylstannylation of alcohols increases their nucleophilicity. Therefore a good yield of methyl 2,3,4,6-tetra-*O*-acetyl-β-D-glucopyranoside was formed when the corresponding α-D-glucosyl bromide was treated with tributylstannyl methiodide

[12] R. Eby and C. Schuerch, *Carbohydrate Res.*, 1976, **50**, 203.
[13] J. Leroux and A. S. Perlin, *Carbohydrate Res.*, 1976, **47**, C8.
[14] P. Di Cesare and B. Gross, *Carbohydrate Res.*, 1977, **58**, C1.
[15] S. Hanessian and J. Banoub, *Tetrahedron Letters*, 1976, 661.

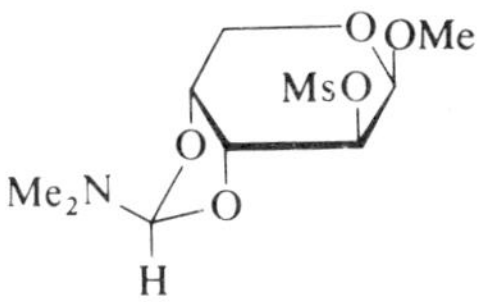

(17)

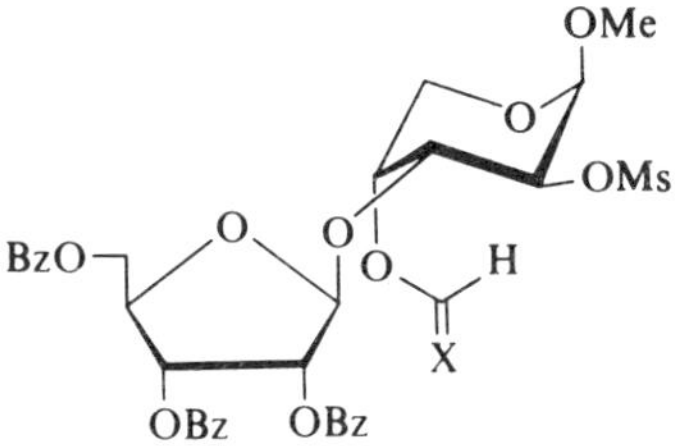

(18) X = $=\overset{+}{N}Me_2\ Cl^-$
(19) X = =O

and stannic chloride in dichloromethane.[16] Other glycosides were formed by treatment with the appropriate stannyl alkoxide. Alkoxides derived from primary alcohols gave β-glycosides, whereas those derived from secondary alcohols gave α-anomers. If the Lewis acid was omitted from the condensation and tetraethylammonium bromide used in its place, orthoesters were formed by participation from the 2-*O*-acetyl group upon the anomerized bromide. In this way several orthoesters were produced in excellent yields. Thus, acetobromoglucose gave (10) almost quantitatively when treated with the tributylstannyl alkoxide of methyl 2,3-*O*-isopropylidene-β-D-ribopyranoside under these reaction conditions.

Simple aglycones have been produced[17] from a variety of dialkyl acetals of dimethylformamide, *e.g.* $Me_2NCH(OR)_2$, in which R = Me, $CHMe_2$, $PhCH_2$, CH_2CMe_3, or cyclohexyl, since these have been shown to react with peracetylated sugars in the presence of stannic chloride to give almost quantitative yields of 1,2-*trans*-glycosides.

Tritylated sugars have been glycosylated.[18] For example, in the presence of trityl perchlorate the anomeric carbon atom in the orthothioacetate (8) was nucleophilically attacked by O-6 of 1,2,3,4-tetra-*O*-acetyl-6-*O*-trityl-β-D-glucopyranose to give a $\beta(1 \rightarrow 6)$-linked disaccharide derivative. The reaction is thought to occur *via* the sulphonium ion intermediate (9).

Interest in the solid-phase method for the synthesis of oligosaccharides continues. For example, 2,3,4-tri-*O*-benzyl-1-thio-β-D-glucopyranose has been attached to a polystyrene support by a thioglycosidic linkage and the free hydroxy-group at C-6 has been glycosylated with 6-*O*-acetyl-2,3,4-tri-*O*-benzyl-α-D-glucopyranosyl bromide.[19] The disaccharide was detached from the resin by sequential methylation and cleavage of the sulphonium salt with benzyl alcohol to give fully benzylated 6′-*O*-acetyl isomaltoside. The presence of the 6-*O*-acetyl residue in the resin-attached dimer offers future scope for selective deblocking and thence oligomer formation.

The hydrolysis of *O*-glycosides remains their most studied reaction, and many investigations have been reported. One of interest[20] monitored the hydrolysis of *p*-nitrophenyl tetrahydropyran-2-yl ether as a model for glycosides with cyclohepta-amylose 2-, 3-, and 6-monophosphoric acids to determine their effectiveness

[16] T. Ogawa and M. Matsui, *Carbohydrate Res.*, 1976, **51**, C13.
[17] S. Hanessian and J. Banoub, *Tetrahedron Letters*, 1976, 657.
[18] N. N. Kochetkov, L. V. Backinowsky, and Y. E. Tsvetkov, *Tetrahedron Letters*, 1977, 3681.
[19] S.-H. L. Chiu and L. Anderson, *Carbohydrate Res.*, 1976, **50**, 227.
[20] B. Siegel, A. Painter, and R. Breslow, *J. Amer. Chem. Soc.*, 1977, **99**, 2309.

as general acid catalysts for the hydrolysis of glycosides. Only the 3-phosphoric acid showed any net catalysis. The 2- and 6-phosphoric acids were not able to overcome the catalytic suppressing effect which binding to the β-cyclodextrin causes.

The synthesis of *C*-glycosides continues to attract much attention because of their potential as precursors for *C*-nucleosides. Buchanan and co-workers[21] report that ethynylmagnesium bromide reacts with 2,3:5,6-di-*O*-isopropylidene-D-mannofuranose (12) to give a 65% yield of 1,2-dideoxy-4,5:7,8-di-*O*-isopropylidene-D-*glycero*-D-*talo*-oct-1-ynitol (20); this efficiently ring-closed to the β-D-mannofuranosylethyne derivative (14) upon monotosylation *via* the 3-sulphonate (21). On the other hand sulphonation of the 3-benzoate (22) followed by methoxide-catalysed debenzoylation caused ring closure in the opposite sense, yielding the talosylethyne derivative (23).

Derivatives of this type are of value because they can be readily modified by cycloadditions to the acetylene residue. For example, two groups[22,23] have prepared the ribofuranosyl-pyrazole derivative (25) by diazomethane addition to the ribosylethyne (24), and another group[24] prepared the closely related derivative (27) by addition of the reagent to (26).

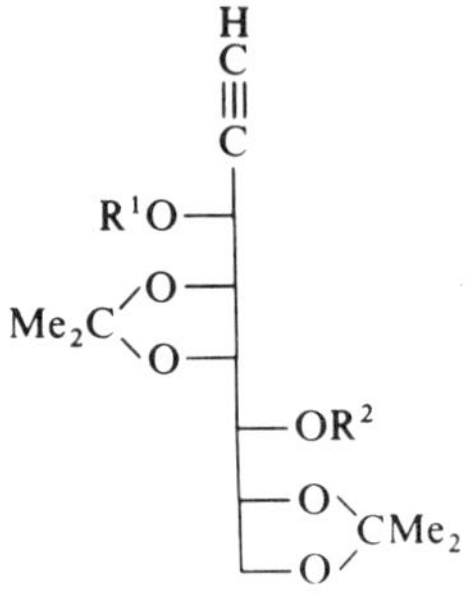

(20) $R^1 = R^2 = H$
(21) $R^1 = Ts, R^2 = H$
(22) $R^1 = Bz, R^2 = H$

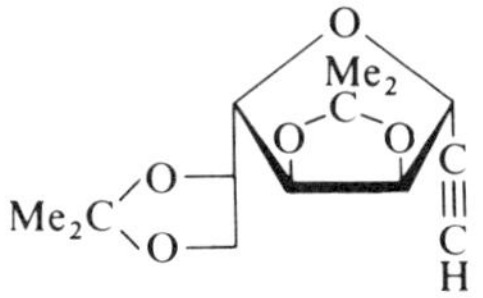

(23)

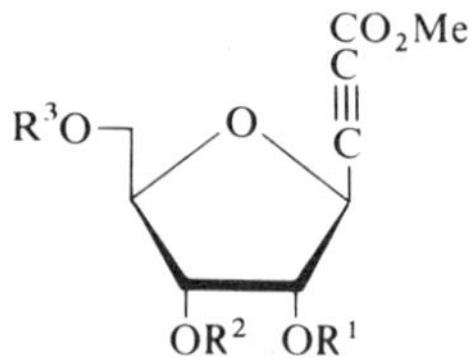

(24) $R^1 = R^2 = R^3 = CH_2Ph$
(26) $R^1R^2 = CMe_2, R^3 = CPh_3$

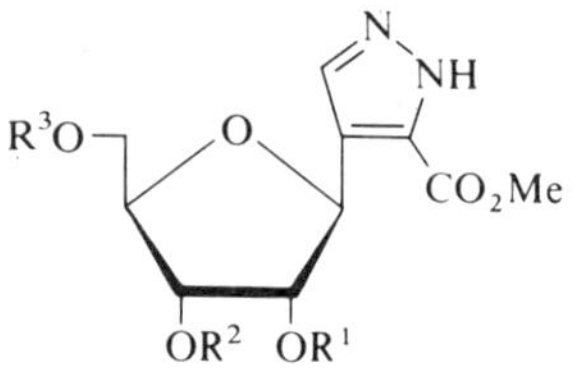

(25) $R^1 = R^2 = R^3 = CH_2Ph$
(27) $R^1R^2 = CMe_2, R^3 = CPh_3$

Formylaminomethylenation of lactones provides *C*-glycosyl precursors. Thus[25] the mannonolactone derivative (28) condenses with ethyl isocyanoacetate in the

21 J. G. Buchanan, A. D. Dunn, and A. R. Edgar, *J.C.S. Perkin I*, 1976, 68.
22 C. M. Gupta, G. H. Jones, and J. G. Moffatt, *J. Org. Chem.*, 1976, **41**, 3000.
23 J. G. Buchanan, A. R. Edar, M. J. Power, and G. C. Williams, *Carbohydrate Res.*, 1977, **55**, 225.
24 F. G. de Las Heras, S. Y.-K. Tam, R. S. Klein, and J. J. Fox, *J. Org. Chem.*, 1976, **41**, 84.
25 R. H. Hall, K. Bischofberger, S. J. Eitelman, and A. Jordaan, *J.C.S. Perkin I*, 1977, 743.

presence of potassium hydride to yield the manno-octenoate (29). This can be hydrogenated to give the amino-acid derivative (30), which is related to the nucleoside amino-acids found in the polyoxin complex of antifungal agents. If, in the condensation, the potassium hydride is omitted and 1,5-diazabicyclo-[4,3,0]non-5-ene is used in its place, then the oxazole (31) is produced.[26] Both condensation products are formed from the oxazolide anion (32), which rearranges either by route (*a*) to give, after protonation, (29) or by route (*b*) to give (31).

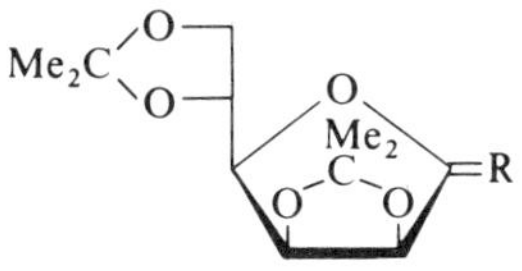

(28) R = O=

(29) R = EtO_2C(OHCNH)C=

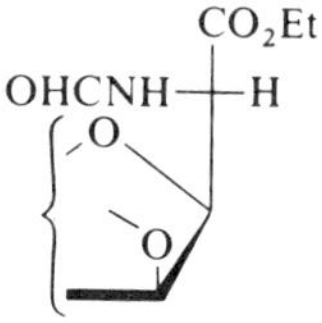

(30)

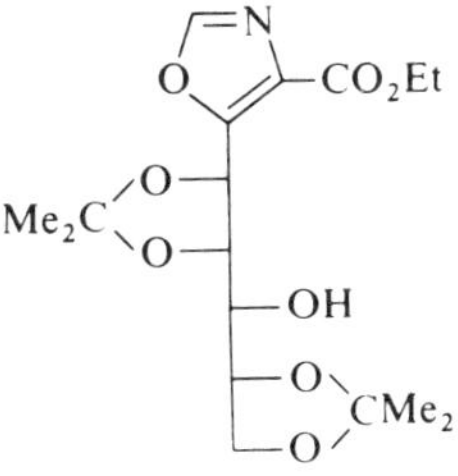

(31)

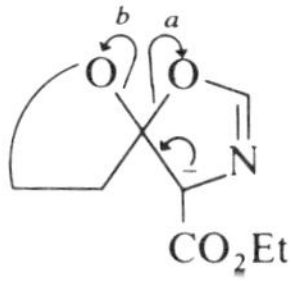

(32)

3 Unsaturated Sugars

2,3-Unsaturated hexopyranose derivatives are of value as synthetic intermediates and there is active interest in their synthesis. Fraser-Reid's group[27] has recommended a high-yielding route to ethyl 2,3,6-trideoxy-α-D-*erythro*-hex-2-enopyranoside (34) in which the 6-deoxy position was formed after the introduction of the double bond. Thus ethyl α-D-*erythro*-hex-2-enopyranoside (33) was deoxygenated at C-6 *via* its primary tosylate, which was nucleophilically displaced with sodium iodide in ethyl methyl ketone containing pyridine; the product was reduced to (34) with Raney nickel without detriment to the double bond. Pyridine was essential during the iodide displacement since in its absence the product rearranged to 2-(2-iodo-1-hydroxyethyl)furan. The enone (35) could be readily prepared from (34), but when it was the compound of interest it proved more economical to oxidize the allylic hydroxy-group in (33), iodinate the enone so formed directly at C-6 with triphenoxyphosphonium methodide, and follow this by selective hydrogenolysis.

A new method which augurs well for the preparation of this class of compounds has recently been introduced by Barton's group.[28] They showed that vicinal diol

[26] R. H. Hall, K. Bischofberger, S. J. Eitelman, and A. Jordaan, *J.C.S. Perkin I*, 1977, 2236.
[27] M. B. Yunker, S. Y.-K. Tam, D. R. Hicks, and B. Fraser-Reid, *Canad. J. Chem.*, 1976, **54**, 2411.
[28] A. G. M. Barrett, D. H. R. Barton, R. Bielski, and S. W. McCombie, *J.C.S. Chem. Comm.*, 1977, 866.

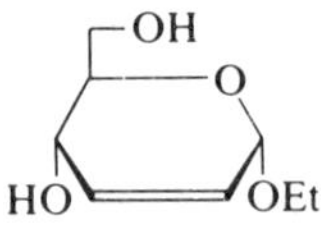

(33)

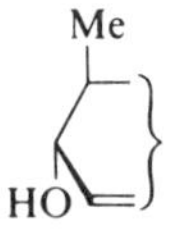

(34)

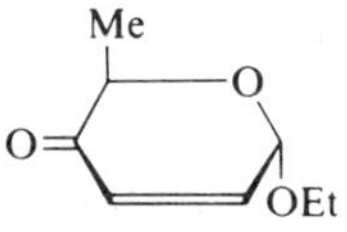

(35)

bisdithiocarbonates reacted with tributylstannane in refluxing toluene to give olefins. Thus the glycopyranoside derivative (36) gave, in 59% yield, the hex-2-enoside derivative (37). The corresponding mannoside also gave (37) in 79% yield. The reaction, which appears to be of general applicability, occurs by the free-radical mechanism outlined in (36).

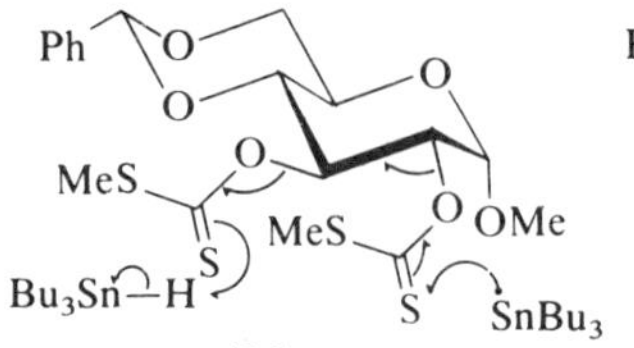

(36)

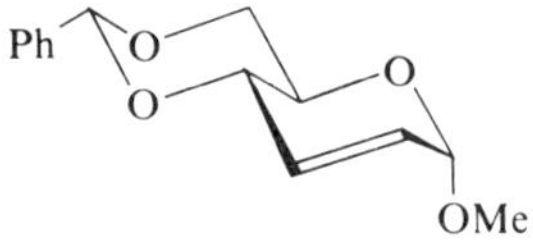

(37)

Nucleosides containing a 2-enopyranosyl unit occur in antibiotics such as blasticidin S and they also serve as potential synthetic intermediates for amicetin and plicacetin. Consequently the reported[29] preparation of the 2,3-dideoxyglyc-2-enopyranosylthymine derivative (40) from the 2,3-di-*O*-mesylthymine derivative (38) is of interest. The best overall yield of 56% was obtained when the 2,2′-anhydride (39), prepared from the dimesylate (38) by the action of sodium benzoate in DMF, was isolated and heated with sodium iodide in DMF. Treatment of either (39) or (38) under the more usual Tipson–Cohen conditions (*e.g.* with added zinc) gave (40) in lower yield.

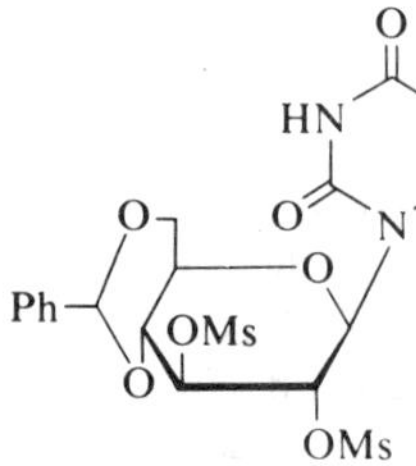

(38)

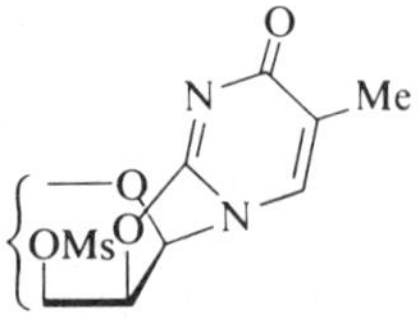

(39)

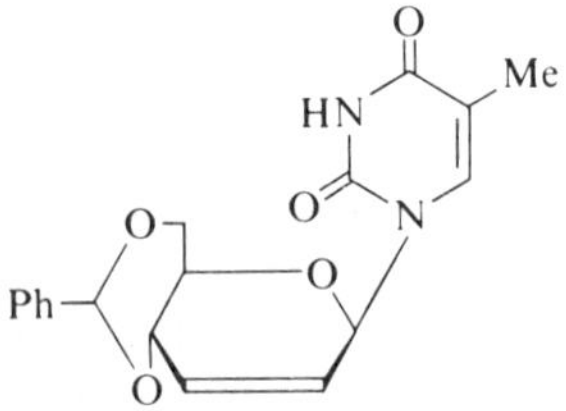

(40)

Methods based upon the Diels–Alder addition of n-butyl glyoxylate (BuO_2CCHO) to dienes have been reported for the preparation of 2,3-unsaturated uronate derivatives.[30,31] If mono-buta-1,3-dienyl ethers derived from sugars are used as dienes in this reaction then disaccharide derivatives containing a uronate are formed. By reducing the ester group and functionalizing the double bond in

[29] T. Yamazaki, K. Matsuda, H. Sugiyama, S. Seto, and N. Yamaoka, *J.C.S. Perkin I*, 1977, 1654.
[30] R. R. Schmidt and R. Angerbauer, *Angew. Chem. Internat. Edn.*, 1977, **10**, 783.
[31] S. David, A. Lubineau, and J.-M. Vatèle, *J.C.S. Perkin I*, 1976, 1831.

these adducts, David *et al.*[31] have developed this reaction into a route to unusual disaccharides.

Furanoid glycal derivatives are not easily produced by zinc reduction of furanosyl halides. Therefore the reported[32] conversion of 2,3 : 5,6-di-*O*-isopropylidene-α-D-mannofuranosyl bromide into the unsaturated disaccharide derivative (42) in 59% yield by sodium in tetrahydrofuran could prove of value. The anion (41), which would be formed from the mannosyl halide by bromine atom abstraction, presumably rearranges with elimination of acetone to an O-3 alkoxide derivative, which attacks another mannosyl bromide molecule.

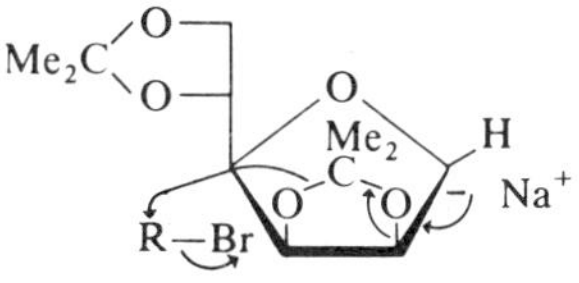

(41)

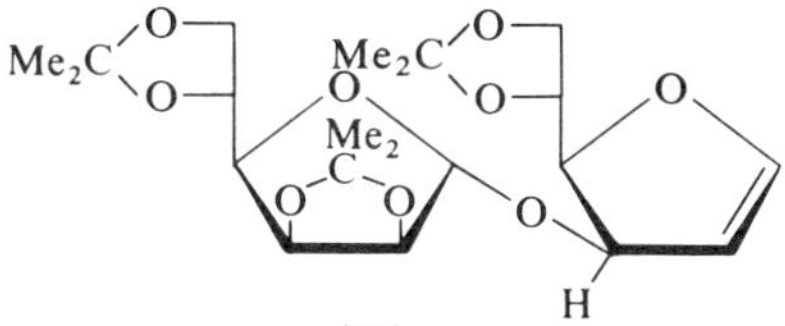

(42)

An unusual reaction, which leads to 1,2-unsaturated pyranoid derivatives, has been observed[33] to occur between phenyl tetra-*O*-benzoyl-1-thio-β-D-glucopyranoside (43) and an excess of *N*-bromosuccinimide in refluxing carbon tetrachloride under visible light; it gives phenyl 2,4,6-tri-*O*-benzoyl-1-thio-D-*erythro*-hex-1-enopyranosid-3-ulose (47) in 76% yield. The *galacto*-isomer of (43) gave the *threo*-epimer of (47) in 83% yield when similarly treated. The reaction is thought to occur *via* a halide intermediate (45), which is reasonable, since free-radical bromination at C-1 would be facilitated by the enhanced radical-stabilizing character of sulphur. Elimination of hydrogen bromide from (45) to give the 1,2-unsaturated derivative (46) followed by its allylic bromination gives, after loss of acyl bromide from C-3, the enone (47). The α-anomer of the glucoside (43) also gave enone (47), but a ten-fold increase in reaction time was required. The orientation of the anomeric hydrogen atoms in the thioglucosides is responsible for this rate difference, since radical attack on axial anomeric hydrogen atoms is stereoelectronically favoured. The tetra-*O*-acetylated analogue (44) gave the corresponding enone (48) in lower yield, accompanied by the 2-bromoacetyl derivative (49).

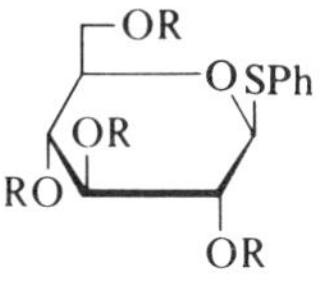

(43) R = Bz
(44) R = Ac

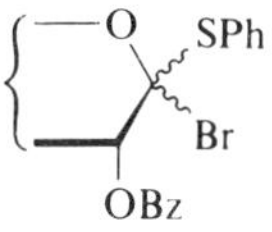

(45)

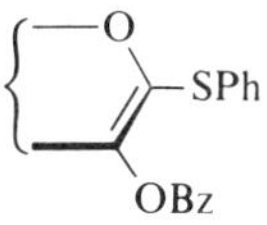

(46)

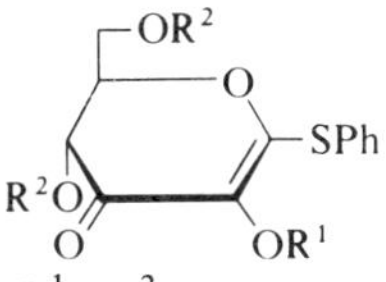

(47) $R^1 = R^2 = Bz$
(48) $R^1 = R^2 = Ac$
(49) $R^1 = COCH_2Br$, $R^2 = Ac$

4 Dicarbonyl Sugars

A mild photochemical method for the preparation of keto- and aldehydo-dicarbonyl sugars from pyruvate esters of secondary or primary hydroxy-groups of

[32] S. J. Eitelman and A. Jordaan, *J.C.S. Chem. Comm.*, 1977, 552.
[33] R. J. Ferrier and R. H. Furneaux, *J.C.S. Perkin I*, 1977, 1993.

suitably blocked sugars has been reported.[34,35] For example, 1,2:5,6-di-*O*-isopropylidene-α-D-glucofuranose was esterified with pyruvoyl chloride in benzene containing pyridine and the product irradiated in benzene with u.v. light either directly or after isolation, which gave the furanos-3-ulose derivative in 71% yield. Several other isopropylidenated sugars have been similarly oxidized, as have suitably blocked sugar acetates. Thus, 1,3,4,6-tetra-*O*-acetyl-α-D-glucopyranose 2-pyruvate (50) gave, upon irradiation, the very unstable fully acetylated pyranos-2-ulose derivative (51), which readily rearranged, with loss of acetic acid, to kojic acid diacetate (52). Thus this oxidation method appears to be one of the few mild enough to oxidize hydroxy-groups in partially acylated sugars. The method has also been applied with success to nucleosides, the 5′-*O*-trityl and 5′-*O*-benzoyl derivatives of 3′-*O*-pyruvoylthymidine being converted in good yields into the corresponding 3′-ulose derivatives. These ulose derivatives were also unstable, since in pyridine or on silica gel the heterocyclic base was eliminated and a 1-ene-3-one formed. The oxidation reaction appears to be an example of the Norish Type II reaction as shown in (50).

Microbiological oxidation of methyl α-D-xylopyranoside to the pentopyranosid-4-ulose (53) has been reported[36] to occur with *Acetobacter suboxydans* during four days in 76% yield. The β-anomer, on the other hand, was 28% oxidized during 30 days. Other pentosides were oxidized only to a small extent and hexopyranosides showed little oxidation.

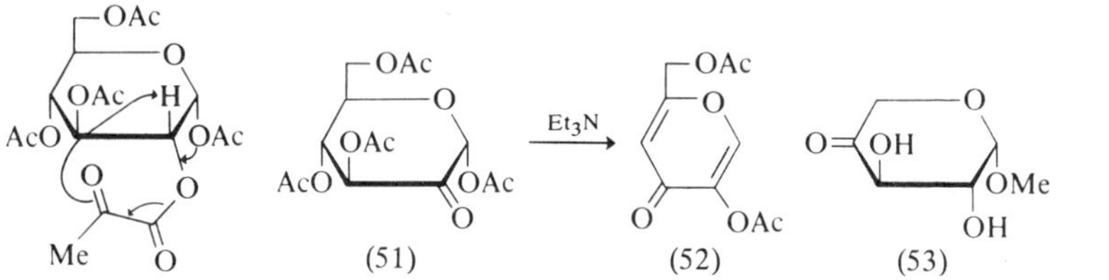

(50) (51) (52) (53)

A base-catalysed degradation of the 3-ulose derivative (54) has been developed[37] into a useful synthesis of the rare deoxyketose derivative (56). A two-phase system of diethyl ether and water containing lithium hydroxide was found to bring the conversion about in 55% yield. The transformation is thought to occur by initial opening of the pyranose ring, *via* (57), rather than by elimination of methanol and hence *via* methyl 4,6-*O*-benzylidene-α-D-*erythro*-hex-1-en-3-ulose.

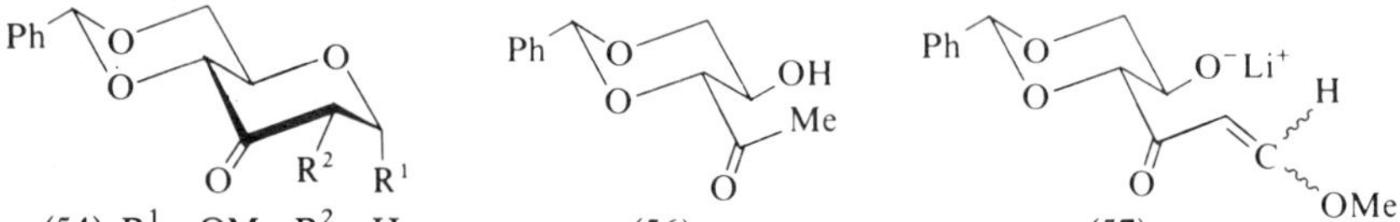

(54) R^1 = OMe, R^2 = H
(55) R^1 = OMe, R^2 = OAc
(56)
(57)

[34] R. W. Binkley, *J. Org. Chem.*, 1977, **42**, 1216.
[35] R. W. Binkley, D. G. Hehemann, and W. W. Binkley, *Carbohydrate Res.*, 1977, **58**, C10.
[36] W. A. Szarek and G. W. Schnarr, *Carbohydrate Res.*, 1977, **55**, C5.
[37] J.-C. Fischner, D. Horton, and W. Weckerle, *Canad. J. Chem.*, 1977, **55**, 4078.

Several groups have investigated the chemistry of α-keto-oxirans. Methyl 3,4-anhydro-6-deoxy-α-L-*lyxo*-hexopyranosid-2-ulose was found to give the chloro-enone (58) when treated with lithium chloride in tetrahydrofuran[38] or with hydrochloric acid in acetone.[39] Catalytic hydrogenation over palladium on charcoal selectively opened the anhydro-ring, yielding methyl 3,6-dideoxy-α-L-*threo*-hexopyranosid-2-ulose.

The sequential oxidation–reduction of partially protected sugars has been adopted as a standard procedure for epimerization. However, a recent report[40] indicates that the reduction step is not always straightforward, as illustrated by methyl 2-*O*-acetyl-4,6-*O*-benzylidene-α-D-*ribo*-hexopyranosid-3-ulose (55), which with sodium borodeuteride in moist methanol or propan-2-ol gave the expected alloside derivative (59) labelled with deuterium at C-3, whereas borodeuteride reduction in dry propan-2-ol gave a 1 : 1 mixture of (59) and the [2-^{2}H]glucopyranoside (60). The formation of (60) may involve an unusual nucleophilic attack upon C-2 of a 2,3-enediol intermediate.

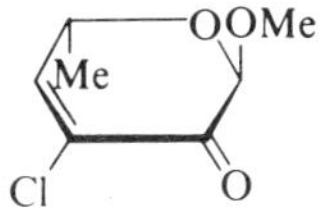

(58)

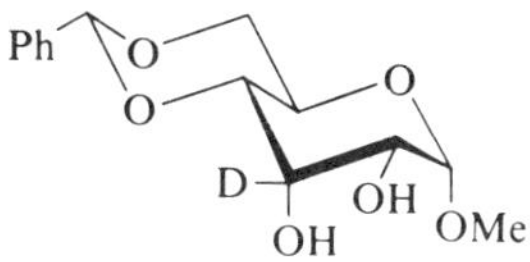

(59)

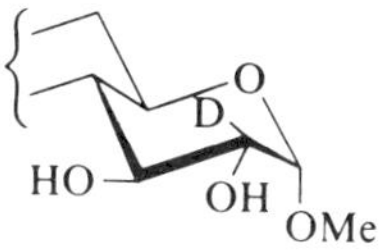

(60)

Deoxygenation of hexosulose derivatives by the Wolff–Kishner reaction has been examined again.[41] 1,5-Anhydro-4,6-*O*-benzylidene-2-deoxy-D-*erythro*-hex-3-ulose was smoothly converted into 1,5-anhydro-4,6-*O*-benzylidene-2,3-dideoxy-*erythro*-hexitol in 69% yield. However the reaction was less satisfactory with glycosidulose derivatives, since extensive degradation occurred.

Deoxygenation of sugar ketones has also been achieved by reduction of their *gem*-dithio and related derivatives, which are formed by the action of phosphorus pentasulphide in pyridine upon protected keto-sugars.[42] Based upon these findings 1,6-anhydro-3,4-*O*-isopropylidene-β-D-*lyxo*-hexopyranosulose gave the bridged derivative (61) whereas 1,2:5,6-di-*O*-isopropylidene-α-D-*ribo*-hexofuranos-3-ulose gave the disulphide (62). Reductive desulphurization of these products gave, respectively, the 2-deoxy derivative corresponding to (61) and 3-deoxy-1,2:5,6-di-*O*-isopropylidene-α-D-*xylo*-hexofuranose. An overall inversion of configuration at C-4 occurred in the transformation of the 3-ulose to the 3-deoxy-derivative,

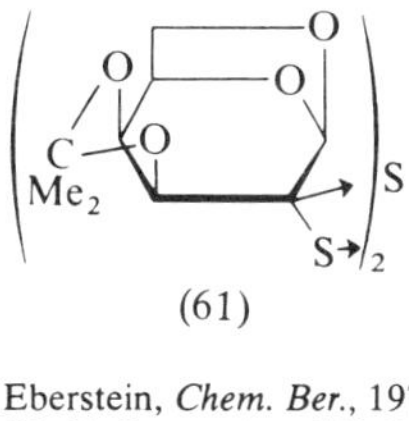

(61)

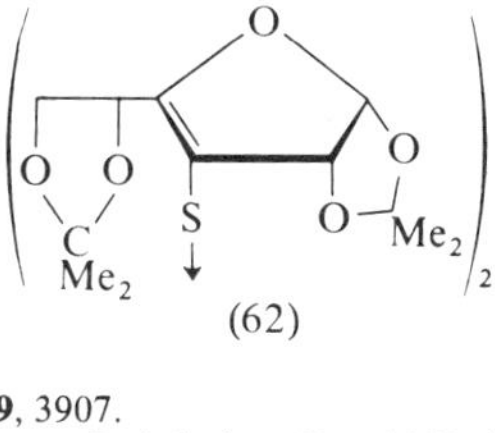

(62)

[38] H. Paulsen and K. Eberstein, *Chem. Ber.*, 1976, **109**, 3907.
[39] G. S. Hajivarnava, W. G. Overend, and N. R. Williams, *Carbohydrate Res.*, 1976, **49**, 93.
[40] D. C. Baker, J. Defaye, A. Gadelle, and D. Horton, *J. Org. Chem.*, 1976, **41**, 3834.
[41] D. Horton, W. Weckerle, L. Odier, and R. J. Sorenson, *J.C.S. Perkin I*, 1977, 1564.
[42] P. Köll, R.-W. Rennecke, and K. Heyns, *Chem. Ber.*, 1976, **109**, 2537.

presumably because hydrogen added to the double bond in (62) from its least hindered face.

5 Esters

Acylations which do not occur readily in pyridine have been shown[43] to proceed efficiently in tetrahydrofuran containing tetra-n-butylammonium fluoride. Thus when 5′-*O*-*p*-methoxytritylthymidine was treated with pivalic anhydride in pyridine the 3′-pivalate was formed in only 9% yield, whereas 95% of the ester was produced from a reaction carried out under the recommended conditions.

Improved selectivity in tosylations of diols under phase-transfer conditions has been achieved,[44] as shown by methyl 4,6-*O*-benzylidene-α-D-glucopyranoside, which gave the 2-tosylate derivative in 78% yield when treated with tosyl chloride in a mixture of dichloromethane and aqueous sodium hydroxide containing tetrabutylammonium hydrogen sulphate. The mannose derivative exhibited even greater selectivity, yielding the 2-tosylate in 95% yield.

The use of di-n-butyltin oxide, first introduced by Moffatt[45] in nucleoside chemistry, has been used to selectively esterify pyranosides and their derivatives.[46] With this reagent in methanol methyl α-D-glucopyranoside gave a quantitative yield of the 2,3-*O*-dibutylstannylene derivative (63), which was converted into methyl 2-*O*-benzoyl-α-D-glucopyranoside in 80–90% yield by subsequent treatment with benzoyl chloride. Similarly, (63) and myristoyl chloride or tosyl chloride gave the corresponding 2-esters. In contrast, the β-glucoside was esterified only at its primary hydroxy-group. The method also gave 2-esters with the methyl 4,6-*O*-benzylidene pyranosides of α-D-glucose, α-D-galactose, and α-D-allose. It failed with the corresponding derivatives of β-D-glucose and α-D-mannose. This has led to the proposal that the success of this reaction depends, in part, upon a stabilizing interaction between the tin atom and a suitably positioned methoxy-group as shown in (63).

In a related process, further selective acylations of secondary positions in hexopyranosides have been achieved through the regioselective enhanced nucleophilicity of one hydroxy-group by trialkylstannylation.[47] In this way methyl α-D-glucopyranoside gave, after sequential treatment with bis(tributylstannyl) oxide and benzoyl chloride, its 2,6-dibenzoate in 82% yield. The selectivity is thought to arise because stannylation occurred preferentially at an equatorial hydroxy-group which has an alkoxy- (or hydroxy-) group *cis* to it. Stabilization of the type depicted in (64) is thought to accrue from such an arrangement. The methyl pyranosides of α-D-mannose and β-D-galactose give, after similar treatment, excellent yields of their 3,6-dibenzoates. The 2,3- and 3-4-*cis*-hydroxy-groups in these pyranosides would be respectively involved in the formation of the stannyl derivative. It is, however, not yet clear why the equatorially disposed 3-hydroxy-group should be esterified from both arrangements.

[43] S. L. Beaucage and K. K. Ogilvie, *Tetrehedron Letters*, 1977, 1691.
[44] P. J. Garegg, T. Iversen, and S. Oscarson, *Carbohydrate Res.*, 1977, **53**, C5.
[45] D. Wagner, J. P. H. Verheyden, and J. G. Moffatt, *J. Org. Chem.*, 1974, **39**, 24.
[46] R. M. Munavu and H. H. Szmant, *J. Org. Chem.*, 1976, **41**, 1832.
[47] T. Ogawa and M. Matsui, *Carbohydrate Res.*, 1977, **56**, C1.

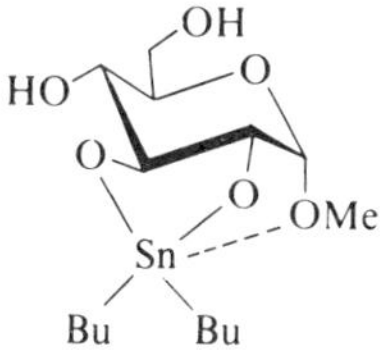

(63)

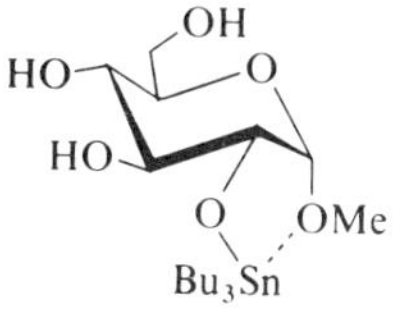

(64)

The order of reactivity of hydroxy-groups in carbohydrate derivatives towards acylation continues to attract attention. It has been found[48] that the hydroxy-groups in methyl β-lactoside react with benzoyl chloride in pyridine in the order 6′ > 3′ > 6 > 2 > 2′,4′ > 3.

Nucleophilic displacement remains a major interest in carbohydrate chemistry. Accordingly there have been several studies using trifluoromethanesulphonates (triflates) as good leaving groups.[49,50] It was found for example that the 3-triflate (65) underwent ready nucleophilic displacement with sodium benzenethiolate at 5 °C with inversion of configuration. Reduction of the 3-phenylthioalloside (66) so formed gave the 3-deoxy-sugar derivative (67).

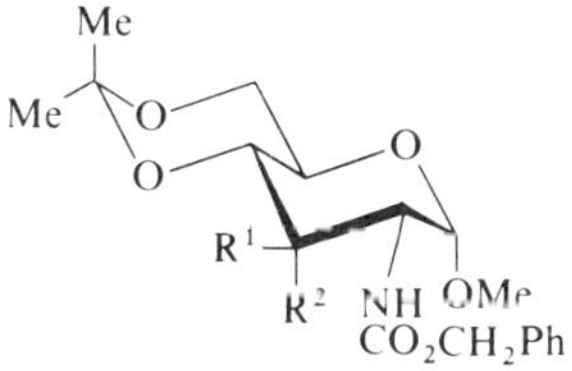

(65) $R^1 = OSO_2CF_3$, $R^2 = H$
(66) $R^1 = H$, $R^2 = SPh$
(67) $R^1 = R^2 = H$

A series of αβ-unsaturated esters have been prepared[51] as temporary blocking groups for the 5′-position in nucleosides which can be removed by treatment with hydrazine under mild conditions. Deblocking occurs *via* a hydrazino addition product which cyclizes to a pyrazolidin-3-one with release of the deblocked nucleoside.

Carbohydrate ester derivatives provide excellent models for examining the importance of stereochemistry in acyloxonium ion chemistry. Paulsen and his co-workers[52] have reported examples of five-membered acetoxonium ions fused *trans* to conformationally locked pyranose rings. For example, treatment of 2,3,4-tri-*O*-acetyl-1,6-anhydro-β-D-idopyranose with trifluoromethanesulphonic acid gave two such ions, fused at the 2,3- and 3,4-positions.

The same group have shown[53] that 1,3-rearrangement processes of acetoxonium ions compete with the more common 1,2-rearrangements. They found this by

[48] R. S. Bhatt, L. Hough, and A. C. Richardson, *J.C.S. Perkin I*, 1977, 2001.
[49] L. D. Hall and D. C. Miller, *Carbohydrate Res.*, 1976, **47**, 299.
[50] T. H. Haskell, P. W. K. Woo, and D. R. Watson, *J. Org. Chem.*, 1977, **42**, 1302.
[51] R. Arentzen and C. B. Reese, *J.C.S. Chem. Comm.*, 1977, 270.
[52] H. Paulsen, H. Höhne, and P. L. Durette, *Chem. Ber.*, 1976, **109**, 597.
[53] H. Paulsen and O. Brauer, *Chem. Ber.*, 1977, **110**, 331.

observing the racemization of 2,3,4-tri-*O*-acetyl-1,5-anhydro-D-arabinitol in hydrogen fluoride containing trifluoromethanesulphonic acid. Racemization can only occur *via* the two 1,3-rearrangements shown in Scheme 1.

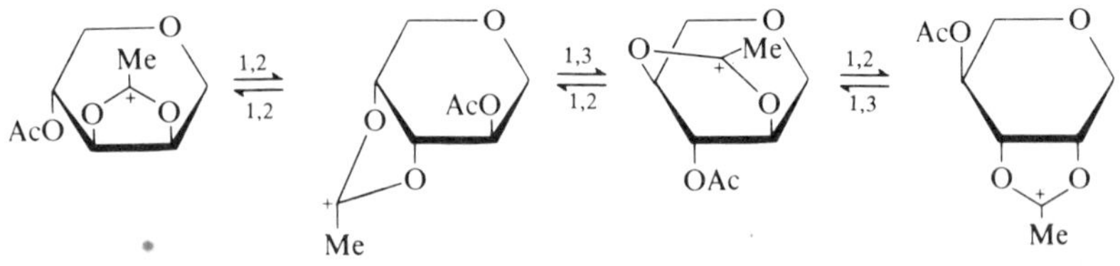

Scheme 1

6 Ethers and Acetals

New methods are being applied to carry out traditional carbohydrate etherifications. Methylations[54,55] and benzylations[55] of partially protected sugars have been achieved with methyl and benzyl trifluoromethanesulphonate, using the sterically hindered weak base 2,6-di-t-butylpyridine. The former reaction was carried out in refluxing dichloromethane, whereas the latter was accomplished at −70 °C, the benzyl triflate being prepared *in situ* from benzyl alcohol and the sulphonic acid anhydride.

Sugars have been built into macrocyclic ethers as a source of asymmetry. Several chiral asymmetric 18-crown-6 macrocycles incorporating D-glucose and D-galactose residues have been prepared[56] by the condensation of disodio derivatives of the methyl 4,6-*O*-benzylidene-2,3-di-*O*-(2-hydroxyethyl)-*α*-D-hexopyranosides with triethylene glycol bis(tosylate). The ability of these crowns to form diastereoisomeric complexes with primary and secondary alkylammonium cations in solution has been studied.[57] One unexpected observation was that the *galacto*-crowns, which have one face more sterically hindered on account of the *cis*-ring junction at C-4 and C-5, formed stronger complexes than the corresponding *gluco*-crowns. The reason for this is that in, for example, the diastereoisomeric complex (68), the O-4 can participate with the macrocyclic ether oxygen atoms in hydrogen bonding and/or electrostatic stabilization of the quaternary ammonium hydrogens.

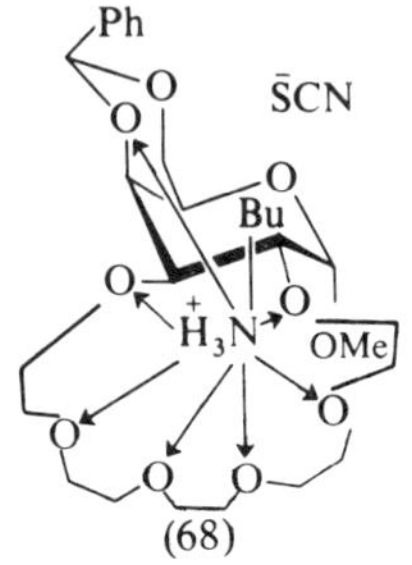

(68)

[54] J. Arnarp, L. Kenne, B. Lindberg, and J. Lönngren, *Carbohydrate Res.*, 1975, **44**, C5.
[55] J. M. Berry and L. D. Hall, *Carbohydrate Res.*, 1976, **47**, 307.
[56] D. A. Laidler and J. F. Stoddart, *Carbohydrate Res.*, 1977, **55**, C1.
[57] D. A. Laidler and J. F. Stoddart, *J.C.S. Chem. Comm.*, 1977, 481.

The role of acetals in carbohydrate chemistry has been changing during the past decade because synthetically useful reactions have been introduced which open cyclic acetals to give partially blocked sugars. Developments in this area have continued. For example, Klemer and her co-workers[58] have extended their studies of the reaction of butyl-lithium on *O*-benzylidene-sugar derivatives to various mono- and di-*O*-isopropylidene derivatives. Their general findings are illustrated by the conversion of 1,2:3,4-di-*O*-isopropylidene-6-*O*-methyl-α-D-galacto-pyranose into the (*E*)- and (*Z*)-isomers of the unsaturated polyol (72). Abstraction of the anomeric proton is thought to initiate this transformation and the anion (69) so formed subsequently rearranges with loss of acetone to (70), which then eliminates another acetone molecule to give, after protonation at O-4, the unsaturated lactone (71). Further attack on (71) gives the isomeric products (72) in low yield.

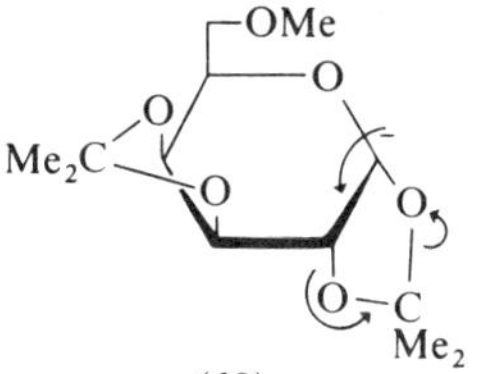

(69)

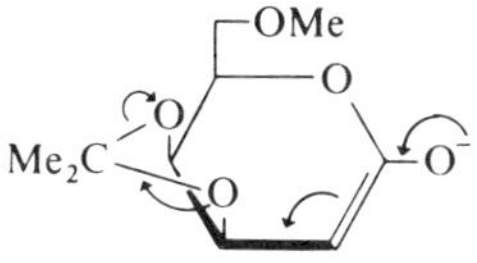

(70)

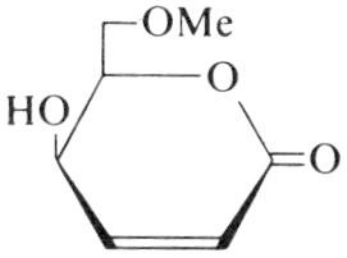

(71)

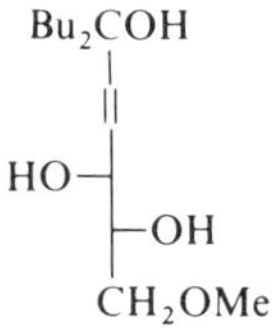

(72)

The synthetic value of this reaction has been studied[59] with methyl 2,3-*O*-benzylidene-α-L-rhamnopyranoside (73) and its 4-*O*-methyl derivative (74). Butyl-lithium smoothly converted (74) at $-30\,°C$ into the 3-ulose derivative (75), whereas the unmethylated compound (73) reacted only at $0\,°C$ to give a mixture of elimination products. The vigorous reaction conditions needed to abstract a carbon-bonded proton from the O-4 oxyanion of (73) were thought to be responsible for the difference in this case.

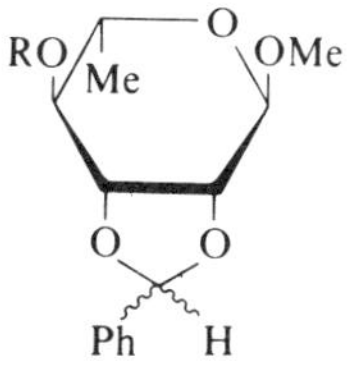

(73) R = H
(74) R = Me

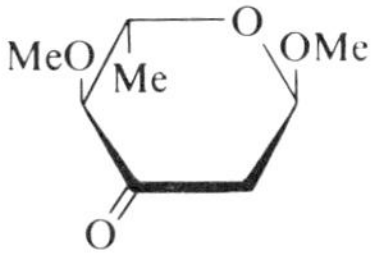

(75)

[58] G. Rodemeyer and A. Klemer, *Chem. Ber.*, 1976, **109**, 1708.
[59] D. M. Clode, D. Horton, and W. Weckerle, *Carbohydrate Res.*, 1976, **49**, 305.

Liptak's studies with the lithium aluminium hydride–aluminium trichloride reducing agent have revealed that the regiospecificity observed in the hydrogenolytic ring cleavage of 2,3-[60] and 3,4-*O*-benzylidene[61] derivatives is dependent upon the stereochemistry at the acetal centre. Accordingly the *exo*-isomer of (73) afforded methyl 3-*O*-benzyl-α-L-rhamnopyranoside, whereas the *endo*-isomer of (73) gave the 2-*O*-benzyl derivative. The high degree of stereoselectivity does not appear to arise from steric interaction with the phenyl group in the *endo*-isomer of (73) since the spin–lattice relaxation times for all carbon atoms in *endo*- and *exo*-(73) were found[62] to be similar, which indicates that rotation of the phenyl groups occurs to the same extent in both isomers.

By use of this reagent *O*-benzylidene residues have been employed as temporary blocking groups in oligosaccharide syntheses.[63]

Cleavage of the 1,3-dioxan ring in 4,6-*O*-benzylidene derivatives with this reagent is less predictable.[60] This was also unfortunately found to be the case with photogenerated ketyl radicals.[64] Thus u.v. irradiation of acetophenone, for example, in an oxygenated solution of 1,2,3-tri-*O*-acetyl-4,6-*O*-benzylidene-β-D-glucopyranose gave a mixture of the 4-*O*-benzoyl and 6-*O*-benzoyl derivatives of 1,2,3-tri-*O*-acetyl-β-D-glucopyranose, plus an adduct formed by ketyl radical addition to the benzylidene acetal carbon atom.

7 Halo- and Thio-sugars, and other Inorganic Sugar Derivatives

Two new approaches[65,66] to the synthesis of halo-sugars have been introduced which are particularly satisfactory for primary iodo- and chloro-derivatives. The precursors for the iodo-compounds are the readily available cyclic thiocarbonates, which have been found to undergo opening of the thiocarbonate ring when treated with methyl iodide.[65,66] Hence the 4,6-thiocarbonate (76), when treated with methyl iodide, gave a high yield of 6-deoxy-6-iodo-4-*O*-methylthiocarbonyl glucoside (78).[66] The ring opening occurred *via* the intermediate (77), which underwent regiospecific attack by iodide at the least hindered primary position. As would be expected from this mechanism, methyl 4,6-*O*-benzylidene-α-D-glucopyranoside 2,3-thiocarbonate opened non-regiospecifically[65] to give a mixture of the 2- and 3-iodo-glucoside derivatives with presumed *manno*- and *allo*-configurations respectively.

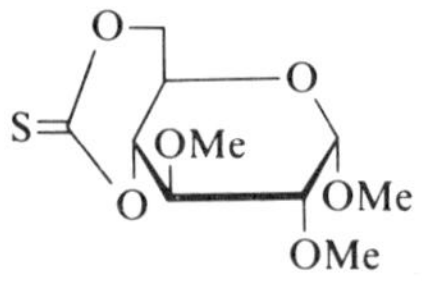

(76)

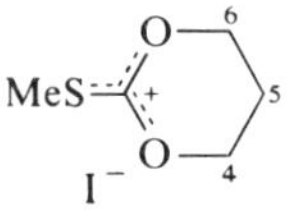

(77)

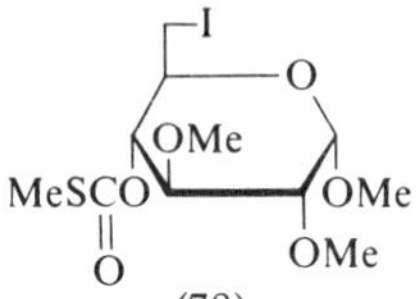

(78)

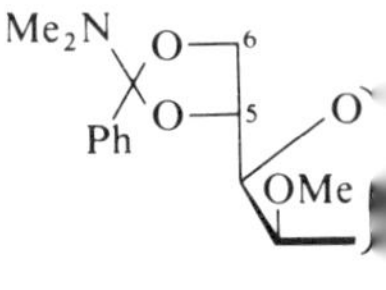

(79)

[60] A. Lipták, P. Fügedi, and P. Nánási, *Carbohydrate Res.*, 1976, **51**, C19.
[61] A. Lipták, *Tetrahedron Letters*, 1976, 3551.
[62] A. Neszmélyi, A. Lipták, and P. Nánási, *Carbohydrate Res.*, 1977, **58**, C7.
[63] A. Lipták and P. Nánási, *Tetrahedron Letters*, 1977, 921.
[64] M. Suzuki, T. Inai, and R. Matsushima, *Bull. Chem. Soc. Japan*, 1976, **49**, 1585.
[65] D. H. R. Barton and R. Subramanian, *J.C.S. Perkin I*, 1977, 1718.
[66] D. H. R. Barton and R. V. Stick, *J.C.S. Perkin I*, 1975, 1773.

The chloro-sugars were prepared[67] by treating partially protected sugar derivatives possessing an unblocked 1,2- or 1,3-diol structural unit with *NN*-dimethyl-α-chlorobenzylideneammonium chloride [$Ph(Cl)C{=}\overset{+}{N}Me_2\,Cl^-$]. In this way, the pair of hydroxy-groups at C-5 and C-6 in 1,2-*O*-isopropylidene-3-*O*-methyl-α-D-glucofuranose react with this reagent to give in almost quantitative yield 5-*O*-benzoyl-6-chloro-6-deoxy-1,2-*O*-isopropylidene-3-*O*-methyl-α-D-glucofuranose. The reaction occurs *via* the cyclic intermediate (79), with ring opening taking place by chloride ion attack at C-6. Chloro-derivatives were not obtained from methyl 4,6-*O*-benzylidene-α-D-glucopyranoside and this reagent.

6-Bromo-6-deoxy-sugars are often prepared by opening dioxan rings in 4,6-*O*-benzylidene-pyranosides with *N*-bromosuccinimide, but there are fewer cases of this reagent being used to cleave dioxolan rings in *O*-benzylidene-sugar derivatives. However, two recent reports[68,69] showed the reaction to be regiospecific, since methyl 2,3-*O*-benzylidene-4-*O*-methyl-α-L-rhamnoside (74) gave[68] a 3-bromo-3-deoxy 2-benzoate, and methyl 3,4-*O*-benzylidene-β-D-fucopyranoside gave[69] a 3-bromo-3-deoxy 4-benzoate.

The biological properties of fluorinated saccharides have led to continued interest in this class of compounds, and new fluorinating reagents are frequently applied to sugar derivatives. One such compound, diethylaminosulphur trifluoride, converted the primary hydroxy-group in 1,2,3,4-tetra-*O*-acetyl-β-D-glucopyranose into a 6-deoxy-6-fluoro-group in high yield.[70] Free keto- or aldehydo-carbonyl groups in otherwise fully blocked sugars are converted[71] into *gem*-difluorides by this reagent, as shown by methyl 2,3-*O*-isopropylidene-β-D-*erythro*-pentopyranosid-4-ulose, which gave the 4-deoxy-4,4-difluoro-derivative (80)[71] when so treated.

Ferrier and Furneaux[72] have reported radical bromination at C-5 in some methyl hexopyranuronates, which yield glycopyranosyl bromides of unusual structure. Thus methyl glycopyranuronate (81) and its 2,6-anhydro analogue (83) gave, upon treatment with *N*-bromosuccinimide in refluxing carbon tetrachloride under bright light, the monobromo-derivatives (82) and (84) respectively. Acetolysis of (84) followed by deacylation and subsequent treatment with hydrogen chloride gave L-ascorbic acid.

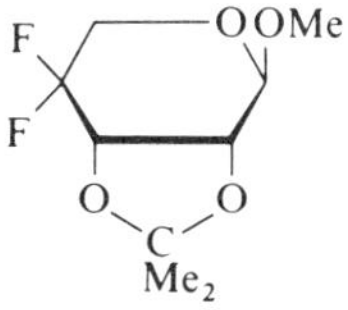

(80)

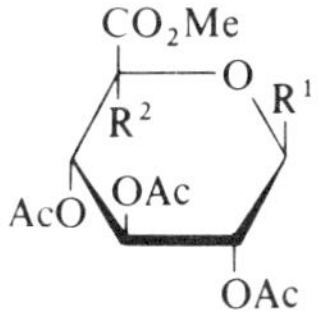

(81) $R^1 = OAc, R^2 = H$
(82) $R^1 = OAc, R^2 = Br$
(83) $R^1 = R^2 = H$,
(84) $R^1 = H, R^2 = Br$

[67] T. G. Black, D. H. R. Barton, and B. L. Rao, *J.C.S. Perkin I*, 1977, 1715.
[68] C. Monneret, J.-C. Florent, N. Gladieux, and Q. Khuong-Huu, *Carbohydrate Res.*, 1976, **50**, 35.
[69] K. Eklind, P. J. Garegg, and B. Gotthammer, *Acta Chem. Scand.* 1975, **B29**, 633.
[70] M. Sharma and W. Korytnyk, *Tetrahedron Letters*, 1977, 573.
[71] R. A. Sharma, I. Kavai, Y. L. Fu, and M. Bobek, *Tetrahedron Letters*, 1977, 3433.
[72] R. J. Ferrier and R. H. Furneaux, *J.C.S. Perkin I*, 1977, 1996.

N.m.r. studies[73] of relevance to glycopyranosyl chlorides have been carried out on chloromethyl methyl ether. At −182 °C the methylene protons give rise to a distinct symmetrical doublet, consistent with a *gauche* conformation about the carbon–oxygen bond. Line-shape calculations on the coalescing methylene proton resonance at −180 °C suggested that two chiral *gauche* forms were exchanging with a rate constant of 250 s^{-1}, from which a value of 2–3 kcal mol^{-1} for the anomeric effect in α-chloro-ethers was calculated.

1-Thioglycosides are usually produced by the action of thiols on 1,2-*trans*-glycosyl acetates, but dithioacetals, formed by opening of the glycosidic ring, often contaminate the products. It has been reported[74] that formation of dithioacetal derivatives can be suppressed by the use of tributylstannyl sulphide (TBSS). Based upon this finding, 2,3,4,6-tetra-*O*-acetyl-1-thio-β-D-glycopyranoside was formed in good yield when the corresponding α-D-glucosyl 1-bromide was treated with this reagent. Unfortunately only 45% of the bromo-derivative reacted under the usual reaction conditions, because the nucleophilicity of the sulphur in TBSS is attenuated. However, upon addition of stannic chloride to the reaction, a quantitative yield of an anomeric mixture of the thioglucoside derivatives was produced. The isomers were formed by the stannic-chloride-catalysed anomerization of the β-isomer.

The propensity of sulphur to participate in displacement reactions and in ring closures is well known, and interest in both continues. For example, 5-thio-D-ribose and 5-thio-D-xylose both gave, upon acid-catalysed acetonation, 1,2:3,4-di-*O*-isopropylidene-pyranoses.[75] Furanose products (oxygen in the ring) were not detected, thus demonstrating the strong preference for those diacetals with a six-membered sulphur-containing ring, despite the strain engendered by a cyclic acetal derived from a vicinal *trans*-diol in the xylopyranose compound and by a *cis–syn–cis*-arrangement of cyclic acetal rings in the ribopyranose derivative.

Neighbouring group participation by a thioether residue has been utilized[76] in the key step in the conversion of tubercidin (85) into 2′-deoxytubercidin (87). Thus when the 3′-*S*-benzyl 2′-mesylate (88), which was obtained from 2′,3′-anhydrotubercidin, was heated with sodium benzoate in DMF, the thiobenzyl group migrated to the 2′-position to give, after deblocking, (86). Reductive desulphurization of this product gave (87).

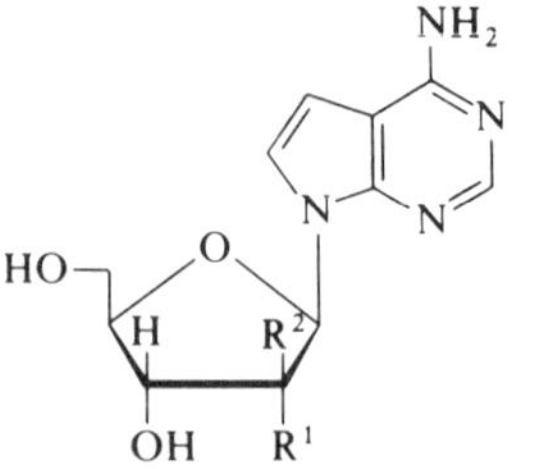

(85) $R^1 = OH, R^2 = H$
(86) $R^1 = H, R^2 = SCH_2Ph$
(87) $R^1 = R^2 = H$

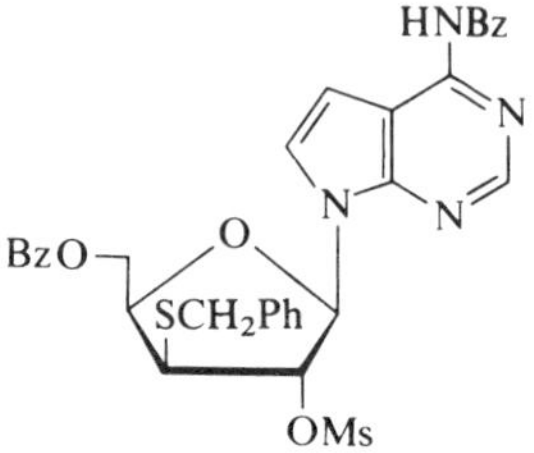

(88)

[73] F. A. L. Anet and I. Yauari, *J. Amer. Chem. Soc.* 1977, **99**, 6752.
[74] T. Ogawa and M. Matsui, *Carbohydrate Res.*, 1977, **54**, C17.
[75] N. A. Hughes and C. J. Wood, *Carbohydrate Res.*, 1976, **49**, 225.
[76] M. J. Robins and W. H. Muhs, *J.C.S. Chem. Comm.*, 1976, 269.

Reports of sugar rings containing phosphorus have appeared. Inokawa's group[77] have, for example, synthesized 5-ethylphosphinyl-5-deoxy-D-xylopyranose (89), the phosphorus being introduced by displacement at C-5 in 3-*O*-acetyl-5-deoxy-5-iodo-1,2-*O*-isopropylidene-α-D-xylofuranose with *OO*-triethylphosphorite.

Compound (90), containing both phosphorus and oxygen in the six-membered ring, was obtained when 2,3:5,6-di-*O*-isopropylidene-D-mannofuranose was subjected to the Abramov reaction with dimethyl phosphite.[78]

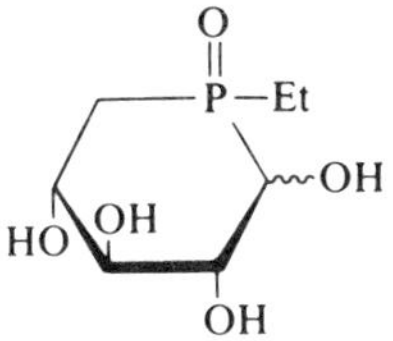

(89)

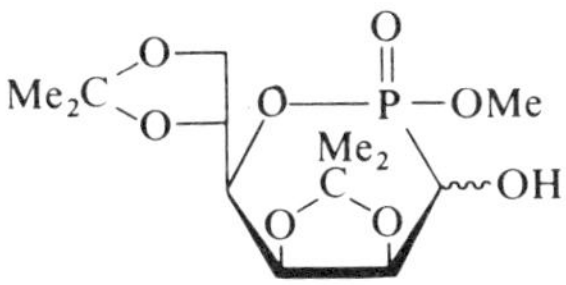

(90)

Application of tin-containing reagents to carbohydrates has now become quite common, so it was inevitable that stannane carbohydrate derivatives would eventually appear. This new class of sugar derivative, containing a carbon—tin bond, has been prepared[79] by triphenyltin lithium displacements in sugar tosylates or opening of oxiran rings in sugar epoxides. For example, with methyl 2,3-anhydro-4,6-*O*-benzylidene-α-D-allopyranoside this reagent gave (methyl 4,6-*O*-benzylidene-α-D-altropyranosid-2-yl)triphenylstannane in 75% yield.

8 Deoxy-sugars and Branched-chain Sugars

Several new methods for preparing deoxy-sugars have recently been reported.

Primary and secondary iodo-sugar derivatives have been reduced[65,66] in high yields by the Cr^{2+}–thiol procedure. For example, the 6-deoxy-6-iodo-glucoside derivative (78) was reduced with chromium(II) acetate in DMF containing butane-1-thiol to give, after de-esterification, methyl 6-deoxy-2,3-di-*O*-methyl-α-D-glucopyranoside in excellent yield. A mild method for deoxygenation of secondary alcohols has been successfully applied to sugars.[80] In this method *NN*-dimethyl-α-chlorobenzylideneammonium chloride reacts with one free hydroxy-group in an otherwise fully blocked sugar derivative to give an intermediate, which is converted into a thionobenzoate by hydrogen sulphide. Reduction of the thiono-ester with tributylstannane gives the corresponding deoxy-sugar in high yield.

In addition to preparing iodo-sugars from cyclic thiocarbonates, Barton and co-workers[65] have reductively ring-cleaved them. Thus, the 5,6-thiocarbonate (91),

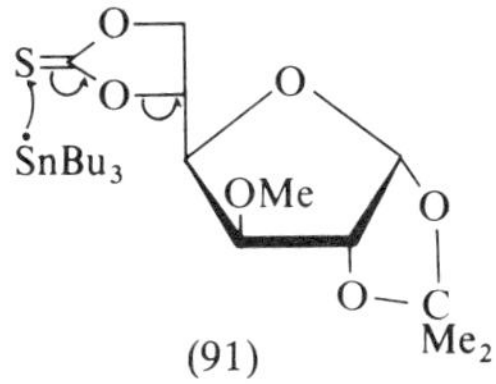

(91)

[77] F. Seo and S. Inokawa, *Bull. Chem. Soc. Japan*, 1975, **48**, 1237.
[78] J. Thiem, M. Günther, H. Paulsen, and J. Kopf, *Chem. Ber.*, 1977, **110**, 3190.
[79] L. D. Hall, D. C. Miller, and P. R. Steiner, *Carbohydrate Res.*, 1976, **52**, C1.
[80] D. H. R. Barton and S. W. McCombie, *J.C.S. Perkin I*, 1975, 1574.

upon refluxing with tributylstannane in toluene containing $\alpha\alpha'$-azoisobutyronitrile, gave, after alkaline hydrolysis, the corresponding derivative of the 5-deoxy-hexose in 67% yield as the only deoxy-sugar product. The regiospecificity arises because the radical-initiated ring cleavage leads to the secondary radical at C-5, as shown in (91). The 4,6-thiocarbonate (76) similarly gave the corresponding derivative of the 4-deoxy-hexose. Specificity was absent in the reaction of methyl 4,6-*O*-benzylidene-α-D-glucopyranoside 2,3-thiocarbonate with the tin hydride, since this gave 3-deoxy- and 2-deoxy-glycoside derivatives in 60 and 30% yields respectively.

Common primary and secondary esters of sugars have been photochemically reduced in aqueous hexamethylphosphortriamide (HMPT). Accordingly 1,2:5,6-di-*O*-isopropylidene-α-D-glucofuranose 3-acetate[81,82] or 3-pivalate,[82] upon u.v. irradiation in aqueous HMPT, gave the corresponding 3-deoxy-hexose derivative in 70% yield. Di- and tri-deoxy-sugars were obtained from the corresponding di- and tri-esters in one step, and in this way the readily available 2,3,6-tri-*O*-acetyl-α-D-glucopyranoside was converted into methyl α-D-amicetoside in reasonable yield.[82]

The branched-chain sugar L-dendroketose (96) has been synthesized, using the well-established aldol-condensation procedure to generate the branch point[83] from the *aldehydo*-arabinose derivative (92) and formaldehyde. The branched alditol derivative (93) so formed was deblocked and oxidized, by *Acetobacter suboxydans*, at the C-2 position, as expected from the Bertrand–Hudson rule, to give (96). Acetonation of (96) gave 1,2:3,4- and 2,3:4,4′-di-*O*-isopropylidene-L-dendroketose derivatives, which were used in ^{13}C n.m.r. spectroscopic studies to assign the configuration at C-4.[84]

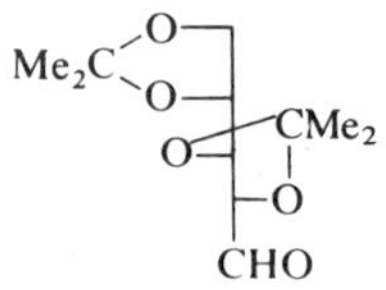

(92)

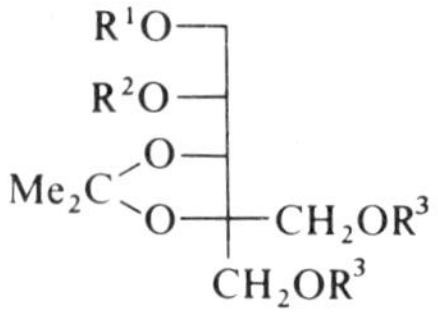

(93) $R^1R^2 = CMe_2$, $R^3 = H$
(94) $R^1 = R^3 = Bz$, $R^2 = H$
(95) $R^1 = CPh_3$, $R^2 = H$, $R^3 = CH_2Ph$

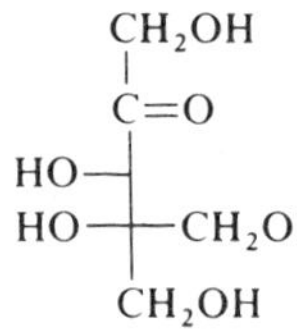

(96)

In an earlier attempt[85] to prepare dendroketose, compound (94) was oxidized with a chromium trioxide–pyridine complex to give the fully protected branched-chain sugar. However, base-catalysed debenzoylation of this material was accompanied by isomerization at C-3. This problem has been overcome[86] by use of compound (95), which upon oxidation yielded a dendroketose derivative from which the blocking groups could be removed without recourse to base.

[81] J.-P. Pete, C. Portella, C. Monneret, J.-C. Florent, and Q. Khuong-Huu, *Synthesis*, 1977, 774.
[82] P. M. Collins and V. R. N. Munasinghe, *J.C.S. Chem. Comm.*, 1977, 927.
[83] W. A. Szarek, G. W. Schnarr, H. C. Jarrell, and K. J. N. Jones, *Carbohydrate Res.*, 1977, **53**, 101.
[84] D. M. Vyas, H. C. Jarrell, and W. A. Szarek, *Canad. J. Chem.*, 1975, **53**, 2748.
[85] H. C. Jarrell, W. A. Szarek, J. K. N. Jones, A. Dmytraczenko, and E. B. Rathbone, *Carbohydrate Res.*, 1975, **45**, 151.
[86] E. B. Rathbone and G. R. Woolard, *Carbohydrate Res.*, 1976, **46**, 183.

A new route to branched-chain sugar derivatives has been reported[87] which involves the condensation shown in (97) between 2,3-*O*-isopropylidene-D-glyceraldehyde and 4,5-dimethyldioxaphospholen to give the corresponding dioxophospholan. The addition is remarkable because the two new chiral centres are generated stereospecifically. Hydrolysis of the dioxophospholan followed by Fischer glycosidation gave a methyl pyranoside and furanoside of 1-deoxy-3-*C*-methyl-*β*-D-*ribo*-hexulose in 20% and 30% yields respectively.

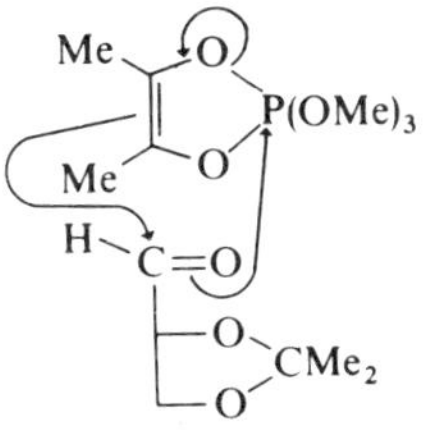

(97)

Anions derived from 1,3-dithians are frequently added to ulose derivatives as the key step in syntheses of branched-chain sugars. The monobenzoyl derivative of pillarose (99) has been prepared in this way.[88] Thus the dianion of 1,3-dithian-2-methanol reacted with methyl 2,3,6-trideoxy-*α*-L-*glycero*-hexopyranosid-4-ulose to give the addition product (98), which was monobenzoylated at the primary hydroxy-group and deblocked at the carbonyl group to give the pillarose derivative (99) in overall 10% yield.

Although additions to ulose derivatives remain one of the most common routes to branched-chain sugars, 1,4-additions to carbohydrate enones are beginning to attract attention. One such example is the addition of lithium dimethylcuprate to the hex-2-enopyranosid-4-ulose (100) to give (in 84% yield) the 2-*C*-methyl derivative (102).[89] The addition occurred to the least hindered side of the pyranose ring, *trans* to the aglycone.

Lithium dimethylcuprate, vinylmagnesium bromide, and 2-lithio-2-ethoxycarbonyl-1,3-dithiolan have been similarly added[90] to methyl hex-2-enopyranosid-4-ulose (101) to give the 2-*C*-branched derivatives shown in (104).

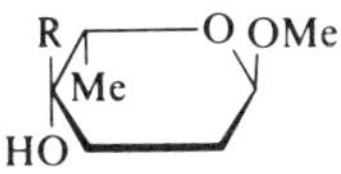

(98) R = 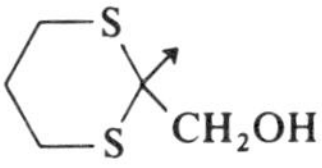

(99) R = $BzOCH_2CO$

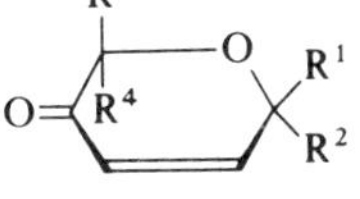

(100) $R^1 = R^4 = H$, $R^2 = OEt$, $R^3 = CH_2OCPh_3$
(101) $R^1 = OMe$, $R^2 = R^3 = H$, $R^4 = Me$

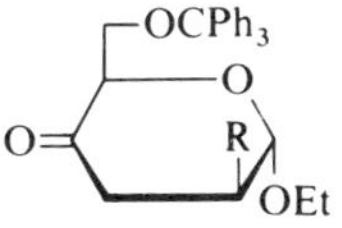

(102) R = Me
(103) $R = CH_2OH$

[87] S. David, M.-C. Lépine, G. Aranda, and G. Vass, *J.C.S. Chem. Comm.*, 1976, 747.
[88] H. Paulsen, K. Roden, V. Sinnewell, and W. Koebernick, *Chem. Ber.*, 1977, **110**, 2146.
[89] M. B. Yunker, D. E. Plaumann, and B. Fraser-Reid, *Canad. J. Chem.*, 1977, **55**, 4002.
[90] H. Paulsen, W. Koebernick, and H. Koebernick, *Tetrahedron Letters*, 1976, 2297.

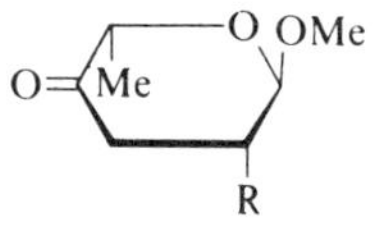

(104) R = Me or $CH_2{=}CH$

or

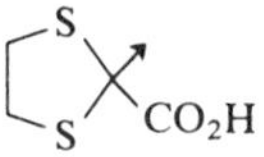

Photochemically induced conjugate additions of simple alcohols,[91] acetals,[92] and aldehydes to this class of carbohydrate derivative have been achieved. Thus u.v. irradiation of benzophenone in methanol containing the enone (100) gave (103), presumably by the 1,4-addition of photogenerated hydroxymethyl radicals. The addition was stereospecific and the configuration at C-2 in (103) was verified by ready formation of a 2,3-cyclopropane derivative with the D-*lyxo*-structure.

Photoexcited 1,2:4,6-diacetalated hexopyranosid-3-uloses have been found[93] to react with methanol to give isomeric mixtures of the corresponding derivatives of 3-*C*-hydroxymethyl branched-chain sugars, the configurations of which were determined by ^{13}C n.m.r. spectroscopy.

A photochemical synthesis of DL-apiose (108) has been achieved[94] by application of the Patterno–Buchii reaction to 1,3-dihydroxy-2-propanone diacetate (105) and 1,3-dioxolen-2-one (106), which gave the oxetan (107), and, upon subsequent alkaline hydrolysis, DL-apiose.

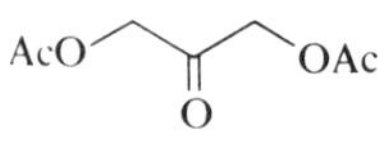

(105)

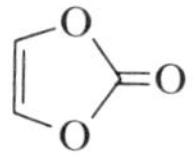

(106)

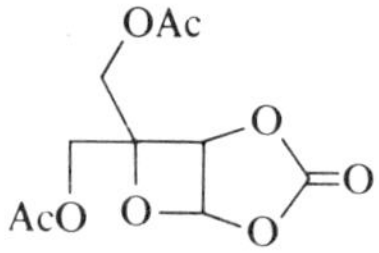

(107)

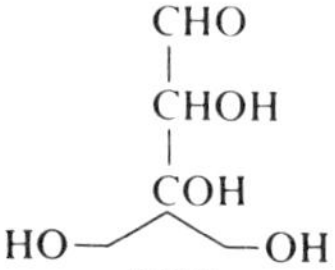

(108)

9 Amino-sugars

The occurrence of amino-sugars that possess unusual structures in antibiotics has been responsible for the continuing interest in the synthesis of this class of compounds. Two synthetic approaches[95,96] to the aminodeoxy-octose lincosamine, which is a constituent of lincomycin, have been reported. One highly stereoselective route[95] commences with the unsaturated nitro-sugar (109), epoxidation of which gave the key intermediate (110). This was opened with benzylamine to give (111) and subsequently reduced and converted into the *N*-acetyl derivative of lincosamine (112). The other route[96] to (112), which uses the diazoketone derivative (113) as the starting material, contained steps that were not stereospecific.

[91] B. Fraser-Reid, N. L. Holder, D. R. Hicks, and D. L. Walker, *Canad. J. Chem.*, 1977, **55**, 3978.
[92] B. Fraser-Reid, R. C. Anderson, D. R. Hicks, and D. L. Walker, *Canad. J. Chem.*, 1977, **55**, 3986.
[93] P. M. Collins, V. R. N. Munasinghe, and N. N. Operaeche, *J.C.S. Perkin I*, 1977, 2423.
[94] Y. Araki, J. Nagasawa, and Y. Ishido, *Carbohydrate Res.*, 1977, **58**, C4.
[95] G. R. Woolard, E. B. Rathbone, W. A. Szarek, and J. K. N. Jones, *J.C.S. Perkin I*, 1976, 950.
[96] S. M. David and J.-C. Fischer, *Carbohydrate Res.*, 1976, **50**, 239.

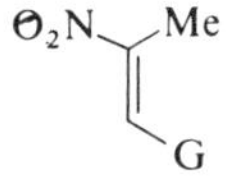

(109)

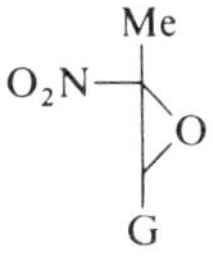

(110)

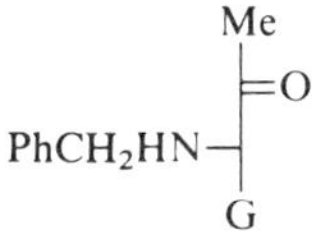

(111)

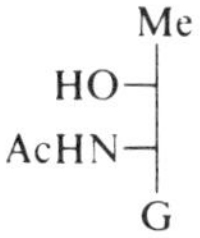

(112)

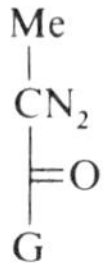

(113)

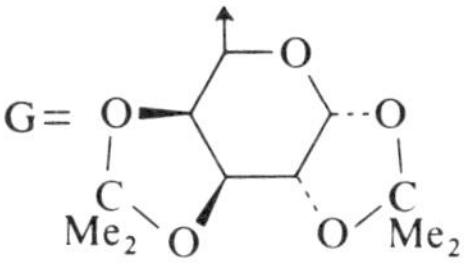

A reaction which could be of value for selectively reducing an azido-group in an unsaturated nucleoside has been reported.[97] The unsaturated azide is treated with hydrogen sulphide in aqueous pyridine at room temperature for 3 h. The sulphur is filtered off and the unsaturated amino-nucleoside isolated.

A method of synthesis of amino-sugar derivatives which has attracted the attention of three groups[98,99] is the addition of the Sharpless reagent[100] to unsaturated sugars. Thus chloramine T (TsNClNa), in the presence of osmium tetroxide, was added[99] to the diacetylated derivative of the pseudo-glucal (33) to give, in moderate yield, a mixture of the 2- and 3-tosylamino-mannosides (114) and (115). A higher yield of the products (116) and (117) was obtained[98] from the corresponding methyl hexenopyranoside after acetylation of the hydroxy-groups in the initially formed adducts.

Attempts have been made to improve the potency and antibacterial spectra, particularly against resistant organisms, of aminoglycoside antibiotics by chemically modifying their structures. For example, gentamicin X_2 (118), which is produced only as a minor component by *Micromonospora purpurea*, has been prepared[101] semi-synthetically. The key step in the synthesis is the α-glycosylation of the 1,3,3′-tris-*N*-benzoyloxycarbonyl derivatives of the semi-synthetic pseudodisaccharide garamine (119)[102] by the Lemieux–Nagabhushan reaction, using the nitrosyl chloride adduct of triacetyl glucal.

The total syntheses of streptomycin and its 3″-deoxydihydro analogue have been reported by Umezawa's group.[103,104]

Labelling studies that have revealed the biosynthetic pathways by which D-glucose is converted into the amino-cyclitol components of streptomycin and spectinomycin have been reviewed.[105]

[97] I. Adachi, Y. Yamada, and I. Inoue, *Synthesis*, 1977, 45.
[98] K. Heyns and J. Feldmann, *Tetrahedron Letters*, 1977, 2789.
[99] I. Dyong, Q. Lam-Chi, G. Schulte, B. Fraser-Reid, and J. L. Primeau, *Angew. Chem. Internat. Edn.*, 1977, **16**, 553.
[100] K. B. Sharpless, A. O. Chong, and K. Oshima, *J. Org. Chem.*, 1976, **41**, 177.
[101] M. Kugelman, A. K. Mallams, H. F. Vernay, D. F. Crowe, G. Detre, M. Tanabe, and D. M. Yasuda, *J.C.S. Perkin I*, 1976, 1097.
[102] M. Kugelman, A. K. Mallmas, H. F. Vernay, D. F. Crowe, and M. Tanabe, *J.C.S. Perkin I*, 1976, 1088.
[103] S. Umezawa, Y. Takahashi, T. Usui, and T. Tsuchiya, *J. Antibiotics*, 1974, **27**, 997.
[104] H. Sano, T. Tsuchiya, S. Kobayashi, M. Hamada, S. Umezawa, and H. Umezawa, *J. Antibiotics*, 1976, **29**, 978.
[105] K. L. Rinehart, jun. and R. M. Stroshane, *J. Antibiotics*, 1976, 319.

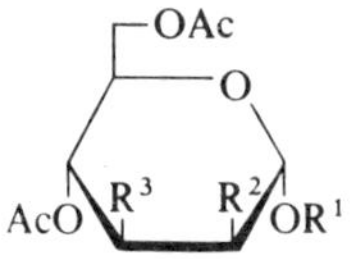

(114) $R^1 = Et, R^2 = NHTs, R^3 = OH$
(115) $R^1 = Et, R^2 = OH, R^3 = NHTs$
(116) $R^1 = Me, R^2 = NHTs, R^3 = OH$
(117) $R^1 = Me, R^2 = OH, R^3 = NHTs$

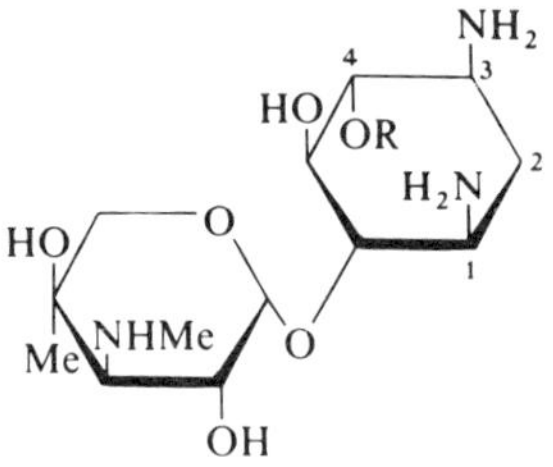

(118) R =

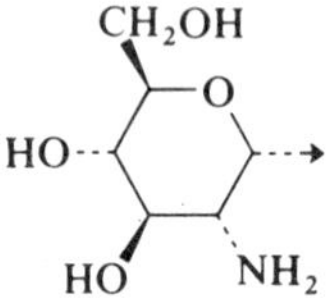

(119) R = H

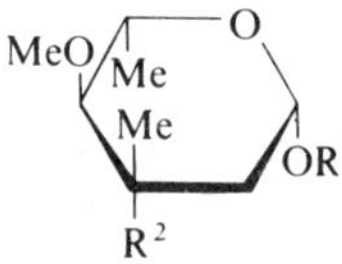

(120) $R^1 = H, R^2 = NO_2$
(121) $R^1 = Me, R^2 = AcNH$

The branched-chain nitro-sugar evernitrose has been shown to possess the L-*arabino*-configuration (120) from an *X*-ray crystallographic analysis of the 3-acetamidopyranoside (121), which was prepared from it in three steps.[106] Thus the complete structure of the octasaccharide antibiotic everninomicin D, with its uncommon orthoester linkages, is now known.[107]

10 Synthesis of Non-carbohydrate Compounds

There is increasing use of carbohydrate derivatives as precursors for the synthesis of optically active non-carbohydrate compounds. Some examples of this approach that have been applied to the synthesis of natural products this year include the preparation of the antifungal metabolite (−)-isoavenaciolide from diacetone-D-glucose[108] and the conversion of 1,6-anhydro-β-D-glucose into the growth factor (+)-biotin[109] in a biomimetic synthesis which involved a stereospecific ring closure to the tetrahydrothiophen. Prostaglandin and related substances have also been prepared from carbohydrate sources. Thus prostaglandin PGE_1 has been prepared from D-glyceraldehyde[110] and thromboxane B_2 has been synthesized by two routes[111,112] from methyl α-D-glucopyranoside.

[106] A. K. Ganguly, O. Z. Sarre, A. T. McPhail, and K. D. Onan, *J.C.S. Chem. Comm.*, 1977, 313.
[107] A. K. Ganguly, O. Z. Sarre, D. Greeves, and J. Morton, *J. Amer. Chem. Soc.*, 1975, **97**, 1982.
[108] R. C. Anderson and B. Fraser-Reid, *Tetrahedron Letters*, 1977, 2865.
[109] T. Ogawa, T. Kawano, and M. Matsui, *Carbohydrate Res.*, 1977, **57**, C31.
[110] G. Stork and T. Takahashi, *J. Amer. Chem. Soc.*, 1977, **99**, 1275.
[111] S. Hanessian and P. Lavallee, *Canad. J. Chem.*, 1977, **55**, 562.
[112] E. J. Corey, M. Shibasaki, and J. Knolle, *Tetrahedron Letters*, 1977, 1625.

14 Biological Chemistry
Part (ii) Insect Chemistry

By R. BAKER and D. A. EVANS

Department of Chemistry, University of Southampton, Southampton SO9 5NH

1 Introduction

Since the last *Annual Report* on this topic,[1] it has become increasingly apparent that the role of chemicals in the mediation of insect behaviour is immeasurably more complex than imagined earlier, and that current knowledge allows us to construct only a superficial understanding of insect life. This Report selectively reviews some major advances reported in the past two years, and the scope of coverage remains broadly similar to the 1975 Report. Several books, articles, and reviews have appeared in the meantime, particularly in the areas concerned with pest management methods and plant–insect interactions.[2]

2 Sex Attraction and Related Phenomena

The sex pheromones of moths and butterflies (Lepidoptera) have again proven to be a fruitful area of study,[3] and the Table lists structures from several detailed examinations. All are C_{12}, C_{14}, C_{16}, or C_{18} straight-chain alcohols, acetates, or aldehydes with the exception of (1a) and (1b) from the Peach Fruit Moth, *Carposina niponensis*,[4] and (2) and (3) from the Potato Tuberworm Moth, *Phthorimaea operculella*[5] (species C in the Table).

In addition to a pyrrolizidine alkaloid (4) reported earlier, the hairpencil secretion of the butterfly *Amauris ochlea* contains, *inter alia*, methyl salicylate, eugenol, and *cis*-jasmone (5).[6] In *Danaus* species, pyrrolizidine alkaloids in food

[1] R. Baker and D. A. Evans, *Ann. Reports (B)*, 1975, **72**, 347.

[2] 'Animal Communication by Pheromones', H. H. Shorey, Academic Press, London, 1976; 'Chemical Control of Insect Behaviour; Theory and Application', H. H. Shorey and J. H. McKelvey, jun., Wiley, New York, 1977; *ACS Symp. Ser.*, 1976, No. 23, 'Pest Management with Insect Sex Attractants and Other Behaviour Controlling Chemicals' (Symposium 1975); *ACS Symp. Ser.*, 1977, No. 37, 'Pesticide Chemistry in the 20th Century' (Symposium 1976); *Recent Adv. Phytochem.*, 1976, **10**, 'Biochemical Interaction of Plants and Insects'; J. W. Wheeler, *Lloydia*, 1976, **39**, 53; B. Tursch, J. C. Braekman, and D. Daloze, *Experientia*, 1976, **32**, 401.

[3] W. L. Roelofs and R. T. Cardé, *Ann. Rev. Entomol.*, 1977, **22**, 377; J. Weatherston and J. Percy, *Endeavour*, 1977, 83.

[4] Y. Tamaki, K. Honma, and K. Kawasaki, *Appl. Entomol. Zool.*, 1977, **12**, 60.

[5] C. J. Persoons, S. Voerman, P. E. J. Verwiel, F. J. Ritter, W. J. Nooijen, and A. K. Minks, *Entomol. Exp. Appl.*, 1976, **20**, 289; R. Yamaoka, H. Fukami, and S. Ishii, *Agric. and Biol. Chem. (Japan)*, 1976, **40**, 1971; W. L. Roelofs, J. P. Kochansky, R. T. Cardé, G. G. Kennedy, C. A. Henrick, J. N. Labovitz, and V. L. Corbin, *Life Sci.*, 1975, **17**, 699.

[6] R. L. Petty, M. Boppre, D. Schneider, and J. Meinwald, *Experientia*, 1977, **33**, 1324.

Table *Some sex pheromone attractants of Lepidoptera species*

Parent chain length	*Terminal functional group*	*Unsaturation*	*Species*	*Ref.*
C_{12}	Acetate	(*Z*)-9	A	*a*
		(*E*,*E*)-8,10	B	*b*
C_{13}	Acetate	(*E*,*Z*)-4,7	C	*c*
		(*E*,*Z*,*Z*)-4,7,10	C	*c*
C_{14}	Alcohol	—	D(?)	*d*
		(*E*)-11	D	*d*
		(*Z*)-11	D, E	*d, e*
		(*Z*,*E*)-9,12	F	*f*
	Acetate	—	D(?), G	*d, g*
		(*E*)-9	G	*g*
		(*Z*)-9	F, H, I, J	*f, h, i, j*
		(*E*)-11	D, G, K, L	*d, g, k, l*
		(*Z*)-11	D, E, G, H, I, L	*d, e, g, h, i, l*
		(*E*,*E*)-9,11	K	*k*
		(*Z*,*E*)-9,12	F	*f*
	Aldehyde	(*Z*)-7	M	*m*
		(*E*)-11	N	*m*
		(*Z*)-11	N	*n*
C_{16}	Alcohol	(*Z*)-11	O	*o*
	Acetate	(*Z*)-9	J	*j*
		(*E*)-11	P	*p*
		(*Z*)-11	J, O, Q	*j, o, q*
		(*E*,*Z*)-6,11	R	*r*
	Aldehyde	(*Z*)-11	S,T	*s, t*
		(*E*,*Z*)-6,11	R	*r*
C_{18}	Aldehyde	(*Z*)-13	S	*s*

Species: A, *Eupoecilia ambiguella*; B, *Rhyacionia rigidana*; C, *Phthorimaea operculella*; D, *Platynota flavedana*; E, *Archips rosana*; F, *Ephestia cautella*; G, *Ostrinia nubilalis* (Iowa); H, *Pandemis limitata*; I, *Pandemis pyrusana*; J, *Naranga aenescens*; K, *Epiphyas postvittana*; L, *Archippus breviplicanus*; M, *Prays citri*; N, *Choristoneura fumiferana*; O, *Scotogramma trifolii*; P, *Brachmia macroscopa*; Q, *Sesamia inferens*; R, *Antheraea polyphemus*; S, *Chilo suppressalis*; T, *Heliothis armigera*.

[a] H. Arn, S. Rauscher, H. R. Buser, and W. L. Roelofs, *Z. Naturforsch.*, 1976, **31c**, 9; [b] A. S. Hill, C. W. Berisford, U. E. Brady, and W. L. Roelofs, *Environ. Entomol.*, 1976, **5**, 959; [c] C. J. Persoons, S. Voerman, P. E. J. Verwiel, F. J. Ritter, W. J. Nooijen, and A. K. Minks, *Entomol. Exp. Appl.*, 1976, **20**, 289; see also R. Yamaoka, H. Fukami, and S. Ishii, *Agric. and Biol. Chem.* (*Japan*), 1976, **40**, 1971 and W. L. Roelofs, J. P. Kochansky, R. T. Cardé, G. G. Kennedy, C. A. Henrick, J. N. Labovitz, and V. L. Corbin, *Life Sci.*, 1975, **17**, 699; [d] A. S. Hill, R. T. Cardé, W. M. Bode, and W. L. Roelofs, *J. Chem. Ecol.*, 1977, **3**, 369; [e] W. L. Roelofs, A. S. Hill, A. Cardé, R. T. Cardé, J. Vakenti, and H. Madsen, *Environ. Entomol.*, 1976, **5**, 362; [f] J. S. Read and C. P. Haines, *J. Stored Product Res.*, 1976, **12**, 49; see also *ibid.* p. 55; [g] J. A. Klun and G. A. Junk, *J. Chem. Ecol.*, 1977, **3**, 447; [h] W. L. Roelofs, A. Cardé, A. S. Hill, and R. T. Cardé, *Environ. Entomol.*, 1976, **5**, 649; [i] W. L. Roelofs, R. F. Lagier, and S. C. Hoyt, *ibid.*, 1977, **6**, 353; [j] T. Ando, K. Kishino, S. Tatsuki, H. Nakajima, S. Yoshida, and N. Takahashi, *Agric. and Biol. Chem.* (*Japan*), 1977, **41**, 1819; [k] R. J. Bartell and L. A. Lawrence, *Physiol. Entomol.*, 1977, **2**, 89; [l] H. Sugie, K. Yaginuma, and Y. Tamaki, *Appl. Entomol. Zool.*, 1977, **12**, 69; [m] B. F. Nesbitt, P. S. Beevor, D. R. Hall, R. Lester, M. Sternlicht, and S. Goldenberg, *Insect Biochem.*, 1977, **7**, 355; [n] C. J. Sanders and J. Weatherston, *Canad. Entomol.*, 1976, **108**, 1285; see also *Compt. rend.*, 1975, **281**, *D*, 1111; [o] E. W. Underhill, W. F. Steck, and M. D. Chisholm, *Environ. Entomol.*, 1976, **5**, 307; [p] C. Hirano, H. Muramoto, and M. Horiike, *Naturwiss.*, 1976, **63**, 439; [q] B. F. Nesbitt, P. S. Beevor, D. R. Hall, R. Lester, and V. A. Dyck, *Insect Biochem.*, 1976, **6**, 105; [r] J. Kochansky, J. Tette, E. F. Taschenberg, K. E. Kaissling, and W. L. Roelofs, *J. Insect Physiol.*, 1975, **21**, 1977; [s] B. F. Nesbitt, P. S. Beevor, D. R. Hall, R. Lester, and V. A. Dyck, *J. Insect Physiol.*, 1975, **21**, 1883; K. Ohta, S. Tatsuki, K. Uchiumi, M. Kurihara, and J. Fukami, *Agric. and Biol. Chem.* (*Japan*), 1976, **40**, 1897; [t] P. Piccardi, A. Capizza, G. Cassani, P. Spinelli, E. Arsura, and P. Massardo, *J. Insect Physiol.*, 1977, **23**, 1443.

plants are postulated to contribute to the unpalatability of these butterflies to potential predators.[7] Specialized organs on the hindwings of male *Ithomiinae* butterflies secrete the lactone (6) which is structually related to the esterifying acids of pyrollizidine alkaloids.[8]

(1) a; R = n-C_9H_{19}
b; R = n-C_8H_{17}

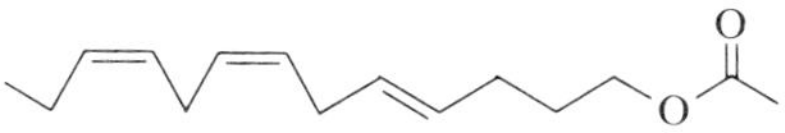

(2)

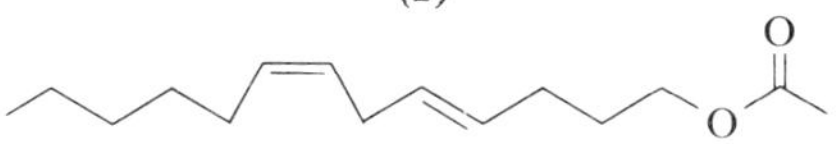

(3)

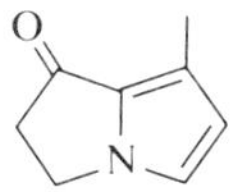

(4)

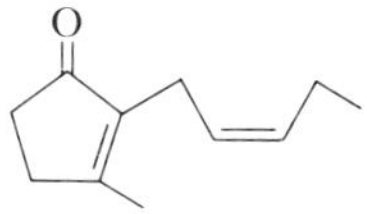

(5)

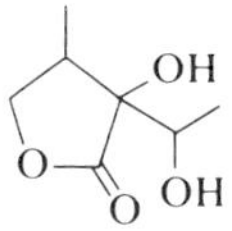

(6)

The differences of sex attractant responses of *Grapholitha molesta* male moths to subtle changes in component ratio of a binary pheromone have been attributed to a normal distribution of responses about an optimum mixture rather than to genetic variation within a single species.[9] Furthermore, two *Archips* sibling species maintain reproductive isolation by employing different blends of the same chemicals as sex pheromones.[10] The effects of enantiomeric[11] and geometric[12] composition on response have also been investigated. Attractants have been discovered for many economically important moth species as a result of screening programmes.[13]

The Eastern Tent Caterpillar *Malacosoma americanum* forages by preception of chemicals present in trails of silk leading from the communal tent to distant food sources.[14] Many male Noctuid moths possess eversible glandular structures, and the genital scent brushes of several species release oxygenated aromatic compounds (*e.g.* 2-phenylethanol) which elicit sexual behaviour in the females of those species.[15]

[7] J. A. Edgar, P. A. Cockrum, and J. L. Frahn, *Experientia*, 1976, **32**, 1535.
[8] J. A. Edgar, C. C. J. Culvenor, and T. E. Pliske, *J. Chem. Ecol.*, 1976, **2**, 263.
[9] R. T. Cardé, T. C. Baker, and W. L. Roelofs, *Experientia*, 1976, **32**, 1406.
[10] R. T. Cardé, A. M. Cardé, A. S. Hill, and W. L. Roelofs, *J. Chem. Ecol.*, 1977, **3**, 71.
[11] D. Klimetzek, G. Loskant, J. P. Vite, and K. Mori, *Naturwiss.*, 1976, **63**, 581.
[12] R. T. Cardé and W. L. Roelofs, *J. Chem. Ecol.*, 1977, **3**, 143.
[13] E. W. Underhill, M. D. Chisholm, and W. Steck, *Environ. Entomol.*, 1977, **6**, 333; K. Yaginuma, M. Kumakura, Y. Tamaki, T. Yushima, and J. H. Tumlinson, *Appl. Entomol. Zool.*, 1976, **11**, 266; L. I. Butler, J. E. Halfhill, L. M. McDonough, and B. A. Butt, *J. Chem. Ecol.*, 1977, **3**, 65; W. Steck, E. W. Underhill, B. K. Bailey, and M. D. Chisholm, *Environ. Entomol.*, 1977, **6**, 270; V. E. Adler, M. Jacobson, J. F. Edmiston, and M. H. Fleming, *J. Econ. Entomol.*, 1976, **69**, 706; A. K. Minks, S. Voerman, and M. Van de Vrie, *Entomol. Exp. Appl.*, 1976, **19**, 301; W. F. Steck, B. K. Bailey, E. W. Underhill, and M. D. Chisholm, *Environ. Entomol.*, 1976, **5**, 523; W. L. Roelofs, W. H. Reissig, and R. W. Weires, *ibid.*, 1977, **6**, 373; R. E. Doolittle, W. L. Roelofs, J. D. Solomon, R. T. Cardé, and M. Beroza, *J. Chem. Ecol.*, 1976, **2**, 399.
[14] T. D. Fitzgerald and E. M. Gallagher, *J. Chem. Ecol.*, 1976, **2**, 187.
[15] M. Jacobson, V. E. Adler, A. N. Kishaba, and E. Priesner, *Experientia*, 1976, **32**, 964; H. J. Bestmann, O. Vostrowsky, and H. Platz, *ibid.*, 1977, **33**, 874.

In the Coleoptera, alcohols and hydrocarbons are employed by female boll weevils (*Anthonomus grandis*) as sex attractants for males.[16] Further investigations of several species of *Trogoderma* Dermestid beetles have shown the importance of (7a) and (7b) in addition to compounds reported earlier in mating behaviour,[17,18] and an informative summary of the chemical basis for interspecific and intraspecific responses has been presented.[19] A major sex-attracting component produced by virgin female *Attagenus elongatus* beetles is reported as (8),[20] whereas the female Japanese Beetle *Popillia japonica* employs (9) as a sex pheromone.[21] It is reported in this context that racemic (9) is completely inactive whereas the (*S*)-enantiomer is inhibitory. Aromatic derivatives including (10) are male-specific natural products in the bug *Leptoglossus phyllopus*.[22]

(7a) (7b has *E*-geometry)

(8)

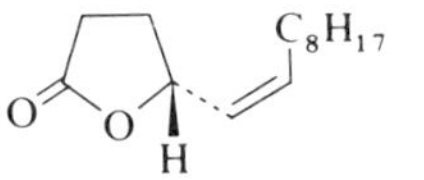

(9)

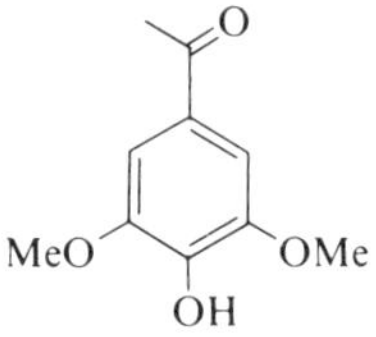

(10)

Pest species of flies have again attracted considerable attention. Females of the stable fly species *Stomoxys calcitrans* produce a mating stimulant consisting of methyl-branched and 1,5-dimethyl-branched hydrocarbons, of which 15-methyltritriacontane and 15,19-dimethyltritriacontane are most active in behavioural bioassays.[23] In this context, it is interesting to note that (*Z*,*Z*)-pentacosa-1,7,13-triene is a major cuticular wax component for males of this species, but is absent in females.[24] The mating-stimulant pheromones and cuticular lipid constituents of the Little House Fly, *Fannia canicularis* have been investigated. Newly emerged

16 G. H. McKibben, P. A. Hedin, W. L. McGovern, N. M. Wilson, and E. B. Mitchell, *J. Chem. Ecol.*, 1977, **3**, 331.

17 J. H. Cross, R. C. Byler, R. F. Cassidy, jun., R. M. Silverstein, R. E. Greenblatt, W. E. Burkholder, A. R. Levinson, and H. Z. Levinson, *J. Chem. Ecol.*, 1976, **2**, 457.

18 R. E. Greenblatt, W. E. Burkholder, J. H. Cross, R. C. Byler and R. M. Silverstein, *J. Chem. Ecol.*, 1976, **2**, 285; J. H. Cross, R. C. Byler, R. E. Greenblatt, J. E. Gorman, and W. E. Burkholder, *ibid.*, 1977, **3**, 115.

19 R. E. Greenblatt, W. E. Burkholder, J. H. Cross, R. F. Cassidy, jun., R. M. Silverstein, A. R. Levinson, and H. Z. Levinson, *J. Chem. Ecol.*, 1977, **3**, 337.

20 H. Fukui, F. Matsumura, A. V. Barak, and W. E. Burkholder, *J. Chem. Ecol.*, 1977, **3**, 539.

21 J. H. Tumlinson, M. G. Klein, R. E. Doolittle, T. L. Ladd, and A. T. Proveaux, *Science*, 1977, **197**, 789.

22 J. R. Aldrich, M. S. Blum, S. S. Duffey, and H. M. Fales, *J. Insect Physiol.*, 1976, **22**, 1201.

23 P. E. Sonnet, E. C. Uebel, R. L. Harris, and R. W. Miller, *J. Chem. Ecol.*, 1977, **3**, 245.

24 P. E. Sonnet, E. C. Uebel, and R. W. Miller, *J. Chem. Ecol.*, 1977, **3**, 251.

adults of both sexes have identical lipid constituents, but considerable differences are apparent after five days. (*Z*)-9-Pentacosene constitutes 66% of female lipids, compared to 1% in males, and this component is suggested as a sex pheromone.[25]

The long-running saga of the sex pheromone of the American Cockroach *Periplaneta americana* has been advanced by the proposal of (11) as a tentative structure.[26] Interestingly, the structurally related germacrene-D (12) has been found to be a phytochemical sex-stimulant for this species.[27]

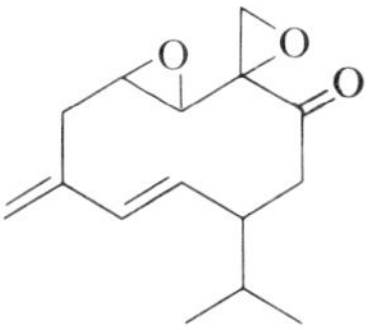

(11)

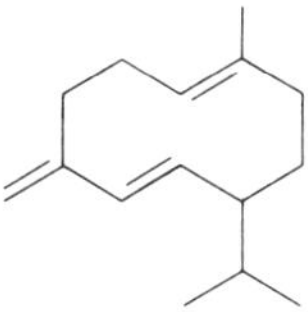

(12)

The chemistry of the metasternal gland of the Eucalypt Longhorn *Phorocantha synonoma* has been studied and the existence of several lactones [*e.g.* (13)] has

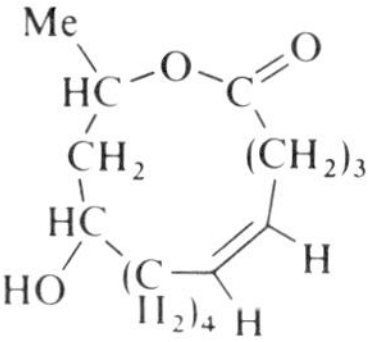

(13)

been demonstrated.[28] A second component of the sex pheromone of the German Cockroach *Blattella germanica* has been shown to be (14).[29] 3,7-Dimethyl-

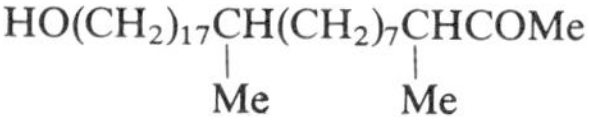

(14)

pentadecan-2-ol (15) plays a key role in the sexual behaviour of three sawfly species, and inter-species specificity is based upon pheromone blends of (15) and the corresponding acetate and propionate esters.[30]

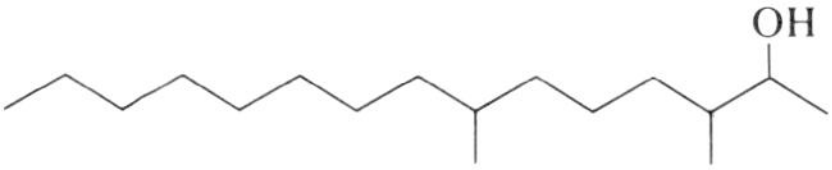

(15)

[25] E. C. Uebel, P. E. Sonnet, R. E. Menzer, R. W. Miller, and W. R. Lusby, *J. Chem. Ecol.*, 1977, **3**, 269.
[26] C. J. Persoons, P. E. J. Verwiel, F. J. Ritter, E. Talman, P. J. F. Nooijen, and W. J. Nooijen, *Tetrahedron Letters*, 1976, 2055.
[27] C. Kitamura, S. Takahashi, S. Tahara, and J. Mizutani, *Agric. and Biol. Chem. (Japan)*, 1976, **40**, 1965; C. Nishino, T. R. Tobin, and W. S. Bowers, *J. Insect Physiol.*, 1977, **23**, 415.
[28] B. P. Moore and W. V. Brown, *Austral. J. Chem.*, 1976, **29**, 1365.
[29] R. Nishida, T. Sato, Y. Kuwahara, H. Fukami, and S. Ishii, *J. Chem. Ecol.*, 1976, **2**, 449.
[30] D. M. Jewett, F. Matsumura, and H. C. Coppel, *Science*, 1976, **192**, 51.

3 Aggregation Pheromones and Population Attractants

This complex area of interplay between host phytochemicals and pheromones has continued to attract considerable attention.[31,32] The aggregation pheromone complex reported earlier[1] for the European Elm Beetle *Scolytus multistriatus* is also associated with the Larger European Elm Beetle *S. scolytus*.[33] The availability by synthesis (Section 13) of enantiomers and isomers of bark-beetle pheromones has enabled testing of response specificity.[34] The dioxaspirononane (16) has been

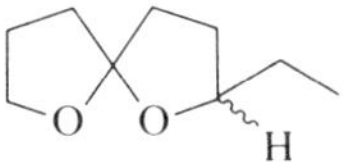

(16)

reported as an aggregation pheromone of *Pityogenes chalcographus*, a pest of Norway spruce.[35]

4 Pheromones of Social Insects and Related Species

The complexity of chemosensory communication is amply illustrated by detailed investigations of the behaviour of social insects.[36] Chemical and ethological studies of the Dufour's gland secretion of the ant *Myrmica rubra* have related specific behavioural responses to individual components of the secretion.[37] The male mandibular gland secretion of the ant *Camponotus clavithorax* is a complex mixture containing (17),[38] whereas the volatiles associated with two *Formica* species are largely monoterpenoids together with methyl 3-isopropylpentanoate (18).[39] Alarm pheromone components have been reported for several ant species,[40] including (19) for eight *Formica* species,[41] and (20) for *Hypoponera opacior* and *Ponera pennsylvanica*.[42] The lactone (21) occurs together with terpenoids in the mandibular gland secretion of carpenter bees.[43] A number of terpenoid alcohols and esters have been implicated in flight-marking in bumblebees,[44] and neral and

[31] J. P. Vite and W. Francke, *Naturwiss.*, 1976, **63**, 550.

[32] C. M. Harring, J. P. Vite, and P. R. Hughes, *Naturwiss.*, 1975, **62**, 488; J. A. A. Renwick, G. B. Pitman, and J. P. Vite, *ibid.*, 1976, **63**, 198; J. A. A. Renwick, J. P. Vite, and R. F. Billings, *ibid.*, 1977, **64**, 226; M. C. Birch, P. E. Tilden, D. L. Wood, L. E. Browne, J. C. Young, and R. M. Silverstein, *J. Insect Physiol.*, 1977, **23**, 1349.

[33] M. M. Blight, F. A. Mellon, L. J. Wadhams, and M. J. Wenham, *Experientia*, 1977, **33**, 845.

[34] D. L. Wood, L. E. Browne, B. Ewing, K. Lindahl, W. D. Bedard, P. E. Tilden, K. Mori, G. B. Pitman, and P. R. Hughes, *Science*, 1976, **192**, 896; J. H. Borden, L. Chong, J. A. McLean, K. N. Slessor, and K. Mori, *ibid.*, p. 894; G. N. Lanier, W. E. Gore, G. T. Pearce, J. W. Peacock, and R. M. Silverstein, *J. Chem. Ecol.*, 1977, **3**, 1.

[35] W. Francke, V. Heeman, B. Gerken, J. A. A. Renwick, and J. P. Vite, *Naturwiss.*, 1977, **64**, 590.

[36] Proceedings of the 8th International Conference of the International Union for the Study of Social Insects, Wageningen, Netherlands, 1977; published by the Centre for Agricultural Publishing and Documentation, Wageningen.

[37] M. C. Cammaerts-Tricot, E. D. Morgan, R. C. Tyler, and J. C. Braekman, *J. Insect Physiol.*, 1976, **22**, 927; see also *ibid.*, 1977, **23**, 511.

[38] H. A. Lloyd, M. S. Blum, and R. M. Duffield, *Insect Biochem.*, 1975, **5**, 489.

[39] M. Buehring, W. Francke, and V. Heeman, *Z. Naturforsch.*, 1976, **31c**, 11.

[40] J. Lofquist, *J. Insect Physiol.*, 1976, **22**, 1331; N. Hayashi and H. Komae, *Experientia*, 1977, **33**, 424.

[41] R. M. Duffield, J. M. Brand, and M. S. Blum, *Ann. Entomol. Soc. Amer.*, 1977, **70**, 309.

[42] R. M. Duffield, M. S. Blum, and J. W. Wheeler, *Comp. Biochem. Physiol.* (*B*), 1976, **54**, 439.

[43] J. W. Wheeler, S. L. Evans, M. S. Blum, H. H. V. Velthuis, and J. M. F. de Camargo, *Tetrahedron Letters*, 1976, 4029; see also *Ann. Entomol. Soc. Amer.*, 1977, **70**, 635.

[44] B. G. Svensson and G. Bergstrom, *Insectes Sociaux*, 1977, **24**, 213.

geranial have been suggested as components of the sex pheromone of the parasitic wasp *Itoplectis conquisitor.*[45]

Faranal (22), reminiscent of a juvenoid, is proposed as a trail pheromone of Pharaoh's Ant, *Monomorium pharaonis.*[46] The configuration of the two adjacent methyl groups was established to be as shown (or its antipode) by comparison of the ozonolysis product of faranal with that of *trans*-3,4-dimethylcyclohexene.

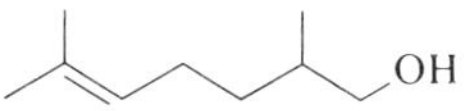

(17)

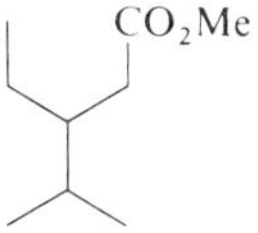

(18)

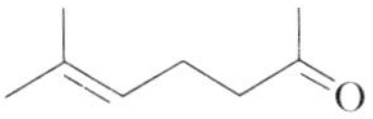

(19)

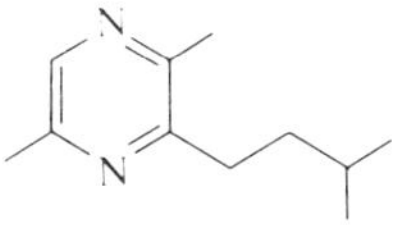

(20)

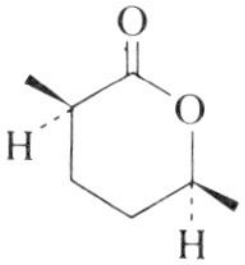

(21)

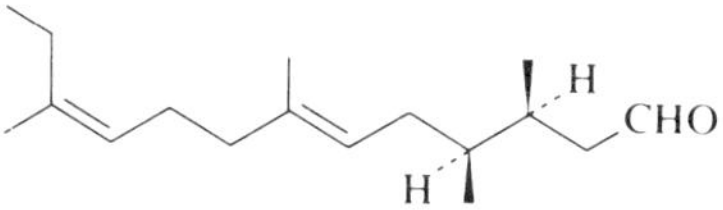

(22)

5 Alarm Behaviour in Aphids

Alarm behaviour in aphid species has been extensively studied. When attacked by predators, aphids of the *Therioaphis* genus secrete droplets containing (−)-germacrene-A (23).[47] Sesquiterpenoid hydrocarbons such as (*E*)-β-farnesene (24) function as alarm pheromones in aphids, and norfarnesenes also elicit this response.[48] It is considered unlikely that such compounds are sufficiently stable for field use in control methods.

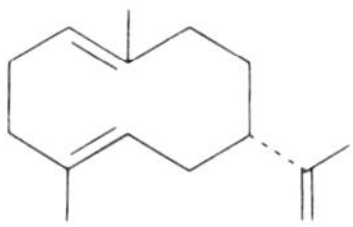

(23)

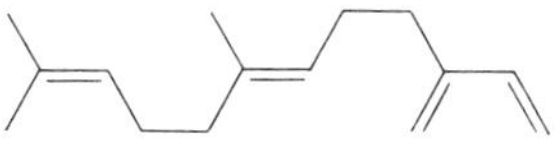

(24)

6 Defence Secretions

Defence secretions are generally produced by insects in much larger quantities than sex pheromones. This consideration, together with the impetus provided by the possibility of uncovering novel insecticides, accounts for the relatively large number of intricate and diverse structures elucidated in this area. An interesting review on arthropod protective agents has appeared.[49]

[45] D. C. Robacker and L. B. Hendry, *J. Chem. Ecol.*, 1977, **3**, 563.

[46] F. J. Ritter, I. E. M. Bruggemann-Rotgans, P. E. J. Verwiel, C. J. Persoons, and E. Talman, *Tetrahedron Letters*, 1977, 2617.

[47] W. S. Bowers, C. Nishino, M. E. Montgomery, L. R. Nault, and M. W. Nielson, *Science*, 1977, **196**, 4290; see also *J. Chem. Ecol.*, 1977, **3**, 349.

[48] C. Nishino, W. S. Bowers, M. E. Montgomery, and L. R. Nault, *Agric. and Biol. Chem. (Japan)*, 1976, **40**, 2303; see also *J. Insect Physiol.*, 1977, **23**, 697; *Appl. Entomol. Zool.*, 1976, **11**, 340.

[49] H. Schildknecht, *Angew. Chem. Internat. Edn.*, 1976, **15**, 214.

The defensive substances of the Opilionids *Leiobunum ventricosum* and *Hadrobunus maculosus* are reported as (25a) and (25b) in addition to 4-methylheptan-3-one.[50] In ants, components of defence secretions range from simple aliphatic compounds in *Formica* species[51] to iridoid monoterpenoids[52] such as iridomyrmecin (26) and dolichodial (27).[53] The powerful alkaloidal venoms of fire ants have

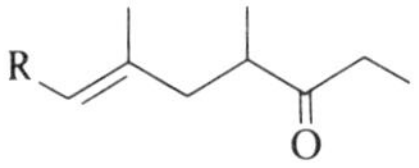

(25) a; R = Me
b; R = Et

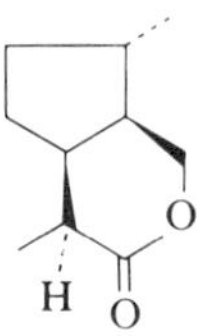

(26)

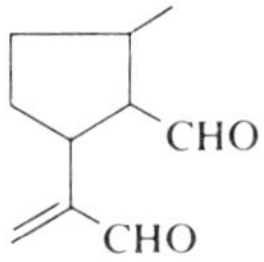

(27)

been inter-related chemotaxonomically,[54] and a South African species (*Solenopsis punctaticeps*) has been shown to contain a novel series of 2,5-dialkyl-pyrrolidines and -pyrrolines, (28a—d) and (29a, b, and c).[55] Ladybugs indigenous to Canada

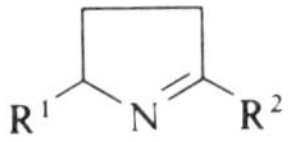

(28) a; $R^1 = C_2, R^2 = C_5$
b; $R^1 = C_5, R^2 = C_2$
c; $R^1 = C_2, R^2 = C_7$
d; $R^1 = C_7, R^2 = C_2$

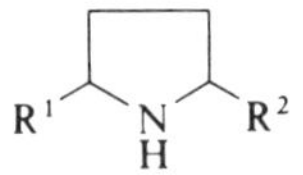

(29) a; $R^1 = C_4, R^2 = C_5$
b; $R^1 = C_2, R^2 = C_7$
c; $R^1 = C_4, R^2 = C_7$

Numbers refer to length of n-alkyl chain

(*Hippodamia caseyi*) have been found to utilize a series of defensive alkaloids [*e.g.* (30), (31)], closely related to structures discovered earlier.[56] Quinone and hydrocarbon defence chemicals from five genera of bombardier beetles have been

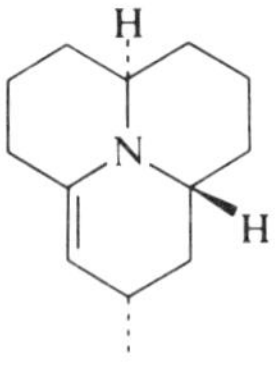

(30)

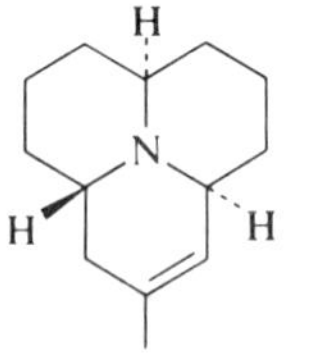

(31)

listed,[57] and the cyclopentanoid monoterpenoids chrysomelidial (32) and plagiolactone (33) are components of the larval defensive secretion of the chry-

[50] T. H. Jones, W. E. Conner, A. F. Kluge, T. Eisner, and J. Meinwald, *Experientia*, 1976, **32**, 1234.
[51] J. Lofquist, *Oikos*, 1977, **28**, 137.
[52] J. W. Wheeler, T. Olagbemiro, A. Nash, and M. S. Blum, *J. Chem. Ecol.*, 1977, **3**, 241.
[53] G. W. K. Cavill, E. Houghton, F. J. McDonald, and P. J. Williams, *Insect Biochem.*, 1976, **6**, 483.
[54] J. G. MacConnell, M. S. Blum, W. F. Buren, R. N. Williams, and H. M. Fales, *Toxicon*, 1976, **14**, 69.
[55] D. J. Pedder, H. M. Fales, T. Jaouni, M. S. Blum, J. G. MacConnell, and R. M. Crewe, *Tetrahedron*, 1976, **32**, 2275.
[56] W. A. Ayer, M. J. Bennett, L. M. Browne, and J. T. Purdham, *Canad. J. Chem.*, 1976, **54**, 1807.
[57] T. Eisner, T. H. Jones, D. J. Aneshansley, W. R. Tschinkel, R. E. Silberglied, and J. Meinwald, *J. Insect. Physiol.*, 1977, **23**, 1383.

somelid beetle *Plagiodera versicolora*.[58] The water beetle *Platambus maculatus* uses platambin (34) in a defensive role;[59] a full report of the defensive behaviour of *Stenus comma* has also appeared.[60] (*E*)-Oct-2-enal and (*E*)-hex-2-enal are present in the defensive scent of the Black Stink Roach, a New Zealand cockroach species.[61]

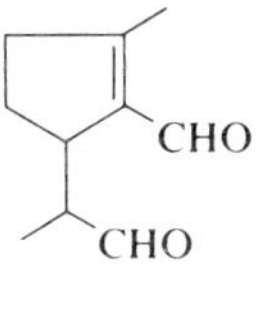

(32)

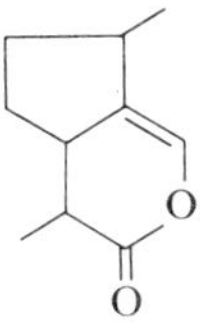

(33) (*S*, *S*-?)

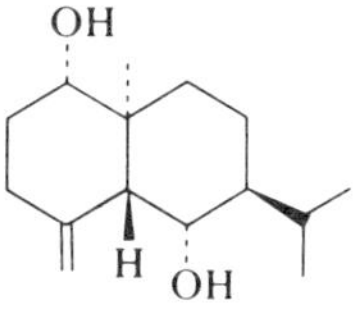

(34)

It has long been recognized that the defensive secretions of many termite species contain complex oxygenated diterpenoids. In particular, the sprays ejected at attackers by the nozzle-like heads of nasute soldiers frequently contain such substances. Several of these structures have been elucidated by *X*-ray methods in conjunction with microchemical studies. The soldier caste of the termite *Trinervitermes gratiosus* contains, in addition to five monoterpenoids,[62] an intriguing series of novel diterpenoids [*e.g.* (35)] based on the 'trinervitene' skeleton.[63] The soldiers of *Nasutitermes kempae* produce[64] the structurally related isomers of (36). The frontal gland defence secretion of *Macrotermes subhyalinus* contains a mixture of alkanes and alkenes,[65] whereas *Schedorhinotermes* species produce long chain enones [*e.g.* (37)] and ketones.[66]

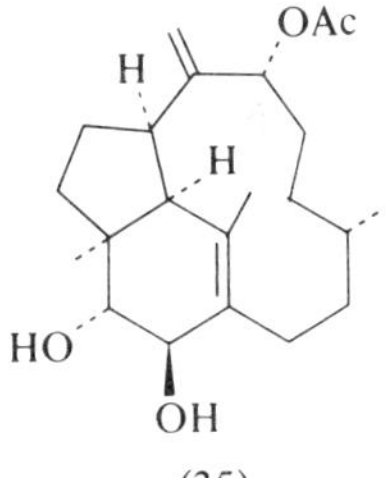

(35)

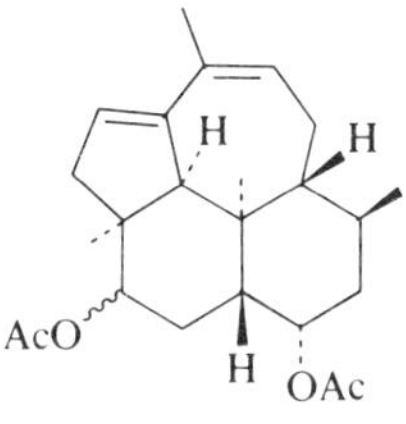

(36)

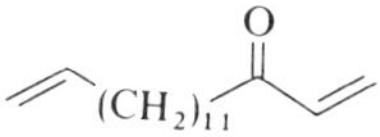

(37)

7 Host Food, Host Prey, and Oviposition Attractants

On the whole, the investigations of chemical signs which are used by phytophagous insects in host selection have advanced surprisingly slowly over the past few years.

[58] J. Meinwald, T. H. Jones, T. Eisner, and K. Hicks, *Proc. Nat. Acad. Sci. U.S.A.*, 1977, **74**, 2189.
[59] H. Schildknecht, H. Holtkotte, and D. Krauss, *Ann. Chem.*, 1975, 1850.
[60] H. Schildknecht, D. Burger, D. Krauss, and J. Connert, J. Gehlhaus, and H. Essenbreis, *J. Chem. Ecol.*, 1976, **2**, 1.
[61] M. H. Benn, R. F.N. Hutchins, and R. Folwell, *J. Insect Physiol.*, 1977, **23**, 1281.
[62] G. D. Prestwich, *Insect Biochem.*, 1977, **7**, 91.
[63] G. D. Prestwich, S. P. Tanis, J. P. Springer, and J. Clardy, *J. Amer. Chem. Soc.*, 1976, **98**, 6061; see also *ibid.*, p. 6062.
[64] G. D. Prestwich, B. A. Solheim, J. Clardy, F. G. Pilkiewicz, Z. Miura, S. P. Tanis, and K. Nakanishi, *J. Amer. Chem. Soc.*, 1977, **99**, 8082.
[65] G. D. Prestwich, B. A. Bierl, E. D. Devilbiss, and M, F. B. Chaudhury, *J. Chem. Ecol.*, 1977, **3**, 579.
[66] G. D. Prestwich, M. Kaib, W. F. Wood, and J. Meinwald, *Tetrahedron Letters*, 1975, 4701.

It has been reported that the attraction of adult Cabbage Maggots, *Hylemya brassicae*, is mediated by allyl isothiocyanate, whereas oviposition is induced by plant glucosinolates.[67] Dipropyl disulphide released by leek leaves stimulates significant egg-laying behaviour in the leek moth *Acrolepiopsis assectella*.[68] A detailed study of Indian Calamus Root oil has unearthed a number of compounds which are attractants of fruit-fly species [*e.g.* acroagermacrone (38)].[69]

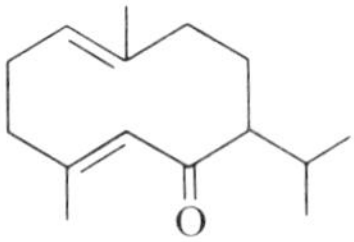

(38)

Chemical cues used by parasitoids of insects have received considerable attention (*e.g.* ref. 70). The parasitoid *Biosteres longicaudatus* locates its fruit-fly host by perception of fermentation products of rotting fruit.[71] Kairomones used by parasitoids have been found in food plants of the host insects, and it has been proposed on the basis of radiotracer evidence that such kairomones are ingested, concentrated, and then released unaltered by the host.[72]

A case of aggressive chemical mimicry has been reported for a female *Mastophora* spider, which attracts prey with a volatile substance which apparently mimics the female-produced sex attractant of the Fall Armyworm, *Spodoptera frugiperda*. This hypothesis is corroborated by the observation that only males of the latter species are trapped, and that these are attracted from downwind, both points being characteristic of mating behaviour.[73]

8 Antifeedants and Repellants

The interactions between plants and insects have been reviewed.[74] There have been numerous reports of antifeedant and insecticidal activity of plant constituents, but coverage here is limited to compounds which have pronounced effects supported by bioassay results. Quinones [*e.g.* (39)] are constituents of pest-resistant strains of cotton, and are associated with insecticidal and larval growth-inhibitory

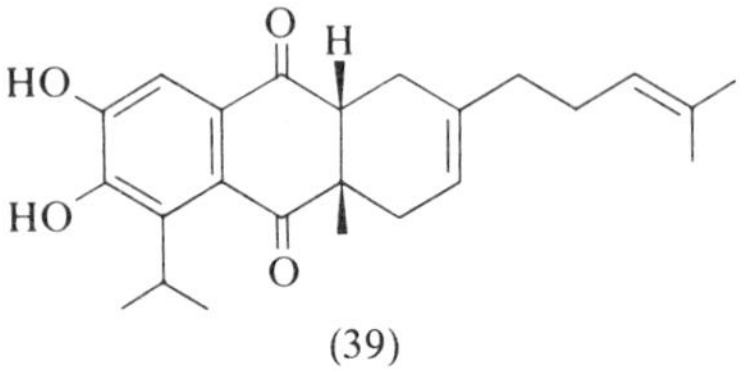

(39)

[67] K. S. S. Nair and F. L. McEwen, *Canad. Entomol.*, 1976, **108**, 1021.
[68] J. Boscher, *Compt. rend.*, 1977, **284**, *D*, 635.
[69] M. Jacobson, I. Keiser, D. Miyashita, and E. J. Harris, *Lloydia*, 1976, **39**, 412.
[70] R. D. Henson, S. B. Vinson, and C. S. Barfield, *J. Chem. Ecol.*, 1977, **3**, 151.
[71] P. D. Greany, J. H. Tumlinson, D. L. Chambers, and G. M. Bousch, *J. Chem. Ecol.*, 1977, **3**, 189.
[72] L. B. Hendry, J. K. Wichmann, D. M. Hindenlang, K. M. Weaver, and S. H. Korzeniowski, *J. Chem. Ecol.*, 1976, **2**, 271.
[73] W. G. Eberhard, *Science*, 1977, **198**, 1173.
[74] H. Z. Levinson, *Experientia*, 1976, **32**, 408.

properties.[75] Four dialdehydes [*e.g.* muzigadial (40)] have been reported as potent armyworm antifeedants.[76] Nic-1 (41) is one of a number of oxygenated steroids from the plant *Nicandra physaloides* which has an inhibitory effect on larval feeding,[77] and harrisonin (42) is reported to have similar properties.[78] Diter-

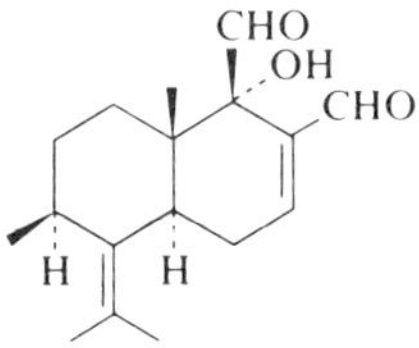

(40)

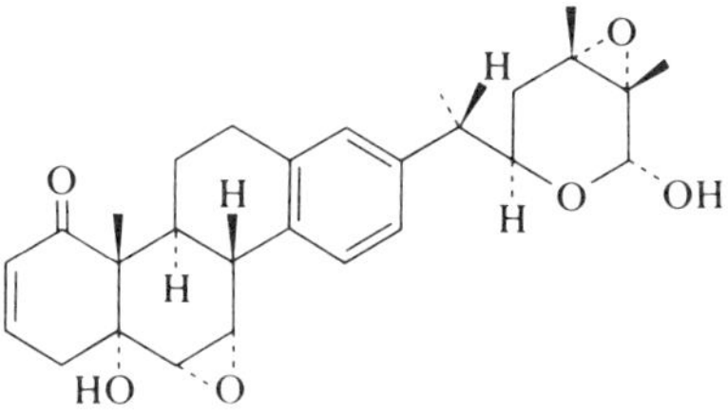

(41)

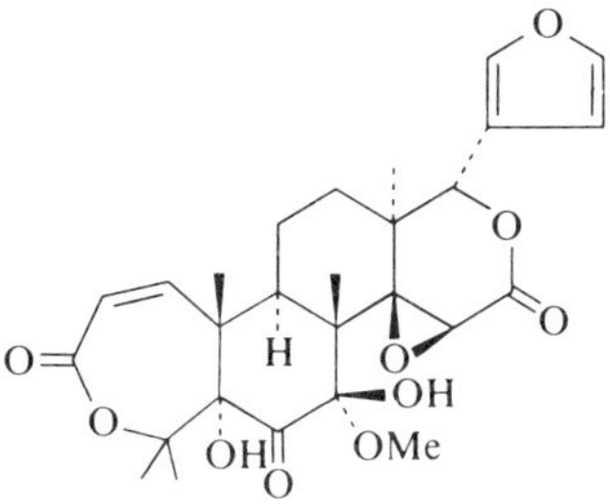

(42)

penoids have also been implicated in host-plant resistance and antifeedant behaviour.[79] Oviposition deterrents[80] and inhibitors[81] represent relatively new areas of study and it is felt that such responses might be important in developing novel control methods.

9 Biosynthesis and Biotransformation

A major point of controversy has centred on the possibility that pheromones of the Lepidoptera may be derived from the diet.[82] Subsequent examination of this postulate established that the 67:33, (*E*)-11-:(*Z*)-11-tetradecenyl acetate pheromone composition of the Oak Leaf Roller did not vary irrespective of whether the insect had been reared on oak leaves or an artificial diet, thus suggesting specific biosynthesis.[83] Indeed, a re-examination of previous results now accords to the latter view.[84] The biosynthesis of the hairpencil pheromone

[75] R. D. Stipanovic, A. A. Bell, D. H. O'Brien, and M. J. Lukefahr, *Tetrahedron Letters*, 1977, 567; J. R. Gray, T. J. Mabry, A. A. Bell, R. D. Stipanovic, and M. J. Lukefahr, *J.C.S. Chem. Comm.*, 1976, 109.
[76] I. Kubo, Z. Miura, M. J. Pettei, Y. W. Lee, F. Pilkiewicz, and K. Nakanishi, *Tetrahedron Letters*, 1977, 4553; I. Kubo, Y. W. Lee, M. J. Pettei, F. Pilkiewicz, and K. Nakanishi, *J.C.S. Chem. Comm.*, 1976, 1013.
[77] M. J. Begley, L. Crombie, P. J. Ham, and D. A. Whiting, *J.C.S. Perkin I*, 1976, 296, 304.
[78] I. Kubo, S. P. Tanis, Y. W. Lee, I. Miura, K. Nakanishi, and A. Chapya, *Heterocycles*, 1976, **5**, 485.
[79] I. Kubo, I. Miura, K. Nakanishi, T. Kamikawa, T. Isobe, and T. Kubota, *J.C.S. Chem. Comm.*, 1977, 555; A. C. Waiss jun., B. G. Chan, C. A. Elliger, V. H. Garrett, E. C. Carlson, and B. Beard, *Naturwiss.*, 1977, **64**, 341; T. Ikeda, F. Matsumura, and D. M. Benjamin, *Science*, 1977, **197**, 497.
[80] J. Hurter, B. Katsoyannos, E. F. Boller, and P. Wirz, *Z. angew. Entomol.*, 1976, **80**, 50.
[81] H. M. Flint, R. L. Smith, J. G. Pomonis, D. E. Forey, and B. R. Horn, *J. Econ. Entomol.*, 1977, **70**, 547.
[82] L. B. Hendry, *Science*, 1976, **192**, 143; *ibid.*, 1975, **188**, 59.
[83] J. R. Miller, T. C. Baker, R. T. Cardé, and W. L. Roelofs, *Science*, 1976, **192**, 140.
[84] D. M. Hindenlang and J. K. Wichmann, *Science*, 1977, **195**, 86.

secretion of 2-phenylethanol in males of *Mamestra configurata* has been studied both *in vivo* and *in vitro* with crude homogenates, and a detailed pathway from phenylalanine has been mapped out. It has been noted that this pheromone is stored in Stobbe's gland as 2-phenethyl-β-glycoside which is hydrolysed on requirement by enzymes contained in hairpencil cap cells.[85] Oleic acid appears to be the biosynthetic precursor of n-nonanal, a pheromone of the Greater Waxmoth, *Galleria melonella*.[86]

The production of pheromones in bark beetles by allylic oxidation of unsaturated terpenoid hydrocarbons has attracted much attention.[87] The intriguing problem of cantharadin (43) biosynthesis in male Spanish flies has been studied

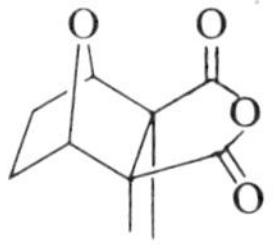

(43)

by feeding and successful incorporation of radiolabelled mevalonate, farnesol, and methyl farnesoate.[88] It is interesting to note that the female is unable to produce this defensive material, but acquires it by transfer from the male during copulation.

10 Perception of Stimuli

The availability of synthetic samples of enantiomerically pure pheromones has much aided studies of olfactory specificity (*e.g.* refs. 89 and 90). The monitoring of olfactory responses by electroantennography (EAG) and single-cell recording has been applied to investigations of many types of stimulus *e.g.* host-plant attractants,[91] cockroach sex attractants,[92] and Lepidopterous pheromones.[93] A good correlation between the EAG method and behavioural bioassays was obtained for the repellancy towards *Reticulitermes* termites of series of quinones and terpenoids of plant origin.[94] The roles of thiol and disulphide groups in nerve receptors for

[85] J. Weatherston and J. E. Percy, *Insect Biochem.*, 1976, **6**, 413; *ibid.*, 1975, **5**, 737.
[86] S. P. Schmidt and R. E. Monroe, *Insect Biochem.*, 1976, **6**, 377.
[87] J. A. A. Renwick, P. R. Hughes, G. B. Pitman, and J. P. Vite, *J. Insect Physiol.*, 1976, **22**, 725; J. A. A. Renwick, P. R. Hughes, and I. S. Krull, *Science*, 1976, **191**, 199; J. M. Brand, J. W. Bracke, L. N. Britton, A. J. Markovetz, and S. J. Barras, *J. Chem. Ecol.*, 1976, **2**, 195; P. A. Hedin, *ibid.*, 1977, **3**, 279.
[88] J. R. Sierra, W. D. Woggon, and H. Schmid, *Experientia*, 1976, **32**, 142; *Helv. Chim. Acta*, 1977, **60**, 2288.
[89] J. P. Vite, D. Klimetzek, G. Loskant, R. Hedden, and K. Mori, *Naturwiss.*, 1976, **63**, 582.
[90] M. Yamada, T. Saito, K. Katagiri, S. Iwaki, and S. Marumo, *J. Insect Physiol.*, 1976, **22**, 755.
[91] R. F. Simpson and R. M. McQuilkin, *Entomol. Exp. Appl.*, 1976, **19**, 205.
[92] H. Washio and C. Nichino, *J. Insect Physiol.*, 1976, **22**, 735; C. Nishino, H. Washio, K. Tsuzuki, W. S. Bowers, and T. R. Tobin, *Agric. and Biol. Chem. (Japan)*, 1977, **41**, 405.
[93] T. C. Baker and W. L. Roelofs, *J. Insect Physiol.*, 1976, **22**, 1357; Y. Aihara and T. Shibuya, *ibid.*, 1977, **23**, 779; T. L. Payne and W. E. Finn, *ibid.*, p. 879.
[94] M. A. Floyd, D. A. Evans, and P. E. Howse, *J. Insect Physiol.*, 1976, **22**, 697.

quinone repellants in the American Cockroach have been chemically investigated.[95]

11 Techniques of Microscale Structure Elucidation

The past two years have seen further improvements in the power and sensitivity of both separative and spectroscopic techniques.[96] Attention has been paid to efficient methods of selective extraction of behaviourally active insect chemicals, *e.g.* collection on Porapak Q of volatiles being produced by living insects.[17] This method has particular relevance to cases where pheromones are stored as precursors prior to release. Another useful technique consists of milking pheromone-producing glands with micro-capillaries followed by direct 'solid sample' injection on to a gas chromatograph (g.c.).[97] A number of adaptations of gas chromatography to insect chemistry have been reported,[98] the most spectacular of which is the monitoring of the response potentials of excised antennae placed post-column in an emergent g.c. gas stream as a selective pheromone detector.[99]

An efficient splitter-trapping system for micropreparative g.c. has been described,[100] and the use of g.c. combined with ^{13}C n.m.r. spectroscopy has been reported for the separation and determination of purity of geometric isomers of dienes.[101] The enantiomeric composition of several insect pheromone alcohols and acetals has been determined by microscale n.m.r. methods using chiral shift reagents and by formation of diastereisomeric Mosher derivatives.[102]

12 Behaviour-modifying Chemicals in Pest Control

The largest increase in activity in the area of chemical ecology has been observed in the field evaluation of behaviour-modifying chemicals in potential pest control programmes, particularly in terms of population survey, behaviour disruption, and trap baiting.[103] The importance of such factors as enantiomeric purity[104] and the correct design of traps[105] has been emphasized. As expected, reports range from significant successes (*e.g.* ref. 106) to disappointment (*e.g.* ref. 107). In other cases, methods based on chemical ecology offer a comparable alternative to conventional methods.[108]

[95] D. M. Norris, J. M. Rozental, and G. Samberg, *Comp. Biochem. Physiol. (C)*, 1977, **57**, 55.
[96] J. H. Tumlinson and R. R. Heath, *J. Chem. Ecol.*, 1976, **2**, 87.
[97] E. D. Morgan and R. C. Tyler, *J. Chromatog.*, 1977, **134**, 174.
[98] E. Talman, P. E. J. Verwiel, and A. C. Lakwijk, *Adv. Mass. Spectrom. Biochem. Med.*, 1977, **2**, 215; L. M. MacDonald and J. Weatherston, *J. Chromatog.*, 1976, **118**, 195.
[99] H. Arn, E. Staedler, and S. Rauscher, *Z. Naturforsch.*, 1975, **30c**, 11.
[100] R. Baker, J. W. S. Bradshaw, D. A. Evans, M. D. Higgs, and L. J. Wadhams, *J. Chromatog. Sci.*, 1976, **14**, 425.
[101] H. Disselkoetter, K. Eiter, W. Karl, and D. Wendisch, *Tetrahedron*, 1976, **32**, 1591.
[102] T. E. Stewart, E. L. Plummer, L. L. McCandless, jun., J. R. West, and R. M. Silverstein, *J. Chem. Ecol.*, 1977, **3**, 27; E. L. Plummer, T. E. Stewart, K. Byrne, G. T. Pearce, and R. M. Silverstein, *ibid.*, 1976, **2**, 307.
[103] P. A. Hedin, *Chemtech.*, 1976, **6**, 444.
[104] J. R. Miller, K. Mori, and W. L. Roelofs, *J. Insect Physiol.*, 1977, **23**, 1447.
[105] E. D. M. Macauley and T. Lewis, *Ecol. Entomol.*, 1977, **2**, 279.
[106] H. Arn, B. Delley, M. Baggiolini, and P. J. Charmillot, *Entomol. Exp. Appl.*, 1976, **19**, 139.
[107] M. D. Proverbs, D. M. Logan, and J. R. Newton, *Canad. Entomol.*, 1975, **107**, 1265.
[108] L. K. Gaston, R. S. Kaae, H. H. Shorey, and D. Sellers, *Science*, 1977, **196**, 904; P. H. Westigard and K. L. Graves, *Canad. Entomol.*, 1976, **108**, 379.

13 Synthetic Studies

Valuable reviews on the methodology of pheromone synthesis,[109—111] the synthesis of optically active pheromones,[112] and the synthesis of insect sex attractants on solid phases[113,114] have appeared.

Acyclic derivatives.—Two methods, which have general applicability, have been used in the synthesis of the queen substance, 9-oxo-(*E*)-dec-2-enoic acid. The first involves the sulphenylation and dehydrosulphenylation of esters or ketones,[115] whereas the key step in the second is an oxidative cleavage which is accompanied by rearrangement of a 2,2-dithiocycloalkan-1-ol (44) (Scheme 1).[116] Conversion of (45) into the queen substance was then achieved by addition of methylmagnesium

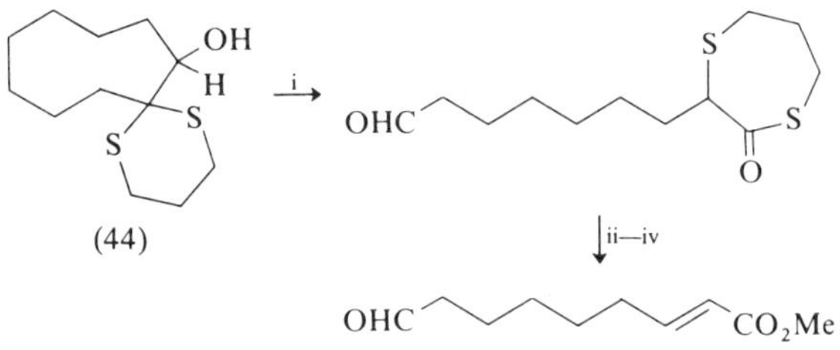

(44)

(45)

Reagents: i, $Pb(OAc)_4$; ii, $MeOH–I_2$; iii, $NaIO_4$; iv, pyrolysis

Scheme 1

chloride followed by Jones oxidation. The queen substance has also been synthesized by a route beginning with reaction of butadiene with diethyl malonate catalysed by palladium acetate and triphenylphosphine.[117] In a similar way, the same reaction, but in the presence of carbon monoxide and t-butyl alcohol, yields (46) (Scheme 2). Further carbonylation in presence of octacarbonyldicobalt yields,

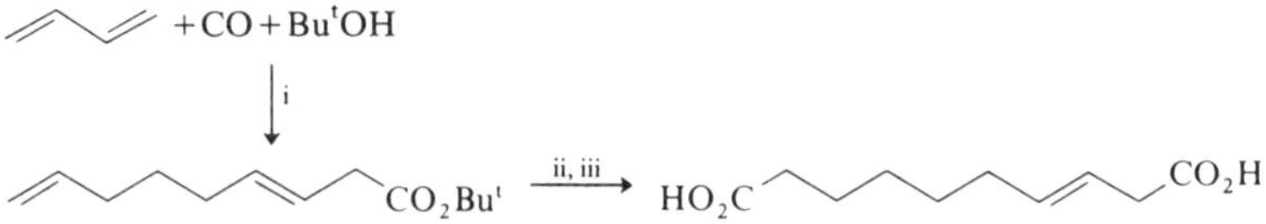

(46)

Reagents: i, $Pd(OAc)_2–PPh_3$; ii, $Co_2(CO)_8–MeOH$; iii, H_3O^+

Scheme 2

[109] C. A. Henrick, *Tetrahedron*, 1977, **33**, 1845.
[110] J. A. Katzenellenbogen, *Science*, 1976, **194**, 139.
[111] R. Rossi, *Synthesis*, 1977, 817.
[112] K. Mori, *Gendai Kagaku*, 1976, **61**, 51.
[113] C. C. Leznoff and T. M. Fyles, *J.C.S. Chem. Comm.*, 1976, 251.
[114] T. M. Fyles, C. C. Leznoff, and J. Weatherston, *Canad. J. Chem.*, 1977, **55**, 1143.
[115] B. M. Trost, T. N. Salzmann, and K. Hiroi, *J. Amer. Chem. Soc.*, 1976, **98**, 4887.
[116] B. M. Trost and K. Hiroi, *J. Amer. Chem. Soc.* 1976, **98**, 4313.
[117] J. Tsuji, K. Masaoka, and T. Takahashi, *Tetrahedron Letters*, 1977, 2267.

after hydrolysis, dec-2-enedioic acid which is one of the components of royal jelly, the sole nutrient of queen bee larvae.[118]

The absolute configuration of (−)-4-methylheptan-3-ol, a pheromone of the Smaller European Elm Bark Beetle has been demonstrated to be (3*S*,4*S*).[119] (*R*)-(+)-Citronellic acid was converted into a mixture of (3*R*,4*R*)-(+)-*threo*- and (3*S*,4*R*)-(+)-*erythro*-4-methylheptan-3-ol which were separable by gas chromatography.

Manicone, (*E*)-4,6-dimethyloct-4-en-3-one (48), the principal alarm pheromone of certain species of Manica ants, has been prepared from a route (Scheme 3) in which the $\alpha\beta$-unsaturated carbonyl moiety was introduced by reaction of a 4-chloro-2-pyrazolin-5-one (47) with aqueous sodium hydroxide.[120] Unfortunately

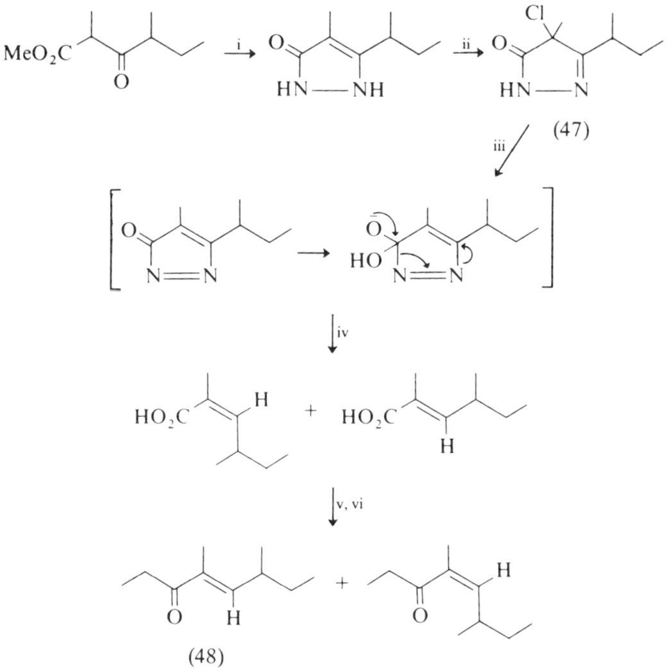

Reagents: i, NH_2NH_2–EtOH; ii, Cl_2–CH_2Cl_2; iii, NaOH; iv, H_3O^+; v, $SOCl_2$–C_6H_6; vi, Et_2Cd–benzene

Scheme 3

the opening of the pyrazolinone gave a mixture of *Z*- and *E*-isomers and further separation was required. Optically active manicone has been prepared from the titanium tetrachloride-promoted reaction of the appropriately substituted vinyl trimethylsilyl ether and (+)-(2*S*)- or (−)-(2*R*)-2-methylbutyraldehyde.[121]

[118] J. Tsuji and H. Yasuda, *J. Organometallic Chem.*, 1977, **131**, 133.
[119] K. Mori, *Tetrahedron*, 1977, **33**, 289.
[120] P. J. Kocienski, J. M. Ansell, and R. W. Ostrow, *J. Org. Chem.* 1976, **41**, 3625.
[121] K. Banno and T. Mukiayama, *Chem. Letters*, 1976, 279.

Further studies on the Ips bark beetle pheromones have appeared. Pyrolysis of the esters obtained from 3-methylbut-3-en-1-yl acetate and senecioic anhydride gave predominantly a ketone which on reduction yielded a racemic mixture of 2-methyl-6-methyleneocta-2,7-dien-4-ol, the main aggregation pheromone of *Ips paraconfusus*.[122] A small amount of asymmetric induction was found in the formation of 2-methyl-6-methyleneoct-7-en-4-ol and 2-methyl-6-methyleneocta-2,7-dien-4-ol by the coupling of Grignard reagents with the appropriate aldehydes in the presence of (+)-(2*S*,3*S*)-*NNN'N'*-tetramethyl-2,3-dimethoxybutane-1,4-diamine.[123] Both enantiomers of the former compound have been synthesized from (*S*)-(+)-leucine and its antipode.[124] The (*R*)-(−)-enantiomer of the dienol has been synthesized from (*R*)-(+)-glyceraldehyde acetonide, obtainable from D-mannitol.[125] The natural pheromone was demonstrated to have (*S*)-configuration.

The sex pheromone of the female Red Bollworm Moth, *Diparopsis castanea*, (*E*)-dodeca-9,11-dien-1-ol has been prepared from undec-10-en-1-ol.[126] The route involved formation of (*E*)-11-hydroxyundec-2-enal by allylic bromination followed by displacement and oxidation. Finally, the synthesis was completed by reaction of methylenetriphenylphosphorane and the aldehyde.

A four-step synthesis (Scheme 4) of the sex pheromone of the European Grape Vine Moth, *Lobesia botrana*, (*E*,*Z*)-dodeca-7,9-dien-1-yl acetate (50) has been reported *via* a functionalized organoborate.[127] Oct-7-ynyl acetate (49) was prepared from hept-1-yne *via* the acetylene 'zipper' reaction of oct-2-yn-1-ol with

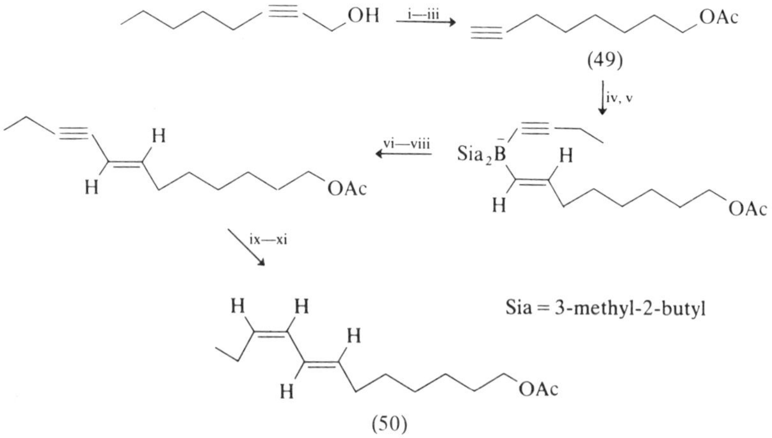

Reagents: i, $KNH(CH_2)_3NH_2$; ii, H_2O; iii, Ac_2O; iv, Sia_2BH; v, ≡—Li; vi, I_2; vii, NaOAc; viii, H_2O_2–NaOAc; ix, Sia_2BH; x, AcOH; xi, H_2O_2–NaOAc

Scheme 4

[122] C. F. Garbers and F. Scott, *Tetrahedron Letters*, 1976, 1625.
[123] S. Karlsen, P. Froyen, and L. Skattebol, *Acta Chem. Scand.*, 1976, **B30**, 664.
[124] K. Mori, *Tetrahedron*, 1976, **32**, 1101.
[125] K. Mori, *Tetrahedron Letters*, 1976, 1609.
[126] J. H. Babler and M. J. Martin, *J. Org. Chem.*, 1977, **42**, 1799.
[127] E. Negishi and A. Abramovitch, *Tetrahedron Letters*, 1977, 411.

potassium 3-aminopropylamide, followed by acetylation. The subsequent reaction depended on the fact that triorganoboranes are much more reactive toward alkynyl-lithiums than the acetoxy-group. The pheromone (50) has also been synthesized by a route involving acid-catalysed ring opening of 1-cyclopropylpent-2-yn-1-ol.[128] Stereoselectivity of the reaction towards the conjugated (*E*)-enyne was obtained by complexion of the triple bond with octacarbonyldicobalts.

(±)-Methyl n-tetradeca-*trans*-2,4,5-trienoate, produced by the male Dried Bean Beetle, *Acanthoscelides obtectus*, has been prepared in which the initial stage was a modified Claisen rearrangement on (51) to produce the allenic ester (52).[129] A

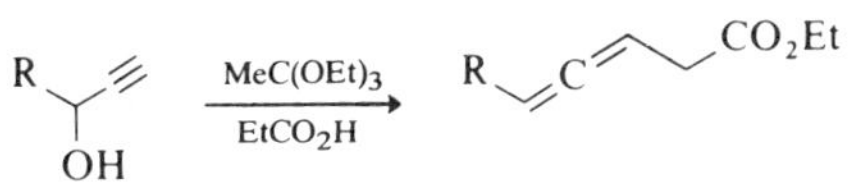

(51) R = octyl (52)

Beckmann fragmentation of the oxime (53) to the isomeric nitriles (54) provided the important stage in the synthesis of 3,7-dimethylpentadec-2-yl acetate, the sex pheromone of the Pine Sawfly *Neodiprion lecontei* (55).[130] The preparation of

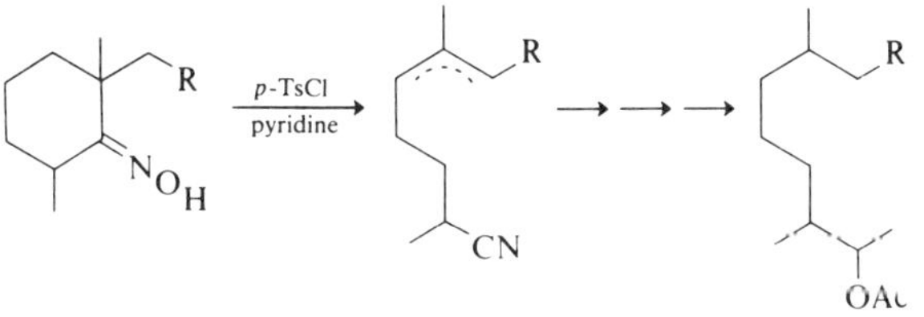

(53) R = heptyl (54) (55)

erythro-3,7-dimethylpentadecan-2-acetate (58), the sex attractant of several species of pine sawflies, has been prepared by a method which has more general applicability.[131] Beginning with *cis*-2,3-dimethylcyclohexanone, Baeyer–Villiger oxidation gave the lactone (56) which on reaction with octyllithium yielded the hydroxy-ketone (57). The acetate (58) was then obtained following a Wittig reaction, reduction, and acetylation.

Further stereospecific syntheses have been reported by Bestmann and co-workers. The base sodium bis-trimethylsilylamide[132] has been employed to produce (*Z*)-olefin formation in the Wittig reaction and this procedure was used in the synthesis of (*Z,Z*)-7,11- and (*Z,E*)-7,11-hexadecadienyl acetate, the sex pheromone of *Pectinophora gossypiella*.[133] The synthesis of a series of (*E,Z*)-conjugated dienes has also been reported.[134] A selective procedure for the general synthesis of conjugated (*E,E*)- and (*E,Z*)-dienes involving the reaction of (*E*)-

[128] D. Descoins, D. Samain, B. Lalanne-Cassou, and M. Gallois, *Bull. Chem. soc. France*, 1977, 941.
[129] P. J. Kocienski, G. Cernigliaro, and G. Felstein, *J. Org. Chem.*, 1977, **42**, 353.
[130] P. J. Kocienski and J. M. Ansell, *J. Org. Chem.*, 1977, **42**, 1102.
[131] G. Magnusson, *Tetrahedron Letters*, 1977, 2713.
[132] H. J. Bestmann, W. Stransky, and O. Vostrowsky, *Chem. Ber.*, 1976, **109**, 1694.
[133] H. J. Bestmann, K. H. Koschatsky, W. Stransky, and O. Vostrowsky, *Tetrahedron Letters*, 1976, 353.
[134] H. J. Bestmann, O. Vostrowsky, H. Paulus, W. Billman, and W. Stransky, *Tetrahedron Letters*, 1977, 121.

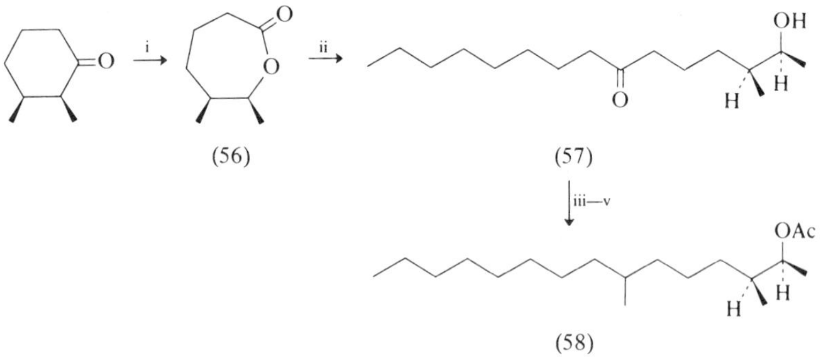

Reagents: i, m-$ClC_6H_4CO_3H$–CH_2Cl_2; ii, n-$C_8H_{17}Li$; iii, $Ph_3P{=}CH_2$–HMPA–ether; iv, H_2–Pd/C–EtOH; v, Ac_2O–pyridine

Scheme 5

alkenylalanes (59), readily obtainable *via* hydroalumination of alkynes, with alkenyl halides in the presence of palladium or nickel complexes has also been reported.[135]

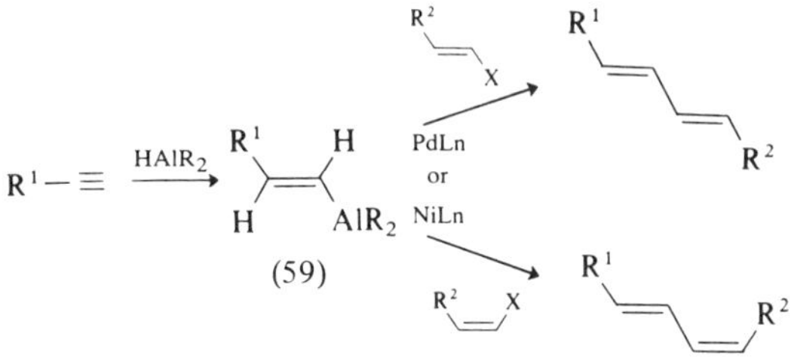

A magnesium iodide rearrangement of α,β-epoxysilane has been used to form β-ketosilanes (60). Subsequent treatment with methyl-lithium and sodium acetate in acetic acid gave (61) which was then transformed into 7-methyl-3-propyldodeca-2(Z),6(Z)-dien-1-ol, previously obtained from the codling moth.[136] A synthesis (Scheme 7) of (62) the pheromone of the Parasitic Bean Weevil, has involved the reaction of an allenic lithium cuprate with an acetylenic ester.[137]

Insoluble polymers have been used as supports in the synthesis of (Z)-dodec-7-enyl, (Z)-tetradec-9-enyl, and (Z)-tetradec-11-enyl acetates.[113,138] Stereoselective synthesis of the isomeric pentadeca-5,10-dienals has been reported.[139] The (S)-enantiomers of (Z)- and (E)-14-methylhexadec-8-en-1-ol and the corresponding aldehydes, which are sex pheromone components of dermestid beetles, have been synthesized.[140] In this work, a new method for the stereospecific lithium aluminium hydride reduction of ω-alkynols to the corresponding (E)-alkenols is described.[141]

[135] S. Baba and E. Negishi, *J. Amer. Chem. Soc.*, 1976, **98**, 6729.
[136] M. Obayashi, K. Utimoto, and H. Nozaki, *Tetrahedron Letters*, 1977, 1807.
[137] D. Michelot and G. Linstrumelle, *Tetrahedron Letters*, 1976, 275.
[138] C. C. Leznoff, T. M. Fyles, and J. Weatherston, *Canad. J. Chem.*, 1977, **55**, 4135.
[139] G. Ohloff, C. Vial, F. Näf, and M. Pawlak, *Helv. Chim. Acta*, 1977, **60**, 1161.
[140] R. Rossi and A. Carpita, *Tetrahedron*, 1977, **33**, 2447.
[141] R. Rossi and A. Carpita, *Synthesis*, 1977, 561.

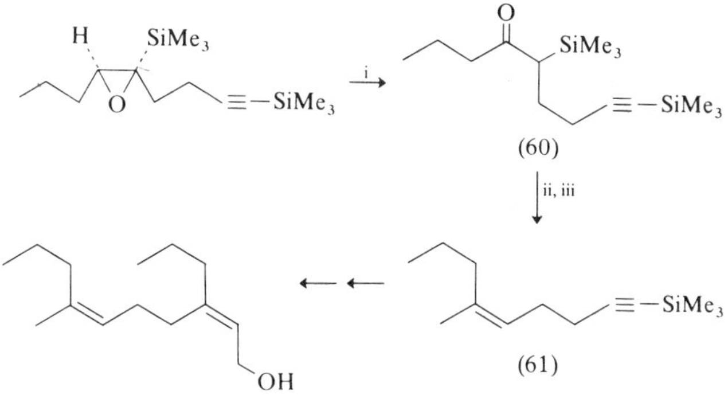

(60)

(61)

Reagents: i, MgI_2–ether; ii, MeLi; iii, AcONa–AcOH

Scheme 6

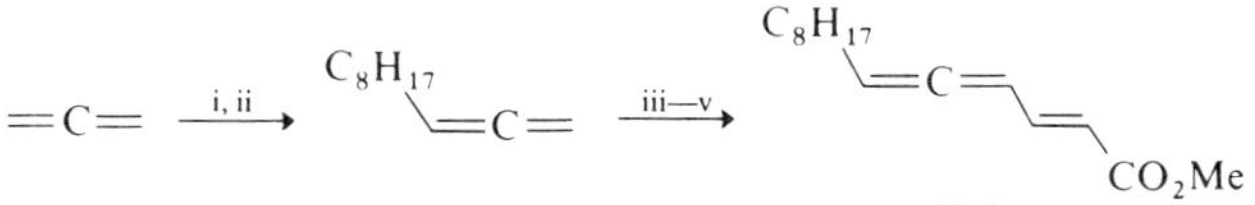

(62)

Reagents: i, BuLi; ii, n-$C_8H_{17}Br$; iii, BuLi; iv, CuI; v, ≡—CO_2Me

Scheme 7

The sex pheromone of the Gypsy Moth, (*Z*)-7,8-epoxy-2-methyloctadecane (65) has been synthesized by a Wittig reaction in presence of the base sodium bis-trimethylsilylamide.[142] The stereospecific reaction of (63) with the vinylsilane (64) has been used in the synthesis of (65) (Scheme 8).[143] Both enantiomers of this

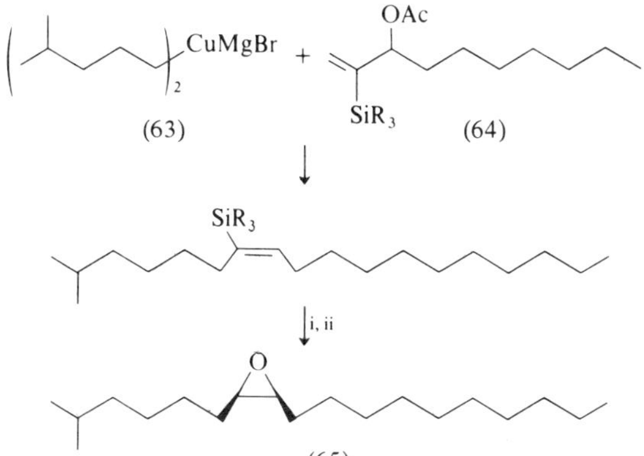

(63) (64)

(65)

Reagents: i, H^+; ii, m-$ClC_6H_4CO_3H$

Scheme 8

[142] H. J. Bestmann, O. Vostrowsky, and W. Stransky, *Chem. Ber.*, 1976, **109**, 3375.
[143] W. Mychajlowskij and T. H. Chan, *Tetrahedron Letters*, 1976, 4439.

compound have been synthesized from the (2*S*,3*S*)-*threo* configuration of L-(+)-tartaric acid[144] and the (7*R*,8*S*)-enantiomer has also been prepared from 1-menthyltoluene-*p*-sulphinate.[145] Further studies on the olefin metathesis reaction have appeared. The pheromones of the Housefly, Face Fly, and Gypsy Moth have been formed from suitable alkenes with homogeneous and heterogeneous catalysts.[146] A synthesis of muscalure, (*Z*)-9-tricosene, in which the first step was reaction of an ester of oleic acid with pentylmagnesium bromide, has been described.[147]

Two interesting syntheses of (*Z*)-6-heneicosen-11-one (68), the sex pheromone of the Douglas Fir Tussock Moth, both involving tosylhydrazone fragmentation, have been reported.[148,149] In one of these (Scheme 9) the epoxide (66) yields the acetylenic ketone (67) which is then converted into (68).[148] The synthesis of racemic[150] and optically active stereoisomers[151] of 3,11-dimethylnonacosan-2-one and 29-hydroxy-3,11-dimethylnonacosan-2-one,[152,153] two compounds isolated from virgin females of *Blattela germanica*, have been reported. Phase-transfer methylation of benzyl 3-oxobutanoate as a route to 3-methyl-2-alkanones has also been shown to be valuable for the synthesis of these two compounds.[154]

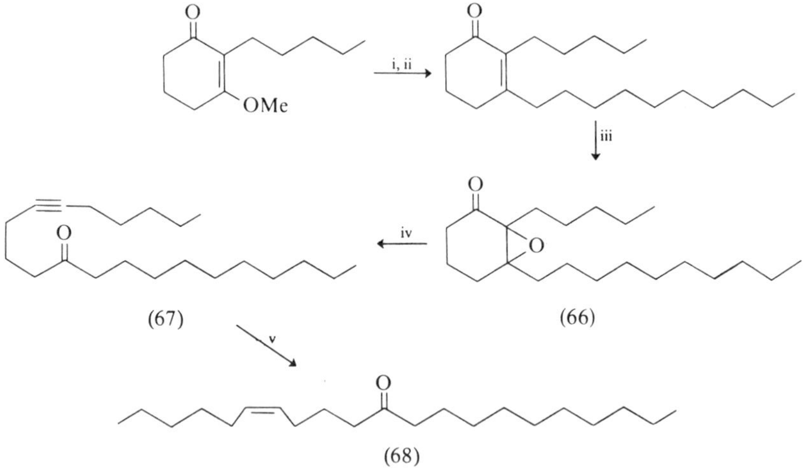

Reagents: i, n-$C_{10}H_{21}MgBr$; ii, H_3O^+; iii, H_2O_2–NaOH–MeOH; iv, *p*-$TsNHNH_2$; v, H_2–Pd/$BaSO_4$

Scheme 9

144 K. Mori, T. Takigawa, and M. Matsui, *Tetrahedron Letters*, 1976, 3953.
145 D. G. Farnum, T. Veysoglu, A. M. Cardé, B. Duhl-Emswiler, T. A. Pancoast, T. J. Reifz, and R. T. Cardé, *Tetrahedron Letters*, 1977, 4009.
146 F. W. Kuepper and R. Streck, *Z. Naturforch.*, 1976, **31b**, 1256.
147 K. Abe, T. Yamasaki, N. Nakamura, and T. Sakan, *Bull. Chem. Soc. Japan*, 1977, **50**, 2792.
148 P. J. Kocienski and G. J. Cernigliaro, *J. Org. Chem.*, 1976, **41**, 2927.
149 K. Mori, M. Uchida, and M. Matsui, *Tetrahedron*, 1977, **33**, 385.
150 L. D. Rosenblum, R. J. Anderson, and C. A. Henrick, *Tetrahedron Letters*, 1976, 419.
151 C. Conti, A. Niccoli, and R. Rossi, *Chimica e Industria*, 1976, **58**, 877.
152 A. W. Burgstahler, L. O. Weigel, M. E. Sanders, C. G. Schaefer, W. J. Bell, and S. B. Vuturo, *J. Org. Chem.*, 1977, **42**, 566.
153 R. Nishida, T. Sato, and Y. Kuwahara, *Agric. and Biol. Chem. (Japan)*, 1976, **40**, 1407.
154 A. W. Burgstahler, M. E. Sanders, C. G. Schaefer, and L. O. Weigel, *Synthesis*, 1977, 405.

Alicyclic Derivatives.—Full details of the synthesis of optically active grandisol (71), one of the four synergistic components of the male Boll Weevil pheromone, has been published.[155] Photolysis of (69), obtainable in a series of steps from (−)-β-pinene, gave (70) which on decarbonylation using chlorotris(triphenylphosphine)rhodium gave grandisol acetate and subsequently (+)-grandisol, (+)-(1*R*,2*S*)-1-methyl-1-(2-hydroxyethyl)-2-isopropenylcyclobutane. A process of geminal alkylation involving 1-lithiocyclopropyl sulphide has also been used in a synthesis of grandisol.[156]

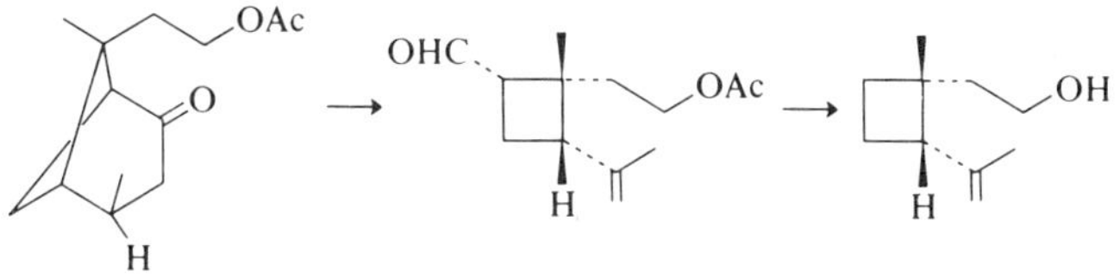

(69) (70) (71)

A two-step synthesis of a mixture of the two (3,3-dimethylcyclohexylidine)acetaldehydes (73), two other components of the boll weevil pheromone, has been reported.[157] In this, Vilsmeier formylation of isophorone afforded (3-chloro-5,5-dimethylcyclohex-2-enylidene)acetaldehyde (72) in 80% yield. Catalytic hydrogenation of (72) over palladium on charcoal poisoned with sulphur and quinoline gave a 2:1 mixture of the *E*/*Z*-isomers of (73). Treatment

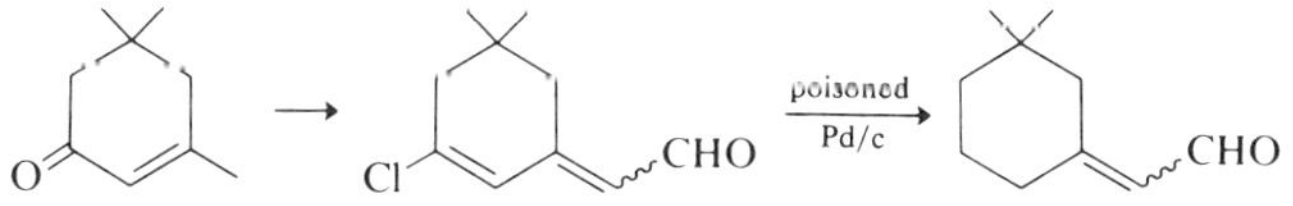

(72) (73)

of the reaction product with a catalytic amount of toluene-*p*-sulphonic acid yielded a 1:1 mixture of stereoisomers. These two compounds together with the other cyclohexyl component (77) have also been prepared from 3-methylcyclohex-2-enone by way of (74)—(76) (Scheme 10).[158] One problem in this synthesis is that (77), the desired alcohol, is obtained as a 1:1 mixture with (78); a difficult separation is necessary here. The formation of α β-unsaturated aldehydes by the bishomologation of ketones has also been applied to the synthesis of the aldehyde components of the pheromone mixture.[159]

Syntheses of *endo*- and *exo*-brevicomin have attracted much attention. Both of them were synthesized by a route involving reaction of the tosylhydrazone formed from the epoxide of 1-ethyl-2-methylcyclohex-1-enone.[160] The acetylenic ketone (80) obtained from this reaction has also been obtained by the reaction of (79) with methyl-lithium.[161] The acetylenic ketone was reduced to the *cis*-olefin (81),

155 P. D. Hobbs and P. D. Magnus, *J. Amer. Chem. Soc.*, 1976, **98**, 4594.
156 B. M. Trost, D. E. Keeley, H. C. Arndt, and M. J. Bogdanowicz, *J. Amer. Chem. Soc.*, 1977, **99**, 3088.
157 P. C. Traas, H. Boelens, and H. J. Takken, *Rec. Trav. chim.*, 1976, **95**, 308.
158 S. W. Pelletier and N. V. Mody, *J. Org. Chem.*, 1976, **41**, 1069.
159 J. H. Babler and J. M. Coghlan, *Synth. Comm.*, 1976, **6**, 469.
160 P. J. Kocienski and R. W. Ostrow, *J. Org. Chem.*, 1976, **41**, 398.
161 J. L. Coke, H. J. Williams, and S. Natarajan, *J. Org. Chem.*, 1977, **42**, 2380.

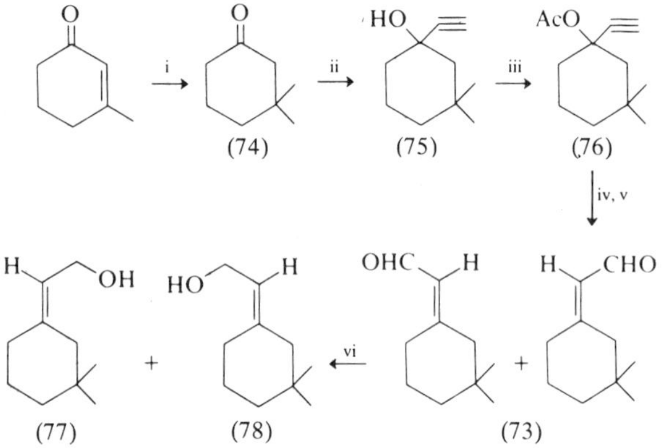

Reagents: i, Me_2CuLi; ii, $LiC{\equiv}CH–NH_2(CH_2)_2NH_2$; iii, $Ac_2O–H^+$; iv, 80% $AcOH–Ag_2CO_3$; v, H_2O; vi, 9-BBN or $NaBH_4$

Scheme 10

whence epoxidation and thermolysis gave *exo*-brevicomin (82), the pheromone from *Dendroctonus brevicomis* (Scheme 11). The selective irradiation of the

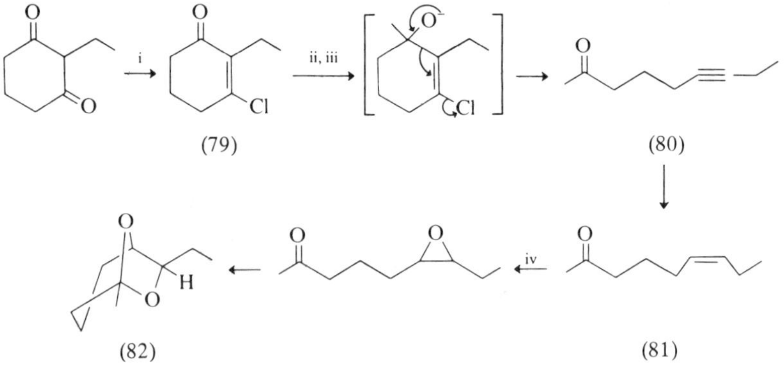

Reagents: i, PCl_3; ii, MeLi; iii, heat; iv, m-$ClC_6H_4CO_3H$

Scheme 11

carbonyl group of 2-propionyl-6-methyl-2,3-dihydro-4*H*-pyran has also been shown to be, after hydrogenation, a selective and elegant route to *exo*-brevicomin.[162] A further synthesis of (−)-(1*S*,7*S*)-*exo*-brevicomin is reported.[163] *endo*-Brevicomin (83), a pheromone inhibitor, has been synthesized by a palladium(II)-catalysed intramolecular cyclization of a terminal olefin containing a suitably located vicinal diol group (84).[164]

[162] D. Chaquin, J. P. Morizur, and J. Kossanyi, *J. Amer. Chem. Soc.*, 1977, **99**, 903.
[163] H. H. Meyer, *Annalen*, 1977, 732.
[164] N. T. Byrom, R. Grigg, and B. Kongkathip, *J.C.S. Chem. Comm.*, 1976, 216.

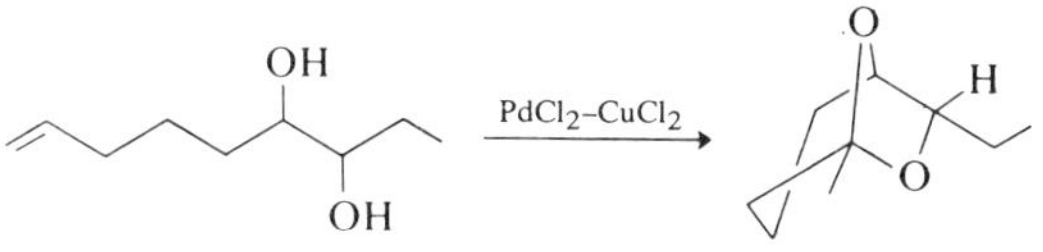

(83) (84)

Both enantiomers of frontalin have been synthesized from the ketone (85) derivable from methyl α-D-glucopyranoside.[165] Thus (1*R*,5*S*)-frontalin (90) was prepared through the intermediate compounds (86)—(89).

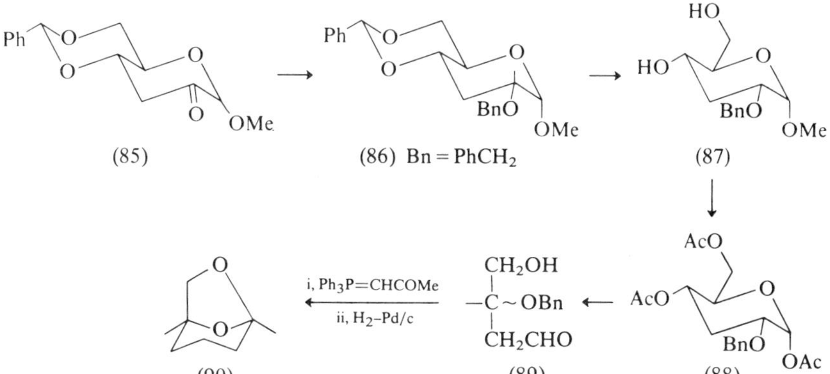

(85) (86) Bn = $PhCH_2$ (87)

(90) (89) (88)

α-Multistriatin (92), one of the compounds of the aggregation pheromone of *Scolytus multistriatus*, has been synthesized from (91), which was available from (*Z*)-but-2-ene-1,4-diol.[166] A similar route, except that (91) was prepared from

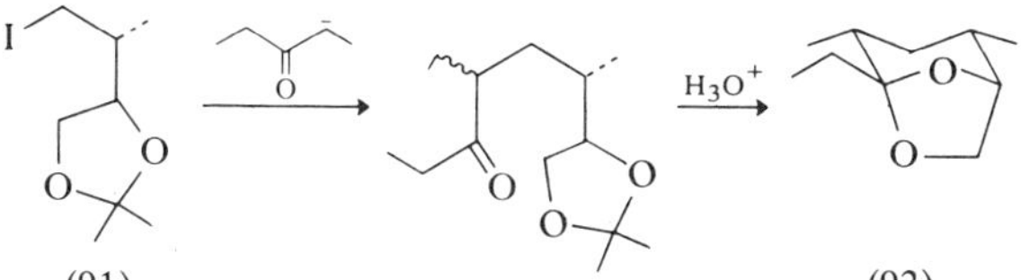

(91) (92)

(*R*)-(+)-glyceraldehyde obtainable from D-mannitol, established the absolute stereochemistry of the natural pheromone to be (1*S*,2*R*,4*S*,5*R*).[167] The same conclusion was reached as a result of a synthesis beginning with (*S*)-(+)-2-methylbut-3-enoic acid.[168] α-Multistriatin has also been synthesized by a route beginning with (+)-(3*R*)-citronellol although the final product was a mixture of α-γ-isomers and gas-chromatographic separation was required.[169]

The oxidation of pinene was used to prepare optically pure (+)-*trans*-verbenol, the pheromone of *Dendroctonus* bark beetles.[170] Both enantiomers of *cis*-verbenol

[165] D. R. Hicks and B. Fraser-Reid, *J.C.S. Chem. Comm.*, 1976, 869.
[166] W. J. Elliot and J. Fried, *J. Org. Chem.*, 1976, **41**, 2475.
[167] K. Mori, *Tetrahedron*, 1976, **32**, 1979.
[168] G. T. Pearce, W. E. Gore, and R. M. Silverstein, *J. Org. Chem.*, 1976, **41**, 2797.
[169] G. J. Cernigliaro and P. J. Kocienski, *J. Org. Chem.*, 1977, **42**, 3622.
[170] K. Mori, *Agric. and Biol. Chem. (Japan)*, 1976, **40**, 415.

were synthesized from (+)- and (−)-verbenone.[171] It was further pointed out that the pheromone of the *Ips* bark beetle should be described as (1*S*,4*S*,5*S*)-pin-2-en-4-ol rather than (+)-*cis*-verbenol since the sign of rotation depends on the solvent enclosed.

A general synthesis of optically active 4-alkyl (or 4-alkenyl)-γ-lactones from commercially available enantiomers of glutamic acid is reported.[172] Both enantiomers of δ-n-hexadecalactone, the pheromone suggested to be responsible for social behaviour of the queens and workers of *Vespa orientalis*, have been prepared from (*R*)-(+) and (*S*)-(−)-1,2-epoxytetradecane.[173] The epoxides were prepared by resolution of 1-dimethylaminotridecan-2-ol with the enantiomers of dibenzoyltartaric acid, followed by Hofmann elimination.

A termiticidal norsesquiterpenoid, *dl*-chamaecynone (93), has been prepared by a route involving a Diels–Alder reaction of (94) and (95).[174] A defensive secretion

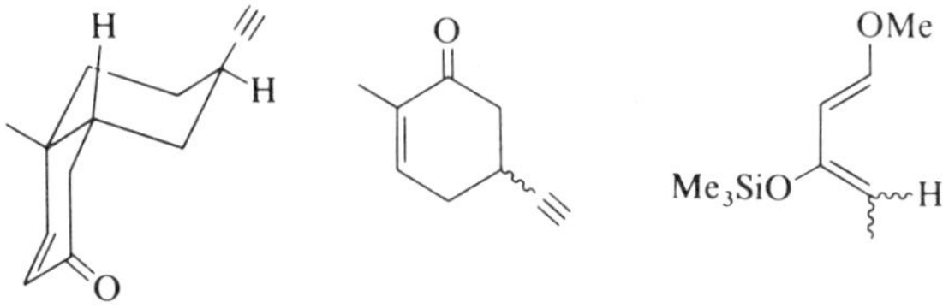

(93) (94) (95)

of the termite *Amitermes evuncifer*, 4,11-epoxy-*cis*-eudesmane (99), has been synthesized stereospecifically from (−)-carvone (96) in which the key steps were annulation to (97) and oxymercuration to (98).[175] Other reported synthesis include those of 2-deoxyecdysones[176] and *dl*-neocembrene.[177]

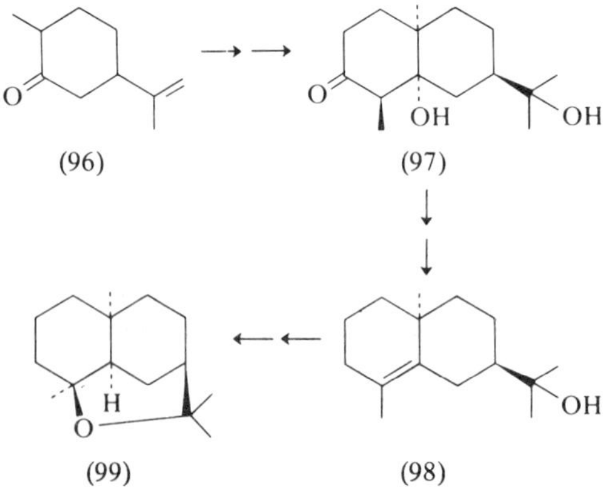

[171] K. Mori, N. Mizumachi, and M. Matsui, *Agric. and Biol. Chem. (Japan)*, 1976, **40**, 1611.
[172] U. Ravid and R. M. Silverstein, *Tetrahedron Letters*, 1977, 423.
[173] J. L. Coke and A. B. Richon, *J. Org. Chem.*, 1976, **41**, 3516.
[174] T. Harayama, H. Cho, and Y. Inubushi, *Tetrahedron Letters*, 1977, 3273.
[175] R. Baker, D. A. Evans, and P. McDowell, *J.C.S. Chem. Comm.*, 1977, 111.
[176] J. F. Kinnear, M. D. Martin, D. H. S. Horn, E. J. Middleton, J. S. Wilkie, M. N. Galbraith, and R. I. Willing, *Austral. J. Chem.*, 1976, **29**, 1815.
[177] Y. Kitahara, T. Kato, and T. Kobayashi, *Chem. Letters*, 1976, **3**, 219.

A synthesis of furanoterpenoids involving a 1,4-cycloaddition reaction of singlet oxygen appears to be generally useful.[178] The 2,5-dialkylpyrrolidines, which are constituents of the South African Fire Ant *Solenopsis punctaticeps*, have been prepared by stepwise alkylation of 2-lithio-1-nitrosopyrrolidines,[179] and by photolysis of the appropriate *N*-chloroamines.[55]

[178] K. Kondo and M. Matsumoto, *Tetrahedron Letters*, 1976, 390.
[179] R. R. Fraser and S. Passannanti, *Synthesis*, 1976, 540.

14 Biological Chemistry
Part (iii) Tetrapyrroles and their Biosynthesis

By D. G. BUCKLEY
Chemistry Department, Queen Mary College, Mile End Road, London E1 4NS

1 Introduction

In the three years since porphyrins and related compounds were last reviewed in *Annual Reports*,[1] a great deal of new work has been published, much of which relates to the biosynthesis of porphyrins, chlorins, and corrins and to details of their biological activity. Most of this work relates directly to results published since the late 1960's, and where appropriate a brief summary of this earlier work will be included.* The publication of the authorititative 'Porphyrins and Metalloporphyrins', edited by K. M. Smith,[2] and based on Falk's original work,[3] was most welcome, as was the appearance of a full account[4] of the important Royal Society discussion on the biosynthesis of porphyrins, chlorophyll, and vitamin B_{12} held in London in February 1975. The informal 'Tetrapyrrole Discussion Group' has been founded recently as a result of the increased activity in this area.†

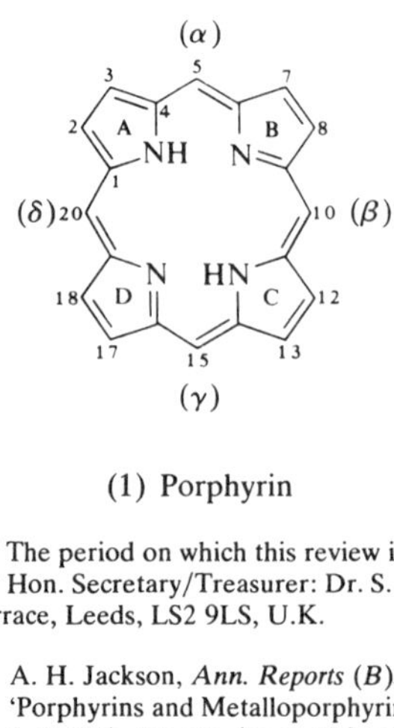

(1) Porphyrin

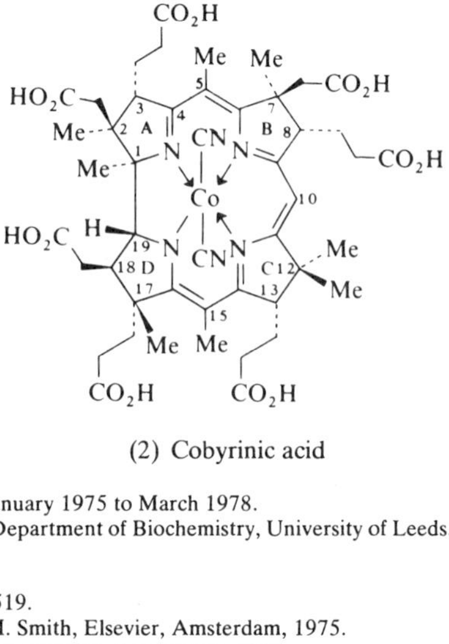

(2) Cobyrinic acid

* The period on which this review is based is January 1975 to March 1978.
† Hon. Secretary/Treasurer: Dr. S. B. Brown, Department of Biochemistry, University of Leeds, 9 Hyde Terrace, Leeds, LS2 9LS, U.K.

[1] A. H. Jackson, *Ann. Reports (B)*, 1974, **71**, 519.
[2] 'Porphyrins and Metalloporphyrins', ed. K. M. Smith, Elsevier, Amsterdam, 1975.
[3] J. E. Falk, 'Porphyrins and Metalloporphyrins', Elsevier, Amsterdam, 1964.
[4] *Phil. Trans. Roy. Soc.*, 1976, **B273**, pp. 75–357.

Both porphyrins and chlorins are based on the porphyrin (porphin) ring system (1), while the cobalamins are corrin derivatives, *e.g.* cobyrinic acid (2). The newer and more systematic numbering system will be used in this article; the older descriptions for the four meso carbon atoms (α, β, γ, and δ) are given for reference in (1).

2 Biosynthesis of Tetrapyrroles: General

Remarkable progress has been made in our understanding of the biosynthesis of the haems, chlorophylls, and corrins over the past thirty years, and a summary of what was known[5] at the end of 1974 is given in Scheme 1. Porphobilinogen (PBG),

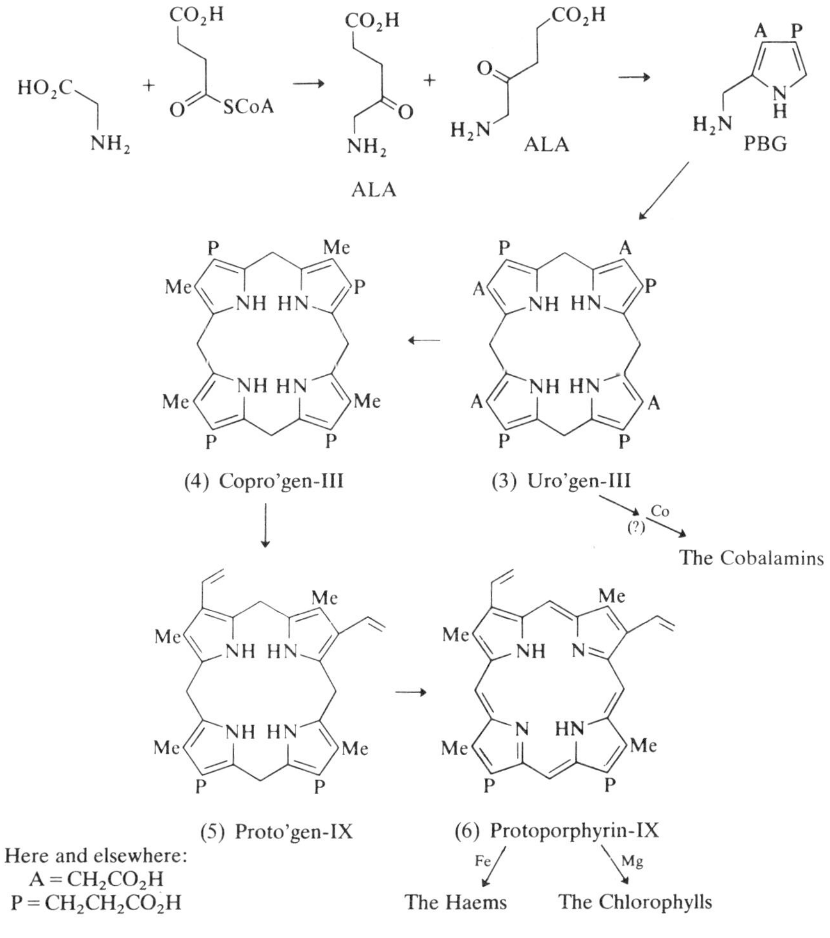

Scheme 1

[5] Reviewed by A. R. Battersby and E. McDonald in ref. 2, pp. 61–122.

normally derived from glycine and succinyl coenzyme A by way of 5-aminolaevulinic acid (ALA), is condensed to give uroporphyrinogen-III (abbreviated to "uro'gen-III") (3) as the next discrete (*i.e.* detectable) intermediate. Modification of the side-chains then occurs to give protoporphyrinogen-IX ("proto'gen-IX") (5) by way of coproporphyrinogen-III ("copro'gen-III") (4); oxidation of the porphyrinogen system follows to give protoporphyrin-IX (6), which either leads directly to the iron-containing haems or else to the chlorophylls, after insertion of magnesium followed by further modification. The pathway to the cobalamins is thought to branch at the uro'gen-III stage; one of the bridging methylene groups in uro'gen-III (3) must be removed to give the corrin ring system, *cf.* (2).

Recently,[4] efforts have been directed towards elucidating the details of various steps in the above biosynthetic pathway; for clarity, this work will be divided into four main sections: (*a*) the specific formation of uro'gen-III; (*b*) modification of side-chains and subsequent aromatization of protoporphyrin-IX; (*c*) the iron and magnesium branches; and (*d*) the biosynthesis of vitamin B_{12}.

3 Biosynthesis of Uro'gen-III: the 'Type-III Problem'

Early work established the crucial role of PBG in the biosynthesis of protohaem, and in turn PBG was shown to be synthesized from glycine and succinyl coenzyme A by way of ALA.[5] Recent work[6] has shown that ALA synthetase catalyses the condensation of glycine and succinyl-CoA by specific removal of the *pro*-2*R*-hydrogen of glycine; the *pro*-2*S*-hydrogen of glycine was shown to occupy the *pro*-5*S*-position in the derived ALA (Scheme 2).

Further studies[7,8] have now revealed the fate of the two hydrogen atoms at C-5 in ALA during its conversion into PBG by ALA dehydratase. C-5 of one ALA molecule provides the aminomethyl side-chain of PBG, and the substituents at C-5 are incorporated intact. Position 2 of PBG is derived from C-5 of the second ALA molecule; specific loss of the *pro*-5*R*-hydrogen of ALA occurs during the aromatization step to yield PBG that has the *pro*-5*S*-hydrogen of ALA at C-2 (Scheme 2).[6–8] These results suggest that the removal of the *pro*-*R*-hydrogen atoms from an intermediate such as (7) occurs while the species is enzyme-bound, and not after its release into the medium [(7) → (7a) → PBG], as is implicated in the otherwise excellent mechanistic sequence proposed earlier by Nandi and Shemin.[9]

The pioneering studies of Shemin, Granich, Bogorad, Neuberger, and Rimington[5] established that the co-operative action of *two enzymes*, porphobilinogen deaminase ('deaminase') and uroporphyrinogen-III cosynthetase ('cosynthetase') was required to catalyse the conversion of four molecules of the monopyrrole PBG into uro'gen-III (3) and ammonia (Scheme 3). In the absence of cosynthetase, deaminase catalyses the conversion of PBG into uro'gen-I (8), the isomer that would be expected from single head-to-tail combination of four PBG units. However, uro'gen-I (8) is *not transformed* into uro'gen-III (3) by cosynthetase alone, nor by the complete deaminase–cosynthetase system. These results sum-

[6] M. M. Abboud, P. M. Jordan, and M. Akhtar, *J.C.S. Chem. Comm.*, 1974, 643.
[7] M. Akhtar, M. M. Abboud, G. Barnard, P. Jordan, and Z. Zaman in ref. 4, pp. 117–136.
[8] M. M. Abboud and M. Akhtar, *J.C.S. Chem. Comm.*, 1976, 1007.
[9] D. L. Nandi and D. Shemin, *J. Biol. Chem.*, 1968, **243**, 1236.

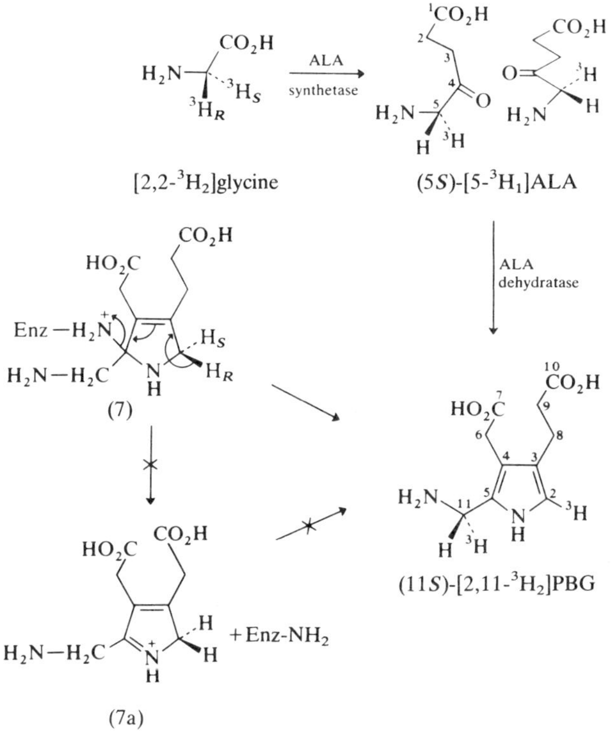

Scheme 2

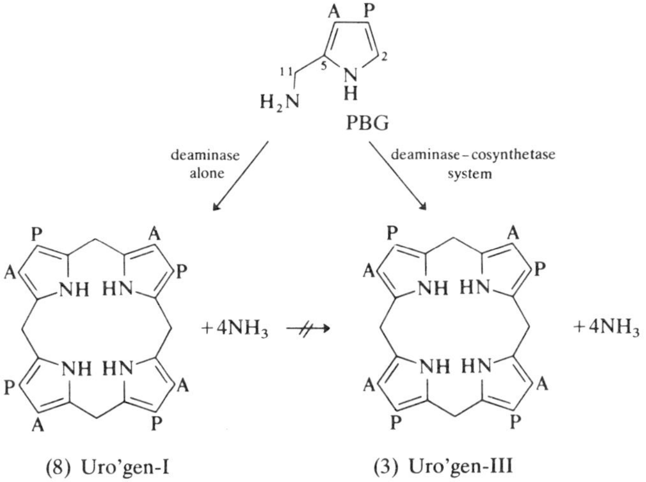

Scheme 3

marize the 'type-III problem': how does the complete enzyme system bring about the molecular rearrangement required to produce the *unexpected type-III isomer*, and what are the intermediates in this intriguing transformation?

The Nature of the Rearrangement.—Until 1973 there were no answers to these questions, although some 25 hypothetical mechanisms had been proposed.[5,10] However, in a series of ^{13}C n.m.r. experiments,[11] Battersby and McDonald were able to deduce *what* happened during the biosynthesis of uro'gen-III (3) from PBG, *i.e.* they were able to deduce the origin of each of the carbon atoms which make up the porphyrin macrocycle. Full details of this work have now been published[12] and are summarized below.

[2,11-$^{13}C_2$]Porphobilinogen (9) was prepared from 90 atom % ^{13}C-labelled ALA (10),[13] using the enzyme ALA-dehydratase. No dilutions were made throughout, so 81% of the PBG molecules were doubly labelled, as in (9). This PBG (1 part) was diluted with normal unenriched PBG (4 parts) and the resultant PBG sample was incubated with a PBG-free enzyme system from the alga *Euglena gracilis.* The product was protoporphyrin-IX [labelled (6)]; this was formed because the biological system contained all the enzymes necessary to effect the conversion of the uro'gen-III [labelled (3)], as it was formed, through the steps in Scheme 1 as far as [labelled (6)]. After ^{13}C n.m.r. analysis, this product was converted *chemically* into the labelled diketone (11) as shown in Scheme 4.

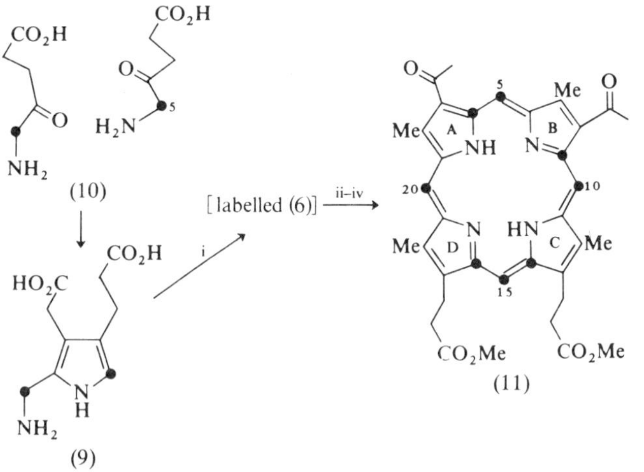

Reagents: i, Enzyme preparation from *E. gracilis*; ii, HBr–HOAc, H_2O; iii, CH_2N_2; iv, $Na_2Cr_2O_7$–H_2SO_4–H_2O.

Scheme 4

[10] Ref. 5, pp. 85–87. See also: E. Margoliash, *Ann. Rev. Biochem.*, 1961, **30**, 551; J. H. Matheson and A. H. Corwin, *J. Amer. Chem. Soc.*, 1961, **83**, 135; E. Bullock, *Nature*, 1965, **205**, 70; E. B. C. Llambias and A. M. del C. Batlle, *Biochem. J.*, 1971, **121**, 327; R. Radmer and L. Bogorad, *Biochemistry*, 1972, **11**, 904.

[11] A. R. Battersby, E. Hunt, and E. McDonald, *J.C.S. Chem. Comm.*, 1973, 442.

[12] A. R. Battersby, G. L. Hodgson, E. Hunt, E. McDonald, and J. Saunders, *J.C.S. Perkin I*, 1976, 273.

[13] A. R. Battersby, E. Hunt, E. McDonald, and J. Moron, *J.C.S. Perkin I*, 1973, 2918.

The low-field part of the [1]H-decoupled [13]C n.m.r. spectrum of the labelled protoporphyrin-IX (6) contained signals from the *meso*-carbon atoms C-5, C-10, and C-20 as 5.5 Hz doublets while that for C-15 was a 72 Hz doublet. The chemical shifts for C-5, C-10, C-15, and C-20 were assigned[14] by rational total synthesis of (6) that was specifically labelled at C-10, C-15, and C-20 respectively. Also, the couplings of 72 Hz and 5.5 Hz were proved to correspond, respectively, to direct coupling of two *adjacent* ^{13}C atoms and to ^{13}C–^{13}C coupling through three bonds, as in Scheme 5, by the unambiguous preparation[12] of the appropriate doubly labelled porphyrins. The spectrum of the derived diketone (11), in which the ^{13}C n.m.r. signals for the four *meso*-carbon atoms are even more clearly differentiated, showed similar features. Thus it was clearly established that the major components in the labelled sample of protoporphyrin-IX (6) that was derived from the incubation experiment were the four doubly labelled species shown in Scheme 5.

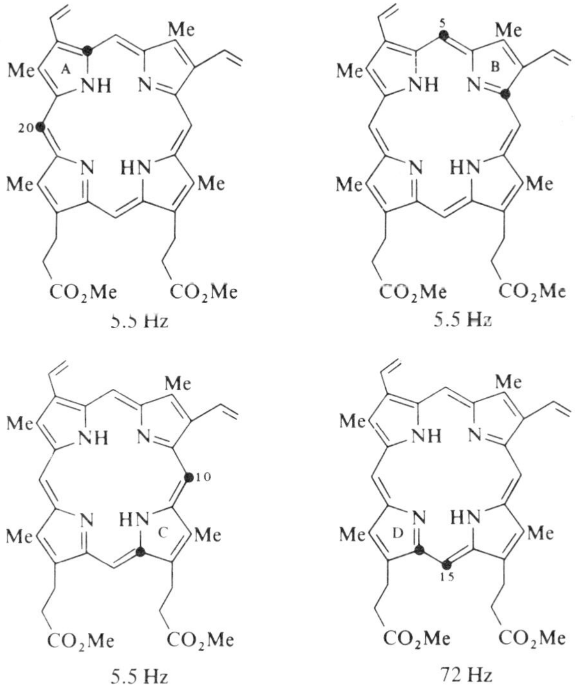

Scheme 5

Exactly the same conclusion was derived from studies[12] of the incorporation of the same diluted [2,11-$^{13}C_2$]PBG (9) into protoporphyrin-IX [labelled (6)] by a mixed enzyme preparation from chicken blood and beef mitochondria. Furthermore, it has been shown[15] that, during enzymic conversion of uro'gen-III (3) into

[14] A. R. Battersby, G. L. Hodgson, M. Ihara, E. McDonald, and J. Saunders, *J.C.S. Perkin I*, 1973, 2923; A. R. Battersby, M. Ihara, E. McDonald, J. Saunders, and R. J. Wells, *ibid.*, 1976, 283.

[15] B. Franck, D. Gantz, F.-P. Montforts, and F. Schmidtchen, *Angew. Chem. Internat. Edn.*, 1972, **11**, 421; A. R. Battersby, J. Staunton, and R. H. Wightman, *J.C.S. Chem. Comm.*, 1972, 1118; A. R. Battersby, E. McDonald, J. R. Redfern, J. Staunton, and R. H. Wightman, *J.C.S. Perkin I*, 1976, 266.

protoporphyrin-IX (6) (see Scheme 1), the macrocycle remains intact and unscrambled. All the results obtained for protoporphyrin-IX (6) therefore hold good for uro'gen-III (3).

From these experiments, the nature of the rearrangement process by which type-III porphyrins are biosynthesized has been defined, and it is characterized by the following features: (*a*) the three PBG units which form ring A and its attached C-20 bridge, ring B and the C-5 bridge, and ring C with its C-10 bridge are all incorporated intact, without rearrangement; (*b*), the PBG unit which forms ring D is built in with rearrangement which is intramolecular with respect to that PBG unit; and (*c*) the rearranged carbon atom forms the bridge at C-15. It is of considerable interest that exactly the same features characterize the biosynthetic process which generates type-III porphyrins in such widely different organisms as an alga (*Euglena gracilis*), a chicken, and a bacterium (*Propionibacterium shermanii*).[16]

The Role of Cosynthetase: Timing of the Rearrangement.—The known[5] deaminase-catalysed conversion of PBG into uro'gen-I (8) is usually considered[4,5,12] to occur by simple head-to-tail combination of four PBG units without rearrangement. Recent studies[16] with [2,11-$^{13}C_2$]PBG (9) and a purified preparation of deaminase from *E. gracilis* have confirmed this view; ^{13}C n.m.r. analysis of the recovered ^{13}C-labelled uroporphyrin-I octamethyl ester showed that each of the four PBG units had been incorporated intact into uro'gen-I (8). A plausible sequence for the interaction of PBG with deaminase is given in Scheme 6.

The fact that deaminase alone catalyses head-to-tail coupling of four PBG units to produce uro'gen-I (8), while deaminase and cosynthetase together catalyse the formation of uro'gen-III (3) from three intact PBG units and one intramolecularly rearranged unit, suggests that cosynthetase must bring about this rearrangement either by operating on an intermediate produced by deaminase or by modifying the way in which deaminase brings about *one* of the coupling steps. Cosynthetase might operate at any stage in the overall process, *i.e.* at the monopyrrole, dipyrrole, tripyrrole, or tetrapyrrole level, and it should be possible to make a distinction by testing the hypothetical intermediates (12)—(15) of Scheme 6 (and rearranged isomers) as precursors of uro'gen-III (3).

Early work[5,17] suggested that rearrangement does not occur as the first step, although incorporation studies have not been reported.

Dipyrrolic Intermediates in Uro'gen-III Biosynthesis.—Pyrromethanes, *e.g.* (16), have been prepared[18–20] in several laboratories and tested for incorporation into type-III porphyrins in the presence of the deaminase–cosynthetase system. The

[16] A. R. Battersby, E. McDonald, R. Hollenstein, and D. C. Williams, unpublished work, Cambridge, 1976; personal communication.

[17] A. T. Carpenter and J. J. Scott, *Biochem. Biophys. Acta*, 1961, **52**, 195.

[18] A. R. Battersby, D. A. Evans, K. H. Gibson, E. McDonald, and L. N. Nixon, *J.C.S. Perkin I*, 1973, 1546; A. R. Battersby, J. F. Beck, and E. McDonald, *ibid.*, 1974, 160.

[19] B. Frydman and R. B. Frydman, *Accounts Chem. Res.*, 1975, **8**, 201.

[20] *e.g.* J. Bausch and G. Müller, *Enzyme*, 1974, **17**, 47; J. M. Osgerby, J. Pluscec, Y. C. Kim, F. Boyer, N. Stojanac, H. D. Mah, and S. F. MacDonald, *Canad. J. Chem.*, 1972, **50**, 2652.

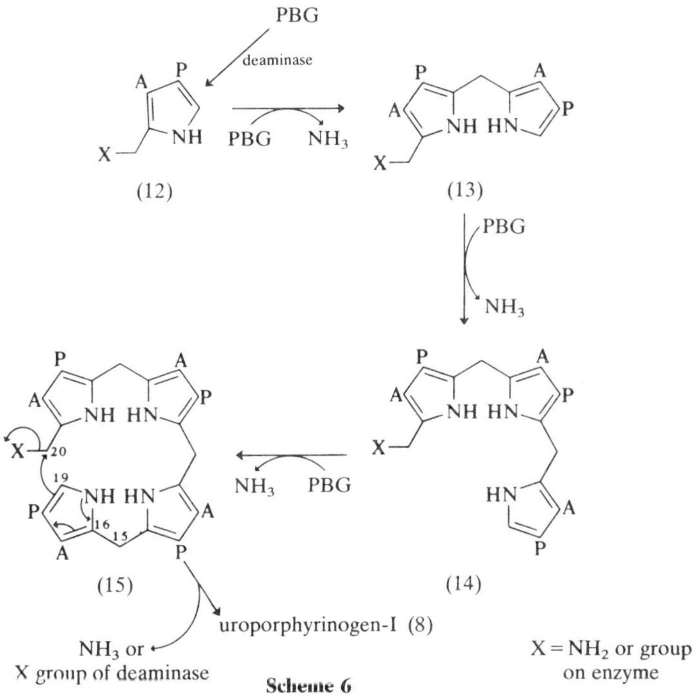

Scheme 6

initial results of such studies[4,5] were confusing, and led to some disagreement between the groups at Buenos Aires,[19,21] Cambridge,[5,22,23] and (later) Yale.[24]

All researches on the enzymic incorporation of PBG and pyrromethanes into porphyrinogens are complicated by concomitant *non-enzymic* conversion of substrates into porphyrinogens (often accompanied by some prior rearrangement),[25] and this results in a troublesome blank. This is a major difficulty encountered by all workers in the area, and undoubtedly is an important factor in the differences in results indicated above. The interested non-expert must remember that a ^{14}C-labelled precursor has not been proved to be incorporated until: (*a*) the precursor is shown to be chemically (*including* isomerically) pure; (*b*) the isolated products are proved to be chemically (*including* isomerically) *and* radiochemically pure; and (*c*) the incorporations observed are proved to be *specific* and the site(s) of labelling are determined.

Recent results from the Cambridge group[26] with both ^{13}C- and ^{14}C-labelled aminomethylpyrromethanes, and with ^{13}C-labelled bilanes (see later), have

[21] B. Frydman, R. B. Frydman, A. Valasinas, E. S. Levy, and G. Feinstein, in ref. 4, pp. 137–160.
[22] A. R. Battersby, D. A. Evans, K. H. Gibson, E. McDonald, L. N. Mander, and J. Moron, *J.C.S. Chem. Comm.*, 1973, 768.
[23] A. R. Battersby and E. McDonald, in ref. 4, pp. 161–180.
[24] A. I. Scott, K. S. Ho, M. Kajiwara, and T. Takahashi, *J. Amer. Chem. Soc.*, 1976, **98**, 1589.
[25] G. W. Kenner, A. H. Jackson, and D. Warburton, *J. Chem. Soc.*, 1965, 1328.
[26] (*a*) Reviewed by A. R. Battersby in the Paul Karrer Lecture, University of Zurich (July 1977), *Experientia*, 1978, **34**, 1; (*b*) See also A. R. Battersby and E. McDonald, *Accounts Chem. Res.*, 1978, **11**, in the press.

clarified the role of pyrromethanes in uro'gen-III (3) biosynthesis, and their results are presented first.

Preliminary work[5,23] had suggested that only the pyrromethane AP.AP (16) was involved in uro'gen-III biosynthesis. When the intact enzyme systems from *E. gracilis* or avian erythrocytes were used, radioactivity was found in the isolated protoporphyrin-IX (6) from $[^{14}C]$AP.AP (16) (range 1.6—9.0% with duck erythrocytes and 0.2—0.5% with *E. gracilis*) and from $[^{14}C]$PA.AP (18) (range 0.06—0.12% from duck erythrocytes and 0.01–0.015% from *E. gracilis*); even lower maximum incorporations were obtained from incorporation studies with $[^{14}C]$AP.PA (17) (0.05%) and $[^{14}C]$PA.PA (19) (0.09%). The incorporations of the rearranged pyrromethanes (17), (18), and (19) were so low that the labelling pattern could not be checked by degradation.

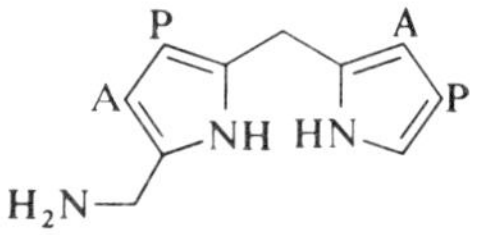

(16) AP.AP

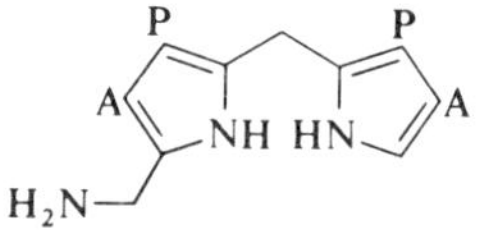

(17) AP.PA

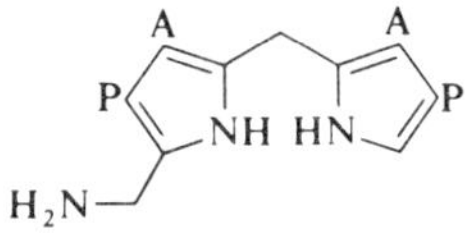

(18) PA.AP

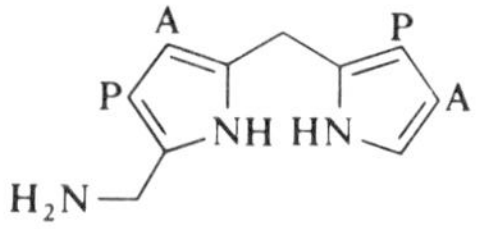

(19) PA.PA

In contrast (Scheme 7), the relatively high level of incorporation of radioactivity from the unrearranged isomer (16a) into the isolated protoporphyrin-IX dimethyl ester (21a) and haemin ester (20a), using the enzyme system from duck erythrocytes, allowed oxidative degradation of the latter to the four biliverdins-IXα, -IXβ, -IXγ, and -IXδ [the IXα isomer (22) is illustrated in Scheme 7].[27] The haemin ester isolated after incubation of $[^{14}C]$AP.AP (16a) with enzymes but without PBG was degraded to the four biliverdins, which showed that the ^{14}C activity of the haemin ester (20) was distributed equally between C-5 and C-15. It followed that the $[^{14}C]$AP.AP (16a) had been *specifically incorporated* into uro'gen-III (3) to label C-5 and C-15 equally.[22,28]

This specific incorporation of AP.AP (16) without PBG into uro'gen-III (3) was also found[22] with the enzymes from *E. gracilis.* The initially formed uro'gen-III (3) was again enzymically converted *in situ* into protoporphyrin-IX (6); incubation of [*bridge-methylene*-^{13}C]AP.AP (16b) led to recovery of [5,15-$^{13}C_2$]protoporphyrin-IX ester (21b) whereas [*aminomethylene*-^{13}C]AP.AP (16c) gave [10,20-$^{13}C_2$]protoporphyrin-IX ester (21c), as in Scheme 7. The sites of labelling were read directly from the ^{13}C n.m.r. spectra, using the earlier unambiguous assignments.[14]

The Frydmans have suggested[29] that the above incorporations with AP.AP (16) may not be leading to protoporphyrin-IX (6) as described but to an unnatural

[27] R. Bonnett and A. F. McDonagh, *Chem. Comm.*, 1970, 237.
[28] A. R. Battersby, Special Lecture, *23rd Internat. Congr. Pure Appl. Chem.*, 1971, **5**, 1.
[29] R. B. Frydman and B. Frydman, *F.E.B.S. Letters*, 1975, **52**, 317.

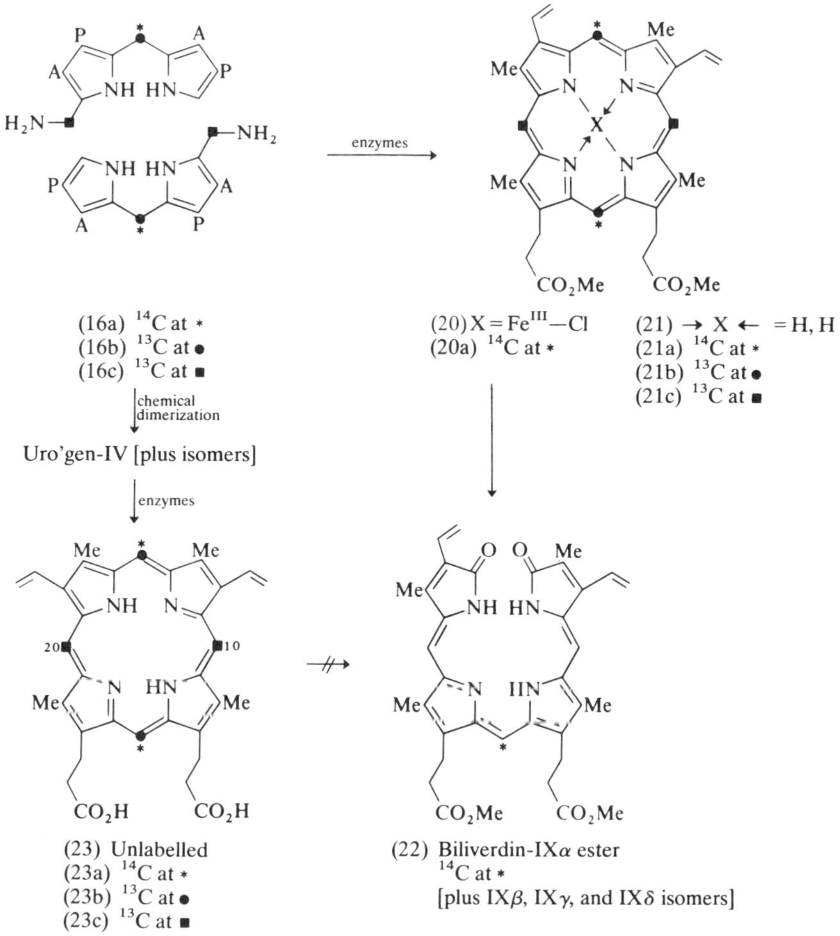

Scheme 7

isomer, protoporphyrin-XIII (23). Their view was based on the known *chemical* formation (by rearrangement) of some uro'gen-IV from AP.AP (16),[22,29] which could then be converted enzymically *via* copro'gen-IV into protoporphyrin-XIII (23) (Scheme 7); the terminal enzymes of protoporphyrin-IX (6) biosynthesis (see Scheme 1) are known[5] to act on these unnatural isomers.

This suggestion cannot be correct. It is clear that the product obtained by the Cambridge group is unquestionably protoporphyrin-IX (6) for the following reasons. First, the recently proved structure of protoporphyrin-XIII (23)[30] has a plane of symmetry which renders C-10 and C-20 equivalent [*cf.* (21)]; the *four*

[30] H. M. G. Al-Hazimi, A. H. Jackson, D. J. Ryder, G. H. Elder, and S. G. Smith, *J.C.S. Chem. Comm.*, 1976, 188; L. Mombelli, E. McDonald, and A. R. Battersby, *Tetrahedron Letters*, 1976, 1037; G. Buldain, J. Hurst, R. B. Frydman, and B. Frydman, *J. Org. Chem.*, 1977, **42**, 2953.

different ^{13}C n.m.r. signals observed[22] from the bridge carbon atoms of the labelled biosynthetic products (21b) and (21c) could not have arisen from protoporphyrin-XIII (23). Second, the haemin ester derived from a protoporphyrin of structure (23) could not be degraded[27] to any of the known biliverdin-IX isomers obtained[22,28] on degradation of the protoporphyrin recovered from the incubation experiments (see Scheme 7). Third, protoporphyrin-IX and protoporphyrin-XIII dimethyl esters are separable by h.p.l.c., and the product from the above enzymic experiments was distinguished in this way from isomer XIII and identified with isomer IX.[26b]

These experiments prove conclusively that two molecules of unrearranged AP.AP (16) can combine and rearrange in a highly specific manner to generate uro'gen-III (3) in the presence of the deaminase–cosynthetase system. The problem of determining the extent of chemical *versus* enzymic formation of uro'gen-III (3) in the incubation experiment[26] was solved by the development of methods for isolating the derived uroporphyrins from highly reactive pyrromethanes[31] and for quantitative separation[32] of the recovered porphyrin isomers. The isolated mixture of uroporphyrin esters was decarboxylated with hot acid[33] to the corresponding mixture of coproporphyrin isomers; a two-stage h.p.l.c. separation of the tetramethyl esters allowed the four coproporphyrin isomers to be collected and assayed.[32]

Incubation[31] of [^{14}C]AP.AP (16a) at pH 7.2 for 16 h *without enzyme* gave a *ca.* 30% yield of uroporphyrins having the isomer composition shown in Table 1. When a strictly parallel experiment was run, but with the addition of purified deaminase–cosynthetase from *E. gracilis*, the isomer ratio was dramatically altered (see Table 1), type-III formation being vastly increased (3%→54%) at the expense of type-I formation (68%→15%). The competing chemical formation of the type-IV isomer was essentially unaffected by deaminase–cosynthetase.

Table 1 *Radioactivity of the uro'gen isomers formed from AP.AP* (16*a*)

	Uro'gens formed (% *of total*)			
	Type-I	*Type-II*	*Type-III*	*Type-IV*
Blank run	68 ± 2	1 ± 0.5	3 ± 2	28 ± 2
Enzyme run	15 ± 1	2 ± 1	54 ± 2	29 ± 2

In addition, the doubly $^{13}C_2$-labelled form of AP.AP (16d), synthesized[18,20] from [2,11-$^{13}C_2$]PBG (9), was incubated[34] with purified deaminase–cosynthetase and then worked up as in the ^{14}C series[31] to give coproporphyrin-III tetramethyl ester. The ^{13}C n.m.r. signal that arises unambiguously from C-15 was a 72 Hz doublet, centred on a smaller singlet. It followed that the labelling pattern around C-15 was that shown in Scheme 8 for uro'gen-III [(3d) and (3e)], and the signal from C-5 was consistent with the illustrated arrangement around ring B. The ratio of split to

[31] A. R. Battersby, D. G. Buckley, E. McDonald, and D. C. Williams, *J.C.S. Chem. Comm.*, 1977, 115.
[32] A. R. Battersby, D. G. Buckley, G. L. Hodgson, R. E. Markwell, and E. McDonald, in 'High Pressure Liquid Chromatography in Clinical Chemistry', ed. P. F. Dixon, C. H. Gray, C. K. Lim, and M. S. Stoll, Academic Press, London, 1976, pp. 63–70.
[33] Ref. 2, p. 825.
[34] A. R. Battersby, D. W. Johnson, E. McDonald, and D. C. Williams, *J.C.S. Chem. Comm.*, 1977, 117.

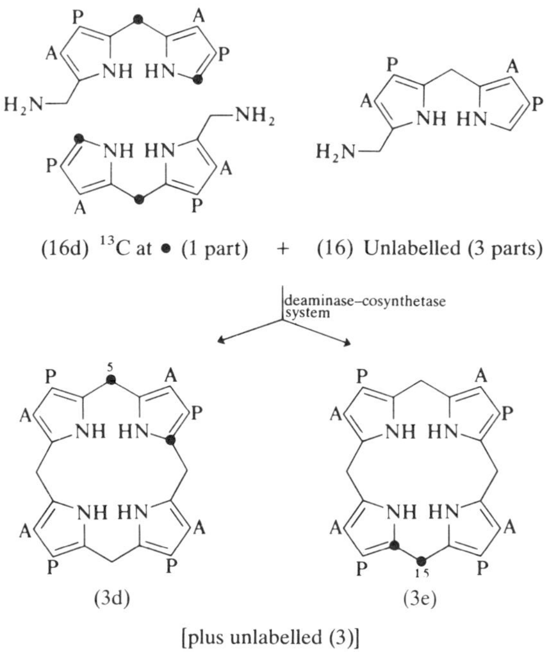

Scheme 8

unsplit signals from C-15 was in agreement with an *intramolecular* rearrangement of ring D, *exactly as had been found earlier for PBG itself.*[11,12]

These results show clearly that deaminase–cosynthetase *does* catalyse the formation of uro'gen-III (3) when AP.AP (16) is provided, and further, that the enzymic conversion involves an intramolecular rearrangement. Nevertheless, they do not prove that free AP.AP (16) is a biosynthetic precursor which is normally dimerized by the enzymes. Eventually it became clear[26] that two molecules of AP.AP (16) might be reacting *chemically* to produce the unrearranged tetrapyrrole system (26), which, as it is formed, could be transformed *enzymically* into uro'gen-III, as in Scheme 9 (see below).

At this stage it is appropriate to consider the results of other workers. In their work with pyrromethanes the Frydmans[19,35] employed enzymes from wheatgerm, and used paper chromatography to separate the isomeric porphyrins prior to radioassay. They found that AP.AP (16), in the presence of PBG, was converted (*ca.* 1%) into uro'gen-I (8) by deaminase alone, but they did not observe incorporation of AP.AP (16) into uro'gen-III (3) with deaminase–cosynthetase when they used a short incubation (1 h). In the light of the results quoted previously, it seems probable that the decisive differences between the experiments of the groups at Cambridge and Buenos Aires were the duration of the incubation and

[35] B. Frydman, S. Reil, A. Valasinas, R. B. Frydman, and H. Rapoport, *J. Amer. Chem. Soc.*, 1971, **93**, 2738; R. B. Frydman, A. Valasinas, and B. Frydman, *Biochemistry*, 1973, **12**, 80; R. B. Frydman, A. Valasinas, H. Rapoport, and B. Frydman, *F.E.B.S. Letters*, 1972, **25**, 309.

the concentration of pyrromethane. When the rearranged [^{14}C]PA.AP [labelled (18)] was incubated with deaminase–cosynthetase, together with PBG, radioactivity (*ca.* 0.7%) appeared in the uro'gen-III (3).

Later, Scott[24] incubated the 'headless' pyrromethane (24) with PBG and deaminase–cosynthetase and found that there was 0.15—0.89% yield of radioactivity in the type-III porphyrins. Both sets of results were thought to prove that the

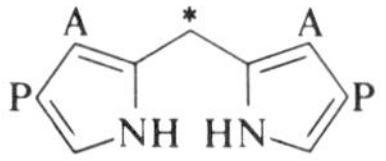

(24) ^{14}C at *

synthesis of type-III porphyrin is controlled at the pyrromethane level. However, the evidence given above (and some of what follows) is so overwhelming that this cannot be correct. The small transfers of radioactivity from rearranged pyrromethanes into uro'gen-III (3) could reasonably be explained as arising from various well-known minor *chemical* side-reactions.[25] The amounts of radioactivity found in the type-III porphyrins in both of these studies with rearranged pyrromethanes are of the same order (< 1%) as were found by the Cambridge group[23] for the three rearranged aminomethylpyrromethanes (17), (18), and (19). Furthermore, none of the foregoing weakly radioactive type-III porphyrins from experiments with rearranged pyrromethanes has been degraded, so as to locate the labels.*

Proof that Rearrangement Occurs at the Tetrapyrrole Level.—Groups at Stuttgart[36] and Cambridge[37] have reported the involvement of the bilane (26) in uro'gen-III biosynthesis. Müller *et al.*[36] prepared the bilane (26) and reported that it was transformed into a uro'gen mixture having a type-I : type-III ratio of 84 : 16 by deaminase–cosynthetase from *P. shermanii.* From a different synthesis, Battersby and McDonald *et al.*[37] prepared the crystalline lactam ester (25), which was hydrolysed to the same bilane (26), corresponding to unrearranged, head-to-tail joining of four PBG units. This cyclized chemically (without enzyme) at pH 7.2 to give virtually pure (>95%) uro'gen-I (8), isolated as uroporphyrin-I and analysed by h.p.l.c. as the coproporphyrin isomer, as described previously.[32] However, incubation of the bilane (26) with purified deaminase–cosynthetase gave a markedly different result: the products after aromatization were uroporphyrin-III (70%), derived from (3), and uroporphyrin-I (30%), derived from (8).

The differences in the isomer ratios from the two experiments were probably due to the different concentrations of purified enzymes used,[36,37] but both sets of results strongly suggested that the unrearranged bilane (26) is a key precursor in the biosynthesis of uro'gen-III (3). The Cambridge group have now established beyond doubt that this is the case.[26]

* The Reporter has attempted to set out clearly the relevant work in this area, but must declare his interest in having been involved in the Cambridge effort.

[36] H.-O. Dauner, G. Gunzer, I. Heger, and G. Müller, *Z. physiol. Chem.*, 1976, **357**, 147.
[37] A. R. Battersby, E. McDonald, D. C. Williams, and H. K. W. Wurziger, *J.C.S. Chem. Comm.*, 1977, 113.

[15-^{13}C]Bilane (26a)[37] was converted into a product consisting of 80% uro'gen-III (3a) and 20% uro'gen-I (8a) by the deaminase–cosynthetase system, as shown by h.p.l.c. analysis of the derived coproporphyrin esters. The crystalline coproporphyrin-III ester was shown by ^{13}C n.m.r. to be labelled specifically at C-15 by comparison with the spectrum of unambiguously synthesized [15-^{13}C]coproporphyrin-III tetramethyl ester. This result[37] established *specific* incorporation of the bilane (26a) into uro'gen-III (3a). The two [$^{13}C_2$]bilanes (26b) and (26c) were prepared[38] in such a way that the enrichment was 90 atom% at each labelled site, and so 81% of the molecules of each bilane sample were doubly labelled, as illustrated in Scheme 9. After dilution with unlabelled bilane the two mixtures were incubated separately with purified deaminase–cosynthetase. The locations of labels in the resultant samples of uro'gen-III (3b) and (3c) were shown to be as illustrated in Scheme 9 by the usual isolation, decarboxylation, and ^{13}C n.m.r. analysis; [20-^{13}C]coproporphyrin-III tetramethyl ester was synthesized, to allow

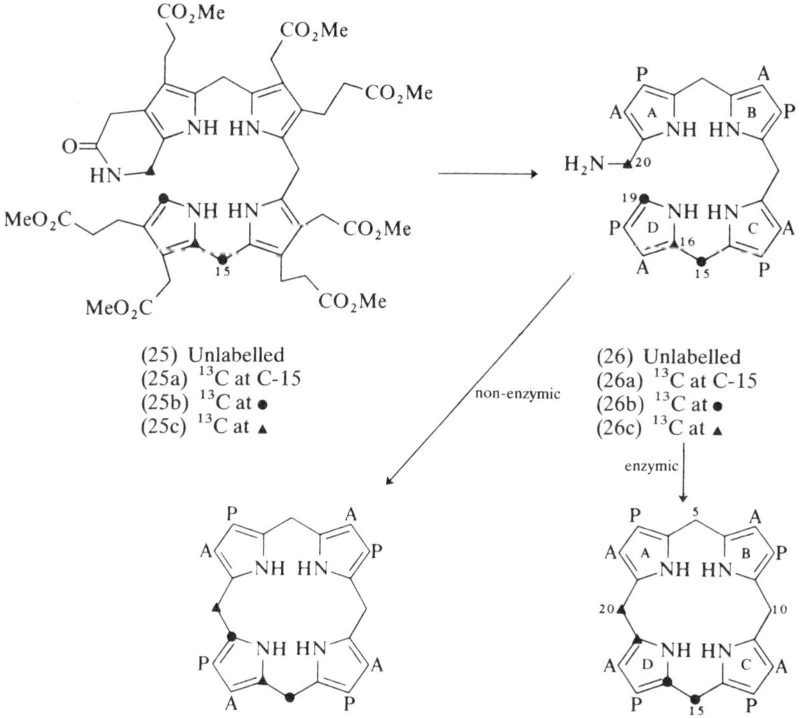

(25) Unlabelled
(25a) ^{13}C at C-15
(25b) ^{13}C at ●
(25c) ^{13}C at ▲

(26) Unlabelled
(26a) ^{13}C at C-15
(26b) ^{13}C at ●
(26c) ^{13}C at ▲

(8) Unlabelled
(8a) ^{13}C at one bridge
(8b) ^{13}C at ●
(8c) ^{13}C at ▲

(3) Unlabelled
(3a) ^{13}C at C-15
(3b) ^{13}C at ●
(3c) ^{13}C at ▲

Scheme 9

[38] A. R. Battersby, C. J. R. Fookes, E. McDonald, and M. J. Meegan, *J.C.S. Chem. Comm.*, 1978, 185.

the assignment of C-20 to be confirmed.[26] Most importantly, both ^{13}C n.m.r. spectra showed a *72 Hz doublet* for the major resonance from the labelled bridge carbon atom;[38] the two ^{13}C atoms in each case have therefore become directly bonded, as illustrated for (3b) and (3c), *via an intermolecular reaction.*

The foregoing combined results prove that intact incorporation of unrearranged bilane (26) *into uro'gen-III* (3) *occurs with inversion of ring* D *by an intramolecular process, exactly as was found earlier*[11,12] *for PBG.* Thus it is now certain that the biosynthesis of the natural type-III porphyrins, chlorins, and corrins involves the following steps. Four PBG units are joined together head-to-tail, and the resultant bilane, bound covalently or by physical forces to the enzyme system, is then converted into uro'gen-III (3) by an intramolecular rearrangement which directly affects only ring D and the two carbon atoms which become C-15 and C-20.[26b]

Battersby and McDonald have suggested[26] that, in the absence of cosynthetase, deaminase catalyses the cyclization of the bilane to join C-20 and C-19 to give uro'gen-I (8) as shown in Scheme 6, but that in the presence of cosynthetase the combined enzyme system causes cyclization between C-20 and C-16 rather than C-20 and C-19 (Scheme 10). This postulated attack would produce the spiro-intermediate (27); the labelling arising from (25b) and (25c) is shown. Fragmentation of the C-15—C-16 bond, followed by cyclization as shown, would generate uro'gen-III (3) (Scheme 10). The spiro-system (27) is related to that proposed by Mathewson and Corwin[39] in 1961, and its intermediacy is consistent with all the evidence available.

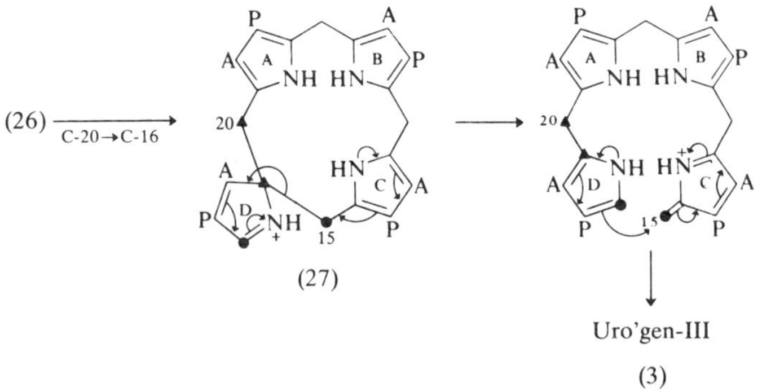

Scheme 10

4 The Pathway from Uro'gen-III to Protoporphyrin-IX

Extensive studies[5] over the past 25 years had indicated that the enzymic decarboxylation of the four acetic acid side-chains of uro'gen-III (3) to give copro'gen-III (4) is a stepwise process, and intermediates with seven, six, and five carboxy-groups had been detected in several laboratories. However, it seemed that a single enzyme (or a highly organized enzyme complex) is able to effect the four successive decarboxylations, although it must have a rather low substrate specificity.

[39] J. H. Mathewson and A. H. Corwin, *J. Amer. Chem. Soc.*, 1961, **83**, 135.

Pure heptacarboxylic acid porphyrins have been isolated from various sources, including haemolysed avian erythrocytes,[40,41] the urine of patients with porphyria,[42] and rat faeces.[42] It now seems very probable that these porphyrins are identical, and they have been variously named phyriaporphyrin, porphyrin-208, and pseudouroporphyrin. From a series of spectroscopic analyses and by synthesis of three of the four possible isomers, Battersby and McDonald showed that the heptacarboxylic acid porphyrin isolated from chicken erythrocytes had the structure (28);[40] this assignment was confirmed subsequently by synthesis,[23,41] using the pyrromethene method.[43]

The heptacarboxylic acid porphyrin isolated from the faeces of rats poisoned with hexachlorobenzene was shown by Jackson *et al.*[42] to have the same structure (28) by synthesis, in this case using the *b*-oxobilane route.[44] This work was part of the Cardiff group's investigation of the porphyrins excreted during both normal and abnormal metabolism,[42] details of which are given in the following two sections.

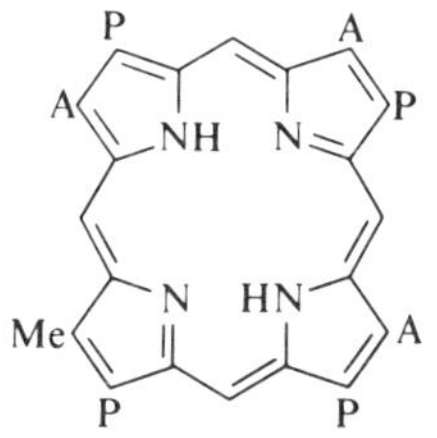

(28) Phyriaporphyrin-III

Intermediates between Uro'gen-III and Copro'gen-III.—Four heptacarboxylic, six hexacarboxylic, and four pentacarboxylic acid porphyrins might be formed by successive decarboxylations of the acetic acid side-chains during the conversion of uro'gen-III (3) into copro'gen-III (4), and a major aim of the Cardiff group was to discover whether or not there was a specific (or preferred) route and whether this was the same in both normal and abnormal metabolism.[42] After extensive preliminary work on isolation procedures and analytical techniques it was possible to carry out preparative-scale separations of the porphyrins excreted in the faeces of rats that had been poisoned with hexachlorobenzene. In this way 10—20 mg of each of the octa-, hepta-, hexa-, and penta-carboxylic fractions were obtained, all of which were shown to be type-III porphyrins by decarboxylation to coproporphyrin-III. The octacarboxylic acid was readily shown to be uroporphyrin-III; the other fractions were essentially homogeneous, and their structures were then deduced and confirmed by synthesis.

The heptacarboxylic acid had the structure (28) (see above), which is now generally termed phyriaporphyrin-III.[41,42] Of the six possible isomeric hexacar-

[40] A. R. Battersby, E. Hunt, M. Ihara, E. McDonald, J. B. Paine, III, F. Satoh, and J. Saunders, *J.C.S. Chem. Comm.*, 1974, 994.
[41] A. R. Battersby, E. Hunt, E. McDonald, J. B. Paine, III, and J. Saunders, *J.C.S. Perkin I*, 1976, 1008.
[42] A. H. Jackson, H. A. Sancovich, A. M. Ferramola, N. Evans, D. E. Games, S. A. Matlin, G. H. Elder, and S. G. Smith, in ref. 4, pp. 191–206.
[43] R. L. N. Harris, A. W. Johnson, and I. T. Kay, *J. Chem. Soc. (C)*, 1966, 22.
[44] *cf.* A. H. Jackson and K. M. Smith in 'The Total Synthesis of Natural Products', ed. J. Apsimon, Wiley, London and New York, 1974, p. 144.

boxylic acid porphyrins which can be derived from uro'gen-III (3), only structure (29) was compatible with the findings of a detailed spectroscopic analysis. This assignment was confirmed[42] by synthesis of the hexamethyl ester by the MacDonald route,[44] and by comparison with the ester derived from the natural compound. The pentacarboxylic acid was deduced to have the structure (30) by a combination of spectroscopic and biosynthetic arguments; other porphyrins derivable from structure (30) had been identified earlier.[45] Synthesis of the pentamethyl ester of (30) and comparison with the ester of the natural compound confirmed this assignment.

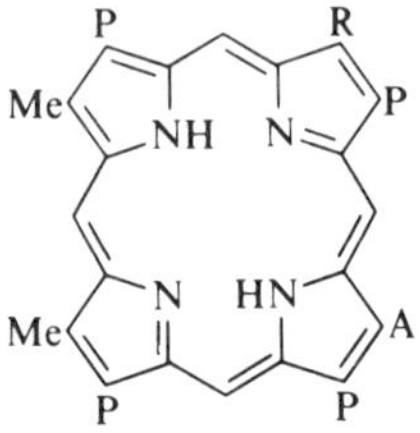

(29) R = A
(30) R = Me

Most of the possible isomeric penta-, hexa-, and hepta-carboxylic acid porphyrins have been synthesized in Cardiff,[42] and some have been synthesized independently by other groups.[41,46] Biosynthetic studies with the porphyrinogens derived from the isomeric penta-, hexa-, and hepta-carboxylic acid porphyrins[42] showed that the porphyrinogens corresponding to the isolated porphyrins (28), (29), and (30) were metabolized at least as fast as those from the synthetic isomers. The fact that each of the isolated porphyrins was homogeneous strongly suggests that the parent porphyrinogens did not arise by a diversion from the normal metabolic pathway; the low specificity of the enzyme suggests that this would be most unlikely.

All these results,[42] combined with the fact[41,42] that the heptacarboxylic acid porphyrins from haemolysed chicken erythrocytes and from humans with porphyria are the same as that from rat faeces, strongly suggest that the porphyrinogens (31), (32), and (33) are the major intermediates in both the normal and abnormal metabolic conversion of uro'gen-III (3) into copro'gen-III (4). The major effect of hexachlorobenzene, which is well known to induce porphyria in animals and humans,[47] seems to be simply to slow down the overall rate of decarboxylation by reducing the activity of the enzyme. Thus, the biosynthesis of copro'gen-III (4) from uro'gen-III (3) does involve a series of discrete steps, as predicted by previous investigators,[5,47] and the preferred (and possibly specific) pathway is shown in Scheme 11.

One striking feature of the preferred pathway [(3) → (31) → (32) → (33) → (4)] is that decarboxylation occurs in a clockwise fashion, starting with the acetic acid side-chain on ring D of uro'gen-III (3) and proceeding by successive decarboxyl-

[45] M. S. Stoll, G. H. Elder, D. E. Games, P. O'Hanlon, D. S. Millington, and A. H. Jackson, *Biochem. J.*, 1973, **131**, 429.
[46] P. S. Clezy, T. T. Hai, and P. C. Gupta, *Austral. J. Chem.*, 1976, **29**, 393.
[47] Reviewed briefly in ref. 42.

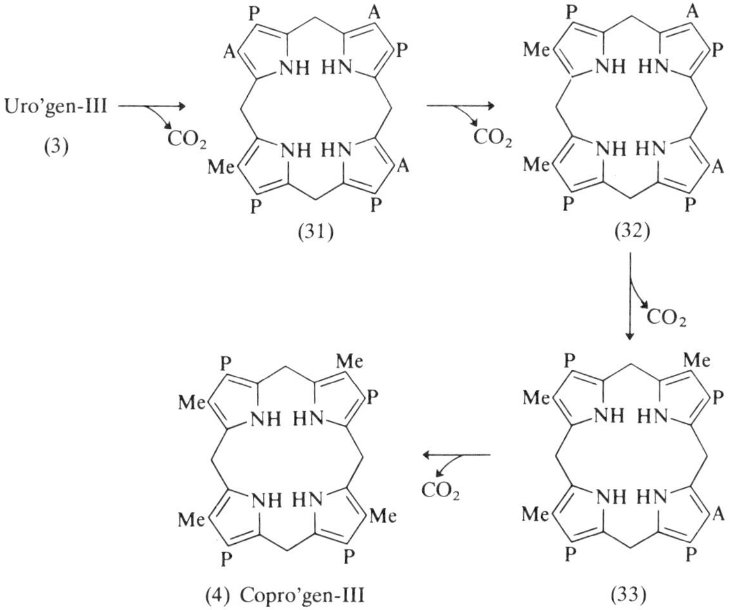

Scheme 11

ations of the acetic acid residues on rings A, B, and C to form copro'gen-III (4) (Scheme 11). It has been suggested that the uro'gen-III (3) in effect 'performs a cartwheel' on the enzyme surface as the side-chains are decarboxylated.[42] However, this intriguing suggestion is largely speculative, and as yet the nature of the enzyme (or group of enzymes) involved is undefined.

In contrast to the above results,[41,42] which show that decarboxylation of the acetic acid side-chains of uro'gen-III (3) starts with ring D and finishes with ring C during its converion into copro'gen-III (4), only the acetic acid group on ring C of uro'gen-III (3) is decarboxylated during the formation of the corrin nucleus of vitamin B_{12}, *cf.* (2). Thus, the pathway to the corrins diverges from that to the chlorins and natural porphyrins immediately after the formation of uro'gen-III (3).[48]

Akhtar's group[49] have defined the stereochemistry of the decarboxylation of the acetate side-chains of the C and D rings of uro'gen-III. (2*R*)-[2-$^2H_1$,2-$^3H_1$]Succinic acid (34)[49] was incubated with a haemolysed chicken erythrocyte preparation and was incorporated into the acetate residues of uro'gen-III (35) (Scheme 12); this, on decarboxylation, generated chiral methyl groups in copro'gen-III (36) and then in haemin (37). The labelled haemin (37) was degraded[5] to ethyl methyl maleimide and haematinic acid (38), which are derived from rings A and B, and rings C and D, respectively. Ozonolysis of (38) gave labelled acetic acid (39), which was converted

[48] For recent reviews, see (*a*) ref. 26*a*, and (*b*) A. I. Scott, *Accounts Chem. Res.*, 1978, **11**, 29.
[49] G. F. Barnard and M. Akhtar, *J.C.S. Chem. Comm.*, 1975, 494.

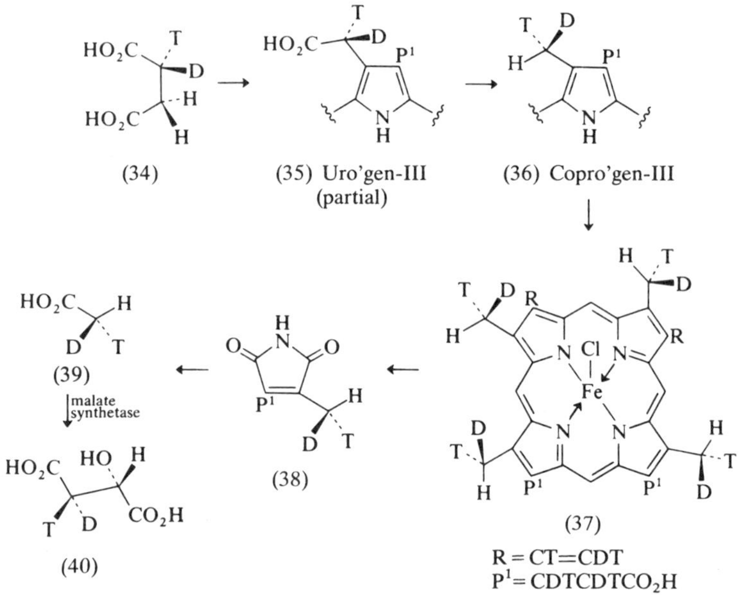

Scheme 12

into labelled malic acid (40).[50a] Incubation of (40) with fumarase and analysis of the results, using the method[50] of Cornforth and Arigoni, showed that the labelled acetic acid (39) had the (*S*) configuration, as illustrated in Scheme 12. Thus the decarboxylation reaction converting the uro'gen-III (3) into copro'gen-III (4) must have occurred as in (35)→(36) (Scheme 12), *i.e.* with retention of configuration.

It is highly likely that this result holds good for decarboxylation of all four acetic acid side-chains, although the above result only refers to acetic acid residues of rings C and D. Akhtar suggests[49] that the mechanism of the decarboxylation reaction involves enzymic protonation of an intermediate such as (35b), formed by enzyme-catalysed decarboxylation of the enzymically protonated porphyrinogen (35a) (Scheme 13).

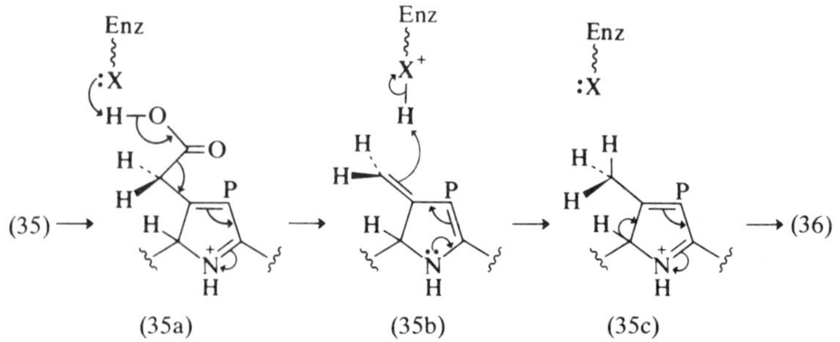

Scheme 13

[50] (*a*) A. I. Rose, *J. Biol. Chem.*, 1970, **245**, 6052; (*b*) J. W. Cornforth, J. W. Redmond, H. Eggerer, W. Buckel, and C. Gutschow, *Nature*, 1969, **221**, 1212; J. Luthy, J. Rétey, and D. Arigoni, *ibid.*, p. 1213.

Biosynthesis of Protoporphyrin-IX from Copro'gen-III.—As was noted earlier, the enzymes involved in the transformation of uro'gen-III (3) into protoporphyrin-IX (6) are not highly specific,[5] and thus results from the testing of putative precursors must be interpreted carefully. Initial studies[5,51] had suggested that hardero'gen-III (41) was a natural intermediate between copro'gen-III (4) and proto'gen-IX (5), although (41) was incorporated only 6—10 times more efficiently than its isomer (42) using an enzyme preparation from *E. gracilis*, and the absolute incorporations were low.

A recent investigation by the Cardiff group has established[52] that hardero'gen-III (41) is indeed an intermediate on the pathway to proto'gen-IX (5). The isomeric porphyrins (43) and (44) were prepared labelled with tritium in the *meso*-positions and were reduced with sodium amalgam to the corresponding porphyrinogens (41) and (42). The latter were incubated separately with haemolysates of mature chicken erythrocytes, and after work-up [which included dilution with unlabelled (6)], the isolated protoporphyrin-IX dimethyl ester was purified by h.p.l.c. The specific incorporations of (41) and (42) were found to be 34 and 0.8% respectively, but, after allowing for the stereospecificity of the later enzyme-mediated aromatization reaction (5)→(6),[53] it is clear that the absolute incorporations of hardero'gen-III (41) and its isomer (42) must have been *ca.* 70 and 1.5% respectively.

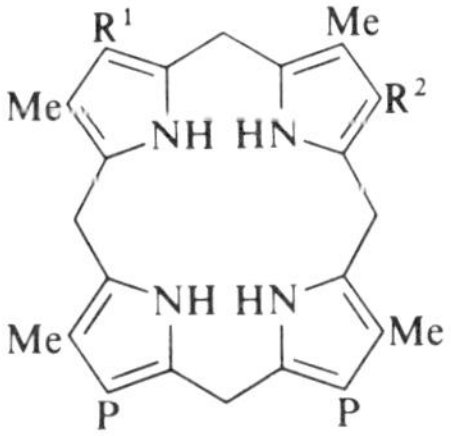

(4) $R^1 = R^2 = P$
(5) $R^1 = R^2 = CH{=}CH_2$
(41) $R^1 = CH{=}CH_2, R^2 = P$
(42) $R^1 = P, R^2 = CH{=}CH_2$

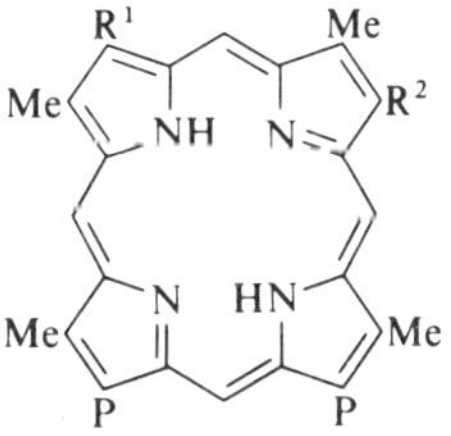

(6) $R^1 = R^2 = CH{=}CH_2$
(43) $R^1 = CH{=}CH_2, R^2 = P$
(44) $R^1 = P, R^2 = CH{=}CH_2$

Further support for the intermediacy of (41) came from a kinetic study of the conversion of (4) into (5). In either the same avian system or in experiments with rat liver homogenates (using h.p.l.c. analysis of the porphyrin esters obtained on work-up), only hardero'gen-III (41) was formed as an intermediate, and none of the isomer (42) could be detected. Thus, the normal biosynthetic route to proto'gen-IX (5) involves formation of the vinyl side-chain on ring A before modification of the propionic acid residue on ring B.

Oxidative Decarboxylation of the Propionic Acid Side-chains. Early work in Cambridge and Southampton showed[5] that when the propionic acid residues on

[51] J. A. S. Cavaliero, G. W. Kenner, and K. M. Smith, *J.C.S. Chem. Comm.*, 1973, 183; *J.C.S. Perkin I*, 1974, 1188.

[52] D. E. Games, A. H. Jackson, J. R. Jackson, R. V. Belcher, and S. G. Smith, *J.C.S. Chem. Comm.*, 1976, 187.

[53] (*a*) A. H. Jackson, D. E. Games, P. Couch, J. R. Jackson, R. V. Belcher, and S. G. Smith, *Enzyme*, 1974, **17**, 81, 88; (*b*) R. Poulson and W. J. Polglase, *J. Biol. Chem.*, 1975, **250**, 1269; (*c*) A. R. Battersby, E. McDonald, J. R. Redfern, J. Staunton, and R. H. Wightman, *J.C.S. Perkin I*, 1976, 266.

both rings A and B of copro'gen-III (4) were oxidatively decarboxylated to give proto'gen-IX (5), only the *pro-S*-hydrogen atom from the methylene group adjacent to the macrocycle was removed, and that each of the two vinyl groups was derived from the remaining three hydrogen and two carbon atoms without rearrangement.

The original work by Akhtar[54] which had defined the stereospecificity of hydrogen removal was based on incubation studies with labelled succinic acid. Recent work by the Cambridge group, which has confirmed the previous findings, was based on incubation of PBG stereospecifically labelled at C-8. This latter approach circumvented the difficulties encountered in the earlier work due to the incorporation of label into each of the eight side-chains of protoporphyrin-IX (6).

(8*S*)-[8-$^3H_1$]PBG (44) was synthesized[26a] as shown in Scheme 14. The initial hydrogenation step was accomplished using the new chiral homogenous catalyst

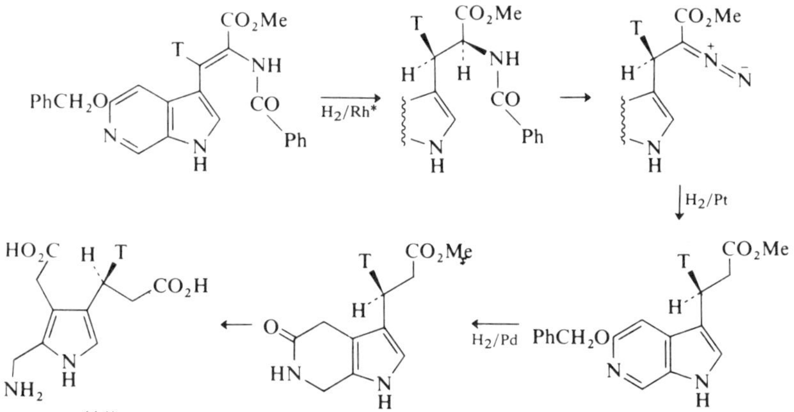

(44)

Scheme 14

developed by Knowles;[55] the stereoselectivity was determined to be as shown by using model compounds, and *ca.* 92% of the derived amido-ester had the (2*S*) configuration. The labelled PBG (44) was mixed with ^{14}C-labelled material to give a $^3H:{}^{14}C$ ratio of 8.7:1. This sample was converted enzymically into protoporphyrin-IX (6a) which had a $^3H:{}^{14}C$ ratio of almost one half of that of the PBG. Further degradation as illustrated in Scheme 15 (after hydrogenation of the vinyl groups) clearly showed that tritium had largely been lost during biosynthesis of the vinyl groups. The values found corresponded closely to those expected (shown in brackets) for *ca.* 92% configurational purity. This shows that it is the *pro*-8*S*-hydrogen atom of PBG which is removed during the formation of the vinyl groups of protoporphyrin-IX (6).

Experiments using PBG specifically deuteriated in the propionic acid side-chain allowed the Cambridge group to determine the overall stereochemistry of the

[54] Z. Zaman, M. M. Abboud, and M. Akhtar, *J.C.S. Chem. Comm.*, 1972, 1263.
[55] W. S. Knowles, M. J. Sabacky, B. D. Vineyard, and D. J. Weinkauff, *J. Amer. Chem. Soc.*, 1975, **97**, 2567.

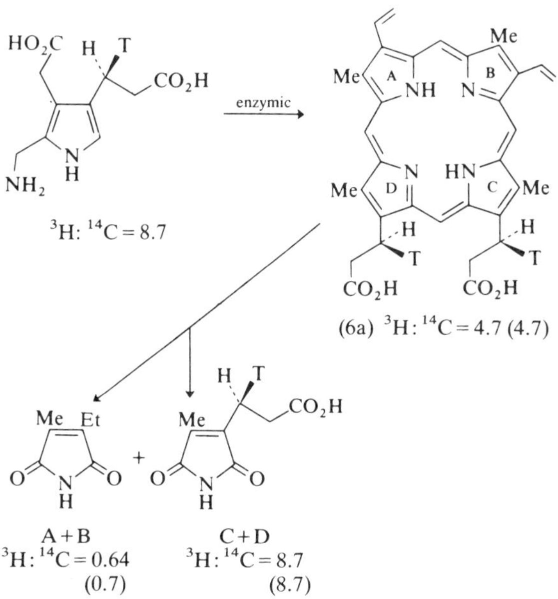

Scheme 15

oxidative decarboxylation.[56] The required labelled PBG was prepared by reduction of the dideuterio-acrylic ester (45) with di-imide, which by its *syn*-stereospecificity fixes the relative configuration at centres X and Y. This arrangement was not affected by the subsequent reactions which produced the racemate (46) + (47). An analysis of the possible outcome of overall *anti*- or *syn*-elimination (irrespective of the precise mechanism) from the molecules (46) and (47) is given in Scheme 16; note that it is the *pro-S*-hydrogen atom which is lost from position X. By using a ^{1}H n.m.r. analysis of the hydrogen atom at centre X in the derived protoporphyrin-IX [structures (48)—(51)], neither product (50) nor (51) from enantiomer (47) will register, and a clear ^{1}H–^{1}H coupling pattern should be observed from products (48) or (49). Under suitable conditions, separate signals could be seen from H_X on each of the vinyl groups of the protoporphyrin-IX dimethyl ester. So this ester, derived enzymically from the labelled PBG [(46) and (47)], was examined by ^{1}H n.m.r. in this way, and two doublets, with *trans*-coupling (18 Hz), were observed from the two H_X [see (48)]. Thus, both vinyl groups of protoporphyrin-IX (6) are formed by an overall *anti*-elimination, and the absolute stereochemistry of this process is as shown in Scheme 17. Also illustrated there is the important reaction which generates the vinyl group of isopentenyl pyrophosphate;[57] the correspondence is striking.[26a]

Of all the mechanisms which have been proposed for formation of the vinyl groups of proto'gen-IX (5), only the two given in Scheme 18 are compatible with

[56] A. R. Battersby, E. McDonald, H. K. W. Wurziger, and (in part) K. J. James, *J.C.S. Chem. Comm.*, 1975, 493.

[57] J. W. Cornforth, R. H. Cornforth, G. Popják, and L. Yengoyan, *J. Biol. Chem.*, 1966, **241**, 3970.

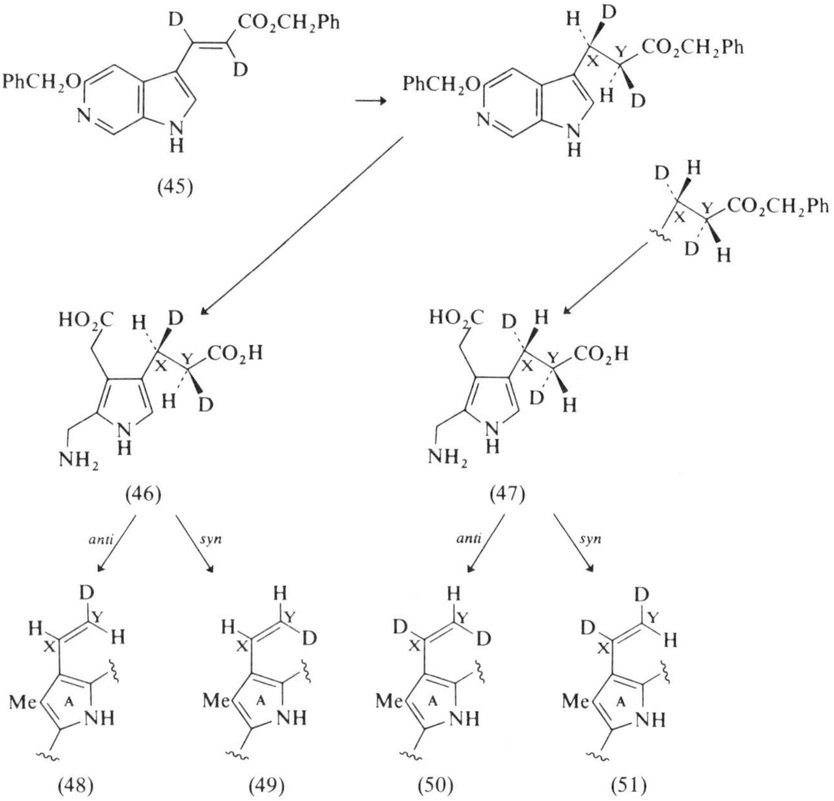

Scheme 16

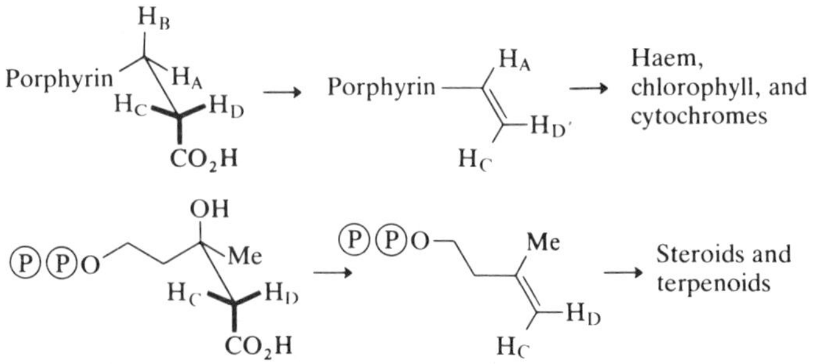

Scheme 17

the above data.* These represent possible mechanisms for the two known types of enzymes, one of which requires oxygen while the other operates in anaerobic conditions. For this latter enzyme, mechanisms involving hydride abstraction have

* Variations of the mechanisms given in Scheme 18 are possible (see ref. 5, pp. 102–104), but those given represent the two main types.

been proposed, but there is no experimental evidence in this area. It has been suggested that the oxygen-requiring enzyme, coproporphyrinogenase, acts by first hydroxylating the propionic acid side-chain and then by catalysing the elimination of carbon dioxide and water; antiperiplanar elimination would require retention of configuration in the hydroxylation step (Schemes 17 and 18).

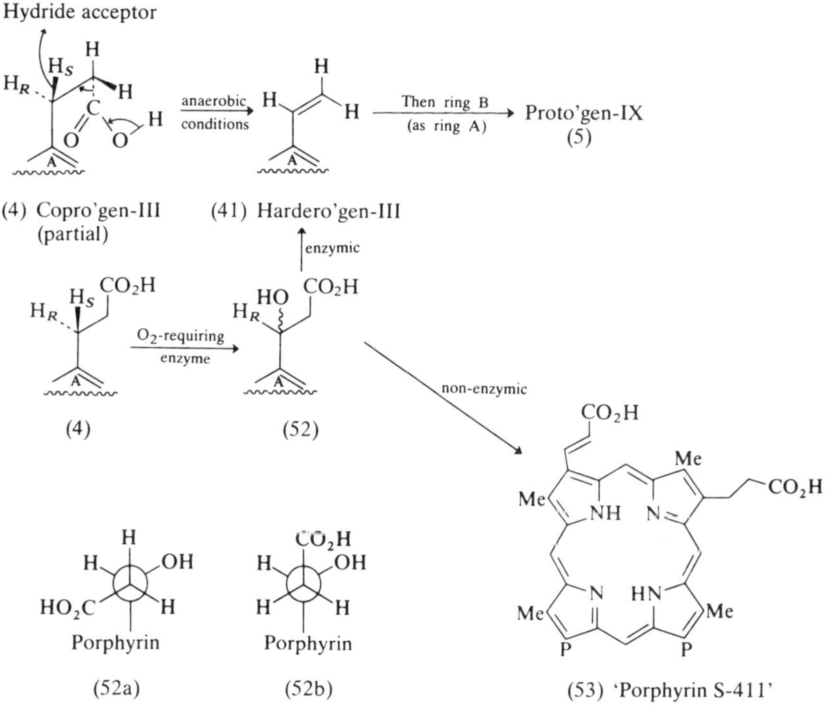

Scheme 18

Incubation studies with derivatives of copro'gen-III (4) hydroxylated in the side-chain of either ring A or ring B were inconclusive.[5] However, indirect evidence for the involvement of the ring-A-hydroxylated porphyrinogen [as (52)] has been presented by both Jackson[42,58] and Clezy.[59] They independently deduced the structure of 'porphyrin S-411' from meconium to be (53), and Jackson suggested that this probably arises by the more favoured *chemical* elimination of water [from (52b)] rather than carbon dioxide and water [from (52a)]. A consideration of the Newman projections (52a) and (52b) indicates strong steric interaction between the porphyrin ring and the carboxy-group in (52a) which is absent in the conformation (52b) required for *anti*-elimination of water to give the *trans*-acrylic acid side-chain of (53). It thus seems that the normal metabolic process requires enzymic assistance to hold the side-chain in the correct conformation for fragmentation to the

[58] P. W. Couch, D. E. Games, and A. H. Jackson, *J.C.S. Perkin I*, 1976, 2492.
[59] I. A. Chaudhry, P. S. Clezy, and V. Diakiw, *Austral. J. Chem.*, 1977, **30**, 879.

vinyl group, whereas the formation of porphyrin S-411 in meconium may be due to non-enzymic loss of water. Alternatively, porphyrin S-411 may be an artefact, produced on work-up from the hydroxypropionate porphyrin (52). In either case, the involvement of the hydroxylated derivative (52) is likely.[58]

Aromatization of Proto'gen-IX (5). Although the autoxidation of porphyrinogens occurs readily in daylight to give the corresponding porphyrins, clear evidence has been obtained for enzymic aromatization of proto'gen-IX (5) *in vivo.*[53] The product, protoporphyrin-IX (6), is the last intermediate common to the biosyntheses of both the haems and the chlorophylls. As a result of the work described above, its biosynthesis has now been defined, and it is of considerable interest that the same pathway is followed in organisms as diverse as bacteria, higher plants, birds, and mammals.

5 The Iron and Magnesium Branches

The Haemoproteins.—An enzyme, ferrochelatase, is known to catalyse the chelation of Fe^{2+} by protoporphyrin-IX (6) to give protohaem (54). Protohaem (54) itself is the prosthetic group of haemoglobin, myoglobin, and the cytochromes-*b*. Protohaem is also the precursor of the cytochromes-*c*, in which the protein is covalently bound to the haem by the addition of an —SH residue to each vinyl side-chain. The cytochromes-*a* have as their prosthetic group a considerably modified protohaem carrying a long hydrocarbon side-chain, and they have been shown to be derived from protohaem (54) in *Staphylococcus.*[5]

The structure of porphyrin C, derived from the cytochrome-*c* of yeast and of horse heart, has been shown to be 3,8-di(*α*-*S*-cysteinylethyl)deuteroporphyrin-IX (55) by a series of degradative, synthetic, and n.m.r. studies.[60] A recent synthesis of haemin C (56) has been achieved[61] from haemin (20), as shown in Scheme 19. The

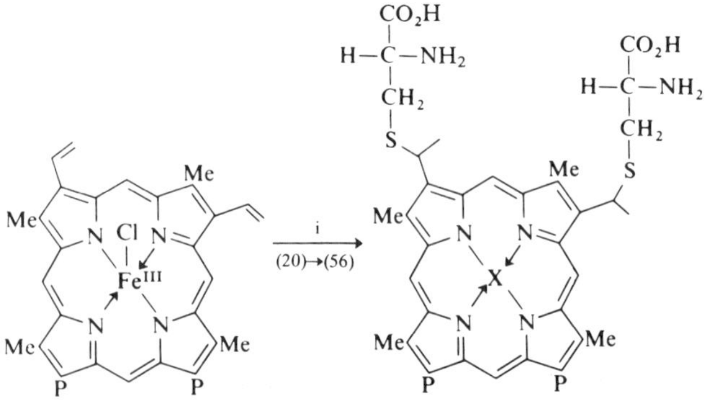

(20) Haemin

(55) → X ← = H, H Porphyrin C
(56) X = Fe^{III}—Cl Haemin C

Reagents: i, Cysteine, n-$C_{16}H_{33}\overset{+}{N}Me_3$ Br^-, room temp., O_2, $NaBH_4$, pH 8.1

Scheme 19

[60] J. T. Slama, H. W. Smith, G. C. Willson, and H. Rapoport, *J. Amer. Chem. Soc.*, 1975, **97**, 6556.
[61] S. Kojo and S. Sano, *J.C.S. Chem. Comm.*, 1977, 249.

iron atom of haemin was found to play an important role in the synthesis, but other observations suggested that the reactive intermediate was not a simple Fe^{II}-protoporphyrin [as (54)].

Cytochrome-*c* oxidase is a key respiratory enzyme, and its prosthetic group(s) have attracted many studies since Warburg's isolation[62] of the first crude metalloporphyrin preparation called haem A. A block to further progress in this area had been the difficulty of demonstrating homogeneity for haem A preparations or, more importantly, of isolating the corresponding metal-free porphyrin(s) A. New work by groups in Australia and Cambridge has shown[63] that the 'porphyrin A' dimethyl ester from beef heart (whether extracted[64] under acidic or basic conditions) was a mixture of two similar compounds in roughly equal amounts. Investigation of the more polar compound has not been reported, but the crystalline, less polar ester (designated porphyrin A dimethyl ester) has been shown to have the structure (57).[63] This confirms that the structure of haem A is (58), as was suggested earlier.[64] The synthesis[63] of porphyrin A dimethyl ester (57) was part of Clezy's investigation of biologically significant porphyrins, much of which has been reported recently.[65]

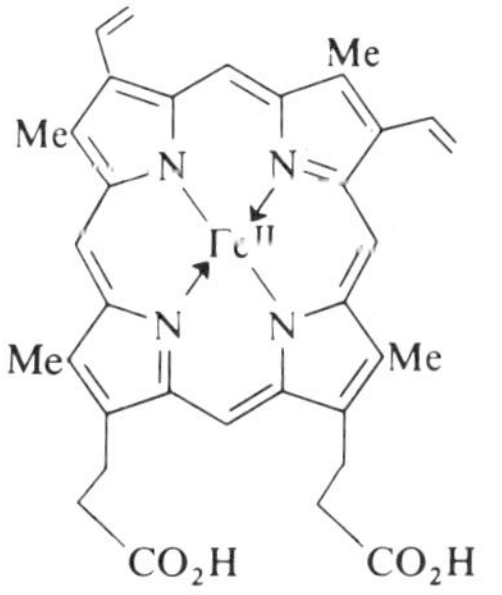

(54) Protohaem

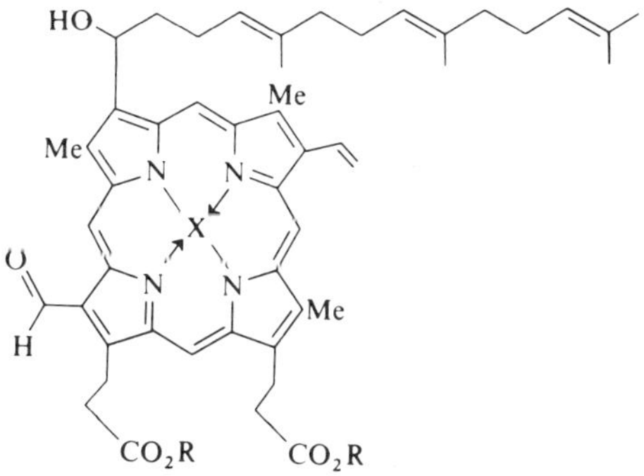

(57) R = Me, →X← = H,H
(58) R = H, X = Fe^{II} Haem A

The Chlorophylls.—Research on the late stages of the biosynthesis of chlorophylls has been hindered by the insolubility of intermediates and relevant enzymes. The pathway first outlined by Granick[66] in 1951 has been generally supported by the numerous investigations since then,[5] and recent work by Jones[67] has confirmed that the steps between protoporphyrin-IX (6) and chlorophyllide-*a* (63) in higher plants are as given in Scheme 20.

The enzyme which catalyses the conversion of chlorophyllide-*a* (63) into chlorophyll-*a* (64) has not been characterized. However, chlorophyllase, the

[62] O. Warburg and H.-S. Gewitz, *Z. physiol. Chem.*, 1951, **288**, 1.
[63] M. Thompson, J. Barrett, E. McDonald, A. R. Battersby, C. J. R. Fookes, I. A. Chaudhry, P. S. Clezy, and H. R. Morris, *J.C.S. Chem. Comm.*, 1977, 278.
[64] See ref. 63, and refs. 2–4 cited therein.
[65] See P. S. Clezy and V. Diakiw, *Austral. J. Chem.*, 1975, **28**, 2703; P. S. Clezy and C. J. R. Fookes, *ibid.*, 1977, **30**, 1799; and refs. therein.
[66] S. Granick, *Ann. Rev. Plant Physiol.*, 1951, **2**, 115.
[67] O. T. G. Jones, in ref. 4, pp. 207–225.

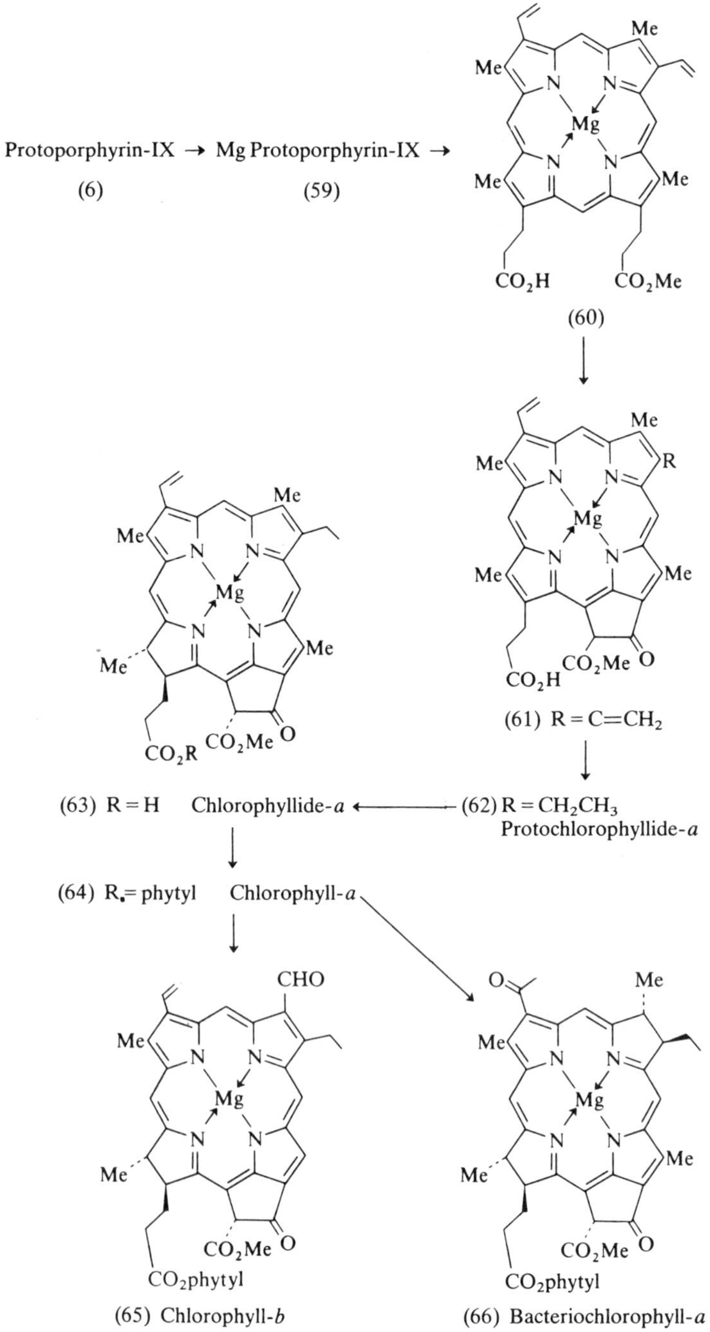

Protoporphyrin-IX → Mg Protoporphyrin-IX →
(6)
(59)
Me
Me
Mg
Me
Me
CO2H
CO2Me
(60)
Me
Me
R
Mg
Me
Me
CO2H
CO2Me
O
(61) R = C=CH2
Me
Me
Mg
Me
Me
CO2R
CO2Me
O
(63) R = H Chlorophyllide-a
(62) R = CH2CH3
Protochlorophyllide-a
(64) R = phytyl Chlorophyll-a
CHO
Me
Mg
Me
CO2Me
O
CO2phytyl
(65) Chlorophyll-b
O
Me
Me
Mg
Me
Me
CO2Me
O
CO2phytyl
(66) Bacteriochlorophyll-a

Scheme 20

enzyme which catalyses the reverse (hydrolysis) reaction of chlorophyll-*a* (64) to give (63) and phytol (67), may also catalyse the biosynthetic esterification which generates chlorophyll-*a* (64); this would probably occur in a lipid environment within the cell.[5] An alternative explanation[67] may be that phytol (67) is not the natural substrate for the synthetic reaction catalysed by chlorophyllase, and that late stages in phytol (67) biosynthesis occur after some other C_{20} alcohol [*e.g.* all-*trans*-geranylgeraniol (69)[68]] has been esterified.

Although the biosynthesis of the chlorophyll macrocycle has been studied extensively,[5,67,69] much less was known about the pathway to phytol (67) until recently.[70] Reasons for this lack of information included difficulties in both obtaining pure phytol (67) and devising specific degradations of the labelled alcohol (67) obtained from feeding experiments.[71]

Battersby *et al.*[70] have now defined conditions for the isolation of pure phytol (67) from cells of *Euglena gracilis.* The method involves separation of the phytol carbamate (68) from the corresponding derivative of geranylgeraniol, which gives both C_{20} alcohols as their corresponding crystalline biphenyl-4-ylcarbamates (68) and (70). This allowed them to study the incorporation of [1-^{13}C]acetate into

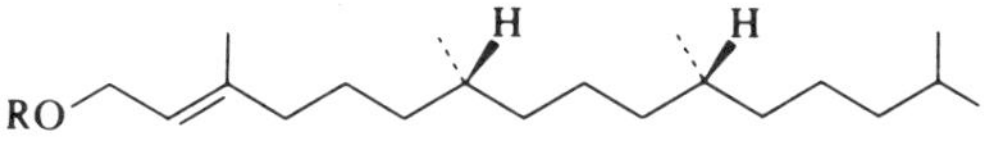

(67) R = H Phytol

(68) R = CONH—

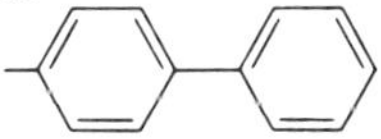

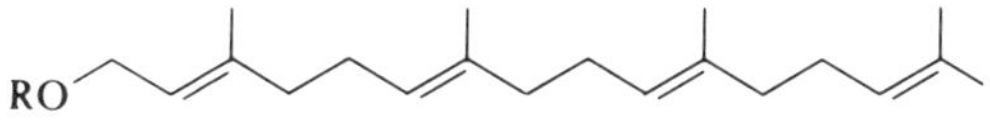

(69) R = H Geranylgeraniol

(70) R = CONH—

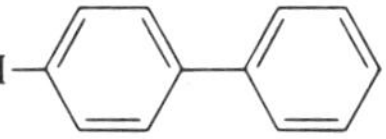

phytol (67), and their results are summarized in Scheme 21. The labelling pattern found supports the operation of the normal terpenoid pathway to geranylgeranyl pyrophosphate[72] with subsequent reduction.

These results do not distinguish between initial formation of the geranylgeranyl ester of protochlorophyllide-*a* (62) or chlorophyllide-*a* (63) followed by reduction, or prior reduction of (69) to phytol (67) before esterification.[68] However, the methods developed should facilitate further studies of this problem.[70]

The Bacteriochlorophylls.—Unlike the higher plants, in which only chlorophylls-*a* (64) and -*b* (65) are normally found, the green photosynthetic bacteria use a wide

[68] J. J. Katz, H. H. Strain, A. L. Harkness, M. H. Studier, W. A. Svec, T. R. Janson, and B. T. Cope, *J. Amer. Chem. Soc.*, 1972, **94**, 7938; C. Liljenberg, *Physiol Plant.*, 1974, **32**, 208 and refs. cited there.

[69] G. W. Kenner, J. Rimmer, K. M. Smith, and J. F. Unsworth, in ref. 4, pp. 255–276.

[70] E. H. Ahrens, jun., D. C. Williams, and A. R. Battersby, *J.C.S. Perkin I*, 1977, 2540.

[71] See ref. 70, and refs. 7–9 cited therein.

[72] G. Popják and J. W. Cornforth, *Biochem. J.*, 1966, **101**, 553, and refs. cited therein; G.P. Moss in 'Terpenoids and Steroids', ed. K. H. Overton (Specialist Periodical Reports), The Chemical Society, London, 1971, Vol. 1, p. 221, and refs. cited therein.

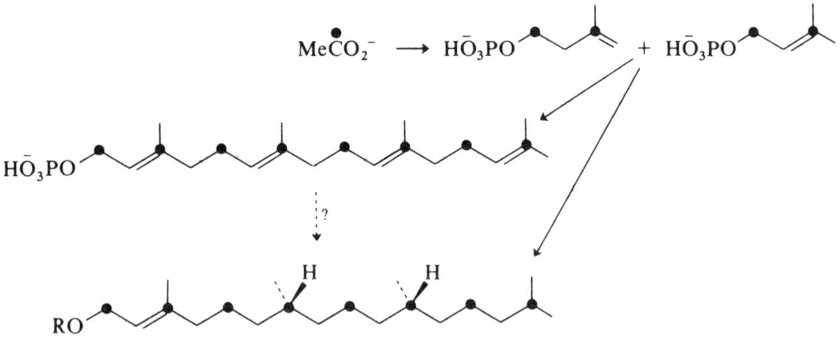

(64) R = chlorophyllide-*a* residue

● = ^{13}C

Scheme 21

range of different chlorophylls for their energy-gathering activities. These bacteriochlorophylls differ in the substituents around the periphery of the macrocycle, and although it appears that bacteriochlorophyll-*a* (65) arises by modification of chlorophyllide-*a* (63) in *Rhodopseudomonas spheroides*,[5] recent work has shown that (63) is not an intermediate in the biosynthesis of other bacteriochlorophylls.[69]

Kenner *et al.*[69] have studied the structures of the various *Chlorobium* chlorophylls (650) and (660) from certain *Chloropseudomonas* species; the designations (650) and (660) refer to the wavelength (in nm) of the red electronic absorption band. Distribution between hydrochloric acid and ether on Celite columns separates both the (650) and (660) chlorophylls into six components (or 'bands') in each case. The structures of the six components of the (650) chlorophylls had been determined beyond doubt[73] (see Table 2), and although the structures of the (660) series were less certain,[69] bands 5 and 6 of the chlorophylls (660) had been shown to be as in Table 2. Many of the differences between the *Chlorobium* chlorophylls and chlorophylls-*a* and -*b* involve simple modifications around the periphery of the macrocycle which are common to all of the former group.

There has been considerable confusion as to the substituent on the δ-*meso*-carbon (C-20) in bands 1—4 of the (660) series. It is now clear that all of the chlorophylls (660) carry a methyl substituent at C-20,[69] but the structures given in Table 2 for bands 1—4 of the (660) series are less certain. Kenner *et al.* have suggested these structures on the basis of a series of synthetic, degradative, and biosynthetic studies. Their proposals[69] for the biosynthesis of the (660) chlorophylls are outlined in Scheme 22. They suggest that the pentacarboxylic acid porphyrinogen (33) is the branch from normal chlorophyll metabolism (Scheme 20) for bands 2, 4, and 5 of chlorophylls (660); this suggestion was made before it was known[42] that (33) is a normal intermediate in porphyrin biosynthesis (see above).

[73] J. L. Archibald, D. M. Walker, K. B. Shaw, A. Markovac, and S. F. MacDonald, *Canad. J. Chem.*, 1966, **44**, 345; R. A. Chapman, M. W. Roomi, T. C. Morton, D. T. Krajcarski, and S. F. MacDonald, *ibid.*, 1971, **49**, 3544.

Table 2 *Structures of the* Chlorobium *chlorophylls*

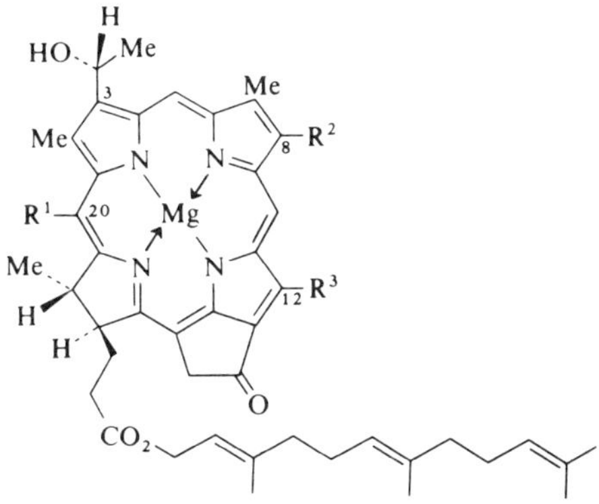

(650) series, $R^1 = H$			(660) series, $R^1 = Me$	
R^2	R^3	band	R^2	R^3
Bu^i	Et	1	$Pr^n(Bu^i)$?	Me
Pr^n	Et	2	Bu^i	Et
Bu^i	Me	3	$Bu^i(Pr^n)$?	Me
Et	Et	4	Pr^n	Et
Pr^n	Me	5	Et	Et
Et	Me	6	Et	Me

The carboxylic acid groups on the substituents at C-8 and C-12 would activate the side-chain for methylation by *S*-adenosylmethionine. This could occur first on the substituent at C-12 to give (71) after decarboxylation of the C-12 residue, which would lead to (660) band 5. Alternatively, methylation of (72) on the propionic acid side-chain at C-8, either once or twice, would lead to (660) band 4, or (660) band 2, respectively.

Scheme 22 also shows the favoured proposal[69] for the formation of chlorophylls (660) bands 1, 3, and 6; it was this proposal that led to the structures given for bands 1 and 3 in Table 2. Methylation of (73) once or twice would give, after decarboxylation, either n-propyl or isobutyl substituents at C-8, whereas simple decarboxylation of (73) would lead to the 8-ethyl substituent known[73] for chlorophyll (660) band 6. Kenner *et al.* have also pointed out that the (650) chlorophylls, which differ from the (660) series in that *meso*-methylation does not occur, contain fractions which have 8-n-propyl-12-methyl (band 5) and 8-isobutyl-12-methyl (band 3) substituent arrays; *meso*-methylation of these would lead to the structures proposed for chlorophylls (660) bands 1 and 3.

Another bacteriochlorophyll has been isolated from both *Chlorobium phaeobacteroides* and *C. phaeovibrioides*, and has been designated bacteriochlorophyll-*e*.[74] This was shown to be a mixture of the three structures shown in (74). These are clearly related to the *Chlorobium* chlorophylls (660),[69] the only significant difference being the formyl group at C-7 in (74), and now the *Chlorobium* chlorophylls (650) and (660) are termed bacteriochlorophylls-*d* and -*c* respectively.[69,73] It

[74] H. Brockmann, jun., in ref. 4, pp. 277–285.

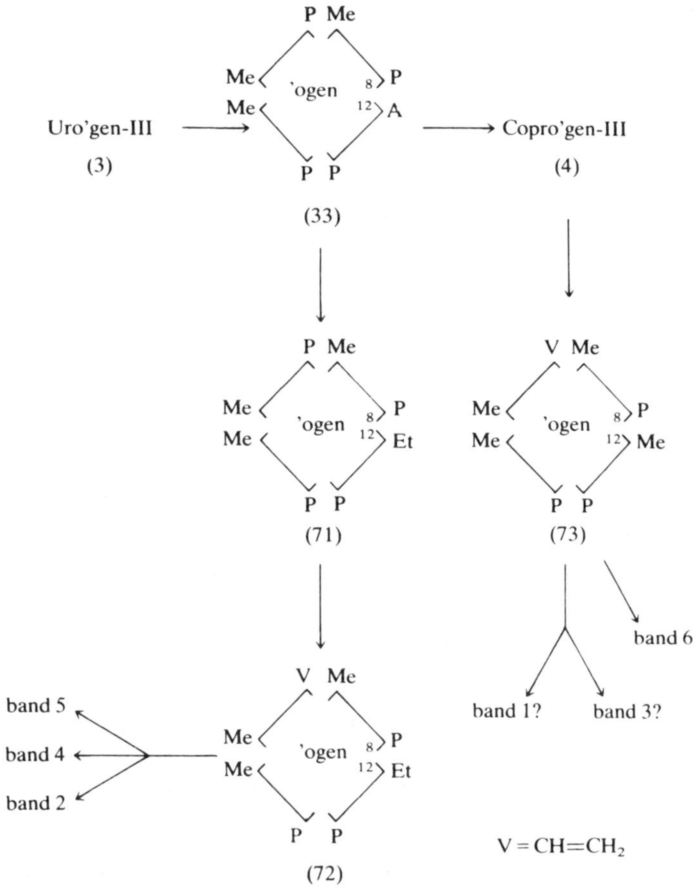

Scheme 22

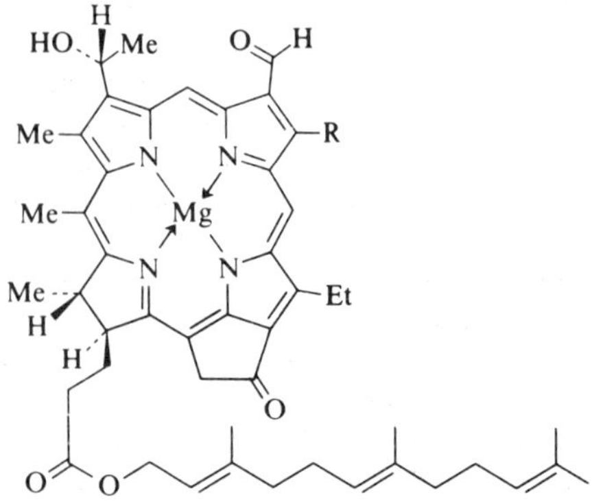

(74) R = Et, Pr^n, or Bu^i

is of interest that the various bacteriochlorophylls-*c*, -*d*, and -*e* all have the (*R*)-configuration at the hydroxyethyl side-chain at C-3.[74,75]

6 The Biosynthesis of Vitamin B_{12}

Coenzyme B_{12} (75) is the biologically active form of vitamin B_{12} (76), and this complex molecule presents formidable challenges to chemists. Determination of the structure[76] and the total synthesis[77] of the vitamin are among the great achievements in organic chemistry. The elucidation of the biosynthesis of vitamin B_{12} and the determination of its role and mechanism of action in biology[78] are two further challenges which are under active investigation.[5,48]

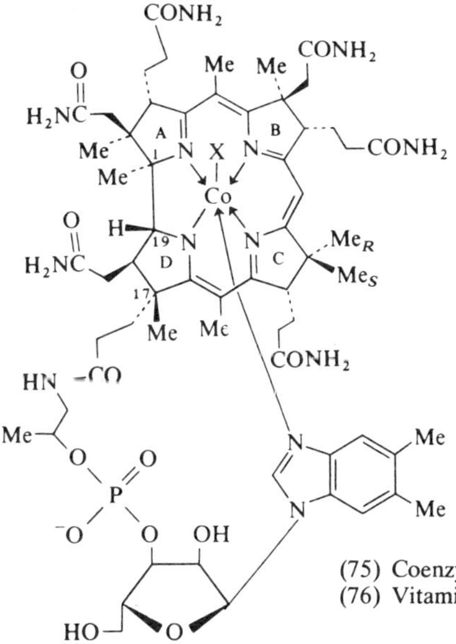

(75) Coenzyme B_{12} X = CH_2

(76) Vitamin B_{12} X = CN

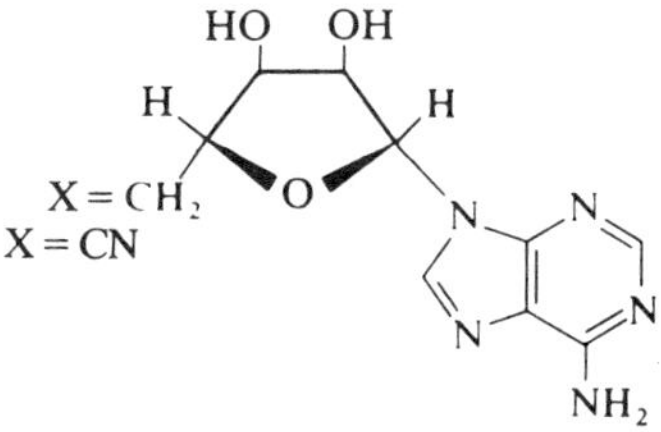

The origin of the complex side-chain attached at C-17 has been studied,[79] but this Report will be concerned only with the biosynthesis of cobyrinic acid (2), the natural corrin nucleus. In common with the porphyrin nucleus, this consists of four pyrrole rings built into a macrocycle, with the important difference of a direct C-1 to C-19 link in the corrins. Furthermore, the acetate and propionate side-chains of

[75] H. Brockmann, jun., and N. Risch, *Angew. Chem. Internat. Edn.*, 1974, **13**, 664.

[76] D. C. Hodgkin, J. Pickworth, J. H. Robertson, K. N. Trueblood, R. J. Prosen, and J. G. White, *Nature*, 1955, **176**, 325; R. Bonnett, J. R. Cannon, A. W. Johnson, I. Sutherland A. R. Todd, and E. L. Smith, *ibid.*, 1955, **176**, 328.

[77] (*a*) R. B. Woodward, *Pure Appl. Chem.*, 1973, **33**, 145; (*b*) A. Eschenmoser, XXIIIrd IUPAC Congress, Boston, *Pure Appl. Chem. Suppl.*, 1971, **2**, 69; (*c*) A. Eschenmoser, *Naturwiss.*, 1974, **61**, 513; (*d*) A. Pfaltz, B. Hardegger, P. M. Muller, S. Farook, B. Krautler, and A. Eschenmoser, *Helv. Chim. Acta*, 1975, **58**, 1444.

[78] T. C. Stadtmin, *Science*, 1971, **171**, 859.

[79] P. Renz and R. Weyhenmeyer, *F.E.B.S. Letters*, 1972, **22**, 124; S. H. Lu and W. L. Alworth, *Biochemistry*, 1972, **11**, 608 and refs. cited therein.

cobyrinic acid (2) are clearly arranged as in the natural type-III porphyrins, with the characteristic 'inversion' in ring D.

By late 1974, the existence of a shared biosynthetic pathway to porphyrins and corrins had been demonstrated by the incorporation of ^{13}C- and ^{14}C-labelled ALA and PBG into cobyrinic acid.[5] It was also known that methionine acted as the source of seven of the eight methyl groups of cobyrinic acid (2), including that at C-1; the eighth, the *pro*-12*S*-methyl group, was known to arise by decarboxylation of the C-12 acetic acid side-chain.[5,48b,80] However, even after several sets of experiments in various laboratories, it was not certain that uro'gen-III (3) was an intermediate in the biosynthesis of cobyrinic acid (2).[5,48]

Proof that vitamin B_{12} (76) is derived from uro'gen-III (3) has now been obtained from various experiments with bacterial whole-cell and cell-free preparations[81] carried out in several laboratories (see Scheme 23).[48,82–84] Scott showed that

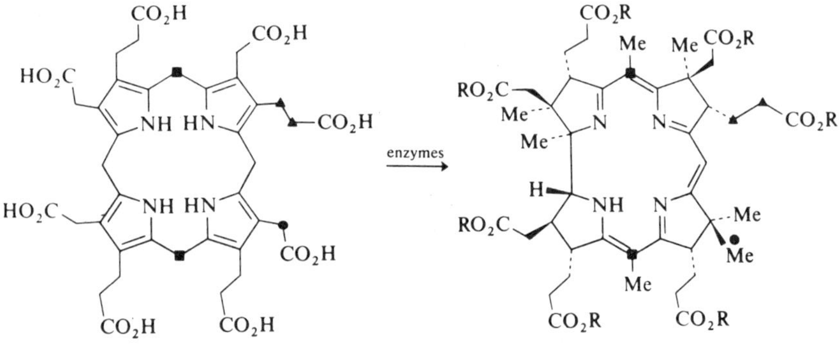

(3) Unlabelled

(3f) ■ = ^{14}C, ▲ = ^{3}H; ▲/■ = 4.10

(3g) ■ = ^{13}C
(3h) ● = ^{14}C

(2) R = H, unlabelled
(77) R = Me, unlabelled

(77f) R = Me, ■ = ^{14}C; ▲/■ = 4.05

(77g) ■ = ^{13}C
(77h) R = Me, ● = ^{14}C

Scheme 23

incubation[82] of the doubly labelled uro'gen-III (3f) (^{3}H : ^{14}C = 4.10 : 1) in the cell-free system from *Propionibacterium shermanii* gave, after dilution with carrier and purification of the recovered heptamethyl cobyrinate (77) ('cobester'), a radiochemically pure sample of cobester (77f) (^{3}H : ^{14}C = 4.05 : 1). Any randomization *via* fragmentation and recombination would have led to a profound change in the ^{3}H : ^{14}C specific activity ratio of this unsymmetrically substituted substrate. Further evidence was obtained on administration[82] of [5, 15-$^{13}C_2$]uro'gen-III (3g) to resting

80 A. I. Scott, in ref. 4, pp. 303–318.
81 A. I. Scott, B. Yagen, and E. Lee, *J. Amer. Chem. Soc.*, 1973, **95**, 5761; A. R. Battersby, European Symposium in Bio-organic Chemistry, Gregynog, Wales, May 1973; see also ref. 83*b*.
82 A. I. Scott, B. Yagen, N. Georgopapadakou, K. S. Ho, S. Klioze, E. Lee, S. L. Lee, G. H. Temme, III, C. A. Townsend, and I. M. Armitage, *J. Amer. Chem. Soc.*, 1975, **97**, 2548; 1976, **98**, 2371.
83 (*a*) A. R. Battersby, M. Ihara, E. McDonald, F. Satoh, and D. C. Williams, *J.C.S. Chem. Comm.*, 1975, 436; (*b*) A. R. Battersby, E. McDonald, R. Hollenstein, M. Ihara, F. Satoh, and D. C. Williams, *J.C.S. Perkin I*, 1977, 166.
84 H.-O. Dauner and G. Müller, *Z. physiol. Chem.*, 1975, **356**, 1353.

whole cells of *P. shermanii*. In this case, pure vitamin B_{12} (76) was recovered; this was shown to be labelled in the corrin nucleus as illustrated in (77g) by a ^{13}C n.m.r. analysis (4.5% specific incorporation).

The Cambridge group used the specifically labelled uro'gen-III (3h)[83] in their studies, which demonstrated 5—8% specific incorporations of (3h) into cobester (77h) using a similar cell-free preparation.[81] Dauner and Müller have reported[84] high incorporations of a mixture of uro'gens-I (8), -II, -III (3), and -IV (labelled equivalently in each pyrrole ring), using a cell-free system from *Clostridium tetanomorphum*. They have also achieved similar incorporations with Battersby's [12-*methylene*-^{14}C]uro'gen-III (3h),[83b] using the same cell-free preparation from *C. tetanomorphum*. These results firmly establish that uro'gen-III (3) is a specific biosynthetic precursor of cobyrinic acid (2) and thus of vitamin B_{12} (76) itself.

By inspection, the biosynthetic conversion of uro'gen-III (3) into cobyrinic acid (2) requires the following structural changes: (*a*) introduction of seven methyl groups from methionine at C-1, C-2, C-5, C-7, C-12 (*pro-R*), C-15, and C-17; (*b*) decarboxylation of the acetic acid side-chain at C-12; (*c*) extrusion of C-20; (*d*) possible adjustments of the oxidation level; and (*e*) insertion of cobalt.

Scott suggested[85] that the first step could be decarboxylation to give the ring-C-methyl heptacarboxylic acid porphyrinogen (78), and subsequently reported[82] a 0.1% incorporation of the ^{14}C-labelled form (78a) into cobester (77), of unknown labelling pattern, using the cell-free system from *P. shermanii*.[81] This result was interpreted[82] as confirmation that uro'gen-III (3) suffers decarboxylation (at C-12) *prior to* the necessary reductive methylation steps. Independent work by the Cambridge group[83] also demonstrated that incubation of the ^{14}C-labelled porphyrinogen (78b) leads to labelled cobester (77) after the usual work-up. However, they concluded that the ring-C-methyl porphyrinogen (78) was *probably not* a normal intermediate in corrin biosynthesis. They reached this conclusion mainly because the porphyrinogen (78) was 30—50 times less effective as a precursor of cobyrinic acid (2) than was uro'gen-III (3), while the ring-C-methyl porphyrinogen (78) and uro'gen-III (3) were almost equally well incorporated into copro'gen-III (4) in the same cell-free system.[83b] The incorporations of ^{14}C-labelled (78) were so low in both cases[82,83] that the recovered cobester (77) could not be degraded to locate the site(s) of labelling.

Groups in Zurich[86] and Cambridge[83b,87] have shown that each of the seven methyl groups in cobyrinic acid (2) derived from methionine is incorporated intact into the corrin nucleus. The Yale group[88] independently determined that the methyl groups at C-12 (*pro-R*) and C-1 of vitamin B_{12} (76) are incorporated intact. Intact incorporation had been demonstrated previously by the Cambridge group for the methyl groups at C-12 (*pro-R*) and C-7,[89] but the most important result concerned the methyl group at C-1.

[85] A. I. Scott, E. Lee, and C. A. Townsend, *Bioorg. Chem.*, 1974, **3**, 229; A. I. Scott, *Tetrahedron*, 1975, **31**, 2639.
[86] M. Imfeld, C. A. Townsend, and D. Arigoni, *J.C.S. Chem. Comm.*, 1976, 541.
[87] A. R. Battersby, R. Hollenstein, E. McDonald, and D. C. Williams, *J.C.S. Chem. Comm.*, 1976, 543; see also ref. 83*b*.
[88] A. I. Scott, M. Kajiwara, T. Takahashi, I. M. Armitage, P. Demou, and D. Petrocire, *J.C.S. Chem. Comm.*, 1976, 544.
[89] A. R. Battersby, M. Ihara, E. McDonald, J. R. Stephenson, and B. T. Golding, *J.C.S. Chem. Comm.*, 1973, 404; 1974, 458; *J.C.S. Perkin I*, 1977, 158.

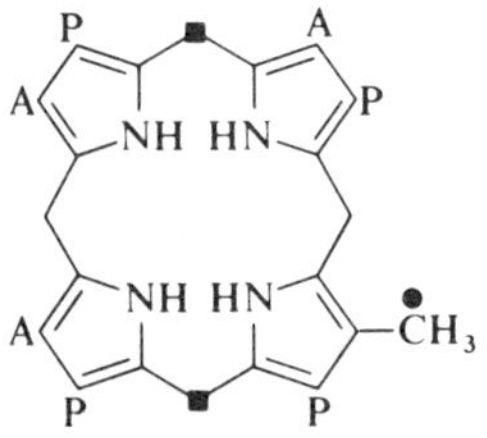

(78) Unlabelled
(78a) ■ = ^{14}C
(78b) ● = ^{14}C

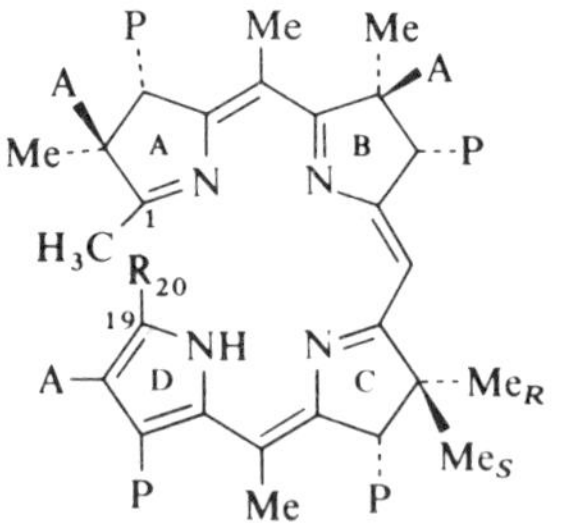

(79)
(79a) R = H
(79b) R = CH_2OH

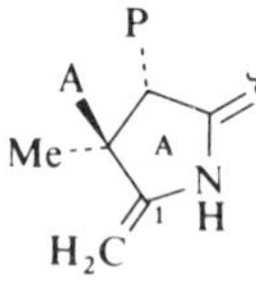

(80)

The question of loss or retention of the methyl hydrogen atoms is of particular importance in the case of this 1-methyl group of vitamin B_{12} (76), as it bears critically on current theories concerning the mechanism of formation of the corrinoid C-1—C-19 bond.[90] Each of the groups used essentially the same methods in their recent studies: they incubated [*Me*-$^{13}CD_3$]methionine (90 atom % ^{13}C, 98 atom % D) with intact cells of *P. shermanii* and examined the recovered vitamin B_{12} (76) by ^{13}C n.m.r. spectroscopy. The Cambridge and Zurich groups used deuterium-noise-decoupled ^{13}C Fourier-transform n.m.r. spectroscopy, which showed that each of the seven methionine-derived methyl groups, including that at C-1, had been incorporated without exchange of D with H of the medium; none of the signals of the ^{13}C-enriched carbon atoms displayed the large splitting (*J ca.* 125 Hz) expected for ^{13}C–^{1}H spin–spin coupling. The Yale group reached the same conclusion about the incorporation of the 1-methyl group, using a different n.m.r. method. These results rule out the intermediacy of any system in which the potential 1-methyl group is temporarily a methylidene residue,[90] *e.g.* (80), although the existence of an intermediate seco-corrin such as (79) would be compatible with the above results.

In principle, C-20 could be lost[5] from the system at any oxidation level between methanol and carbon dioxide, and this could occur at various stages in the formation of the corrin ring, *e.g.* before bond formation between C-1 and C-19 occurs,[85] or as a consequence of it. Recent work by Scott[48b] indicates that C-20 is lost as formaldehyde, possibly from a species such as (79b); electrocyclic ring closure of the product (79a)[90] would lead to a dehydrocorrin, which may then be reduced by insertion of Co^{2+} rather than Co^{3+}.[48b] However, there are many other possibilities, and further experimental work is required.

Sirohydrochlorin.—The prosthetic group of the widespread class of enzymes which catalyse the six-electron reduction of sulphite to sulphide was first characterized in 1973 by Kamen and Siegel[91] and was named sirohaem. Removal of the iron afforded an orange fluorescent compound, sirohydrochlorin (81), which was shown to be an isobacteriochlorin. It was suggested[92] that sirohydrochlorin might arise by two *C*-methylations, at C-12 and C-18 of uro'gen-III (3), and it was also noted that

[90] A. Eschenmoser, *Chem. Soc. Rev.*, 1976, **5**, 377; see also refs. 26*a*, 48*b*, 77*d*, and 85.
[91] L. M. Siegel, M. J. Murphy, and H. Kamen, *J. Biol. Chem.*, 1973, **248**, 251.
[92] M. J. Murphy, L. M. Siegel, H. Kamen, and D. Rosenthal, *J. Biol. Chem.*, 1973, **248**, 2801.

sirohydrochlorin could represent an early intermediate in the biosynthesis of vitamin B_{12} (76).[92]

In 1976, Bykhovsky's group, in Moscow, reported[93] the isolation of a methylated tetrapyrrole which contained cobalt from *P. shermanii*. The metal-free form of this compound appeared to be similar to sirohydrochlorin, incorporated radioactive methionine into its methyl groups, and furthermore increased the rate of vitamin B_{12} production in the organism. Müller's group, in Stuttgart, then reported[94] the isolation of 'Factor II' from *P. shermanii*, and showed that labelled forms of 'Factor II' were incorporated into cobyrinic acid (2) by an enzyme preparation of *C. tetanomorphum*. The Cambridge and Moscow groups, working in collaboration, showed[95,96] that the isobacteriochlorin from *P. shermanii* was the same as the sirohydrochlorin obtained from the sulphite-reducing organism *Desulphovibrio gigas*. Finally, 'Factor II' was shown[26a] to be identical with sirohydrochlorin.

Chemical and spectroscopic work by the Cambridge and Moscow groups led to the postulation of two alternative structures for sirohydrochlorin; the A–B structure (81) and an isomeric B–C structure, although the former was preferred.[95,96] Recent work by the Cambridge group has rigorously established the structure (81) for sirohydrochlorin, and its *specific* incorporation into cobyrinic acid (2) has also been demonstrated.[97]

The origin of the two *C*-methyl groups of sirohydrochlorin was confirmed[97] by incorporation of (2*S*)-[*Me*-^{13}C]methionine, using *D. gigas*, and examination of the product (81a) by ^{13}C n.m.r. spectroscopy. Incubation of (2*S*)-[*Me*-^{14}C]methionine with *P. shermanii* under suitable conditions gave labelled sirohydrochlorin (81b). The purified ester (82) was proved to be radiochemically pure, and the derived labelled acid (81b) was then incubated in the broken cell system[83] from *P. shermanii*; the incorporation into cobyrinic acid (2b), isolated as cobester (77b), was in the range 3—7%. Ozonolysis of the purified cobester (77b) gave the fragments (83), (84), and (85),[98] the radioactivities of which confirmed that the labelling pattern was shown in Scheme 24 (* = ^{14}C).

These results, when combined with those noted above, establish[95–97] that sirohydrochlorin has the structure and absolute configuration (81) and that sirohaem is the corresponding iron complex. The proven[97] efficient enzymic conversion of sirohydrochlorin† (81) into cobyrinic acid (2) confirms the Cambridge group's earlier finding[83b] that the heptacarboxylic acid (78) is not the next intermediate beyond uro'gen-III (3) *en route* to vitamin B_{12} (76), contrary to what had been thought.[82] The middle section of the biosynthetic pathway to vitamin B_{12} can now be defined as uro'gen-III (3) → (81) → (2) → vitamin B_{12} (76).

† They envisage that the biosynthetic intermediate may be a dihydro-derivative of (81), which would be the product expected from simple methylation in rings A and B of uro'gen-III (3) (see refs. 26*a* and 48*b*).

93 V. Ya. Bykhovsky, N. I. Zaitseva, A. V. Umrikhina, and A. N. Yavorskaya, *Priklad. Biokhim. Mikrobiol.*, 1976, **12**, 825, and refs, cited therein.
94 R. Deeg, H.-P. Kriemler, K.-H. Bergmann, and G. Müller, *Z. physiol. Chem.*, 1977, **358**, 339.
95 A. R. Battersby, E. McDonald, H. R. Morris, M. Thompson, D. C. Williams, V. Ya. Bykhovsky, N. I. Zaitseva, and V. N. Bukin, *Tetrahedron Letters*, 1977, 2217.
96 A. R. Battersby, K. Jones, E. McDonald, J. A. Robinson, and H. R. Morris, *Tetrahedron Letters*, 1977, 2213.
97 A. R. Battersby. E. McDonald, M. Thompson, and V. Ya. Bykhovsky, *J.C.S. Chem. Comm.*, 1978, 150.
98 D. Arigoni, E.T.H., Zurich; personal communication to A. R. Battersby, quoted in ref. 97.

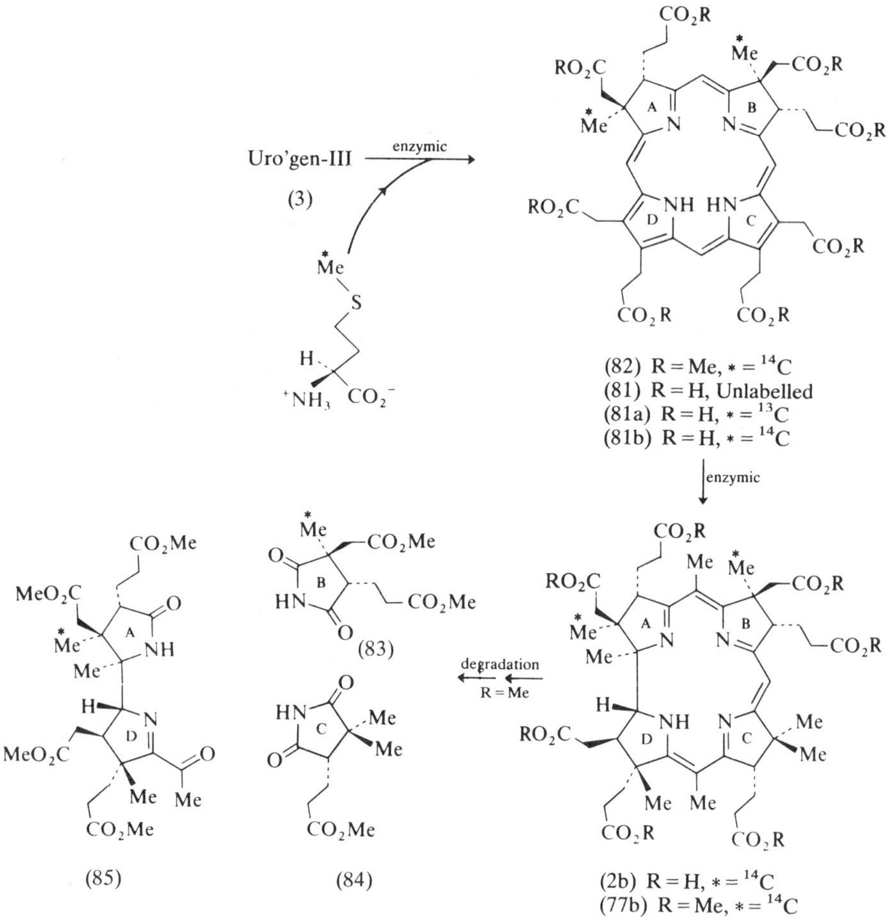

Scheme 24

In his recent review,[48b] Scott describes the work of the Yale group with sirohydrochlorin, which confirms both the structure as (81) and the intermediacy of sirohydrochlorin* in the biosynthesis of vitamin B_{12}. Interestingly, they found that (81) and its dihydro-derivative were both efficiently incorporated into cobyrinic acid (2).

7 Miscellaneous

New Techniques.—The extensive survey of laboratory methods made available early in 1975 by Fuhrhop and Smith in the book edited by K. M. Smith[2] has now been published separately.[99] Of the three most important new techniques

* He also envisages that the intermediate may be a dihydro-derivative of (81); see footnote on p. 427.

99 J.-H. Fuhrhop and K. M. Smith, 'Laboratory Methods in Porphyrin and Metalloporphyrin Research', Elsevier, Amsterdam, 1975.

developed since the early 1970s, only the use of ^{13}C-labelling was sufficiently well advanced to be included in detail in ref. 2.[5] The crucial role played by ^{13}C-labelling, and the associated ^{13}C n.m.r. techniques, in the elucidation of various parts of the biosynthesis of porphyrins, chlorins, and corrins has been published subsequently, and is described above.[26,48b,85]

The Cambridge[11,12,23,41] and Cardiff[42,45] groups have used ^{1}H and ^{13}C n.m.r. spectroscopy of porphyrin esters in the presence of various amounts of shift reagents[100] to show the distribution of the side-chains which flank the four *meso* positions. This technique has both facilitated the identification of various natural porphyrins[23,42] and the confirmation of their structure by comparisons of the esters of the natural and synthetic compounds.

The use of shift reagents and of other new techniques which change n.m.r. spectra as probes for the study of chlorophyll chemistry has been reviewed in a recent Meldola Medal Lecture.[100] Other important new work in the area of chlorophyll chemistry, using the n.m.r. method, has been reported by Katz *et al.*[101]

High-performance liquid chromatography (h.p.l.c.)[102] has been used very successfully by various groups in the past four years for the efficient separation of different porphyrin types[103] and, more importantly, for the otherwise very difficult separation of *isomeric* porphyrins.[31,32,51,52] The *efficient* separation of *all four* coproporphyrin isomers,[31,32] which was not possible until recently,[104] was one of the crucial developments by the Cambridge group in their successful work on the 'type-III problem'.[26] Much of the recent work by the Cambridge and Cardiff groups has relied heavily on h.p.l.c. for both analytical and preparative separations of porphyrins (see above).

Bile Pigments.—O'Carra has reviewed the biochemistry of haem cleavage and associated chemical analogues.[105] Oxidative cleavage of protohaem (54) at the α-*meso*-carbon (C-5) and loss of the iron atom normally gives biliverdin-IXα (86), although in isolated cases oxidative attack does occur at one of the other *meso* positions. In mammals the biliverdin-IXα (86) is immediately converted into bilirubin (87) (sometimes named bilirubin-IXα) by enzymic reduction (Scheme 25). The structures shown for biliverdin-IXα (86) and bilirubin (87), with the (Z) configuration about each of the three or two exocyclic double bonds, are what might reasonably be expected from their mode of formation; the particular tautomers shown have been accepted on the basis of various spectroscopic studies.[105,106]

The first X-ray analysis of any naturally occurring linear tetrapyrrole has been reported recently.[106] This study of bilirubin confirms that the structure is as shown in (87), thus confirming the gross chemical structure assigned to bilirubin by Fisher

[100] J. K. M. Sanders, *Chem. Soc. Rev.*, 1977, **6**, 467.
[101] H. Scheer and J. J. Katz, in ref. 2, pp. 399–524; J. J. Katz, W. Oettmeier, and J. R. Norris in ref. 4, pp. 227–254; M. J. Wasielewski, U. H. Smith, B. T. Cope, and J. J. Katz, *J. Amer. Chem. Soc.*, 1977, **99**, 4172.
[102] L. R. Snyder and J. J. Kirkland, 'Introduction to Modern Liquid Chromatography', Wiley, New York, 1974.
[103] Ref. 32, and accompanying papers.
[104] Ref. 2, pp. 859–861.
[105] P. O'Carra, in ref. 2, pp. 123–153.
[106] R. Bonnett, J. E. Davies, and M. B. Hursthouse, *Nature*, 1976, **262**, 326.

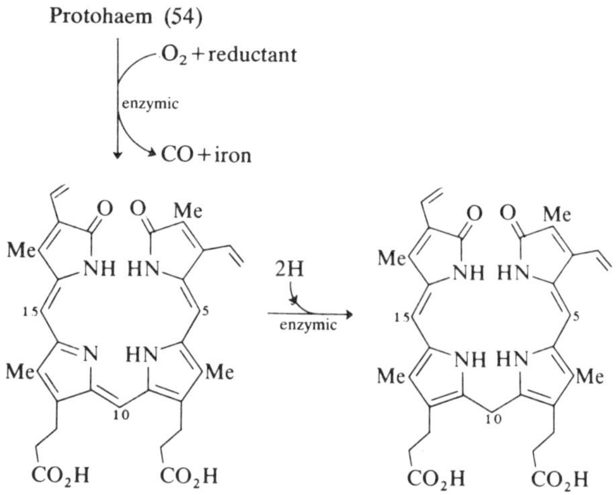

(86) Biliverdin-IXα (87) Bilirubin

Scheme 25

et al.[107] There is considerable intramolecular hydrogen bonding in the crystalline bilirubin (Figure 1); the low solubility of bilirubin in water, which has important biological consequences, can be understood readily in terms of this pattern of hydrogen bonding.[106] Since bilirubin (87) is derived directly from biliverdin-IXα (86), and the former has the same configuration about the two exocyclic double bonds as is found in protohaem (54), it is clear that biliverdin-IXα (86) has the illustrated (*Z*) configuration at the C-5 and C-15 double bonds. The stereochemistry of the C-10 double bond of biliverdin-IXα has not been determined, but presumably it also has the (*Z*) configuration shown in (86).

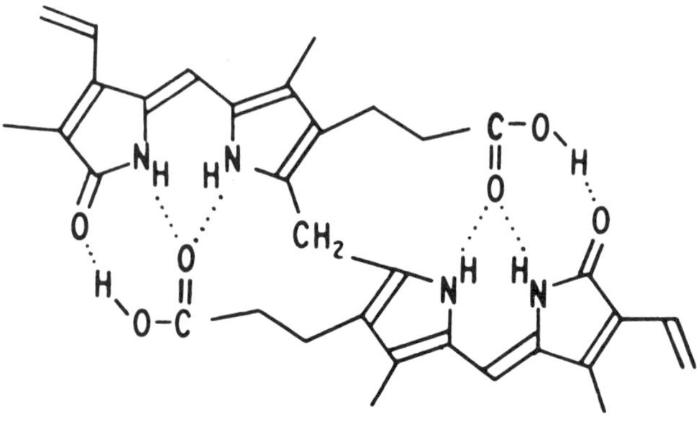

Figure 1 *Structure of bilirubin, in simplified form.*
(Reproduced by permission from *Nature*, 1976, **262**, 326)

[107] H. Fischer, H. Plieninger, and O. Weissbarth, *Z. physiol. Chem.*, 1941, **268**, 231.

Haemoglobin/Myoglobin Model Studies.—The respiratory haemoproteins haemoglobin and myoglobin transport and store molecular oxygen, and thus are essential to the life of all vertebrates.[108] The most significant property of haemoglobin[109] is co-operative oxygen binding: the oxygen affinity of the biologically active tetramer increases with increasing saturation.[109,110] Co-operativity is required for the transfer of oxygen from the carrier haemoglobin to the receptor myoglobin as well as for responses to other physiological requirements.

The way that the protein in haemoglobin and myoglobin regulates oxygen affinity, controls axial ligation of oxygen, and provides kinetic stability to the iron–dioxygen group is under active investigation by the preparation and full characterization of synthetic model compounds. Since Collman's synthesis of his 'picket-fence' porphyrin,[1] which led to the isolation and characterization of a series of *crystalline* dioxygen–iron porphyrin complexes,[111] many different groups have synthesized a wide variety of different iron porphyrins.[112] The different approaches used in these various studies have already helped to define the degree of hindrance required around the iron–dioxygen grouping to allow *reversible* oxygen binding,[112] but further work is necessary before these model studies can make a significant contribution to our understanding of the chemistry of the binding of oxygen to haemoglobin and myoglobin *in vivo.*

[108] For recent reviews, see: N. W. Makinen, in 'Techniques and Topics in Bioinorganic Chemistry', ed. C. A. McAuliffe, Wiley, New York, 1975, Part 1, Chapter 2; J. H. Prat, *ibid.*, Part 2, Chapter 7.
[109] M. F. Perutz, *Brit. Med. Bull.*, 1976, **32**, 193.
[110] J. M. Baldwin, *Brit. Med. Bull.*, 1976, **32**, 213.
[111] J. P. Collman, R. R. Gagne, T. R. Halbert, T.-C. Marchon, and C. A. Reed, *J. Amer. Chem. Soc.*, 1973, **95**, 7868; J. P. Collman, R. R. Gagne, C. A. Reed, T. R. Halbert, G. Lang, and W. T. Robinson, *ibid.*, 1975, **97**, 1427.
[112] J. P. Collman, *Accounts Chem. Res.*, 1977, **10**, 265.

14 Biological Chemistry
Part (iv) Enzyme Chemistry

By M. C. SUMMERS
Department of Biochemistry, University of Cambridge, Tennis Court Road, Cambridge, CB2 1QW
and
D. C. WILLIAMS
Department of Biochemistry, Trinity College, Dublin 2, Ireland

1 Irreversible Enzyme Inhibitors

Irreversible enzyme inhibitors constitute a specific type of affinity labelling in which the inhibitor, because of its similar structure to the natural substrate, binds to the enzyme and subsequently becomes covalently attached to it.[1]

One class of irreversible inhibitor is the substrate analogue which carries a reactive functional group, and by virtue of its structural similarities to the natural substrate binds at the enzyme active site.[2] Accordingly, the effectiveness of this class of inhibitor depends on the binding affinity of the inhibitor for its target enzyme, and the availability of reactive amino-acid residues at the enzyme active site. Another class of irreversible enzyme inhibitor is the k_{cat} inhibitor,[3] where the enzyme substrate(s) carries a masked chemically reactive group and only after enzymic transformation is a reactive molecule produced. The product, depending on its reactivity and availability of a suitable nucleophile, either dissociates from the enzyme or becomes covalently attached. Inhibitors of this class have also been euphemistically referred to as 'Suicide Inhibitors'.[4]

Irreversible inhibitors can provide information on the amino-acid residue(s) involved in the inactivation, and the peptide sequence of, the region enclosing the position of inactivation. In addition, inhibitors of this type may be used to determine the reactive pK of an amino-acid group involved in the inactivation.[5]

1 Methods in Enzymology Vol. XLVI. 'Affinity Labelling', ed. W. B. Jakoby and M. Wilchek, Academic Press, N.Y. and London, 1977; E. Shaw, in 'The Enzymes', ed. P. D. Boyer, Academic Press, N.Y. and London, 3rd ed., 1970, vol. 1, Ch. 2.

2 (*a*) M. Akhtar and D. C. Wilton, *Ann. Reports* (B), 1973, **70**, 98; (*b*) M. Akhtar and D. C. Wilton, *Ann. Reports* (*B*), 1971, **68**, 167.

3 (*a*) R. R. Rando, *Science*, 1974, **185**, 320; (*b*) R. R. Rando, *Biochem. Pharmacol.*, 1975, **24**, 1153; (*c*) R. R. Rando, *Accounts Chem. Res.*, 1975, **8**, 281.

4 R. H. Abeles and A. L. Maycock, *Accounts Chem. Res.*, 1976, **9**, 313.

5 (*a*) D. E. Schmidt and F. H. Westheimer, *Biochemistry*, 1971, **10**, 1249; (*b*) J. R. Knowles, *Crit. Rev. Biochem.*, 1976, **4**, 165; (*c*) K. Brocklehurst and H. B. F. Dixon, *Biochem. J.*, 1977, **167**, 859.

2 Acetylenic Irreversible Enzyme Inhibitors

The original discovery by Bloch *et al.*[6] that acetylenic substrate analogues are effective irreversible inhibitors of their respective target enzymes has led to the synthesis and study of a large number of inhibitors of this type. In recent years, acetylenic substrate analogues of diverse structure have been found to inhibit irreversibly Δ^5-3-keto-steroid isomerase (see later),[7] flavin-linked oxidases,[8] Cu^{2+}-amine oxidases,[9] γ-cystathionase,[10] L-alanine aminotransferase,[11] thiolase,[12] and 4-aminobutyric acid aminotransferase (see below).

Two of the flavin-linked oxidases which are inhibited by acetylenic substrate analogues have been investigated with regard to the structure of the flavin-inhibitor adduct *viz.* L-lactate oxidase (EC.1.13.12.4) and mitochondrial monoamine oxidase (EC.1.4.3.4).

L-Lactate oxidase from *Mycobacterium smegmatis* catalyses the oxidative decarboxylation of L-lactate to give acetic acid and CO_2.[2a] Abeles and co-workers[13] using 2-hydroxybut-3-ynoate (1), which is both a substrate and an irreversible inhibitor of the enzyme, characterized the structure of the isolated flavin-inhibitor adduct as (2) and using a combination of spectroscopic and chemical methods presented evidence for the structure of the initially formed enzyme-bound flavin-inhibitor complex as (3) (Scheme 1).

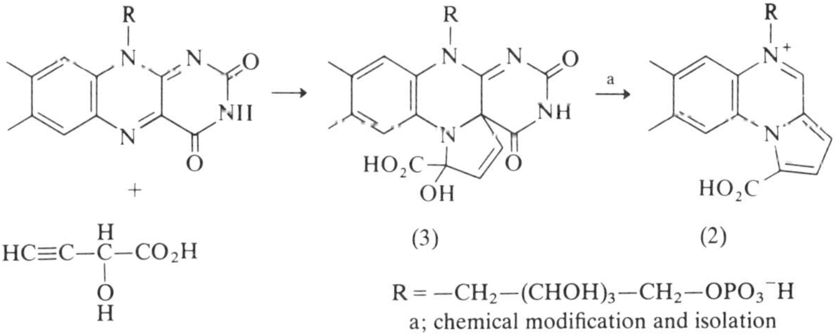

(1) (3) (2)

R = $-CH_2-(CHOH)_3-CH_2-OPO_3^-H$

a; chemical modification and isolation

Scheme 1

[6] (*a*) G. M. Helmkamp, R. R. Rando, D. J. H. Brock, and K. Bloch, *J. Biol. Chem.*, 1968, **243**, 3229; (*b*) K. Bloch, *Accounts Chem. Res.*, 1969, **2**, 193; (*c*) K. Endo and G. M. Helmkamp *J. Biol. Chem.*, 1970, **245**, 4293; (*d*) K. Bloch, in 'The Enzymes', ed. P. D. Boyer, Academic Press, N.Y. and London, 3rd ed., 1971, vol. 5, p. 441.

[7] F. H. Batzold and C. H. Robinson, *J. Amer. Chem. Soc.*, 1975, **97**, 2576.

[8] (*a*) L. Hellerman and V. G. Erwin, *J. Biol. Chem.*, 1968, **243**, 5234; (*b*) A. L. Maycock, R. H. Abeles, J. L. Salach, and T. P. Singer, *Biochemistry*, 1976, **15**, 114; (*c*) J. L. Kraus and J. J. Yaouanc, *Mol. Pharmacol.*, 1977, **13**, 378; (*d*) J. L. Kraus, J. J. Yaouanc, and G. Stratz, *European J. Med. Chem.*, 1975, **5**, 507; (*e*) C. T. Walsh, A. Schonbrunn, O. Lockridge, V. Massey, and R. H. Abeles, *J. Biol. Chem.*, 1972, **247**, 6004 (*f*) T. H. Cromartie, J. Fisher, G. Koczorowski, R. Laura, P. Marcotte, and C. T. Walsh, *Chem. Comm.*, 1974, 597; (*g*) P. Marcotte and C. T. Walsh, *Biochemistry*, 1976, **15**, 3070; (*h*) K. Horiike, Y. Hishina, Y. Migake, and T. Yamano, *J. Biochem. (Japan)*, 1975, **78**, 57; (*i*) T. H. Cromartie and C. T. Walsh, *Biochemistry*, 1975, **14**, 3482.

[9] (*a*) R. C. Hevey, J. Babson, A. L. Maycock, and R. H. Abeles, *J. Amer. Chem. Soc.*, 1973, **95**, 6125; (*b*) R. R. Rando and J. DeMairena, *Biochem. Pharmacol.*, 1974, **23**, 463.

[10] R. H. Abeles and C. T. Walsh, *J. Amer. Chem. Soc.*, 1973, **95**, 6124.

[11] P. Marcotte and C. T. Walsh, *Biochem. Biophys. Res. Comm.* 1975, **62**, 677.

[12] P. C. Hollan, M. G. Clark, and D. P. Bloxham, *Biochemistry*, 1973, **12**, 3309.

[13] A. Schonbrunn, R. H. Abeles, C. T. Walsh, S. Ghisla, H. Ogata, and V. Massey, *Biochemistry*, 1976, **15**, 1798.

In the case of mitochondrial monoamine-oxidase, studies aimed at elucidating the structure of the flavin-inhibitor adduct have been undertaken using model systems[14,15] and partially purified enzyme preparations.[16] In both types of study, when propargylamines, *e.g.* 3-dimethylamino-prop-1-yne (4), are used as inhibitors, the same flavocyanine, flavin-inhibitor adduct has been characterized.

The characterized adduct (5) (Scheme 2) is the result of attachment of the C-1 atom of (4) to N-5 of the flavin nucleus.[17] Several mechanisms were proposed by Abeles *et al.* to account for the structure of the flavocyanine (5): reaction between

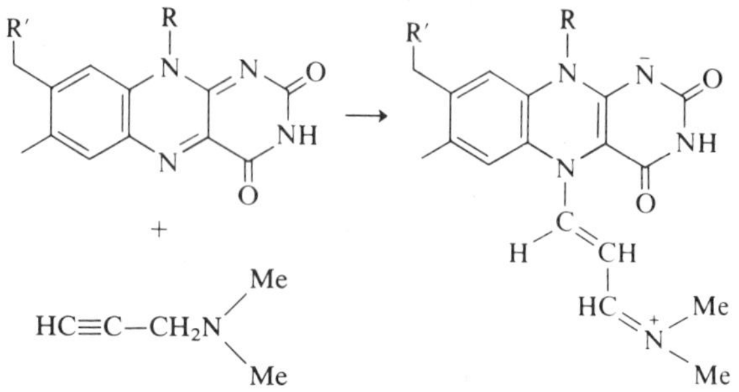

(4) (5)

R = $-CH_2-(CHOH)_3-CH_2-O-(PO_3^-)_2-$adenosine
R′ = 8α-cysteinyl peptide

Scheme 2

oxidized flavin and allenic carbanion; Michael addition of reduced flavin with oxidized substrate; and radical-pair formation of flavin and substrate with subsequent collapse and protonation to give a stable covalent adduct. However, a recent report by Krantz and Lipkowitz[18] indicates that the mechanism of inactivation may be more complex than had previously been proposed.[16] These workers studied the irreversible inhibition of monoamine oxidase using the acetylenic inhibitor (6) and the allenic isomer (7), and observed distinct spectral properties for the respective flavin-inhibitor adducts, with only the former inhibitor (6) giving a spectrum typical of a flavocyanine. Krantz and Lipkowitz[18] argued that if one of

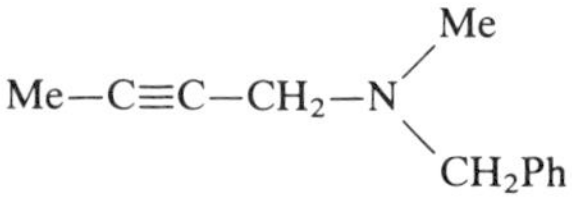

(6)

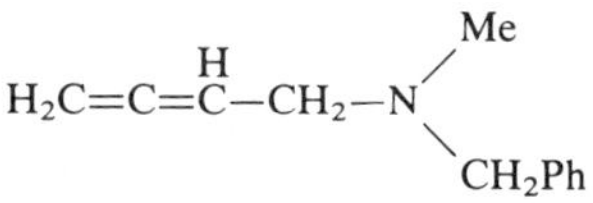

(7)

[14] (*a*) B. Gartner and P. Hemmerich, *Angew. Chem. Internat. Edn.*, 1975, **14**, 110; (*b*) A. L. Maycock, *J. Amer. Chem. Soc.*, 1975, **97**, 2270.
[15] B. Gartner, P. Hemmerich, and E. A. Zeller, *European J. Biochem.*, 1976, **63**, 211.
[16] A. L. Maycock, R. H. Abeles, J. I. Salach, and T. P. Singer, *Biochemistry*, 1976, **15**, 114.
[17] For recent reviews of the chemistry of flavins and flavoenzymes, see, (*a*) P. Hemmerich, *Fortschr. Chem. org. Naturstoffe*, 1976, **33**, 451; (*b*) T. C. Bruice, *Prog. Bioorganic Chem.*, 1976, **4**, 1; (*c*) 'Flavins and Flavoproteins', ed. T. P. Singer Elsevier, Amsterdam, 1976; P. Hemmerich, V. Massey, and H. Fenner, *F.E.B.S. Letters*, 1977, **84**, 5.
[18] A. Krantz and G. S. Lipkowitz, *J. Amer. Chem. Soc.*, 1977, **99**, 4156.

the three mechanisms proposed by Abeles *et al.* is correct then the mechanism of inactivation of inhibitors (6) and (7) should be very similar and give rise to the same flavin-inhibitor adduct. However, this appears not to be the case. At the moment, the structure of the flavin-allene inhibitor adduct is unknown.

Apart from the use of acetylenic substrate analogues to inhibit flavin-linked enzymes they have also been used successfully irreversibly to inhibit pyridoxal phosphate-dependent enzymes. The mechanism of inactivation of a pyridoxal phosphate enzyme is illustrated by the work of Washtien and Abeles with γ-cystathionase.[19] The enzyme is inhibited irreversibly by propargylglycine (8; 2-aminopent-4-ynoic acid); the inactivation displays pseudo-first order kinetics and is accompanied by the covalent attachment of inhibitor to the enzyme. In addition, when radio-labelled (8) was used, it was possible to isolate an inhibitor–enzyme adduct which on treatment with aqueous acid yielded (2*S*)-[2-^{14}C]-2-amino-4-ketopentanoic acid (9). The proposed mechanism of inactivation is shown in Scheme 3. From indirect evidence it was suggested that the active site nucleophile

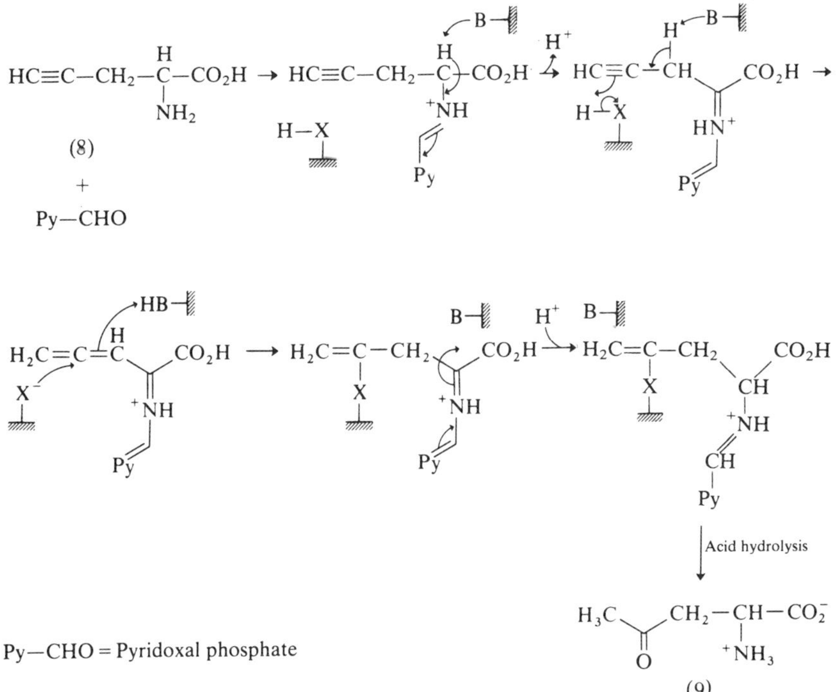

Scheme 3

(X) involved in covalent bond formation is either tyrosine or cysteine. Recently, Washtien *et al.*[20] have shown that γ-cystathionase catalyses exchange of the α- and

[19] W. Washtien and R. H. Abeles, *Biochemistry*, 1977, **16**, 2485.
[20] W. Washtien, A. J. L. Cooper, and R. H. Abeles, *Biochemistry*, 1977, **16**, 460.

β-protons of (2*S*)-amino-acids that are unable to undergo γ-elimination, and yet are competitive inhibitors of the normal enzyme-catalysed reaction.

The high specificity of k_{cat} inhibitors makes them useful agents for the study of the *in vivo* function of enzymes and, in particular, in the investigation of enzyme synthesis and degradation; *e.g.* Abeles and Walsh[10] found that when mice were injected i.p. with propargylglycine a condition similar to the genetic defect cystathionuria was induced. A particularly useful target enzyme in this respect is γ-aminobutyrate-α-ketoglutarate transaminase [GABA-T (EC.2.6.1.19)].

The putative inhibitory neurotransmitter, γ-aminobutyric acid (10; GABA) is degraded, *via* transamination with 2-oxoglutarate, by pyridoxal phosphate-dependent GABA-T. Irreversible inhibitors of the enzyme include ethanolamine-*O*-sulphate,[21] 4-amino-hex-5-ynoic acid,[22] 4-amino-hex-5-enoic acid (11)[23] and the naturally occurring neurotoxin, gabaculine (12; cyclohexa-1,3-dienyl-5-amino-carboxylic acid).[24] The mechanism of inactivation of GABA-T by gabaculine involves Schiff base formation with pyridoxal phosphate, tautomerization and finally aromatization to give *m*-carboxyphenyl pyridoxamine phosphate (13). Chemical studies by Rando and Bangerter[25] support the mechanism outlined above, and in particular that the structure of the inactivator adduct has the structure shown and not the other possibility resulting from Michael addition at C-3. Interestingly, the dihydro analogue (14) does not irreversibly inhibit the enzyme.[25]

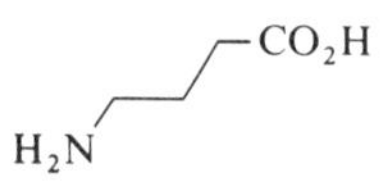

(10)

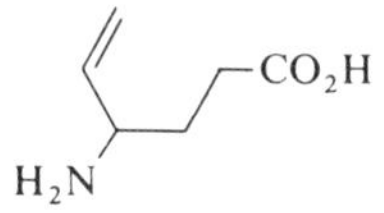

(11)

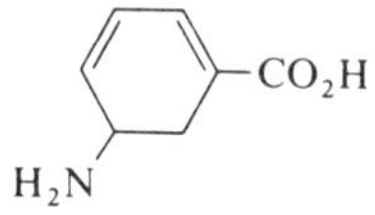

(12)

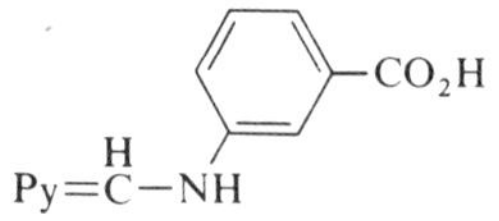

(13)

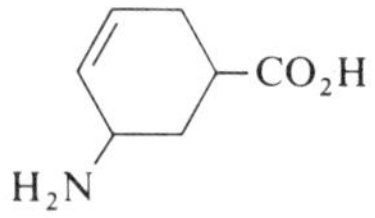

(14)

Both (11) and (12) appear to be very specific k_{cat} inhibitors of GABA-T. Thus, i.p. injection of gabaculine solution leads to a dose-dependent loss of GABA-T activity and an increase in GABA levels in mouse brain. Other pyridoxal phosphate-dependent enzymes such as glutamate decarboxylase, ornithine decarboxylase, aspartate aminotransferase, and alanine aminotransferase are not inhibited irreversibly.[24c] Similarly, 4-amino-hex-5-enoic acid (11) when administered peripherally causes both a dose-dependent irreversible inhibition of GABA-T and an increase of brain GABA in mice.[26] The $t_{1/2}$ of GABA-T in mouse brain was

[21] L. J. Fowler and R. A. John, *Biochem. J.*, 1972, **130**, 569.
[22] M. J. Jung and B. W. Metcalf, *Biochem. Biophys. Res. Comm.*, 1975, **67**, 301.
[23] B. Lippert, B. W. Metcalf, M. J. Jung, and P. Cassara, *European J. Biochem.*, 1977, **74**, 441.
[24] (*a*) R. R. Rando and F. W. Bangerter, *J. Amer. Chem. Soc.*, 1976, **98**, 6762; (*b*) K. Kobayashi, S. Miyazawa, and A. Endo, *F.E.B.S. Letters*, 1977, **76**, 207; (*c*) R. R. Rando and F. W. Bangerter, *Biochem. Biophys. Res. Comm.*, 1977, **76**, 1276.
[25] R. R. Rando and F. W. Bangerter, *J. Amer. Chem. Soc.*, 1977, **99**, 5141.
[26] M. J. Jung, B. Lippert, B. W. Metcalf, P. Bohlen, and P. J. Schechter, *J. Neurochem.*, 1977, **29**, 797.

estimated to be 3.4 days. Also, the inhibitor has little or no effect on L-glutamate decarboxylase (*cf.* 4-amino-hex-5-ynoic acid, ref. 27), succinic semialdehyde dehydrogenase, aspartate aminotransferase and alanine aminotransferase.[26]

3 Steroid Isomerases

Rose,[28] in a review of aldose-ketose isomerases emphasized the importance of investigating the mechanism of action of enzymes catalysing α-hydroxyketone-α-hydroxyaldehyde interconversions which do not use sugar phosphates as substrates.

Monder *et al.*[29] have described an enzyme from hamster liver that catalyses the isomerization of the ketol side-chain of 11-deoxycorticosterone (DOC) to the corresponding hydroxy aldehyde (isoDOC). The proposed mechanism for the isomerization is shown in Scheme 4. Initial results obtained by Monder's group indicate that the enzyme has both similar and different properties to the more extensively studied sugar phosphate isomerases.[28,30] The enzyme catalyses intramolecular proton transfer between C-21 and C-20 of the steroid side-chain; this is a common feature of other isomerases, *e.g.* glucosamine-6-phosphate isomerase,[31] glucose-6-phosphate isomerase,[32] and triose phosphate isomerase.[33]

The generally accepted mechanism for intramolecular proton transfer to and from adjacent carbon atoms is that a single base (—B; Scheme 4) removes a proton

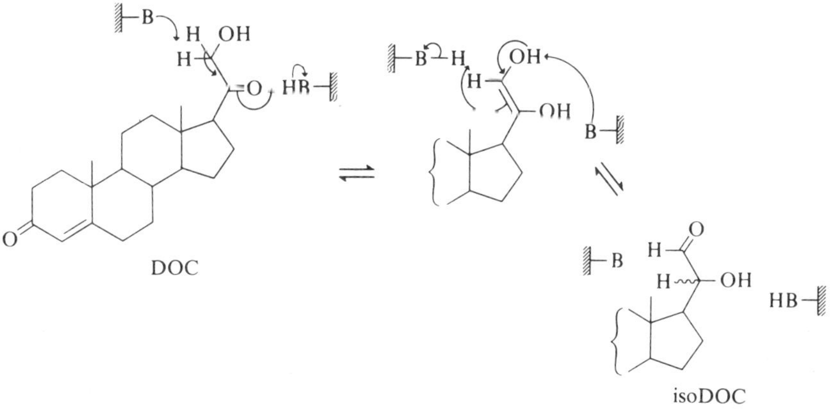

Scheme 4

from the substrate to give a *cis*-enediol intermediate and protonated enzyme (B—H), this is followed by protonation of the *cis*-enediol on the adjacent carbon

[27] M. J. Jung, B. Lippert, B. W. Metcalf, P. J. Schechter, P. Bohlen, and A. Sjoerdsma, *J. Neurochem.*, 1977, **28**, 717.

[28] I. A. Rose, *Adv. Enzymol.*, 1975, **43**, 491.

[29] (*a*) A. K. Willingham and C. Monder, *Endocrin. Res. Comm.*, 1974, **1**; 145; (*b*) C. Monder, B. Zumoff, H. L. Bradlow, and L. Hellman. *J. Clin. Endocrin. Metabolism*, 1975, **40**, 86; (*c*) K. O. Martin, S. W. Oh, H. J. Lee and C. Monder, *Biochemistry*, 1977, **16**, 3803.

[30] I. A. Rose, *Brookhaven Symp. Biol.*, 1962, **15**, 293.

[31] C. F. Midelfort and I. A. Rose, *Biochemistry*, 1977, **16**, 1590.

[32] I. A. Rose and E. L. O'Connell, *J. Biol. Chem.*, 1961, **236**, 3086.

[33] J. M. Herliky, S. G. Maister, W. J. Albery, and J. R. Knowles, *Biochemistry*, 1976, **15**, 5601.

atom. There appears to be a time-dependent stereoselective exchange of the C-21 protons. However, Lee and Monder[34] have evidence of a C-20 epimerase in their isomerase perparation and it is possible that the stereochemistry at C-20 determines the stereochemistry of proton removal at C-21.

Another isomerase, which uses a steroid as substrate, is Δ^5-3-ketosteroid isomerase, and in this case the reaction catalysed is the isomerization of a Δ^5-3-ketosteroid to the corresponding Δ^4-3-ketosteroid, *e.g.* (15) → (16) in Scheme 5.

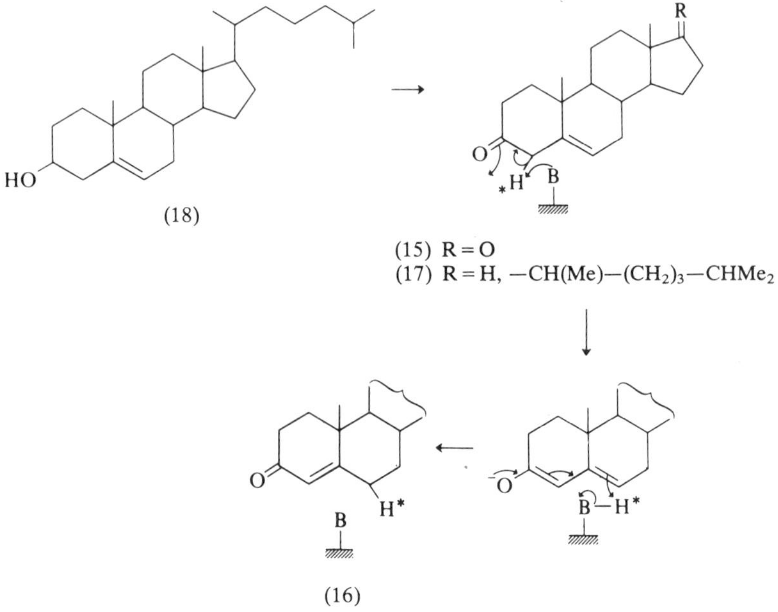

Scheme 5

The enzyme from *Pseudomonas testosteroni* has been studied in detail and, in fact, was the first example found of an enzyme catalysing an intramolecular proton transfer.[35] However, reinvestigation of the mechanism of proton transfer for the *P. testosteroni* enzyme indicates that the proton transfer may not be as specific as had previously been assumed[36] Viger and Marquet have now shown that during enzyme-catalysed isomerization of Δ^5-androstendione (15) there is not only $4\beta \rightarrow 6\beta$ hydrogen transfer, but also exchange of the 4β hydrogen. It appears, therefore, that the assumed dienol intermediate may be generated by removal of either the α or β proton at C-4; at the moment there are no accurate data on the relative contribution of the various exchange processes in the overall mechanism, *i.e.* $4\beta \rightarrow 6\beta$ and $4\alpha \rightarrow 6\alpha$ proton migration; 4β and 4α proton exchange. It was mentioned earlier that *P. testosteroni* Δ^5-3-ketosteroid isomerase is inhibited by

[34] H. J. Lee and C. Monder, *Biochemistry*, 1977, **16**, 3810.
[35] P. Talalay and V. S. Wang, *Biochim. Biophys. Acta*, 1955, **18**, 300.
[36] A. Viger and A. Marquet, *Biochim. Biophys. Acta*, 1977, **485**, 482.

acetylenic substrate analogues such as (19) in Scheme 6. In 1973, Martyr and Benisek[37] reported that the enzyme is also inactivated irreversibly by light, at wavelengths greater than 300 nm, in the presence of Δ^4-3-ketosteroids. It has now been shown that under the conditions of photoinactivation aspartic acid-38 in the polypeptide chain is decarboxylated to alanine.[38]

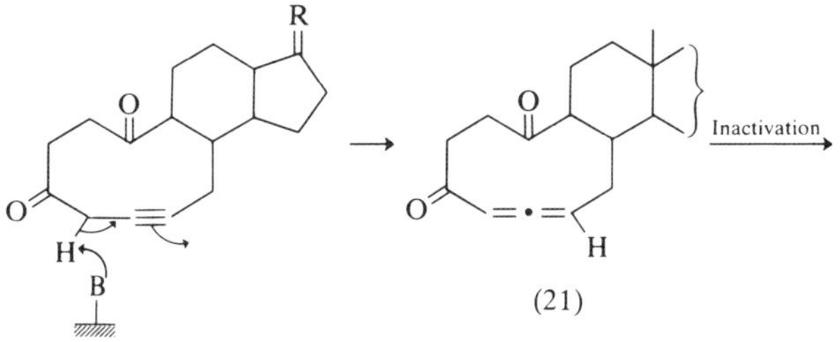

(21)

(19) R = O
(20) R = H, $-CH(Me)-(CH_2)_3-CH(Me)_2$

Scheme 6

Cholesterol oxidase from *Nocardia erythropolis* has mechanistic similarities to the Δ^5-3-ketosteroid isomerase described above. The enzyme oxidizes cholesterol (18) to cholest-4-en-3-one (16) *via* cholest-5-en-3-one (17) with hydrogen peroxide as the other product. Smith and Brooks, using cholesterol stereospecifically deuteriated at C-4, showed that the enzyme catalysed $4\beta \rightarrow 6\beta$ proton transfer and 4α-proton exchange.[39] In addition, the enzyme is inhibited irreversibly by the acetylenic substrate analogue (20).[40] The mechanism of inactivation of the Δ^5-3-ketosteroid isomerase and cholesterol oxidase is most probably the same and involves proton loss from the α-carbon atom to give an allene (21) which reacts with an active-site nucleophile. This is outlined in Scheme 6. Whereas the bacterial Δ^5-ketosteroid isomerase has been studied extensively the same cannot be said of the corresponding enzyme-catalyzed process of animal origin Benson and Talalay[41] observed that human and rat liver cytoplasm contains Δ^5-ketosteroid isomerase activity which was stimulated specifically by reduced glutathione (GSH). They also noted that the molecular and physical properties of the partially purified isomerase were similar to a group of basic proteins isolated from rat (6 proteins designated AA, A, B, C, D, and E) and human (5 proteins designated α, β, γ, δ, and ε) liver, and referred to as glutathione-S-transferases.[42] The GSH-transferases catalyse the transfer of GSH to a wide range of small organic molecules such as epoxides and halobenzenes to give the corresponding thioether. Also, it transpires that one of the GSH-transferases from rat liver is identical to ligandin (GSH-transferase B), an intracellular binding protein, so termed because of its ability to

[37] (*a*) R. J. Martyr and W. F. Benisek, *Biochemistry*, 1973, **12**, 2172; (*b*) R. J. Martyr and W. F. Benisek, *J. Biol. Chem.*, 1975, **250**, 1218.
[38] J. R. Ogez, W. F. Tivol, and W. F. Benisek, *J. Biol. Chem.*, 1977, **252**, 6151.
[39] A. G. Smith and C. J. W. Brooks, *Biochem. Soc. Trans.*, 1977, **5**, 1088.
[40] A. G. Smith and C. J. W. Brooks, *Biochem. J.*, 1977, **167**, 121.
[41] A. M. Benson and P. Talalay, *Biochem. Biophys. Res. Comm.*, 1976, **69**, 1073.
[42] W. B. Jakoby and J. H. Keen, *Trends Biochem. Sci.*, 1977, 229.

bind hydrophobic substrates such as bilirubin, bromosulphophthalein, drugs, cortisol metabolites, and azocarcinogens.[43,44] Remarkably, Benson *et al.*[45] have now shown that the same rat and human liver proteins are responsible for the GSH-activated Δ^5-3-ketosteroid isomerase activity. In rat liver the isomerase activity is associated mainly with ligandin (GSH-transferase B), while in humans all five GSH-transferases have isomerase activity, but transferase γ, δ, and ε are considerably more active than α and β.

4 Biotin-requiring Enzymes

Biotin acts as a CO_2 carrier in a number of enzyme reactions where it functions either to accept (decarboxylases), to donate (carboxylases), or to transfer (trans-carboxylases) CO_2.[46,47] There are nine known biotin-requiring enzymes and the list includes transcarboxylase,[48] pyruvate carboxylase,[49] and acetyl-CoA carboxylase.[50] The biotin is covalently linked to the ε-amino group of a lysine residue to give a biotin carboxyl carrier protein (22) and this subunit transfers CO_2 between two

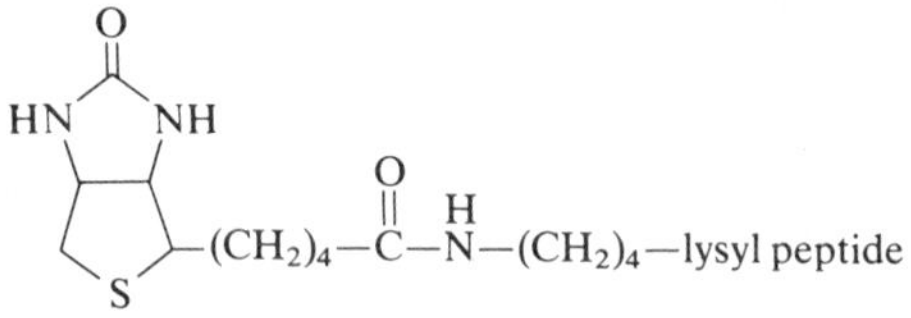

(22)

active sites. Amino-acid sequence data of the biotin carboxyl carrier-protein of transcarboxylase, acetyl-CoA carboxylase and several pyruvate carboxylases shows a high degree of conservation, particularly around the biotin attachment-site.[46,51] However, whether the amino-acid sequence serves as a recognition factor for the attachment of biotin to the apoenzyme, or a conformational, or catalytic role is not known.

In general, the reaction catalysed by a biotin enzyme is shown in Scheme 7. The overall reaction consists of two partial reactions, first, carboxylation of biotin by

[43] (*a*) G. Litwack, B. Ketterer, and I. M. Arias, *Nature*, 1971, **234**, 466; (*b*) G. J. Smith, K. Huebrer, and G. Litwack, *Biochem. Biophys. Res. Comm.*, 1977, **76**, 1174; (*c*) A. Grahnen and I. Sjoholm, *European J. Biochem.*, 1977, **80**, 573.

[44] (*a*) A. J. Levi, Z. Gatmaritan, and I. M. Arias, *J. Clin. Invest.*, 1969, **48**, 2156; (*b*) B. Ketterer, P. Ross-Mansell, and J. K. Whitehead, *Biochem. J.*, 1967, **103**, 316; (*c*) K. S. Morey and G. Litwack, *Biochemistry*, 1969, **8**, 4813.

[45] A. M. Benson, P. Talalay, J. H. Keen, and W. B. Jakoby, *Proc. Nat. Acad. Sci. U.S.A.*, 1977, **74**, 158.

[46] (*a*) H. G. Wood and R. E. Barden, *Ann. Rev. Biochem.*, 1977, **46**, 385; (*b*) J. Kaappe, *Ann. Rev. Biochem.*, 1970, **39**, 757.

[47] J. Moss and M. D. Lane, *Adv. Enzymol.*, 1971, **35**, 757.

[48] (*a*) H. G. Wood and G. K. Zwolinski, *Crit. Rev. Biochem.*, 1976, **4**, 47; (*b*) H. G. Wood, in 'The Enzymes', ed. P. D. Boyer, Academic Press, N.Y. and London, 3rd ed., 1972, vol 6, p. 83.

[49] (*a*) M. Scrutton and M. R. Yount, in 'The Enzymes', ed. P. D. Boyer, Academic press, N.Y. and London, 3rd ed., 1972, vol **6**, p. 1; (*b*) M. F. Utter, R. E. Barden, and L. B. Taylor, *Adv. Enzymol.*, 1975, **42**, 1.

[50] (*a*) A. W. Alberts and P. R. Vagelos, in 'The Enzymes', ed. P. D. Boyer, Academic Press, N.Y. and London, 3rd ed., 1972, vol 6, p. 37; (*b*) M. D. Lane, J. Moss, and S. E. Palakis, *Current Topics Cell Regul.*, 1974, **8**, 139.

[51] (*a*) D. B. Rylatt, D. B. Keech, and J. C. Wallace, *Biochem. Soc. Trans.*, 1977, **5**, 1544; (*b*) M. R. Sutton, R. R. Fall, A. M. Nervi, A. W. Alberts, P. R. Vagelos, and R. A. Bradshaw, *J. Biol. Chem.*, 1977, **252**, 3934.

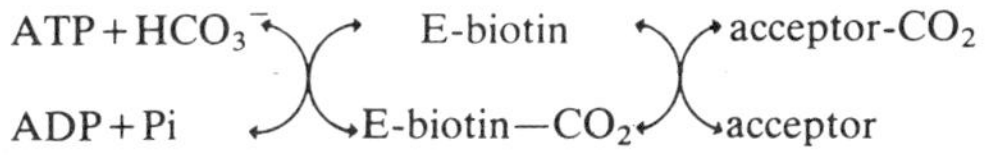

Scheme 7

HCO_3^- and ATP, and second, transfer of CO_2 to an acceptor to give carboxylated product. The decarboxylases carry out the second partial-reaction in reverse.

The following discussion concerns recent studies on transcarboxylase (EC.2.1.3.1) which catalyses the transfer of CO_2 from methylmalonyl-CoA to biotin to give carboxybiotin. Consequently, the enzyme does not require HCO_3^- and ATP. The second partial-reaction is the carboxylation of pyruvate to oxaloacetate by carboxybiotin (Scheme 8).

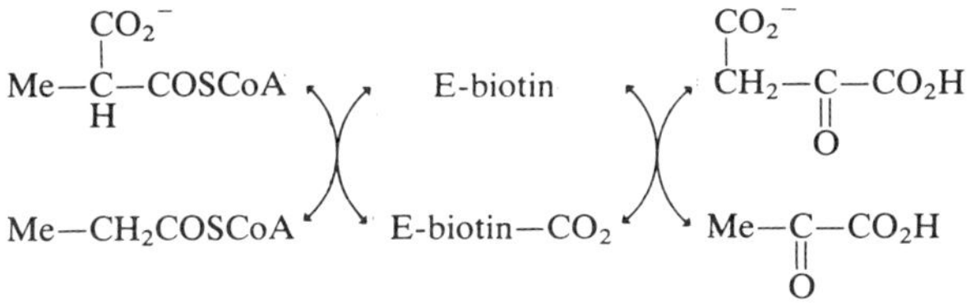

Scheme 8

The enzyme is found in the propionic acid bacteria and it provides the mechanism by which propionate is formed in fermentation. Studies on the quaternary structure of the 26*S* form of the enzyme have been published recently; the early studies being carried out on smaller, more stable forms of the enzyme complex (18*S* form). The 26*S* form of transcarboxylase is composed of a central hexameric subunit of MW 360 000 (12*S*) and to this are attached two sets of three subunits on opposite sides of the central subunit. Each outside subunit is a dimer of MW 120 000 (5.8*S*) and each dimer is attached to the central subunit by two biotin carboxyl carrier proteins of MW 12 000 (1.3*S*). The 26*S* form of the enzyme is therefore made up of thirty individual proteins to give a complex of MW 1 210 000. At acid pH the 26*S* form is stable, but as the pH is raised selective dissociation occurs and as the pH is raised even higher only the central hexameric subunit remains intact.[52] The methylmalonyl-CoA and pyruvate binding sites are on different subunits and as the enzyme displays two-site Ping-Pong kinetics the biotin 'carries' the CO_2 from one binding site and delivers it to the other.[53]

The actual mechanism of CO_2 transfer from methylmalonyl-CoA to biotin and from biotin-CO_2 to pyruvate is not completely understood although a plethora of data has been obtained from kinetic,[53] stereochemical,[54] subunit replacement,[55]

52 (*a*) H. G. Wood, *Fed. Proc.*, 1976, **35**, 1899; (*b*) H. G. Wood, J. P. Chiao, and E. M. Poto, *J. Biol. Chem.*, 1977, **252**, 1490; (*c*) N. G. Wrigley, J. P. Chiao, and H. G. Wood, *J. Biol. Chem.*, 1977, **252**, 1500; (*d*) E. M. Poto and H. G. Wood, *Biochemistry*, 1977, **16**, 1949.

53 (*a*) D. B. Northrop, *J. Biol. Chem.*, 1969, **244**, 5808; (*b*) D. B. Northrop and H. G. Wood, *J. Biol. Chem.*, 1969, **244**, 5820.

54 Y. F. Cheung, C. H. Fung, and C. T. Walsh, *Biochemistry*, 1975, **14**, 2981.

55 (*a*) H. G. Wood, H. Lochmueller, C. Riepertinger, and F. Lynen, *Biochem. Z.*, 1963, **337**, 247; (*b*) M. Chueng, F. Ahman, B. Jacobson, and H. G. Wood, *Biochemistry*, 1975, **14**, 1611.

isotope,[56] model,[57] and n.m.r. studies.[58] A plausible mechanism for CO_2 transfer is shown in Scheme 9. The mechanism incorporates a pertinent observation by Rose *et al.*[56] who observed that transcarboxylase not only catalyses CO_2 transfer between methylmalonyl-CoA and pyruvate, but also proton transfer. When [3-$^3H_1$]pyruvate and unlabelled methylmalonyl-CoA were incubated together with transcarboxylase a small amount of pyruvate-derived tritium was located in the *pro*-2*R* position of propionyl-CoA. Conversely, incubation of (2*R*)-[2-$^3H_1$]propionyl-CoA and unlabelled oxaloacetate resulted in the transfer of a small amount of tritium to pyruvate. Rose *et al.* considered biotin as the most likely proton transfer agent. Thus, during transfer of CO_2 to [3-$^3H_1$]pyruvate, *via* a cyclic mechanism as shown in Scheme 9, some tritium is transferred to the ureido oxygen of biotin which

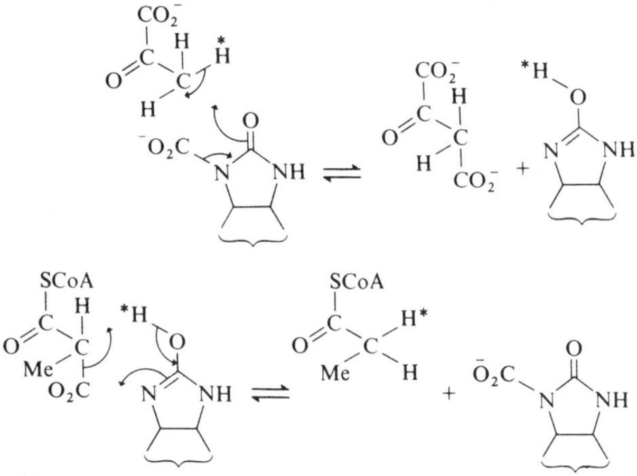

Scheme 9

is then converted to carboxybiotin by another molecule of methylmalonyl-CoA. The carboxylation of biotin by methylmalonyl-CoA must be sufficiently fast that complete exchange of the tritium with solvent does not occur and some of it ends up in propionyl-CoA. A modified mechanism to the one above has been proposed by Cleland.[59] However, the problem of a concerted or stepwise mechanism for biotin-dependent carboxylation reactions is unresolved.[60]

5 Adenylate Cyclase: Receptor Complexes

Adenylate cyclase catalyses the conversion of adenosine-5′-triphosphate into cyclic 3′,5′-adenosine monophosphate, and plays an important role in the cellular modulation of regulatory signals. It is generally accepted that the extracellular messengers, hormones, bind to a receptor protein on the external surface of the cell membrane and this binding process activates the adenylate cyclase.[61] Fundamental

[56] I. A. Rose, E. L. O'Connell, and F. Solomon, *J. Biol. Chem.*, 1976, **251**, 902.
[57] H. G. Wood, *Trends Biochem. Sci.*, 1976, 4.
[58] A. S. Mildvan, *Accounts Chem. Res.*, 1977, **10**, 246.
[59] W. W. Cleland, *Adv. Enzymol.*, 1977, **45**, 273.
[60] J. Stubbe and R. H. Abeles, *J. Biol. Chem.*, 1977, **252**, 8338.
[61] See M. F. Greaves, *Nature*, 1977, **265**, 681.

to the understanding of the mechanism of adenylate cyclase is the elucidation of the structural and functional organization of the receptor and cyclase molecules.

Using the technique of radiation-enhanced inactivation using an electron beam, the component sizes of the glucagon receptor-cyclase system from rat liver plasma membranes have been determined[62] in the absence and presence of glucagon. The technique neatly demonstrated that in the absence of hormone, the receptor and cyclase have molecular weights corresponding to their individual molecular weights, and thus are probably unlinked. In the presence of glucagon, association of receptor and cyclase takes place, and the complex appears to be in the dimeric form with two molecules each of receptor and cyclase.

The interaction between receptors and adenylate cyclase from different cells using the technique of cell fusion has also been reported.[63] The catecholamine receptor from turkey erythrocytes has been shown to activate, and thus interact with, adenylate cyclase from mouse erythroleukaemia cells within minutes of cell fusion. Recently it was demonstrated[64] that the catecholamine receptor can be donated and coupled to adenylate cyclase from a number of diverse cell types, *e.g.*, erythrocytes containing receptor have been fused with mouse adrenal tumour cells; rat glioma cells containing receptor have been fused with mouse erythroleukaemia cells.

Thus evidence has been produced which supports the 'mobile-receptor' hypothesis, that receptor and cyclase are freely dissociable in cell membranes. On agonist-receptor binding, receptor-cyclase interaction is locked causing activation of cyclase and thus formation of *c*AMP to perform its regulatory functions within the cell.

6 Multienzyme Complexes

Multi-enzyme complexes are aggregates of enzymes which catalyse two or more steps in a metabolic sequence, the most-widely known example being fatty acid synthase, a cluster of seven enzymes which catalyses the biosynthesis of fatty acids from acetyl- and malonyl-CoA. Many enzymes have been implicated in multi-enzyme complexes and the advantages of association of enzymes for metabolic pathways has been discussed.[65]

An extreme example of a multienzyme complex is a multifunctional polypeptide, a single protein chain having more than one catalytic activity. Multifunctional polypeptides are common in the biosynthetic pathway to the aromatic amino-acids, tyrosine, phenylalanine, and tryptophan; examples being; chorismate mutase-prephenate dehydrogenase of *Escherichia coli*,[66] tryptophan synthase of *Neurospora crassa*,[67] indol-3-yl glycerol phosphate synthase of *E. coli*,[68] and 3-deoxy-D-arabino-heptulosonate 7-phosphate synthase–chorismate mutase of *Bacillus*

[62] M. D. Housley, J. C. Ellery, G. A. Smith, T. R. Hesketh, J. M. Stein, G. B. Warren, and J. C. Metcalfe, *Biochim. Biophys. Acta*, 1977, **467**, 208.
[63] J. Orly and M. Schramm, *Proc. Nat. Acad. Sci. U.S.A.*, 1976, **73**, 4410.
[64] M. Schramm, J. Orly, S. Eimerl, and M. Korner, *Nature*, 1977, **268**, 310.
[65] See for instance, L. J. Reed and D. J. Cox, in 'The Enzymes' (Student Edition), ed. P. D. Boyer, Academic Press, New York, 1970, vol. 1, Ch. 4.
[66] B. E. Davidson, E. H. Blackburn, and T. A. A. Dopheide, *J. Biol. Chem.*, 1972, **247**, 4441.
[67] W. H. Matchett and J. A. DeMoss, *J. Biol. Chem.*, 1975, **250**, 2941.
[68] T. E. Creighton and C. Yanofsky, *J. Biol. Chem.*, 1966, **241**, 4616.

subtilis,[69] which all contain bi-functional polypeptides. A trifunctional polypeptide, anthranilate synthase from *N. crassa* has also been described.[70]

The *arom* multienzyme complex of *N. crassa* has been shown[71] to catalyse five consecutive steps in the aromatic amino-acid biosynthetic pathway, the conversion of 3-deoxy-D-arabino-heptulosonate-7-phosphate (23) into 5-enoylpyruvoyl-shikimate-3-phosphate (28), the immediate precursor of chorismic acid which is itself the common precursor of tyrosine, phenylalanine, and tryptophan. The pathway proceeds *via* 3-dehydroquinate (24), 3-dehydroshikimate (25), D-shikimate (26), and 3-phospho-shikimate (27), as shown in Scheme 10.

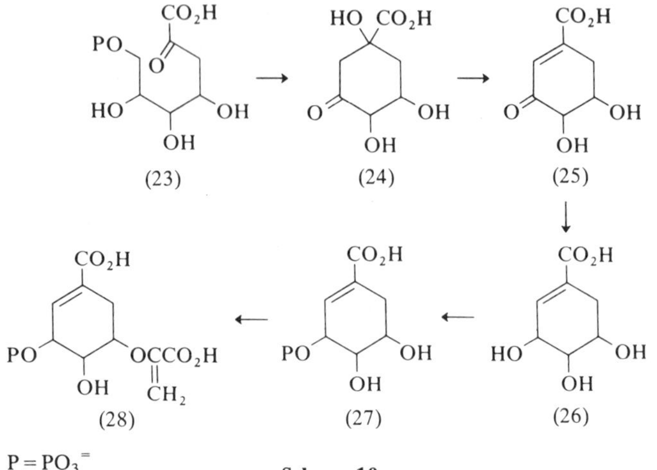

Scheme 10

Recently a new procedure for the purification of the *arom* complex from *N. crassa* has been reported[72] which minimized possible proteolytic damage to the enzyme. A homogeneous enzyme was obtained, MW 270 000, which under dissociating conditions produced apparently identical subunits of 165 000. Evidence was presented that proteolysis leads to progressive degradation of the complex into smaller subunits which retain enzymic activity, thus accounting for the confusing subunit compositions previously obtained.

It was concluded that the *arom* complex consists of two identical subunits each composed of a single polypeptide-chain capable of catalysing all five reactions. It would appear that discrete folding domains exist for each catalytic activity and that the domains have become connected as a result of fusion of the five structural genes coding for the enzymes, and indeed, it has been reported,[71,73] that the five genes occur as a cluster. The co-synthesis of these five activities thus provides an economy of synthesis of the enzymes of the biosynthetic pathway.

[69] L. Huang, A. L. Montoya, and E. W. Nester, *J. Biol. Chem.*, 1974, **249**, 4473.
[70] F. M. Hulett and J. A. DeMoss, *J. Biol. Chem.*, 1975, **250**, 6648.
[71] N. H. Giles, M. E. Case, C. W. H. Partridge, and S. I. Ahmed, *Proc. Nat. Acad. Sci. U.S.A.*, 1967, **58**, 1453.
[72] J. Lumsden and J. R. Coggins, *Biochem. J.*, 1977, **161**, 599.
[73] S. R. Gross and A. Fein, *Genetics*, 1960, **45**, 885.

Pyruvate dehydrogenase multienzyme complex[74] from *E. coli* catalyses the reaction according to the generally accepted mechanism shown in Scheme 11.

$$\text{Pyruvate} + \text{NAD}^+ + \text{CoA} \rightarrow \text{Acetyl-CoA} + \text{NADH} + \text{H}^+ + \text{CO}_2$$

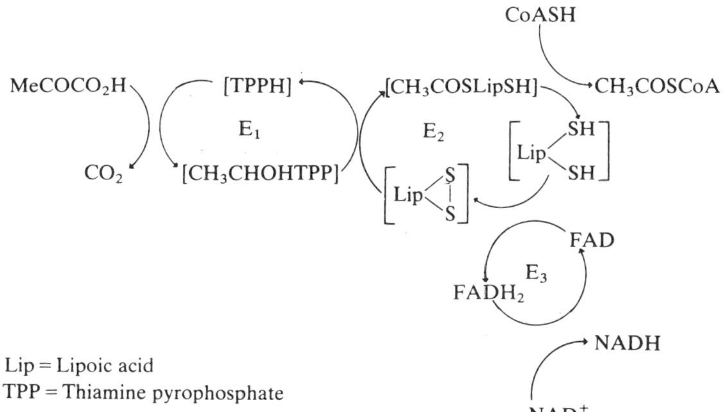

Lip = Lipoic acid
TPP = Thiamine pyrophosphate

Scheme 11

The complex has a large molecular weight ($\sim 8 \times 10^6$) and is composed of multiple copies of three different subunits having pyruvate decarboxylase (E_1), lipoate acetyltransferase (E_2), and lipoamide dehydrogenase (E_3) activity, respectively.

There has been some confusion as to the stoicheiometry of the subunits in the complex and the symmetry of the complex (see[78]). Evidence has been provided[75,76] for the involvement of a lipoyl-lysine 'swinging arm' which can transfer the 'acyl' moiety between subunits, but the mechanism cannot be simple since each E_2 subunit carries two functionally active lipoyl residues.[77]

Recent studies[78] on the self-assembly of the complex have demonstrated a 2:1 stoicheiometry of $E_1 : E_2$ and shown that the overall catalytic activity is proportional to the state of assembly of the complex. In the presence of [2-^{14}C]pyruvate, and absence of CoA, the lipoic acid residues of the complex become acetylated and the reaction ceases. Thus a partial reaction involving E_1 and E_2 subunits can be studied, while excluding that involving E_3. It has been shown[78] that the dependance of acetylation of the E_2 subunit on $E_1 : E_2$ ratio (partially re-assembled complexes) is not, as predicted, by a simple $E_1 : E_2$ ratio of 2:1, but rather by an $E_1 : E_2$ ratio of $24 \pm 2 : 1$.

Thus it would appear that each E_2 subunit can be 'catalytically serviced' by 12 E_1 dimers. On this basis a complex sub-structure is postulated consisting of a central E_2 core, a cube with octahedral symmetry comprising 24 polypeptide chains. Since each E_1 dimer can acetylate half of the total E_2 subunits and must perform this

[74] L. J. Reed, *Accounts Chem. Res.*, 1974, **7**, 40.
[75] M. C. Ambrose and R. N. Perham, *Biochem. J.*, 1976, **155**, 429.
[76] H. J. Grande, H. J. Van Telgen, and C. Veeger, *European J. Biochem.*, 1976, **71**, 509.
[77] M. J. Danson and R. N. Perham, *Biochem. J.*, 1976, **159**, 677.
[78] D. L. Bates, M. J. Danson, G. Hale, E. A. Hooper, and R. N. Perham, *Nature*, 1977, **268**, 313.

reaction indirectly (for direct acetylation, large distances would have to be spanned) transacetylation reactions between E_2 subunits have been postulated. An E_1 dimer bound to one of the E_2 subunits could cause direct acetylation of either 3 (or 4) E_2 subunits. Indirect acetylation (transacetylation) could then occur between a directly-acetylated E_2 subunit and either 3 (or 2) other E_2 subunits. Thus only 12 E_2 subunits could be serviced by each E_1 dimer.

This 'transacetylation' reaction provides a novel functional connection between active sites; however, the significance of the reaction is not yet apparent.[78]

The possibility of interaction between the glycolytic enzymes and the advantages of such interactions have been frequently discussed.[79] Recently direct evidence has been produced in favour of the presence of such glycolytic complexes. Glycolytic enzymes from rat skeletal muscle have been shown[80] to exist as a complex with myosin. However, this situation could be regarded as an exclusive case, being a consequence of accommodating rapid anaerobic respiration during muscle action.

Recently a large aggregate has been isolated from *E. coli* spheroplasts[81,82] which demonstrates all the enzyme activities of glycolysis. Gel chromatography demonstrated the presence of a fraction with high mol. wt. (1.6×10^6) containing all the enzyme activities, as well as low mol. wt. fractions (corresponding to the individual enzymes). Reassembly of the individual enzymes was demonstrated on concentration of the fractions. Total flux through the glycolytic pathway from [*U*-^{14}C]glucose to pyruvate was demonstrated. The presence of unlabelled glycolytic intermediates reduced the specific radioactivity of the pyruvate (isolated as alanine) by a much smaller extent than that expected for free mixing of intermediates, and thus a functional organization within the complex is indicated.

Another glycolytic complex has been reported[83] from the parasitic protozoan, *Trypanosoma brucei*, where all the glycolytic enzymes are associated with a rapidly sedimentable particle on sucrose density gradient centrifugation. Latency studies have indicated that the enzymes were contained in a microbody which has been named the 'glycosome'. These organisms, when in the morphological form found in the host bloodstream, rely completely on a modified glycolysis scheme for energy production, the respiratory chain and citric-acid cycle enzymes being absent. Glucose is converted to pyruvate and NADH reoxidized *via* coupled glycerol-3-phosphate dehydrogenase : glycerol-3-phosphate oxidase. It would appear that the glycosome has developed to optimize conditions for glycolysis, with high substrate and enzyme concentrations being maintained inside the microbody.

[79] See, for instance, C. de Duve, in 'Structure and Function of Oxidation-Reduction Enzymes', ed. A. Akeson and A. Ehrenberg, Pergamon Press, Oxford and N.Y., 1972, pp. 715-728.
[80] F. M. Clarke and C. J. Masters, *Biochim. Biophys. Acta*, 1973, **327**, 233; 1974, **358**, 193.
[81] J. Mowbray and V. Moses, *European J. Biochem.*, 1976, **66**, 25.
[82] D. M. Gorringe and V. Moses, *Biochem. Soc. Trans.*, 1978, **6**, 167.
[83] F. R. Opperdoes and P. Borst, *F.E.B.S. Letters*, 1977, **80**, 360.

Author Index

Aalbersberg, W. G. L., 63
Aasen, S. M., 166
Abbayes, H. D., 127
Abbond, J.-L. M., 7
Abboud, M. M., 394, 412
Abe, K., 386
Abeles, R. H., 432, 433, 434, 435, 442
Abramovitch, A., 136, 176, 382
Abramovitch, R. A., 108, 109, 274
Achari, A., 17
Achiwa, K., 121, 205
Acton, N., 303
Adachi, I., 365
Adam, W., 169, 170, 258, 282
Adams, D. R., 184
Adcock, W., 151, 216
Adembri, G., 263
Adler, A. E., 20
Adler, V. E., 369
Adolin, G., 278
Adzima, L. J., 261
Agopian, G. K., 239
Agranat, I., 230
Aguiar, A., 158
Ahlers, H., 151
Ahlrichs, R., 148
Ahman, F., 441
Ahmed, S. I., 444
Ahr, A.-J., 114
Ahrens, E. H., jun., 419
Aihara, J.-i., 215
Aihara, Y., 378
Airey, J. E., 265
Aizenshtat, Z., 125
Akhtar, M., 394, 409, 412, 432
Akimoto, T., 273
Akiyama, S., 247, 249, 291
Alange, G. G., 270
Alberti, F., 300
Alberts, A. W., 440
Albery, W. J., 437
Albinati, A., 13
Albizati, K. F., 140
Albright, T. A., 265
Alder, R. W., 281
Aldrich, J. R., 370
Alewood, P. F., 108
Al-Hazimi, H. M. G., 401
Aliwi, S. M., 90
Allegra, G., 13
Allen, L. C., 297
Allinger, N. L., 297
Allison, D. A., 216
Alonso, M. E., 201
Alper, H., 109, 127, 129, 243
Alvernhe, G., 274
Alworth, W. L., 423
Alzerreca, A., 258
Ambrose, M. C., 445
Ambrosius, P. M. M., 269
Amemiya, T., 245
Amin, S. G., 252
Ammon, H. L., 13
Amouroux, R., 187
Anastassiou, A. G., 276, 285
Anbar, M., 10
Anderson, A. G., 227
Anderson, L., 347
Anderson, R. C., 364, 366
Anderson, R. D., 39
Anderson, R. J., 386
Anderson, W. G., 173
Ando, T., 368
Ando, W., 114, 117, 148
Andrade, J. G., 147
Andreeva, M. A., 261
Andrews, G. D., 69, 172, 305, 306
Andrews, L. J., 226
Andrieux, C. P., 161
Andrieux, J., 124
Andrus, W. A., 325
Aneshansley, D. J., 374
Anet, F. A. L., 272, 296, 360
Anex, B. G., 22, 25, 32, 37, 39
Angerbauer, R., 350
Angres, I., 100
Angyal, S. J., 343
Anh, N. T., 190
Anjo, D. M., 144
Ankner, K., 161
Annis, G. D., 127, 335
Ansell, H. V., 230, 240
Ansell, J. M., 381, 383
Anselme, J.-P., 109
Apeloig, Y., 136
Appel, R., 258
Arai, H., 273
Arai, M., 327
Arakawa, S., 253
Araki, Y., 364
Aranda, G., 363
Aratani, M., 242
Aratani, T., 299
Archibald, J. L., 420
Arentzen, R., 355
Arias, I. M., 440
Arigoni, D., 410, 425, 427
Ari-Izumi, A., 327
Armand, J., 162
Armarego, W. L. F., 282
Armitage, I. M., 424, 425
Arn, H., 368, 379
Arnarp, J., 356
Arndt, H. C., 287, 387
Arnett, E. M., 48, 85, 202
Aronowitz, Y. J., 22
Arsura, E., 368
Asao, T., 244
Asaoka, M., 330
Ashby, E. C., 141, 146, 309
Ashe, A. J., 278
Ashkenazi, P., 246
Ashworth, B., 100
Ashworth, R. W., 218
Asirvatham, M. R., 154
Askari, M., 295, 297
Atkins, P. W., 90
Atkins, R. J., 171
Atkinson, R. S., 107, 109
Atsumi, K., 260
Attia, S. Y., 189
Attinà, M., 223
Atzmüller, M., 235
Aue, D. H., 86, 297
Aumann, R., 127
Ausloos, P., 7
Au-Yeung, B.-W., 136
Awad, S. B., 109
Ayalon-Chass, D., 331
Ayediran, D., 228
Ayer, W. A., 374
Ayscough, P. B., 103
Azami, T., 7
Azegami, I., 260
Azogu, C. I., 109

Baasner, B., 63, 249
Baba, H., 23, 30
Baba, S., 384
Babiarz, J. E., 258
Babler, J. H., 382, 387
Babsch, H., 218, 279
Babson, J., 433
Bacher, A., 39
Bachler, V., 60
Bachman, G. L., 122
Back, T. G., 336
Backinowsky, L. V., 347
Backlund, S. J., 144, 149, 332
Bäckvall, J.-E., 192
Baggiolini, M., 379

Bagus, P. S., 110
Bahl, J. J., 300
Bailey, B. K., 369
Bailey, F. C., 319
Bakalik, D. P., 102
Baker, D. C., 353
Baker, E. W., 19
Baker, R., 134, 185, 367, 379, 390
Baker, T. C., 369, 377, 378
Balchunis, R. J., 235
Baldwin, J. E., 64, 69, 172, 260, 285, 305, 306, 340
Baldwin, J. M., 431
Balke, D. E., 31
Balthazor, T. M., 261
Balyeat, J. R., 117
Bamford, C. H., 90
Bamkole, T. O., 228
Banasiak, D. S., 147
Banda, F. M., 154
Bandura, A. V., 45
Banerji, A., 265
Banet, D. M., 311
Bangerter, F. W., 436
Banister, A. J., 270
Banno, K., 326, 381
Banoub, J., 344, 346, 347
Banwell, T., 225
Barak, A. V., 370
Barbadoro, S., 259
Barber, G. N., 285
Barber, W., 295
Bard, A. J., 162
Barden, R. E., 440
Barfield, C. S., 376
Barker, D. A., 344
Barker, J. M., 282
Barlos, K., 143
Barltrop, J. A., 173
Barnard, G. F., 394, 409
Barnett, J. W., 224
Barras, S. J., 378
Barrett, A. G. M., 312, 349
Barrett, J., 417
Barron, P. F., 142
Barsotti, R., 43
Bartell, R. J., 368
Bartels-Keith, J. R., 271
Bartlett, P. A., 200, 324
Bartmess, J. E., 85, 86, 140, 202
Bartoli, G., 227, 264
Barton, D. H. R., 124, 173, 213, 229, 270, 312, 314, 325, 341, 349, 358, 359, 361
Barton, T. J., 147
Basha, A., 208
Basilevsky, M. V., 59
Basolo, F., 109
Bass, R. G., 116
Basset, J. M., 119, 133, 185, 304
Bassindale, A. R., 151
Bates, D. L., 445
Bates, G. S., 336
Bates, R. B., 300
Batlle, A. M. del C., 396
Bats, J. W., 12
Battersby, A. R., 393, 396, 397, 398, 399, 400, 401, 402, 404, 405, 407, 411, 413, 417, 419, 424, 425, 427
Batzold, F. H., 433
Baudoir, J. L., 13
Baudoug, R., 336
Bauld, N. L., 68, 298
Baumgärtel, O., 292
Baumstark, A. L., 341
Bausch, J., 398
Bauschlicher, C. W., jun., 110
Bayles, R., 236
Beak, P., 139, 282
Bean, G. P., 282
Beard, B., 377
Beaucage, S. L., 354
Beauchamp, J. L., 85, 88
Beavers, W. A., 300
Beck, J. F., 398
Becker, H. D., 279
Becker, K. B., 310
Becker, R. S., 31, 32
Beckhaus, H.-D., 180
Beckmann, B. G., 222
Bedard, W. D., 372
Beer, R. J. S., 268
Beevor, P. S., 368
Begley, M. J., 294, 334, 377
Beheshti, I., 169, 193
Behrens, U., 253
Beier, B. F., 135
Belcher, R. V., 411
Belkind, B. A., 262
Bell, A. A., 190, 377
Bell, B., 270
Bell, L., 48
Bell, W. J., 386
Bellas, M., 220
Bellus, D., 65, 290
Belzecki, C., 197, 198
Benati, L., 112, 253
Bendig, J., 237
Benedetti, E., 122
Benisek, W. F., 439
Benitez, F. M., 112
Benjamin, D. M., 377
Benn, F. R., 68, 192
Benn, M., 199
Benn, M. H., 375
Bennett, J. A., 30
Bennett, J. E., 101
Bennett, M. J., 374
Bensel, N., 202
Benson, A. M., 439, 440
Benson, B. W., 108
Benson, S. W., 101
Bente, P. F., 5
Bentley, T. W., 71
Bentor, Y., 230
Bentrude, W. G., 100
Benu, M., 108
Bercaw, J. E., 220
Berchtold, G. A., 218
Bergbreiter, D. E., 141, 204
Berger, D., 201
Bergman, R. G., 63, 65, 113, 217, 220, 297, 308
Bergmann, K.-H., 427
Berg-Nielsen, K., 110
Bergstrom, G., 372
Berisford, C. W., 368
Berk, H. C., 216, 294
Berkovitch-Yellin, Z., 12
Berlin, K. D., 282
Berliner, E., 178
Berman, H. M., 17
Bernard, D., 131
Bernasconi, C. F., 228
Bernauer, K., 255
Berndt, A., 298
Bernhardt, J. C., 175, 229
Berning, W., 170
Bernstein, H. J., 33
Beroza, M., 369
Berry, J. M., 356
Berry, J. P., 189
Berthod, H., 50
Bertrand, M., 184, 318
Bertsch, K., 182
Best, G., 268
Bestmann, H. J., 181, 369, 383, 385
Bethell, D., 99
Bettolo, R. M., 159
Beugelmans, R., 282
Bewick, A., 156
Beyermann, M., 237
Beynon, J. H., 3, 7, 9
Bhatnagar, S. P., 184
Bhatt, R. S., 355
Bhattacharyya, S., 336
Bianchi, M., 122
Bickart, P., 137
Bickelhaupt, F., 218, 233
Bidan, G., 168
Bieber, W., 234
Bielski, R., 312, 349
Bieri, J. H., 255
Bierl, B. A., 375
Biggs, I. D., 237
Bilhou, J. L., 133, 185, 304
Billingham, N. C., 180
Billingham, R. C., 309
Billings, R. F., 372
Billman, W., 383
Billups, C., 34
Billups, W. E., 114, 286, 287
Binger, P., 301
Binkley, J. S., 136
Binkley, R. W., 352
Binkley, W. W., 352
bin Samsudin, M. W., 171

Birch, D. J. S., 28
Birch, M. C., 372
Birge, R. R., 30, 31
Birks, J. B., 28
Birnbaum, G. I., 296
Bischof, P., 299, 301
Bischofberger, K., 348, 349
Bishop, S. W., 223
Black, T. G., 359
Blackburn, E. H., 443
Blackburn, T. F., 322
Blacklock, T. J., 166, 167
Blair, A. S., 5
Blair, L. K., 85
Blakeney, A. J., 286, 287
Blankenship, R. M., 294
Blauer, G., 25
Blight, M. M., 372
Bloch, K., 433
Blomberg, C., 141
Blount, J. F., 194, 265
Bloxham, D. P., 433
Blum, J., 106, 125
Blum, L., 229
Blum, M. S., 370, 372, 374
Blum, R. B., 326
Blumenfeld, J., 122
Blunt, J. W., 296
Boatman, R. J., 138, 222
Bobbitt, J. M., 157
Bobek, M., 359
Bobkova, R. G., 282
Boche, G., 69
Bochmann, M., 262
Bocian, D. F., 296
Bock, H., 268
Bockhoff, F. M., 5
Bodart-Gilmont, J., 254
Bode, W. M., 368
Bodkin, C. L., 343
Bodoev, N. V., 237
Boeckman, R. K., jun., 111, 331
Böhme, P., 199
Boekelheide, V., 234
Boelens, H., 338, 387
Boelhouwer, C., 133, 185
Bogdanowicz, M. J., 287, 387
Bogorad, L., 396
Bohandy, J., 19, 22
Bohlen, P., 436, 437
Bohm, H., 116
Bohme, D. K., 85, 87
Bolewska, K., 14
Bolin, D. R., 173
Bolis, G., 44
Boller, E. F., 377
Bolton, R., 230
Bon, M., 271
Bonaccorsi, R., 46
Bonini, B. F., 269
Bonneau, R., 165
Bonnett, R., 11, 25, 400, 423, 429
Boon-Keng Teo, 163
Boontanonda, P., 302
Booth, B. L., 99
Boppre, M., 367
Borchers, F., 5, 6
Borden, J. H., 372
Borden, W. T., 68, 293
Bordwell, F. G., 86
Borer, R., 207
Borhani, K. J., 162
Borromeo, P. S., 186, 203, 313
Borst, P., 446
Boscher, J., 376
Bose, A. K., 202, 252
Bosnich, B., 120
Boswell, R. F., 116
Botta, M., 196
Botteghi, C., 122
Boulares, L., 162
Bouma, W. J., 63
Bousch, G. M., 376
Bovill, M. J., 180
Bowden, B. F., 239
Bowen, R. D., 4, 5
Bowers, M. T., 86, 297
Bowers, W. S., 371, 373, 378
Bowman, D. F., 98
Boxer, S. G., 20
Boyd, D. R., 174, 197
Boyd, G. V., 270
Boyd, R. J., 113, 297
Boyd, R. K., 9
Boyer, F., 398
Boyer, J. H., 265
Bracke, J. W., 378
Bradford, K. C., 173
Bradlow, H. L., 437
Bradshaw, J. W. S., 379
Bradshaw, R. A., 440
Bradsher, C. K., 60
Brady, P. A., 240
Brady, U. E., 368
Braekman, J. C., 367, 372
Bräutigam, I., 200
Bräutigam, K.-H., 200
Brand, J. M., 372, 378
Branz, S. E., 64
Braslavsky, S., 110, 282
Braterman, P. S., 127
Brauer, O., 355
Brauman, J. I., 85, 87
Bregman, R., 226
Breslow, R., 242, 347
Brestkin, P. A., 177
Brettle, R., 154
Breuer, A., 168
Breuer, E., 311
Brewer, A., 239
Brewster, P. D., 222
Briat, B., 25
Bridges, A. J., 289
Briggs, N. H., 226
Brinker, U. H., 249, 291
Brinkmann, A., 301
Brinkmeyer, R. S., 205
Britton, L. N., 378
Brock, D. J. H., 433
Brocklehurst, K., 432
Brockmann, H., jun., 421, 423
Broekhof, N., 277
Broer, W. J., 5
Bromilow, J., 217
Brooks, C. J. W., 439
Brooks, D. W., 246, 290
Brown, C. A., 143, 176, 316
Brown, F. J., 9
Brown, H. C., 71, 72, 81, 142, 143, 188, 190, 282, 320
Brown, J. N., 37
Brown, R. D., 60
Brown, R. F. C., 118
Brown, W. V., 371
Brownbridge, P., 211
Browne, A. R., 238
Browne, L. E., 372
Browne, L. J., 289
Browne, L. M., 374
Brownlee, R. T. C., 217
Brownstein, S., 180
Bruce, M. I., 119
Bruck, D., 216
Bruder, H., 126
Brückner, J., 13
Brüntrup, G., 312
Bruggemann-Rotgans, I. E. M., 373
Bruice, T. C., 434
Bruins, A. P., 8
Brunet, J. J., 179
Brunsvold, W. R., 130
Bruza, K. J., 331
Bryce-Smith, D., 171, 219, 220
Brylikowska-Piotrowicz, J., 321
Bryson, T. A., 211, 221, 331
Bubb, W. A., 270
Bucciarelli, M., 197
Buchanan, G. W., 296
Buchanan, J. G., 348
Buchardt, O., 174, 273
Buckel, W., 410
Buckley, D. G., 402
Buckwalter, B. L., 201
Budzikiewicz, H., 239
Büchel, K. H., 264
Büchi, G., 187
Buehring, M., 372
Buenker, R. J., 27
Bug, J. E., 280
Buhl, H., 112
Bukin, V. N., 427
Buldain, G., 401
Bullock, E., 396
Bullpit, M., 216
Buncel, E., 228
Bundle, D. R., 344
Bunes, L., 196
Bunnell, C. A., 335

Bunnett, J. F., 227
Burden, I. J., 281
Buren, W. F., 374
Burford, C., 149, 324
Burgada, R., 271
Burger, D., 375
Burger, J. J., 313
Burger, K., 266
Burger, U., 218, 302
Burgess, M. T., 271
Burgstahler, A. W., 386
Burke, L. A., 59
Burkholder, W. E., 370
Burley, R. E. M., 230
Burlingame, A. L., 3
Burns, P. A., 258
Burns, W., 302
Bursey, M. M., 3, 9
Burton, D. J. 111
Busch, J. H., 55
Buse, C. T., 202, 326
Buser, H. R., 368
Bush, C. A., 36
Bushweller, C. H., 255
Butler, L. I., 369
Butler, R. N., 282
Butt, B. A., 369
Bykhovsky, V. Ya., 427
Byler, R. C., 370
Byrne, K., 379
Byrom, N. T., 388

Cabrino, R., 243
Cacace, F., 82, 223, 226
Caddy, P., 220
Cadiot, P., 182
Cadogan, J. I. G., 98
Cahiez, G., 131, 319
Caillean, H., 13
Caillet, J., 13
Cairncross, A., 306
Calas, R., 192
Calderazzo, F., 119
Callis, P. R., 37
Cambie, R. C., 189
Cameron, D. W., 239
Cammaerts-Tricot, M. C., 372
Campagna, F., 208
Camparini, A., 263
Campbell, D. E., 285
Camus, A., 123
Cannon, J. R., 152, 423
Capizza, A., 368
Capon, B., 75
Caporusso, A. M., 126
Capozzi, G., 179, 253
Caramella, P., 59, 61, 170
Card, P., 238
Cardé, A. M., 212, 368, 369, 386
Cardé, R. T., 212, 367, 368, 369, 377, 386
Carey, P. R., 33
Cariou, M., 77
Carlini, C., 126
Carlsen, P. H. J., 61, 170
Carlson, E. C., 377
Carmody, M. J., 246, 300
Carnduff, J., 240
Carotenuto, A., 269
Carotti, A., 208
Carpenter, A. T., 398
Carpita, A., 384
Carr, D. B., 131, 146
Carr, R. M., 274
Carrell, H. L., 17
Carrié, R., 254
Carroll, S. E., 107
Caruthers, W., 232
Case, M. E., 444
Casey, C. P., 130
Casini, G., 208
Cass, J. C., 265
Cassani, G., 368
Cassar, L., 127
Cassara, P., 436
Cassidy, R. F., jun., 370
Caubere, P., 105, 179
Caulton, K. G., 135
Cauquis, G., 275
Cava, M. P., 115
Cavaliero, J. A. S., 411
Cavallone, F., 43
Cavazza, M., 243
Cave, R. J., 65
Cavell, R. G., 216
Cavill, G. W. K., 374
Cerfontain, H., 225, 226
Cernigliaro, G. J., 383, 386, 389
Cessac, J., 68, 298
Chabaud, B., 275
Chadwick, D. J., 260
Chae, Q., 25, 30, 32
Chaloupka, S., 255
Chamberlain, W. T., 286
Chambers, D., 189
Chambers, D. L., 376
Chambers, J. Q., 157
Chambers, O. R., 109
Chambers, R. D., 174
Chan, B. G., 377
Chan, K.-K., 202
Chan, T. H., 117, 187, 385
Chandross, E. A., 239
Chang, C.-S., 73
Chang, Y.-H. 285
Chantrapromma, K., 66
Chapdelaine, M. J., 319
Chapman, D., 32
Chapman, O. L., 92
Chapman, R. A., 420
Chappell, I., 68, 192
Chapya, A., 377
Chaquin, D., 388
Chaquin, P., 168
Charbonneau, G. P., 13
Charles, G., 196
Charles, H. C., 239
Charmillot, P. J., 379
Chatt, J., 109
Chattha, M. S., 158
Chau, M. M., 261, 277
Chaudhry, I. A., 415, 417
Chaudhury, M. F. B., 375
Chawla, B., 48
Chawla, H. P. S., 252
Chellathurai, T., 108, 109
Chen, B., 137
Chen, H. H., 35
Chen, S. J., 262
Chen, T. B. R. A., 313
Chen, W. Y., 265
Cheng, L. D., 22
Cheng, P.-T., 75
Cheng, Y. Y., 161
Cheong, K. K., 37
Chern, C.-I., 322
Cherry, R. J., 32
Chertkov, V. A., 327
Cheung, Y. F., 441
Chheda, G. B., 17
Chiang, C. C., 14
Chiang, C.-S., 222
Chiang, W., 171
Chiao, J. P., 441
Chiasson, B. A., 218
Chiba, T., 155
Chiellini, E., 110
Childs, R. F., 215, 290
Chin, C. A., 23, 24
Chisholm, M. D., 368, 369
Chittattu, G., 206, 335
Chiu, S.-H. L., 347
Chivois, A. B., 19
Chmielarz, B., 234
Cho, H., 390
Chong, A. O., 365
Chong, L., 372
Choplin, F., 64
Chou, K. J., 201
Chow, M.-F., 170
Christensen, J. J., 174, 273
Christensen, R. L., 27, 28, 30
Christoffersen, R. E., 19
Christoph, G. G., 216, 238
Chuankamnerdkarn, M., 276
Chuche, J., 69, 254
Chudek, J. A., 246
Chueng, M., 441
Chuisoli, G. P., 125
Chuit, C., 189
Chwang, T. L., 177
Chwang, W. K., 189
Ciardelli, F., 126
Cimiraglia, R., 68
Ciranni, G., 223
Clardy, J., 61, 68, 216, 261, 375
Clark, D. T., 52
Clark, G., 130
Clark, J. H., 204, 244, 339
Clark, L. B., 35

Clark, M. G., 433
Clarke, F. M., 446
Clarke, M. T., 220
Clarke, T. C., 63
Claveni, P., 13
Clayton, R. K., 24
Cleland, W. W., 442
Clemens, A. H., 225
Clementi, E., 43, 44
Clennan, E. L., 163
Clezy, P. S., 408, 415, 417
Clive, D. L. J., 206, 312, 335
Clode, D. M., 357
Closs, G. L., 20
Coates, R. M., 235
Cockrum, P. A., 369
Coggins, J. R., 444
Coghlan, J. M., 387
Cohen, B. J., 139
Cohen, T., 223
Coke, J. L., 317, 387, 390
Cole, T. E., 323
Collins, J., 302
Collins, P. M., 362, 364
Collman, J. P., 431
Collonges, F., 313
Comins, D. L., 276
Comisarow, M., 89
Commerçon, A., 141
Cone, C., 7, 245
Conlon, M., 98
Connelly, N. G., 126
Conner, W. E., 374
Connert, J., 375
Connon, H. A., 302
Consiglio, G., 228
Conti, C., 386
Cook, D., 82
Cook, M. J., 263, 273
Cooke, F., 149, 324
Cooke, M. P., jun., 128, 139, 330, 340
Cooks, R. G., 3, 7, 8, 9, 10
Cookson, R. F., 265
Coombes, R. G., 224
Cooper, A. J. L., 435
Cooper, C. M., 74
Cooper, J. W., 97, 99, 100
Cope, B. T., 419, 429
Copenhafer, R. A., 158
Copland, D., 245, 260
Coppel, H. C., 371
Coppens, P., 12
Corbin, V. L., 367, 368
Corcoran, J. W., 336
Corey, E. J., 321, 326, 366
Cornforth, J. W., 410, 413, 419
Cornforth, R. H., 413
Cornu, H., 3
Corrigan, D. A., 163
Corriu, R. J. P., 149
Corwin, A. H., 19, 396, 406
Cory, R. M., 112, 341
Costa, L. C., 182
Costopoulos, M. G., 78, 241
Cottingham, A. B., 66
Cotton, T. M., 23
Couch, P. W., 411, 415
Coughlin, D. J., 169
Cousseau, J., 178
Coutrot, P., 314
Cowherd, F. G., 331
Cox, B. G., 208
Cox, D. J., 443
Cox, D. P., 151
Coxon, A. C., 281
Coxon, J. M., 296
Crabbé, P., 181, 336
Crabtree, R. H., 123
Craik, D. J., 217
Craine, L., 223
Cram, D. J., 281
Crampton, M. R., 227
Crandall, J. K., 313
Craven, B. M., 16
Creaser, I. I., 281
Creighton, T. E., 443
Cremer, D., 298
Cresp, T. M., 247, 248, 249, 291
Crewe, R. M., 374
Crimmin, M. J., 134, 185
Cromartie, T. H., 433
Cromarty, B. J., 52
Crombie, L., 377
Crosby, G. A., 139
Cross, J. H., 370
Crossland, N. M., 168, 341
Crowe, D. F., 365
Crumbie, R. L., 320
Csacsko, B., 305
Cu, A., 236
Cueto, O., 258
Cuingnet, E., 234
Culvenor, C. C. J., 369
Cunico, R. F., 63
Cunningham, A. J., 85
Cuppen, T. J. H. M., 174, 240
Curtin, D. Y., 14, 15
Curtis, M. D., 134
Curtis, W. D., 281
Cusmano, G., 268
Cussans, N. J., 325
Cuvigny, T., 203
Czarny, M. R., 108

Dach, R., 195
Dagan, A., 216
Dahl, L. F., 163
Dali, H. M., 238
Dallatomasina, F., 125
Daloze, D., 367
Dalton, J., 87
Daly, R. F., 324
Damji, S. W. H., 227
Dan, N., 242
d'Angelo, J., 126, 140, 221, 325
Danilov, V. I., 36
Danishefsky, S., 232, 289
Dannenberg, J. J., 49
Danson, M. J., 445
Dao, L. H., 269
Daray, R., 98
Darby, N., 247
Das, K. G., 7
Das, P. K., 31
Dattaguysta, J. K., 14
Daub, J., 245
Daum, H., 174
Dauner, H.-O., 404, 424
Dauplaise, D., 261
David, S., 350, 363
David, S. M., 364
Davidson, A., 20
Davidson, A. H., 186
Davidson, B. E., 443
Davidson, N., 35
Davidson, W. R., 297
Davies, J. E., 11, 429
Davies, L. B., 272
Davies, S. G., 218, 296
Davis, J. H., 65, 106, 113, 118, 217, 297, 308
Davis, R. E., 168, 239
Davis, W. H., jun., 222
Day, A. C., 173
Day, P., 24
Day, R. A., 324
Deakyne, C. A., 297
De Anngelis, F., 196
De Bruin, K. E., 210
de Camargo, J. M. F., 372
Declerq, J. P., 240, 254
de Duve, C., 446
Deeg, R., 427
Defay, N., 240, 241
Defaye, J., 353
de Fonseka, K. K., 64, 231
De Frees, D. J., 216
Degenhardt, C. R., 149, 187, 216, 294, 312
Degrand, C., 158, 159
Dehmlow, E. V., 110
Deininger, D., 49, 138
De Jeso, B., 196
De Jongh, D. C., 105
Dekkers, H. P. J. M., 169, 258
Del Alley, W., 100
de la Mare, P. B. D., 226
De La Mater, M. R., 223
de Las Heras, F. G., 248
Delaunay, L., 273
de la Vega, J. R., 55
Del Bene, J. E., 57
Déléris, G., 192
Dellaria, J. F., jun., 223
Dell'Erba, C., 240
Delley, B., 379
De Lue, N. R., 143
Delugeard, Y., 13
De Mairena, J., 433
de Maria, P., 208

de Mayo, P., 167
de Meijere, A., 305, 323
De Meyer, C., 4, 7
Demitras, G. C., 134
De Moss, J. A., 443, 444
Demou, P., 425
Dennis, R. W., 99
Denny, C. T., 174, 273
Deno, N. C., 181
Deol, B. S., 332
Derguini-Boumechal, F., 141
De Rossi, R. H., 228
Derrick, P. J., 3
Deschamps, B., 311
Descoins, D., 383
Desimoni, G., 60
Desirajin, G. R., 15
Destro, R., 12, 253
Desucher, J., 13
De Titta, G. T., 16
Detre, G., 365
Detsina, A. N., 225
Detty, M. R., 219
Devaquet, A., 68
Devilbiss, E. D., 375
Deville, C. G., 170
Deville, C. S., 61
Devlin, J. L., tert., 216
De Vries, R. A., 120
de Waard, E. R., 313
Dewar, M. J. S., 7, 65, 68, 145, 169, 241, 245, 298
de Wolf, W. H., 218, 233
De Young, S., 178
Deyrup, J. A., 262
Diakiw, V., 415, 417
Diamond, J., 28
Diaz, A. F., 161, 163
Di Cesare, P., 346
Dietz, A. G., jun., 223
Dietz, R., 130
Di Gregorio, S., 99
Dill, J. D., 136
Dillon, P. W., 68
Dilworth, J. R., 109
Dirks, G. W., 282
Disselkoetter, H., 379
Dittmer, D. C., 258
Dixit, D. M., 213
Dixon, H. B. F., 432
Dixon, W. T., 101
Djerassi, C., 9, 25
Dmytraczenko, A., 362
Dobashi, S., 168
Dobbie, R. C., 99
Dobbin, C. J. B., 269
Doddrell, D., 142, 216
Dolak, T. M., 211, 331
Dolbier, W. R., jun., 231
Doleschall, G., 323
Dolgushin, M. D., 47, 52
Dollat, J.-M., 181
Dollimore, D., 94
Dolphin, D. H., 20
Domelsmith, L. N., 59
Dondoni, A., 252
Donelson, D. M., 221
Donovan, D. J., 81
Donovan, S. F., 336
Doolittle, R. E., 369, 370
Dopheide, T. A. A., 443
Doria, M.-C., 331
Dorie, J., 208
Dotz, K. H., 130
Dougherty, D. A., 150
Dougherty, R. C., 87
Dowle, M. D., 208
Doyle, M. P., 188, 223, 313
Draggett, P. T., 126
Draxl, K., 3
Driguez, H., 344
Drozd, V. N., 227
Drucker, G. E., 86
Druey, J., 66
Dryhurst, G., 156
Dubois, J.-E., 84, 177, 206
Duboudin, J. G., 105
Duchet, D., 345
Dueber, T. E., 76
Dünnbier, K., 269
Dürr, H., 267, 268
Duesler, E. N., 261
Duffey, S. S., 370
Duffield, R. M., 372
Duhamel, L., 140
Duhl-Emswiler, B., 212, 386
Dumas-Bouchiai, J. M., 161
Dumont, W., 212
Dunaway-Mariano, D., 276
Dunbar, R. C., 7
Duncan, D. P., 148
Dunford, J. A., 176, 319
Dunkelblum, E., 165
Dunkin, I. R., 174, 273
Dunn, A. D., 348
Dunn, L. C., 245, 260
Dunogues, J., 180, 192
Du Priest, M. T., 195
Duréault, A., 243
Durette, P. L., 355
Durner, G., 314
Durr, H., 111, 114
Dussauge, A., 113
Dwyer, J., 68, 192
Dyck, V. A., 368
Dymerski, P. P., 5
Dyong, I., 365

Ealick, S. E., 14
Earnshaw, C., 186, 187, 321
Eastwood, F. W., 118
Eaton, W. A., 21, 38, 39
Eberbach, W., 69
Eberhard, W. G., 376
Eberson, L., 155
Eberstein, K., 353
Ebine, S., 244
Ebrey, T. G., 24
Eby, R., 346
Echegoyen, L., 163
Echigo, Y., 251
Echigoya, E., 133
Eckell, A., 61
Eckert, R., 25
Eckhardt, G. J., 25
Edgar, A. R., 348
Edgar, J. A., 369
Edge, D. J., 95, 99, 101
Edmiston, J. F., 369
Edmonds, J. S., 152
Edmondson, D. E., 39
Edwards, G. J., 156
Edwards, J. H., 192
Edwards, L., 20
Eggelte, H. J., 169
Eggerer, H., 410
Eggimann, W., 228
Eguchi, S., 116
Ehlinger, E., 149, 324, 331
Ehrlich, S., 178
Eichenauer, H., 204, 338
Eimerl, S., 443
Eisch, J. J., 145
Eisenstein, O., 60, 190
Eisner, T., 374, 375
Eitelman, S. J., 348, 349, 351
Eiter, K., 379
Eklind, K., 359
El-Abbady, S., 273
El-Deek, M., 282
Elder, G. H., 401, 407, 408
Elder, J. F., 7, 9
Elgavi, A., 252
Eliel, E. L., 272
Ellery, J. C., 443
Elliger, C. A., 377
Elliot, A. J., 90, 103
Elliot, J. D., 213
Elliot, R. M., 9
Elliot, W. J., 389
Elliott, I. W., jun., 155
Elliott, R. C., 223
Ellis, B. S., 183, 211
Ellison, G. B., 300
El-Sayed, M. A., 29
Elson, I. H., 99
El'tsov, A. V., 230
Elving, P. J., 153
Enders, D., 195, 204, 338
Endo, A., 436
Endo, K., 433
Endo, Y., 229
Engel, J., 16
English, T. H., 103
Enkaku, M., 248
Ephritikhine, M., 303
Erdtman, H., 280
Erman, M., 16
Ernest, I., 117
Erwin, V. G., 433

Eschenmoser, A., 423, 426
Essenbreis, H., 375
Evans, D. A., 325, 367, 378, 379, 390, 398, 399
Evans, D. H., 160, 163
Evans, G. T., 90
Evans, N., 407
Evans, R. H., 194
Evans, S., 6, 9, 265, 266
Evans, S. L., 372
Evaresto, R. A., 45
Eweiss, N. F., 252
Ewing, B., 372
Ewing, D. F., 216
Ewing, G. D., 215, 216, 294
Eyring, H., 37

Faehl, L. G., 295
Fahrni, H.-P., 243
Fairhurst, S. A., 99
Fales, H. M., 370, 374
Falk, J. E., 392
Fall, R. R., 440
Falou, S., 126, 140, 221
Falsig, M., 162
Fanconi, B., 37
Faragher, R., 266
Farcasiu, D., 116, 223
Farnum, D. G., 212, 280, 386
Farook, S., 423
Fedeli, W., 13
Fedorynski, M., 110
Fein, A., 444
Feinstein, G., 399
Feldmann, J., 365
Feldstein, A. C., 191
Felkin, H., 66, 189
Felman, S. W., 197, 338
Felstein, G., 383
Felton, R. H., 22
Fenner, H., 434
Ferguson, G., 302
Ferguson, J. A., 163
Feringa, B., 185
Ferramola, A. M., 407
Ferrer-Correia, A. J. V., 10
Ferrier, R. J., 351, 359
Ficini, J., 126, 140, 221, 243
Field, D. J., 64
Filippo, J. S., 322
Filler, R., 282
Filosa, M. P., 286
Finn, W. E., 378
Fiorentino, M., 227
Fioshin, M. Ya, 153
Firestone, R. A., 59, 110
Fischer, H., 101, 151, 167, 430
Fischer, J.-C., 352, 364
Fischer, P., 147
Fisher, C., 121
Fisher, J., 433
Fishwick, B. R., 183, 211
Fitjer, L., 294
Fitzgerald, T. D., 369
Fitzsimmons, B., 150
Flammang, R., 4, 7
Flatow, A., 233
Fleet, G. W. J., 322
Fleming, I., 136, 186
Fleming, M. H., 369
Fleming, M. P., 185
Flicker, W. M., 29
Flint, H. M., 377
Flood, T. C., 324
Florent, J.-C., 359, 362
Floyd, M. A., 378
Foa, M., 127
Förster, H., 232
Förster, S., 39
Folwell, R., 375
Fomum, Z. T., 184
Fonken, G. J., 307
Fookes, C. J. R., 405, 417
Foos, J. S., 64, 307
Ford, G. P., 65, 68
Ford, W. T., 61, 69
Forey, D. E., 377
Forni, A., 197
Forster, L. S., 36, 38
Forsyth, D. A., 80
Forte, P. A., 227
Fortier, S., 11
Forster, M. S., 85
Foster, R., 246
Foti, F., 264
Foucaud, A., 107
Fox, D. P., 118
Fox, J. J., 348
Fox, J. L., 38, 39
Fox, M. A., 306
Foxall, J., 101
Fowler, L. J., 436
Frackowiak, D., 39
Frahn, J. L., 369
Frajerman, C., 66, 189
Francesconi, K. A., 152
Franck, B., 397
Franck, R. W., 238
Francke, W., 372
Franck-Neumann, M., 62, 113, 267
Francotte, E., 254
Frandsen, E. G., 268
Frank, A., 236
Fraser, R. R., 391
Fraser-Reid, B., 349, 363, 364, 365, 366, 389
Fray, G. I., 171
Fredericksen, S., 4
Frediani, P., 122
Freeman, J. P., 266, 267
Freeman, P. K., 110, 117
Freer, A. A., 11
Freidlina, R. Kh., 90
Frey, H. M., 61
Fried, J., 389
Friedman, H. S., 278
Friedrich, E., 327
Friesen, M. D., 6
Fringuelli, F., 282
Fritchie, C. J., 39
Fritz, H., 218, 244, 279
Frölich, S., 111, 268
Fromm, J., 43
Froyen, P., 382
Fruehey, O. S., 318
Frydman, B., 398, 399, 400, 401, 403
Frydman, R. B., 398, 399, 400, 401, 403
Fryer, R. I., 265
Fryzuk, M. D., 120
Fu, P. P., 238, 297
Fu, Y. C., 37
Fu, Y. L., 359
Fucaloro, A. F., 36, 37, 39
Fuchikami, T., 126, 148
Fuchs, P. L., 335
Fügedi, P., 358
Fühlhuber, H. D., 279
Fünfschilling, J., 22
Fueno, T.,155
Fugate, R. D., 24
Fuhrhop, J.-H., 428
Fuji, K., 331
Fujimori, M., 63, 245
Fujimoto, H., 64
Fujino, S., 278
Fujise, Y., 244
Fujita, E., 236, 331
Fujita, K., 112
Fujita, M., 30
Fujiwara, K., 224
Fukami, H., 367, 368, 371
Fukami, J., 368
Fukazawa, N., 246
Fukui, H., 370
Fukui, K., 55, 64, 107, 247
Fukumoto, K., 10, 190, 314
Fukumura, M., 108
Fukunaga, T., 300
Fukuyama, Y., 336
Fukuzumi, S., 92, 93
Fung, C. H., 441
Fung, M. M., 24
Funk, R. L., 126, 314
Furakawa, J., 126
Furakawa, N., 108
Furneaux, R. H., 351, 359
Furusaki, F., 282
Fusco, F., 236
Fyfe, C. A., 226, 227
Fyles, T. M., 238, 380, 384

Gabhe, S. Y., 223
Gadelle, A., 353
Gagne, R. R., 431
Gais, H.-J., 205
Gal, J. Y., 8, 163
Galbraith, M. N., 390

Gale, R., 20
Galeazzi, E., 331
Gallagher, E. M., 369
Galle, J. E., 145
Gallenkamp, B., 70
Galli, C., 205
Gallois, M., 383
Games, D. E., 407, 408, 411, 415
Gamliel, A., 247
Gan, T. H., 217
Gandour, R. W., 61, 170
Ganguly, A. K., 366
Gantz, D., 397
Gara, W. B., 97
Garbers, C. F., 382
Garcia, B. A., 318
Gardlik, J. M., 215
Gardy, E. M., 101, 102
Garegg, P. J., 345, 354, 359
Gargano, M., 123
Garnett, J. L., 94
Garrett, V. H., 377
Garrison, P. J., 327
Gartner, B., 434
Gartshore, D., 216
Gase, R. A., 194
Gassman, P. G., 133, 185, 207, 235, 332
Gaston, L. K., 379
Gatmaritan, Z., 440
Gautier, A., 123
Gavezzotti, A., 46
Gavin, R. M., jun., 27
Gavrilov, L. D., 180
Gawley, R. E., 285
Gearhart, R. C., 266
Gedge, D. R., 334, 336
Gehlhaus, J., 375
Geiss, K.-H., 212
Geittner, J., 60
Gell, K. I., 243
Gensler, A., 248
Gentric, E., 270
George, B., 252
George, M. V., 61
Georgopapadakou, N., 424
Gerber, U., 255
Geresh, S., 122
Gerken, B., 372
Gerlach, H., 336
Germa, H., 271
Germain, G., 240, 254
Gessner, M., 238
Getman, D., 113, 166
Geue, R. J., 281
Gewitz, H.-S., 417
Ghisla, S., 433
Ghriofa, S. N., 98
Giacomelli, G., 126
Giacomello, P., 82, 223, 226
Giacovazzo, G., 11
Giam, C. S., 128
Giannoccaro, P., 123
Gibson, H. W., 319
Gibson, K. H., 398, 399
Giese, B., 180, 217
Giffney, J. C., 224
Giglio, E., 13
Gil, G., 184, 318
Gilb, W., 234
Gilbert, A., 171, 219, 220
Gilbert, B. C., 91, 92, 95, 96, 97, 100, 101
Gilbert, K. E., 305, 306
Gilbert, S. G., 17
Gilchrist, T. L., 266
Giles, N. H., 444
Gilgen, P., 172, 282
Gill, G. B., 68, 192, 314
Gillan, T., 98
Gilman, S., 320, 335
Gilmore, C. J., 11
Ginebreda, A., 110
Gingrich, H. L., 262
Ginsberg, R. J., 149
Ginsburg, D., 246
Giordano, C., 79, 123
Giovannini, E., 221
Girgenti, S. J., 276
Gladiali, S., 122
Gladieux, N., 359
Gladysz, J. A., 233
Glaser, R., 122
Gleiter, R., 299, 301
Glidewell, C., 150
Glover, S. A., 229
Glusker, J. P., 17
Glyde, E., 223
Goddard, W. A., 106, 113, 118, 256, 297
Godfrey, J. M., 25
Godfrey, M., 217
Goel, A. B., 141
Goerdeler, J., 269
Goering, H. L., 73
Goffin, E., 208
Gohee, Y., 275
Gokel, G. W., 229
Goldenberg, S., 368
Golding, B. T., 425
Golding, J. G., 224
Goldschmidt, E., 235
Goldstein, M. J., 69, 307
Goldwasser, J. M., 213
Golob, N. F., 287
Gonnermann, J., 200
Gordon, M. D., 300
Gordon, M. S., 241
Gore, W. E., 372, 389
Gorelik, M. V., 239
Gorman, J. E., 370
Gorringe, D. M., 446
Goswami, R., 139, 340
Gotoh, H., 109
Gotoh, N., 280
Gotthammer, B., 359
Gouesnard, J. P., 208
Gouin, L., 178
Gouterman, M., 19, 20, 21, 22
Gouverneur, P., 223
Grabe, B., 40
Grabley, F. F., 321
Grabley, S., 255
Gradushko, A. T., 20
Gräf, R., 147
Gragerov, I. P., 98
Graham, R., 6
Grahnen, A., 440
Grande, H. J., 445
Granick, S., 417
Grassi, G., 264
Graves, K. L., 379
Gray, J. R., 377
Graydon, W. F., 133, 185, 304
Grayson, J. I., 186
Graziano, M. L., 269
Greany, P. D., 376
Greaves, M. F., 442
Grebenkina, V. M., 230
Greé, R., 254
Green, F. R., III, 200, 324
Green, M., 126, 220
Green, M. L. H., 303
Green, M. M., 9
Greenberg, A., 285
Greenblatt, R. E., 370
Greene, A. E., 336
Greene, J., 134
Greenhouse, R., 68, 293
Greeves, D., 366
Gregory, B. H., 281
Greif, N., 67, 149
Gresh, N., 50
Gribble, G. W., 237
Grieco, P. A., 320, 335, 338
Grierson, J. R., 339
Griffin, A. C., 169
Griffin, G. W., 239
Griffith, R. C., 276
Griffiths, G., 183, 211
Grigg, R., 302, 388
Grima, J. P., 64
Grimm, K. G., 325
Gröbel, B. T., 140, 331
Gross, B., 346
Gross, S. R., 444
Grout, A., 83
Grovenstein, E., 66
Grubbs, R. H., 120, 132
Grudyushko, A. T., 20
Grütze, J., 233, 235
Gruntz, U., 196, 252, 334
Grunwell, J. R., 112
Gruska, R., 238
Guanti, G., 240
Günther, H. J., 289
Günther, M., 361
Günthert, P., 247
Guerin, C., 149
Guest, M. F., 149
Guggenberger, L. J., 266

Guinot, A., 182
Gunn, B. C., 276
Gunzer, G., 404
Gupta, B. G. B., 189, 324
Gupta, C. M., 348
Gupta, P. C., 408
Gurke, A., 233
Gustafsson, K., 279
Gutman, A. L., 246
Gutman, I., 215
Gutschow, C., 410
Guzier, F. S., 321

Habraken, C. L., 266
Hache, K., 314
Hacker, N. P., 241
Haddon, R. C., 246
Haenel, M. W., 233, 234
Hafner, K., 62, 243
Hagenbruch, B., 111
Hager, D. C., 282
Haggerty, J. G., 232
Hahnfeld, J. L., 111
Hai, T. T., 408
Haider, R., 301
Haiduc, I., 137
Haines, A. H., 281
Haines, C. P., 368
Hájek, J., 339
Hajivarnava, G. S., 353
Halbert, T. R., 431
Hale, G., 445
Halevi, E. A., 110, 299
Haley, N. F., 263
Halfhill, J. E., 369
Hall, C. R., 210
Hall, D., 226
Hall, D. R., 368
Hall, J. A., 61, 170
Hall, L. D., 355, 356, 361
Hall, R. H., 348, 349
Halton, B., 238
Ham, P. J., 377
Hamada, M., 365
Hamaguchi, H., 323
Hamana, H., 278
Hamer, G. K., 151
Hamilton, D. C., 167
Hammons, J. H., 299
Hanack, M., 75
Hanafusa, T., 225
Handoo, K. L., 99
Hanes, R. M., 323
Hanessian, S., 344, 346, 347, 366
Hansen, H.-J., 65, 235, 236, 282
Hansen, P.-E., 213
Hanson, A. W., 234
Hanson, L. K., 21
Hanzawa, Y., 278
Harada, T., 266, 330
Harayama, T., 390
Hardegger, B., 423
Harders, H., 39
Harding, L. B., 118
Harding, T. A., 110
Hardy, T. A., 117
Harger, M. J. P., 210
Hargrove, R. J., 76
Harkness, A. L., 419
Harless, J. M., 293
Harman, M. E., 109
Harman, P. J., 217
Harms, R., 199
Harnden, R. M., 157
Harnisch, J., 292
Harpool, R. D., 42
Harring, C. M., 372
Harris, E. J., 376
Harris, J. M., 75
Harris, R. L. N., 370, 407
Harrowfield, J. M., 281
Hart, H., 137, 165, 238, 239, 253
Hart, I., 268
Hartke, K., 208
Hartog, F. A., 141
Hartshorn, M. P., 179, 225
Hartzell, G. E., 63
Harvey, R. G., 238, 297
Hase, H.-L., 242
Hashimoto, S., 207, 342
Haskell, T. H., 355
Hass, J. R., 9
Hassner, A., 149
Haszeldine, R. N., 99
Hatanaka, N., 166
Hatch, R. P., 287
Hathaway, B., 62
Hauptmann, H., 11
Hauser, J. J., 261
Hausmann, M., 125
Hawley, M. D., 162
Hayashi, H., 275
Hayashi, J., 153
Hayashi, K., 142
Hayashi, N., 372
Hayashi, T., 119
Hayes, K. S., 277
Heath, R. R., 379
Heathcock, C. H., 202, 326
Heckert, D. C., 92
Hedden, R., 378
Hedin, P. A., 370, 378, 379
Heeman, V., 372
Hegedus, L., 119
Heger, I., 404
Hehemann, D. G., 352
Hehre, W. J., 7, 48, 59, 140, 216, 298
Heicklen, J., 110, 282
Heider, J., 110
Heidrich, D., 138
Heimbach, H., 6, 8
Heimbach, P., 126
Heimgartner, H., 255, 282
Heinzelmann, W., 172, 255
Heinzmann, R., 148
Heitzmann, M., 222
Hekman, M., 290
Heldeweg, R. F., 79
Helgée, B., 155
Helgeson, R. C., 281
Hellerman, L., 433
Hellman, L., 437
Hellwinkel, D., 138
Helmhold, R. B., 12
Helmkamp, G. M., 433
Helquist, P., 222, 313
Hemetsberger, H., 107
Hemingson, J. A., 230
Hemley, R., 32
Hemmerich, P., 434
Hemsworth, R. S., 85
Henderson, G. N., 70, 236
Hendriks, K. B., 343
Hendry, L. B., 373, 376, 377
Hengesbach, J., 262
Henne, A., 167
Henneke, K.-W., 199
Henning, R., 200
Henrici-Olivé, G., 119, 126, 185
Henrick, C. A., 185, 367, 368, 380, 386
Henriquez, R., 222
Henson, R. D., 376
Hentchova Hemo, J., 196
Herberich, G. E., 262
Herbert, J. A. L., 264
Hergrueter, C. A., 222
Herliky, J. M., 437
Herlt, A. J., 281
Hermann, H., 253
Herndon, W. C., 60
Herron, J. T., 3, 256
Herron, N. R., 222
Hersh, W. H., 133, 185
Herz, C. P., 233
Hesketh, T. R., 443
Hess, B. A., jun., 246
Hess, J., 258
Hetherington, W. M., 29
Hevey, R. C., 433
Heyns, K., 353, 365
Hiatt, R., 101
Hibbert, D. B., 102
Hiberty, P. C., 298
Hicks, D. R., 349, 364, 389
Hicks, K., 375
Hidaka, A., 127, 289
Higa, T., 10
Higgins, R. H., 258
Higgs, M. D., 379
Hill, A. S., 368, 369
Hillard, R. L., 105, 126, 220, 222
Hindenlang, D. M., 376, 377
Hinton, J. F., 42

Hirabayashi, Y., 215
Hirai, Y., 190, 314
Hirako, Y., 116
Hirano, C., 368
Hiraoka, K., 85, 89
Hiroi, K., 380
Hirotsu, K., 61, 68
Hirschler, M. M., 240
Hirsekorn, F. J., 135
Hirst, J., 228
Hisaoka, M., 239
Hishina, Y., 433
Hitchcock, P. B., 279
Hiyama, T., 112
Ho, C.-T., 111, 189, 202
Ho, K. S., 399, 424
Ho, T.-L., 323, 324, 332
Hoard, L. G., 16
Hobbs, P. D., 387
Hobolth, E., 158
Hochstrasser, R. M., 21, 174, 273
Hodgkin, D. C., 423
Hodgson, G. L., 396, 397, 402
Höfle, G., 199, 263, 269
Högberg, H. E., 234, 280
Höhne, H., 355
Hoekstra, M. S., 287
Hönig, H., 297
Hörster, H.-G., 243
Hoffman, M. K., 6
Hoffmann, H. M. R., 192, 329
Hoffmann, R., 208
Hoffmann, R. W., 111, 114
Hofmann, G., 138
Hofmann, H., 278
Hofrichter, J., 39
Hogeveen, H., 79
Hogg, A. M., 10
Hogrefe, F., 63, 249
Hohl-Blumer, M., 270
Holcomb, W. D., 108, 109
Holder, N. L., 364
Hollman, P. C., 433
Hollenstein, R., 398, 424, 425
Holm, K. H., 112
Holmes, J. L., 5
Holmes, R. G. G., 91, 92, 95, 99
Holtkotte, H., 375
Holtom, G. R., 29
Holubowitch, E. J., 237
Holy, N. L., 203
Hong, P., 130
Honig, B., 30
Honma, K., 367
Hooper, E. A., 445
Hopf, H., 233, 234
Hoppe, B., 257
Hopper, S. P., 149
Hoppin, C. R., 132
Horak, D. V., 231
Hori, A., 209
Horiike, K., 433
Horiike, M., 368
Horikawa, H., 154
Horita, H., 235
Horn, B. R., 377
Horn, D. H. S., 390
Horner, M., 162, 241, 291, 300
Horton, D., 352, 353, 357
Horwitz, J., 34
Hosaka, K., 230
Hoshino, M., 232
Hosomi, A., 230, 340
Houalla, D., 272
Hough, L., 355
Houghton, E., 374
Houk, K. N., 59, 61, 62, 170, 245, 260
Hounshell, W. D., 150
Housley, M. D., 443
Houssier, C., 23
Houts, J. J., 161
Howard, J. A., 94
Howes, P. D., 183, 211
Howie, J. K., 161
Howse, P. E., 378
Hoyt, S. C., 368
Hoz, S., 227
Hrib, N. J., 233
Huang, B.-S., 174
Huang, F., 262
Huang, L., 444
Hudrlik, A. M., 149, 319
Hudrlik, P. F., 149, 319
Hudson, B. S., 27, 28, 29
Hudson, C. W., 116
Hudson, R. F., 60, 190
Hudspeth, J. P., 288
Huebrer, K., 440
Hünig, S., 162, 170, 241, 291, 300
Huffman, J. C., 135
Hug, G., 31
Hug, P., 218, 279
Hug, W., 35
Hugelin, B., 126
Huggins, M.-A., 166
Hughes, L., 144
Hughes, N. A., 360
Hughes, P. R., 372, 378
Huie, R. E., 256
Huisgen, R., 59, 60, 61, 69, 184, 253, 282, 285
Hulett, F. M., 444
Hulshof, L. A., 173, 341
Hummel, J. P., 150
Hunt, E., 396, 407
Hunter, D. H., 70, 334
Huntress, W. T., 85
Hurst, J., 401
Hursthouse, M. B., 11, 429
Hurter, J., 377
Hutchins, R. F. N., 375
Hutton, R. S., 90
Huynh, C., 117
Huys-Francotte, M., 208
Hvistendahl, G., 4, 5
Ibers, J. A., 126
Içli, S., 248
Ide, T., 244
Iesce, M. R., 269
Igeta, H., 273
Ignatova, N. P., 282
Ihara, M., 10, 397, 407, 424, 425
Iitaka, Y., 273
Iizuka, K., 61
Ikeda, M., 275
Ikeda, T., 377
Ikeno, M., 148
Illuminati, G., 205
Imafuku, K., 244
Imagawa, T., 239
Imanaka, T., 126
Imfeld, M., 425
Imsgard, F., 32
Inaba, S., 345
Inagaki, S., 64, 215
Inai, T., 358
Inamoto, N., 217
Inazu, T., 233
Inbasekaran, M. N., 274
Inch, T. D., 210
Inesi, A., 159
Ingold, K. U., 94, 98, 102, 180
Ingraham, L. L., 24
Inokawa, S., 361
Inomata, M., 92
Inoue, I., 326, 365
Inoue, Y., 117, 165, 290
Inubushi, Y., 390
Inuzuka, K., 31
Ioki, Y., 92
Irngartinger, H., 235, 242
Iroff, L. D., 218
Isabelle, M. E., 221
Isagawa, K., 110
Iseli, R., 243
Ishida, T., 336
Ishido, Y., 345, 364
Ishii, S., 367, 368, 371
Ishii, Y., 126, 141
Ishikawa, K., 239
Ishikawa, M., 126, 148
Ishikawa, N., 229
Ishiyama, M., 248
Isobe, T., 377
Isomura, K., 107, 126
Ito, I., 273
Ito, S., 60, 244, 275
Ito, Y., 170, 260, 330
Itoh, A., 207, 342
Itoh, K., 126, 141
Itoh, M., 335
Ittah, Y., 106
Iversen, P. E., 162
Iversen, T., 354
Ives, J. L., 329
Iwaki, S., 378
Iwasaki, T., 154
Iwata, S., 54

Iwatani, F., 247
Iyoda, M., 291
Izawa, Y., 116
Izumichi, N., 61

Jackman, G. P., 118
Jackman, L. M., 137, 325
Jackson, A. H., 392, 399, 401, 407, 408, 411, 415
Jackson, J.-A. A., 7
Jackson, J. R., 411
Jackson, R. A., 180, 309
Jacob, P., 188
Jacobson, B. M., 191, 441
Jacobson, M., 369, 376
Jacquinet, J.-C., 345
Jäger, V., 289
Jahngen, E. G. E., 323
Jain, K. M., 30
Jain, S. C., 17
Jakabouski, A. A., 321
Jakoby, W. B., 439, 440
James, K. J., 343, 413
Janiga, E. R., 266, 267
Janousek, Z., 208, 252
Janson, T. R., 419
January, J. R., 162
Janzen, E. G., 102
Jaouni, T., 374
Jaques, B., 105
Jarrell, H. C., 362
Jastrzebski, J. T. B. H., 229
Jean, Y., 68, 111
Jedziniak, E. J., 181
Jefferson, A. M., 264
Jeffrey, G. A., 52
Jemmis, E. D., 140, 177
Jen, C. K., 19, 22
Jenner, P. J., 74
Jennings, K. R., 9, 10
Jenny, E. F., 66
Jensen, R. B., 4
Jentsch, R., 199
Jespers, J., 241
Jew, S.-S., 332
Jewett, D. M., 371
Jiang, J. B. C., 253
Jochims, J. C., 181, 182
Jørgensen, S. E., 4
Johansen, H., 24
Johansen, J. E., 32
John, R. A., 436
Johnson, A. W., 407, 423
Johnson, D. W., 402
Johnson, M., 327
Johnson, M. C., 236
Johnson, N. A., 100
Johnson, R. A., 102
Johnson, T. H., 133, 185
Johnson, W. L., 318
Jolidon, S., 65, 235
Joly, M., 240
Jona, R. J., 319
Jones, A., 205
Jones, A. J., 108, 180
Jones, D. W., 61, 64, 231
Jones, G. H., 281, 348
Jones, J. H., 213
Jones, J. K. N., 362, 364
Jones, K., 427
Jones, M., jun., 218
Jones, O. T. G., 417
Jones, P. G., 15
Jones, P. R., 137, 147
Jones, R. A., 282
Jones, S. R., 156
Jones, T. H., 374, 375
Jones, W. M., 110, 219, 249
Jordaan, A., 348, 349, 351
Jordan, P. M., 394
Jorgensen, W. L., 48, 300
Jousseaume, B., 105
Joussen, R., 151
Joussot-Dubien, J., 165
Julia, S., 110, 117
Jung, M. E., 137, 202, 204, 288, 318, 320, 322, 325, 326, 332
Jung, M. J., 436, 437
Junino, A., 184
Junk, G. A., 368
Jurd, L., 232
Jutzi, P., 145

Kaae, R. S., 379
Kaappe, J., 440
Kachinski, J. L. C., 302, 328
Kai, Y., 13
Kaib, M., 375
Kaiser, E. M., 137
Kaissling, K. E.,
Kaita, S., 275
Kaji, K., 257
Kajigaeshi, S., 245
Kajiwara, M., 399, 425
Kakehi, A., 275
Kakitani, H., 34
Kakitani, T., 34
Kalabin, G. A., 180
Kalinowski, H.-O., 136, 195, 336
Kalman, A., 277
Kalman, J. R., 5
Kamada, M., 157
Kamen, H., 426
Kametani, T., 10, 190, 314
Kamigata, N., 61, 170
Kamikawa, T., 377
Kamiya, Y., 102
Kamiyama, Y., 148, 257
Kammula, S. L., 218
Kanamaru, H., 189, 309
Kanbe, T., 155
Kande, A. S., 265
Kane, J., 262
Kaneda, K., 126
Kanemasa, S., 245
Kanematsu, K., 61
Kanghae, W., 327
Kanoktanaporn, S., 274
Kanzawa, H., 245
Kapnang, H., 196
Kapoor, V. M., 205
Kapteijn, F., 133, 185
Karich, G., 181, 182
Karl, W., 379
Karle, I. L., 15
Karlsen, S., 382
Karlsson, B., 280
Karntiang, P., 281
Karplus, M., 27, 30
Kasahara, T., 143
Kasai, N., 13
Kasha, M., 38
Kashdan, D. S., 323
Katada, T., 209
Katagiri, K., 378
Kato, K., 55
Kato, M., 139
Kato, S., 55, 156, 209
Kato, T., 342, 390
Katritzky, A. R., 196, 252, 263, 265, 272, 273, 275, 334
Katsoyannos, B., 377
Katz, J. J., 20, 23, 419, 429
Katz, T. J., 132, 133, 185, 303, 308
Katzenellenbogen, J. A., 380
Kauffmann, T., 151, 201, 280, 312
Kaufmann, G., 64
Kaupp, G., 166
Kausch, E., 255
Kausch, M., 111
Kavai, I., 359
Kawai, S., 142
Kawalek, B., 221
Kawamura, M., 257
Kawamura, T., 96
Kawanisi, M., 239
Kawano, T., 366
Kawasaki, K., 367
Kay, I. T., 407
Kazarians-Moghaddam, H., 101
Ke, B., 32
Kebarle, P., 85, 89
Keech, D. B., 440
Keefer, R. M., 226
Keehn, P. M., 235
Keeley, D. E., 287, 387
Keen, J. H., 439, 440
Kees, F., 278
Kees, K. L., 341
Keinan, E., 200, 324
Keiser, I., 376
Keller, L., 130
Kemp, D. L., 9
Kempe, T., 216
Kende, A. S., 61
Kennard, O., 15

Kenne, L., 356
Kennedy, G. G., 367, 368
Kenner, G. W., 399, 411, 419
Kenny, D. H., 196, 252, 334
Kent, J. E., 217
Kern, C. W., 53
Ketterer, B., 440
Kevan, L., 101
Khalil, G. E., 21
Khan, T., 123
Khan, Z. U., 107
Khanna, P. L., 242
Khattak, R. K., 262
Khor, T.-C., 216
Khuong-Huu, Q., 359, 362
Khurshudyan, S. A., 327
Kice, J. L., 261, 277
Kiji, J., 126
Kikuchi, K., 319
Killian, L., 144
Kim, B. F., 19, 22
Kim, C. S., 265
Kim, J. B., 19
Kim, J. H., 115, 222
Kim, M. G., 262
Kim, S. C., 143
Kim, Y. C., 398
Kimble, B. J., 3
Kimura, A., 248
Kimura, Y., 110
Kinberger, K., 262
King, A. O., 141, 175, 229, 317
King, D. S., 174, 273
King, J., 31
King, R. W., 249, 291
Kingston, D. G. I., 9
Kinkemo, C. L., 336
Kinnear, J. F., 390
Kirby, G. W., 109
Kirk, C. M., 96, 97
Kirkland, J. J., 429
Kirmse, W., 64
Kirrstetter, R. G. H., 235
Kishaba, A. N., 369
Kishino, K., 368
Kita, J., 107
Kitagawa, Y., 207, 342
Kitahara, Y, 246, 342, 390
Kitamura, C., 371
Kitamura, T., 239
Kitasawa, M., 233
Kitaura, K., 51
Kitching, W., 142, 151, 216
Kito, T., 240
Kitschke, B., 62, 265
Kjøsen, H., 32
Kleijn, H., 141, 176, 313
Klein, A. J., 160
Klein, J., 106
Klein, M. G., 370
Klein, R. S., 348
Klemer, A., 357
Kleschick, W. A., 202, 326
Kleveland, K., 111
Kliger, D. S., 29
Klimetzek, D., 369, 378
Klink, J. R., 237
Klinkmüller, K.-D., 202
Klioze, S., 424
Kluge, A. F., 374
Klun, J. A., 368
Klunkin, G., 220
Knapps, S., 280
Kneen, G., 64, 231
Knittel, P., 189
Knözinger, H., 124, 193
Knoll, F., 258
Knolle, J., 366
Knothe, L., 243
Knowles, J. R., 432, 437
Knowles, W. S., 122, 412
Knust, E. J., 224
Knutson, P. L. A., 137
Knyazev, V. N., 227
Ko, E. C. F., 77
Kobayashi, H., 21
Kobayashi, K., 260, 436
Kobayashi, S., 365
Kobayashi, T., 390
Kobayashi, Y., 278, 330
Kobrina, L. S., 222
Koch, T. H., 258
Kochansky, J. P., 367, 368
Kochetkov, N. N., 347
Kochi, J. K., 96
Kochloefl, K., 124, 193
Kocienski, P. J., 381, 383, 386, 387, 389
Koczorowski, G., 433
Kodama, H., 146
Kodama, M., 92
Koeberg-Telder, A., 225
Koebernick, H., 363
Koebernick, W., 363
Köll, P., 353
Kölle, O., 262
Köppel, C., 8
Koetzle, T. F., 12
Koge, M., 225
Kogure, T., 121, 205
Kohler, B. E., 27, 28, 30, 31, 32
Koizumi, M., 10
Kojo, S., 416
Koka, P., 23, 32
Koll, A., 227
Kollman, P., 57, 58
Kollmar, H., 241
Kolotilo, N. V., 278
Komae, H., 372
Komamura, T., 112
Komatsu, K., 63, 241, 245
Komornicki, A., 241, 298
Komura, H., 345
Kondo, K., 63, 275, 288, 312, 391
Kondrat, R. W., 10
Kongkathip, B., 388
Konishi, H., 241
Konno, Y., 275
Konoike, T., 330
Kopf, J., 361
Koptyug, V. A., 225, 237
Kornblum, N., 322
Korner, M., 443
Kornuta, P. P., 278
Korytnyk, W., 359
Korzeniowski, S. H., 229, 376
Kosbahn, W., 277
Koschatsky, K. H., 383
Koshy, K. M., 189
Kossanyi, J., 168, 388
Kosugi, M., 151
Kovner, M., 28
Kowalski, J., 192
Koyama, K., 108
Krajcarski, D. T., 420
Kramer, R., 313
Krantz, A., 174, 253, 257, 434
Krapcho, A. P., 323
Kraus, G., 288
Kraus, G. A., 320
Kraus, J. L., 433
Kraus, M. A., 139
Krause, J. G., 265
Krauss, D., 375
Krautler, B., 423
Krawczyk, Z., 197
Krepski, L. R., 185
Kresge, C. T., 299
Krestonosich, S., 220
Kretschmer, G., 258
Kreysig, D., 237
Krief, A., 212
Krieger, C., 235
Kriegesmann, R., 312
Kriemler, H.-P., 427
Krishnamurthy, S., 143, 320
Krolikiewicz, K., 336
Kropp, P. J., 167
Kroto, H. W., 149
Krueger, D. S., 326, 332
Kruger, A., 163
Kruger, T. L., 8, 10
Kruglyak, Y. A., 47
Krull, I. S., 378
Kruse, C., 277
Kruse, L. I., 285, 340
Krusell, W. C., 135
Krusic, P. J., 96
Kryger, R. G., 222
Krysin, A. P., 237
Ksander, G. M., 327
Ku, A., 61, 170
Kubo, I., 377
Kubota, T., 377
Kucherov, V. F., 327
Kudo, Y., 273
Kueh, J. S. H., 342
Kuepper, F. W., 386
Küsefoglu, S. H., 233
Kugelman, M., 365
Kuhla, D. E., 282

Kuhn, H., 25
Kukla, M. J., 216, 293
Kulakowska, I., 14
Kulkarni, S. U., 190
Kumada, M., 126, 148, 229
Kumadaki, I., 278
Kumagae, K., 233
Kumakura, M., 369
Kuper, D. G., 110
Kupperman, A., 29
Kurihara, M., 368
Kurita, J., 279
Kuroda, K., 168
Kuroda, S., 244
Kurtin, W. E., 34, 38
Kurzawa, J., 83
Kuster, T., 9
Kuwahara, Y., 371, 386
Kuwajima, I., 203, 209, 260, 326, 327, 333
Kuznetsov, S. G., 177
Kvarnström, I., 345
Kventsel, G. V., 26
Kwiatkowski, J. S., 38
Kyba, E. P., 116, 117

Laarhoven, W. H., 174, 240
Labaw, C. S., 265
L'abbé, G., 256, 257
Laberge, S. P., 38
Labhart, H., 37
Labovitz, J. N., 367, 368
Ladd, T. L., 370
Lämmerzahl, F., 138
Lagier, R. F., 368
La Grange, J., 58
Lai, C. C., 265
Laidler, D. A., 281, 356
Laing, G., 78
Lake, D. H., 221
Lakwijk, A. C., 379
Lal, B., 202
Lalanne-Cassou, B., 383
La Mattina, J. L., 277, 332
Lambert, G., 103
Lambert, J. B., 3, 272
Lambert, R. L., jun., 149
Lam-Chi, Q., 365
Lamm, B., 161
Lampin, J. P., 311
Landells, R. G. M., 98
Landman, D., 7, 245
Landor, P. D., 182, 184, 317
Landor, S., 118
Landor, S. R., 182, 184, 317
Lane, G. A., 296
Lane, M. D., 440
Lang, E., 295
Lang, G., 431
Lange, B., 199, 263
Lange, B. C., 137, 325
Lange, G. L., 166
Langs, D. A., 16
Lanier, G. N., 372
Larcheveque, M., 203
Lardicci, L., 126
Larock, R. C., 175, 229
Latham, D. W. S., 264
Lattes, A., 133, 185
Lau, P. W. K., 187
Laube, B. L., 154
Laue, H. A. H., 92, 96
Launer, C. R., 300
Laura, R., 433
Lauransan, J., 270
Laurenco, C., 314
Laureni, J., 174, 253
Laurie, V. W., 297
Laval, J.-P., 133, 185
Lavallee, P., 366
Lavrik, P. B., 216
Law, K. Y., 167
Lawrence, L. A., 368
Leardini, R., 264
Leary, G., 230
Leaver, D., 245, 260
le Bot, Y., 276
Le Breton, P., 85
Lechleiter, J. C., 166
Ledford, N. D., 295
Ledlie, D. B., 78, 241, 295
Le Drian, C., 336
Lee, C. B., 183
Lee, C. C., 74, 77
Lee, E., 424, 425
Lee, H. J., 437, 438
Lee, H. M., 63
Lee, S. J., 303
Lee, S. L., 424
Lee, T. S., 231
Lee, Y. W., 377
Leenstra, W. R., 21
Leffek, K. T., 83, 227
Lefferts, J. L., 149
Lefour, J. M., 60, 190
Legrand, Y., 208
Le Guillanton, G., 77
Lehman, T. A., 3
Lehmann, H., 243
Lehr, F., 200
Leight, R. S., 69, 307
Leiserowitz, L., 12
Lelli, M., 264
Lemal, D. M., 255
Lemieux, R. U., 343, 344
Lenoir, D., 185
Lentz, C. M., 138, 340
Leonard, J. J., 19
Lépine, M.-C., 363
Leroux, J., 346
Leroy, G., 58, 59
Lessinger, L., 12
Lester, D. J., 325
Lester, R., 368
Leung, H. W., 228
Leung, T., 144
Leung, Y.-K., 189
Levi, A. J., 440
Levinson, A. R., 370
Levinson, H. Z., 370, 376
Levit, A. F., 98
Levitt, L. S., 87
Levitt, T. E., 205
Levkoeva, E. I., 282
Levsen, K., 3, 4, 5, 6, 8
Levy, A. B., 176, 319
Levy, E. S., 399
Lewars, E., 253
Lewellyn, M., 166, 186, 292, 312
Lewis, N. J., 223
Lewis, T., 379
Lewis, T. P., 38
Lex, J., 63, 249
Ley, S. V., 127, 216, 229, 293, 325, 335
Leznoff, C. C., 213, 310, 380, 384
Liaaen-Jensen, S., 32
Liang, G., 241, 243, 246, 300
Lias, S. G., 7
Libman, N. M., 177
Liebelt, W., 124, 193
Liebl, R., 277
Liebman, J. F., 285
Lieder, C. A., 87
Liepa, A. J., 238
Liles, D. C., 150
Lilienblum, W., 111
Liljenberg, C., 419
Lim, T. F. O., 147, 148
Limbach, H.-H., 243
Lin, C. T., 14
Lin, J. J., 141, 309
Lindahl, K., 372
Lindberg, B., 356
Linden, H. W., 269
Lindley, J. M., 109
Lindley, N. B., 296
Lindley, P. F., 270
Lindner, H. J., 62, 265
Lindsay, D., 180
Lines, R., 162
Ling, A. C., 101
Link, H., 255
Linke, S., 108
Linstrumelle, G., 141, 384
Liotta, D., 207, 332
Liotta, R., 190
Lipkowitz, G. S., 434
Lipkowitz, K. B., 282
Lippert, B., 436, 437
Lipscomb, W. N., 54
Lipták, A., 358
Lipton, M. S., 307
Lissel, M., 110
Little, W., 322
Litton, J. F., 8, 10
Litwack, G., 440
Liu, J. C., 258
Liu, L.-K., 168, 239
Llambias, E. B. C., 396

Llort, F. M., 137
Lloyd, H. A., 372
Loc, C. V., 10
Lochmann, R., 56
Lochmueller, H., 441
Lockridge, O., 433
Lönngren, J., 356
Lofquist, J., 372, 374
Logan, D. M., 379
Lohmann, J.-J., 113, 267
Lohse, C., 174, 273
Lombardino, J. G., 282
Longo, F. R., 19
Longone, D. T., 233
Loof, I. H., 114
Loots, M. J., 177, 313
Lorand, J. P., 222
Lorenc, J. F., 266, 267
Lorne, R., 141
Loskant, G., 369, 378
Lotts, K. D., 285
Loubinoux, B., 179
Loudon, A. G., 4
Louterman-Leloup, G., 58
Lovas, F. J., 256
Lo Vecchio, G., 264
Lovey, A. J., 323
Loza, R., 113
Lu, S. H., 423
Luanglath, K., 107
Lubineau, A., 350
Lubosch, W., 136, 336
Lucchese, R. R., 110
Lucchini, V., 179, 253
Luche, J.-L., 181
Ludwig, M. L., 39
Lüddecke, E., 267
Lüdemann, H.-D., 295
Luger, P., 270
Lukáč, J., 255
Lukefahr, M. J., 190, 377
Lumsden, J., 444
Lunazzi, L., 98
Lund, H., 158, 159
Lusby, W. R., 371
Luteri, G. F., 61
Luthy, J., 410
Lutz, M., 19
Lutz, W., 327
Lwowski, W., 108
Ly, N. D., 260
Lyda, M., 95
Lyle, R. E., 276
Lynch, B. M., 217
Lynd, R. A., 317
Lynen, F., 441
Lysenko, Z., 206, 335
Lyster, M. A., 202, 320, 332
Lythgoe, B., 65, 312
Lyznicki, E. P., 189

Maarsen, P., K., 226
Mabry, T. J., 377
McAdoo, D. J., 9
McAlister, D. R., 220
Macaluso, G., 268
Macauley, E. D. M., 379
MacBride, J. A. H., 274
McBride, J. M., 110
McCaffery, L. K., 21
Maccagnani, G., 269
McCandless, L. L., jun., 379
McCapra, F., 169, 193
McCaskie, J. E., 258
McClain, W. M., 29
McClelland, R. A., 77
McCloskey, C. J., 341
McCombie, S. W., 312, 349, 361
MacConnell, J. G., 374
McConwick, J., 11
McCullough, J. J., 64, 231
McCurry, P. M., jun., 289
McDaniel, C. R., jun., 295
MacDiarmid, A. G., 163
McDonach, A. F., 400
McDonald, E., 393, 396, 397, 398, 399, 401, 402, 404, 405, 407, 411, 413, 417, 424, 425, 427
McDonald, F. J., 374
MacDonald, L. M., 379
McDonald, R. N., 162
MacDonald, S. F., 398, 420
MacDonell, G. D., 282
McDonough, L. M., 369
McDowell, P., 390
McEwen, F. L., 376
McGarvey, G., 319
McGinnis, J., 132, 185, 303
McGovern, W. L., 370
McGuirk, P. R., 313
Machiguchi, T., 242, 299
McHugh, A. J., 19
McIver, R. T., jun., 85, 140, 202, 216
Mackay, D., 269
Mackay, G. I., 87
McKean, D. R., 318
McKee, M. L., 65, 145
McKelvey, J. H., jun., 367
McKervey, M. A., 302
McKibben, G. H., 370
McKillop, A., 147, 222, 229
McKinnie, B. G., 139
Mackinnon, J. W. M., 109
McLafferty, F. W., 5, 7, 8, 9, 10
McLaren, F. R., 112, 341
McLaughlan, K. A., 104
McLaughlin, J. L., 10
McLean, J. A., 372
McLennan, D. J., 82, 83
McMahon, T. B., 85
McManus, S. P., 47, 75
McMaster, I. T., 108, 109
McMeeking, J., 301
McMurray, J. E., 185, 327, 336, 341
McOmie, J. F. W., 241
McOsker, C. C., 188, 313
McPhail, A. T., 366
MacPhee, J. A., 206
McQuilkin, R. M., 378
McQuillan, F. J., 192
McRobbie, I. M., 109
Madan, P. B., 265
Madawinata, K., 199
Madhavarao, M., 129
Madsen, H., 368
Märkl, G., 277, 278
Märky, M., 172
Magee, W. L., 113
Maggiora, G. M., 19, 20
Magnus, P., 149, 324, 331, 387
Magnusson, G., 383
Mah, H. D., 398
Mahaffy, P. G., 231
Maier, G., 242, 257, 274, 299
Main, P., 11, 12
Maisey, R. F., 236
Maister, S. G., 437
Maitlis, P. M., 123
Mak, N., 10
Maki, Y., 142, 257, 319
Makinen, M. W., 39, 431
Makino, T., 16
Malek, F., 180, 309
Malek, J., 339
Malkiewich, C. D., 227
Malkowski, J., 270
Mallams, A. K., 365
Mallik, B., 30
Mallon, C. B., 111
Malpass, J. R., 107
Mamatyuk, V. I., 225
Mammarella, R. E., 151, 312
Mandai, T., 189
Mandal, A. K., 190
Mandal, K., 30
Mandell, L., 324
Mander, L. N., 399
Mandolini, L., 205
Mangner, T. J., 9
Mango, F. G., 133
Mangum, M. G., 118
Manhas, M. S., 202, 252
Manisse, N., 254
Mann, C. K., 154
Manning, C., 64, 220, 231
Manning, M. J., 232
Mant, D. M., 11
Mantulin, W. W., 23
Maquestiau, A., 4, 7
Marchon, T.-C., 431
Marcinal, P., 234
Marcotte, P., 433
Marfat, A., 313
Margoliash, E., 396
Margolis, H., 261
Margolis, Z., 86
Mariano, P. S., 171, 276
Marino, G., 282
Marino, J. P., 289

Mark, F., 60
Mark, H. B., jun., 163
Markiewicz, W., 207
Markovac, A., 420
Markovetz, A. J., 378
Markovskii, L. N., 278
Markowski, V., 69, 253
Markwell, R. E., 402
Marquet, A., 438
Marrero, R., 332
Marschall, H., 202
Marsden, J. R., 118
Marshall, J. A., 166, 186, 292, 312
Marshall, J. H., 163
Marshall, J. L., 262, 276, 295
Marshall, P. D. R., 95
Marten, D. F., 129
Martigny, P., 161
Martin, G. J., 208
Martin, H.-D., 290
Martin, J. C., 261
Martin, K. O., 437
Martin, M. D., 390
Martin, M. J., 382
Martin, M. L., 208
Martin, R. H., 240, 241
Martin, S. F., 195, 327
Martina, D., 62
Martineau, A., 105
Martinez, R. I., 256
Martyr, R. J., 439
Marumo, S., 378
Maruoka, K., 207
Maruthamuthu, P., 99
Maruyama, K., 61, 313
Maryanoff, C. A., 277
Marziano, N. C., 224
Masaka, K., 189
Masaki, M., 107
Masamune, S., 242, 246, 252, 290, 299, 336
Masamune, T., 317, 330
Masaoka, K., 380
Mashenkov, V. A., 21
Masilamani, D., 192
Massardo, P., 368
Massey, V., 433, 434
Massot, R., 3
Massuda, D., 117
Masters, C. J., 446
Masuda, S., 217
Masumoto, K., 241
Matchett, W. H., 443
Math, V. B., 25
Matheson, J. H., 396
Mathewson, J. H., 406
Mathey, F., 311
Mathis, F., 271
Matinopoulos-Scordou, A. E., 227
Matjeka, E. R., 232, 260
Matlin, S. A., 105, 407
Matsen, F. A., 299
Matsuda, K., 350
Matsue, H., 317
Matsui, K., 231
Matsui, M., 347, 354, 360, 366, 386, 390
Matsumoto, K., 154
Matsumoto, M., 63, 168, 250, 391
Matsumura, F., 370, 371, 377
Matsumura, H., 244
Matsumura, T., 93
Matsumura, Y., 153, 323
Matsuno, A., 345
Matsuo, M., 200
Matsushima, R., 358
Matsuura, T., 230, 231
Mattay, J., 171, 220
Matteoli, U., 122
Matthews, W. G., 86
Maurin, R., 184
Maycock, A. L., 432, 433, 434
Mayr, H., 80
Mazenod, F., 218, 302
Mazid, R. M., 318
Mazur, Y., 200, 324
Mazza, F., 13
Mazzanti, G., 269
Medimagh, M. S., 69
Meegan, M. J., 405
Meesters, A. C. M., 199
Meguro, S., 124
Meier, H., 63, 112, 115, 198, 279, 282
Meijer, J., 141, 149, 176, 313
Meiler, V. W., 49
Meinwald, J., 261, 280, 367, 374, 375
Meissner, U. E., 248
Meister, J., 180
Melchior, M. T., 223
Melega, W. P., 149, 187, 246, 300, 312
Mellon, F. A., 372
Mellor, J. M., 156, 281
Mellor, M., 342
Mel'nikov, G. D., 177
Mel'nikov, N. N., 282
Mel'nikov, S. P., 177
Menchen, S. M., 312
Mendelowitz, P. C., 149
Mendelsohn, R., 33
Menke, K., 233, 234
Menzer, R. E., 371
Menzies, W. B., 245, 260
Merenyi, R., 254
Meritt, M. V., 102
Merrifield, D. L., 295
Merz, A., 159
Messer, L. A., 181
Mestroni, G., 123
Metcalf, B. W., 436, 437, 443
Metcalfe, D. A., 65
Metcalfe, J., 110
Meth-Cohn, O., 109, 264
Metzger, J., 270
Meyer, H. H., 388
Meyer, L.-U., 305
Meyer, M. D., 181
Meyer, R., 199
Meyer, T. J., 163
Meyer, V., 168
Michalski, J., 210
Michel, D., 49
Michelot, D., 384
Michl, J., 231
Middleton, E. J., 390
Middleton, R., 174
Midelfort, C. F., 437
Midland, M. M., 143, 181, 317
Migake, Y., 433
Migita, T., 151
Miki, Y., 275
Milat, M.-L., 345
Mildvan, A. S., 442
Millard, A. A., 335
Millen, P. W., 270
Miller, B., 141, 232, 260
Miller, D. C., 355, 361
Miller, I. J., 230
Miller, J. M., 204, 244, 339
Miller, J. R., 377, 379
Miller, J. S., 85
Miller, R. B., 319
Miller, R. W., 370, 371
Millington, D. S., 408
Mills, J. A., 343
Milun, M., 215
Mimura, T., 327
Minami, N., 209, 333
Minato, T., 64
Minks, A. K., 367, 368, 369
Minot, C., 60, 190
Minov, V. M., 227
Minter, D. E., 307
Mioskowski, C., 212
Mirbach, M. F., 171, 220
Mirbach, M. J., 171, 220
Mirek, J., 221
Miser, J. R., 223
Mislow, K., 137, 150, 277
Misra, R. N., 149, 319
Misra, T. N., 30
Misumi, S., 13, 233, 235, 250
Mitchell, E. B., 370
Mitchell, R. H., 238
Mitchell, T. R. B., 302
Mitsuhashi, T., 238
Miura, I., 377
Miura, Y., 233, 241
Miura, Z., 375, 377
Miyashita, D., 376
Miyaura, N., 335
Miyazawa, S., 436
Miyoshi, M., 154
Miyoshi, N., 189
Mizoroki, T., 124
Mizumachi, N., 390
Mizuno, H., 13
Mizuta, M., 209
Mizutani, J., 371
Mo, F., 13

Mock, W. L., 63
Modena, G., 178, 179, 253
Modro, A., 84, 178
Mody, N. V., 387
Moffatt, J. G., 348, 354
Mol, J. C., 133, 185
Molander, G. A., 143
Molnar, A., 278
Mombelli, L., 401
Monaghan, F., 240
Monder, C., 437, 438
Monneret, C., 359, 362
Monroe, R. E., 378
Monstrey, J., 6
Montevecchi, P. C., 112, 253
Montforts, F.-P., 397
Montgomery, L. K., 231
Montgomery, M. E., 373
Monti, S. A., 293
Montoya, A. L., 444
Moodie, R. B., 224
Moore, B. P., 371
Moore, T. A., 22, 28, 30, 32, 33, 38, 39
Morales, O., 63
Mordenti, L., 179
Moreau-Hochu, M. F., 105
Moretti, I., 197
Morey, K. S., 440
Morgan, E. D., 372, 379
Mori, K., 369, 372, 378, 379, 380, 381, 382, 386, 389, 390
Morimoto, N., 13
Morin, C., 282
Morin, F. G., 296
Morio, K., 290
Morita, T., 245
Morizur, J.-P., 168, 388
Morokuma, K., 44, 51, 54, 57
Moron, J., 396, 399
Morris, H. R., 417, 427
Morrison, A., 32
Morrison, G., 253
Morrison, H. A., 173
Morse, C. S., 225
Morton, J., 366
Morton, T. C., 420
Moses, P. R., 157
Moses, V., 446
Mosher, O. A., 29
Mosher, H. S., 121
Moss, G. P., 419
Moss, J., 440
Moss, R. A., 111, 200
Mostowicz, D., 197
Mott, R. C., 326
Mount, D. L., 75
Moutsokapas, A. A., 113
Mowbray, J., 446
Mpango, G. M., 184
Muchowski, J. M., 331
Müllen, K., 249
Müller, G., 398, 404, 424, 427
Mues, P., 249
Muetterties, E. L., 123, 134, 135
Muhs, W. H., 360
Mukaiyama, T., 203, 251, 318, 326, 381
Mulder, F., 46
Mulder, J. J. C., 218
Muller, P., 114
Muller, P. M., 423
Mullins, M. J., 280
Mulzer, J., 312
Munasinghe, V. R. N., 362, 364
Munavu, R. M., 354
Mundy, P. P., 282
Munoz, A., 271
Munroe, J. E., 48
Murahashi, S.-I., 110, 143, 189, 309, 313, 338
Murai, A., 317, 330
Murai, S., 127, 142, 189, 289, 312
Murakami, M., 100
Muramoto, H., 368
Murase, H., 317
Murata, I., 245
Murata, N., 108
Murawski, H.-R., 64
Murayama, K., 143
Murphy, M. J., 426
Murphy, W. S., 62
Murray, R. E., 317
Murrell, J. N., 149
Mutin, R., 133, 185, 304
Mychajlowskij, W., 187, 385
Myhre, P. C., 225

Nabbefeld, E. F., 126
Nader, F., 12
Näf, F., 384
Nagai, S. I., 273
Nagasawa, J., 364
Nagata, C., 92
Naiman, A., 126
Nair, K. S. S., 376
Nakadaira, Y., 148, 257
Nakagawa, K.-I., 148
Nakagawa, M., 246, 247, 249, 291
Nakagawa, Y., 108
Nakai, T., 327
Nakajima, H., 368
Nakamura, C. Y., 325
Nakamura, E., 203, 326
Nakamura, K., 195, 230
Nakamura, N., 386
Nakamura, R., 133
Nakanishi, K., 375, 377
Nakatsuji, S., 247, 249
Nakayama, J., 232
Nakazaki, M., 165, 233
Nakazawa, T., 245
Nakazima, Y., 342
Nánási, P., 358
Nandi, D. L., 394
Nash, A., 374
Nassr, M., 345
Natarajan, S., 317, 387
Nau, H., 3
Nault, L. R., 373
Nay, B., 107
Neckers, D. C., 172, 219
Negishi, A., 63
Negishi, E.-I., 132, 136, 141, 142, 143, 175, 176, 229, 282, 313, 316, 317, 382, 384
Neidert, E., 166
Neidle, S., 17
Neill, D. C., 174, 197
Nelander, B., 262
Nelsen, S. F., 163
Nelson, D. J., 103
Nervi, A. M., 440
Nesbitt, B. F., 368
Nesbitt, S. L., 325
Nester, E. W., 444
Neszmélyi, A., 358
Neuenschwander, M., 243
Newcomb, M., 69, 204
Newkome, G. R., 274, 282
Newman, M. S., 238
Newton, J. R., 379
Newton, M. D., 52
Newton, P. F., 296
Newton, R. F., 168, 341
Nibbering, N. M. M., 6, 7, 8, 273
Nicolaou, K. C., 206, 335, 336
Nicoletti, R., 196
Nie, P. L., 252
Nielsen, J. U. R., 4
Nielson, M. W., 373
Niki, E., 102
Ning, R. Y., 265
Nishida, R., 371, 386
Nishiguchi, I., 112
Nishiguchi, K., 13
Nishikawa, Y., 275
Nishikida, K., 100
Nishimoto, K., 19, 38
Nishinaga, A., 230, 231
Nishino, C., 371, 373, 378
Nishio, T., 108
Nishishita, T., 5, 8, 10
Nishiyama, K., 109
Nishizawa, K., 231
Nishizawa, M., 320
Nixon, L. N., 398
Noall, W. I., 189
Nocci, R., 126
Noding, S. A., 146
Noe, E. A., 205
Noell, J. O., 44
Nöth, H., 143
Noltes, J. G., 177, 229
Nomoto, T., 247
Nonhebel, D. C., 222

Nooijen, P. J. F., 371
Nooijen, W. J., 367, 368, 371
Nord, L., 262
Norden, B., 20
Nordman, C. E., 16
Norman, R. O. C., 91, 92, 95, 96, 97, 100, 101, 274
Normant, J. F., 131, 140, 141, 314, 319
Norris, D. M., 379
Norris, J. R., 23, 429
Norris, T., 270
Northrop, D. B., 441
Noto, R., 228
Nov, E., 236
Novi, M., 240
Novoselov, N. P., 45
Nowak, R. J., 163
Nowlan, V. J., 84, 184, 189, 248
Nowotnik, D. P., 265
Noyori, R., 203, 326
Nozaki, H., 112, 149, 176, 188, 207, 342, 384
Numan, H., 169, 258
Nunokawa, O., 327
Nutakul, W., 221
Nyburg, S. C., 75

Oae, S., 108, 282
Oakley, R. T., 278
Obayashi, M., 149, 176, 384
Oberhänsli, W. E., 255
O'Brien, D. H., 190, 377
O'Brien, E., 220
O'Carra, P., 429
Ochoa, M., 161
O'Connell, E. L., 437, 442
O'Connor, J. G., 235
O'Connor, J. P., 135
Oda, M., 246
Oda, N., 273
Odier, L., 353
Odyek, O., 182, 317
Oehling, H., 100
Oei, A. T. T., 94
Oei, H.-A., 174
Oertle, K., 336
Oettmeier, W., 429
Offenhartz, B. H., 24
Offenhartz, P. O., 24
Ogata, H., 433
Ogata, I., 119
Ogata, Y., 237
Ogawa, S., 280
Ogawa, T., 347, 354, 360, 366
Oh, S. W., 437
O'Hanlon, P., 408
Ohfune, Y., 335
Ohloff, G., 384
Ohnishi, Y., 194
Ohno, A., 195
Ohno, T., 142
Orht, J. M., 17
Ohsawa, A., 273
Ohta, K., 368
Ohta, T., 229
Ohto, N., 102
Oikawa, H., 246
Ojima, I., 121, 205
Ojima, J., 248
Oka, S., 195
Okamoto, H., 257
Okamoto, K., 63, 241, 245
Okamoto, T., 229
Okano, M., 179, 189
Okazaki, H., 179
Ōki, M., 238
Okinoshima, H., 148
Okude, Y., 112
Okukado, N., 141, 175, 229, 317
Okumura, K., 110
Olagbemiro, T., 374
Olah, G., 7, 78, 80, 81, 181, 185, 189, 202, 241, 243, 246, 298, 299, 300, 323, 324, 332
Olivé, S., 119, 126, 185
Oliver, J. P., 137
Olivers, W. L., 272
Ollis, W. D., 66, 270
Olmstead, W. N., 87
Olofson, R. A., 196, 285
Omoto, H., 153
Onan, K. D., 366
O'Neill, P., 96
Ono, M., 330
Ono, Y., 92, 93
Onoe, A., 179
Onyido, I., 228
Operaeche, N. N., 364
Opperdoes, F. R., 446
Oppolzer, W., 285
Orere, D. M., 338
Orger, B. H., 171
Orly, J., 443
Ornstein, P. L., 202, 318, 325
Orr, A. F., 336
Ors, J. A., 220
Osawa, E., 302
Osawa, Y., 16
Osborn, M. E., 276
Oscarson, S., 354
Oschmann, W., 262
Osella, D., 135
Osgerby, J. M., 398
O'Shea, D. C., 22
Oshima, K., 365
Osman, F. H., 272
Ostlund, N. S., 297
Ostrow, R. W., 381, 387
Osuka, M., 247, 249
Ota, K., 240
Otani, S., 142
Oth, J. F. M., 249
Othen, D. G., 150
Otsubo, T., 13, 233, 234, 235, 250
Otsuji, Y., 110
Otsuka, S., 238
Ovchinnikov, A., 26
Overend, W. G., 353
Overman, L. E., 140
Owicki, J. C., 43
Oyama, K., 189
Ozaki, A., 124

Packard, B. S., 272
Paddock, N. L., 278
Paddon-Row, M. N., 243, 258
Padwa, A., 61, 113, 166, 167, 170
Page, A. D., 263
Pahor, B., 220
Paige, J. N., 63
Paik, H. N., 129, 243
Paine, A. J., 77
Paine, J. B., III, 407
Painter, A., 347
Palakis, S. E., 440
Palensky, F. J., 173
Pancoast, T. A., 212, 386
Pandit, U. K., 194
Panfil, I., 197
Papadopoulos, E. P., 252, 265
Paquette, L. A., 61, 116, 149, 187, 215, 216, 219, 246, 293, 294, 299, 300, 312
Parfitt, R. T., 265
Parkanyi, L., 277
Parker, D. G., 181
Parker, D. H., 29
Parker, V. D., 162
Parkes, D. A., 101
Parkhurst, L. J., 32
Parlman, R. M., 128, 330
Parr, W. J. E., 295
Parriot, P., 140
Parrish, F. W., 343
Parrott, M. J., 99
Parthasarathy, R., 16, 17
Partridge, C. W. H., 444
Paschal, J. W., 295
Passannanti, S., 391
Passerini, R., 224
Patchornik, A., 139
Patel, R. C., 272
Patin, H., 124
Patnaik, P., 123
Paton, R. M., 98, 101
Patrick, T. B., 118
Pattenden, G., 334, 336, 342
Paul, I. C., 14, 15
Paulsen, H., 344, 353, 355, 361, 363
Paulus, H., 383
Pavlik, J. W., 173
Pawlak, M., 384
Payne, T. L., 378
Payzant, J. D., 85, 87
Peacock, J. W., 372

Peacock, R. D., 20
Pearce, A., 186
Pearce, G. T., 372, 379, 389
Pecul, K., 54
Pedder, D. J., 374
Pedersen, E. B., 137
Peel, J. B., 217
Pellacani, L., 109
Pelletier, S. W., 387
Pellicciari, R., 287
Pelter, A., 144, 205, 268
Peng, C., 137
Penney, M. R., 269
Pepe, J. P., 196
Perahia, D., 49
Percy, J. E., 367, 378
Perevalov, V. P., 261
Pereyre, J., 165
Perham, R. N., 445
Periasamy, M. P., 298
Peries, R., 140
Perkins, M. J., 102
Perlin, A. S., 346
Perrin, C. L., 224
Perry, N., 74
Perry, R. A., 334
Persico, M., 68
Persin, M., 163
Person, H., 107
Persoons, C. J., 367, 368, 371, 373
Perutz, M. F., 431
Pete, J.-P., 362
Peters, E. N., 71, 81
Peters, K. S., 256, 300
Peterson, M., 28
Petke, J. D., 19
Petrocire, D., 425
Petrov, A. A., 178
Pettei, M. J., 377
Pettersen, R. C., 276
Pettit, R., 323
Pettit, W. A., 64, 231
Petty, J. D., 137
Petty, R. L., 367
Pfaltz, A., 423
Phadungkul, N., 276
Philips, J. C., 63
Piacenti, F., 122
Piancatelli, G., 259
Picavet, J. P., 271
Piccardi, P., 368
Picker, K., 213
Pickholtz, Y., 125
Pickworth, J., 423
Pienta, N. J. 167
Pierce, B. M., 30
Piers, E., 339
Pieter, R., 195
Pietra, F., 243
Pikulik, I., 215, 290
Pilati, T., 12, 249, 253
Pilkiewicz, F. G., 375, 377
Pilotti, A. M., 280
Pinchuk, V. M., 47, 52
Pincock, J. A., 113, 297
Pine, S. H., 236
Pinet, M., 105
Pinnick, H. W., 285
Pitman, G. B., 372, 378
Pittman, C. U., jun., 135, 323
Placucci, G., 98
Platz, H., 369
Plaumann, D. E., 363
Plieninger, H., 430
Pliske, T. E., 369
Plomp, R., 258
Plummer, E. L., 379
Plummer, E. W., 193
Plus, R., 19
Pluscec, J., 398
Podslawczyńska, I., 213, 321
Poehlmann, H., 260
Poels, E. K., 266
Pohmakotr, M., 139, 205
Poindexter, G. S., 167
Poirier, J.-M., 140
Pojer, P. M., 343
Polglase, W. J., 411
Pollack, S. K., 216
Pommer, H., 184, 310
Pommier, J.-C., 196
Pomonis, J. G., 377
Pong, R. G. S., 174
Ponticelli, F., 263
Poor, R. W., 19
Popa, V., 137
Pople, J. A., 53, 136, 140, 177
Popják, G., 413, 419
Poppinger, D., 63, 140, 177
Portella, C., 362
Porter, N. A., 60
Posner, G. H., 138, 319, 340
Post, D. E., 29
Poto, E. M., 441
Potter, G. J., 189
Potts, K. T., 262
Pougny, J.-R., 345
Poulain, E., 55
Poulson, R., 411
Poulter, C. D., 186, 203, 313
Power, M. J., 348
Prakash, G. K. S., 185, 243, 324
Pramanik, B. N., 202
Preiner, G., 147
Preston, P. N., 270
Prestwich, G. D., 375
Pretzer, W. R., 135
Prickett, J. E., 109
Priesner, E., 369
Priester, W., 177
Primeau, J. L., 365
Prinstein, R. M., 5
Prinzbach, H., 70, 218, 243, 244, 279
Pritt, J. R., 75
Proidakov, A. G., 180
Prokseh, E., 323
Prosen, R. J., 423
Pross, A., 217
Protschuk, G., 287
Proveaux, A. T., 370
Proverbs, M. D., 379
Pryor, W. A., 222
Puckett, S. A., 14
Pullman, A., 41, 47, 49, 50
Pullman, B., 13, 38, 49, 50
Purdham, J. T., 374
Purdum, W. R., 282

Quagliata, C., 13
Quast, H., 256, 276
Quinkert, G., 314
Quinn, C. P., 101

Rabek, J. F., 90
Rabideau, P. W., 295, 297
Rabiller, C., 208
Rabinovich, D., 14
Rabinovitz, M., 216, 246, 247, 249
Radchenko, S. I., 178
Radmer, R., 396
Radom, L., 7
Rae, I. D., 216
Raghavan, R. S., 110
Rahimi, P. M., 248
Rajan, S., 227
Rakowski, M. C., 123
Ralph, L. M., 238
Ramaiah, M., 331
Raman, K. P., 62
Rampal, J. B., 277, 278
Rampazzo, L., 159
Ramsden, C. A., 252, 270, 272, 282
Ramsey, B. G., 144
Ramunni, G., 59
Rånby, B., 90
Randić, M., 215
Rando, R. R., 432, 433, 436
Ranieri, R. L., 10
Rao, B. L., 359
Rao, C. G., 71
Rao, Y. S., 282
Raoult, E., 162
Rapoport, H., 403, 416
Rapp, K. M., 245
Rapp, U., 8
Rappoport, Z., 75, 77, 78
Rasheed, K., 280
Rasmussen, J. K., 325
Raston, C. L., 152
Rathbone, E. B., 362, 364
Rathke, M. W., 335
Ratovelomanana, V., 117
Radom, L., 63
Rau, H., 267
Raucher, S., 318
Rauchschwalbe, R., 295

Rauscher, S., 368, 379
Ravid, U., 390
Ravindran, N., 142
Ravindranathan, M., 71, 72
Ravindranathan, T., 68, 293
Rawlins, M. F., 281
Rayez, J. C., 49
Raynolds, P. W., 232
Read, D. M., 272
Read, J. S., 368
Reasoner, J. W., 92
Reddy, G. S., 7
Redfern, J. R., 397, 411
Redmond, J. W., 410
Reed, C. A., 431
Reed, L. E., 114
Reed, L. J., 443, 445
Rees, J. H., 224
Reese, C. B., 338, 355
Reetz, M. T., 67, 83, 149
Regel, E., 264, 265
Regen, S. L., 110
Reich, H. J., 211
Reichert, O. M., 141
Reichle, R., 74
Reichmanis, E., 276
Reifz, T. J., 386
Reil, S., 403
Reinehr, D., 126
Reinhoudt, D. N., 282
Reissenweber, G., 279
Reissig, W. H., 369
Reiter, S. E., 245, 260
Reitz, D. B., 139
Reitz, T. J., 212
Renga, J. M., 280
Renger, B., 136, 195, 336
Rennecke, R.-W., 353
Renner, C. A., 308
Renwick, J. A. A., 372, 378
Renz, P., 423
Rétey, J., 410
Reuter, J. M., 328
Reutrakul, V., 327
Reynolds, W. F., 151, 216
Rezende, M. C., 196, 252, 334
Rho, M. M., 71
Rhyne, L. D., 241
Rice, S. A., 27
Rich, A., 38
Richards, K. E., 225
Richardson, A. C., 355
Richmond, G., 6
Richon, A. B., 390
Ricker, A., 279
Ridd, J. H., 224
Ridley, D. D., 320, 332
Rieke, R. D., 137, 158, 241
Riemann, J. M., 222, 274
Riepertinger, C., 441
Riganti, V., 60
Righetti, P. P., 60
Rimmer, J., 419
Rinehart, K. L., jun., 365
Ring, H., 127
Ripoll, J. L., 181
Ris, C., 225
Risch, N., 423
Risitano, F., 264
Rist, G., 290
Ritchie, C. D., 217
Ritter, F. J., 367, 368, 371, 373
Rivetti, F., 178
Robacker, D. C., 373
Robbiani, R., 9
Roberts, B. P., 97, 99, 100
Roberts, B. W., 128
Roberts, J. D., 87
Roberts, J. L., 186, 203, 313
Roberts, P., 302
Roberts, S. M., 168, 341
Roberts, T. D., 112, 253
Robertson, A. J. B., 102
Robertson, J. H., 423
Robertson, T. H., 127
Robey, M. J., 36
Robins, M. J., 360
Robins, R. K., 37
Robinson, C. H., 433
Robinson, J. A., 427
Robinson, W. T., 431
Rodemeyer, G., 357
Roden, K., 363
Rodewald, H., 235
Rodionov, V. Ya., 261
Rodriquez, O., 170
Roelofs, W. L., 367, 368, 369, 377, 378, 379
Rogers, V., 118
Rogić, M. M., 192
Roloff, A., 126
Rommel, E., 168
Rona, R. J., 149
Ronchi, N., 98
Roomi, M. W., 420
Rooney, J. J., 119, 302
Roos, B. O., 110
Roper, J. M., 282
Rose, I. A., 410, 437, 442
Rosenblum, L. D., 386
Rosenblum, M., 129
Rosenstock, H. M., 3
Rosenthal, D., 426
Rosini, G., 264
Rosmus, P., 268
Ross, I. G., 36
Ross, J. A., 255
Ross, K. H., 256
Rossa, L., 233
Rossi, M., 123
Rossi, R., 380, 384, 386
Ross-Mansell, P., 440
Roth, M. D., 110
Rothenberg, S., 57
Rothermel, W., 262
Rothschild, A. J., 114
Roussel, C., 270
Roussi, G., 189
Roustan, J. L., 182
Rowe, M. D., 21
Rozental, J. M., 379
Rozhkov, I. N., 153
Ruasse, M.-F., 84, 177
Rubottom, G. M., 326, 332
Ruchiwarat, S., 276
Ruder, J.-P., 243
Rüchardt, C., 69, 180
Ruge, B., 238
Ruggeri, M. V., 258
Rundle, J. W., 85
Runquist, A. W., 319
Runzheimer, H.-V., 63, 249
Russell, M. J., 123
Russell, R. A., 231
Russell-King, J., 74
Russiello, A. B., 265
Rutledge, P. S., 189
Ryan, R. C., 135
Rycroft, D. S., 109
Ryder, D. J., 401
Rylatt, D. B., 440
Ryu, I., 142
Rzepa, H. S., 7, 65, 68

Sabacky, M. J., 122, 412
Saegusa, T., 260, 330
Saenger, W., 14
Said, I. M., 235
Saiki, K., 273
Sainsbury, M., 155
St. John, G. A., 10
Saito, I., 60, 244, 378
Saito, M., 209
Saji, T., 162
Sakaino, M., 244
Sakan, K., 244
Sakan, T., 386
Sakata, J., 203, 326
Sakata, Y., 250
Sakore, T. D., 17
Sakurai, H., 117, 148, 165, 230, 257, 290, 340
Salach, J. L., 433, 434
Salares, V. R., 33
Salem, L., 59
Salerno, G., 125
Salmond, W. G., 316
Salomon, M. F., 150, 302
Salomon, R. G., 150, 169, 302, 328
Salzmann, T. N., 380
Samain, D., 383
Samberg, G., 379
Samman, N. G., 302
Sammes, M. P., 275
Sammes, P. G., 272
Samori, B., 20
Samuel, C. J., 173
Sanchez, M., 272
Sancovich, H. A., 407
Sanders, C. J., 368

Sanders, J. K. M., 429
Sanders, M. E., 386
Sanechika, K.-I., 245
Sannicolo, F., 236
Sano, H., 365
Sano, S., 416
Santiesteban, H., 207, 332
Sargent, F. P., 101, 102
Sargeson, A. M., 281
Sarhangi, A., 324
Sarkar, S., 123
Sarre, O. Z., 366
Sasaki, N., 335
Sasaki, T., 61, 116
Sasaoka, M., 253
Sato, F., 146
Sato, M., 146, 244
Sato, S., 146
Sato, T., 157, 327, 371, 386
Satoh, F., 190, 314, 407, 424
Satoh, J. Y., 325
Saucy, G., 202, 207
Sauer, J. D., 274, 279, 282
Sauer, K., 23
Sauer, W., 242, 299
Saunders, J., 396, 397, 407
Saus, A., 171, 220
Sauter, H., 70, 243, 290
Saveant, J. M., 161
Savoia, D., 137
Sawada, M., 217
Sawada, S.-I., 241
Sawyer, D. T., 161
Scaiano, J. C., 167
Scamehorn, R. G., 227
Scarborough, R. M., 207
Scarcelli, N., 13
Scarpati, R., 269
Scettri, A., 259
Schaad, L. J., 246
Schaap, A. P., 258
Schaefer, C. G., 386
Schäfer, H., 277
Schaefer, H. F., 7, 58, 110
Schäfer, L., 295, 297
Schäfer, P., 256
Schaefer, T., 295
Schäfer, U., 257
Schaffer, A. M., 31, 32
Schallner, O., 294
Scharf, H.-D., 171, 220
Schaumann, E., 253, 255, 321
Schechter, P. J., 436
Schechtman, B. H., 20
Scheele, J. J., 160
Scheer, H., 429
Scheerer, B., 201
Schegoler, A. A., 327
Scheiner, S., 53, 54
Schenk, W. N., 207, 332
Schenkluhn, H., 126
Scheraga, H. A., 36, 43
Scherowsky, G., 269
Schickaneder, H., 266
Schiess, P., 222
Schiff, H. I., 85
Schildknecht, H., 373, 375
Schlecker, R., 139
Schley, B., 268
Schleyer, P. von R., 71, 116, 140, 177, 302
Schlosser, M., 260
Schmid, G., 262
Schmid, G. H., 84, 178, 282
Schmid, H., 65, 255, 280, 282, 378
Schmid, P., 102, 228, 236
Schmid, R., 280
Schmidbaur, H., 149
Schmidt, D. E., 432
Schmidt, D. J., 118
Schmidt, R. R., 198, 350
Schmidt, S. P., 378
Schmidtchen, F., 397
Schmieder, K. R., 314
Schmitz, H., 267
Schnarr, G. W., 352, 362
Schneider, D., 367
Schneider, D. F., 294
Schneider, D. R., 69
Schneider, H. P., 279
Schneider, M. P., 305
Schnur, R. C., 196
Schoeller, W. W., 64, 306
Schöllkopf, U., 137, 199
Schönholzer, P., 255
Schofield, K., 224
Schonbrunn, A., 433
Schonfelder, W., 107
Schoppee, C. W., 70
Schramm, M., 443
Schreimer, K., 100
Schreiner, J. L., 332
Schriener, K., 298
Schroll, G., 4
Schuchardt, V., 301
Schuda, P. F., 232
Schuerch, C., 346
Schug, R., 285
Schuh, H., 101
Schulte, G., 365
Schulte, K.-W., 242
Schulte-Frohlinde, D., 96
Schulten, K., 27, 30
Schulze, J., 262
Schurvell, H. F., 3
Schuster, H. F., 258
Schuttler, R., 114
Schwalm, W. A., 94
Schwartz, J., 131, 134, 146, 177, 313, 322
Schwarz, H., 3, 4, 5, 8
Schweig, A., 242
Schweizer, E. E., 265, 266
Scopes, P. M., 25
Scordamaglia, R., 43
Scorrano, G., 178
Scott, A., 322
Scott, A. I., 399, 409, 424, 425
Scott, F., 382
Scott, J. J., 398
Scott, L. T., 239
Scrimin, P., 179, 253
Scriven, E. F. V., 107
Scrocco, E., 46
Scrutton, M., 440
Sealy, R. C., 104
Seebach, D., 136, 139, 140, 174, 195, 200, 205, 212, 287, 331, 336
Segal, G., 59
Segden-Penne, J., 311
Seguchi, K., 61
Seibl, J., 9
Seibold, K., 37
Seidel, K. D., 255
Seitz, B., 112
Seki, K., 124
Seki, Y., 127, 289
Sekiguchi, A., 114, 148
Sekiya, A., 229
Sellers, D., 379
Selva, E., 60
Semmelhack, M. F., 64, 130, 307
Seo, K., 361
Sepiol, J., 221
Seppelt, W., 244
Sera, A., 61
Sergheraert, S., 234
Sessions, R. B., 281
Seto, S., 350
Seufert, A., 145
Seuring, B., 136, 212, 336
Severin, T., 200, 260
Sevrin, M., 212
Seyferth, D., 148, 149, 150, 151, 312
Shaden, G., 238
Shah, S. K., 211
Shahak, I., 106
Shakked, Z., 14
Shamov, A. G., 59
Shannon, M. L., 148
Sharma, D. K. S., 10
Sharma, M., 359
Sharma, R. A., 359
Sharma, R. P., 109
Sharp, D. W. A., 109
Sharp, J. T., 98
Sharpless, K. B., 192, 320, 365
Shaw, J., 7
Shaw, K. B., 420
Shaw, T. J., 137, 204
Shea, K. J., 65, 217, 308
Shearer, H. M. M., 150
Shechter, H., 113, 238
Sheikh, H., 196, 252, 334
Sheldrick, G. M., 15
Shelnutt, J. A., 22
Shemin, D., 394
Sheng, S. J., 29

Sheradsky, T., 236
Shibanuma, T., 251
Shibasaki, M., 366
Shibuya, T., 378
Shieh, H. S., 16
Shih, Y.-S., 230
Shimizu, K., 342
Shimizu, M., 203, 326
Shimizu, Y., 151
Shin, H. J., 247
Shine, H. J., 236
Shinmyozu, T., 233
Shinoda, M., 225
Shiono, M., 251
Shipman, L. L., 19, 22, 23
Shoda, S.-I., 251, 318
Shoer, L. I., 134, 146
Shono, T., 112, 153, 323
Shorey, H. H., 367, 379
Shudo, K., 229
Shvetsov-Shilovskii, N. I., 282
Siebert, W., 262
Siegbahn, P. M., 110
Siegel, B., 347
Siegel, L. M., 426
Siegfried, B., 223
Sierra, J. R., 378
Silberglied, R. E., 374
Silberman, L., 340
Silvers, J. H., 85
Silverstein, R. M., 370, 372, 379, 389, 390
Sim, S. K., 70
Simmons, H. E., 300
Simon, K., 126
Simonet, J., 161
Simonetta, M., 12, 46, 249, 253
Simoni, R. D., 29
Simons, J., 45
Simpson, G. W., 320, 332
Simpson, R. F., 378
Simpson, W. T., 37
Sinaÿ, P., 345
Sinclair, J. A., 143
Singer, S. P., 280
Singer, T. P., 433, 434
Singh, T. D., 273
Sinnewell, V., 363
Siodmiak, J., 39
Sjoerdsma, A., 437
Sjoholm, I., 440
Skattebol, L., 111, 112, 382
Skell, P. S., 289
Skelton, B. W., 152
Šket, B., 172, 219, 220
Sklar, L. A., 28, 29
Skrabal, P., 270
Skrzypczynski, Z., 210
Slama, J. T., 416
Slessor, K. N., 372
Sliwa, H., 271, 273, 274, 276
Slobbe, J. 153
Small, L. E., 85, 202
Small, R. D., 167
Smallwood, J. I., 191
Smart, L. E., 220
Smit, C. J., 160
Smit, W. A., 327
Smith, A. B., 174, 207, 273
Smith, A. G., 439
Smith, A. K., 119
Smith, D. M., 111
Smith, E. L., 423
Smith, G. A., 15, 443
Smith, G. J., 440
Smith, H. W., 416
Smith, J. J., 36
Smith, K., 136, 137, 205
Smith, K. M., 407, 411, 419, 428
Smith, M. R., 75
Smith, R. A., 61
Smith, R. L., 377
Smith, S. G., 401, 407, 411
Smith, T. W., 278
Smith, U. H., 429
Snider, B. B., 233
Snieckus, V., 279
Snow, M. R., 281
Snowden, R. L., 186
Snyder, L. R., 429
Sobczak, R. L., 290
Sobell, H. M., 17
Soderquist, J. A., 149
Söderholm, A. C., 280
Sohar, P., 277
Solaro, R., 110
Solheim, B. A., 375
Solladié, G., 212
Solomon, F., 442
Solomon, J. D., 369
Solonki, B., 268
Solov'ev, K. N., 20, 21
Solyom, S., 277
Somfai, I., 234
Sommerfeld, C.-D., 168, 239
Sondengam, B. L., 196
Sondheimer, F., 247, 248, 249, 291
Song, P. S., 22, 23, 24, 25, 28, 30, 32, 34, 38, 39
Sonnet, P. E., 370, 371
Sonoda, A., 143, 189, 309, 313
Sonoda, N., 127, 142, 189, 289, 312
Sood, H. R., 118
Sorenson, R. J., 353
Sotheeswaran, S., 216
Soumillion, J.P., 223
Souto-Bachiller, F. A., 252
Soysa, H. S. D., 148
Spackman, I. H., 83
Spangler, D., 19
Spangler, R. J., 115, 222
Spear, R. J., 80, 298
Specchiarella, M., 60
Speltz, L. M., 322
Spencer, T. A., 108
Speranza, M., 223
Spernol, A., 101
Spicer, W. E., 20
Spiegel, E., 256, 276
Spinelli, D., 228
Spinelli, P., 368
Spira, I., 107
Spiro, T. G., 22
Springborg, J., 281
Springer, J. P., 216, 375
Sridaran, P., 227
Srikrishnan, T., 16
Srinivasan, R., 220
Staab, H. A., 233, 235, 247, 248
Stackhouse, J., 150
Stadtmin, T. C., 423
Staedler, E., 379
Staemmler, V., 241
Stafforst, D., 199
Staires, S. K., 274
Staley, R. H., 88
Staley, S. W., 306
Stallings, W. C., 17
Stang, P. J., 118, 227
Stanislawski, D., 241
Staral, J. S., 246, 299, 300
Stark, M., 166
Starr, T. L., 13
Starukhin, A. S., 20
Staunton, J., 397, 411
Stavinoha, J. L., 171
Steck, W. F., 368, 369
Steenken, S., 95, 96
Steevens, J. B., 194
Stegk, A., 314
Steigerwald, H., 123
Stein, J. M., 443
Steiner, B. W., 3
Steiner, P. R., 70, 361
Stenzel, W., 344
Stepanov, B. I., 261
Stephan, W., 83
Stephenson, J. R., 425
Stern, R. L., 265
Sternbach, L. H., 265
Sternlicht, M., 368
Stevens, N. R., 222
Stevenson, J. M., 271
Stewart, A., 119
Stewart, R. F., 35
Stewart, T. E., 379
Stezowski, J. J., 147
Stick, R. V., 343, 358
Stiles, P. J., 180
Still, W. C., 150, 324
Stipanovic, R. D., 190, 377
Stirling, C. J. M., 183, 211
Stock, L. M., 225
Stodala, R. K., 17
Stoddart, J. F., 281, 356
Stojanac, N., 398
Stoll, M. S., 408
Stolle, W. T., 310

Stolze, R., 239
Stone, J. G., 135
Storesund, H., 74
Stork, G., 288, 366
Stout, G. H., 3
Strain, H. H., 419
Strand, A., 248
Strang, P. J., 76
Stransky, W., 383, 385
Stratz, G., 433
Strauss, H. L., 296
Streck, R., 386
Strickland, E. H., 34
Strohmeier, W., 123
Stroshane, R. M., 365
Stroud, S. G., 181
Struckmeier, G., 16
Stubbe, J., 442
Studier, M. H., 419
Studzinskii, O. P., 230
Subramanian, R., 358
Suck, S. H., 39
Suda, M., 62, 243
Süess, R., 209
Suemitsu, R., 128, 329
Suenram, R. D., 256
Sugie, H., 368
Sugimura, N., 330
Sugino, K., 209
Sugiyama, H., 350
Sullivan, M. J., 31
Sumengen, D., 268
Sumida, Y., 275
Summerhays, K. D., 216
Summers, R., 101
Summerville, R. H., 216
Summons, R. E., 238
Sun, M., 22, 34, 39
Sundbom, M., 20
Suschitzky, H., 107, 109, 264
Sustmann, R., 60, 69
Sutcliffe, L. H., 99
Sutherland, I., 423
Sutherland, I. O., 66
Sutton, M. R., 440
Suzuki, A., 335
Suzuki, H., 123, 126, 225
Suzuki, M., 257, 342, 358
Svec, W. A., 419
Svensson, B. G., 372
Swanson, R., 171
Swenton, J. S., 232
Swierczewski, G., 189
Switzer, F., 295
Swofford, R. L., 29
Sydnes, L. V., 111
Symanyk, W., 310
Syriopoulus, G. T., 341
Szarek, W. A., 352, 362, 364
Szeimies, G., 292
Szmant, H. H., 354

Taagepera, M., 48, 273
Taber, D. F., 289
Tacconi, G., 60
Taft, R. W., 7, 48, 87, 216, 273
Tagashira, Y., 92
Tagat, J., 222
Taglieber, V., 233
Taguchi, H., 107
Taguchi, T., 330
Tahara, S., 371
Tajima, S., 7
Takagi, K., 237
Takagi, T., 209
Takahashi, N., 368
Takahashi, S., 371
Takahashi, T., 189, 325, 366, 380, 399, 425
Takahashi, Y., 365
Takahatake, Y., 288
Takamuku, S., 117, 165, 290
Takase, K., 245
Takata, Y., 155
Takaya, T., 109
Takegami, Y., 320
Takehira, Y., 242
Takei, H., 330
Takemasa, T., 232
Takemura, T., 31
Takeshita, H., 244
Takeuchi, H., 108
Takeuchi, K., 72, 245
Takeuchi, Y., 282
Takigawa, T., 386
Takken, H. J., 338, 387
Tal, D., 125
Talalay, P., 438, 439, 440
Talbiersky, J., 198
Talley, P., 176, 319
Talman, E., 371, 373, 379
Tam, S. Y.-K., 348, 349
Tamaki, Y., 367, 368, 369
Tamao, K., 229
Tamaru, Y., 266
Tamborra, P., 205
Tamura, Y., 275
Tan, H.-W., 100
Tanabe, M., 365
Tanaka, J., 35, 38
Tanaka, K., 87, 242
Tanaka, M., 38, 119
Tanaka, T., 107
Tanigawa, Y., 189, 309
Taniguchi, H., 99, 107, 126
Tanimoto, S., 106
Tanis, S. P., 375, 377
Tapia, O., 55
Tardella, P. A., 109
Tarnowski, T. L., 281
Tartar, A., 274
Taschenberg, E. F., 368
Taschner, M. J., 320
Ta-Shma, R., 78
Tassi, D., 252
Taticchi, A., 282
Tatsuki, S., 368
Tatsuoka, T., 280
Taylor, D. R., 183
Taylor, E. C., 147, 222, 229, 277, 332
Taylor, G., 171, 220
Taylor, G. L., 17
Taylor, L. B., 440
Taylor, R., 223, 230, 240
Taylor, R. T., 116, 149, 187, 219, 312
Tebby, J. C., 311
Tedeschi, P., 263
Teitell, M. F., 39
Temme, G. H., 424
ten Hoedt, R. W. M., 177
Teranishi, A. Y., 192
Teranishi, S., 126
Terasawa, M., 126
Terashima, S., 332
Terent'ev, A. B. 90
Terlouw, J. K., 5
Tette, J., 368
Thackaberry, S. P., 92
Thalmann, A., 336
Thankachan, C., 75, 248
Thankwerden, B., 182
Thea, S., 240
Thebtaranonth, C., 276
Thebtaranonth, Y., 130
Theissling, C. B., 6, 273
Thewalt, U., 16
Thiel, W., 169
Thiem, J., 361
Thijs, L., 269
Thomas, D. R., 107
Thomas, J. K., 102
Thomas, L. L., 210
Thomas, M. G., 135
Thomas, P. N., 224
Thomas, R. N., 25
Thomas, T. M., 30
Thomas, W., 230
Thompson, A. J., 25
Thompson, J. T., 237
Thompson, M., 417, 427
Thompson, R. C., 340
Thompson, R. S., 179
Thuillier, A., 181
Thulin, B., 234, 248
Thummell, R. P., 221
Tidwell, T. T., 75, 84, 184, 189, 248
Tigler, D., 280
Tikhomirov, V. A., 59
Tilden, P. E., 372
Tilhard, H.-J., 151
Timko, J. M., 281
Timm, U., 112
Timms, P. L., 137
Tinnemans, A. H. A., 172, 219
Tinoco, I., jun., 35
Tishler, M., 321
Titova, S. P., 239
Tivol, W. F., 439
Tobin, G. D., 224
Tobin, T. R., 371, 378
Toda, F., 242

Todd, A. R., 423
Todesco, P. E., 227
Tokanou, H., 109
Tokumaru, K., 239
Tokuno, E., 330
Toldy, L., 277
Tollin, G., 39
Tolson, T. L., 341
Tomasi, J., 46, 68
Tomasik, W., 198
tom Dieck, H., 126
Tomezsko, E. S., 181
Tomioka, H., 116
Tomita, H., 231
Tomono, K., 19
Tonellato, U., 178
Tonge, K. H., 94
Topol, A., 77
Toppett, S., 256, 257
Tori, K., 217
Torre, G., 197
Toshimitsu, A., 189
Toube, T. P., 3
Touzin, A.-M., 140
Towarowski, A. J., 29
Townsend, C. A., 424, 425
Townsend, R. E., 59
Toyne, K. J., 216
Toyoda, T., 13
Traas, P. C., 338, 387
Trahanovsky, W. S., 222, 274
Tramontano, A., 143
Trampe, S., 233
Traynor, S. G., 215, 216, 293
Trdatyan, V. A., 239
Trefonas, L. M., 37
Tremelling, M. J., 149
Trenary, M., 58
Trickes, G., 198
Trifunac, A. D., 103
Trinajstić, N., 215
Trinh-Toan, 163
Trombini, C., 137
Trong Anh, N., 60
Trost, B. M., 185, 207, 287, 289, 336, 380, 387
Trostman, U., 69
Troxler, E., 126
Trueblood, K. N., 423
Truesdale, L. K., 325
Trullinger, D. P., 207
Tsai, C. C., 17
Tschinkel, W. R., 374
Tseng, K. L., 231
Tsuchida, T., 108
Tsuchiya, T., 7, 279, 365
Tsue Rima, T., 329
Tsuji, A., 273
Tsuji, J., 123, 189, 327, 380, 381
Tsunetsugu, J., 244
Tsuchima, T., 192
Tsuzuki, K., 378
Tsvetkov, Y. E., 347
Tsvirko, M. P., 20
Tubault, M., 162
Tumlinson, J. H., 369, 370, 376, 379
Tunemoto, D., 288
Turchi, I. J., 169
Turnbull, K., 270
Turner, J. J., 174, 273
Turner, R. W., 236
Turner, S. P., 9
Turney, T. W., 137
Turrell, A. G., 229
Turro, N. J., 170, 308
Tursch, B., 367
Tyler, R. C., 372, 379
Tyrrell, H. M., 171

Uchida, K., 149, 188
Uchida, M., 386
Uchikama, K., 275
Uchimaru, T., 326
Uchiumi, K., 368
Uebel, E. C., 370, 371
Ueda, M., 331
Uemura, H., 109, 179, 189
Ueno, K., 128
Ukrainskii, I. I., 26
Umani-Ronchi, A., 137
Umans, R. S., 22
Umbreit, M. A., 192, 320
Umemoto, T., 288
Umeyama, H., 51, 57
Umezawa, H., 365
Umezawa, S., 365
Umrikhina, A. V., 427
Underhill, E. W., 368, 369
Underwood, G. R., 68
Unsworth, J. F., 419
Usui, T., 365
Utimoto, K., 149, 176, 188, 384
Uto, K., 126
Utter, M. F., 440
Uwai, C., 248

Vagelos, P. R., 440
Vaglio, G. A., 135
Vakenti, J., 368
Valasinas, A., 399, 403
Valette, G., 203
Valeur, B., 34
Valle, M., 135
Valter, W., 255
van Dam, H., 194
van de Graaf, B., 5
Van Der Avoird, A., 46
van der Gen, A., 277
Vander Helm, R., 14
Vanderpool, D. P., 108, 109
Van de Sande, C. C., 6, 8
Van de Vrie, M., 369
van Dijck, L. A., 182
Van Gaever, F., 6
van Haverbeke, Y., 7
Vanhooren, M., 6
Van Horn, D. E., 132, 313
van Houte, J. J., 4
van Koten, G., 177, 229
van Leusen, A. M., 329
van Leusen, D., 329
Van Meerssche, M., 254
van Straten, J. W., 218, 233
Van Telgen, H. J., 445
van Thuijl, J., 4
van Tilborg, W. J. M., 160, 258
Van Veen, L., jun., 226
Vass, G., 363
Vatèle, J.-M., 350
Vaughan, J., 179
Vedejs, E., 280, 310
Veeger, C., 445
Veillard, H., 47
Velthuis, H. H. V., 372
Venema, A., 7
Venkataramu, S. D., 282
Venuti, M. C., 237
Verbit, L., 3
Vereschchagin, L. I., 180
Vergamini, P. G., 122
Verhelst, G., 257
Verheyden, J. P. H., 354
Verhoeven, T. R., 207, 336
Verkuijlen, E., 133, 185
Vermeer, J. G. C. M., 182
Vermeer, P., 141, 149, 176, 313
Vermeulen, G., 257
Vernay, H. F., 365
Vernon, J. M., 274
Verwiel, P. E. J., 367, 368, 371, 373, 379
Veysoglu, T., 212, 386
Vial, C., 384
Viala, J., 318
Vick, S. C., 148, 150
Victor, R., 222
Vidal, M., 113
Viehe, H. G., 208, 252, 254
Viger, A., 438
Villiéras, J., 140, 141
Vincens, M., 113
Vineyard, B. D., 122, 412
Vinson, S. B., 376
Vishveshwara, S., 53
Vite, J. P., 369, 372, 378
Vittorelli, P., 255
Vivona, N., 268
Vögeli, R., 243
Vögtle, F., 232, 233, 234, 235
Voelter, W., 39
Voerman, S., 367, 368, 369
Vogel, E., 63, 168, 239, 246, 249
Vogel, F., 143, 202, 320
Voigt, E., 115
Voinova, S. E., 261
Volkov, V. B., 47
Vollhardt, K. P. C., 63, 105, 119, 126, 175, 220, 222, 242, 314
Vollmar, A., 225

Volz, W. E., 299
Von Dreele, R. B., 15
von Schnering, H. G., 256
Vorbruggen, H., 336
Vorburger, T. V., 193
Vos, A., 12, 258
Vostrowsky, O., 369, 383, 385
Vuilleumier, H., 221
Vuturo, S. B., 386
Vyas, D. M., 362

Wachi, H., 239
Waclawski, B. J., 193
Wada, K., 336
Waddell, W. H., 31, 32
Wade, K., 150
Wade, L. E., 65
Wadhams, L. J., 372, 379
Wadt, W. R., 256
Wagemann, W., 249
Wagenknecht, J. H., 158
Wagner, D., 354
Wagniere, G., 25
Wah, H. K., 275
Waiss, A. C., jun., 377
Wakabayashi, T., 209
Walborsky, H. M., 298
Walker, B. S., 127
Walker, D. L., 364
Walker, D. M., 420
Walker, J. A., 64
Wallace, B., 68, 192, 314
Wallace, J. C., 440
Wallace, R. G., 105
Wallis, C. J., 187, 321
Wallis, T. G., 60, 61, 216
Walsh, C. T., 433, 441
Wan, J. K. S., 90
Wancowicz, D. J., 196
Wand, A. H.-J., 14
Wang, V. S., 438
Wang, Y. F., 203
Warburg, O., 417
Warburton, D., 399
Warren, G. B., 443
Warren, S., 186, 187, 211, 321
Warrener, R. N., 231, 243, 258
Warshel, A., 30
Washburne, S. S., 185
Washio, H., 378
Washtien, W., 435
Wasielewski, M. J., 429
Wasserman, E., 90
Wasserman, H. H., 329
Watanabe, Y., 244, 251, 318
Waterhouse, I., 65, 312
Watson, D. R., 355
Watt, R. A., 272
Watts, R. O., 43
Weatherston, J., 367, 368, 378, 379, 380, 384
Weaver, K. M., 376
Webb, H. M., 86
Webb, J. G. K., 228
Webb, K. S., 4
Webb, T. R., 200, 324
Weber, D., 163
Weber, G., 34
Weber, R. U., 98, 101
Weber, W. P., 148
Weckerle, W., 352, 353, 357
Weeks, C. M., 16
Weese, G. W., 5
Wehner, W., 235
Wehrli, R., 65
Weidlein, J., 147
Weigel, L. O., 386
Weimann, L. J., 20
Weinkam, R. J., 8
Weinkauff, D. J., 122, 412
Weinreb, S. M., 208
Weires, R. W., 369
Weisgerber, H., 268
Weiss, C. jun., 19, 22
Weissbarth, O., 430
Welch, J., 202
Weller, H. N., 64, 307
Weller, T., 56
Wells, R. J., 397
Welsher, T. L., 299
Wender, P. A., 166, 286, 338
Wendisch, D., 379
Wendling, L. A., 63
Wenham, M. J., 372
Wenkert, D., 287
Wenkert, E., 201, 287
Wennerström, O., 234, 248
Wentrup, C., 114
Wentrup-Byrne, E., 114
Weringa, W. D., 5
Wescott, L. D., jun., 117
Wesdemiotis, C., 5, 8
West, J. R., 379
West, P. R., 95
West, R., 177, 241
Westerman, I. J., 60
Westheimer, F. H., 432
Westigard, P. H., 379
Westley, J. W., 194
Westmijze, H., 141, 149, 176, 313
Weston, A. F., 9
Weston, J. B., 224
Weyerstahl, P., 202
Weyhenmeyer, R., 423
Wheatley, C. M., 281
Wheeler, J. W., 367, 372, 374
White, A. H., 152
White, C. K., 123, 158, 241
White, D. N. J., 180
White, J. D., 262, 336
White, J. G., 238, 423
White, W. N., 237
Whitehead, G., 150
Whitehead, J. K., 440
Whiter, P. F., 182
Whitesell, J. K., 197
Whitesell, M. A., 197, 338
Whitesides, G. M., 182
Whitham, G. H., 218, 296
Whiting, D. A., 377
Whiting, M. C., 74, 75
Whitlock, B. J., 138, 222
Whitlock, H. W., jun., 138, 222
Whitten, D. G., 19
Whitten, J. L., 27
Wiberg, K. B., 300, 302
Wiberg, N., 147
Wichmann, J. K., 376, 377
Wieringa, J. H., 258
Wiersig, J. R., 9
Wieser, J. D., 231
Wiessler, M., 232, 274
Wieting, R. D., 88
Wightman, R. H., 221, 397, 411
Wihler, H. D., 258
Wijsman, A., 277
Wilding, H. F., 87
Wilkie, J. S., 390
Willbe, C., 260
Willett, G. D., 217
Williams, D. C., 398, 402, 404, 419, 424, 425, 427
Williams, D. E., 13
Williams, D. F., 22
Williams, D. H., 3, 4, 5, 15
Williams, D. L. H., 237
Williams, D. R., 321, 336
Williams, G. C., 348
Williams, G. H., 101
Williams, H. J., 317, 387
Williams, J. C., jun., 158
Williams, N. R., 353
Williams, P. J., 374
Williams, R. N., 374
Williams, S. B., 188, 313
Williamson, A. D., 85
Willing, R. I., 390
Willingham, A. K., 437
Willis, J. P., 157
Willner, I., 246, 247, 249
Willson, G. C., 416
Willy, W. E., 318
Wilson, G. E., 262
Wilson, I. F., 311
Wilson, J. M., 3
Wilson, J. W., 64, 231
Wilson, N. M., 370
Wilson, S., 220
Wilton, D. C., 432
Wiltshire, J. F., 228
Windel, B., 8
Winfield, J. M., 109
Winkler, H. U., 5
Winkler, J., 7
Winkler, T., 65
Winnik, M. A., 180
Winter, J. N., 99
Wirtz, K., 101
Wirz, J., 168

Wirz, P., 377
Withers, G. P., 149, 319
Witiak, D. N., 9
Wittman, J. M., 104
Woggon, W. D., 378
Wolf, R., 272
Wolfschütz, R., 5
Wollenberg, R. H., 140, 336
Wollowitz, S., 109
Woltermann, A., 151, 201, 280
Wong, J., 108, 128
Wong, P. K., 129
Wong, R. J., 83
Wong, S. K., 167
Woo, P. W. K., 355
Wood, C. J., 360
Wood, D. E., 99
Wood, D. L., 372
Wood, G. W., 10
Wood, H. G., 440, 441, 442
Wood, W. F., 375
Woodgate, P. D., 189
Woodward, P., 220
Woodward, R. B., 62, 423
Woolard, G. R., 362, 364
Wooldridge, T., 112, 253
Woolfson, M. M., 12
Woolhouse, A. D., 230
Worley, S. D., 47
Wormer, P. E. S., 46
Wothermann, A., 312
Wrackmeyer, B., 144
Wright, G. J., 225
Wright, T. L., 225
Wrigley, N. G., 441
Wudl, F., 163, 261
Wüest, H., 187
Wunderli, A., 65
Wursthorn, K. R., 151, 312
Wurziger, H. K. W., 404, 413
Wyatt, J., 155
Wynberg, H., 169, 258
Wynberg, M., 185
Wyvratt, M. J., 294

Yagen, B., 424
Yaginuma, K., 368, 369
Yakakoto, K., 327
Yakhontov, L. N., 282
Yamabe, S., 57
Yamada, M., 209, 378
Yamada, S., 237
Yamada, Y., 230, 365
Yamaguchi, T., 280
Yamamoto, H., 207, 342
Yamamoto, I., 109, 127
Yamamoto, K., 165, 233, 245
Yamamoto, N., 31
Yamamoto, Y., 35, 143, 196, 313
Yamamura, K., 338
Yamano, T., 433
Yamaoka, N., 350
Yamaoka, R., 367, 368
Yamasaki, T., 386
Yamashita, K., 232
Yamashita, M., 128, 320, 329
Yamaya, M., 188
Yamazaki, H., 130
Yamazaki, I., 23
Yamazaki, T., 350
Yamdagni, R., 85
Yan, C. F., 289
Yanagawa, Y., 21
Yanagi, J., 165
Yang, N. C., 171
Yanofsky, C., 443
Yany, F., 170, 258
Yaouanc, J. J., 433
Yarwood, A. J., 64, 231
Yarwood, J., 165
Yasuda, D. M., 365
Yasuda, H., 381
Yasunami, M., 245
Yasuoka, N., 13
Yatagai, H., 313
Yates, K., 84, 178
Yavari, I., 272, 296, 360
Yavorskaya, A. N., 427
Yee, L. S., 242
Yegorova, G. D., 21
Yen, C. Y., 272
Yengoyan, L., 413
Yeo, A. N., 9
Yim, M. B., 99
Yogo, T., 141
Yokomichi, Y., 245
Yokoyama, K., 203, 326
Yokoyama, Y., 248, 320, 335, 338
Yoneda, N., 181
Yoshida, J.-I., 229
Yoshida, K., 155, 230, 323
Yoshida, S., 368
Yoshida, Z.-I., 241, 266
Yoshifuji, M., 177, 313
Yoshikawa, S., 126
Yoshikawa, Y., 247, 249
Yoshimura, Y., 217
Yoshino, T., 233
Yoshioka, T., 242
Young, J. C., 372
Young, N. M., 33
Yount, M. R., 440
Yu, C. C., 256
Yu, M. W., 39
Yu, N. Y., 22
Yunker, M. B., 349, 363
Yushima, T., 369
Yushkevich, N. A., 21

Zaitseva, N. I., 427
Zaklika, K. A., 258
Zaman, Z., 394, 412
Zanardi, G., 112, 253
Zander, G., 278
Zapf, U., 233
Zaretskii, Z. V., 3
Zassinovich, G., 123
Zderic, S. A., 143
Zehavi, U., 322
Zeller, E. A., 434
Zeller, K.-P., 63, 111, 112, 282
Zembayashi, M., 229
Zeppa, A., 159
Zettl, C., 266
Zheltovskii, N. V., 36
Zhogolev, D. A., 47
Ziegler, F. E., 338
Ziegler, H. E., 100
Zimmer-Gasser, B., 149
Zimmerman, H. E., 166
Zumoff, B., 437
Zuniga Juaristi, G., 272
Zupan, M., 172, 219, 220
Zwanenburg, B., 269
Zweifel, G., 144, 149, 317, 332
Zwierzak, A., 213, 321
Zwolinski, G. K., 440